ISOTOPE GEOLOGY

Radiogenic and stable isotopes are used widely in the Earth sciences to determine the ages of rocks, meteorites, and archeological objects, and as tracers to understand geological and environmental processes. Isotope methods determine the age of the Earth, help reconstruct the climate of the past, and explain the formation of the chemical elements in the Universe. This textbook provides a comprehensive introduction to both radiogenic and stable isotope techniques. An understanding of the basic principles of isotope geology is important in a wide range of the sciences: geology, astronomy, paleontology, geophysics, climatology, archeology, and others.

Claude Allègre is one of the world's most respected and best-known geochemists, and this textbook has been developed from his many years of teaching and research experience.

Isotope Geology is tailored for all undergraduate and graduate courses on the topic, and is also an excellent reference text for all Earth scientists.

CLAUDE ALLÈGRE is extremely well known globally in the Earth science research and teaching community. He is currently an Emeritus Professor at the Institut Universitaire de France, Université Denis Diderot, and the Institut de Physique du Globe de Paris, and has had a long and illustrious career in science. He is a former Director of the Department of Earth Sciences, Université Paris VII, former Director of the Institut de Physique du Globe de Paris, past President of the French Bureau of Geological and Mining Research (BRGM), and former National Education Minister for Research and Technology for the French government. In his career he has won most of the available honours and awards in the geosciences, including the Crafoord Prize from the Swedish Royal Academy of Sciences, the Goldschmidt Medal from the Geochemical Society of America, the Wollaston Medal from the Geological Society of London, the Arthur Day Gold Medal from the Geological Society of America, the Médaille d'Or du CNRS, the Holmes Medal from the European Union Geosciences, and the Bowie Medal from the American Geophysical Union. He is member of several academics: Foreign Associate of the National Academy of Sciences (USA), Foreign Member of the Academy of Art and Science, Foreign Member of the Philosophical Society, Foreign Member of the Royal Society, Foreign Member of the National Academy of India, and Membre de l'Académie des Sciences de Paris. He is also a Commandeur de la Légion d'Honneur, a past President of the European Union of Geosciences, past President of the NATO Earth Sciences Committee, and former editor of the journals *Physics of the Earth and Planetary Interiors* and *Chemical Geology*. He has written hundreds of research articles, and 25 books in French.

Isotope Geology

CLAUDE J. ALLÈGRE

Institut de Physique du Globe de Paris
and
Université Denis Diderot

CAMBRIDGE UNIVERSITY PRESS
Cambridge, New York, Melbourne, Madrid, Cape Town, Singapore, São Paulo

Cambridge University Press
The Edinburgh Building, Cambridge CB2 8RU, UK

Published in the United States of America by Cambridge University Press, New York

www.cambridge.org
Information on this title: www.cambridge.org/9780521862288

Original edition:
Géologie isotopique
© Éditions Belin, Paris, 2005

First published 2008

Printed in the United Kingdom at the University Press, Cambridge

A catalog record for this publication is available from the British Library

Library of Congress Cataloging in Publication data
Allhgre, Claude J.
Introduction to isotope geology / Claude J. Allegre.
p. cm.
Includes bibliographical references and index.
ISBN 978-0-521-86228-8
1. Isotope geology. I. Title.
QE501.4.N9A45 2008
551.9–dc22 2008029603

ISBN 978-0-521-86228-8 hardback

Dedication

I dedicate this book to all those who have helped me to take part in the extraordinary adventure of developing isotope geology.

To my family, who have probably suffered from my scientific hyperactivity.

To those who were paragons for me and have become very dear friends: Jerry Wasserburg, Paul Gast, George Wetherill, Al Nier, John Reynolds, Mitsunobu Tatsumoto, Clair Patterson, George Tilton, Harmon Craig, Samuel Epstein, Karl Turekian, Paul Damon, Pat Hurley, Edgar Picciotto, Wally Broecker, and Devendra Lal. I have tried to stand on their shoulders.

To my colleagues and friends with whom I have shared the intense joy of international scientific competition: Stan Hart, Keith O'Nions, Al Hofmann, Marc Javoy, Don DePaolo, Charles Langmuir, Jean Guy Schilling, Chris Hawkesworth, and many others.

To my undergraduate and graduate students, and postdoctoral fellows, to my laboratory staff and first and foremost to those who have participated in almost all of this adventure: Jean-Louis Birck, Gérard Manhès, Françoise Capmas, Lydia Zerbib, and the sorely missed Dominique Rousseau. Without them, none of this would have been possible, because modern research is primarily teamwork in the full sense of the word.

CONTENTS

The color plates are situated between pages 220 and 221.

PREFACE

Isotope geology is the offspring of geology on one hand and of the concepts and methods of nuclear physics on the other. It was initially known as "nuclear geology" and then as "isotope geochemistry" before its current name of isotope geology came to be preferred because it is based on the measurement and interpretation of the isotopic compositions of chemical elements making up the various natural systems. Variations in these isotope compositions yield useful information for the geological sciences (in the broad sense). The first breakthrough for isotope geology was the age determination of rocks and minerals, which at a stroke transformed geology into a quantitative science. Next came the measurement of past temperatures and the birth of paleoclimatology. Then horizons broadened with the emergence of the concept of isotopic tracers to encompass not only questions of the Earth's structures and internal dynamics, of erosion, and of the transport of material, but also problems of cosmochemistry, including those relating to the origins of the chemical elements. And so isotope geology has not only extended across the entire domain of the earth sciences but has also expanded that domain, opening up many new areas, from astrophysics to environmental studies.

This book is designed to provide an introduction to the methods, techniques, and main findings of isotope geology. The general character of the subject defines its potential readership: final-year undergraduates and postgraduates in the earth sciences (or environmental sciences), geologists, geophysicists, or climatologists wanting an overview of the field.

This is an educational textbook. To my mind, an educational textbook must set out its subject matter and explain it, but it must also involve readers in the various stages in the reasoning. One cannot understand the development and the spirit of a science passively. The reader must be active. This book therefore makes constant use of questions, exercises, and problems. I have sought to write a book on isotope geology in the vein of Turcotte and Schubert's *Geodynamics* (Cambridge University Press) or Arthur Beiser's *Concepts of Modern Physics* (McGraw-Hill), which to my mind are exemplary.

As it is an educational textbook, information is sometimes repeated in different places. As modern research in the neurosciences shows, learning is based on repetition, and so I have adopted this approach. This is why, for example, although numerical constants are often given in the main text, many of them are listed again in tables at the end. In other cases, I have deliberately not given values so that readers will have to look them up for themselves, because information one has to seek out is remembered better than information served up on a plate.

Readers must therefore work through the exercises, failing which they may not fully understand how the ideas follow on from one another. I have given solutions as we go along, sometimes in detail, sometimes more summarily. At the end of each chapter, I have set a number of problems whose solutions can be found at the end of the book.

Another message I want to get across to students of isotope geology is that this is not an isolated discipline. It is immersed both in the physical sciences and in the earth sciences. Hence the deliberate use here and there of concepts from physics, from chemistry (Boltzmann distribution, Arrhenius equation, etc.), or from geology (plate tectonics, petrography, etc.) to encourage study of these essential disciplines and, where need be, to make readers look up information in basic textbooks. Isotope geology is the outcome of an encounter between nuclear physics and geology; this multidisciplinary outlook must be maintained.

This book does not set out to review all the results of isotope geology but to bring readers to a point where they can consult the original literature directly and without difficulty. Among current literature on the same topics, this book could be placed in the same category as Gunter Faure's *Isotope Geology* (Wiley), to be read in preparation for Alan Dickin's excellent *Radiogenic Isotope Geology* (Cambridge University Press).

The guideline I have opted to follow has been to leave aside axiomatic exposition and to take instead a didactic, stepwise approach. The final chapter alone takes a more synthetic perspective, while giving pointers for future developments.

I have to give a warning about the references. Since this is a book primarily directed towards teaching I have not given a full set of references for each topic. I have endeavored to give due credit to the significant contributors with the proper order of priority (which is not always the case in modern scientific journals). Because it is what I am most familiar with, I have made extensive use of work done in my laboratory. This leads to excessive emphasis on my own laboratory's contributions in some chapters. I feel sure my colleagues will forgive me for this. The references at the end of each chapter are supplemented by a list of suggestions for further reading at the end of the book.

ACKNOWLEDGMENTS

I would like to thank all those who have helped me in writing this book.

My colleagues Bernard Dupré, Bruno Hamelin, Éric Lewin, Gérard Manhès, and Laure Meynadier made many suggestions and remarks right from the outset. Didier Bourles, Serge Fourcade, Claude Jaupart, and Manuel Moreira actively reread parts of the manuscript.

I am grateful too to those who helped in producing the book: Sandra Jeunet, who word-processed a difficult manuscript, Les Éditions Belin, and above all Joël Dyon, who did the graphics. Christopher Sutcliffe has been a most co-operative translator.

My very sincere thanks to all.

CHAPTER ONE

Isotopes and radioactivity

1.1 Reminders about the atomic nucleus

In the model first developed by **Niels Bohr** and **Ernest Rutherford** and extended by **Arnold Sommerfeld**, the atom is composed of two entities: a central nucleus, containing most of the mass, and an array of orbiting electrons.[1] The nucleus carries a positive charge of $+Ze$, which is balanced by the electron cloud's negative charge of $-Ze$. The number of protons, Z, is matched in an electrically neutral atom by the number of electrons. Each of these particles carries a negative electric charge e.

As a rough description, the nucleus of any element is made up of two types of particle, neutrons and protons. A neutron is slightly heavier than a proton with a mass of $m_n = 1.674\,95 \cdot 10^{-27}$ kg compared with $m_p = 1.672\,65 \cdot 10^{-27}$ kg for the proton. While of similar masses, then, the two particles differ above all in their charges. The proton has a positive charge ($+e$) while the neutron is electrically neutral. The number of protons (Z) is the **atomic number**. The sum $A = N + Z$ of the number of neutrons (N) plus the number of protons (Z) gives the **mass number**. This provides a measure of the mass of the nuclide in question if we take as our unit the approximate mass of the neutron or proton. **Thomson** (1914) and **Aston** (1919) showed that, for a given atomic number Z, that is, for a given position in Mendeleyev's periodic table, there are atoms with different mass numbers A, and therefore nuclei which differ in the number of neutrons they contain (see Plate 1). Such nuclides are known as the **isotopes** of an element.

Thus there is one form of hydrogen whose nucleus is composed of just a single proton and another form of hydrogen (deuterium) whose nucleus comprises both a proton and a neutron; these are the two stable isotopes of hydrogen. Most elements have several naturally occurring isotopes. However, some, including sodium (Na), aluminum (Al), manganese (Mn), and niobium (Nb), have just one natural, stable isotope.

The existence of isotopes has given rise to a special form of notation for nuclides. The symbol of the element – H, He, Li, etc. – is completed by the atomic number and the mass number – ${}^1_1H$, ${}^2_1H$, ${}^6_3Li$, ${}^7_3Li$, etc. This notation leaves the right-hand side of the symbol free for chemical notations used for molecular or crystalline compounds such as ${}^2_1H_2\ {}^{16}_8O_2$. The notation at the lower left can be omitted as it duplicates the letter symbol of the chemical element.

[1] For the basic concepts of modern physics the exact references of original papers by prominent figures (Einstein, Fermi, etc.) are not cited. Readers should consult a standard textbook, for example Leighton (1959) or Beiser (1973).

The discovery of isotopes led immediately to that of **isobars**. These are atoms with the same mass numbers but with slightly different numbers of protons. The isobars rubidium $^{87}_{37}Rb$ and strontium $^{87}_{38}Sr$ or alternatively rhenium $^{187}_{75}Re$ and osmium $^{187}_{76}Os$ do not belong in the same slots in the periodic table and so are chemically distinct. It is important to know of isobars because, unless they are separated chemically beforehand, they "interfere" with one another when isotope abundances are measured with a mass spectrometer.

1.2 The mass spectrometer

Just as there would be no crystallography without x-rays nor astronomy without telescopes, so there would be no isotope geology without the invention of the mass spectrometer. This was the major contribution of **Thomson** (1914) and **Aston** (1918). Aston won the 1922 Nobel Prize for chemistry for developing this instrument and for the discoveries it enabled him to make.[2] Subsequent improvements were made by **Bainbridge** and **Jordan** (1936), **Nier** (1940), and **Inghram** and **Chupka** (1953). Major improvements have been made using advances in electronics and computing. A decisive step was taken by **Arriens** and **Compston** (1968) and **Wasserburg** *et al.* (1969) in connection with Moon exploration with the development of automated machines. More recent commercial machines have improved quality, performance, and reliability tenfold!

1.2.1 The principle of the mass spectrometer

The principle is straightforward enough. Atoms of the chemical element whose isotopic composition is to be measured are ionized in a vacuum chamber. The ions produced are then accelerated by using a potential difference of 3–20 kV. This produces a stream of ions, and so an electric current, which is passed through a magnetic field. The magnetic field exerts a force perpendicular to the "ionic current" and so bends the beam of ions. The lighter ions are deflected more than the heavier ones and so the ions can be sorted according to their masses. The relative abundance of each isotope can be measured from the relative values of the electron currents produced by each stream of ions separated out in this way.

Let us put this mathematically. Suppose atoms of the element in question have been ionized. The ion acceleration phase is:

$$eV = \frac{1}{2} mv^2$$

where eV is the electrical energy, $\frac{1}{2}mv^2$ is the kinetic energy, e is the ion's charge, v its speed, m its mass, and V the potential difference. It can be deduced that:

$$v = \left(\frac{2eV}{m}\right)^{\frac{1}{2}}.$$

[2] The other inventor of the mass spectrometer, J. J. Thomson, had already been awarded the 1906 Nobel Prize for physics for his discovery of the electron.

Magnetic deflection is given by equating the magnetic force Bev to centripetal acceleration (v^2/R) multiplied by mass m, where B is the magnetic field and R the radius of curvature of the deflected path:

$$Bev = m\left(\frac{v^2}{R}\right).$$

It can be deduced that:

$$v = \frac{BeR}{m}.$$

Making the two values of v equal, which is tantamount to removing speed from the equation, gives:

$$\frac{m}{e} = \frac{B^2 R^2}{2V}.$$

Therefore R is proportional to $\sqrt{m}$, for an ion of a given charge. Allowing for electron charge, elemental mass,[3] and differences in units, we can write:

$$m = \frac{B^2 R^2}{20\,721\ V} \times 10^{12}$$

in which B is in teslas, m in atomic mass units, R in meters, and V in volts.

Exercise

A mass spectrometer has a radius of 0.3 m and an acceleration voltage of 10 000 V. The magnetic field is adjusted to the various masses to be measured. Calculate the atomic mass corresponding to a field of 0.5 T.

Answer

Just apply the formula with suitable units:

$$m = \frac{(0.5)^2 \times (0.3)^2}{20\,721 \times (10^4)} \times 10^{12} = 108.58.$$

Exercise

If hydrogen ions (mass number = 1) are accelerated with a voltage of 10 kV, at what speed are they emitted from the source?

Answer

Just apply the formula $v = (2eV/m)^{\frac{1}{2}}$. The electron charge is $1.60219 \cdot 10^{-19}$ coulombs and the atomic mass unit is $1.660\,540\,2 \cdot 10^{-27}$ kg.

$$v = \sqrt{1.9272 \cdot 10^{12}} = 1388 \text{ km s}^{-1}$$

[3] Atomic mass unit: $m = 1.660\,540\,2 \cdot 10^{-27}$ kg, electron charge: $e = 1.602\,19 \cdot 10^{-19}$ C, therefore $e/m = 0.964 \cdot 10^8$ C kg^{-1} (C = coulomb).

which is about 5 million km per hour. That is fast, admittedly, but still well below the speed of light, which is close to 1 billion km per hour! Heavy ions travel more slowly. For example ions of $m = 100$ would move at just a tenth of the hydrogen speed.

1.2.2 The components of a mass spectrometer

The principal components of a mass spectrometer are the source, the magnet, and the collector and measurement systems, all of which are maintained under vacuum.

The source

The source has three functions:

- To generate ions from atoms. The ions may be positive or negative.
- To accelerate the ion by potential differences created by plates at different potentials (from ground to 20 KeV, and in accelerator mass spectrometers to several MeV).
- To shape the beam, through calibrated slits in the high-voltage plates. The beam from the source slit is usually rectangular.

The magnet

The magnet has two functions. It deviates the ions and this deflection separates them by mass. At the same time it treats the various components of the ion beam or a single mass as an optical instrument would. It handles both colors (masses) and also beam geometry. One of its properties is to focus each ion beam for each mass on the collector. The characteristics of focusing vary with the shape of the magnet and the shape of the pole face, which may be curved in various ways (Figure 1.1 and Plates 2 and 3). A further recent improvement, using computer simulation, has been to focus the beam not only in the x and y directions but in the z direction too. In modern solid-source mass spectrometers, the angular dispersion of ions is fully corrected and almost all the ions leaving the source end up in the collectors which are arranged in a focal plane.

The collectors

The collectors collect and integrate the ion charges so generating an electric current. The collector may be a Faraday bucket, which collects the charges and converts them into a current that is conducted along a wire to an electrical resistor. By measuring the potential difference across the resistor terminals, the current can be calculated from Ohm's law, $V = IR$. The advantage is that it is easy to amplify a potential difference electronically. It is convenient to work with voltages of about 1 V. As the currents to be measured range from 10^{-11} to 10^{-9} A, by Ohm's law, the resistors commonly used range from 10^{11} to 10^{9} Ω. This conversion may be made for small ion fluxes of **electron multipliers** or **ion counters**.[4] In all cases the results are obtained by collecting the ion charges and measuring them. Just ahead of the collector system is a slit that isolates the ion beam. This is explained below.

[4] Each ion pulse is either counted (ion counter) or multiplied by a technical trick of the trade to give a measurable current (electron multiplier).

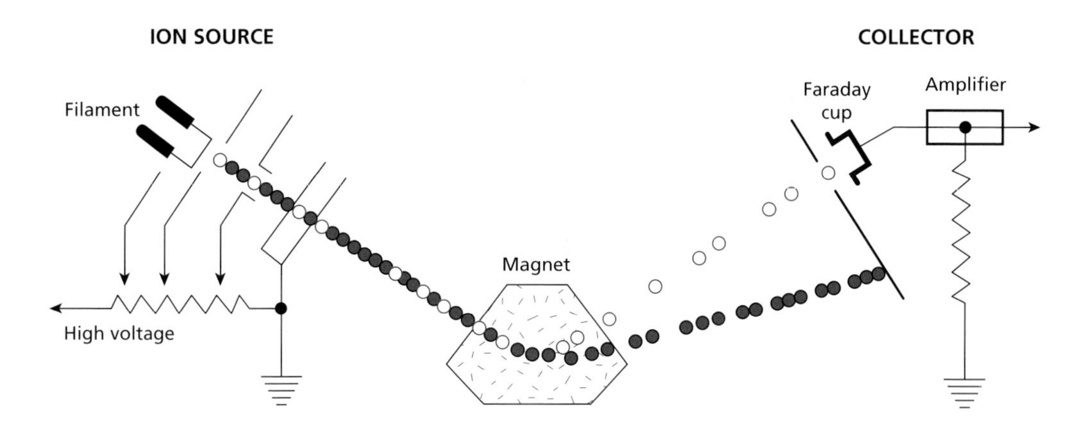

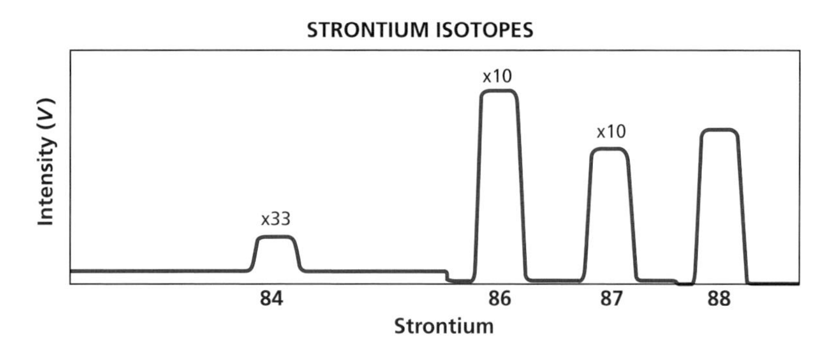

Figure 1.1 A thermal-ionization mass spectrometer. Top: in the center is the electromagnet whose field is perpendicular to the figure and directed downwards (through the page). By Fleming's rules, the force is directed upwards (towards the top of the page) since the stream of ions is coming from the left. An array of Faraday cups may be used for multicollection, that is, for simultaneous measurement of the current produced by each isotope. One important feature has been omitted: the whole arrangement is held in a vacuum of 10^{-7}–10^{-9} mm Hg.[5] Bottom: the mass spectrum of strontium.

The vacuum

A fourth important component is the vacuum. Ions can travel from the source to the collector only if they are in a vacuum. Otherwise they will lose their charge by collision with air molecules and return to the atom state. The whole system is built, then, in a tube where a vacuum can be maintained. In general, a vacuum of 10^{-7} millibars is produced near the source and another vacuum of 10^{-9} millibars or better near the collector. Even so, some air

[5] The SI unit of pressure is the pascal (Pa) but for a time it was measured by the height expressed in centimeters (cm) of a column of mercury in reference to Torricelli's experiment. This unit has been used ever since for measuring extreme vacuums. "Standard" atmospheric pressure of 10^5 Pa corresponds to 76 cm of mercury.

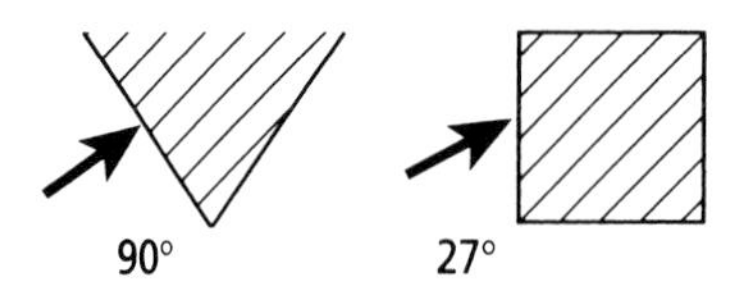

Figure 1.2 (a) Incident beam in the focal plane. (b) Magnet focalization. The beam from the source has a certain aperture. The trajectories of some ions that are not strictly perpendicular to the source are refocused by the magnetic field. The refocusing surface for the various masses at the collector end is curved if the magnet faces are plane, but may be plane if a curved magnet face is used. The figure shows schematically how the magnet separates three isotopes in both configurations.

molecules remain inside the mass spectrometer and collide with the beams, partially disrupting their initial rectangular section.

All of these components contribute to the quality of the data obtained. Mass spectrometer quality is characterized by a number of features.

Efficiency

This is given by the ratio

$$\frac{\text{Number of ions collected}}{\text{Number of atoms in source}} = E$$

$$\text{Number of ions} = \text{intensity} \times \text{duration} \times \frac{6.24 \times 10^{18}}{Z \leftarrow \text{ion charge}}$$

$$\text{Number of atoms} = \frac{\text{mass}}{\text{atomic mass}} \times \text{Avogadro's number.}$$

Efficiency varies with atomic mass.

Ionization efficiency of the source (I) and transmission efficiency of the total ion optics (T):

$$I \times T = E$$

The value of T is variable: 1% for ICP-MS, 25% for ion probes.

The values of I have been greatly improved but vary with the nature of the element and the ionization process. The range is 5‱ to 100% (ICP-MS)!

Power of resolution

The question is, what is the smallest difference in mass that can be separated and then measured using a mass spectrometer? A formal definition is:

$$\text{resolving power } RP = \frac{M_1}{\Delta M}$$

where M_1 is the mass. ΔM is defined as $M_2 = M_1 + \Delta M$, where M_2 is the closest mass to M_1 that does not overlap by more than 50% in the collector.

We can also define a resolving power at 1%.

The distance Δx between two beams in the focal plane is written:

$$\Delta x = K \frac{\Delta m}{m}.$$

Depending on the angle of the incident beam to the magnet, $K = R$ for an angle of incidence of 90°; $K = 2R$ for an angle of 27°.

From the formula above:

$$RP = C \frac{R}{\Delta x},$$

R being the radius of curvature and Δx the distance between two beams of M and $M + \Delta M$.

This is just to show that when one wants to separate two masses more efficiently, the radius has to be increased and then the voltage adjusted accordingly. Suppose we want to separate ^{87}Rb and ^{87}Sr by the difference in mass of neutrons and protons alone. A "monster" mass spectrometer would be required. However, interferences between two masses can be avoided when separating isotopes of an element from contaminating molecules. (Methane $^{12}CH_4$ has the same mass as ^{16}O and benzene C_6H_6 interferes at mass 78 with krypton.)

Abundance sensitivity

Another important characteristic is the Δx distance (in millimeters) between the beams. We have to come back to the slits in the collectors. The problem is easy enough to understand. At first, the beam is rectangular. Collisions with residual air molecules means that, when it reaches the collector slit, the beam is wider and trapezoid-shaped with long tails. Collector slits are open so that they can receive one mass but no contribution from the adjacent mass. When the abundances of two adjacent isotopes are very different, the tail of the more abundant isotope forms background noise for the less abundant one. Measuring the less abundant isotope involves reconstructing the tail of the more abundant one mathematically. This is possible only if the tail is not too big. Narrowing the collector slit brings about a rapid decline in sensitivity.

Abundance sensitivity is the measurement of the contribution of the tail of one isotope to the signal of the neighboring isotope. It is given as a signal/noise ratio multiplied by the mass ratios. Special instruments have been developed for measuring abundance sensitivity in extreme cases, such as measuring ^{14}C close to the massively more abundant ^{12}C. Abundance sensitivity is related to resolving power but also to the quality of the ion optics.

Exercise

The isotopic composition of the element rubidium (Rb) is measured, giving a current $i = 10^{-11}$ A for the mass of ^{87}Rb. How many ions per second is that? If the measurement lasts 1 hour how much Rb has been used if the ionization yield is 1%?

Answer

The intensity of an electrical current is defined by $i = dq/dt$, where dq is the quantity of electrical charge and dt the unit of time. Electrical current is therefore the quantity of electrical charge flowing per unit time. The ampere, the unit of electrical current, is defined as being 1 coulomb per second, the coulomb being the unit of electrical charge. The charge of an electron is $-1.6 \cdot 10^{-19}$ coulombs. The positive charge is identical but with the opposite sign. An intensity of 10^{-11} amps therefore corresponds to 10^{-11} coulombs per second / $1.6 \cdot 10^{-19}$ coulombs $= 62.5 \cdot 10^{6}$ ions per second.

If this intensity is maintained for 1 hour: $6.25 \cdot 10^{7} \times 3600 = 2.2464 \cdot 10^{11}$ ions of $^{87}Rb^{+}$. As the ionization is 1%, this corresponds to $2.2464 \cdot 10^{13}$ atoms of 87^{Rb} placed on the emitter filament. As $^{85}Rb/^{87}Rb = 2.5933$, Rb_{total} (in atoms) $= {}^{87}Rb$ (in atoms) $(1 + 2.5933)$.

So a total number of $8.0719 \cdot 10^{13}$ atoms of Rb is placed on the filament. As the atomic mass of Rb is 85.468 g, the total weight of Rb is 11 ng.

Exercise

How much rock is needed to determine the isotopic composition of Rb by measuring a sample for 20 minutes at 10^{-11} A if its concentration in Rb is 10 ppm (parts per million)?

Answer

If 11 ng of Rb are needed for 1 hour, for 20 minutes we need $(11 \times 20)/60 = 3.66$ ng, that is $3.66 \cdot 10^{-9}/10^{-5} = 0.36$ mg of rock or mineral.

It can be seen, then, that mass spectrometry is a very sensitive technique.

1.2.3 Various developments in mass spectrometers

Mass spectrometers have come a long way since the first instruments developed by J. J. Thomson and F. Aston. To give some idea of the advances made, when Al Nier was measuring lead isotopes as a postdoctoral fellow at Harvard in 1939 (more about this later), he used a galvanometer projecting a beam of light onto the wall and measured the peak with a ruler! Nowadays everything takes the form of a computer output.

Ionization

The first technique was to use the element to be measured as a gaseous compound. When bombarded by electrons, atoms of the gas lose electrons and so become ionized (Nier, 1935, 1938, 1939). Later came the thermal-ionization technique (TIMS) (Inghram and Chupka, 1953). In the so-called solid-source mass spectrometer, a salt of the element is deposited on a metal filament (Ta, W, Pt). The filament is heated by the Joule effect of an increasing electric current. At a certain temperature, the element ionizes (generally as positive ions [Sr, Rb, Sm, Nd, U, Pb] but also as negative ions [Os]). Ionization became a fundamental characteristic of mass spectrometry.

Nowadays, as an alternative, plasma is used for optimal ionization in instruments named ICP-MS.

Ion optics

Substantial effort has been put into optics combining various geometries and assemblies. **Bainbridge** and **Jordan** (1936) used a magnetic field to turn the beam through 180°. **Mattauch** and **Herzog** (1934) combined electric and magnetic fields to separate ions and focus beams. Magnet shapes were modified to improve transmission efficiency.

Computerized numerical simulation has allowed tremendous advances in ion optics design. All of the techniques used tended to maximize ionization and transmission, to increase resolution power and abundance sensitivity, and to minimize the high voltage requirement and the size of the magnet, which are both big factors in cost. However, when the ionization process created a wide dispersion in ion energies, more sophisticated ion optics were required to refocus the ion beam in a narrow band on the collectors. So ICP-MS, ion probe, and AMS instruments have become large and more expensive.

Collectors are another important issue. Early mass spectrometers had a single collector. By scanning the magnetic field, the ion beam passed in sequence through the collector and a spectrum of ion abundance was recorded (Figure 1.1). Nowadays most mass spectrometers use simultaneous ion collection with an array of collectors side by side, each collector corresponding to a distinct mass. This seems an obvious technique to use as it eliminates fluctuations between the recordings of one mass (peak) to another. However, it is technically extremely difficult to achieve, both mechanically, accurately installing several collectors in a small space, and electronically, controlling drifting of the electronic circuits with time. These problems have now been virtually eradicated. It is worth noting that, unlike in most areas of science, all advances since 1980 have been made by industrial engineers rather than by academic scientists. However, because of electronic "noise" and electrical instabilities, all isotopic measurements are statistical. On each run, thousands of spectra are recorded and statistically processed. Only since microcomputers have been available have such techniques become feasible.

1.2.4 Preparatory chemistry and final accuracy

In most mass spectrometry techniques (except for ion probes) chemical separation is used before measurement to purify the element whose isotopic composition is under study. Since the elements to be measured are present as traces, they have to be separated from the major elements which would otherwise prevent any ionization as the major elements rather than the trace elements would give out their electrons. For example, an excess of K inhibits any Rb ionization. Chemical separation also prevents isobaric interference between, say, ^{87}Rb and ^{87}Sr or ^{187}Re and ^{187}Os.

Chemical separation can be done in gaseous form in purification lines, as for rare gases or for oxygen or hydrogen measurement, or in liquid solution for most elements. The basic technique in the latter case is the ion-exchange column as introduced by Aldrich *et al.* (1953). All these operations have to be performed in very clean conditions, otherwise sample contamination will ruin measurement. The greater the accuracy of mass spectrometry, the cleaner the chemistry required. The chemistry is carried out in a clean room with special equipment using specially prepared ultra-clean reagents, that are far cleaner than any commercial versions.

When judging measurement reliability, investigators have to state the level of their blanks. The blank is the amount of the target elements measured in a chemical process done without any sample. The blank has to be negligible or very small compared with the amount of material to be measured. So increases in accuracy are linked not only with the improvement of the mass spectrometer but also of the blanks.

Although this is not the place to give full technical details about conditions for preparing and measuring samples, as these can only be learned in the laboratory and not from textbooks, a few general remarks may be made.

Modern techniques allow isotope ratios to be measured with a degree of precision of 10^{-5} or 10^{-6} (a few ppm!) on samples weighing just a few nanograms (10^{-9} g) or even a few picograms (10^{-12} g). For example, if a rock contains 10 ppm of strontium, its isotope composition can be measured on 10^{-9} g with a degree of precision of 30 ppm. Therefore just 10^{-4} g, that is, 0.1 mg would be needed to make the measurement. This method can be used for studying precious rocks, such as samples of moon rock or meteorites, or minor or rare minerals, that is, minerals that are difficult to separate and concentrate. What do such levels of precision mean? They mean we can readily tell apart two isotope ratios of strontium, say 0.702 21 and 0.702 23, that is, to within 0.000 03, even where low concentrations are involved. To achieve such precision the measurement must be "internally calibrated." When measuring the abundance ratio (A_1/A_2) of two isotopes, the electrical current ratio (I_1/I_2) detected is slightly different from (A_1/A_2). The difference is engendered by the measurement itself. This is termed **mass discrimination**.[6] Either of two methods is used for calibrating measurements.

The first is the internal standard method. If the element has three or more isotopes one particular ratio is chosen as the reference ratio and correction is made for mass discrimination. So if the abundances are A_1, A_2, A_3, we take $(A_1/A_3) = R$. The measurement (I_1/I_2) is written $R(1 + \delta\Delta m)$, where Δm is the difference in mass between A_1 and A_3. The fractionation coefficient δ is calculated and then applied to the measurement of the ratio (A_1/A_2).[7]

The second method is to measure a standard sample periodically and to express the values measured in terms of that standard.

The extraordinary precision the mass spectrometer can achieve must not be jeopardized by accidental contamination when preparing samples. To this end ultra-clean preparatory chemistry is developed using ultra-pure chemical reagents in clean rooms (Plate 3 bottom).

1.2.5 Ionization techniques and the corresponding spectrometers

Four major ionization techniques are used depending on the characteristics of the various chemical elements (ionization potential).

Thermal-ionization mass spectrometry (TIMS)

The element to be analyzed is first purified chemically (especially to separate any isobars) and deposited on a refractory filament. Heating the filament in a vacuum by the Joule effect

[6] Such discrimination depends on the type of mass spectrometer used. It decreases with mass for any given type.

[7] In high-precision mass spectrometry an exponential law rather than a linear one is used to correct mass fractionation.

ionizes the elements, which either lose an electron becoming positive ions, as in the cases of Rb^+, Sr^+, and Pb^+, or gain an electron becoming negative ions as with OsO_3^- and WO_3^-. Instrumental mass fractionation is of the order of 1% by mass deviation for light elements (Li) and 0.1% by mass deviation for heavy elements (Pb, U).

Electronic bombardment

Light elements such as hydrogen (H), carbon (C), nitrogen (N), and oxygen (O) or rare gases are analyzed as gases (H_2, CO_2, N_2, O_2, or atoms of He, Ne, At, or Xe) bombarded in a vacuum by an electron beam. Positive ions are thus formed by stripping an electron from such molecules or atoms. The ions are then accelerated and sorted magnetically as in TIMS. Substances are prepared for analysis in gaseous form by extracting the gas from the sample under vacuum either by fusion or by chemical reaction. The gas is then purified in vacuum lines where other gases are captured either by adsorption or by manipulating their liquefaction temperatures.

Inductively coupled plasma mass spectrometry (ICPMS) in an argon plasma

The sample is ionized in an argon plasma induced by a high-frequency electrical field (plasma torch). The high temperature of the plasma, about 10 000 K, means elements like hafnium or thorium, which are difficult to ionize by thermal emission, can be completely ionized. The element to be analyzed is atomized and then ionized. It is sprayed into the plasma from a solution as a liquid aerosol. Or it may be released from a solid sample by laser ablation. The solid aerosol so formed is injected into the plasma torch. Mass fractionation is between a twentieth of 1% for a light element like boron and 1% for heavy elements. Fractionation is corrected for by using the isotope ratios of other similar elements as internal standards, because, at the temperature of the plasma, fractionation is due to mass alone and is not affected by the element's chemical characteristics.

Ionic bombardment in secondary-ion mass spectrometry (SIMS)

The solid sample (rock, mineral) containing the chemical element for analysis is cut, polished, and put into the vacuum chamber where it is bombarded by a "primary" beam of ions (argon, oxygen, or cesium). This bombardment creates a very-high-temperature plasma at about 40 000 K in which the element is atomized and ionized. The development of high resolution secondary-ion mass spectrometers (**ion microprobes**) means *in-situ* isotope measurements can be made on very small samples and, above all, on tiny grains. This is essential for studying, say, the few grains of interstellar material contained in meteorites.

Remark

All of the big fundamental advances in isotope geology have been the result of improved sensitivity or precision in mass spectrometry or of improved chemical separation reducing contamination (chromatographic separation using highly selective resins, use of high-purity materials such as teflon). These techniques have recently become automated and automation will be more systematic in the future.

1.3 Isotopy

As said, each chemical element is defined by the number of protons Z in its atomic structure. It is the number of protons Z that defines the element's position in the periodic table. But in

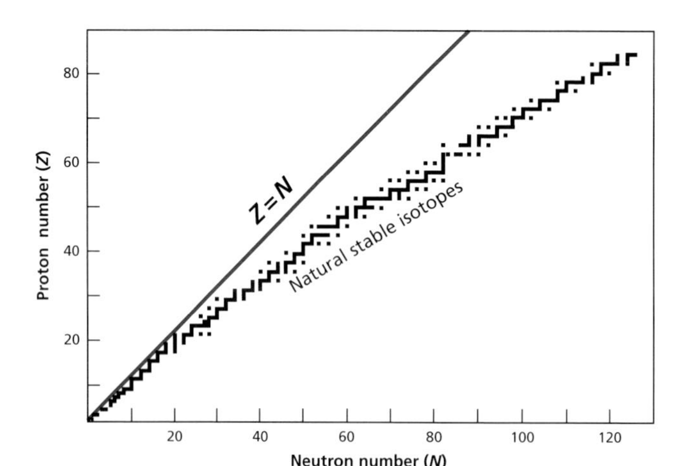

Figure 1.3 The distribution of natural stable isotopes in the neutron–proton diagram. The diagram is stippled because natural or artificial radioactive isotopes lie between the stable isotopes. After $N = 20$, the zone of stable nuclei moves away from the diagonal for which the number of neutrons equals the number of protons. For $N > 20$, the number of neutrons then exceeds the number of protons. This zone is called the valley of stability as it corresponds to a minimum energy level of the nuclides.

each position there are several isotopes which differ by the number of neutrons N they contain, that is, by their mass. These isotopes are created during nuclear processes which are collectively referred to as **nucleosynthesis** and which have been taking place in the stars throughout the history of the Universe (see Chapter 4).

The isotopic composition of a chemical element is expressed either as a percentage or more conveniently as a ratio. A reference isotope is chosen relative to which the quantities of other isotopes are expressed. **Isotope ratios** are expressed in terms of numbers of atoms and not of mass. For example, to study variations in the isotopic composition of the element strontium brought about by the radioactive decay of the isotope ^{87}Rb, we choose the ^{87}Sr/^{86}Sr isotope ratio. To study the isotopic variations of lead, we consider the ^{206}Pb/^{204}Pb, ^{207}Pb/^{204}Pb, and ^{208}Pb/^{204}Pb ratios.

1.3.1 The chart of the nuclides

The isotopic composition of all the naturally occurring chemical elements has been determined, that is, the number of isotopes and their proportions have been identified. The findings have been plotted as a (Z, N) graph, that is, the number of protons against the number of neutrons. Figure 1.3, details of which are given in the Appendix, prompts a few remarks.

First of all, the stable isotopes fall into a clearly defined zone, known as the **valley of stability** because it corresponds to the minimum energy levels of nuclides. Initially this energy valley follows the diagonal $Z = N$. Then, after $N = 20$, the valley falls away from the diagonal on the side of a surplus of neutrons. It is as if, as Z increases, an even greater number of neutrons is needed to prevent the electrically charged protons from repelling each other and breaking the nucleus apart. (Things are actually more complicated than this simplistic image suggests!)

A second remark relates to parity. Elements for which Z is an even number have far more isotopes than elements for which Z is an odd number. Fluorine ($Z=9$), sodium ($Z=11$), phosphorus ($Z=15$), and scandium ($Z=21$) have just a single isotope.

Lastly, and not least importantly, the heaviest element with stable isotopes is lead.[8]

1.3.2 Isotopic homogenization and isotopic exchange

As the isotopes of any given chemical element all have the same electron suite, they all have pretty much the same chemical properties. But in all chemical, physical, or biological processes, isotopes of any given element behave slightly differently from each other, thus giving rise to **isotopic fractionation**. Such fractionation is very weak and is apparent above all in light elements. It is also exploited in isotope geology as shall be seen in Chapter 7.

Initially we shall ignore such fractionation, except where allowance has to be made for it as with ^{14}C or when making measurements with a mass spectrometer where, as has been seen, correction must be made for mass discrimination. This virtually identical behavior of chemical isotopes entails a fundamental consequence in the tendency for **isotopic homogenization** to occur. Where two or more geochemical objects (minerals within the same rock, rocks in solution, etc.) are in thermodynamic equilibrium, the isotope ratios of the chemical elements present are generally equal. If they are unequal initially, they exchange some atoms until they equalize them. It is important to understand that **isotopic homogenization** occurs through isotopic exchange without **chemical homogenization**. Each chemical component retains its chemical identity, of course. This property of isotopic homogenization "across" chemical diversity is one of the fundamentals of isotope geochemistry. A simple way of observing this phenomenon is to put calcium carbonate powder in the presence of a solution of carbonate in water in proportions corresponding to thermodynamic equilibrium. Therefore no chemical reaction occurs. Repeat the experiment but with radioactive ^{14}C in solution in the form of carbonate. If after 10 days or so the solid calcium carbonate is isolated, it will be found to have become radioactive. It will have exchanged some of its carbon-14 with the carbonate of mass 12 and 13 which were in the solution.

Exercise

A liter of water saturated in $CaCO_3$ whose Ca^{2+} content is $1 \cdot 10^{-2}$ moles per liter is put in the presence of 1 g of $CaCO_3$ in solid form. The isotopic ratio of the solid $CaCO_3$ is $^{40}Ca/^{42}Ca = 151$. The isotopic ratio of the dissolved Ca^{2+} has been artificially enriched in ^{42}Ca such that $^{40}Ca/^{42}Ca = 50$. What is the common isotopic composition of the calcium when isotopic equilibrium is achieved?

Answer

$^{40}Ca/^{42}Ca = 121.2$.

As said, when two or more geochemical objects with different isotopic ratios are in each other's presence, atom exchange (which occurs in all chemical reactions, including at

[8] Until recently it was believed to be bismuth ($Z=83$), whose only isotope is ^{209}Bi. In 2003 it was discovered to be radioactive with a half-life of $1.9 \cdot 10^{19}$ years!

equilibrium) tends to homogenize the whole in terms of isotopes. This is known as **isotopic exchange**. It is a kinetic phenomenon, depending therefore on the temperature and physical state of the phases present. Simplifying, isotope exchange is fast at high temperatures and slow at low temperatures like all chemical reactions which are accelerated by temperature increase. In liquids and gases, diffusion is fast and so isotope exchange is fast too. In solids, diffusion is slow and so isotope exchange is slow too. In magmas (high-temperature liquids), then, both trends are compounded and isotope homogenization occurs quickly. The same is true of supercritical fluids, that is, fluids deep within the Earth's crust. Conversely, in solids at ordinary temperatures, exchange occurs very slowly and isotope heterogeneities persist. Two important consequences follow from these two properties. The first is that a magma has the same isotope composition as the solid source from which it has issued by fusion, but not the same chemical composition. The second is that, conversely, a solid at ordinary temperatures retains its isotopic composition over time without becoming homogeneous with its surroundings. This is why **rocks are reliable isotope records**. This property is a direct consequence of the diffusion properties of natural isotopes in liquids and solids.

The theory of diffusion, that is, the spontaneous motion of atoms influenced by differences in concentration, provides an approximate but adequate formula:

$$x \approx \sqrt{Dt}$$

where x is the distance traveled by the element, t is time in seconds, and D the diffusion coefficient ($cm^2\,s^{-1}$).

Exercise

In a liquid silicate at 1200 K the diffusion coefficient for elements like Rb, Sr, or K is $D = 10^{-6}\,cm^2\,s^{-1}$. In solid silicates heated to 1200 K, $D = 10^{-11}\,cm^2\,s^{-1}$.

How long does it take for two adjacent domains of 1 cm diameter to become homogeneous:

(1) within a silicate magma?

(2) between a silicate magma and a solid, which occurs during partial melting when 10% of the magma coexists with the residual solid?

Answer

(1) In a silicate magma if $D = 10^{-6}\,cm^2\,s^{-1}$, $t \approx x^2/D = 10^6$ s, or about 11 days.
If it takes 11 days for the magma to homogenize on a scale of 1 cm, on a 1-km scale ($= 10^5$ cm), it will take $t \approx x^2/D = 10^{10} = 10^{16}$ s, or close to $3 \cdot 10^8$ years, given that 1 year $\approx 3 \cdot 10^7$ s.
In fact, homogenization at this scale would not occur by diffusion but by advection or convection, that is, a general motion of matter, and so would be much faster.

(2) In the case of a magma impregnating a residual solid with crystals of the same dimensions (1 cm), $t \approx x^2/D = 10^{11}$ s or about $3 \cdot 10^5$ years, or 300 000 years, which is rather fast in geological terms. For 1-mm crystals, which is more realistic, $t \approx 10^{-2}/D = 3 \cdot 10^3$ years, or 3000 years. So isotope equilibrium is established quite quickly where a magma is in the presence of mineral phases.

A second important question is whether rocks at ordinary temperatures can retain their isotope compositions without being modified and without being re-homogenized. To answer this, it must be remembered that the diffusion coefficient varies with temperature by the Arrhenius law:

$$D = D_0 \exp\left(\frac{-\Delta H}{RT}\right)$$

where ΔH is the activation energy, which is about 40 kcal per mole, R is the ideal gas constant (1.987 cal per K per mole), and T the absolute temperature (K). If $D = 10^{-11}\,\text{cm}^2\,\text{s}^{-1}$ in solids at 1300 °C, what is the diffusion coefficient D_{or} at ordinary temperatures?

$$\frac{D_{\text{or}}}{D_{1300}} = \exp\left[\frac{-\Delta H}{R}\left(\frac{I}{T_{\text{or}}} - \frac{I}{T_{1300}}\right)\right] = \langle B \rangle$$

from which $D_{\text{or}} = \langle B \rangle\, D_{1300}$.

Calculate $\langle B \rangle$ to find that it gives $1.86 \cdot 10^{-25}$, therefore $D_{\text{or}} = 2 \cdot 10^{-36}\,\text{cm}^2\,\text{s}^{-1}$.

To homogenize a 1-mm grain at ordinary temperatures takes

$$t = \frac{10^{-2}}{2 \cdot 10^{-36}} = 0.5 \cdot 10^{34}\,\text{s} \approx 1.5 \cdot 10^{26} \text{ years},$$

which is infinitely long to all intents and purposes.

Important remark

Rocks therefore retain the memory of their history acquired at high temperatures. This is the prime reason isotope geology is so incredibly successful and is the physical and chemical basis of isotope memory. The phenomenon might be termed isotopic quenching, by analogy with metal which, if it is immersed when hot in cold water, permanently retains the properties it acquired at high temperature.

1.3.3 A practical application of isotopic exchange: isotopic dilution

Suppose we wish to measure the rubidium content of a rock. Rubidium has two isotopes, of mass 85 and 87, in the proportion $^{85}\text{Rb}/^{87}\text{Rb} = 2.5933$. (This is the value found when measuring the Rb isotope composition of natural rocks.) The rock is dissolved with a mixture of acids. To the solution obtained, we add a solution with a known Rb content which has been artificially enriched in ^{85}Rb (spike), whose $^{85}\text{Rb}/^{87}\text{Rb}$ ratio in the spike is known. The two solutions mix and become isotopically balanced. Once equilibrium is reached, the **absolute** Rb content of the rock can be determined by simply measuring the **isotope composition** of a **fraction** of the mixture.

Writing $\left(\frac{^{85}\text{Rb}}{^{87}\text{Rb}}\right) = \left(\frac{85}{87}\right)$ gives:

$$\left(\frac{85}{87}\right)_{\text{mixture}} = \frac{(^{85}\text{Rb})_{\text{rock}} + (^{85}\text{Rb})_{\text{spike}}}{(^{87}\text{Rb})_{\text{rock}} + (^{87}\text{Rb})_{\text{spike}}} = \frac{\left(\frac{85}{87}\right)_{\text{spike}} + \left(\frac{85}{87}\right)_{\text{rock}}\left[\frac{^{87}\text{Rb}_{\text{rock}}}{^{87}\text{Rb}_{\text{spike}}}\right]}{\left[1 + \frac{^{87}\text{Rb}_{\text{rock}}}{^{87}\text{Rb}_{\text{spike}}}\right]}.$$

Top and bottom are divided by ^{87}Rb, bringing out $\left(\frac{85}{87}\right)$.

A little manipulation gives:

$$^{87}\text{Rb}_{\text{rock}} = {}^{87}\text{Rb}_{\text{spike}}\left(\frac{85}{87}\right)_{\text{spike}}\left[\frac{\left(\frac{85}{87}\right)_{\text{spike}} - \left(\frac{85}{87}\right)_{\text{mixture}}}{\left(\frac{85}{87}\right)_{\text{mixture}} - \left(\frac{85}{87}\right)_{\text{rock}}}\right].$$

Because we know the ^{87}Rb content of the spike, the Rb content of the rock can be obtained by simply measuring the isotope ratio of the mixture, without having to recover all the Rb by chemical separation.

Suppose, say, the isotopic composition of the spike of Rb is $^{85}\mathrm{Rb}/^{87}\mathrm{Rb} = 0.12$. The naturally occurring $^{85}\mathrm{Rb}/^{87}\mathrm{Rb}$ ratio is 2.5933. We dissolve 1 g of rock and add to the solution 1 g of spike containing 3 ng of ^{87}Rb per cubic centimeter. After thoroughly mixing the solution containing the dissolved rock and the spike solution, measurement of a fraction of the mixture yields a ratio of $^{85}\mathrm{Rb}/^{87}\mathrm{Rb} = 1.5$. The Rb content of the rock can be calculated. We simply apply the formula:

$$^{87}\mathrm{Rb}_{\mathrm{rock}} = {}^{87}\mathrm{Rb}_{\mathrm{spike}}\left[\frac{0.12 - 1.5}{1.5 - 2.59}\right] = 1.266,$$

or

$$^{87}\mathrm{Rb}_{\mathrm{rock}} = 1.266 \times {}^{87}\mathrm{Rb}_{\mathrm{spike}} = 1.266 \times 3 \cdot 10^{-9}\,\mathrm{g} = 3.798 \cdot 10^{-9}\,\mathrm{g}.$$

As we took a sample of 1 g of rock, $C_{^{87}\mathrm{Rb}} = 3.798\ \mathrm{ng\ g^{-1}} = 3.798\,\mathrm{ppb}$. Therefore $C_{87_{\mathrm{total}}} = C_{^{87}\mathrm{Rb}}(1 + 2.5933) = 13.42\,\mathrm{ppb}$.

This method can be used for all chemical elements with several stable isotopes for which spikes have been prepared that are artificially enriched in one or more isotopes and for elements with a single isotope, provided it is acceptable to use a radioactive isotope (and so potentially dangerous for whoever conducts the experiment) as a spike. The method has three advantages.

First, after mixing with the spike, chemical separation methods need not be entirely quantitative. (The yields of the various chemical operations during analysis do not count.) Isotope ratios alone matter, as well as any contamination, which distorts the measurement, of course.

Then, as the mass spectrometer makes very sensitive and very precise measurements of isotope ratios, isotopic dilution may be used to measure the amounts of trace elements, even the tiniest traces, with great precision.

Isotope dilution was invented for the needs of laboratory analysis but may be extended to natural processes. As shall be seen, variable isotope ratios occur in nature. Mixes of them can be used to calculate proportions by mass of the geochemical elements involved just by simple measurements of isotope ratios.

As can be seen, isotope dilution is an essential method in isotope geochemistry. But just how precise is it? This exercise will allow us to specify the error (uncertainty) in isotope dilution measurement.

Exercise

The isotope ratios of the spike, sample, and mixture are denoted R_T, R_S, and R_M, respectively. We wish to find out the quantity X of the reference isotope C_j in the sample. To do this, quantity Y of spike has been mixed and the isotope ratios $(C_i/C_j) = R$ of the spike, sample, and mix have been measured. What is the uncertainty of the measurement?

Answer

Let us begin with the formula

$$X = Y\left|\frac{R_T - R_M}{R_M - R_S}\right|$$

which may be written $R_M X - R_S X = R_T Y - R_M Y$ or $R_M (X + Y) = R_T Y + R_S X$, or alternatively

$$R_M - R_T \frac{Y}{X+Y} + R_S \frac{X}{X+Y}.$$

We posit $\frac{X}{X+Y} = W$ and $Y = 1 - W$. Let us calculate the logarithmic derivative and switch to Δ (finite difference):

$$\frac{\Delta X}{X} = \frac{\Delta Y}{Y} + \frac{\Delta(R_T - R_M)}{(R_T - R_M)} + \frac{\Delta(R_M - R_S)}{(R_M - R_S)}.$$

We can transform $(R_T - R_M)$ and $(R_M - R_S)$ as a function of $(R_T - R_S)$, from which:

$$\frac{\Delta X}{X} = \frac{\Delta Y}{Y} + \frac{2\Delta R_M}{(R_T - R_S)}\left[\frac{1}{W} + \frac{1}{1-W}\right] = \frac{\Delta Y}{Y} + \frac{2\Delta R_M}{(R_T - R_S)}\left[\frac{1}{W(1-W)}\right].$$

Neglecting the uncertainty on R_T and R_S, which are assumed to be fully known, uncertainty is minimum when:

- $R_T - R_S$ is maximum. A spike must therefore be prepared whose composition is as remote as possible from the sample composition.
- $1/[w(1-w)]$ is maximum for given values of R_T and R_S, that is, when $W = 0.5$, in other words when the samples and spike are in equal proportions.

By way of illustration, let us plot the curve of relative error $\Delta X/X$ as a function of W. It is assumed that $\Delta R_M/(R_T - R_M) = 10^{-4}$ and $\Delta Y/Y = 10^{-4}$.

The curve is shown in Figure 1.4.

Conversely, the formulae for isotope dilution show how contamination of a sample by reagents used in preparatory chemistry modifies the isotope composition of a sample to be

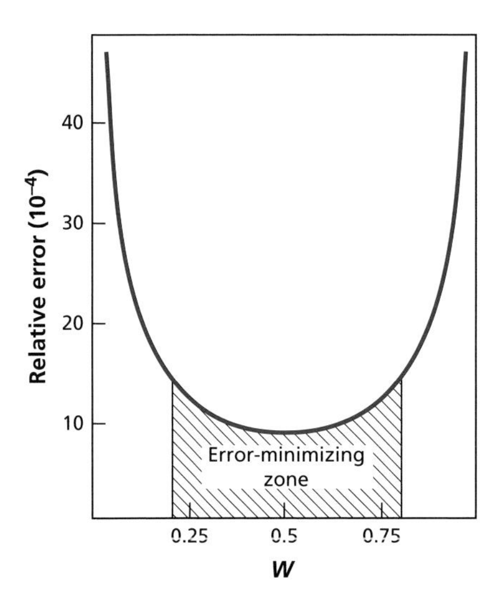

Figure 1.4 Relative error due to isotope dilution. Relative error is plotted as a function of the ratio W, which is the proportion of the isotope from the sample in the sample–spike mixture. The greatest precision is achieved with comparable amounts of spike and sample, but with a relatively large tolerance for this condition.

measured. To evaluate this uncertainty (error) (or better still to make it negligible), isotope dilution is used to gauge the quantity of the element to be measured that has been introduced accidentally during preparation. To do this, a blank measurement is made with no sample. The blank is the quantity of contamination from the preparatory chemistry. A good blank has a negligible influence on measurement. See Problem 3 at the end of the chapter for more on this.

1.4 Radioactivity

Radioactivity was discovered and studied by **Henri Becquerel** and then **Pierre** and **Marie Curie** from 1896 to 1902. In 1902 **Pierre Curie** (1902a) and independently **Ernest Rutherford** and **Frederick Soddy** (1902a, b, c) proposed an extremely simple mathematical formalization for it.

1.4.1 Basic principles

Radioactivity is the phenomenon by which certain nuclei transform (transmute) spontaneously into other nuclei and in so doing give off particles or radiation to satisfy the laws of conservation of energy and mass described by **Albert Einstein**. The **Curie–Rutherford–Soddy (CRS) law** says that the number of nuclei that disintegrate per unit time is a constant fraction of the number of nuclei present, regardless of the temperature, pressure, chemical form, or other conditions of the environment. It is written:

$$\frac{dN}{dt} = -\lambda N$$

where N is the number of nuclei and λ is a proportionality constant called the **decay constant**. It is the probability that any given nucleus will disintegrate in the interval of time dt. It is expressed in yr^{-1} (reciprocal of time).

The expression λN is called the **activity** and is the number of disintegrations per unit time. Activity is measured in **curies** ($1\ Ci = 3.7 \cdot 10^{10}$ disintegrations per second, which is the activity of 1 g of ^{226}Ra). A value of 1 Ci is a very high level of activity, which is why the millicurie or microcurie are more generally used. The international unit is now the **becquerel**, corresponding to 1 disintegration per second. 1 Ci = 37 gigabecquerels.

This law is quite strange a priori because it seems to indicate that the nuclei "communicate" with each other to draw by lots those to be "sacrificed" at each instant at an unchanging rate. And yet it has been shown to be valid for nuclei with very short (a few thousandths of a second) or very long (several billion years, or more than 10^{20} s) lifespans. It holds whatever the conditions of the medium. Whether the radioactive isotope is in a liquid, solid, or gas medium, whether heated or cooled, at high pressure or in a vacuum, the law of decay remains unchanged. For a given radioactive nucleus, λ remains the same over the course of time. Integrating the Curie–Rutherford–Soddy law gives:

$$N = N_0 e^{-\lambda t}$$

where N is the number of radioactive nuclei now remaining, N_0 the initial number of radioactive atoms, and t the interval of time measuring the length of the experiment. Thus the

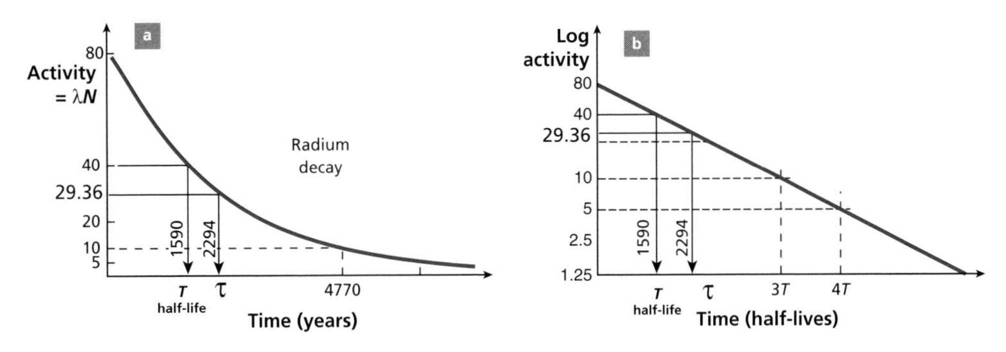

Figure 1.5 Curves of the radioactive decay of radium established by Pierre and Marie Curie. The activity curve is shown with normal coordinates (a) and semi-logarithmic coordinates (b). Both plots show the half-life, that is, the time taken for half of the atoms to disintegrate, and the mean life, that is, the reciprocal of λ.

number of radioactive atoms remaining is a function of just the initial number of radioactive atoms and of time.

Each radioactive isotope is characterized by its **decay constant** λ. We also speak of the **mean life** $\tau = 1/\lambda$. The equation is then written:

$$N = N_0 e^{-(t/\tau)}.$$

Radioactivity is therefore a stopwatch, a natural clock, which, like an hourglass, measures the passage of time unperturbed. The phenomenon can be displayed graphically in two forms.

On an (N, t) graph, the negative exponential decreases becoming tangential to the x-axis at infinity (Figure 1.5a). On a semi-log $(\ln N, t)$ graph, as $\ln N = N_0 - \lambda t$, the curve describing decay is a straight line of slope $-\lambda$ (Figure 1.5b).

To characterize the speed with which the "nuclear hourglass" empties in a less abstract way than by the decay constant λ, the **half-life** (T) (also written $T_{\frac{1}{2}}$) of a radioactive element is defined by the time it takes for half the radioactive isotope to disintegrate. From the fundamental equation of radioactivity we have: $\ln (N_0/N) = \ln 2 = \lambda T$, from which:

$$T_{\frac{1}{2}} = \ln 2/\lambda = 0.693\ \tau,$$

where $T_{\frac{1}{2}}$ is the half-life, λ the radioactive constant, and τ the mean life.

The half-life (like the mean life) is expressed in units of time, in thousands, millions, or billions of years.[9] It can be used to evaluate, in a simple way, the speed at which any radioactive isotope decays. Reviewing these half-lives, it is observed that while some are very brief, a millionth of a second (or even less), others are measured in thousands and in billions of years. This is the case of ^{238}U or ^{87}Rb and other isotopes we shall be using. This observation immediately prompted **Pierre Curie** in 1902 and independently **Ernest Rutherford** and **Frank Soddy** to think that geological time could be measured using radioactivity. This was

[9] Care is required because tables may give either the half-life or the mean life.

probably the most important discovery in geology since **Hutton** in 1798 had laid down its foundations from field observations.

Exercise

Given that the decay constant of ^{87}Rb is $\lambda = 1.42 \cdot 10^{-11}$ yr^{-1} and that there are 10 ppm of Rb in a rock, how much Rb was there 2 billion years ago?

Answer

We have seen that Rb is composed of two isotopes of masses 85 and 87 in the ratio $^{85}\text{Rb}/^{87}\text{Rb} = 2.5933$. The atomic mass of Rb is therefore:

$$\frac{85 \times 2.5933 + 87 \times 1}{3.5933} = 85.556.$$

In 10 ppm, that is, $10 \cdot 10^{-6}$ by mass, there is $\frac{10 \times 10^{-6}}{85.556} = 0.116 \cdot 10^{-6}$ mole of Rb. There is therefore $\frac{0.116 \cdot 10^{-6} \times 1}{3.5933} = 0.032\,282 \cdot 10^{-6}$ mole of ^{87}Rb and $\frac{0.116 \times 10^{-6} \times 2.5933}{3.5933} = 0.083\,717 \cdot 10^{-6}$ atom g^{-1} of ^{85}Rb.

$N = N_0\, e^{-\lambda t}$, therefore with $\lambda = 1.42 \cdot 10^{-11}$ yr^{-1} and $t = 2 \cdot 10^9$ yr, $e^{-\lambda t} = 0.97199$.

Therefore, 2 billion years ago there was 0.032 282 / 0.971 99 = $0.033\,212\,2 \cdot 10^{-6}$ mole of ^{87}Rb.

The isotopic composition of ^{87}Rb was 85/87 = 2.5206, or a variation of 2.8% relative to the current value in isotopic ratio, which is not negligible.

Exercise

The ^{14}C method is undoubtedly the most famous method of radioactive dating. Let us look at a few of its features that will be useful later. It is a radioactive isotope whose half-life is 5730 years. For a system where, at time $t = 0$, there are 10^{-11} g of ^{14}C, how much ^{14}C is left after 2000 years? After 1 million years? What are the corresponding activity rates?

Answer

We use the fundamental formula for radioactivity $N = N_0\, e^{-\lambda t}$. Let us first, then, calculate N_0 and λ. The atomic mass of ^{14}C is 14. In 10^{-11} g of ^{14}C there are therefore $10^{-11}/14 = 7.1 \cdot 10^{-13}$ atoms per gram (moles) of ^{14}C. From the equation $\lambda T = \ln 2$ we can calculate $\lambda_C = 1.283 \cdot 10^{-4}$ yr^{-1}.[10]

By applying the fundamental formula, we can write:

$N = 7.1 \cdot 10^{-13} \exp(-1.283 \cdot 10^4 \times 2000) = 5.492 \cdot 10^{-13}$ moles.

After 2000 years there will be $5.492 \cdot 10^{-13}$ moles of ^{14}C and so $7.688 \cdot 10^{-12}$ g of ^{14}C.

After 10^6 years there will remain $1.271 \cdot 10^{-68}$ moles of ^{14}C and so $1.7827 \cdot 10^{-67}$ g, which is next to nothing.

In fact there will be no atoms left because $\frac{1.271 \cdot 10^{-68}\ \text{moles}}{6.02 \cdot 10^{23}} \approx 2 \cdot 10^{-44}$ atoms!

The number of disintegrations per unit time dN/dt is equal to λN.

The number of atoms is calculated by multiplying the number of moles by Avogadro's number $6.023 \cdot 10^{23}$. This gives, for 2000 years, $5.4921 \cdot 10^{-13} \times 6.023 \cdot 10^{23} \times 1.283 \cdot 10^{-4} = 4.24 \cdot 10^7$ disintegrations per year. If 1 year $\approx 3 \cdot 10^7$ s, that corresponds to 1.4 disintegrations per second (dps), which is measurable.

[10] This value is not quite exact (see Chapter 4) but was the one used when the method was first introduced.

For 10^6 years, $1.27 \cdot 10^{-68} \times 6.023 \cdot 10^{23} \times 1.283 \cdot 10^{-4} = 9.7 \cdot 10^{-49}$ disintegrations per year. This figure shows one would have to wait for an unimaginable length of time to observe a single disintegration! (10^{48} years for a possible disintegration, which is absurd.)

This calculation shows that the geochronometer has its limits in practice! Even if the ^{14}C content was initially 1 g (which is a substantial amount) no decay could possibly have been detected after 10^6 years!

This means that if radioactivity is to be used for dating purposes, the half-life of the chosen form of radioactivity must be appropriate for the time to be measured.

Exercise

We wish to measure the age of the Earth with ^{14}C, the mean life of which is 5700 years. Can it be done? Why?

Answer

No. The surviving quantity of ^{14}C would be too small. Calculate that quantity.

1.4.2 Types of radioactivity

Four types of radioactivity are known. Their laws of decay all obey the **Curie–Rutherford–Soddy** formula.

Beta-minus radioactivity

The nucleus emits an electron spontaneously. As **Enrico Fermi** suggested in 1934, the neutron disintegrates spontaneously into a proton and an electron. To satisfy the law of conservation of energy and mass, it is assumed that the nucleus emits an antineutrino along with the electron. The decay equation is written:

$$\mathrm{n} \rightarrow \mathrm{p} + \beta^- + \bar{\nu}$$

neutron → proton + electron + antineutrino

To offset the (+) charge created in the nucleus, the atom captures an electron and so "moves forwards" in the periodic table:

$$^{A}_{Z}\mathrm{A} \rightarrow {}_{Z+1}^{\;\;A}\mathrm{B} + \mathrm{e}^- + \bar{\nu}.$$

In the (Z, N) diagram, the transformation corresponds to a diagonal shift up and to the left. For example, ^{87}Rb decays into ^{87}Sr by this mechanism (see Figure 1.6).

We write, then:

$$^{87}\mathrm{Rb} \rightarrow {}^{87}\mathrm{Sr} + \beta^- + \bar{\nu}.$$

This long-lived radioactivity is very important in geochemistry. Its decay constant is $\lambda = 1.42 \cdot 10^{-11}\,\mathrm{yr}^{-1}$. Its half-life is $T_{\frac{1}{2}} = 49 \cdot 10^9$ years.

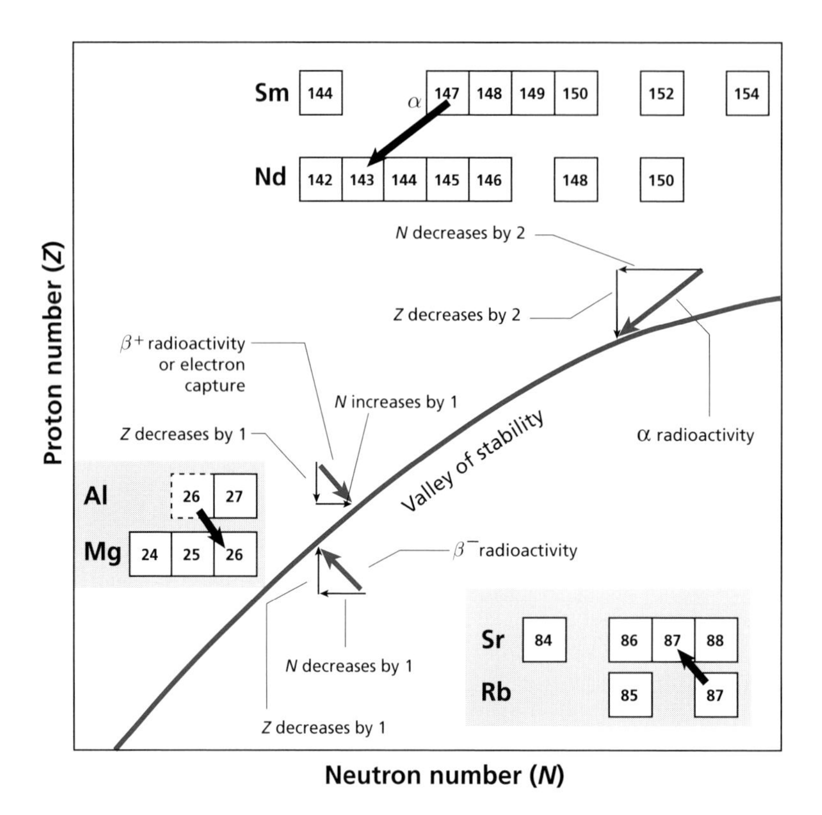

Figure 1.6 The various types of radioactivity in the neutron–proton diagram. Notice that all forms of disintegration shift the decay products towards the valley of stability. Radioactivity seems to restore the nuclear equilibrium of nuclides lying outside the valley of stability and so in disequilibrium.

Beta-plus radioactivity and electron capture

The nucleus emits a positron (anti-electron) at the same time as a neutrino. A proton disintegrates into a neutron. A similar but different process is **electron capture** by a proton.

$$p + e^- \rightarrow n + \nu$$

proton + electron → neutron + neutrino

The atom emits a peripheral electron to ensure the nuclide remains neutral.

$${}^{A}_{Z}A \rightarrow {}_{Z-1}^{A}B + e^+ + \nu \qquad \beta^+ \text{ radioactivity}$$

or

$${}^{A}_{Z}A + e^- \rightarrow {}_{Z-1}^{A}B + \nu \qquad \text{electron capture.}$$

This is represented in the (Z, N) diagram by a diagonal shift down and to the right. Notice that neither of these forms of radioactivity involves a change in mass number. It is said to be **isobaric radioactivity**. For example, potassium-40 (^{40}K) decays into argon-40 (^{40}Ar):

$$^{40}K + e^- \rightarrow {}^{40}Ar + \nu.$$

This is a very important form of radioactivity for geologists and geochemists. Its radioactive constant is $\lambda^{40}{}_K = 0.581 \cdot 10^{-10}\ yr^{-1}$ and its half-life $T_{\frac{1}{2}} = 1.19 \cdot 10^{10}$ years.[11] We shall be looking at it again.

Alpha radioactivity

The radioactive nucleus expels a helium nucleus ^4_2He (in the form of He^+ ions) and heat is given off. The radiogenic isotope does not have the same mass as the parent nucleus. By conservation of mass and charge, the decay equation can be written:

$$^A_Z A \rightarrow {}^{A-4}_{Z-2}B + ^4_2 He.$$

In the (N, Z) diagram, the path is the diagonal of slope 1 shifting down to the left. For example, samarium-147 (^{147}Sm) decays into neodymium-143 (^{143}Nd) by the decay scheme:

$$^{147}Sm \rightarrow {}^{143}Nd + {}^4_2He$$

with $\lambda = 6.54 \cdot 10^{-12}\ yr^{-1}$ and $T_{\frac{1}{2}} = 1.059 \cdot 10^{11}$ years.

This form of decay has played an important historical role in the development of isotope geology and we shall be using it on many occasions.

Spontaneous fission

Fission is a chain reaction caused by neutrons when they have sufficient energy. The elementary reaction splits a uranium nucleus into two unequal parts – for example a krypton nucleus and a xenon nucleus, a bromine nucleus and an iodine nucleus – and many neutrons. These neutrons in turn strike other uranium atoms and cause new fission reactions, and neutron reactions on the nuclei formed by fission. This is "statistical break-up" of uranium atoms into two parts of unequal masses. The nucleus that splits does not always produce the same nuclei but a whole series of pairs. Figure 1.7 shows the abundance of the various isotopes produced by spontaneous fission of ^{238}U.

Notice that the last two types of radioactivity (α and fission) break up the nucleus. They are called partition radioactivity. Remember that spontaneous fission too obeys the mathematical (CRS) law of radioactivity.

EXAMPLE

The Oklo natural reactor

The isotope ^{238}U undergoes spontaneous fission while ^{235}U is subject to fission induced by the impact of neutrons. Both these forms of fission occur naturally.

[11] This is for disintegration of ^{40}K into ^{40}Ar. ^{40}K also disintegrates giving ^{40}Ca, as shall be seen later.

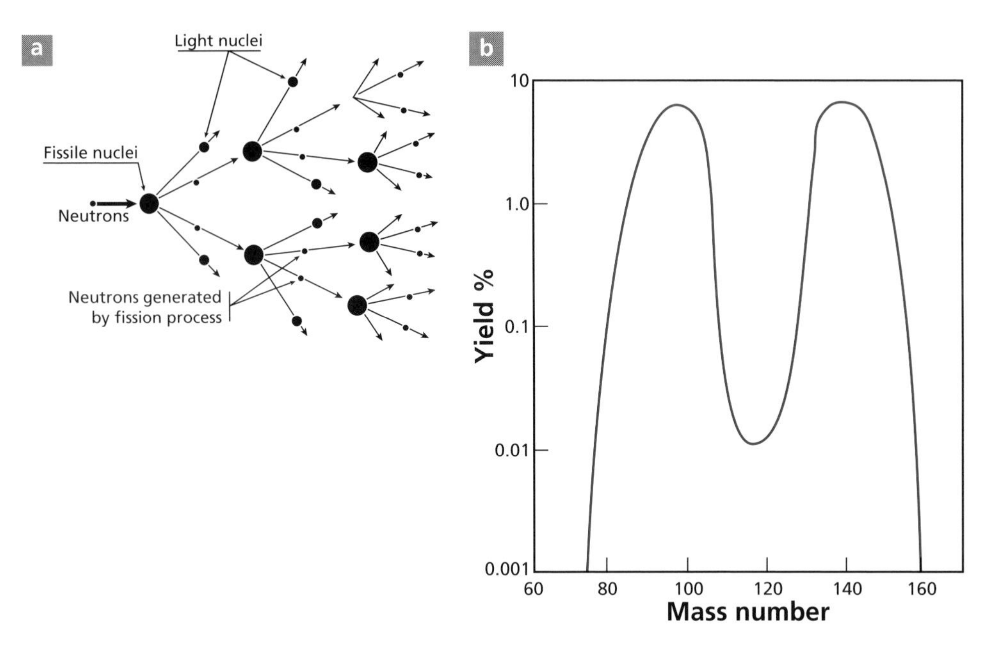

Figure 1.7 Spontaneous fission: (a) chain reactions multiply the number of neutrons as the reaction unfolds; (b) the curve of the distribution of fission products as a function of their mass number.

Spontaneous fission of ^{238}U has an extremely low decay constant $\lambda = 8.62 \cdot 10^{-17}$ yr^{-1}. Induced fission of ^{235}U is a reaction produced in the laboratory or in nuclear reactors by bombarding uranium with neutrons.

In 1973, a natural nuclear reactor some 2 billion years old was discovered in the Oklo uranium mine in Gabon. This uranium deposit worked like an atomic pile, that is, with induced fission of ^{235}U. Apart from a negative anomaly in the abundance of ^{235}U, the whole series of fission-induced products corresponding to this isotope was detected. This fission of ^{235}U, which was believed to be confined to laboratories or industrial nuclear reactors, therefore occurred naturally, probably triggered by α disintegration of ^{235}U, which was much more abundant at the time. Nature had discovered nuclear chain reactions and atomic piles some 2 billion years before we did! Oklo is a unique example to date.

Exercise

Given that the $^{238}U/^{235}U$ ratio nowadays is 137.8, what was the activity level of ^{235}U per gram of ore 2 billion years ago for a uranium ore that today contains 30% uranium?

The decay constants are $\lambda_{238} = 0.155\,125 \cdot 10^{-9}$ yr^{-1} and $\lambda_{235} = 0.9885 \cdot 10^{-9}$ yr^{-1}.

Answer

The activity of ^{235}U was 1247 disintegrations per second per gram (dsg). Today the activity of ^{235}U is 172 dsg.

Exercise

What types of radioactivity are involved in the following very important reactions in cosmochronology and geochronology: $^{146}Sm \rightarrow {}^{142}Nd$, $^{53}Mn \rightarrow {}^{53}Cr$, $^{230}Th \rightarrow {}^{226}Ra$?

Answer

See Chapter 2, Section 2.4.3.

1.4.3 Radioactivity and heat

Each form of radioactive decay is associated with the emission of particles or γ electromagnetic radiation. Interaction of this radiation with the material surrounding the radioactive isotope creates heat, as **Pierre Curie** and **Albert Laborde** realized in 1903, just 7 years after **Becquerel**'s discovery. This heat is exploited in nuclear reactors to generate electricity. Inside the Earth, the radioactivity of ^{40}K, ^{238}U, ^{235}U, and ^{232}Th is one of the main sources of internal energy, giving rise to plate tectonics and volcanism and to the heat flow measured at the surface. In the early stages of the Earth's history, this radioactive heat was greater than today because the radioactive elements ^{40}K, ^{238}U, ^{235}U, and ^{232}Th were more abundant.[12]

A LITTLE HISTORY

The age of the Earth

In the mid nineteenth century, when Joseph Fourier had just developed the theory of heat propagation, the great British physicist William Thomson (Lord Kelvin)[13] had been studying how the Earth cooled from measures of heat flow from its interior. He had come to the conclusion that the Earth, which was assumed to have been hot when it first formed, could not be more than 40–100 million years old. That seemed intuitively too short to many geologists, particularly to Charles Lyell, one of the founders of geology, and also to an obscure naturalist by the name of Charles Darwin. Lyell had argued for the existence of an unknown heat source inside the Earth, which Kelvin, of course, dismissed as unscientific reasoning! It was more than 50 years before Pierre Curie and Laborde in 1903 measured the heat given off by the recently discovered radioactivity and Rutherford could redo Kelvin's calculations and prove Lyell right by confirming his intuition. See Chapter 5 for more historical information on the age of the Earth.

Exercise

Heat emissions in calories per gram and per second of some isotopes are:[14]

^{238}U	^{235}U	^{40}K	^{232}Th
$2.24 \cdot 10^{-8}$	$1.36 \cdot 10^{-7}$	$6.68 \cdot 10^{-9}$	$6.44 \cdot 10^{-9}$

[12] At the time there were other radioactive elements such as ^{26}Al which have now disappeared but whose effects compounded those listed.

[13] See Burchfield (1975) for an account of Kelvin's work on the age of the Earth.

[14] These values include heat given off by all isotopes of radioactive chains associated with ^{238}U, ^{235}U, and ^{232}Th (see Chapter 2).

Calculate how much heat is given off by 1 g of peridotite of the mantle and 1 g of granite given that $^{40}K = 1.16 \cdot 10^{-4}\ K_{total}$; $^{238}U/^{235}U = 137.8$; Th/U = 4 for both materials; and that the mantle contains 21 ppb U and 260 ppm K and that the granite contains 1.2 ppm U and $1.2 \cdot 10^{-2}$ K.

Answer

Calculation of heat given off by 1 g of natural uranium:

$$0.992\,79 \times 2.24 \cdot 10^{-8} + 0.007\,20 \times 1.36 \cdot 10^{-7} = 2.32 \cdot 10^{-8}\ \text{cal g}^{-1}\ \text{s}^{-1}.$$

Calculation of the heat given off by 1 g of potassium:

$$6.68 \cdot 10^{-9} + 1.16 \cdot 10^{-4} = 7.74 \cdot 10^{-13}\ \text{cal g}^{-1}\ \text{s}^{-1}.$$

Calculation for the mantle:

$$\underbrace{(2.32 \times 10^{-8} \times 21 \times 10^{-9})}_{\text{uranium}} + \underbrace{(6.44 \times 10^{-9} \times 84 \times 10^{-9})}_{\text{thorium}} + \underbrace{(7.74 \times 10^{-13} \times 2.60 \times 10^{-4})}_{\text{potassium}} = (48.7 + 54 + 20) \cdot 10^{-17}$$
$$= 0.1227 \cdot 10^{-14}\ \text{cal g}^{-1}\text{s}^{-1}.$$

To convert this result into SI units, 1 calorie = 4.18 joules and 1 watt = 1 joule per second. Therefore 1 g of peridotite of the mantle gives off $0.512 \cdot 10^{-14}$ W s^{-1}. Calculation for granite:

$$[2.32 \cdot 10^{-8} \times 1.2 \cdot 10^{-6}] + [6.44 \cdot 10^{-9} \times 4.8 \cdot 10^{-6}]$$
$$+ [7.74 \cdot 10^{-13} \times 1.2 \cdot 10^{-2}] = 2.78 \cdot 10^{-14} + 3.09 \cdot 10^{-14} + 0.928 \cdot 10^{-14}$$
$$= 6.79 \cdot 10^{-14}\ \text{cal g}^{-1}\text{s}^{-1}.$$

1 g of granite gives off $28.38 \cdot 10^{-14}$ W.

It can be seen that today the two big contributors are ^{238}U and ^{232}Th; ^{40}K contributes less and ^{235}U is non-existent. The granite produces 55 times more heat than the mantle peridotite.

Exercise

The decay constants of ^{238}U, ^{235}U, ^{232}Th, and ^{40}K are $\lambda_{238} = 0.155\,125 \cdot 10^{-9}\ \text{yr}^{-1}$, $\lambda_{235} = 0.984\,85 \cdot 10^{-9}\ \text{yr}^{-1}$, $\lambda_{232} = 0.049\,47 \cdot 10^{-9}\ \text{yr}^{-1}$, and $\lambda_K = 0.5543 \cdot 10^{-9}\ \text{yr}^{-1}$, respectively. Calculate heat production 4 billion years ago for the peridotite of the mantle and the granite of the continental crust.

Answer

Total heat production H can be written:

$$\begin{aligned} H = {} & 0.9927 \times C_0^U \times P_{238} \exp(0.155\,125\,T) \\ & + 0.007\,20 \times C_0^U \times P_{235} \exp(0.984\,85\,T) \\ & + C_0^{Th} \times P_{232} \exp(0.049\,47\,T) \\ & + 1.16 \cdot 10^{-4}\ C_0^K \times P_{K40}(0.5543\,T). \end{aligned}$$

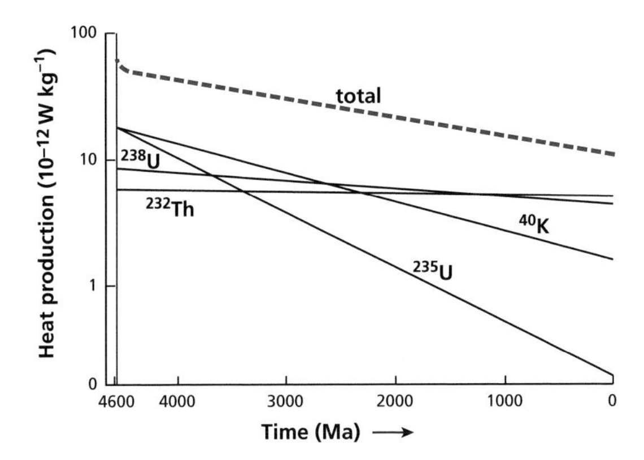

Figure 1.8 Heat production by various forms of radioactivity in the Earth versus geological age.

With $T = 4 \cdot 10^9$ years, the result for the mantle is: $8.68 \cdot 10^{-16} + 10.56 \cdot 10^{-16} + 6.59 \cdot 10^{-16} + 18.4 \cdot 10^{-16} = 44.23 \cdot 10^{-16}$ cal g^{-1} s^{-1} $= 1.84 \cdot 10^{-14}$ W g^{-1} s^{-1}.
For granite of the continental crust: $4.96 \cdot 10^{-14} + 6.03 \cdot 10^{-14} + 3.76 \cdot 10^{-14} + 8.537 \cdot 10^{-14} = 23.28 \cdot 10^{-14}$ cal g^{-1} s^{-1} or $97.3 \cdot 10^{-14}$ W g^{-1} s^{-1}.

Notice that, at the present time, radioactive heat is supplied above all by the disintegration of ^{238}U and to a lesser extent ^{40}K and ^{232}Th. Four billion years ago heat was supplied mainly by ^{40}K and ^{235}U (Figure 1.8). It will be observed, above all, that 4 billion years ago the mantle produced 3.5 times as much heat as it does today. So it may be thought that the Earth was 3.5 times more "active" than today.

Problems

1 Which molecules of simple hydrocarbons may interfere after ionization with the masses of oxygen ^{16}O, ^{17}O, and ^{18}O when measured with a mass spectrometer? How can we make sure they are absent?

2 The lithium content of a rock is to be measured. A sample of 0.1 g of rock is collected. It is dissolved and 2 cm^3 of lithium spike added with a lithium concentration of $5 \cdot 10^{-3}$ moles per liter and whose isotope composition is ^{6}Li/^{7}Li $= 100$. The isotope composition of the mixture is measured as ^{6}Li/^{7}Li $= 10$.

Given that the isotopic composition of natural lithium is ^{6}Li/^{7}Li $= 0.081$, what is the total lithium content of the rock?

3 A sample contains 1 μg of strontium. What must be the maximum acceptable chemical blank, that is, the quantity with which the sample is accidentally contaminated, if precision of measurement with the mass spectrometer of ^{87}Sr/^{86}Sr $\sim$0.7030 bears on $\pm$ 0.0001?

The ^{87}Sr/^{86}Sr ratio of the blank is 0.7090. What must the blank be if precision is 10 times better?

4 We are to construct a mass spectrometer for separating ^{87}Rb from ^{87}Sr. What should its radius be?

5 Suppose that 1 mg of purified uranium has been isolated. It contains two (main) isotopes ^{238}U and ^{235}U in the current proportions of $^{238}U/^{235}U = 137.8$. What was the total activity of this milligram of uranium 4.5 billion years ago and what is its activity today if $\lambda_{238} = 0.155\,125 \cdot 10^{-9}\ yr^{-1}$ and $\lambda_{235} = 0.9875 \cdot 10^{-9}\ yr^{-1}$?

6 The Urey ratio is the ratio of heat from radioactivity to total heat which includes heat from the accretion of the Earth and the formation of its core. The average heat flow measured at the Earth's surface is $4.2 \cdot 10^{13}$ W, which is 42 terawatts.

(i) In a first hypothesis the mantle composition is assumed to be uniform, with:

$U = 21$ ppb $(Th/U)_{mass} = 4$ and $K = 210$ ppm.

Calculate the Urey ratio today.

(ii) In a second hypothesis, it is assumed that the entire mantle is similar to the upper mantle. The upper mantle has a uranium content of 5 ppb, and the Th/U ratio is 2. Calculate the Urey ratio.

CHAPTER TWO

The principles of radioactive dating

It can never be repeated enough that radioactive dating was the greatest revolution in the geological sciences. Geology is an historical science which cannot readily be practised without a precise way of measuring time. It is safe to say that no modern discovery in geology could have been made without radioactive dating: reversals of the magnetic field, plate tectonics, the puzzle of the extinction of the dinosaurs, lunar exploration, the evolution of life, human ancestry, not to mention the age of the Earth or of the Universe!

The ages involved in the earth sciences are very varied. They are measured in years (yr), thousands of years (ka), millions of years (Ma), and billions of years (Ga). Geological clocks must therefore be varied too, with mean lives ranging from a year to a billion years.

2.1 Dating by parent isotopes

Imagine we have a radioactive isotope R and N_R^* is the number of atoms of this isotope. Suppose that geological circumstances (crystallization of a rock or mineral, say) enclose an initial quantity of R, i.e., the number of atoms of R at time zero, written $N_R^*(0)$, in a "box." If the box has remained closed from when it first formed until today, the number of atoms of R remaining is $N_R^*(t) = N_R^*(0)e^{-\lambda t}$, where t is the time elapsed since the box was closed. If we know the quantity $N_R^*(0)$ and the decay constant λ, by measuring $N_R^*(t)$ we can calculate **the age *t* at which the box closed** by using the radioactivity formula "upside down":

$$t = \frac{1}{\lambda} \ln \left(\frac{N_R^*(0)}{N_R^*(t)} \right).$$

Methods where the initial quantities of radioactive isotopes are well enough known are above all those where the radioactive isotope is produced by irradiation by cosmic rays. This is the case of carbon-14 (^{14}C) and beryllium-10 (^{10}Be).

Exercise

The half-life of ^{14}C is 5730 years. The ^{14}C content of the atmosphere is 13.2 disintegrations per minute and per gram (dpm g^{-1}) of carbon (initial activity A_0). We wish to date an Egyptian artefact dating from approximately 2000 BC. What is the approximate activity (A) of this artefact? If our method can measure 1 dpm, what mass of the (probably precious) sample will have to be destroyed?

Answer

If the half-life $T=5730$ years, the decay constant is $\lambda = \ln 2/T = 1.209 \cdot 10^{-4}\ \mathrm{yr}^{-1}$. 2000 BC corresponds to a time 4000 years before the present, therefore since $A_0 = 13.2\ \mathrm{dpm\,g}^{-1}$ and $A = A_0\,e^{-\lambda t}$, $A = 7\ \mathrm{dpm\,g}^{-1}$.

Making the measurement means using at least 1/7 g, or 142 mg of the sample.

As seen in the examples, the abundance of a radioactive isotope is estimated relative to a reference. For ^{14}C we use total carbon. The dating formula is therefore:

$$t = \frac{1}{\lambda}\ln\left[\frac{(^{14}C/C)_0}{(^{14}C/C)_t}\right]$$

where (^{14}C) and (C) represent the concentrations of ^{14}C and total carbon, respectively.

In other cases, a neighboring stable isotope that is not subject to radioactive decay is taken as the reference. So for ^{14}C, we use stable ^{13}C and we write:

$$t = \frac{1}{\lambda}\ln\left[\frac{(^{14}C/^{13}C)_0}{(^{14}C/^{13}C)_t}\right].$$

This formulation has the advantage of bringing out isotopic ratios, that is, the ratios measured directly by mass spectrometry.

2.2 Dating by parent–daughter isotopes

2.2.1 Principle and general equations

The difficulty with dating by the parent isotope is of course knowing $N_R^*(0)$, that is, knowing exactly how many radioactive atoms were trapped in the box at the beginning. This difficulty is overcome by involving the stable daughter isotope produced by the disintegration noted (D) (the asterisk being a reminder of the radioactive origin of the isotope R*). The parent–daughter relation is written:

$$(\mathrm{R})^* \underset{\lambda}{\rightarrow} (\mathrm{D}).$$

From the Curie–Rutherford–Soddy law, we can write:

$$\frac{\mathrm{d}N_R^*(t)}{\mathrm{d}t} = -\lambda\ N_R^*(t)$$

$$\frac{\mathrm{d}N_D(t)}{\mathrm{d}t} = -\frac{\mathrm{d}N_R^*(t)}{\mathrm{d}t} = \lambda\ N_R^*(t).$$

Integrating the first equation yields the decay law, $N_R^*(t) = N_R^*(0)\ e^{-\lambda t}$. The second is therefore written:

$$\frac{\mathrm{d}N_D(t)}{\mathrm{d}t} = \lambda\ N_R^*(0)e^{-\lambda t},$$

which integrates to:

$$N_{\mathrm{D}}(t) = -N^*_{\mathrm{R}}(0)\mathrm{e}^{-\lambda t} + C.$$

The integration constant C is determined by writing in $t=0$, $N_{\mathrm{D}}=N_{\mathrm{D}}(0)$, hence: $C = N_{\mathrm{D}}(0) + N^*_{\mathrm{R}}(0)$. This gives:

$$N_{\mathrm{D}}(t) = N_{\mathrm{D}}(0) + N^*_{\mathrm{R}}(0)(1 - \mathrm{e}^{-\lambda t}).$$

But this expression leaves the troublesome unknown $N^*_{\mathrm{R}}(0)$. This is advantageously replaced by:

$$N^*_{\mathrm{R}}(0) = N^*_{\mathrm{R}}(t)\ \mathrm{e}^{\lambda t}.$$

This gives:

$$N_{\mathrm{D}}(t) = N_{\mathrm{D}}(0) + N^*_{\mathrm{R}}(t)(\mathrm{e}^{\lambda t} - 1).$$

If the box remains closed for both the radioactive isotope and the radiogenic isotope, by measuring the present values $N_{\mathrm{D}}(t)$ and $N_{\mathrm{R}}(t)$, we can calculate t, provided we know $N_{\mathrm{D}}(0)$. This can be plotted as $(N_{\mathrm{D}}(t), t)$. The slope of the curve at each point equals:

$$\frac{\mathrm{d}N_{\mathrm{D}}(t)}{\mathrm{d}t} = -\frac{\mathrm{d}N^*_{\mathrm{R}}(t)}{\mathrm{d}t} = \lambda N^*_{\mathrm{R}}(t) = \lambda N^*_{\mathrm{R}}(0)\mathrm{e}^{-\lambda t}.$$

It therefore equals $\lambda N_{\mathrm{R}}(t)$, at λ times the parent isotope content. As this content decays constantly, the slope of the tangent does likewise, and the curve is concave downwards. To calculate an age, we write the dating equation:

$$t = \frac{1}{\lambda}\ln\left\{\left[\frac{N_{\mathrm{D}}(t) - N_{\mathrm{D}}(0)}{N^*_{R}(t)}\right] + 1\right\}.$$

The values of $N_{\mathrm{D}}(t)$ and $N^*_{\mathrm{R}}(t)$ can be measured, but t can only be calculated if $N_{\mathrm{D}}(0)$ can be estimated or ignored and, of course, if we know the decay constant λ. Figure 2.1 illustrates all these points.

EXAMPLE

Rubidium–strontium dating

Let us take rubidium–strontium dating by way of an example. As seen, ^{87}Rb decays to ^{87}Sr with a decay constant $\lambda = 1.42 \cdot 10^{-11}$ yr^{-1}. The parent–daughter dating equation is written:

$$t = \frac{1}{\lambda_{\mathrm{Rb}}}\ln\left[\frac{^{87}\mathrm{Sr}(t) - {}^{87}\mathrm{Sr}(0)}{^{87}\mathrm{Rb}(t)} + 1\right],$$

where ^{87}Sr(0) is the quantity of ^{87}Sr present at time $t=0$, and ^{87}Rb(t) and ^{87}Sr(t) the quantities of ^{87}Rb and ^{87}Sr present at time t. (The term *quantity* must be understood here as the number of atoms or atom–grams.) Notice that time can be reversed and the present time considered as the starting point, which is more practical. The initial time is then in the past, age (t) such that $T = t$.

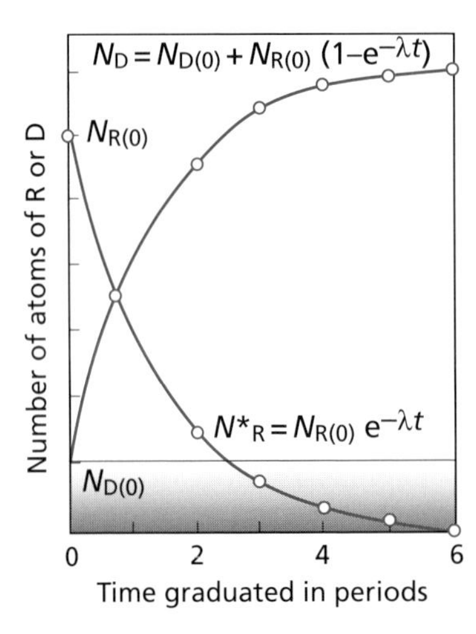

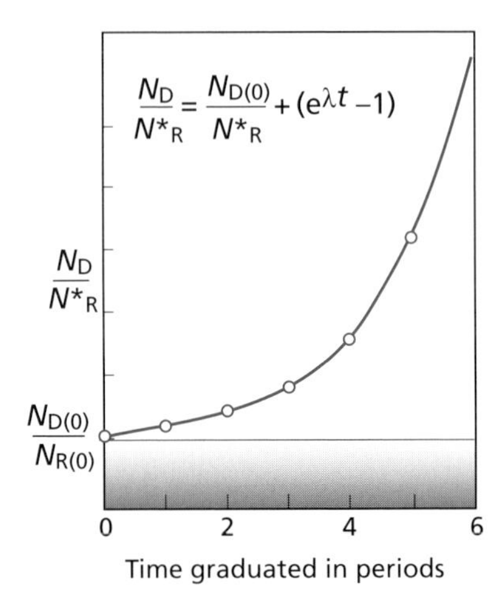

Figure 2.1 Left: the decrease in the radioactive isotope and the increase in the radiogenic isotope. Right: the increase in the radiogenic/radioactive isotope ratio.
N^*_R number of atoms of the radioactive isotope (R)
N_D number of atoms of the radiogenic isotope (D)
$N_R(0)$ number of atoms of R at time $t = 0$
$N_D(0)$ number of atoms of D at time $t = 0$
λ radioactive decay constant

The equation is then written:

$$T = \frac{1}{\lambda} \ln \left[\frac{{}^{87}\mathrm{Sr(p)} - {}^{87}\mathrm{Sr}(T)}{{}^{87}\mathrm{Rb(p)}} + 1 \right],$$

where ^{87}Sr(p) and ^{87}Rb(p) relate to the present-day values (p = present), and ^{87}Sr(T) relates to the initial values at time (T). When dealing with minerals that are very rich in Rb such as biotite and muscovite, the initial ^{87}Sr is negligible relative to the ^{87}Sr produced by decay of ^{87}Rb. For such systems, which are said to be **radiogenically rich** (or just **rich** for short), the dating formula is written:

$$T = \frac{1}{\lambda} \ln \left[\frac{{}^{87}\mathrm{Sr(p)}}{{}^{87}\mathrm{Rb(p)}} + 1 \right].$$

This formula can be extended to rich systems in general:

$$T = \frac{1}{\lambda} \ln \left[\frac{N_D(p)}{N_R(p)} + 1 \right].$$

As seen, if λ is known, the age can be calculated directly from measurements of the present-day abundances of N_D(p) and N_R(p). The only assumption made, but which is crucial, is that the box to be dated, that is, the mineral or rock, has remained **closed** ever since the time it formed and that closure was short compared with the age to be measured. This is indeed the case when a mineral crystallizes or a magma solidifies as with volcanic or plutonic rock.

Remark

Where λ is very small relative to the age, the exponential can be approximated by $e^{\lambda t} \sim 1 + \lambda$. This is the case for the constant λ of rubidium ($\lambda = 1.42 \cdot 10^{-11}\ \mathrm{yr}^{-1}$) and for many others. The dating formula is then:

$$T \approx \frac{1}{\lambda}\left[\frac{^{87}\mathrm{Sr(p)}}{^{87}\mathrm{Rb(p)}}\right].$$

Exercise

Suppose we have a specimen of biotite from Quérigut granite (Pyrénées Orientales, France) whose age is to be determined from $^{87}\mathrm{Rb} \to {}^{87}\mathrm{Sr}$ decay. We measure the total content of Rb = 500 ppm and of Sr = 0.6 ppm. Knowing that Rb is composed of ^{87}Rb and ^{85}Rb in the proportions $^{85}\mathrm{Rb}/^{87}\mathrm{Rb} = 2.5$, that the Sr is "pure" radiogenic ^{87}Sr, and that the decay constant is $\lambda = 1.42 \cdot 10^{-11}\ \mathrm{yr}^{-1}$, calculate the age of the biotite in Quérigut granite.

Answer

The Rb content is written $\mathrm{Rb_{total}} = {}^{85}\mathrm{Rb} + {}^{87}\mathrm{Rb} = (2.5 + 1)\ {}^{87}\mathrm{Rb}$, therefore:

$$^{87}\mathrm{Rb} = \frac{\mathrm{Rb_{total}}}{3.5} \approx 142.8\ \mathrm{ppm}.$$

The ^{87}Sr content is 0.6 ppm. (There is no need to come back to atom–grams since ^{87}Rb and ^{87}Sr have virtually the same atomic mass.) We can therefore write directly:

$$T \approx \frac{1}{1.42 \cdot 10^{-11}}\left(\frac{0.6}{142.8}\right) \approx 298 \cdot 10^{6}\ \mathrm{yr} = 298\ \text{million years}.$$

If we had not adopted the linear approximation, we would have obtained:

$$T \approx \frac{1}{1.42 \cdot 10^{-11}} \ln\left(\frac{0.6}{142.8} + 1\right) \approx 297.36 \cdot 10^{6}\ \mathrm{yr}.$$

As can be seen, the linear approximation is valid for the $^{87}\mathrm{Rb}/^{87}\mathrm{Sr}$ system when the age is not too great.

2.2.2 Special case of multiple decay

Let us consider the case of the long-period ^{40}K radioactive isotope which decays in two different ways, either by electron capture giving ^{40}Ar or by β^- decay giving ^{40}Ca, each with its own decay constant λ_e and λ_{β^-}, respectively:

$$^{40}\mathrm{K} \xrightarrow{e^{-}\mathrm{cap}} {}^{40}\mathrm{Ar}, \qquad ^{40}\mathrm{K} \xrightarrow{\beta^{-}} {}^{40}\mathrm{Ca}$$

What is the dating formula? Let us get back to basics:

$$d/dt[N_{\mathrm{K}}(t)] = -(\lambda_e + \lambda_\beta)N_{\mathrm{K}}(t)$$

where N_K is the number of ^{40}K nuclides and N_{Ar} the number of ^{40}Ar nuclides.

$$d/dt[N_{Ar}(t)] = \lambda_e N_K(0) e^{-(\lambda_e+\lambda_\beta)t}.$$

Integrating using the usual notation with ^{40}K and $^{40}Ar_0$ gives:

$$^{40}Ar(t) = {}^{40}Ar_0 + {}^{40}K(t)\frac{\lambda_e}{\lambda_e+\lambda_\beta}\left(e^{(\lambda_e+\lambda_\beta)t} - 1\right)$$

with $\lambda_e = 0.581 \cdot 10^{-10}\,yr^{-1}$ and $\lambda_\beta = 4.962 \cdot 10^{-10}\,yr^{-1}$.

$$\lambda = \lambda_e + \lambda_\beta = 5.543 \cdot 10^{-10}\,yr^{-1}.$$

The initial ^{40}Ar is usually negligible. We are therefore generally dealing with **rich systems** but not with very young systems where what is known as **excess argon** raises difficulties for accurate age calculations.

Exercise

We analyze 1 g of biotite extracted from Quérigut granite by the ^{40}K–^{40}Ar method. The biotite contains 4% K and $^{40}K = 1.16 \cdot 10^{-4}$ of K_{total}.

The quantity of argon measured at standard temperature and pressure is $4.598 \cdot 10^{-5}\,cm^3$ of ^{40}Ar. What is the radiometric age of this biotite?

Answer

The dating formula to calculate the age is written:

$$T = \frac{1}{\lambda_e + \lambda_b} \ln\left[\frac{^{40}Ar}{^{40}K} \cdot \left(\frac{\lambda_e + \lambda_b}{\lambda_e}\right) + 1\right].$$

We must therefore calculate the $^{40}Ar/^{40}K$ ratio in atoms.

As there are 22 400 cm^3 in a mole at standard temperature and pressure, the value of ^{40}Ar in number of moles is:

$$\left(\frac{4.598 \cdot 10^{-5}}{22\,400}\right) = 2.053 \cdot 10^{-9} \text{ moles of } {}^{40}Ar.$$

The value of ^{40}K is:

$$\frac{0.04 \times 1.16 \cdot 10^{-4}}{40} = 1.16 \cdot 10^{-7} \text{ moles of } {}^{40}K.$$

Therefore:

$$T = \frac{10^9}{0.5543} \ln\left[\left(\frac{2.053 \cdot 10^{-9}}{1.16 \cdot 10^{-7}} \times 0.1048\right) + 1\right] = 280 \text{ million years.}$$

Comparing this with the result obtained previously using the ^{87}Rb–^{87}Sr method, we find about the same age but slightly younger.

There is another branched decay used in geology, that of ^{138}La which decays into ^{138}Ba and ^{138}Ce (Nakaï *et al.*, 1986).

$$^{138}\text{La} \xrightarrow{e^-\text{cap}} {}^{138}\text{Ba} \quad \lambda_{\text{Cap } e^-} = 4.44 \cdot 10^{-12}\,\text{yr}^{-1}$$

$$^{138}\text{La} \xrightarrow{\beta^-} {}^{138}\text{Ce} \quad \lambda_\beta = 2.29 \cdot 10^{-12}\,\text{yr}^{-1}$$

There is also a case where even more intense multiple decay occurs, in the spontaneous fission of ^{238}U, which yields a whole series of isotopes. Luckily, fission decay is negligible compared with α decay. In the dating equation we can consider the constant λ_α alone as the decay constant, but allowance must be made, of course, for the fission products. For example, for ^{136}Xe we write the dating equation:

$$\frac{^{136}\text{Xe}_{\text{radio}}}{^{238}\text{U}} = Y \frac{\lambda_{\text{fission}}}{\lambda_\alpha}\left(e^{\lambda_\alpha t} - 1\right)$$

where Y is the yield of ^{136}Xe produced during fission $\sim$0.0673, with $\lambda_{\text{fission}} = 8.47 \cdot 10^{-17}\,\text{yr}^{-1}$ and $\lambda_\alpha = 1.55 \cdot 10^{-10}\,\text{yr}^{-1}$.

We saw when looking at dating by parent isotopes that it was convenient to express the dating equation by introducing isotope ratios rather than moles of radioactive and radiogenic isotopes. This is called normalization. Thus for ^{87}Rb–^{87}Sr we use a stable strontium isotope, ^{86}Sr. The dating equation is then written:

$$t = \frac{1}{\lambda}\ln\left[\left(\frac{\frac{^{87}\text{Sr}}{^{86}\text{Sr}}(t) - \frac{^{87}\text{Sr}}{^{86}\text{Sr}}(0)}{\frac{^{87}\text{Rb}}{^{86}\text{Sr}}}\right) + 1\right].$$

This is the form in which dating equations will be expressed from now on. A system will be rich when:

$$\frac{^{87}\text{Sr}}{^{86}\text{Sr}}(0) \ll \frac{^{87}\text{Sr}}{^{86}\text{Sr}}(t).$$

2.2.3 Main geochronometers based on simple parent–daughter ratios

- **Rubidium–Strontium** $^{87}\text{Rb}\ \beta^{-}\ {}^{87}\text{Sr}$; $\lambda = 1.42 \cdot 10^{-11}\,\text{yr}^{-1}$. The normalization isotope is ^{86}Sr. Developed in its modern form by **Aldrich** *et al.* (1953).
- **Potassium–Argon** ^{40}K–^{40}Ar, with the constants already given. The reference isotope is ^{36}Ar. Developed in its modern form by **Aldrich** and **Nier** (1948a).
- **Rhenium–Osmium** $^{187}\text{Re}\ \beta^{-}\ {}^{187}\text{Os}$ with $\lambda = 1.5 \cdot 10^{-11}\,\text{yr}^{-1}$. The reference isotope is ^{186}Os and more recently ^{188}Os. Developed by **Luck**, **Birck**, and **Allègre** in 1980 after a first attempt by **Hirt** *et al.* in 1963.

These are the three simple clocks that are commonly found as rich systems in nature. We shall see that other forms of decay can be used but under more difficult circumstances.

- **Samarium–Neodymium** $^{147}\text{Sm}\ \alpha\ {}^{143}\text{Nd}$; $\lambda = 6.54 \cdot 10^{-12}\,\text{yr}^{-1}$. Normalization by ^{144}Nd. Developed by **Lugmair** and **Marti** in 1977 after an attempt by **Notsu** *et al.* in 1973.

- **Lutetium–Hafnium** $^{176}Lu \; \beta^{-\,176}Hf$; $\lambda = 2 \cdot 10^{-11}\ yr^{-1}$. Normalization with ^{177}Hf. Developed by **Patchett** and **Tatsumoto** (1980a, 1980b).

These methods are supplemented by others related to radioactive chains which are to be examined next, by the extinct radioactive methods covered in Section 2.4, and by the induced radioactive methods examined in Chapter 4.

2.3 Radioactive chains

2.3.1 Principle and general equations

When ^{238}U, ^{235}U, and ^{232}Th decay, they give rise to three other radioactive isotopes which, in turn, decay into new radioactive elements, and so on. The process stops when the last isotopes produced are the three lead isotopes ^{206}Pb, ^{207}Pb, and ^{208}Pb, which are stable. It was radioactive chains which allowed both **Pierre** and **Marie Curie** and **Rutherford** and **Soddy** to discover the mechanisms of radioactivity. A radioactive chain can be represented by writing:

$$(1)^* \rightarrow (2)^* \rightarrow (3)^* \rightarrow (4)^* \rightarrow \ldots \rightarrow (n) \text{ stable.}$$

Decay involves α and β radioactivity. Alpha radioactivity gives off helium nuclei. Their path, in the (Z, N) plot, brings the end product into the valley of stability.

Figure 2.2 shows the precise structure of the three chains. Mathematically, as studied by **Bateman** (1910), radioactive chains can be described by the Curie–Rutherford–Soddy laws written one after the other:

$$\frac{dN_1(t)}{dt} = -\lambda_1 N_1(t) \qquad \frac{dN_4(t)}{dt} = \lambda_3 N_3(t) - \lambda_4 N_4(t)$$

$$\frac{dN_2(t)}{dt} = \lambda_1 N_1(t) - \lambda_2 N_2(t) \qquad \vdots$$

$$\frac{dN_3(t)}{dt} = \lambda_2 N_2(t) - \lambda_3 N_3(t) \qquad \frac{dN_n}{dt} = \lambda N_{n-1}(t)$$

where N_i is the number of nuclides of elements i and N_n the number of nuclides of the final stable isotope. Successive integration of these equations presents no real difficulty and is even a good revision exercise for integrating first-order differential equations with constant coefficients. Let us leave that for now and concentrate on a few simple and important limiting cases.

2.3.2 Secular equilibrium: uranium–lead methods

We suppose that, in view of the length of geological time, the radioactive chain reaches a stationary state where the content of all the intermediate radioactive isotopes remains constant (this is known as secular equilibrium):

$$\frac{dN_2}{dt} = \frac{dN_3}{dt} = \cdots = \frac{dN_{n-1}}{dt} = 0.$$

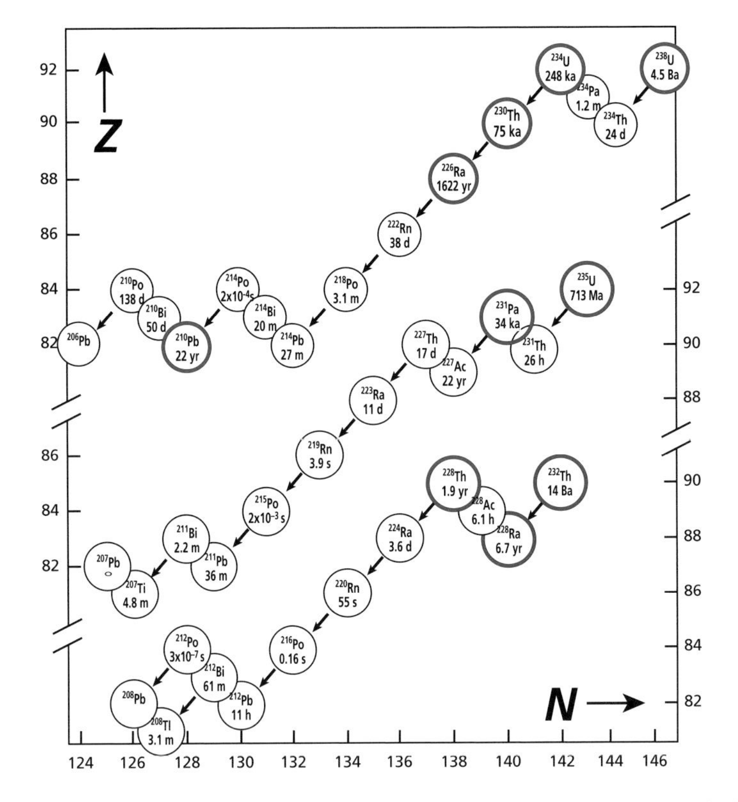

Figure 2.2 Radioactive chains represented on the proton–neutron graph. Each slot contains the symbol of the isotope and its period $T_{\frac{1}{2}}$. Identify the various types of radioactivity of the three chains for yourself.

In this case, we therefore have:

$$\lambda_1 N_1 = \lambda_2 N_2 = \lambda_3 N_3 = \cdots = \lambda_{n-1} N_{n-1},$$

therefore:

$$\frac{\mathrm{d}N_n}{\mathrm{d}t} = \lambda_1 N_1.$$

It is as if there was just a single direct decay reaction $(1) \rightarrow (n)$.

Considering the fact that in the course of geological time natural radioactive chains rapidly reach equilibrium, the initial products give the end products directly. Dating equations can then be written:

$$^{206}\mathrm{Pb} = {}^{238}\mathrm{U}(\mathrm{e}^{\lambda_8 t} - 1) + {}^{206}\mathrm{Pb}_0$$
$$^{207}\mathrm{Pb} = {}^{235}\mathrm{U}(\mathrm{e}^{\lambda_5 t} - 1) + {}^{207}\mathrm{Pb}_0$$
$$^{208}\mathrm{Pb} = {}^{232}\mathrm{Th}(\mathrm{e}^{\lambda_2 t} - 1) + {}^{208}\mathrm{Pb}_0$$

where constants λ_8, λ_5, and λ_2 are those of ^{238}U, ^{235}U, and ^{232}Th.

$$^{238}\text{U} \rightarrow {}^{206}\text{Pb} \qquad \lambda_{238} = \lambda_8 = 1.551\,25 \cdot 10^{-10}\,\text{yr}^{-1}$$
$$^{235}\text{U} \rightarrow {}^{207}\text{Pb} \qquad \lambda_{235} = \lambda_5 = 9.8485 \cdot 10^{-10}\,\text{yr}^{-1}$$
$$^{232}\text{Th} \rightarrow {}^{208}\text{Pb} \qquad \lambda_{232} = \lambda_2 = 4.9475 \cdot 10^{-11}\,\text{yr}^{-1}.$$

This comes down to **direct parent–daughter dating**. Warning! Linear approximation cannot be used with these chronometers. To be convinced of this, let us compare the values of $(e^{\lambda t} - 1)$ with (λt) for time intervals of 1 and 2 billion years for ^{235}U and ^{238}U.

Let us consider the quantity $\left\{[(e^{\lambda t} - 1) - \lambda t]/[e^{\lambda t} - 1]\right\} \times 100$, which is the relative error that can be expressed as a percentage.

With ^{235}U, for 1 billion the error is 41.2% and for 2 billion 68.2%! With ^{238}U, for 1 billion the error is 7.5% and for 2 billion 14.7%. These are highly non-linear systems, then, as can be seen from the shape of the typical curves in Figure 2.3.

2.3.3 The special case of lead–lead methods

We have seen that on the geological timescale radioactive chains attain equilibrium and it can be considered that ^{238}U decays directly to ^{206}Pb, ^{235}U to ^{207}Pb, and ^{232}Th to ^{208}Pb. For uranium-rich minerals like zircon, uraninite, or even monazite, the initial amount of lead can be considered negligible. Assuming the system has remained closed, we can write:

$$^{206}\text{Pb} = {}^{238}\text{U}(e^{\lambda_8 t} - 1)$$
$$^{207}\text{Pb} = {}^{235}\text{U}(e^{\lambda_5 t} - 1).$$

Taking the ratio, remembering that nowadays[1] the ratio $^{238}\text{U}/^{235}\text{U} = 137.8$, we get:

$$\frac{^{207}\text{Pb}}{^{206}\text{Pb}} = \frac{1}{137.8}\left(\frac{e^{\lambda_5 t} - 1}{e^{\lambda_8 t} - 1}\right).$$

It can be seen that the $^{207}\text{Pb}/^{206}\text{Pb}$ isotope ratio of lead alone gives a direct measurement of time. This function is implicit and calculating it requires prior numerical values (Table 2.1).

Exercise

The $^{206}\text{Pb}/^{207}\text{Pb}$ ratio of a uranium ore deposit is found to be 13.50. What is the age of the ore supposing it has remained closed since it crystallized and that common lead can be ignored?

Answer

We shall invert the ratio so Table 2.1 can be used. $^{207}\text{Pb}/^{206}\text{Pb} = 0.074$. The table shows that the ratio measured lies between 1 and $1.2 \cdot 10^9$ years. The result can be improved either by refining the table by calculating the ratio $\left[e^{\lambda_5 t} - 1/e^{\lambda_8 t} - 1\right]$ in the interval 1 and 1.2, or by

[1] The two uranium isotopes decay each at their own rate from the time that they are first formed. Their ratio is constant as there is no isotopic fractionation between them. However, this ratio has varied over geological time.

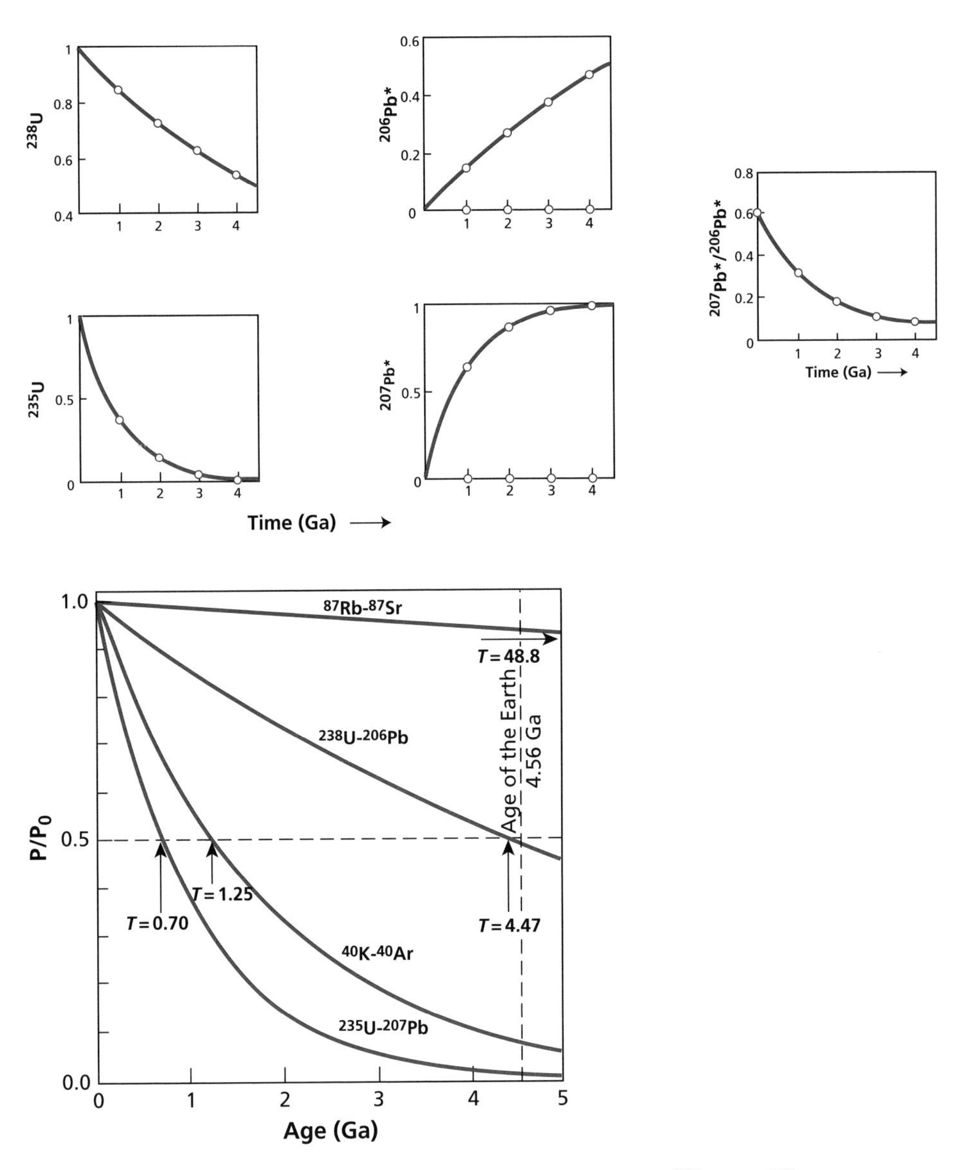

Figure 2.3 Non-linear decay of ^{235}U and ^{238}U and their end products ^{207}Pb and ^{206}Pb. (a) The radioactive constants of these reactions are very different. The curve on the right shows the change in radiogenic ratio $^{207}Pb^*/^{206}Pb^*$ versus time. (b) Comparison in the same figure of the principal radioactive clocks used in geology, emphasizing the behavior of U–Pb systems compared to others.

using a linear approximation between 1 and $1.2 \cdot 10^9$ years. The value $^{207}Pb/^{206}Pb$ for 1.2 billion is 0.080 12 and for 1 billion is 0.0725. The variation is therefore:

$$\frac{0.080\,12 - 0.0725}{200} = 3.8 \cdot 10^{-5} \text{per million years.}$$

The difference is between 0.0740 and 0.0725, that is, $1.5 \cdot 10^{-3}$ million years.

Table 2.1 Numerical values of the radiogenic $^{207}Pb/^{206}Pb$ isotope ratio

Time (Ga)	$e^{\lambda_1 t}-1$	$e^{\lambda_2 t}-1$	$\left(\frac{^{207}Pb}{^{206}Pb}\right)_{radiogenic}$
0	0.0000	0.0000	
0.2	0.0315	0.2177	0.05012
0.4	0.0640	0.4828	0.05471
0.6	0.0975	0.8056	0.05992
0.8	0.1321	1.1987	0.06581
1.0	0.1678	1.6774	0.07250
1.2	0.2046	2.2603	0.08012
1.4	0.2426	2.9701	0.08879
1.6	0.2817	3.8344	0.09872
1.8	0.3221	4.8869	0.11000
2.0	0.3638	6.1685	0.12298
2.2	0.4067	7.7292	0.13783
2.4	0.4511	9.6296	0.15482
2.6	0.4968	11.9437	0.17436
2.8	0.5440	14.7617	0.19680
3.0	0.5926	18.1931	0.22266
3.2	0.6428	22.3716	0.25241
3.4	0.6946	27.4597	0.28672
3.6	0.7480	33.6556	0.32634
3.8	0.8030	41.2004	0.37212
4.0	0.8599	50.3878	0.42498
4.2	0.9185	61.5752	0.48623
4.4	0.9789	75.1984	0.55714
4.6	1.0413	91.7873	0.63930

The approximate age is therefore 1040 Ma. Direct calculation from the $^{207}Pb/^{206}Pb$ ratio for 1040 Ma gives 0.073 98. The age is therefore 1042 Ma, but such precision is illusory because error is far greater (see Chapter 5) than the precision displayed.

2.3.4 The helium method

Natural chains feature many instances of α decay, that is, expulsion of ^4He nuclei. Thus ^{238}U decay ultimately produces eight ^4He, ^{235}U decay produces seven ^4He, and ^{232}Th decay produces six ^4He. We can therefore write:

$$\frac{d\,^4He}{dt} = 8\lambda_8\,^{238}U + 7\lambda_5\,^{235}U + 6\lambda_2\,^{232}Th.$$

Remark

This equation underlies the first helium dating method thought up by Rutherford (1906).

Integrating the equation, assuming $^4He(0) = 0$, gives:

$$^4He = 8\,^{238}U(e^{\lambda_8 t}-1) + 7\,^{235}U(e^{\lambda_5 t}-1) + 6\,^{232}Th(e^{\lambda_2 t}-1).$$

By admitting linear approximations, which is valid only for durations that are not excessive, $e^{\lambda t} - 1 \approx \lambda t$ and observing that $^{235}U/^{238}U = 1/137.8$ and that $^{232}Th/^{238}U \approx 4$ we obtain:

$$^{4}He = {}^{238}U \left[8\,\lambda_8 + \frac{7}{137.8}\lambda_5 + 6 \times 4 \times \lambda_2 \right] t.$$

By noting $\Lambda = \left[8\lambda_8 + \frac{7\lambda_5}{137.8} + 24\lambda_2\right] = 14.889 \cdot 10^{-10}\,yr^{-1}$, we obtain

$$\tau \approx \frac{1}{\Lambda}\left(\frac{^{4}He}{^{238}U}\right).$$

This formula is valid for young ages.

Exercise

Take 1 kg of rock containing 2 ppm of uranium. How many cubic centimeters of ^{4}He will it have produced in 1 billion years?

Answer

At 2 ppm 1 kg of rock represents $10^3\,g \times 2 \cdot 10^{-6} = 2 \cdot 10^{-3}$ g of uranium, and $2 \cdot 10^{-3}$ g of uranium corresponds to $\frac{2 \cdot 10^{-3}}{238} = 8.4 \cdot 10^{-6}$ moles. Using the approximate formula:

$$^{4}He = \Lambda t U = 14.89 \cdot 10^{-10} \times 10^{9} \times 8.4 \cdot 10^{-6}$$

$$^{4}He \approx 125 \cdot 10^{-7}\,\text{moles}.$$

If 1 mole of ^{4}He corresponds to 22.4 liters at standard temperature and pressure, the amount of ^{4}He produced in 1 billion years is $0.28\,cm^3$.

A LITTLE HISTORY

The beginnings of radioactive dating

Rutherford performed the first radioactive age determination in 1906. He calculated the amount of helium produced by uranium and radium per year and per gram of uranium and found $5.2 \cdot 10^{-8}\,cm^3\,yr^{-1}\,g^{-1}$ of uranium.

Ramsay and Soddy had measured the helium content (long confused with nitrogen because it is inert like the noble gases) in a uranium ore known as fergusonite. The fergusonite contained 7% uranium and $1.81\,cm^3$ of helium per gram. Rutherford calculated an age of 500 million years. The following year he found uranium ores more than 1 billion years old! At that time Lord Kelvin was claiming the Earth was less than 100 million years old! (See previous chapter.)

2.3.5 The fission track method

Fission reactions emit heavy atoms. The atoms ejected by fission create flaws in crystals. Such defects – known as tracks – can be shown up on mineral surfaces through acid etching

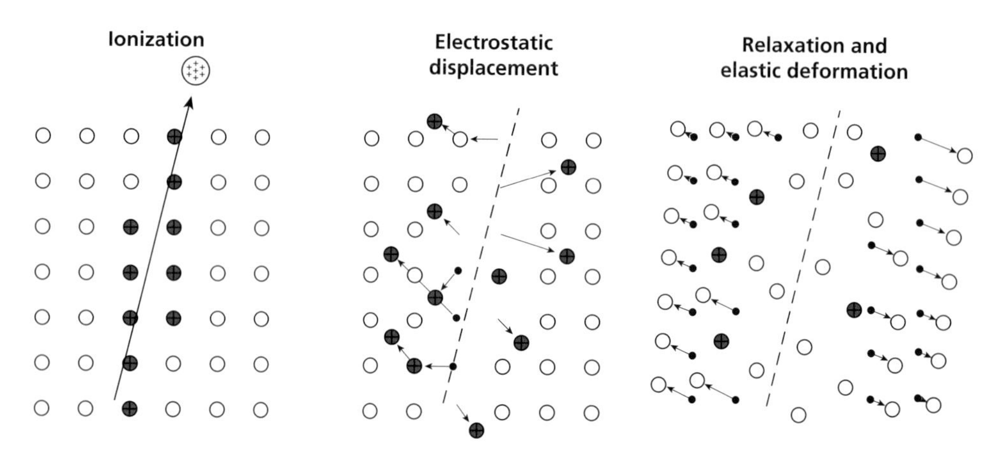

Figure 2.4 Creation of fission tracks in an insulating material.

which preferentially attacks damaged areas, those where atoms have been displaced (Price and Walker, 1962) (Figure 2.4).

The number of fission tracks is written:

$$F_s = \frac{\lambda_{\text{fission}}}{\lambda_\alpha} \; {}^{238}\text{U}(e^{\lambda_\alpha t} - 1)$$

$$\lambda_{\text{fission}} = 7 \cdot 10^{-17}\,\text{yr}^{-1} \quad t_{1/2} = 9.9 \cdot 10^{15}\,\text{yr}^{-1}.$$

In practice, a thin section is cut and the number of surface tracks (and not in a volume) is measured. The visible proportion ρ_s is calculated:

$$\rho_s = q\frac{\lambda_{\text{fission}}}{\lambda_\alpha} \; {}^{238}\text{U}(e^{\lambda_\alpha t} - 1)$$

where q is a geometric factor for switching from a surface to a volume.

To measure ^{238}U, we take advantage of knowledge that $^{238}\text{U}/^{235}\text{U} = 137.8$ and measure ^{235}U by causing **induced fission** (by placing the ore under study in a reactor), which produces tracks that are revealed and counted. We then have the number of tracks before (ρ_s) and after (ρ_i) irradiation. The geometric factor disappears:

$$\frac{\rho_s}{\rho_i} = \frac{\lambda_{\text{fission}}}{\lambda_\alpha}\frac{137.8}{\phi\Gamma}(e^{\lambda_\alpha t} - 1).$$

We calibrate the flux ϕ of neutrons inducing ^{235}U fission and the effective cross-section Γ.[2] By using a standard sample whose uranium content is known and which is irradiated at the same time as the study sample and by counting the fission tracks produced, we can then calculate t.

[2] The effective cross-section is the probability of a reaction occurring, that is, here, the probability of producing a fission with a given flux of neutrons. The characteristics of this will be seen in Chapter 4.

2.3.6 Isolation of a part of the chain and dating young geological periods

The various radioactive isotopes of the radioactive chain belong to different chemical elements so that, under certain geological conditions, one or two isotopes in the chain fractionate chemically and become isolated, thereby breaking the secular equilibrium. Once isolated they create a new partial chain in turn. Two straightforward specific cases are of practical importance.

The radioactive isotope becomes isolated on its own

First, when a radioactive isotope in the chain (but with a long enough period) becomes isolated on its own, it gives rise to a partial chain, but being isolated from the parent it decays according to the law:

$$N_i(t) = N_i(0)e^{-\lambda t}$$

where $N_i(0)$ is the number of nuclides of the intermediate element at time $t = 0$. If this number can be estimated, the decay scheme can be used as a chronometer. This is equivalent, then, to dating by the parent isotope.

EXAMPLE

The ionium method and the rate of sedimentation

Thorium is virtually insoluble in sea water. Thus ^{230}Th (still known as ionium from the terminology of the pioneers of radioactivity), a decay product of ^{234}U with $T_{\frac{1}{2}} = 75$ ka, precipitates on the sea floor, is incorporated in the sediment and so gradually buried. There, now isolated, it decays.

At any depth x of sediment from the surface (Figure 2.5) we can write:

$$^{230}\mathrm{Th}(x) = {}^{230}\mathrm{Th}(0)e^{-\lambda_{230}t}$$

where ^{230}Th(0) is the surface content which is assumed constant over time. If the sedimentation rate is constant, time can be replaced by the ratio $t = x/V_s$ where V_s is the sedimentation rate and x the length (depth):

$$^{230}\mathrm{Th}(x) = {}^{230}\mathrm{Th}(0)\exp([x/V_s]\lambda)$$

or in logarithms:

$$\ln(^{230}\mathrm{Th}(x)) = \ln\left(^{230}\mathrm{Th}(0)\right) - \frac{x}{V_s}\lambda.$$

The slope of the curve (ln ^{230}Th, x) gives a direct measure of sedimentation rate and the ordinate at the origin gives ^{230}Th(0). (Note its order of magnitude of a millimeter per thousand years.)

This method only works, of course, if it is assumed that ^{230}Th(0), that is the thorium content at the sediment surface, is constant and if the sediment has not been disturbed by chemical, physical, or biological phenomena.

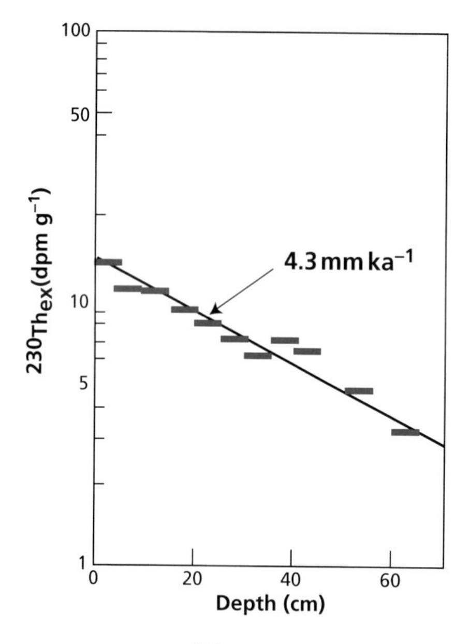

Figure 2.5 Decreasing ^{230}Th content in a core from the sea floor and determination of the sedimentation rate. Th_{ex} is the excess thorium compared with the equilibrium value counted in disintegrations per minute per gram (dpm g^{-1}).

Exercise

The lead isotope ^{210}Pb (as a member of the radioactive chain) is radioactive with a decay constant $\lambda = 3.11 \cdot 10^{-2}$ yr^{-1}. This natural radioactive lead is incorporated into ice deposited in Greenland by forming successive layers of ice which can be studied like sedimentary strata. The activity of ^{210}Pb is measured at four levels in disintegrations per hour per kilogram of ice (dph kg^{-1}). Table 2.2 shows the results.

Calculate the sedimentation rate of the ice. Assuming a constant rate and a compaction factor of 5, how thick will the glacier be in 5000 years? Calculate the ^{210}Pb content of fresh ice.

Answer

The dating equation is written noting activity by square brackets:

$$\left[{}^{210}\mathrm{Pb}\right] = \left[{}^{210}\mathrm{Pb}\right] e^{-\lambda t}.$$

If the rate of deposition is V and height h, we have $t = h/V$.

The equation becomes:

$$\left[{}^{210}\mathrm{Pb}\right] = \left[{}^{210}\mathrm{Pb}\right] \exp(\lambda h/V)$$

or

$$\ln\left[{}^{210}\mathrm{Pb}\right] = \ln\left[{}^{210}\mathrm{Pb}\right]_0 - \frac{\lambda h}{V}.$$

If the ^{210}Pb content has remained constant over time, the data points must be aligned in a (ln [activity, h]) plot. The slope is therefore $-\lambda/V$. The data points are plotted on the graph and the slope determined. This gives $V = 45$ cm yr^{-1}.

Table 2.2 Activity of ^{210}Pb with depth

	Depth			
	0 m	1 m ± 20 cm	1.50 m ± 40 cm	2.50 m ± 50 cm
dph kg^{-1}	75	32 ± 5	24 ± 5	10 ± 3

In 5000 years' time, allowing for compaction, there will be $5000 \times 45/5 = 450$ m of ice. The ^{210}Pb content is calculated: surface activity is 75 dph kg^{-1} ice.

$$\lambda_{210} \times N_{210} = 75 \text{ dph kg}^{-1}$$

hence

$$N_{210} = \frac{75}{3.1110^{-2} \times (8760)^{-1}} = \frac{75}{3.5 \cdot 10^{-7}} = 21.4 \cdot 10^{7} \text{atoms of}^{210}\text{Pb nuclides}$$

per kilogram of ice,

8760 being the number of hours in a year. As a mass that gives:

$$\frac{21.4 \cdot 10^{7} \times 210}{6.023 \cdot 10^{23}} = 7.5 \cdot 10^{-14} \text{ kg of ice,}$$

where Avogadro's number is in the denominator and the previous atomic mass of ^{210}Pb in the numerator.

The ^{210}Pb content is $7.5 \cdot 10^{-17}$, which is very little! This shows the incredible sensitivity of radioactive methods because ^{210}Pb in Greenland's glaciers really can be measured and used for estimating the rate of sedimentation of ice.

The parent isotope is isolated and engenders its daughters

This is what happens with uranium; for example, when certain solid phases like calcium carbonate are precipitated uranium is entrained with calcium and then isolated. For the first radioactive product of any notable half-life, we then have:

$$\lambda_{2-3} N_{2-3}(t) = \lambda_1 \mathrm{U}(t)\left(\mathrm{e}^{-\lambda_1 t} - \mathrm{e}^{-\lambda_{2-3} t}\right)$$

where N_{2-3} is the number of nuclides in the third intermediate product in the chain. Why? Because the chain includes very short-lived radioactive products such as thorium-234 (^{234}Th) or protactinium-234 (^{234}Pa) which reach equilibrium very quickly. Thus, ^{238}U decays to ^{234}Th, an element whose lifespan is 24 days, then ^{234}Pa, whose lifespan is 1.18 minutes: it can be considered, then, that ^{238}U directly gives ^{234}U whose half-life is $2.48 \cdot 10^{5}$ years for all types of usual samples (corals, speleothems, travertines, etc.).

Such a method is applied, for example, to secondary mineralizations of uranium. The soluble uranium migrates and is deposited further away, leaving the insoluble thorium where it is. It then "resumes" its decay giving ^{230}Th and we can write:

$$\lambda_{230}{}^{230}\mathrm{Th} = \lambda_{230}{}^{238}\mathrm{U}\left(\mathrm{e}^{-\lambda_{238} t} - \mathrm{e}^{-\lambda_{230} t}\right).$$

Therefore the age of migration can be determined by measuring ^{230}Th and ^{238}U.

Exercise

Uranium-238 (^{238}U) decays to ^{234}U which is itself radioactive with $\lambda = 2.794 \cdot 10^{-6}\ yr^{-1}$.

Sea water is not in secular equilibrium relative to uranium isotopes as ^{234}U weathers better than ^{238}U from rocks of the continental crust and is enriched in the rivers flowing into the ocean. In activity, noted in square brackets [ ], $[^{234}U/^{238}U]_{seawater} = 1.15$. When limestone forms from sea water, it is isotopically balanced with the sea water and so takes the value 1.15. A fossil mollusk has been found in a Quaternary beach formation and its activity ratio measured as $[^{234}U/^{238}U] = 1.05$. Work out the dating equation. What is the age of the mollusk?

Answer

The dating equation is:

$$\left[\frac{^{234}U}{^{238}U}\right] = \left[\frac{^{234}U}{^{238}U}\right]_0 e^{-\lambda t} + \left(1 - e^{-\lambda_{234} t}\right)$$

because $\lambda_{234} \gg \lambda_{238}$ and $\lambda_{238}\, t \approx 0$ if $t > 10_6$ years.

This gives:

$$T = \frac{1}{\lambda} \ln \left(\frac{\left[\frac{^{234}U}{^{238}U}\right]_0 - 1}{\left[\frac{^{234}U}{^{238}U}\right] - 1} \right) \quad \text{and } T = 393\ 000 \text{ years.}$$

2.3.7 General equation and equilibration time

We shall do an exercise to help understand equilibration time and by the same token give the theoretical answer to the previous exercise.

Exercise

Establish the general equation for evolution of an isotope in a radioactive chain and where the parent has a longer half-life and the daughter a markedly shorter half-life. We shall take the example of $^{234}U \rightarrow {}^{230}Th$ decay to get our ideas straight.

Answer

This is a review exercise for mathematics on integrating a linear differential equation with constant coefficients:

$$\frac{d\,^{234}U}{dt} = \lambda_{238}\ ^{238}U - \lambda_{234}\ ^{234}U$$

$$\frac{d\,^{230}Th}{dt} = \lambda_{234}\ ^{234}U - \lambda_{230}\ ^{230}Th$$

with ^{238}U being considered constant.

Integrating the two previous equations in succession for the example in question gives:

$$\lambda_{234}\ ^{234}U = \lambda_{234}\ ^{234}U_0 e^{-\lambda_{234} t} + \lambda_{238}\ ^{238}U\left(1 - e^{-\lambda_{234} t}\right)$$

$$\lambda_{230}\ ^{230}Th = \lambda_{230}\ ^{230}Th_0 e^{-\lambda_{230} t} + \lambda_{234}\ ^{234}U\left(e^{-\lambda_{234} t} - e^{-\lambda_{230} t}\right).$$

These two equations can be written with the activity notation in square brackets: $\lambda N = [N]$, which simplifies notation and means the λ constants can be dispensed with. This gives:

$$[^{234}U] = [^{234}U_0]e^{-\lambda_{234}t} + [^{238}U](1 - e^{-\lambda_{234}t})$$
$$[^{230}Th] = [^{230}Th_0]e^{-\lambda_{230}t} + [^{234}U](e^{-\lambda_{234}t} - e^{-\lambda_{230}t}).$$

This is the general equation for equilibration of a radioactive chain used for chronological purposes.

From there, without resorting to long and tedions numerical simulation, let us try to answer the question: at what speed does a greatly disturbed radioactive chain return to secular equilibrium? Let us consider the preceding equation with $\lambda_{234} \ll \lambda_{230}$. The equation becomes:

$$\lambda_{230}\ {}^{230}Th = \lambda_{230}\ Th_0 e^{-\lambda_{230}t} + \lambda_{234}\ {}^{234}U(1 - e^{-\lambda_{230}t}).$$

For the chain to achieve equilibrium, it is necessary and sufficient that:

$$\lambda_{230}\ {}^{230}Th \approx \lambda_{234}\ {}^{234}U.$$

For this $e^{-\lambda_{230}t}$ must be virtually zero. So t must be less than to 6–10 periods of ^{230}Th. The chains therefore achieve equilibrium after more than six half-lives of the daughter product (with the smaller radioactive decay constant).

If $^{234}U/^{238}U$ isotope fractionation occurs, which is the case in surface processes, it is ^{234}U which is the "limiting factor" for the ^{238}U chain. The time required is therefore 1.5 million years.

If there is no $^{234}U/^{238}U$ isotope fractionation, the limiting factor is ^{230}Th and the equilibration time is 450 000 years.

For the ^{235}U chain the limiting factor is ^{231}Pa and therefore a time of 194 000 years. Both chains of the two uraniums are equilibrated at about the same time in this case.

The isotopes of radioactive series used as geochronometers are those with decay constants of more than one year.

Uranium-238 chain

- ^{234}U, half-life 248 ka, is used in sedimentary or alteration processes because $^{234}U/^{238}U$ varies during alteration. The radioactive recoil of ^{234}U extracts this uranium from its crystallographic site and so it is easily altered.
- ^{230}Th, half-life 75 ka. This element, named ionium, is certainly the most widely used for surface and volcanic processes.
- ^{226}Ra, half-life 1.622 ka. It is used like ^{230}Th but for shorter-lived surface or volcanic processes.
- ^{210}Pb, half-life 22 years. Used for studying glaciers, oceans or volcanics involving very young phenomena.

Uranium-235 series

- ^{231}Pa, half-life 32.48 ka. This is the sister of ^{230}Th but slightly more difficult to master analytically.
- ^{277}Ac, half-life 22 years. This element is not much used as yet because of difficulties in making precise analyses.

Thorium-232 chain

- ^{228}Ra, half-life 6.7 years. This is complementary to ^{226}Ra to constrain the timescale of Ra fractionation.
- ^{228}Th, half-life 1.9 years. A very good complement for ^{228}Ra and ^{230}Th.

These elements can be measured by alpha and gamma radioactive spectroscopy or by mass spectrometry. The second technique is far more precise but less sensitive than counting. In practice, ^{234}U, ^{230}Th, ^{226}Ra, and ^{231}Pa are measured by mass spectrometry and the others by counting. Pb-210 can be measured by mass spectrometry but this entails great difficulties and requires special precautions. For a general review see Ivanovich (1982).

A LITTLE HISTORY

The polemic that followed the discovery of radioactivity

Henri Becquerel discovered radioactivity almost by chance in 1898 while studying rays from phosphorescent uranium salts excited by sunlight and trying to understand the nature of x-rays discovered by **Röntgen**. But one day, although there was no sunlight, a sample from the Joachimsthal mine in Bohemia spontaneously emitted radiation which blackened a photographic plate, indicating as yet unknown properties of matter (see Barbo, 2003).

Some years later, when measuring the effect of these radioactive substances on an electrometer (the particles ionized the air of the electrometer which then discharged) **Pierre** and **Marie Curie** proposed calling the phenomenon radioactivity (activity created by radiation). For them, it was the property certain substances, including uranium, had of spontaneously emitting radiation.

They immediately came in for harsh criticism from British scientists relying on a crucial observation: when the activity of 1 g of purified uranium was measured with an electrometer, the activity was less than that of 1 g of uranium contained in, say, 100 g of uranium ore. In other words 1 g of uranium "diluted" in 100 g of inert rock was more "active" than 1 g of pure uranium.

How could concentrated uranium be less active than diluted uranium? It smacked of magic. It was **Marie Curie** who came up with the hypothesis of intermediate radioactive products to explain this paradox. The discovery of radium must be set in this polemical context, thus taking on its full significance. It was the second intermediate radioactive product to be found after polonium.

The New Zealander **Ernest Rutherford** brought grist to the Curies' mill and within a few years the mechanisms of successive "cascade" decay was understood. Radioactive chains had been discovered.

Shedding light on a paradox

Nowadays the paradox of diluted uranium being more active than pure uranium can be fully explained. Suppose a radioactive chain is in equilibrium, say the ^{238}U chain:

$$\lambda_1 N_1 = \lambda_2 N_2 = \lambda_3 N_3 = \cdots = \lambda_n N_n$$

noting $N_1, N_2, \ldots, N_n$ the numbers of nuclides in the various isotopes of the chain. Thus in secular equilibrium there is 14 times the $\lambda_1 N_1$ activity of ^{238}U in the chain. (The input from the ^{235}U chain must be added to this although its contribution is small.)

This is what happens in a rock several hundred million years old in which the chain has had time to attain secular equilibrium.

Now, in purified uranium, the activity is merely $\lambda_1 N_1$.

So, if the "activity" of 1 g of uranium is measured, it will be 14 times less than the activity of 100 g of 1% uranium ore. Indeed, the ore would be slightly more active because there must also be at least 0.5 g of thorium whose chain also produces 11 times more activity than pure thorium. However, as the thorium constant is 3.5 times less, the outcome would be an increase in the uranium activity of about 1.5 times.

In all, the rock is about 15.5 times more active than purified uranium. This is the paradox behind the polemic!

2.4 Dating by extinct radioactivity

2.4.1 The historical discovery

Chemical elements – that is, the nuclei that make up most of their mass – have been manufactured in stars ever since the Universe came into being. This is nucleosynthesis. Of the nuclei formed, some are stable and others radioactive. Among the radioactive nuclei some have very short half-lives: they decay quickly giving rise to their stable daughter isotopes. All of this goes on everywhere in interstellar space and is what provides the ordinary matter of the Universe. These isotopes are incorporated into interstellar matter as gases or dust (see Chapter 4).

But suppose that nucleosynthesis of heavy elements (explosion of a supernova) occurred in the vicinity of the place where the Solar System formed giving rise to certain short-lived radioactive isotopes. Suppose too that the solid bodies of the Solar System form while these radioactive isotopes are not yet extinct. These "young" radioactive isotopes will be incorporated with the other interstellar material in these solid bodies (planets or meteorites) and there they will decay and give rise to daughter isotopes. The solid objects having received such inputs will therefore have abnormal isotope abundances for the isotopes produced by decay of the radioactive isotope. Detecting such anomalies is therefore the first step in showing the existence of extinct radioactivity.

In 1960 **John Reynolds** at Berkeley discovered a large excess of the isotope 129 in the isotopic composition of xenon in the Richardton (H4) meteorite (see Reynolds, 1960) (Figure 2.6). Now, this is not any old isotope but the decay product of iodine-129 (radioactive iodine we know how to make in nuclear reactors) whose half-life is 17 million years and which, if it formed before the birth of the Solar System, has disappeared today. And indeed naturally occurring iodine has only a single isotope, ^{127}I.

To prove that the excess ^{129}Xe did come from the ^{129}I, **Peter Jeffrey** and **John Reynolds** came up with a most ingenious experiment combining neutron activation analysis, mass spectrometry, and stepwise outgassing by temperature increments. They irradiated a sample of the Richardton meteorite using a flux of neutrons. The ^{127}I was transformed by nuclear reaction (n, γ) into radioactive ^{128}I which transformed by β^- decay into ^{128}Xe. This reaction is written

$$^{127}\mathrm{I}(\mathrm{n},\gamma)^{128}\mathrm{I} \xrightarrow{\beta^-} {}^{128}\mathrm{Xe}.$$

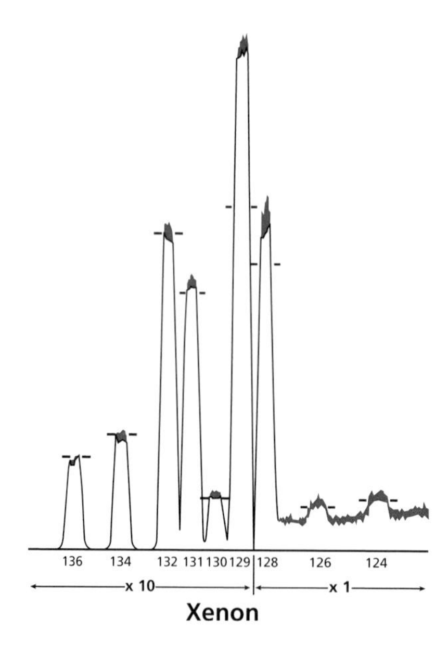

Figure 2.6 Mass spectrum of xenon in the Richardton meteorite as measured by Reynolds. The bars indicate the height of ordinary, say atmospheric, xenon.

They then heated the irradiated sample progressively, after placing it in the vacuum and purification line of a mass spectrometer. They analyzed the isotope composition of the xenon extracted at each temperature increment and observed that the excess ^{129}Xe was extracted at the same temperature as half of the ^{128}Xe, whereas "ordinary" xenon was extracted at a different temperature. Jeffrey and Reynolds (1961) concluded that ^{129}Xe is situated at the same (crystallographic) site as natural iodine and therefore is indeed the daughter of ^{129}I.

This extraordinary discovery has two important consequences. First, it shows that, before the Solar System formed (at a time in the past 5–10 times the half-life of ^{129}I, that is, 85–170 million years), there was a synthesis of heavy chemical elements. In addition, this radioactive decay provided an exceptional and unexpected dating tool for studying the period when meteorites (and also, as we shall see, the Earth) were formed (see Figure 2.7). We shall concentrate on this aspect now.

2.4.2 Iodine–xenon dating

Let $^{129}\mathrm{I}^*(0)$ be the number of nuclides formed by nucleosynthesis at time $t = 0$, defined as the end of the nucleosynthetic process. The radioactive iodine decays by the law:

$$^{129}\mathrm{I}^*(t) = {}^{129}\mathrm{I}^*(0)\mathrm{e}^{-\lambda t}.$$

Suppose now that, at time t_1, some of the iodine is incorporated in a meteorite (A) and at time t_2 some other iodine in a meteorite (B). We can write:

$$^{129}\mathrm{I}_\mathrm{A}^*(t_1) = K_\mathrm{A}\ {}^{129}\mathrm{I}^*(0)\mathrm{e}^{-\lambda t_1}$$
$$^{129}\mathrm{I}_\mathrm{B}^*(t_2) = K_\mathrm{B}\ {}^{129}\mathrm{I}^*(0)\mathrm{e}^{-\lambda t_2}.$$

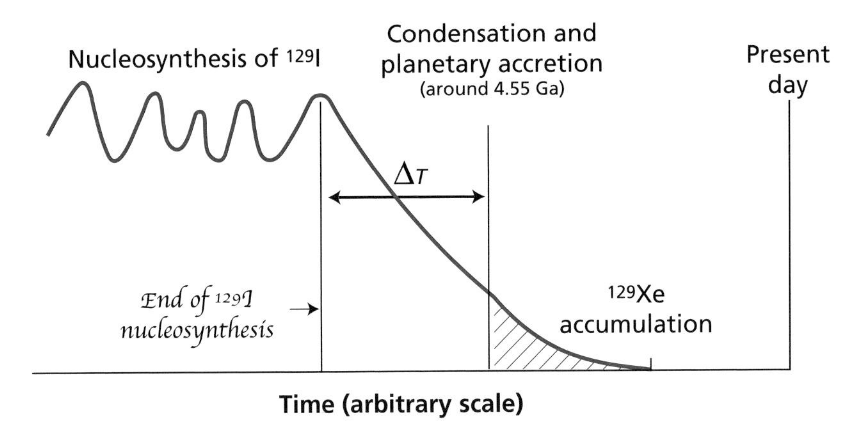

Figure 2.7 Evolution of ^{129}I in the Solar System and its trapping in planetary bodies; ΔT is the time between the end of nucleosynthesis and the accretion of planetary objects.

Here K_A and K_B are the factors of incorporation of iodine between the interstellar cloud and meteorites A and B, which is the ratio between iodine concentration in the interstellar cloud and in the meteorite. For two meteorites of different chemical compositions, K_A and K_B are different. In each meteorite, the ^{129}I decays completely into radiogenic ^{129}Xe, ^{129}Xe* (with an asterisk):

$$^{129}\mathrm{Xe}^*_\mathrm{A} = {}^{129}\mathrm{I}_\mathrm{A}(t_1) \qquad {}^{129}\mathrm{Xe}^*_\mathrm{B} = {}^{129}\mathrm{I}_\mathrm{B}(t_2).$$

How can this be used for dating as we do not know the values of K_A, K_B, and I(0)?

Iodine has a stable isotope ^{127}I (the only one for that matter). We divide the expressions describing ^{129}I decay by ^{127}I. We can assume that the incorporation of iodine by the meteorites obeys chemical laws, and so the same rules apply for the 129 isotope as for the 127 isotope. This means the K values are the same for both isotopes. The K coefficients can therefore be removed from the equations describing the evolution of the **isotopic** ratios. This gives:

$$\left[\frac{^{129}\mathrm{I}}{^{127}\mathrm{I}}(t_1)\right]_\mathrm{A} = \left[\frac{^{129}\mathrm{I}}{^{127}\mathrm{I}}(0)\right]\mathrm{e}^{-\lambda t_1}$$

$$\left[\frac{^{129}\mathrm{I}}{^{127}\mathrm{I}}(t_2)\right]_\mathrm{B} = \left[\frac{^{129}\mathrm{I}}{^{127}\mathrm{I}}(0)\right]\mathrm{e}^{-\lambda t_2}.$$

There remains one unknown in these equations, namely the ratio $^{129}\mathrm{I}^*/^{127}\mathrm{I}(0)$, in other words the (^{129}I/^{127}I) isotope ratio at the end of nucleosynthesis. It is a reasonable assumption that it was identical throughout the Solar System (and so for all meteorites). We can then find the ratio between the two isotope ratios of our two meteorites:

$$\frac{\left[\frac{^{129}\mathrm{I}^*}{^{127}\mathrm{I}}(t_1)\right]_\mathrm{A}}{\left[\frac{^{129}\mathrm{I}^*}{^{127}\mathrm{I}}(t_2)\right]_\mathrm{B}} = \mathrm{e}^{\lambda(t_2-t_1)}.$$

By measuring the $\left({}^{129}\mathrm{I}^{*}/{}^{127}\mathrm{I}(t)\right)$ ratios in both meteorites, we can in principle calculate the time interval between the formation of the two meteorites $(t_2 - t_1)$. This method therefore provides **absolute–relative dating!**

The problem comes down to measuring the $({}^{129}\mathrm{I}^{*}/{}^{127}\mathrm{I})$ isotope ratio at the time the meteorite formed. Total ${}^{129}\mathrm{Xe}$ is the sum of initial ${}^{129}\mathrm{Xe} + {}^{129}\mathrm{I}^{*}(t)$. Today ${}^{129}\mathrm{I}^{*}(t) = {}^{129}\mathrm{Xe}^{*}$, since ${}^{129}\mathrm{I}$ has decayed entirely. Therefore:

$$ {}^{129}\mathrm{Xe}^{*} = {}^{129}\mathrm{Xe}_{\mathrm{total}} - {}^{129}\mathrm{Xe}_{\mathrm{initial}}. $$

The interesting ratio is:

$$ \frac{{}^{129}\mathrm{I}}{{}^{127}\mathrm{I}}(t) = \frac{{}^{129}\mathrm{Xe}^{*}}{{}^{127}\mathrm{I}} = \frac{{}^{129}\mathrm{Xe}^{*}}{\mathrm{Xe}_{\mathrm{total}}} \cdot \left(\frac{\mathrm{Xe}_{\mathrm{total}}}{\mathrm{I}_{\mathrm{total}}}\right) = A\ \mathrm{e}^{-\lambda t} $$

with $A = {}^{129}\mathrm{I}/{}^{127}\mathrm{I}(0)$ at the end of nucleosynthesis.

The problem comes down to measuring the fraction of radiogenic ${}^{129}\mathrm{Xe}^{*}$ in total xenon, and then measuring the chemical (Xe/I) ratio, since the iodine is all ${}^{127}\mathrm{I}$.

We can calculate the age to the nearest coefficient and therefore, by taking the ratio between the values for the two meteorites, determine the relative age of the two meteorites. The beauty of this method lies in its capacity to measure very brief intervals of time between the formation of planetary objects billions of years ago.

Exercise

The ${}^{129}\mathrm{I}/{}^{127}\mathrm{I}$ isotope composition measured on the Karoonda and Saint-Séverin meteorites is $1.3 \cdot 10^{-4}$ and $0.8 \cdot 10^{-4}$, respectively. Given that the half-life of iodine is 17 Ma, what is the age difference between the two meteorites?

Answer

If $T_{\frac{1}{2}} = 17$ Ma

$$ \lambda = \frac{\ln 2}{T} = 4 \cdot 10^{-8}\ \mathrm{yr}^{-1}. $$

The dating formula is applied:

$$ \frac{\left(\frac{{}^{129}\mathrm{I}}{{}^{127}\mathrm{I}}\right)_{\mathrm{A}}}{\left(\frac{{}^{129}\mathrm{I}}{{}^{127}\mathrm{I}}\right)_{\mathrm{B}}} = \frac{1.3 \cdot 10^{-4}}{0.8 \cdot 10^{-4}} = \mathrm{e}^{\lambda\,\Delta t}, $$

hence $\Delta t = \frac{1}{\lambda}\ln(1.3/0.8) = 12.1$ million years. This age is actually the maximum interval measured.

Exercise

Given that the half-life of ${}^{129}\mathrm{I}$ is 17 Ma, what is the shortest interval of time that can be estimated, given the uncertainty in measuring the ${}^{129}\mathrm{Xe}^{*}/{}^{127}\mathrm{I}$ ratio is 2%?

Answer

Suppose we always take the same reference, say, Karoonda. There will no longer be any error relative to Karoonda but everything will be expressed in terms of this reference. Let us take the previous measurement and look at the limits of uncertainty. The value 0.8 has two limits, at 0.816 and 0.784. Let us calculate the age Δt. We obtain 11.64 and 12.64, or 1 million years. The age of Saint-Séverin (relative to Karoonda) is written 12.1 $\pm$ 0.5 Ma.

2.4.3 Discoveries of other forms of extinct radioactivity

Since then many forms of extinct radioactivity have been discovered, which we shall use as required. Each discovery has required the identification by experiment of parent–daughter relations in meteorites. We review them with a few brief comments.

- **Iodine–Xenon** ^{129}I–^{129}Xe, $\tau = 25$ Ma, discovered by **Reynolds** (1960).
- **Plutonium–Xenon** ^{244}Pu–$Xe_{fission}$, $\tau = 84$ Ma. This radioactivity, discovered by **Kuroda** (1960), produces fission tracks and the fission isotopes of xenon ^{131}Xe, ^{132}Xe, ^{134}Xe, and ^{136}Xe. It is an important supplement to the iodine–xenon method and was discovered very shortly after it.
- **Samarium–Neodymium** ^{146}Sm–^{142}Nd, $\tau = 146$ Ma. This form of radioactivity, discovered at San Diego by **Lugmair** *et al.* (1975), is interesting because it is has a long half-life and allows us to connect long-duration phenomena that occurred around the time of 4.5 billion years ago.
- **Aluminum–Magnesium** ^{26}Al–^{29}Mg, $\tau = 1$ Ma. This form of radioactivity was first detected in certain minerals from very ancient meteorites by the team of **Gerald Wasserburg** at the California Institute of Technology (Caltech) (Lee *et al.*, 1977). It was of historical importance but it is probably more important still that aluminum is a decisive constituent of meteorites (2–3%). It is probable that ^{26}Al was instrumental in the very early thermal history of planetesimals and its influence should be added to the calculations already done on this topic in Chapter 1.
- **Palladium–Silver** ^{107}Pd–^{107}Ag, $\tau = 9.4$ Ma. This form of radioactivity was detected in iron meteorites by the Caltech team (**Kelly** and **Wasserburg**, 1978). It has shown how old these meteorites are. This means that metallic iron differentiation is a very ancient phenomenon in the processes of formation of the Solar System.
- **Manganese–Chromium** ^{53}Mn–^{53}Cr, $\tau = 5.3$ Ma. This form of radioactivity, discovered in Paris by **Birck** and **Allègre** in 1985, is interesting because the manganese and chromium fractionate because of their different volatilities.
- **Iron–Nickel** ^{60}Fe–^{60}Ni, $\tau = 2.1$ Ma. This form of radioactivity has been found in just a few basaltic meteorites by **Shukolyukov** and **Lugmair** (1993) at San Diego. It is important because iron is a very abundant element.
- **Calcium–Potassium** ^{41}Ca–^{41}K, $\tau = 0.143$ Ma. This form of extinct radioactivity is important because of its short half-life. It was discovered by **Srinivasan** *et al.* (1994).
- **Hafnium–Tungsten** ^{182}Hf–^{182}W, $\tau = 13$ Ma. This form of extinct radioactivity, discovered by **Harper** and **Jacobsen** (1994) at Harvard and then by **Lee** and **Halliday** (1995) at the University of Michigan, is very important because Hf and W fractionate during metal–silicate separation, allowing this separation to be dated, including in planets (differentiation of the core). We shall use this scheme later on.
- **Niobium–Zirconium** ^{92}Nb–^{92}Zr, $\tau = 36$ Ma. This newcomer to the "club" of forms of extinct radioactivity, discovered in Zurich by **Schonbachler** *et al.* (2002), is yet to be exploited.

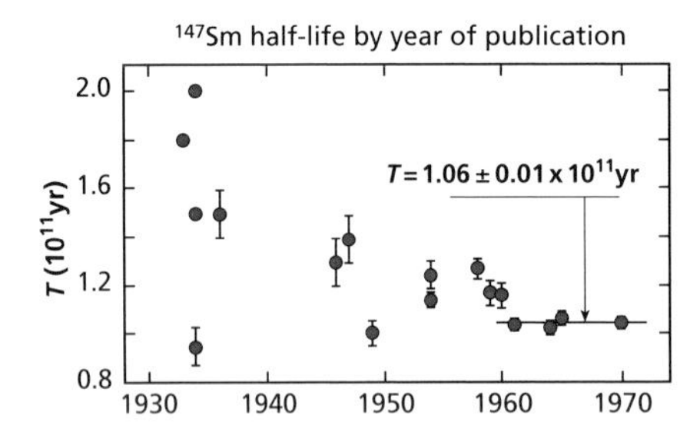

Figure 2.8 Improvement in determination of decay constants over time: the example of ^{147}Sm.

- **Chlorine–Sulfur** ^{36}Cl–98% ^{36}Ar / 2% ^{36}S, $\tau = 0.43$ Ma was discovered recently by a Chinese team, **Lin** *et al.* (2005). No anomaly on ^{36}Ar was found, only on ^{36}S.
- **Beryllium–Boron** ^{10}Be–^{10}B, T = 15 Ma was discovered by **McKeegan**, **Chaussidon**, and **Robert** (2000).
- **Lead–Thallium** ^{205}Pb–^{205}Tl, T = 15.1 Ma was recently discovered by **Neilsen**, **Rehkämper**, and **Halliday** (2006).

Others have not been confirmed and are not listed here.

2.5 Determining geologically useful radioactive decay constants

As just seen, what allows us to calculate age and is the very essence of the radioactive clock is the radioactive constant λ, namely the probability that a nucleus will decay. How can this be determined? This is difficult a priori, given that the constants are generally very small because activity is low (see Figure 2.8). To simplify there are three methods more or less derived from the dating method:

(1) direct measurement of radioactivity by λN activity: if we know N, we can deduce λ;
(2) measurement of accumulation of the daughter isotope (both these series of measurements are done in the laboratory);
(3) geological "comparison" of ages obtained by various methods.

We shall examine these three techniques in the case of ^{87}Rb.

2.5.1 Measurement of activity

We start with the fundamental equation describing decay

$$\frac{\mathrm{d}\,^{87}\mathrm{Rb}}{\mathrm{d}t} = \lambda\,^{87}\mathrm{Rb},$$

that is, the number of β particles emitted by unit time is equal to $\lambda\,^{87}$Rb.

Let us take 1 kg of ^{87}Rb, which corresponds to $10^3/87 \times 6.023 \cdot 10^{23}$g ($6.23 \cdot 10^{23}$ is Avogadro's number), or $6.92 \cdot 10^{24}$ atoms of ^{87}Rb. If $\lambda = 1.42 \cdot 10^{-11}$ yr^{-1}, the number of β particles emitted in 1 year is $6.92 \cdot 10^{24} \times 1.41 \cdot 10^{-11}$, or $9.8264 \cdot 10^{13}$ particles.

Remembering that 1 year $\approx 3 \cdot 10^7$ seconds, that corresponds to $3.275 \cdot 10^6$ disintegrations per second, which is a measurable value: even 10 g of pure rubidium will suffice.

The difficulty lies in measuring β^- particles. Some of these particles are absorbed by the rubidium deposit itself. It is fundamental then to make layers of rubidium of variable thicknesses and to correct what is known as auto-absorption. These are tricky methods to master.

2.5.2 The radiogenic isotope produced

For ^{87}Rb we try to measure the ^{87}Sr accumulated. Take 1 kg of pure ^{87}Rb. How much ^{87}Sr does it produce in 1 year?

$$^{87}\mathrm{Sr} = 10^3\,\mathrm{g} \times 1.42 \cdot 10^{-11} = 1.42 \cdot 10^{-8}\,\mathrm{g},$$

that is 14.2 ng. As strange as it may seem, such a quantity can easily be measured with a mass spectrometer by isotope dilution.

Naturally, in practice we try to use both methods and to compare the results. The question is, of course, how do we obtain 1 kg of pure or almost pure ^{87}Rb? Isotope separation is expensive so we try rather to use natural rubidium, in which there is only a fraction of ^{87}Rb. The difficulty is that there must be very little ^{87}Sr in the rubidium being measured. The rubidium must therefore be very carefully purified by chemical methods, which is difficult for such a large amount of rubidium. The measurement uncertainty stems from this.

2.5.3 The method of geological comparison

If we know the age of certain rocks from methods with decay constants that are easier to determine (such as uranium) we can then calculate the constant λ_{Rb} by measuring the ^{87}Sr/^{87}Rb ratios on a series of rocks or minerals whose (U/Pb) age is known.

This method too is difficult to implement as we must be sure that the various systems have remained closed, as we shall see in the next chapter. Even so, the method is essential for ensuring the geological reliability of the different methods. To avoid geological difficulties, much use is made of cross-calibrations with meteorites and moon rocks. Why so? Because meteorites are rocks dating from the origin of the Solar System (and therefore old) and have not been subjected to major "disruptive" events. We proceed by trial and error combining the different approaches. An international commission makes regular reviews and updates the constants as need be.

Let us give three important geological comparisons that use the constants given in the table.

Moon rocks

We choose Rock 10072, which has come in for particularly close scrutiny. As the U/Pb ratio is low this dating method is not good but it is a good way of comparing Rb–Sr, K–Ar, and Sm–Nd.

	Dating method		
Rock 10072	K–Ar	Rb–Sr	Sm–Nd
Time (Ga)	3.52±0.4	3.57±0.05	3.57±0.03

Eucrites (basaltic achondrites)

These are basalt meteorites, in other words ancient extraterrestrial lava flows. Dating by U–Pb is extremely precise because the U/Pb ratios are high. As said, the uranium decay constants are the references. However, Rb–Sr is not very precise as Rb/Sr ratios are really low. (We are anticipating a little on the next chapter.) The K–Ar, Lu–Hf, and Sm–Nd datings are also relatively precise.

	Dating method				
Eucrite meteorites	U–Pb	Rb–Sr	K–Ar	Sm–Nd	Lu–Hf
Time (Ga)	4.55±0.05	4.50±0.14	4.50±0.1	4.53±0.04	4.57±0.19

Ordinary chondrites

These are the most common meteorites characterized by the presence of chondrules. Comparison of Pb–Pb and Rb–Sr datings is very precise. Similarly, K–Ar and Re–Os are in suitable agreement.

	Dating method			
Ordinary chondrites	U–Pb	Rb–Sr	K–Ar	Re–Os
Time (Ga)	4.55±0.05	4.55±0.08	4.52±0.05	4.54±0.02

Table 2.3 Comparison of methods for determining radioactive decay constants

Isotope	Laboratory counting	Accumulation	Geological comparison
^{238}U, ^{235}U, ^{232}Th, and radioactive chains	This is the reference method and is (relatively) precise	Not used	This is the reference for other methods
^{87}Rb	Difficult because of auto-absorption of β^- rays by the powder being counted	Difficult to show up because of traces of initial ^{87}Sr, which is hard to remove	Comparison between methods and in particular between meteorites is essential
^{187}Re	Impossible: insufficient β energy (high auto-absorption)	This is the best method; we measure accumulated ^{187}Os	Comparison with other methods and particularly meteorites remains essential
^{176}Lu	Poor determination	Difficult because of the importance of initial Hf	Comparisons between methods are useful
^{147}Sm	Poor determination	Difficult because of initial Nd, which is hard to remove	Comparisons between methods are useful
^{40}K	Very precise counting	Should be possible in a well-sealed flask	Comparison is difficult because ^{40}Ar diffuses readily

2.5.4 Conclusions

Laboratory determination of uranium and thorium constants by counting is the most precise method. These decay constants are then taken as references for other measurements.

Table 2.3 shows the pros and cons of the different techniques for determining decay constants for the different chronometers.

Problems

1 Given that the potassium content of the silicate Earth is 250 ppm,[3] how much ^{40}Ar is created in $4.5 \cdot 10^9$ years? $^{40}K = 1.16 \cdot 10^{-4}\ K_{total}$.
Given that ^{40}Ar cannot escape from the Earth and the quantity of ^{40}Ar in the atmosphere is $66 \cdot 10^{18}$ g, what conclusion do you draw?

2 Suppose a series of zircons gives the results in the table below. Calculate the $^{206}Pb/^{238}U$, $^{207}Pb/^{235}U$, $^{207}Pb/^{206}Pb$, and $^{208}Pb/^{232}Th$ ages. Which ages appear most reliable to you? If we know that the samples must be of the same geological age, which age would you recommend?

U (ppm)	Th (ppm)	Radiogenic Pb (ppm)	$^{206}Pb/^{204}Pb$	$^{207}Pb/^{204}Pb$	$^{208}Pb/^{204}Pb$
415	86.4	30.5	1138	84.34	116.2
419	84.0	30.4	1984	130.7	162.5
482	85.0	32.6	2292	147.5	174.1
507	81.8	34.0	3301	206.0	229.4

3 The K–Ar ages of two volcanic rocks from the island of Santa Maria in the Azores are measured as below.

Rock mass (g)	K_2O%	Radiogenic ^{40}Ar (10^{-12} moles g^{-1})
1.81	1.83	13.90
0.64	0.92	5.60

Calculate their ages. Given that the error is $\pm 5\%$, what can you say about these two rocks?

4 To apply the ^{230}Th–^{238}U method to carbonate rocks formed in the ocean, allowance must be made for the fact that $(^{234}U/^{238}U)_0 = 1.15$ in activity and that ^{230}Th derives entirely from ^{234}U decay.
(i) Draw up the complete $^{234}U/^{238}U$ dating equation.
(ii) Draw up the ^{230}Th, ^{234}U, ^{238}U dating equation assuming that $(^{230}Th)_0 = 0$.

5 The isotopes ^{231}Pa and ^{230}Th are both insoluble and isolated while uranium isotopes remain in solution.
(i) Draw up the dating equation based on the $^{231}Pa/^{230}Th$ ratio.
(ii) What is the time-span over which it can be applied?

[3] Previously we used 210 ppm. Readers should be aware of variations in values used by different authors in such determinations.

CHAPTER THREE

Radiometric dating methods

We have so far examined the principles of radioactive dating with simple assumptions, namely that the initial amount of daughter isotope is negligible and that the system (mineral or rock) whose age is to be determined has remained closed since it first formed, that is, it has neither lost nor gained parent or daughter isotopes in the course of its geological past. These two conditions do not usually pertain in nature. So how can these difficulties be overcome?

3.1 General questions

3.1.1 Rich systems, poor systems

The dating equation showed that a distinction has to be made between systems with negligible initial radiogenic isotope and systems with abundant initial radiogenic isotope. In the first instance, that of a **rich system** (understood as radiogenically rich), an age can be calculated in theory from direct measurement of the parent and daughter isotopes. In the second instance, that of a **poor system**, some method must be found for estimating the initial abundance of the radiogenic isotope.

For a system to be considered rich, the radioactive isotope must be abundant compared with the radiogenic element. The chemical composition of the system under study must be such that the ratio of the radioactive isotope to the stable reference isotope[1] is very high. The time for which it has been decaying must be long enough for radioactive disintegration to have produced enough of the radiogenic isotope.

Exercise

The $[^{87}Rb/^{86}Sr]$ ratio in a biotite is 3000. Can a 300-million-year-old biotite be considered a rich system?

Answer

The $[^{87}Sr/^{86}Sr]$ radiogenic ratio is written $[^{87}Sr/^{86}Sr] \approx [^{87}Rb/^{86}Sr]\ \lambda t$. Calculation with $T = 3 \cdot 10^8$ years and $\lambda = 1.42 \cdot 10^{-11}\ \text{yr}^{-1}$ gives:

$$[^{87}Sr/^{86}Sr] = 3 \cdot 10^3 \times 1.42 \cdot 10^{-11} \times 3 \cdot 10^8 = 12.6.$$

[1] The reference isotope is an isotope close to the radiogenic isotope of the same chemical element which is stable and is not itself a product of radioactivity.

The initial $[^{87}Sr/^{86}Sr]$ ratios of common strontium vary from about 0.720 to 0.705. These variations are negligible compared with 12.6. The biotite is indeed therefore a **rich system**.

In practice, the initial daughter isotope ratio is not completely ignored when calculating the age of rich systems but this so-called **normal ratio** is estimated from measurements of various minerals and a conventional value is taken to make the correction. A catalog of rich systems for each radioactive clock has been drawn up by the systematic study of the commonest rocks and minerals. In each case, allowance must be made for the age of the system, which is an essential feature, of course. This catalog is relatively limited.

For the $^{87}Rb-^{87}Sr$ method, it comprises above all micas like biotite $K(Mg,Fe)_3AlSi_3O_{10}(OH)_2$ and muscovite $KAl_2Si_3O_{10}(OH)_2$ as well as some quite old potassium feldspars ($KAlSi_3O_8$).

For U–Th–Pb methods there are uranium ores proper, and then uranium-rich minerals. The commonest is zircon ($ZrSiO_4$) but there are also sphene ($CaTiSiO_5$) and apatite $Ca_5(PO_4)_3(OH)$.

For $^{40}K-^{40}Ar$ and $U-^4He$ methods, all systems may be considered rich since rocks usually have negligible initial argon or helium contents (except the very young ones).

For the $^{187}Re-^{186}Os$ method, the current rich mineral is molybdenite (MoS_2), but also basaltic or granitic rocks (where $[^{187}Re/^{186}Os]$ ratios are frequently 500–20 000) as well as minerals of these rocks such as olivine or the pyroxenes.

Rich systems are very rare for the other chronometers such as $^{147}Sm-^{143}Nd$ or $^{176}Lu-^{176}Hf$ because geochemical systems do not clearly separate samarium from neodymium or lutetium from hafnium.

3.1.2 Closed system, open system

The fundamental assumption in radioactive dating is that the **box** (the rock or mineral) has remained closed, that is, it has neither lost nor gained parent or daughter nuclides through exchange with its environment since it first closed.

Now, while this assumption is accepted for parent elements nestled in suitable crystallographic sites (Rb in place of K, U in place of Zr, etc.) it is less obvious for daughter elements. The daughter isotopes produced by radioactivity are intruders in the crystallographic lattice. They have been introduced "artificially" by radioactive transmutation. Why should they stay there? This is particularly true of rich systems, since these radiogenic isotopes are very abundant. They have been produced in a mineral which is "unfamiliar" to them, and will therefore tend to escape from it.

EXAMPLE

The use of biotite in radiometric dating

Biotite is a black mica containing potassium (K) in its structural formula. Rubidium, which is an alkali similar to potassium, can enter its structure and remain there. Potassium forms a K^+ ion and rubidium a Rb^+ ion. However, when it decays, ^{87}Rb yields ^{87}Sr.

Strontium is an alkaline earth element similar to calcium. In ionic form it is bivalent with formula Sr^{2+}. It is not structurally stable in biotite and so tends to escape by diffusion whenever it can. The biotite system therefore has little chance of remaining closed for the ^{87}Rb–^{87}Sr pair, above all if, in the course of its history, it has been subjected to metamorphic heating or weathering, which provide conditions amenable to the diffusion of ^{87}Sr from biotite.

The question of whether a system is open or closed has been asked since the earliest work in geochronology. Before trying to answer it, let us see what the effects are of any leakage from the system. Let us take the example of rich systems, where the equation is of the form:

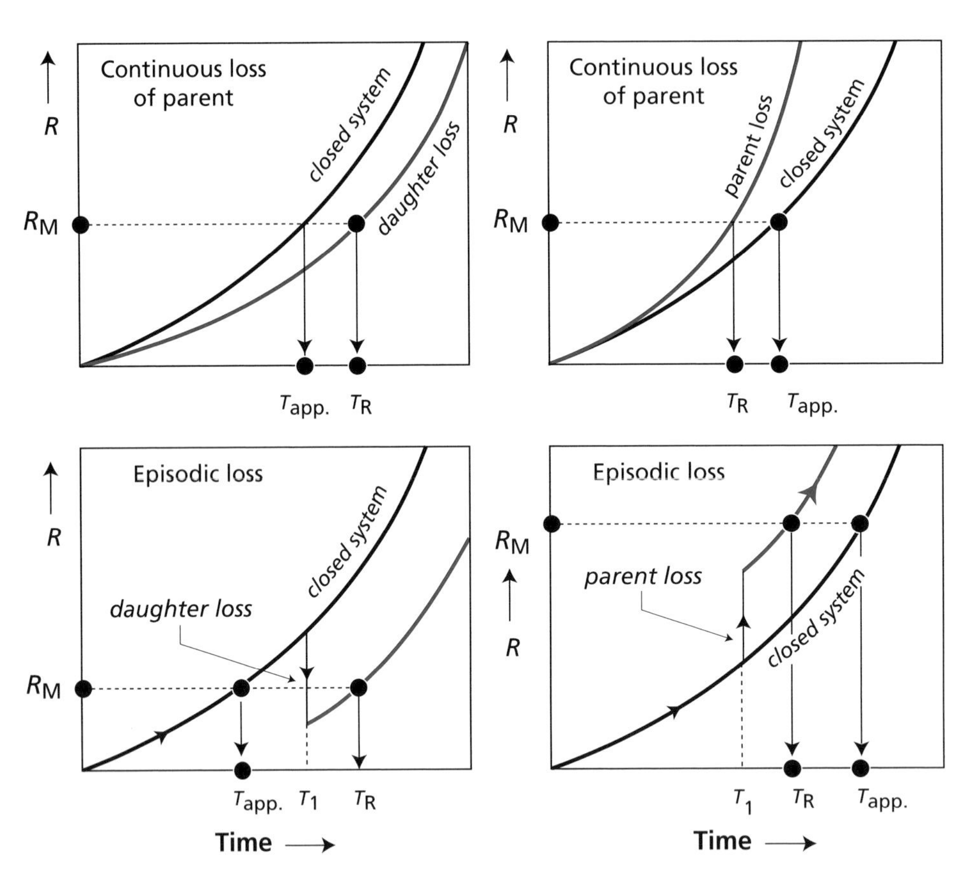

Figure 3.1 Opening of a radiometric dating system: R is the radiogenic/radioactive parent ratio; R_M is the measured ratio. The top two figures show continuous, constant losses over time. The bottom two figures show a sudden episodic loss followed by a long period in a closed system. T_R = real time, $T_{app.}$ = apparent time. Where $T_R > T_{app.}$ the system becomes younger and there is a loss of the radiogenic daughter isotope (left-hand diagram). Where $T_R < T_{app.}$ the system becomes older and there is a preferential loss of the radioactive parent isotope. Where both are lost, the outcome depends on the relative values of the losses. If they are equal, the system is equivalent to a closed system. If more daughter isotope than parent isotope is lost, the left-hand diagram is relevant, and vice versa.

$$R = \frac{I_D}{I_P} = \varphi(t)$$

where φ is an increasing function of time, I_D and I_P the daughter and parent isotopes, and R the parent–daughter ratio. If a little of the daughter isotope is lost, the age calculated will be false because "too young." This is by far the commonest case. If a little of the parent isotope is lost, the age calculated will be false because "too old." A gain in I_D or I_P would, of course, have the opposite effect (Figure 3.1).

3.1.3 Continuous or episodic losses

The question of the closed character of the boxes arises with two very different geological scenarios.

In the first scenario, the system continuously loses (or more rarely gains) radiogenic isotopes over the course of geological time. This process has been evoked for argon, helium, and the other rare gases, which are not chemically bonded within minerals and so tend to escape constantly by diffusion. It may also be true of isotopes of soluble elements exposed to chemical alteration at the Earth's surface (K, Rb, Sr, U, Pb).

The alternative scenario involves sudden events that punctuate geological time: tectono-metamorphic crises, igneous intrusions, volcanic eruptions, folding, etc. Over a short period (compared with the duration of geological time) the system loses or gains the isotopes in question.

There may also be a combination of both circumstances, when the history of a rock is divided into two episodes: one, say, at low temperature when the system is closed and the other at high temperature where continuous loss occurs between moments of "crisis." Naturally, the more complex the scenario the more difficult it is to decipher and above all the more difficult it is to come up with a single interpretation.

Exercise

Suppose we have the rich ^{87}Rb–^{87}Sr system of muscovite (white mica). The simplified dating equation is $[^{87}Sr^*/^{87}Rb] = \lambda t$.

(1) Suppose that recent heating has caused the mineral to lose 30% of its $^{87}Sr^*$.
If the mineral is really 1 billion years old, what will be its apparent age determined by measuring $^{87}Sr^*$ and ^{87}Rb today?

(2) Suppose the muscovite has been weathered by water which has leached 10% of the Sr and 50% of the Rb (the Rb was more soluble than the Sr). What will its apparent age be?

Answer

(1) The simplified expression can be written:

$$\frac{t_{\text{apparent}}}{t_{\text{real}}} = \left[\frac{^{87}Sr_{\text{real}}}{^{87}Sr_{\text{closed}}}\right]$$

giving

$$^{87}Sr_{\text{real}} = (1 - 0.3)\ ^{87}Sr_{\text{closed}} = 0.7\ ^{87}Sr_{\text{closed}}.$$

Therefore, $t_{apparent} = 700$ million years.
(2)

$$\frac{t_{apparent}}{t_{real}} = \left[\frac{^{87}Sr_{apparent}}{^{87}Sr_{closed}}\right] \cdot \left[\frac{^{87}Rb_{closed}}{^{87}Rb_{apparent}}\right] = \frac{0.9}{0.5} = 1.8.$$

The apparent age will be 1.8 billion years.

3.1.4 Concordant and discordant ages

With this concern, how can we be sure that a geological age is reliable? The method unanimously employed for testing whether a result is reliable is the age **concordance** and **discordance** method. Suppose that, with the methods just described, that is, those applicable to rich systems, we wish to test whether a geological age determination is reliable.

Suppose we measure the ages of potassium feldspar, muscovite, and biotite minerals of the same granitic rock by the ^{87}Rb–^{87}Sr method. We can hope they will be identical and yield an age for when the granite crystallized.

If the ages are **concordant**, that is, if they are close to each other, we accept there is a good chance the common age is geologically significant. This is because we admit that a disruptive event will have different effects on different clocks because of their chemical differences. There will then remain the matter of relating this age to a specific geological event (magmatism, metamorphism, sedimentation, etc.). Suppose that, for the Quérigut granite already discussed, the ^{87}Rb–^{87}Sr method were to yield ages of 296 million years for potassium feldspar, 295 million years for muscovite, and 293 million years for biotite. These ages can be considered concordant around 296 million years. Given the geological setting (ages of the terrain they cut across and of the overlying terrains) it is reasonable to accept that this is the age of emplacement (intrusion) of the granite. Suppose, though, that for a rock of the same granite sampled close to a fault we find ages of 110 million, 90 million, and 50 million years for the same three minerals. We would conclude that the system had been disrupted by some secondary phenomenon, probably related to the formation of the fault, and so that proper age determination is not possible. There are two stages in the reasoning, then. In the first, concordance supports the idea that the age is geologically significant. In the second, the geological context allows the age to be identified, that is, to be attributed to some specific geological phenomenon.

If the ages are **discordant**, then we accept that the basic assumptions of the model have been breached and that the ages so determined are not geologically meaningful. From now on, therefore, we shall speak of **apparent age** for a crude date measurement and shall reserve the term "age" for geologically significant dates.

Remark

Apparent age is an isotope ratio converted into time units. Validating it as a geological age is a complex process. The apparent age is chemical and isotopic. It is of temporal and geological significance only if certain conditions are met.

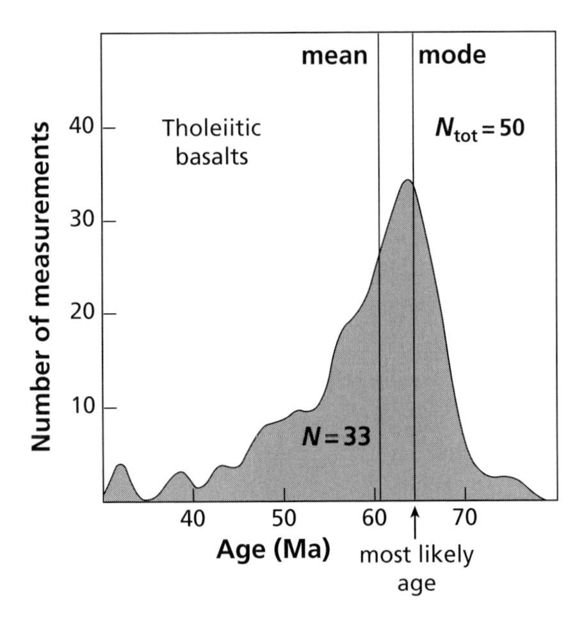

Figure 3.2 Graph of apparent ^{40}K–^{40}Ar ages. Dispersion is quite wide because ages range from 30 to 80 Ma. The most probable age is indicated by the arrow, since the most likely process is that the system lost argon over the course of time.

Suppose we measure the ^{40}K–^{40}Ar age of a series of presumed **cogenetic** basalts (here is an essential word! – cogenetic means derived from the same genetic phenomenon). We draw the graph (Figure 3.2) and the mean indicates an age of 60 million years. The "chemical" statistical test is correct and indicates an age of 60 ± 5 million years. Have we met the concordance criteria? The answer is no.

First, the histogram is asymmetrical and is not a normal distribution. The point is that ^{40}Ar is a gas produced by the decay of ^{40}K and tends to escape from minerals. This loss is a statistical process and makes ages younger. In the case in question, an age of 65 million years is more likely to be the real age than the mean age is, which reflects rather the statistic of argon losses. But this is only a hypothesis as we have no means of calculating the age for certain. The concordance criterion must be based on various sets of measurements:

- Several different sorts of boxes and a single chronometer.
- Several chronometers and a single type of box (Rb–Sr and K–Ar on biotite).
- Several chronometers on several boxes (Rb–Sr and K–Ar on biotite and muscovite). In such cases the boxes have to be closed at the same time during the same geological phenomenon. Such boxes are said to be **cogenetic**.

Cases of perfect concordance are rare with rich systems. The question is how can we go beyond this disillusion.

3.2 Rich systems and solutions to the problem of the open system

3.2.1 The semi-quantitative systematic comparative approach

The basic idea is simple enough. For a given problem in a given geological case, we measure apparent ages by various methods applied to various minerals. We compare the apparent ages and try to establish a **systematics**. For example, we observe that amphibole generally has a greater apparent age than biotite when measured by the K–Ar method. We deduce that amphibole conserves its argon better than biotite does. Similarly, we observe that the apparent Rb–Sr age of biotite is generally greater than its apparent K–Ar age. We deduce that biotite retains the radiogenic $^{87}Sr^*$ better than it does the ^{40}Ar. In this specific instance, these purely comparative studies were first based on field studies that we looked to multiply. Laboratory studies of diffusion were required to supplement this approach.

Laboratory study of diffusion and extrapolation

Diffusion is the process whereby a chemical species propagates in a random walk, just like heat, say, can propagate. Mathematically it obeys Fourier's law, written here for a single dimension:

$$\frac{\partial C}{\partial t} = D\frac{\partial^2 C}{\partial x^2}$$

where C is concentration, t is time, x is distance, and D the coefficient of diffusion. This law is also expressed by a simpler-looking formula:

$$Q = -D\left(\frac{\partial C}{\partial x}\right)$$

where $(\partial C/\partial x)$ is the concentration gradient and Q the flux of matter which diffuses. The diffusion coefficient D is expressed in $cm^2\ s^{-1}$ or in $m^2\ s^{-1}$ (be careful as other units are sometimes found in the literature).

The analytical solutions of this equation are known for simple geometrical cases: plates, spheres, half-planes, etc. Nowadays computers can provide numerical solutions to all problems, or almost all. It is therefore easy to calculate the evolution of a radioactive system in a box of given shape and size by superimposing diffusion on radioactive decay. Figure 3.3 shows the results of such calculations for a K–Ar system supposed to evolve from a sphere in which either argon or potassium can diffuse.

Solutions to the diffusion equations are generally expressed by the approximation $x \approx \sqrt{Dt}$ where x is distance, t time, and D the diffusion coefficient. The relevant parameter for diffusion is (D/a^2), where a is the radius of the sphere. The diffusion coefficient D obeys Arrhenius' equation:

$$D = D_0 \exp\left(\frac{-E}{RT}\right)$$

where T is temperature (in degrees Kelvin), E activation energy, and R the ideal gas constant. D_0 is a constant dependent on the nature of the mineral but is independent of temperature.

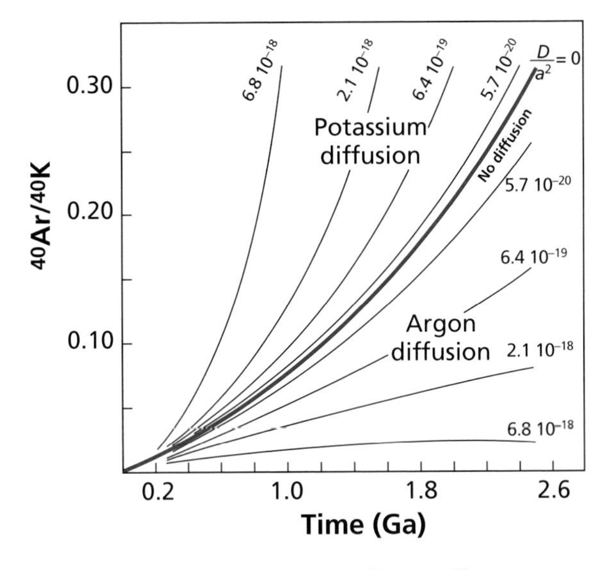

Figure 3.3 Calculation of the diffusion of ^{40}Ar and ^{40}K over geological time out of a spherical mineral and its effect on the chronometric curve. The figures are relative to the parameter D/a^2 expressed in s^{-1}.

In addition, as might be expected, mineral size is essential: the larger it is the better the retention of argon. This is just what is found when we study, say, the K–Ar age versus mineral size (Wasserburg and Hayden, 1955).

Remark

To determine diffusion coefficients in the laboratory we take a mineral whose shape and dimensions have been "measured" under the microscope and we measure the amount of isotopes escaping from it by heating the mineral progressively at different controlled temperatures. By plotting log D in these results versus $(1/T)$, we can calculate the activation energy and estimate D_0 from the formula $D = D_0 \exp(-E/RT)$ or $\log D = \log D_0(-E/RT)$ (Figure 3.4).

To conduct such laboratory experiments it is essential to take natural samples containing "intrusive" radiogenic isotopes, and thus quite old samples, which have not been exposed to secondary phenomena (metamorphism, erosion, etc.). (These are rare and precious samples.) How can the results be applied? A first exercise will help us with this.

Exercise

What is the diffusion coefficient of ^{40}Ar in biotite at 50 °C and at 1000 °C if we know that the activation energy is $E = 21$ kcal per mole and that at 600 °C $D/a^2 = 2 \cdot 10^{-10}$ per second (dividing by a^2 removes the dimension in cm^2)?

Answer

Temperatures are converted into Kelvin, giving 323 K and 1273 K. For the temperatures considered, $E/R \approx 10^4$ since $R = 1.98$ cal mol^{-1} K^{-1}. By using Arrhenius' equation this gives $D_{323}/a^2 = 1.3 \cdot 10^{-16}$ s^{-1} and $D_{1273}/a^2 = 7 \cdot 10^{-7}$ s^{-1}.

Table 3.1 Closure temperatures

Mineral	Closure temperature (°C)	Grain dimension (10^{-6} m)	Reference
Hornblende	685 ± 53	210–840	Berger and York (1981)
Biotite	373 ± 21	500–1410	Berger and York (1981)
K-feldspar	230 ± 18	125–840	Berger and York (1981)
Plagioclase	176 ± 54	125–210	Berger and York (1981)
Microcline (pure K-feldspar)	132 ± 13	125–250	Harrison and McDougall (1982)

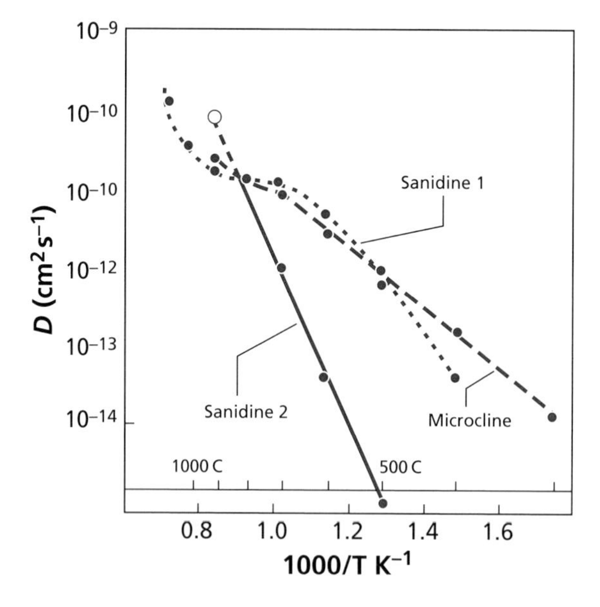

Figure 3.4 Variation of the diffusion coefficient with temperature. Two sanidines (volcanic potassium feldspar) from two different rocks yield very different results. One is very close to the coefficient measured on a microcline crystal (granitic potassium feldspar) while the other is very different. This shows that diffusion coefficients depend largely on mineral history and that it is difficult therefore to use these experiments to make precise geological age determinations.

Clearly, then, **the diffusion coefficient is extremely sensitive to temperature**. Many rocks and minerals that are to be dated by Rb–Sr or K–Ar methods of rich systems (biotites, muscovites, potassium feldspars) crystallize at high temperatures, be they magmatic rocks or metamorphic rocks. Emplacement is generally followed by cooling. This prompted the idea of defining a temperature at which the system begins to retain the radiogenic isotope for each mineral. As the activation energies of the various minerals are different, these **closure temperatures** are different (Table 3.1). Slight differences in apparent age are used to define cooling curves for the rock and therefore for the massif or even for the region to which the minerals belong (Figure 3.5).

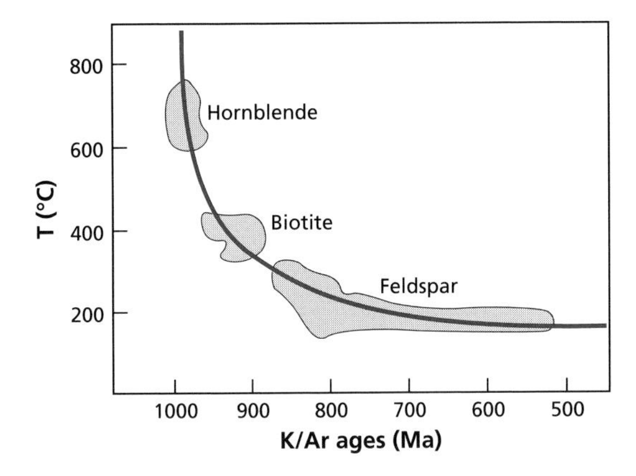

Figure 3.5 Cooling curve of an orogenic segment of Grenville Province, Canada, based on K–Ar apparent ages and on the retentivity of the different minerals. After Berger and York (1981).

Remark

The cases most studied by these methods are of thermal variations during orogeny (mountain building) where estimations are made of how the crustal block cooled as it rose to the surface; at what speed and over what temperature range, etc.

The study of contact metamorphism

The decisive step in this type of semi-quantitative approach was the study by **Stanley Hart** (1964) (Figure 3.6), then a student at the Massachusetts Institute of Technology, confirmed by **Gil Hanson** and **Paul Gast** (1967), then at the University of Minnesota, who studied the ages obtained by the different methods for an instance of contact metamorphism.

Contact metamorphism occurs when granite is intruded into a geological series. Physically, such metamorphism corresponds to sudden heating by a body of defined geometry. The thermal evolution of such a problem can be readily processed mathematically. The isotherms and their variations over time are easily obtained. Hart chose to study the result of a Tertiary intrusion some 60 Ma old (Eldora) in the Precambrian terrain of Colorado. There is a big age difference between the intrusion and the surrounding rock. He studied the variation in Rb–Sr, K–Ar, and U–Pb ages on different rich minerals of the Precambrian formation with distance from the point of contact. The "apparent" age of all the minerals varied with distance from the contact, which was what he expected. At the contact point, it is the same as the age of the intrusion and then it progressively converges towards the age of the surrounding rock at a great distance from the contact. What is interesting is that at any given distance from the contact, the order of apparent ages determined by the various methods on the various minerals obeys a coherent logical schema.

The reliability of "rich" chronometers has been estimated on the basis of such a sequence and a few laboratory experiences. They are ranked here in decreasing order of reliability of the ages obtained.

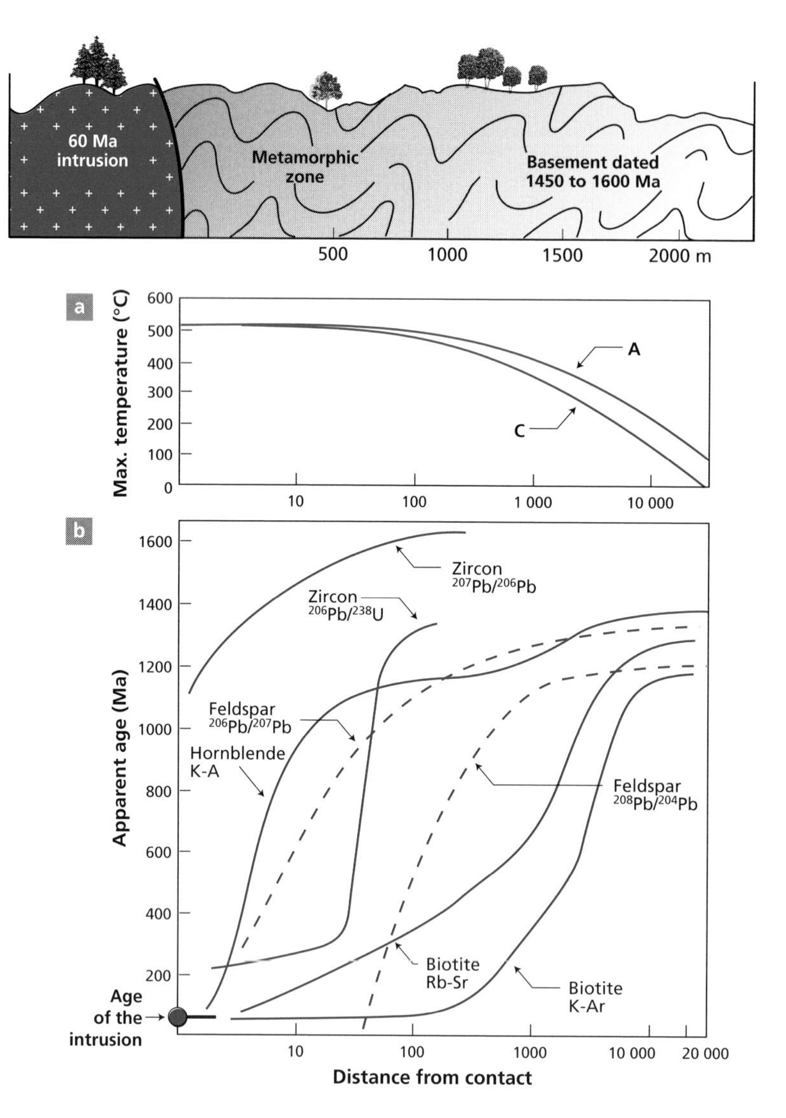

Figure 3.6 Contact metamorphism of the Eldora stock, Colorado, studied by Stanley Hart. This figure is famous. Top: the geological section. The scale of distances in (a) and (b) is logarithmic. (a) The thermal profile, that is, the maximum temperatures reached at each point versus distance. Curves A and C are from two values of thermal diffusivity. (b) Apparent ages versus distance.

$$^{206}Pb-^{207}Pb \text{ zircon} > ^{206}Pb-^{238}U \text{ zircon} > ^{40}Ar-^{40}K \text{ amphibole} > ^{87}Rb-^{87}Sr \text{ muscovite} > ^{87}Rb-^{87}Sr \text{ biotite} > ^{40}Ar-^{40}K \text{ biotite} > ^{40}Ar-^{40}K \text{ feldspar}$$

Although there are many exceptions to this ranking, it allows a series of apparent ages to be gauged rapidly. Thus when, say, $^{207}Pb-^{206}Pb$ on zircon gives the same age as $^{40}Ar-^{40}K$ on biotite there is a good chance the age is correct. When the difference between the $^{40}Ar-^{40}K$ on amphibole and $^{87}Rb-^{87}Sr$ on muscovite ages is slight, we consider we are close to the true age, etc. While this empirical approach is useful, it does not, alas, provide a thoroughgoing solution to the issue of reliability. In particular, how can an age be calculated when the hypotheses of the simple model are not satisfied?

Exercise

Analysis of a granite yields: U–Pb on zircon, 520 Ma; K–Ar on amphibole, 480 Ma; K–Ar on biotite, 400 Ma; Rb–Sr on biotite, 460 Ma; Rb–Sr on muscovite, 470 Ma. What seems the likely age of intrusion of this granite to you? How reliable is the result?

Answer

The true age is probably around 550–540 Ma. The reason being that some disruptive phenomenon has affected the region as there is clearly a discordance in ages. The most robust of the chronometers is U–Pb on zircons, but it is not perfect and this apparent age may be made a little older. But this reasoning is very rough and ready as can be seen!

3.2.2 The concordia method

This method has been devised for calculating a system's true age even though the system is an open one. This is done by exploiting pairings of chemical behavior which may be found with various radiochronometers.

Uranium–lead systems

Uranium–lead systems using $^{238}U \rightarrow {}^{206}Pb$ and $^{235}U \rightarrow {}^{207}Pb$ decay have an interesting feature: both parents and both daughters are of the same chemical nature but have very different decay constants. This pairing is exploited for determining geological ages even when the system is an open one.

Let us consider a uranium-rich material. It may be zircon ($ZrSiO_4$), a common mineral in granite rocks, sphene, uraninite, or apatite, minerals containing little "common" lead (detected by measuring the non-radiogenic isotope ^{204}Pb). The initial lead can therefore be neglected in chronometric equations (rich systems) and we write:

$$\frac{^{206}Pb^*}{^{238}U} = \left(e^{\lambda_{238} t} - 1\right)$$

$$\frac{^{207}Pb^*}{^{235}U} = \left(e^{\lambda_{235} t} - 1\right).$$

Two ages can therefore be calculated by measuring the ^{206}Pb, ^{207}Pb, ^{238}U, and ^{235}U contents of minerals. If everything complied with the assumptions (closed system, rich system, etc.) these two ages should be identical, that is, concordant. The common age would therefore indicate the time the zircon or apatite crystallized in the granite magma. This is sometimes the case, but generally ages are found to be different and so **discordant**. To bring out these age discrepancies we consider a plot:

$$x = \left[\frac{^{207}Pb}{^{235}U}\right]; \; y = \left[\frac{^{206}Pb}{^{238}U}\right].$$

We plot the curve of the parametric equation:

$$y = \left(e^{\lambda_{238} t} - 1\right); \quad x = \left(e^{\lambda_{235} t} - 1\right).$$

This curve, which can be graduated for time, is known as the **concordia** curve. It is the geometric locus of concordant ages. Any concordant age lies on the curve and any discordant age lies off the curve.

The South African **Louis Ahrens** noticed in 1955 that when the $^{206}Pb^*/^{238}U$ and $^{207}Pb^*/^{235}U$ ratios (as before * indicates the isotopes are radiogenic) measured on suites of cogenetic minerals are plotted, they tend to be aligned. These alignments cut the concordia curve at two points, corresponding to two ages (t_0 and t_1). In 1956 **George Wetherill** of the Carnegie Institution in Washington showed that the two "ages" obtained by Ahrens' constructions could be interpreted as the age of uranium ore crystallization and the age of a disruption that affected the minerals and caused uranium and lead exchanges with the environment. Here is a simplified demonstration of Wetherill's model.

The uranium decays in a closed system from the time at which the mineral is crystallized at time $t = 0$ until time t_1.

$$U(t_1) = U_0 e^{-\lambda_{238} t_1}.$$

Suppose that at t_1 some of the uranium was lost. Let us term the lost proportion α. There remains $(1 - \alpha)$. At time $t_1 + \Delta t$, the uranium becomes:

$$U(t_1 + \Delta t) = (1 - \alpha) U_0 e^{-\lambda t_1}.$$

This uranium decays in a closed system until time t when the analysis is conducted:

$$U(t) = U(t + \Delta t) e^{-\lambda(t - t_1)} = (1 - \alpha) U_0 \, e^{-\lambda t_1} \, e^{-\lambda(t - t_1)}$$

$$U(t) = (1 - \alpha) U_0 \, e^{-\lambda t}.$$

Let us see what becomes of the corresponding lead isotope. From t_0 to t_1 the mineral system is closed. We have $Pb(t_1)$ as $(t_1) = U_0(1 - e^{-\lambda t_1})$, as we assume there is no initial lead. At time t_1 the system loses a proportion (β) of lead so at $(t_1 + \Delta t)$ the lead is therefore:

$$^*Pb(t_1 - \Delta t) = (1 - \beta) U_0 (1 - e^{-\lambda t_1}).$$

Between t_1 and t, lead is produced by decay of the uranium remaining at that time.

$$^*Pb(t) = Pb(t_1 + \Delta t) + U(t_1 + \Delta t) \left[1 - e^{-\lambda(t - t_1)}\right].$$

Replacing by their values $Pb(t_1 + \Delta t)$ and $U(t_1 + \Delta t)$ gives:

$$^*Pb(t) = (1 - \beta) U_0 \left[1 - e^{-\lambda t_1}\right] + (1 - \alpha) U_0 \, e^{-\lambda t_1} \left[1 - e^{-\lambda(t - t_1)}\right].$$

We replace U_0 by $U(t)$ to return to the traditional expression:

$$\mathrm{U}_0 = \frac{\mathrm{U}(t)\mathrm{e}^{\lambda t}}{(1-\alpha)}.$$

We have to change variables because we have decided to take the present time as the origin. In the new notation (in capitals), $T = t, T_1 = t - t_1$. This gives:

$$\frac{^*\mathrm{Pb}}{\mathrm{U}} = \left[\frac{1-\beta}{1-\alpha}\right]\left[\mathrm{e}^{\lambda T} - \mathrm{e}^{\lambda T_1}\right] + \left[\mathrm{e}^{\lambda T_1} - 1\right].$$

But for the decay constant the expressions are identical for both $^{206}\mathrm{Pb}$–$^{238}\mathrm{U}$ and $^{207}\mathrm{Pb}$–$^{235}\mathrm{U}$ pairs, because the two lead isotopes and the two uranium isotopes must react chemically in an identical manner to the "crisis" which affected the systems at T_1. We write:

$$\left[\frac{^{206}\mathrm{Pb}}{^{238}\mathrm{U}}\right] = r_1 \qquad \left[\frac{^{207}\mathrm{Pb}}{^{235}\mathrm{U}}\right] = r_2$$

$$r_1 = \left(\frac{1-\beta}{1-\alpha}\right)\left[\mathrm{e}^{\lambda_8 T} - \mathrm{e}^{\lambda_8 T_1}\right] + \left[\mathrm{e}^{\lambda_8 T_1} - 1\right]$$

$$r_2 = \left(\frac{1-\beta}{1-\alpha}\right)\left[\mathrm{e}^{\lambda_5 T} - \mathrm{e}^{\lambda_5 T_1}\right] + \left[\mathrm{e}^{\lambda_5 T_1} - 1\right].$$

We can eliminate $[(1-\beta)/(1-\alpha)]$ between the two equations, so giving:

$$\frac{r_1 - \left[\mathrm{e}^{\lambda_8 T_1} - 1\right]}{r_2 - \left[\mathrm{e}^{\lambda_5 T_1} - 1\right]} = \left[\frac{\mathrm{e}^{\lambda_8 T} - \mathrm{e}^{\lambda_8 T_1}}{\mathrm{e}^{\lambda_5 T} - \mathrm{e}^{\lambda_5 T_1}}\right].$$

For fixed T and T_1, this expression takes the form:

$$\frac{r_1 - y_0}{r_2 - x_0} = constant.$$

This is the equation of a straight line in an (r_1, r_2) plot. When $\beta = \alpha = 0$, that is, when the system has remained closed, $r_1 = \left(\mathrm{e}^{\lambda_8 T} - 1\right)$ and $r_2 = \left(\mathrm{e}^{\lambda_5 T} - 1\right)$. The upper intercept with the concordia therefore corresponds to T, which is the initial age of the mineral population. When $\beta = 1$, that is, when the minerals have lost all their lead at T_1, $r_1 = \left(\mathrm{e}^{\lambda_8 T_1} - 1\right)$ and $r_2 = \left(\mathrm{e}^{\lambda_5 T_1} - 1\right)$.

The lower intercept with the concordia corresponds to T_1. When β and α are between 0 and 1, the data points are on the straight line joining the two points of the concordia T and T_1. When more lead than uranium is lost the points are between T and T_1. When more uranium than lead is lost the points are "above" T and therefore above the concordia (see Figure 3.7).

We can then address the inverse problem and say that when the data points obtained on a series of uranium-rich cogenetic minerals are aligned in the $(^{206}\mathrm{Pb}/^{238}\mathrm{U}, {}^{207}\mathrm{Pb}/^{235}\mathrm{U})$ plot, the intersections of the straight line they define with the concordia curve determine the initial age of the family of minerals and the age of the disruption experienced. We give a simple historical example, that of the Morton gneiss of Minnesota. The U–Pb measurements being made on zircons, we obtain the age of the perturbation (Figure 3.8).

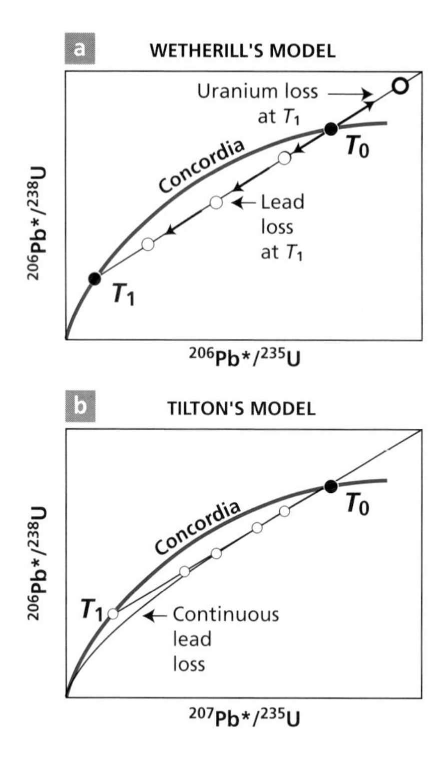

Figure 3.7 Theoretical diagrams of concordia construction for ^{238}U–^{206}Pb and ^{235}U–^{207}Pb systems. (a) The episodic loss model (Wetherill's model) at T_1 for zircons of age T_0; (b) the constant loss model (Tilton's model) by diffusion for the same zircons of age T_0. The parametric curves are plotted on each diagram. The ratios are given in number of atoms.

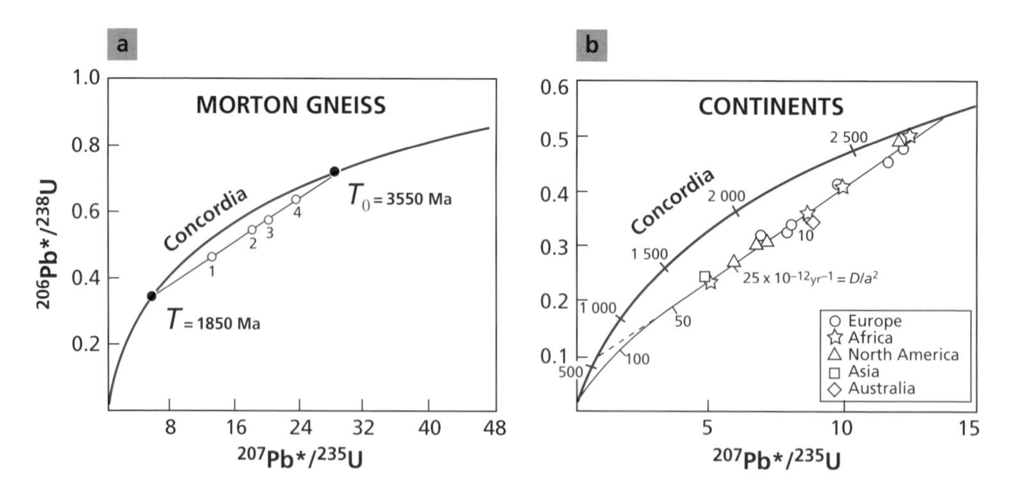

Figure 3.8 Concordia plots. (a) The Morton gneiss of Minnesota; (b) different continental zircons of common age 2.7 Ga. These plots can be interpreted as constant losses although they have different geological histories.

Wetherill 's model has been used abundantly. Generally it is borne out but it has often been observed that the lower intercept is difficult to interpret in geological terms because the age determined corresponds to neither a known tectonic nor a metamorphic episode. Thus, taking zircons of several continents of the same age of 2.7 Ga, and plotting the U–Pb results on the concordia diagram, they seem to define the same lower intercept, while the geological histories of the various parts of the continents and in particular the tectono-metamorphic episodes later than 2.7 Ga are very different. To account for this, in the 1960s **George Tilton**, at the Carnegie Institution, developed another model (Tilton, 1960). He assumed a continuous loss of lead throughout the rock's history. In such a case, the first part of the curve of evolution in the concordia diagram is a straight line starting from the initial age. It then curves towards the origin. From this model, the lower intercept is geologically meaningless and depends solely on the initial age. This can be demonstrated mathematically by using kinetic equations. If we consider a continuous lead loss, the equations are written:

$$\begin{cases} \dfrac{\mathrm{dPb}^*}{\mathrm{d}t} = \lambda \mathrm{U} - G\,\mathrm{Pb} \\ \dfrac{\mathrm{dU}}{\mathrm{d}t} = -\lambda \mathrm{U} \end{cases}$$

where G is a coefficient of lead loss which is assumed constant to simplify matters (we assume there is no uranium loss). Therefore:

$$\frac{\mathrm{d}\dfrac{\mathrm{Pb}^*}{\mathrm{U}}}{\mathrm{d}t} = \lambda + (\lambda - G)\frac{\mathrm{Pb}^*}{\mathrm{U}}.$$

By positing $\mathrm{Pb}^*/\mathrm{U} = r$ and by integrating r_8 and r_5, we obtain:

$$r_8 = \frac{\lambda_8}{\lambda_8 - G}\left(e^{(\lambda_8 - G)t} - 1\right), \quad r_5 = \frac{\lambda_5}{\lambda_5 - G}\left(e^{(\lambda_5 - G)t} - 1\right).$$

Numerically, it can be seen that in an (r_8, r_5) plot supposing G is identical for both lead isotopes, the curve is a straight line in its first part and only curves towards the origin. It therefore has no significant lower intercept. Only the initial age is significant. If a straight line is drawn through the data points, the upper intersection with the concordia indeed gives the initial age but the lower intersection with the concordia is meaningless.

Allègre, Albarède, Grünenfelder, and **Köppel** (1974) showed that we could switch from one model to the other. If several tectono-metamorphic crises are superimposed, they may also generate a **discordia** whose lower intersection with the concordia is meaningless. This is the case of old inherited zircons in the Alps. A complex polymetamorphic history generates regularities similar to a continuous loss.

As seen, interpreting the lower intercept is neither straightforward nor unequivocal! Each case must be carefully examined, that is, one must have sound geological knowledge of the region before reaching any conclusion. However, the upper intercept, when well defined mathematically (spaced data points), seems more robust.

Of course, an "apparent age" can be determined for each discordant mineral by the $^{207}Pb/^{206}Pb$ method alone. These apparent $^{207}P/^{206}P$ ages vary from the age of the lower intersection to the initial age. We therefore have a whole series of apparent ages.[2]

And yet the chronometric equation is misleading a priori since if we consider only the $^{207}Pb-^{206}Pb$ dating equation and suppose that lead is lost without isotopic fractionation, it is easy to believe the system has remained closed for lead! This is what was thought some 50 years ago when some workers asserted that since the same granite had $^{207}Pb/^{206}Pb$ ages ranging from say 200 million to 1 billion years, that was evidence it had formed by diffusion in solid state over hundreds of millions of years! In fact, the dating equation contains a hidden variable – uranium! The true equation of the $^{207}Pb/^{206}Pb$ ratio, in the case of a episodic loss, is written:

$$\frac{^{207}\text{Pb}}{^{206}\text{Pb}} = \frac{1}{137.8}\left[\frac{\left(\dfrac{1-\beta}{1-\alpha}\right)\left(e^{\lambda_5 T} - e^{\lambda_5 T_1}\right) + \left(e^{\lambda_5 T_1 - 1}\right)}{\left(\dfrac{1-\beta}{1-\alpha}\right)\left(e^{\lambda_8 T} - e^{\lambda_8 T_1}\right) + \left(e^{\lambda_8 T_1 - 1}\right)}\right].$$

This equation shows that the ratio depends on the coefficients of loss β and α (unless $T_1 = 0$, that is, if the loss is at the present day). This also illustrates that apparent and true ages must not be mixed without care! Of course, it is tempting to measure the $^{207}Pb/^{206}Pb$ ratio alone because it requires just an isotope measurement of lead without measuring the lead and uranium content, but, as has been seen, its interpretation is ambiguous. It must therefore be used by taking certain precautions and by being familiar with the limits.

Exercise

We assume a continuous loss with $G = 2\lambda_8$. Calculate the apparent $^{206}Pb/^{238}U$, $^{207}Pb/^{235}U$, and $^{207}Pb/^{206}Pb^*$ ages for a zircon whose initial age is 2.7 Ga.

Answer

$^{206}Pb/^{238}U = 1.8$ Ga, $^{207}Pb/^{235}U = 2.18$ Ga, and $^{207}Pb/^{206}Pb^* = 2.480$ Ga.

Generalizing the concordia method

Allègre (1964) and **Allègre** and **Michard** (1964) extended this approach to other pairs of chronometers (Figure 3.9).

- **Uranium/lead–thorium/lead.** This was a natural extension. The big difference was that the Th/U ratio is variable and so the "magical" conditions of the U–Pb system do not hold. The discordia straight line cuts the concordia at the upper intercept only. There is no lower intercept as the discordia is actually a curve (Steiger and Wasserburg, 1966; Allègre, 1967) (Figure 3.9).
- **Rubidium/strontium–potassium/argon.** Here again the discordia cuts the concordia at one point only. This generalization has confirmed the coherence of the two systems and their behavior when disrupted (Allègre and Michard, 1964).

[2] Apparent ages are measured by the slope of the straight line joining the origin and the experimental point on the (r_5, r_8) plots.

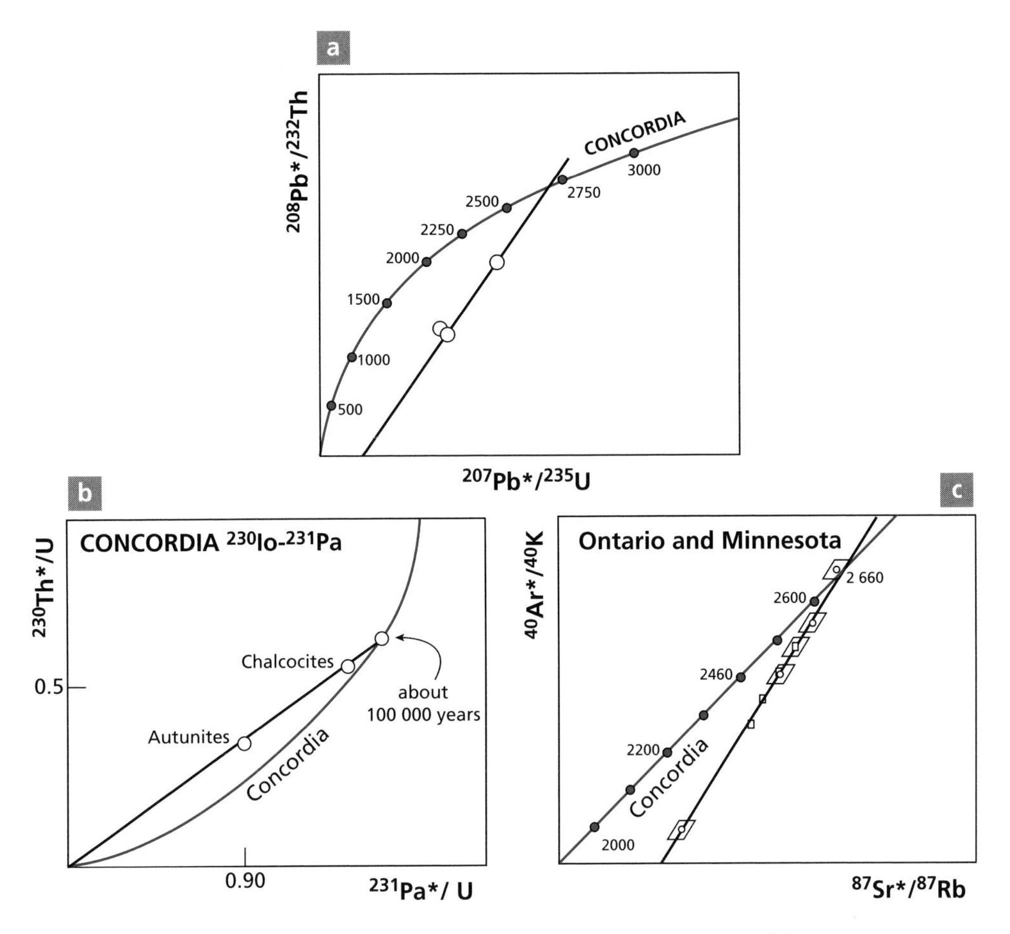

Figure 3.9 Generalization of the concordia method. (a) U–Th–Pb concordia; (b) Concordia applied to methods of radioactive disequilibrium; (c) concordia applied to K–Ar, Rb–Sr methods. After Allègre (1967), Allègre (1964), and Allègre and Michard (1964).

- **Uranium–^{230}Th, U–^{231}Pa.** This approach has been used for dating secondary mineralizations of uranium. It was then extended recently to other examples of dating corals. These are mentioned here but not elaborated on (Allègre, 1964).

Such generalization of the concordia method is above all of fundamental methodological interest. It shows that the behavior of the various radiogenic isotopes, although original, is not "autonomous" and that there is redundancy which explains the regularities observed and allows simple mathematical modeling.

3.2.3 Stepwise thermal extraction

The idea behind this approach is to suppose that, when disruptive phenomena occur, radiogenic isotopes migrate by diffusion towards the mineral boundaries or to cracks and that there are "cores" which have resisted and which can be considered to be closed boxes. We go

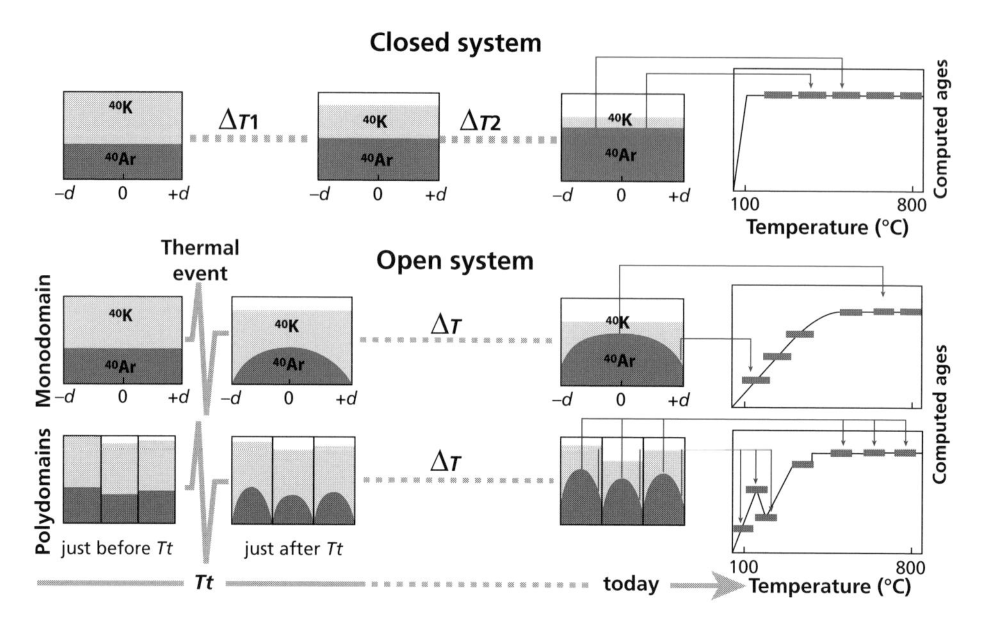

Figure 3.10 The ^{39}Ar–^{40}Ar method. Each small rectangle represents horizontally the size of a mineral (the center has coordinate zero and the edges $+d, -d$). The ordinates show the ^{40}K and ^{40}Ar contents. Top: evolution over time in a closed system. Notice that ^{40}K decreases and ^{40}Ar increases. When the experiment is conducted we find a simple clear plateau indicating the age. Bottom: the evolution of an open system. We suppose a thermal event occurred at T_t, thermal diffusion occurred, and argon escaped from the rims, hence the bell-shaped distribution. Then the system remained closed and ^{40}K decayed producing ^{40}Ar. When the argon is extracted stepwise, it can be seen that a certain temperature is required to reach the plateau because the edges are degassed first. The bottom plot is analogous except we have supposed a mineral with a complex structure with several domains separated by cracks, faults, or grain joints. The first part of the extraction has a complex appearance.

and get the "information" by causing artificial diffusion. Two chronometers have been the subject of such an approach: that of the K–Ar method on various minerals or rocks and that of the U–Pb method on zircon.

The ^{39}Ar–^{40}Ar method

This method first came to light at Berkeley in the laboratory of **John Reynolds** (Merrihue and Turner, 1966) but was above all the product of work by **Grenville Turner** (1968) at Sheffreld University in England. Here is what it involves. First the sample under test is irradiated by a stream of neutrons. A nuclear reaction transforms the ^{39}K into ^{39}Ar. This is a neutron–proton (n, p) reaction. Once this operation has been accomplished, the test sample is heated by temperature stages and the $^{40}Ar/^{39}Ar$ ratio measured at each step (Figure 3.10).

Naturally, at the same time as the sample, we irradiate a control sample containing a known quantity of K whose ^{39}Ar content is to be analyzed. The $^{40}Ar/^{39}Ar$ ratio is equivalent to an $^{40}Ar/^{40}K$ ratio and so the dating formula can be applied to it. We therefore obtain a range of (apparent) ages depending on the outgassing temperature.

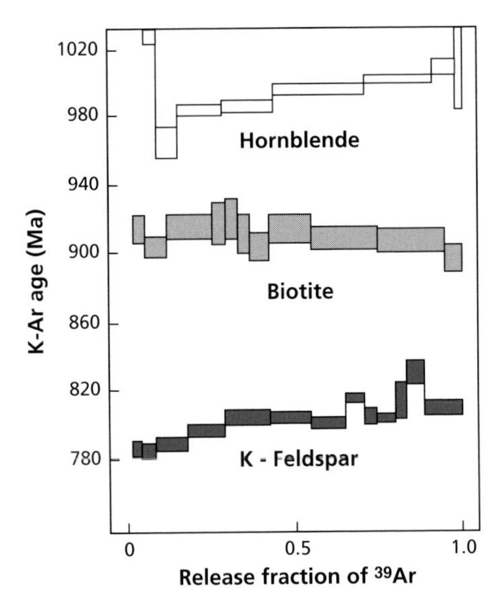

Figure 3.11 Examples of ^{39}Ar–^{40}Ar age spectra for three minerals from a rock aged 1000 Ma, showing sensitivity to the different minerals (from Berger and York, 1981).

The idea underpinning the method is that, since ^{40}Ar diffuses readily, it has departed from the most vulnerable crystallographic sites over the course of geological time but has remained in the more resistant sites (Figure 3.11). By stepwise heating, we drive out successively the fractions of argon located in increasingly retentive sites, finally getting it from "closed" sites. This method became particularly successful with lunar exploration. At the time, the ages reported by Turner concurred so well with those obtained by the Rb–Sr method that the model was "established" in spectacular fashion. Subsequently it was refined. First it was recalled that the various minerals did not close at the same temperature, which is unsurprising. Instead of looking for just the "true" age, attempts were made to take advantage of the differences observed to determine the conditions under which the rock had cooled by using the concept of closure age already mentioned. Later still, a laser was used to degas specific sites of the various minerals and in stages.

An analogous method was proposed for U–Xe dating but it is still little used for the time being (Figure 3.12), which is probably a shame. It may be resurrected. It involves irradiating the sample and producing various xenon isotopes by fission of ^{235}U. As the xenon-induced fission spectrum is different from the spontaneous fission spectrum of ^{238}U, we can measure the ^{136}Xe produced by natural fission and the uranium in the mineral. Then the age is calculated (Shukolyukov *et al.*, 1974; Teitoma *et al.*, 1975).

Direct ^{206}Pb–^{207}Pb analysis

This technique is even more audacious although it stems from the same principle as the previous one. It involves depositing the mineral under analysis by the Pb–Pb method (zircon) directly on the TIMS mass spectrometer filament. It is heated in stages and the Pb diffuses from the zircon and ionizes and the $^{207}Pb/^{206}Pb$ ratio given off is measured (Figure 3.13). This method, devised by **Koztolanyi** at Nancy in France in 1965, was neglected for 20 years

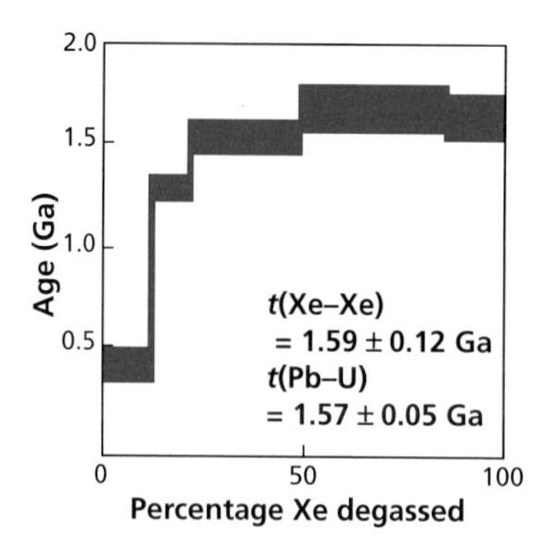

Figure 3.12 Example of a spectrum of the ^{133}Xe–^{136}Xe method imitated from ^{39}Ar–^{40}Ar for a Rapakivi granite from Finland. After Shukolyukov *et al.* (1994).

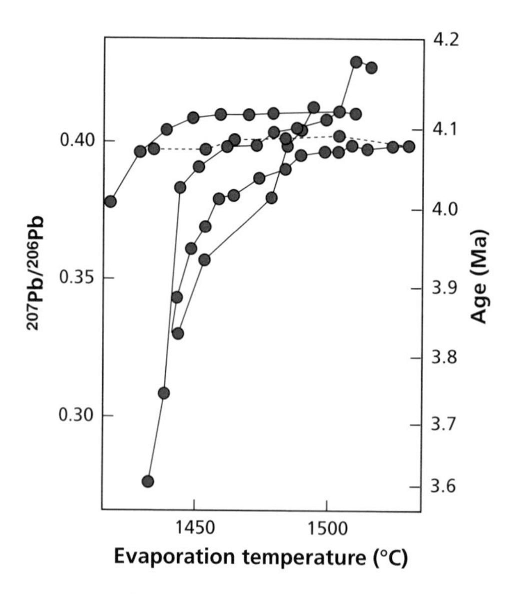

Figure 3.13 The ^{207}Pb–^{206}Pb spectrum obtained by thermo-ionization of lead. This spectrum is obtained by progressively heating the filament after having deposited zircon on it. The example shown is of detrital zircon from Australia, among the oldest recorded (Kober, 1986).

before being taken up by **Kober** in 1986! It yields interesting and sometimes impressive results, even if all criteria of concordance or discordance are lost, preventing any reliability check as the concordia diagram cannot be used.

3.3 Poor systems and the radiometric isotopic correlation diagram

With poor systems, the main problem is not just the question of openness but also that of the initial presence, when the system closes, of a certain amount of radiogenic isotope. How can this be evaluated?

3.3.1 The ^{87}Rb–^{87}Sr system

This method was proposed, after an initial attempt by Australians **William Compston** and **Peter Jeffrey** (1959), then at the University of Western Australia, by **Louis Nicolaysen** (1961) of South Africa's Witwatersrand University for the ^{87}Rb–^{87}Sr system, but it is more general and applies to all parent–daughter systems.

Let us turn back to the fundamental ^{87}Rb–^{87}Sr decay equation in a closed system:

$$^{87}\mathrm{Sr}(t) = {}^{87}\mathrm{Sr}(0) + {}^{87}\mathrm{Rb}(t)(e^{\lambda t} - 1).$$

Let us divide member by member by ^{86}Sr, which is one of the stable Sr isotopes and whose abundance is not dependent on radioactivity and is therefore constant over time. The equation becomes:

$$\left(\frac{^{87}\mathrm{Sr}}{^{86}\mathrm{Sr}}\right)_t = \left(\frac{^{87}\mathrm{Sr}}{^{86}\mathrm{Sr}}\right)_0 + \left(\frac{^{87}\mathrm{Rb}}{^{86}\mathrm{Sr}}\right)_t (e^{\lambda t} - 1).$$

The mathematical operation of ^{87}Sr normalization by ^{86}Sr is not just a simple algebraic manipulation. It brings out the strontium isotope ratios in the equation. This is the magnitude measured by the mass spectrometer. Now, we have seen that isotope ratios obey precise laws, particularly those of isotope exchange.

Let us now consider a series of cogenetic boxes (assumed to be in thermodynamic equilibrium). As a result of isotopic equilibration, they have the same initial $(^{87}\mathrm{Sr}/^{86}\mathrm{Sr})_0$ isotope ratio and the same age. In an $\left[^{87}\mathrm{Sr}/^{86}\mathrm{Sr},\ ^{87}\mathrm{Rb}/^{86}\mathrm{Sr}\right]$ plot they are aligned on a straight line of slope $(e^{\lambda t} - 1)$ and of ordinate at the origin of $\left(^{87}\mathrm{Sr}/^{86}\mathrm{Sr}\right)_0$. This straight line is an **isochron**, which is the locus of boxes of the same age (Figure 3.14). If we wish to break down the way the system evolves over time, we can say that at time $t = 0$, the points representing all the boxes lie on a horizontal line, each box being characterized by its own ^{87}Rb/^{86}Sr ratio but with the same strontium isotope ratios. Then each box evolves along a diagonal of slope -1 with $+d\,^{87}\mathrm{Sr}/dt = -d\,^{87}\mathrm{Rb}/dt$. The straight line rotates around the fixed point $\left(^{87}\mathrm{Sr}/^{86}\mathrm{Sr}\right)_0$. Its slope is constantly $(e^{\lambda t} - 1)$.

Remark

The inverse problem is interesting, of course. Let us take the example of a series of minerals (or total rocks) which are assumed to be cogenetic (minerals from the same rock, or rock from the same massif). The ^{87}Sr/^{86}Sr and ^{87}Rb/^{86}Sr isotope ratios of each of them are measured. If the representative points lie in a straight line, the slope of the straight line is equal to $(e^{\lambda t} - 1)$ and the age of the system of boxes can be calculated.

Exercise

Annie Michard measured the Mont-Louis, granite, Milhas, and Quérigut in the Pyrénées Orientales (France) (Table 3.2). Draw the isochron graph and calculate the age of the series of granite rocks of the Quérigut and the initial ratio (Michard and Allègre, 1979).

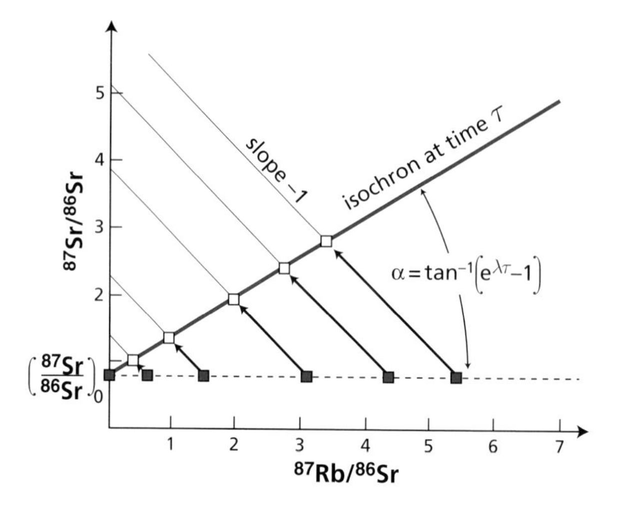

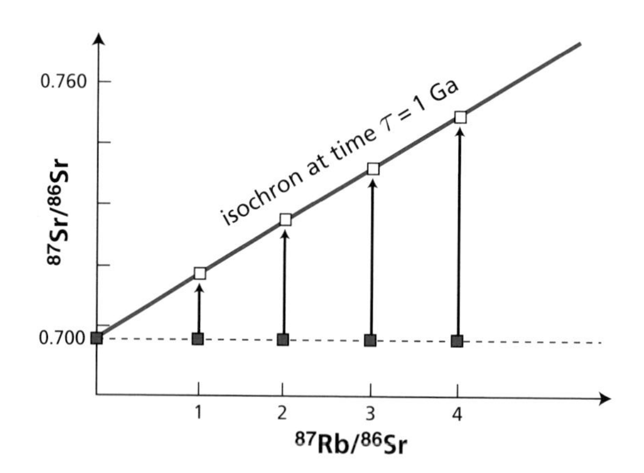

Figure 3.14 Principle of the ^{87}Rb–^{87}Sr isochron method. At time $t=0$ the points have the same strontium isotope ratio and different $^{87}Rb/^{87}Sr$ ratios. They therefore all lie on a horizontal line corresponding to $(^{87}Sr/^{86}Sr)(_0)$. After time T, the points are aligned on a straight line of slope $(e^{\lambda t} - 1) \approx \lambda t$. In practice, in real graphs, the straight lines of evolution do not have a slope of -1 because the $^{87}Rb/^{86}Sr$ ratio varies little and the $^{87}Sr/^{86}Sr$ ratio varies greatly, given the chosen scales. The evolution vectors are very steep.

We calculate the $^{87}Rb/^{86}Sr$ ratios. The experimental points are plotted on the $(^{87}Sr/^{86}Sr$, $^{87}Rb/^{86}Sr)$ graphs not shown here. The straight line through the points "yields" an age of $T = 271 \pm 10$ Ma, with an $(^{87}Sr/^{86}Sr)_{initial}$ ratio $= 0.710 \pm 0.005$.

This method, often called the **isochron method**, is one of the pillars of radiochronology. It both dispenses with the assumption (accepted for rich systems) that the $^{87}Sr(0)$ is negligible or constant and so allows poor systems to be used and **tests the closed system hypothesis** by alignment or non-alignment of the measurements made on the different boxes. If Rb

Table 3.2 Mont-Louis, Milhas, and Quérigut granites

	$[^{87}Sr/^{86}Sr]$	Rb (ppm)	Sr (ppm)
Mont-Louis sample numbers:			
4 B1	0.721	157	181
4 B2	0.719	146	200
Milhas sample numbers:			
V 90	0.723	150	146
V 92	0.717	164	302
Quérigut sample numbers:			
A	0.714	105	241
B	0.719	189	214
C	0.723	161	135
A4	0.720	160	153
X	0.7205	155	170
Q1	0.742	202	88
Q2	0.743	257	90
Q3	0.746	245	74
Ap1	0.780	192	32
Ap2	0.875	257	18

and Sr have migrated differentially, then the representative points of the various boxes will not remain aligned. In the favorable case, this method can be used to measure an age (t) and to determine the initial $^{87}Sr/^{86}Sr$ ratio, which we will see later to be important. Measuring an age therefore involves measuring the Sr isotopic ratio, measuring Rb and Sr concentrations for each box and this is for a series of cogenetic boxes.

3.3.2 Two-stage models

Imagine we have a series of cogenetic boxes evolving in a closed system. All the systems are therefore aligned on an isochronous straight line. Suppose that at t_1 a sudden event occurs which re-homogenizes the boxes isotopically. This event may be metamorphism, for example. The sub-systems will exchange their atoms and become isotopically homogeneous. The representative points of the boxes then lie on a horizontal line whose ordinate at the origin is written $\left(^{87}Sr/^{86}Sr\right)_{m,\,t_1}$ with m, t_1 signifying the mean value at t_1. Then the boxes evolve radiogenically until the present time in a closed system. The alignment observed between the points representing the boxes has a slope of $\left(e^{\lambda t_1}-1\right)$ and an ordinate at the origin of $\left(^{87}Sr/^{86}Sr\right)_{m,\,t_1}$.

Let us explain our model a little to make it realistic. We consider the cogenetic boxes are both whole rocks and minerals. Metamorphism occurs at time t_1 which re-homogenizes the minerals isotopically around the value of the whole rock, but it is not intense enough to re-homogenize the whole rocks at the scale of the entire massif (Figure 3.15).

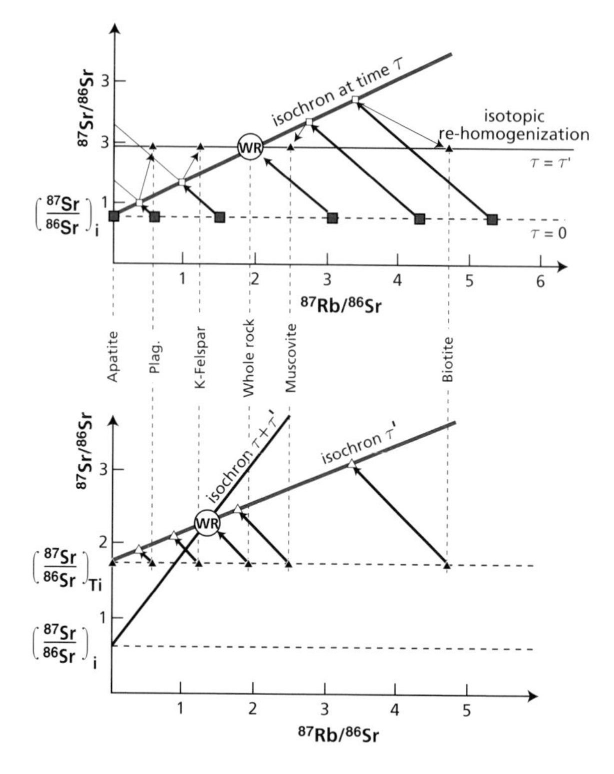

Figure 3.15 Isotopic re-homogenization of minerals at time T_1. Top: the re-homogenization event. Bottom: the evolution of the system after T_1. WR, whole rock. After Lanphere *et al.* (1964).

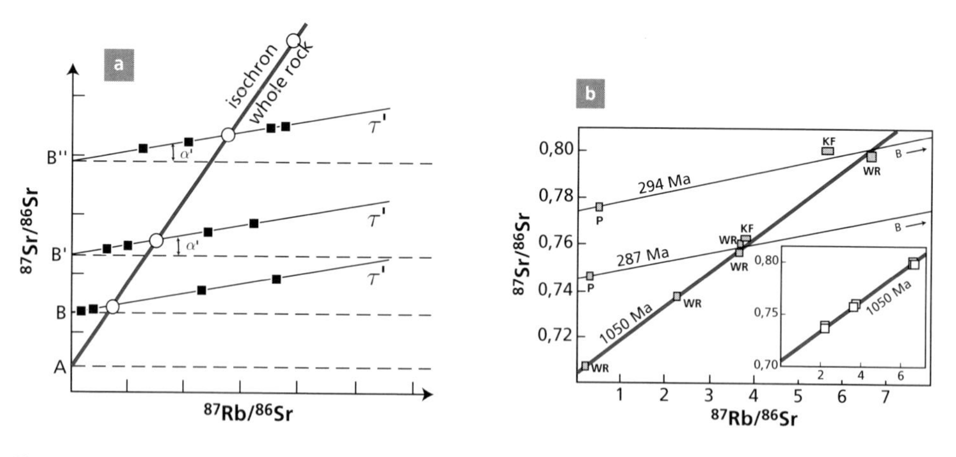

Figure 3.16 Example of two types of isochron. (a) The whole-rock isochron indicates the initial age, that of the minerals dates the age of the disruptive event; (b) classic example of Baltimore gneiss with the whole-rock isochron and the mineral isochrons. WR, whole rock; B, biotite; KF, potassium feldspar; P, phosphate. After Wetherill *et al.* (1968).

The $\left({}^{87}\mathrm{Sr}/{}^{86}\mathrm{Sr},\ {}^{87}\mathrm{Rb}/{}^{86}\mathrm{Sr}\right)$ graph shows two types of straight line: the straight line for whole rock of slope $\left(e^{\lambda t_0}-1\right)$ and the straight lines of the minerals of slope $\left(e^{\lambda t_1}-1\right)$. This method is illustrated with the historical example of the Baltimore gneiss of the United States (Figure 3.16.b) studied by the group at the Carnegie Institution of Washington.

3.3.3 Extension of the isochron graph to other radiochronometers

The method initially developed for ^{87}Rb–^{87}Sr was extended to other chronometers with specific features for each, particularly different sensitivities to disruptive phenomena. Without this isochron method, the ^{147}Sm–^{143}Nd and ^{176}Lu–^{176}Hf methods would not have come to light, as they have no rich systems.

The case of ^{147}Sm–^{144}Nd

This method was introduced by **Notsu** *et al.* (1973) and developed by **Lugmair** and **Marti** (1977). The basic dating equation is written:

$$\frac{^{143}\mathrm{Nd}}{^{144}\mathrm{Nd}} = \left(\frac{^{143}\mathrm{Nd}}{^{144}\mathrm{Nd}}\right)_0 + \left(\frac{^{147}\mathrm{Sm}}{^{143}\mathrm{Nd}}\right) \lambda_{147} t \qquad \lambda_{147} = 6.54 \cdot 10^{-12}\mathrm{yr}^{-1}.$$

The stable reference isotope is ^{144}Nd. This method has the advantage of being applicable to ultrabasic and basic rocks, which cannot be readily dated by ^{87}Rb–^{87}Sr, and to granites (Figure 3.17). With regard to minerals, garnet and pyroxene are useful since geochemical fractionation between Sm and Nd, although slight, is relatively constant across the petrological spectrum.

Exercise

Calculate the initial $^{143}Nd/^{144}Nd$ ratio of a rock of which we know the age (2 Ga), the present-day $^{143}Nd/^{144}Nd$ ratio (0.512 556) and whose neodymium and samarium contents are 36 and 10.4 ppm respectively. What do you think the error is if samarium and neodymium are measured to the nearest 3%, the isotope ratio to 1.10^{-5} and age to $\pm 10\%$?

Answer

$^{143}Nd/^{144}Nd = 0.510\,31$ Absolute error $= \pm 0.000\,35$.

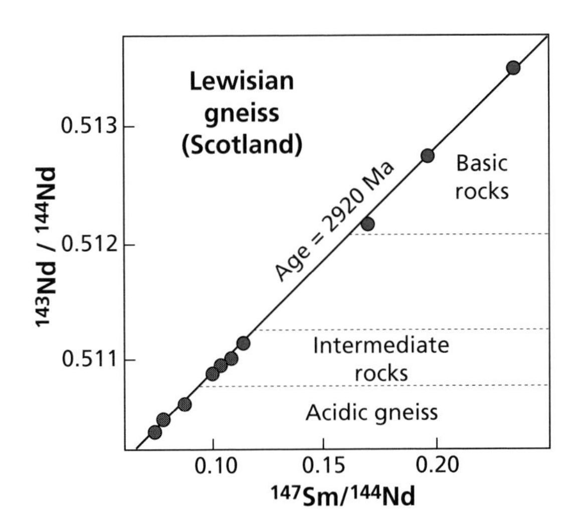

Figure 3.17 Example of a ^{147}Sm–^{143}Nd isochron on rocks of the Lewisian Complex of Scotland. After Hamilton *et al.* (1979).

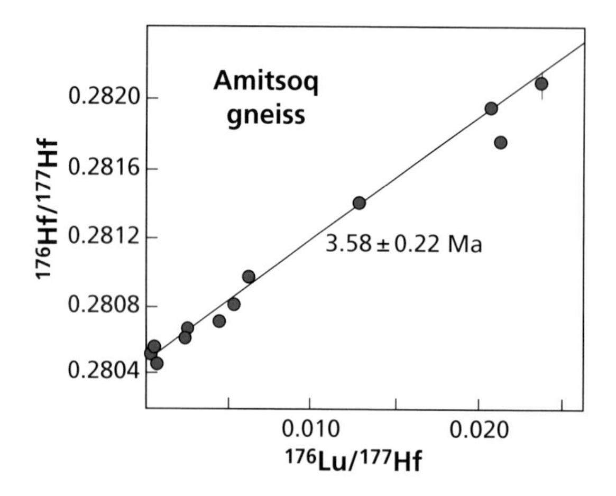

Figure 3.18 The lutetium–hafnium isochron of the Amitsoq gneiss. After Blichert-Toft *et al.* (1999).

The case of ^{176}Lu–^{176}Hf

In this case we take the ^{177}Hf isotope as the stable reference.

$$\frac{^{176}\text{Hf}}{^{177}\text{Hf}} = \left(\frac{^{176}\text{Hf}}{^{177}\text{Hf}}\right)_0 + \left(\frac{^{176}\text{Lu}}{^{177}\text{Hf}}\right)\lambda t \qquad \lambda_{176} = 1.94 \cdot 10^{11}\ \text{yr}^{-1}$$

This method, introduced by **John Patchett** and **Mitsunobu Tatsumoto** (1980) working at the U.S. Geological Survey in Denver was long neglected because of analytical difficulties in measuring hafnium. It has been increasingly used with the advent of ICPMS (Figure 3.18).

Exercise

The question of metamorphism related to the subduction of the Himalayas is an important one for geologists. One way of determining this age is to use eclogite (that is, basalt metamorphosed at high pressure) whose mineral composition is sodium pyroxene termed omphacite and pyrope garnets.

The occurrence of garnet and pyroxene makes lutetium–hafnium dating easier. Table 3.3 gives the results obtained on eclogite rock from Ladakh. Calculate the age of metamorphism (formation of eclogite) (De Sigoyer *et al.*, 2000).

Answer

55 Ma.

The case of ^{187}Re–^{187}Os

This method was introduced by **Hirt** *et al.* (1963) and developed by **Luck** *et al.* (1980). The reference isotope being ^{186}Os and more recently ^{188}Os, we can write:

$$\frac{^{187}\text{Os}}{^{186}\text{Os}} = \left(\frac{^{187}\text{Os}}{^{186}\text{Os}}\right)_0 + \left(\frac{^{187}\text{Re}}{^{186}\text{Os}}\right)\lambda_{187} t \qquad \lambda_{187} = 1.62 \cdot 10^{-11}\text{yr}^{-1}.$$

This method is potentially important but, although meaningful on meteorites (Luck *et al.*, 1980), it is difficult to apply because the system does not seem to remain closed, particularly

Table 3.3 Dating of eclogite rock from Ladakh

	Lu (ppm)	Hf (ppm)	Lu/Hf	$^{176}Hf/^{177}Hf$
Whole rock	0.3630	3.5991	0.1009	0.282996 ± 0.014310
Garnet	0.7324	3.0907	0.2370	0.283012 ± 0.033630
Pyroxene	0.0302	3.0727	0.0098	0.28977 ± 0.001400

Source: After de Sigoyer *et al.* (2000).

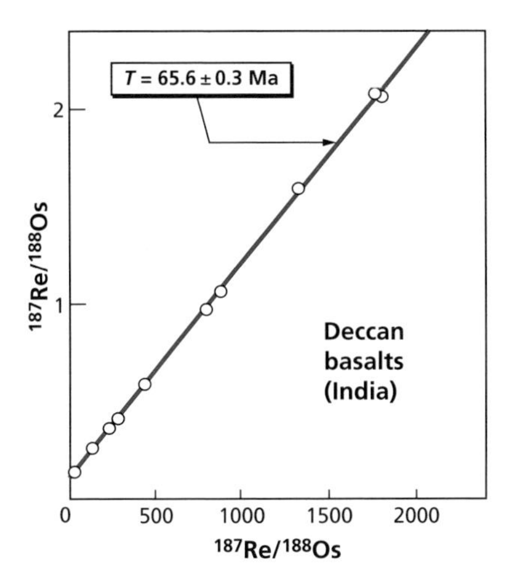

Figure 3.19 The rhenium–osmium isochron of the Deccan Traps. After Allègre *et al.* (1999).

for molybdenite, although containing a large amount of Re, nor for basalt, with a very high Re/Os ratio. Even so the present author obtained isochrons including one for Deccan basalt (Figure 3.19).

The case of uranium–thorium–lead

Taking ^{204}Pb as the reference isotope, the dating equations are written:

$$\frac{^{206}Pb}{^{204}Pb} = \left(\frac{^{206}Pb}{^{204}Pb}\right)_0 + \left(\frac{^{238}U}{^{204}Pb}\right)\left(e^{\lambda_8 t} - 1\right)$$

$$\frac{^{207}Pb}{^{204}Pb} = \left(\frac{^{207}Pb}{^{204}Pb}\right)_0 + \left(\frac{^{235}U}{^{204}Pb}\right)\left(e^{\lambda_5 t} - 1\right)$$

$$\frac{^{208}Pb}{^{204}Pb} = \left(\frac{^{208}Pb}{^{204}Pb}\right)_0 + \left(\frac{^{232}Th}{^{204}Pb}\right)\left(e^{\lambda_2 t} - 1\right).$$

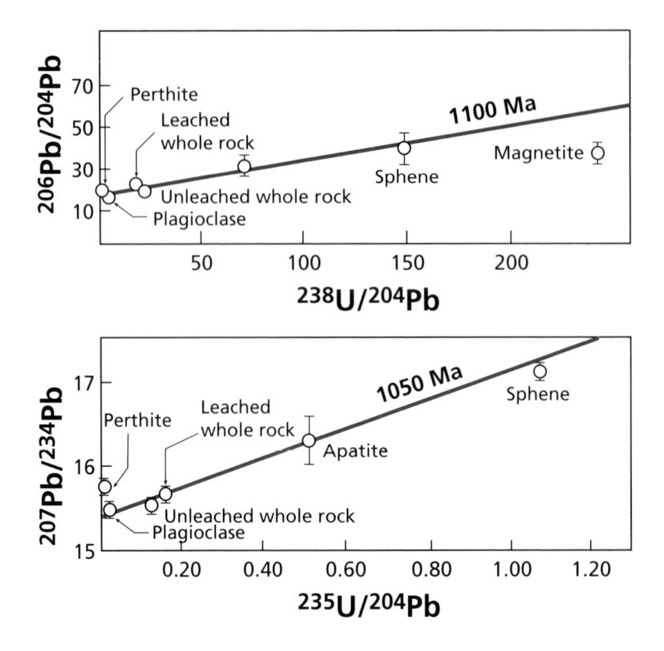

Figure 3.20 Isochrons of lead–uranium minerals of the Essonville granite. After Tilton *et al.* (1958).

These methods have undoubtedly been insufficiently used (Figure 3.20) although they are very powerful as they have an internal concordance criterion (chemically identical parent and daughter isotopes) guaranteeing their reliability. They should combine the advantages of the isochron and concordia methods. Unfortunately this is not yet the case.

Exercise

Given that the $^{238}U/^{204}Pb$ ratio of granite varies from 2 to 30 and that the Th/U ratios vary from 1 to 10, calculate the deviation in the $^{206}Pb/^{204}Pb$, $^{207}Pb/^{204}Pb$, and $^{208}Pb/^{204}Pb$ ratios for granites of 1, 2, and 3 Ga.

In calculating the $^{206}Pb/^{204}Pb$, $^{207}Pb/^{204}Pb$, and $^{208}Pb/^{204}Pb$ ratios of the granite, it is assumed that this granite derives from evolution of a reservoir with ratios $^{238}U/^{204}Pb = 7$ and $Th/U = 4$, having evolved for 4.5 Ga. The initial ratios of these reservoirs are $^{206}Pb/^{204}Pb = 9.30$, $^{207}Pb/^{204}Pb = 10.29$, and $^{208}Pb/^{204}Pb = 29.47$.

Answer

$\Delta(^{206}Pb/^{204}Pb) = 15.534–29.998$, $\Delta(^{207}Pb/^{204}Pb) = 14.44–16.18$, and $\Delta(^{208}Pb/^{204}Pb) = 35.14–79.69$.

3.3.4 The lead–lead method

We have come across the $^{206}Pb/^{207}Pb$ method when the two isotopes were of purely radiogenic origin. Minerals or rocks generally contain some initial lead. The specific feature of the two U–Pb systems can be exploited provided correction is made for the initial lead. Combining the first two equations of the previous paragraphs gives:

$$\frac{\dfrac{^{206}\text{Pb}}{^{204}\text{Pb}} - \left(\dfrac{^{206}\text{Pb}}{^{204}\text{Pb}}\right)_0}{\dfrac{^{207}\text{Pb}}{^{204}\text{Pb}} - \left(\dfrac{^{207}\text{Pb}}{^{204}\text{Pb}}\right)_0} = 137.8\,\frac{(e^{\lambda_8 t} - 1)}{(e^{\lambda_5 t} - 1)} = \frac{^{206}\text{Pb}^*}{^{207}\text{Pb}^*},$$

$^{206}Pb^*$ and $^{207}Pb^*$ being the radiogenic fractions of ^{206}Pb and ^{207}Pb.

In a $(^{206}Pb/^{204}Pb,\ ^{207}Pb/^{204}Pb)$ plot, if a suite of cogenetic boxes becomes isotopically homogenized, the representative points will cluster in a **point**, instead of a straight line in the simple parent–daughter case (Figure 3.21). The radioactive system will evolve

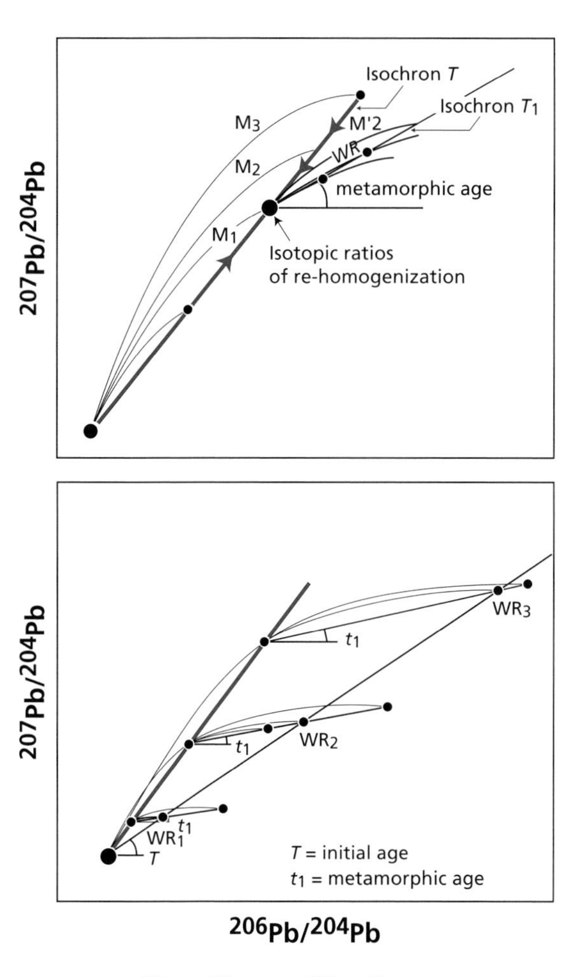

Figure 3.21 Plot of ^{207}Pb–^{204}Pb and ^{206}Pb–^{204}Pb isotope evolution for a complex geological history. (a) The system is assumed to have re-homogenized at T_1 as a result of a secondary event; (b) the evolution of three rocks (WR_1, WR_2, and WR_3) whose minerals re-homogenized at t_1.

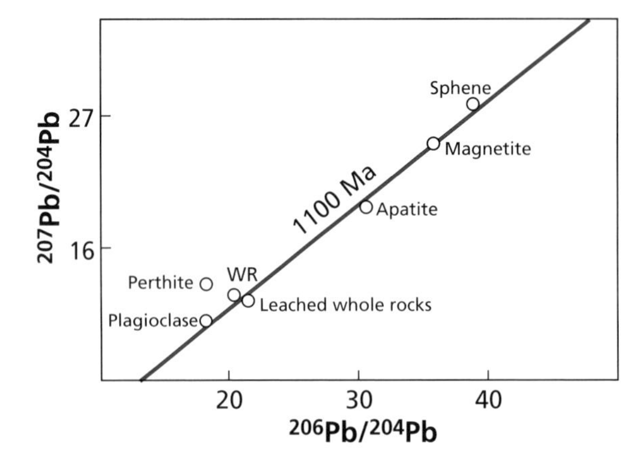

Figure 3.22 Mineral isochron (Pb–Pb) of the Essonville granite. Data after Tilton *et al.* (1958).

subsequently with different trajectories depending on the $^{238}U/^{204}Pb$ ratios. But if the system remains closed, the cogenetic boxes will remain on a straight line (Figure 3.22). The initial age t_0 of the system can be determined by measuring the slope of whole rocks and t_1 can be determined by measuring the isochron of the minerals. As this method involves measuring lead isotope compositions only, with no need to measure U and Pb concentrations, it is far easier to implement experimentally. It is very widely employed.

Exercise

The Pb isotope composition and U and Pb contents of the Muntshe Tundra massif in Russia's Kola Peninsula have been determined (Manhès *et al.*, 1980) (Table 3.4). Determine the Pb–Pb age. What can you say of the U–Pb or Th–Pb ages?

Answer

The Pb–Pb isotope age is 2.13 ± 0.25 Ga. No age can be determined from the U–Pb isotope plot. The Th–Pb isotope plot indicates an age of 2.09 ± 0.28 Ga.

Table 3.4 Lead isotope composition and uranium, thorium, and lead contents of the Muntshe Tundra massif

Sample	$^{206}Pb/^{204}Pb$	$^{207}Pb/^{204}Pb$	$^{208}Pb/^{204}Pb$	Pb (ppm)	U (ppm)	Th (ppm)	$^{232}Th/^{204}Pb$
1B	14.70	14.88	35.07	3.63	0.0391	0.375	6.04
2B	14.83	14.91	35.18	8.93	0.0318	0.928	6.11
3B	14.72	14.89	35.15	0.97	0.0038	0.1194	7.2
4B	15.38	14.98	35.72	1.40	0.0413	0.299	12.2
5B	15.84	15.03	36.29	1.02	0.0625	0.287	16.85
6B	15.12	14.95	35.55	9.85	–	1.55	9.45

Source: After Manhès *et al.* (1980).

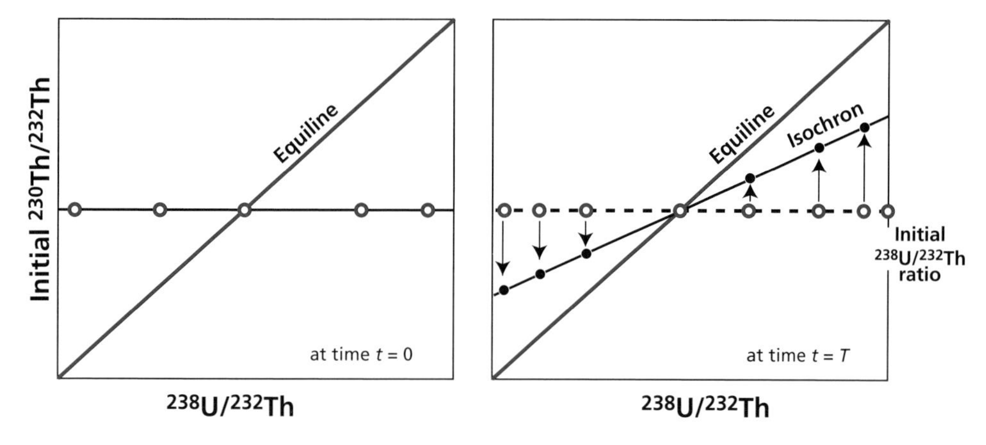

Figure 3.23 Principle of the (^{230}Th–^{232}Th, ^{238}U–^{232}Th) dating method. The sub-systems are isotopically homogeneous in thorium at time $T = 0$. They lie on a horizontal line representing isotopic equilibrium. The system has evolved by isotopic decay of ^{238}U and ^{230}Th. It evolves along an isochronous straight line whose slope gives the age. After about 10 periods the system is once more in radioactive equilibrium defined by the equiline. The initial isotopic ratio is given by the intersection with the equiline since this point is fixed because it is in equilibrium. After Allègre and Condomines (1976).

3.3.5 The ^{230}Th–^{238}U method and disequilibria of radioactive chains

These are methods related to radioactive chains for periods younger than 300 000 years. Radioactive equilibrium is assumed to have been destroyed as a result of some geological phenomenon or other (volcanism, formation of a shell in the ocean, erosion). As seen in the previous chapter, the system then tends to return to equilibrium. Let us deal with the case of ^{230}Th. The general equation is written:

$$\lambda_{230}\ {}^{230}\mathrm{Th} = \lambda_{230}\ {}^{230}\mathrm{Th}_0\ \mathrm{e}^{-\lambda t} + \lambda_8\ {}^{238}\mathrm{U}\ \left(1-\mathrm{e}^{-\lambda_2 t}\right).$$

Noticing that ^{232}Th has a very long period which may be considered constant relative to ^{230}Th (Allègre, 1968), we can write:

$$\frac{\lambda_0\ {}^{230}\mathrm{Th}}{\lambda_2\ {}^{232}\mathrm{Th}} = \left(\frac{\lambda_0\ {}^{230}\mathrm{Th}}{\lambda_2\ {}^{232}\mathrm{Th}}\right)_0 \mathrm{e}^{-\lambda t} + \left(\frac{\lambda_8\ {}^{238}\mathrm{U}}{\lambda_2\ {}^{232}\mathrm{Th}}\right)\ \left(1-\mathrm{e}^{-\lambda_0 t}\right).$$

On a $\left(\lambda_0^{230}\mathrm{Th}/\lambda_2^{232}\mathrm{Th},\ \lambda_8^{238}\mathrm{U}/\lambda_2^{232}\mathrm{Th}\right)$ plot that is in activity $\left[{}^{230}\mathrm{Th}/{}^{232}\mathrm{Th},\ {}^{238}\mathrm{U}/{}^{232}\mathrm{Th}\right]$, the cogenetic boxes lie at time zero (time when the radioactive equilibrium is destroyed) on a horizontal line and then evolve along a straight line whose slope is $\left(1-\mathrm{e}^{-\lambda_1 t}\right)$.

This straight line cuts the first bisector, which is the locus of points in secular equilibrium (**equiline**); the point of intersection (**equipoint**) is a fixed point around which the isochron pivots. It corresponds to the original $[{}^{230}\mathrm{Th}/{}^{232}\mathrm{Th}]_0$ value. This method is used successfully for young volcanic rocks (Figures 3.23 and 3.24).

Table 3.5 Concentrations and isotope ratios measured on andesite rock of the Irazú volcano, Costa Rica

Sample CA12	U (ppm)	Th (ppm)	Th/U	Activity ($^{238}U/^{232}Th$)	Activity ($^{230}Th/^{232}Th$)
Whole rock	5.83	16.4	2.80	1.07	1.13 ± 0.02
Magnetite	0.45	1.34	2.98	1.01	1.08 ± 0.04
Plagioclase	0.30	0.73	2.43	1.22	1.16 ± 0.05
Hypersthene	0.53	1.43	2.70	1.12	1.14 ± 0.04
Glass	7.15	19.4	2.71	1.11	1.15 ± 0.03
Apatite	14.0	53.6	3.83	0.78	0.98 ± 0.04

Source: After Allègre and Condomines (1976).

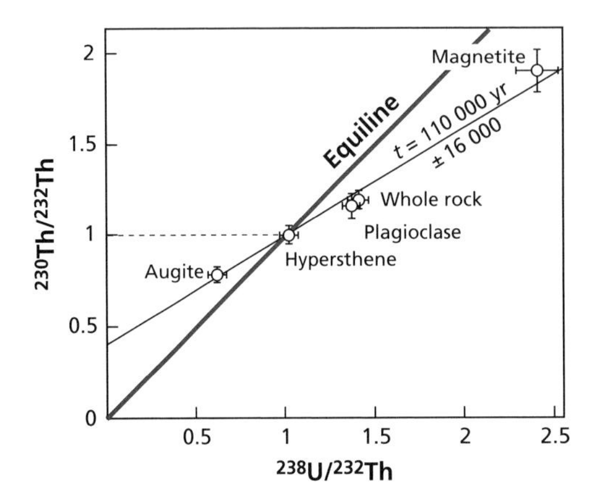

Figure 3.24 Isochron of minerals on a volcanic rock from Irazú, Costa Rica. After Allègre and Condomines (1976).

Exercise

Table 3.5 gives the concentrations and isotope ratios measured on an andesite rock of the Irazú volcano, Costa Rica (Allègre and Condomines, 1976). Calculate the age of the lava. What was its initial $^{230}Th/^{232}Th$ isotope ratio?

Answer

Age: 68 000 years; initial ratio in activity: 1.17.

The same method may be used for other isotopes of radioactive chains. Unfortunately ^{226}Ra has no stable isotope and barium is used for normalization. Similarly, ^{231}Pa has no stable isotope, and niobium is used as the norm. These slight difficulties explain why the ^{230}Th–^{238}U method is the most widely used.

3.3.6 Extinct radioactivities

These are radioactive elements that were present when the Solar System first formed and which have become extinct since. We shall speak of their origin in the next chapter (see also Wasserburg, 1985).

The case of ^{26}Al–^{26}Mg

The aluminum isotope ^{26}Al is radioactive. It decays to ^{26}Mg with a half-life of 0.72 million years. Imagine that ^{26}Al is incorporated in a meteorite at time t. From what we have already said of extinct radioactivity, by noting the initial magnesium $^{26}Mg(i)$ and ^{26}Al incorporated at time t $^{26}Al^*(t)$, we can write:

$$^{26}\mathrm{Mg} = {}^{26}\mathrm{Mg(i)} + {}^{26}\mathrm{Al}^*(t).$$

If the stable and non-radiogenic isotopes ^{27}Al and ^{24}Mg are introduced as references:

$$\frac{^{26}\mathrm{Mg}}{^{24}\mathrm{Mg}} = \left(\frac{^{26}\mathrm{Mg}}{^{24}\mathrm{Mg}}\right)(\mathrm{i}) + \frac{^{26}\mathrm{Al}}{^{27}\mathrm{Al}}(t)\,\frac{^{27}\mathrm{Al}}{^{24}\mathrm{Mg}}.$$

In a $\left(^{26}\mathrm{Mg}/^{24}\mathrm{Mg},\ ^{27}\mathrm{Al}/^{24}\mathrm{Mg}\right)$ plot, the slope of the isochron straight line gives $\left(^{26}\mathrm{Al}^*/^{27}\mathrm{Al}\right)_t$

Now, we know that:

$$\left(\frac{^{26}\mathrm{Al}}{^{27}\mathrm{Al}}\right)_t = \left(\frac{^{26}\mathrm{Al}}{^{27}\mathrm{Al}}\right)_{t=0} \mathrm{e}^{-\lambda t},$$

$t = 0$ being the end of the nucleosynthetic process that engendered ^{26}Al.

If we have two meteorites (I) and (II) we can therefore write:

$$\frac{\left[\left(\frac{^{26}\mathrm{Al}}{^{27}\mathrm{Al}}\right)_{t_1}\right]_{\mathrm{I}}}{\left[\left(\frac{^{26}\mathrm{Al}}{^{27}\mathrm{Al}}\right)_{(t_2)}\right]_{\mathrm{II}}} = \mathrm{e}^{\lambda(t_2 - t_1)}.$$

Therefore, as in the case of iodine–xenon (see below), we can date the difference in age of formation of two meteorites (see Figure 3.25).

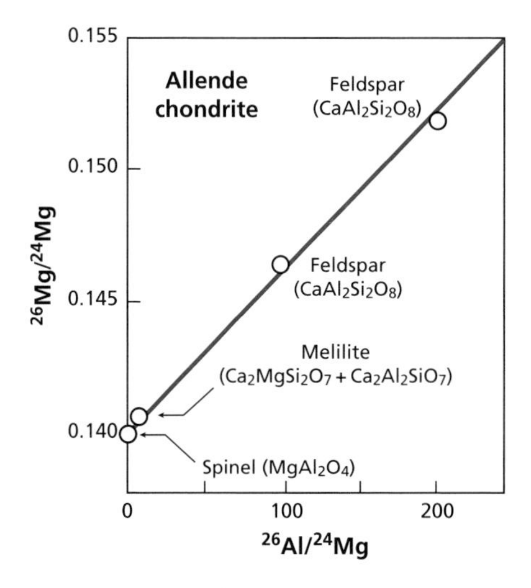

Figure 3.25 Plot of the ^{26}Mg–^{24}Mg, ^{26}Al–^{24}Mg isochron obtained for the Allende meteorite. After Wasserburg (1985).

The case of ^{129}I–^{129}Xe

As said, this technique was pioneered by **Reynolds** (1960). As shown in Section 2.4.2, ^{129}Xe today is the sum of $(^{129}Xe)_0$ plus ^{129}Xe produced by the complete decay of ^{129}J. If we chose to refer to a non-radiogenic xenon isotope ^{130}Xe, introducing the stable iodine isotope ^{127}I, we can write:

$$\left(\frac{^{129}Xe}{^{130}Xe}\right)_{total} = \left(\frac{^{129}Xe}{^{130}Xe}\right)_0 + \frac{^{129}I}{^{130}Xe} = \left(\frac{^{129}Xe}{^{130}Xe}\right)_0 + \left(\frac{^{129}I}{^{127}I}\right)_{t_1} \cdot \left(\frac{^{127}I}{^{130}Xe}\right).$$

As stated above, to measure ^{127}I, we resort to the neutron activation method of analysis hit upon by **Jeffrey** and **Reynolds** consisting in transforming ^{127}I into ^{128}Xe.

$$^{127}I = k^{128}Xe,$$

where k is a proportionality factor relative to the activation analysis. This finally yields the dating equation:

$$\left(\frac{^{129}Xe}{^{130}Xe}\right)_{total} = \left(\frac{^{129}Xe}{^{130}Xe}\right)_0 + \left(\frac{^{129}I}{^{127}I}\right) k \left(\frac{^{128}Xe}{^{130}Xe}\right).$$

In practice, the meteorite is irradiated and then degassed in stages (in a combination between the ^{39}Ar–^{40}Ar and isochron methods). The isotope composition of Xe is measured at each step and a plot made of $^{129}Xe/^{130}Xe$ as a function of $^{128}Xe/^{130}Xe$.

The slope of the straight line is equal to $(^{129}I/^{127}I)$ at the time the meteorite formed. The ordinate to the origin is equal to $(^{129}Xe/^{130}Xe)_0$ at the time the meteorite formed (Figure 3.26).

By taking a given meteorite as a reference (the Bjürbole meteorite) the various "ages" of different meteorites can be calculated step by step. This was done by the team at Berkeley under the supervision of **John Reynolds** (Hohenberg *et al.*, 1967) (Figure 3.26).

The isochron plot therefore works in the same way as for the other pairs. In just the same way, similar plots can be constructed:

$$\left[\frac{^{53}Cr}{^{54}Cr}, \frac{^{55}Mn}{^{54}Cr}\right], \left[\frac{^{182}W}{^{184}W}, \frac{^{180}Hf}{^{184}W}\right], \text{etc.},$$

for other extinct radioactivities: ^{53}Mn–^{53}Cr, ^{182}Hf–^{182}W, etc.

Exercise

Table 3.6 gives the ^{182}Hf–^{182}W experimental results for three meteorites. Plot the isochrons. What are the relative ages of the three meteorites given that the reference for the formation of the Solar System is $^{182}Hf/^{180}Hf = 2.5 \cdot 10^{-4}$?

Answer

The relative ages in respect of the initial reference are 5 Ma, 3 Ma, and 3 Ma. Estimate the errors for yourself.

Table 3.6 Studies of three meteorites

	Hf (ppb)	W (ppb)	($^{180}Hf/^{184}W$) atomic	$^{182}W/^{184}W$
Forest Vale				
M-L	3.61	647	0.00658	0.864716 ± 16
M-B-L	14.8	758	0.0230	0.864690 ± 50
M-B-2	63.0	642	0.1157	0.864737 ± 17
M-R	204	588	0.4087	0.864800 ± 33
WR	172	178	1.14	0.864945 ± 43
NM	176	114	1.82	0.865049 ± 29
Sainte-Marguerite				
M-L	156	761	0.241	0.864751 ± 24
M-B*	9.81	754	0.0154	0.864700 ± 41
M-R	77.6	1523	0.0601	0.864778 ± 28
WR	141	166	1.01	0.864888 ± 31
Richardton				
M-B-1*	56.2	460	0.144	0.864681 ± 44
M-B-2	53.2	582	0.108	0.864680 ± 31
NM	87.1	41.2	2.49	0.865110 ± 43

Source: After Lee and Halliday (2000).

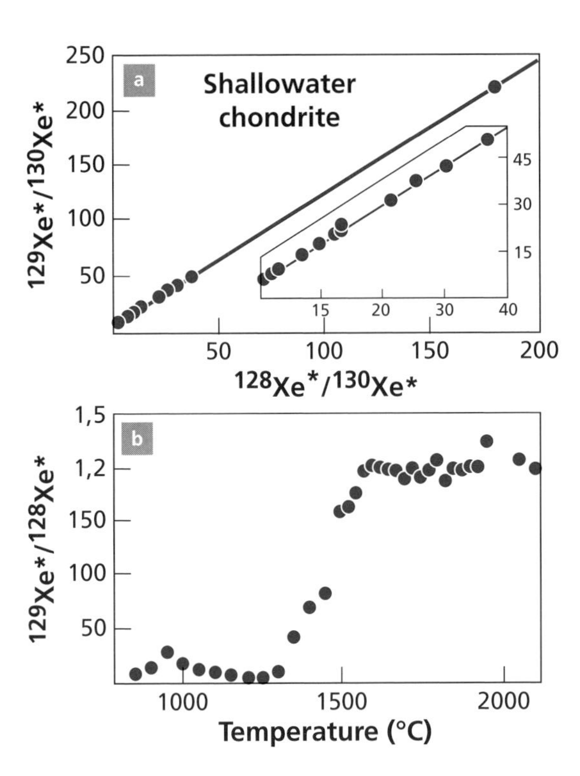

Figure 3.26 Dating by (I–Xe) on the Shallowater meteorite. (a) Isochron representation; (b) step degassing. After Hohenberg *et al.* (1998).

3.3.7 Conditions of use of the isochron diagram

The essential criterion for using the isotopic correlation diagram and the isochron construction is that the experimental points should be aligned. Otherwise one of the conditions for applying the model is no longer met and no straight line can be plotted to make an age calculation. When the data points form a cloud, this criterion is easy enough to apply. But in practice, where each point is affected by experimental error, the alignment is imperfect and it is no simple matter to decide whether or not there is a straight line. We shall come back to this issue when discussing uncertainties.

When dealing with minerals, whether in rocks or in meteorites, there is a phenomenon one must be wary of and which is related to isotopic exchange. Rich minerals (typically biotite and muscovite for $^{87}Rb/^{87}Sr$ and zircon or apatite for U–Pb systems) tend to lose their radiogenic isotopes as those isotopes have been introduced "indirectly" through radioactivity.

But where do these radiogenic isotopes go when they migrate?

The places where they might be accommodated are the usual crystallographic sites of their chemical element, that is, generally in "poor minerals" (apatite and sphene for $^{87}Rb/^{87}Sr$ and feldspar for U–Pb). Thus ^{87}Sr will migrate from biotite to apatite (and even more easily as often apatite is included in biotite). Similarly ^{206}Pb will migrate from zircon to feldspar (and again there are often zircon inclusions in feldspar).

In the isochron diagram, the data point of the recipient mineral will move vertically sometimes greatly and so move off the isochron, and sometimes little and so will tend to be indistinct.

Caution is required, then, with these "poor," purely radiogenic minerals close to the y-axis. One might think they could be used to determine the initial isotopic ratio ($^{87}Sr/^{86}Sr$ or $^{206}Pb/^{204}Pb$) accurately, as often they are more radiogenic than it is! This is the way the donor–acceptor pair works. Failure to allow for it has led to errors, particularly when dating meteorites.

3.4 Mixing and alternative interpretations

3.4.1 Mixing

Using the two methods just exposed (concordia and isochron) it is possible to determine the age of the formation of a cogenetic rocky system and also to obtain some information about its complex geological history. These are valuable methods therefore for anyone wanting to study the geological history of continents, an often complex history made of superpositions of geological events. But the interpretations we have developed are **not unique**!

When we obtain a straight line in the $\left[^{87}Sr/^{86}Sr,\ ^{87}Rb/^{86}Sr\right]$ plot, we may consider it not as an **isochron** but as a straight line of **mixing** (Allègre and Dars, 1966) (Figure 3.27). The same is true of the $\left[^{206}Pb^{*}/^{238}U,\ ^{207}Pb/^{235}U\right]$ concordia diagram.

Let there be two reservoirs, 1 and 2, with distinct $^{87}Sr/^{86}Sr$ and $^{87}Rb/^{86}Sr$ ratios. The mix of the two reservoirs is written:

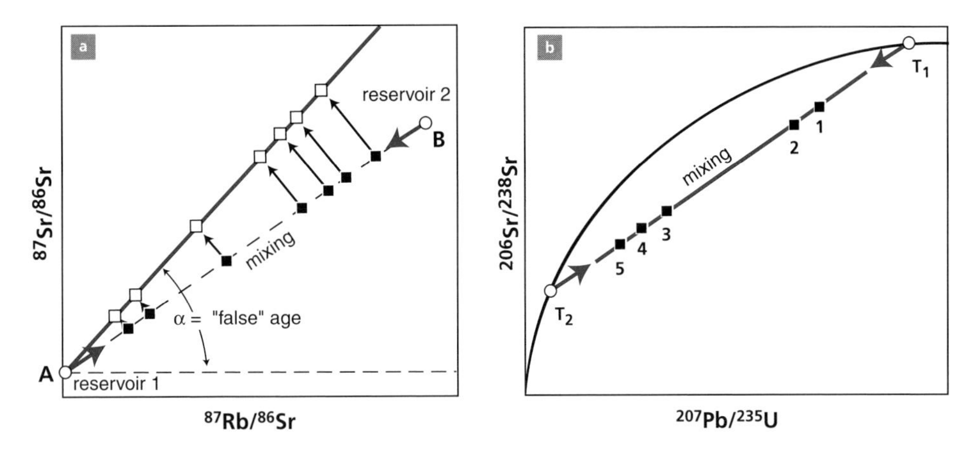

Figure 3.27 Mixing straight lines. (a) The mixing line in the (^{87}Sr/^{86}Sr, ^{87}Rb/^{86}Sr) plot. It represents the straight line homogenized by the mixture in variable proportions between reservoir 1 and reservoir 2. (b) A mixing straight line in the concordia diagram between two populations of zircon of ages T_1 and T_2.

$$\left(\frac{^{87}\mathrm{Sr}}{^{86}\mathrm{Sr}}\right)_{\mathrm{mixture}} = \frac{\left(^{87}\mathrm{Sr}\right)_1 + \left(^{87}\mathrm{Sr}\right)_2}{\left(^{86}\mathrm{Sr}\right)_1 + \left(^{86}\mathrm{Sr}\right)_2} = \frac{\left(\frac{^{87}\mathrm{Sr}}{^{86}\mathrm{Sr}}\right)_1 \left(^{86}\mathrm{Sr}\right)_1 + \left(\frac{^{87}\mathrm{Sr}}{^{86}\mathrm{Sr}}\right)_2 \left(^{86}\mathrm{Sr}\right)_2}{\left(^{86}\mathrm{Sr}\right)_1 + \left(^{86}\mathrm{Sr}\right)_2}.$$

By noting $x = \left(^{86}\mathrm{Sr}\right)_1 / \left[\left(^{86}\mathrm{Sr}\right)_1 + \left(^{86}\mathrm{Sr}\right)_2\right]$ and $R = {}^{87}\mathrm{Sr}/{}^{86}\mathrm{Sr}$, we obtain:

$$R_{\mathrm{mixture}} = R_1 x + R_2(1 - x).$$

If a same mixing equation is written with the "chemical" ratio $r = {}^{87}\mathrm{Rb}/{}^{86}\mathrm{Sr}$, we get:

$$r_{\mathrm{mixture}} = r_1 x + r_2(1 - x).$$

We can eliminate x between the two preceding equations, hence:

$$\frac{R_{\mathrm{mixture}} - R_2}{r_{\mathrm{mixture}} - r_2} = \frac{R_1 - R_2}{r_1 - r_2}.$$

In an (R, r) plot this is the equation of a straight line through the coordinates (R_1, r_1) and (R_2, r_2) and whose slope, equal to $(R_1 - R_2)/(r_1 - r_2)$, is unrelated a priori to λt. Note that these are the same equations as for isotope dilution.

Exercise

Calculate the apparent age of an alignment on a (^{87}Rb/^{86}Sr, ^{87}Sr/^{86}Sr) plot which actually results from mixing of a limestone component for which ^{87}Rb/^{86}Sr = 0.001 and ^{87}Sr/^{86}Sr = 0.708 and a schist component for which ^{87}Rb/^{86}Sr = 1 and ^{87}Sr/^{86}Sr = 0.720.

Answer

857 Ma.

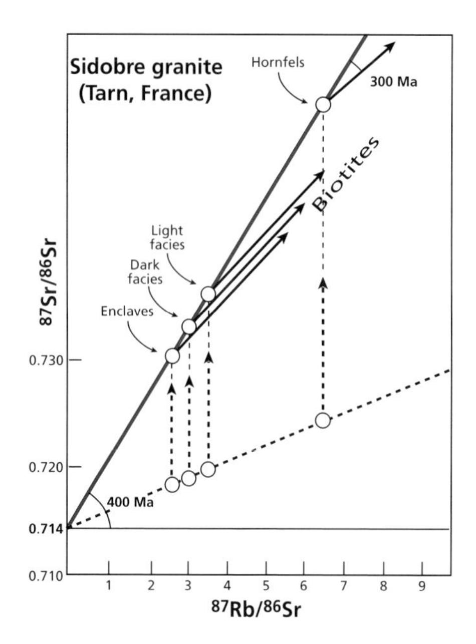

Figure 3.28 The Sidobre granite. The age of emplacement is 300 Ma. The apparent age due to mixing is 400 Ma (from Allègre and Dars, 1966).

EXAMPLE

The Sidobre granite

The Sidobre granite of the Montagne Noire in the south of the Massif Central (France) near Castres is intruded in schists transformed by contact into hornfels (Figure 3.28). It contains schist enclaves but also "basic" enclaves. This granite is composed of two imbricated facies: one is light (and richer in SiO_2) and the other dark. The latter has a high "basic" enclave content.

Analyses of the light facies, dark facies of a "basic" enclave, and hornfels define a straight line whose slope corresponds to 400 Ma. The analysis of biotites gives about 300 Ma.

A subsequent analysis of whole rock from the same massif gives 285 Ma.

The interpretation is that the straight line at 400 Ma is a straight line of mixing between the enclosing rock and the basic magma which penetrated and cannibalized the enclosing rock.

There are more complex cases where the mixture occurs among several components. In principle they do not give a single straight line but in some circumstances they may mimic a pseudo-straight line.

Exercise

(1) At time $t = 1$ Ga, an ultra-basic magma mixes with the enclosing granite. The ultrabasic magma has isotopic ratios $^{147}Sm/^{144}Nd = 0.35$ and $^{143}Nd/^{144}Nd = 0.511\,25$.

Ratios for the surrounding granite are $^{147}Sm/^{144}Nd = 0.1$ and $^{143}Nd/^{144}Nd = 0.510\ 358$, respectively.

Calculate the equation of the straight line of mixing.

(2) Radioactive decay has occurred and today we analyze the rocks produced by the mixing. What apparent age we will obtain? What is the initial $^{143}Nd/^{144}Nd$ ratio? Is it the same as that of the mix?

Answer

(1) Apparent age: 1558 Ma.

(2) Initial ratio: $^{143}Nd/^{144}Nd = 0.5100$. This is the same as the ratio of the mixture as its Sm/Nd ratio is zero and so it cannot evolve over time.

Mixtures are extremely common in geological phenomena: when a magmatic rock is emplaced, it digests the surrounding rock. When a sediment forms it is always a mixture of various detrital components and some chemical component, etc.

When mixing is followed by **isotopic re-homogenization** it does not affect the age calculation but may sometimes be detected by examining the initial isotope ratio. The mixture is troublesome when not followed by homogenization, either at low temperature or because the system cools very quickly (these are the sort of cases we have referred to).

3.4.2 The (1/*C*) test

Let us return to the mixing equation:

$$R_M = R_1 x + R_2 (1 - x),$$

R_M, R_1, and R_2 being the isotopic ratios of the mixture and of the two components:

$$x = \frac{m_1\, C_1}{m\, C_M}$$

where m_1 is the mass of component 1 whose concentration is C_1, m is the mass of the mixture, and C_M the concentration of the mixture. Let us posit:

$$\beta = \frac{m_1}{m} \quad x = \frac{C_1}{C_M} \cdot \beta.$$

If we examine the concentration of the mixture, we have:

$$C_M = C_1 \beta + C_2(1 - \beta) \text{ or } \frac{C_M - C_2}{C_1 - C_2} = \beta.$$

R_M becomes:

$$R_M = R_1 \beta \frac{C_1}{C_M} + R_2 \left(1 - \beta \frac{C_1}{C_M}\right) \text{ or } \frac{R_M - R_2}{R_1 - R_2} = \beta \frac{C_1}{C_M}.$$

We can eliminate β from both equations. This simplifies to:

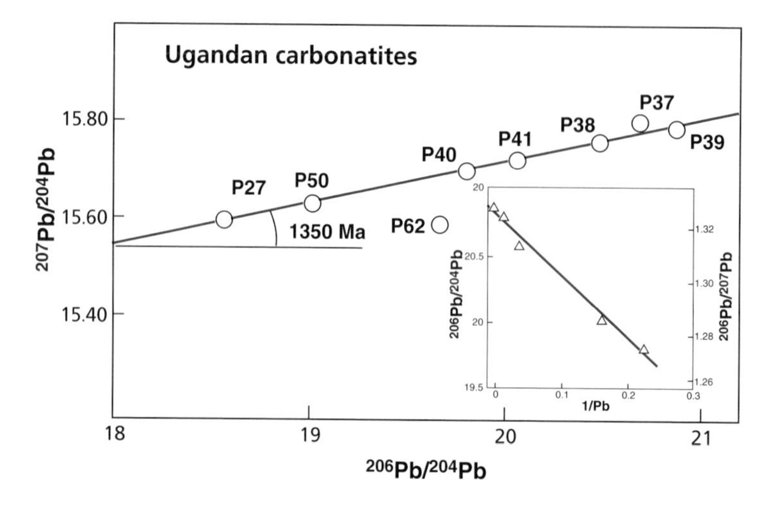

Figure 3.29 Carbonatites of Uganda. After Lancelot and Allègre (1974).

$$R_M = A\left(\frac{1}{C_M}\right) + B.$$

The plot of $R_M = f(1/C_M)$ is a straight line.

EXAMPLE

Carbonatites of East Africa

A series of carbonatites (lava composed of $CaCO_3$) from East Africa has been analyzed for its Pb isotope composition (Lancelot and Allègre, 1974) (Figure 3.29). They define a straight line in the $(^{207}Pb/^{204}Pb\ ,\ ^{206}Pb/^{204}Pb)$ plot corresponding to an apparent age of 1300 Ma. In fact, the true age is very recent!

The $(^{206}Pb/^{204}Pb,\ 1/Pb)$ plot shows it is in fact a mixture of two components. In this case, the isotopic composition of one component can be determined $^{206}Pb/^{204}Pb = 20.8$.

For the $\left(^{206}Pb/^{238}U,\ ^{207}Pb/^{235}U\right)$ concordia diagram, the problem of the mixture is also a fundamental issue. The mathematical demonstration is identical to the foregoing. Here again, it can be easily understood that a population of zircons is a mixture between two populations of different ages. The difference here is that the information given by the concordia construction is chronologically correct. It defines the age of two populations.

An example is the analysis of U–Pb zircon of gneiss from the Alps by **Marc Grünenfelder** of the Eidgenössische Technische Hochschule of Zurich and colleagues (Köppel and Grünenfelder, 1971). The paragneiss was found to have a very wide spread of age distributions such as that shown, whereas the orthogneiss yielded very different and tightly clustered distributions (Figure 3.30).

How can these two types of distribution be accounted for? The Zurich team gave a simple explanation. The orthogneisses have ages of crystallization of granite of 450 Ma and were

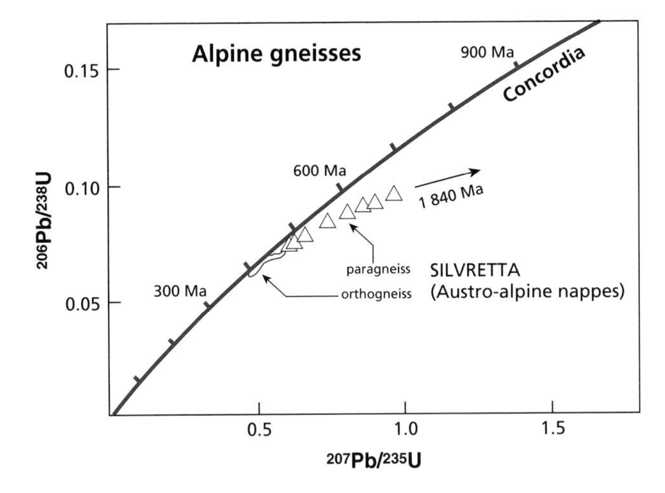

Figure 3.30 Zircon measured in the Alps. The paragneiss results from a mixture of zircon 1590–1840 Ma old and of 380-Ma-old zircon. After Köppel and Grünnenfelder (1971).

gneissified during the Alpine orogeny between 100 and 300 Ma. The ages are somewhat younger but relatively close. The paragneiss contains very old zircon (whose exact geological origin is unknown) inherited from ancient rocks whose debris is found in sediments formed 450 Ma ago. The paragneiss discordia is therefore a mixing straight line of inherited zircon and zircon formed 450 Ma ago, or possibly of zircon which seeded from pieces of inherited zircon around which they grew.

As can be seen, sound geological knowledge is called for when interpreting measurements.

3.5 Towards the geochronology of the future: *in situ* analysis

Over the last 15 years or so, new spot geochronology measuring methods have been developed. Instead of mechanically separating minerals, grinding them, and dissolving them in acids to extract the elements for analysis, the mineral matter is pulverized in one spot and then analyzed without any preparatory chemistry. Various techniques are available. We shall review them succinctly, describing those that are already operational, those under development, and those that are still only at the idea stage.

3.5.1 The ion probe

This is a method already mentioned in Chapter 1 invented by the two French physicists **Castaing** and **Slodzian** (1962) at the University of Orsay, consisting of bombarding the rock sample with a beam of oxygen or cesium ions. The mineral material is thus pulverized and ionized to form an ionized plasma. This plasma, made up of ions of different natures and species, is analyzed using a mass spectrometer, that is, the ions are accelerated and

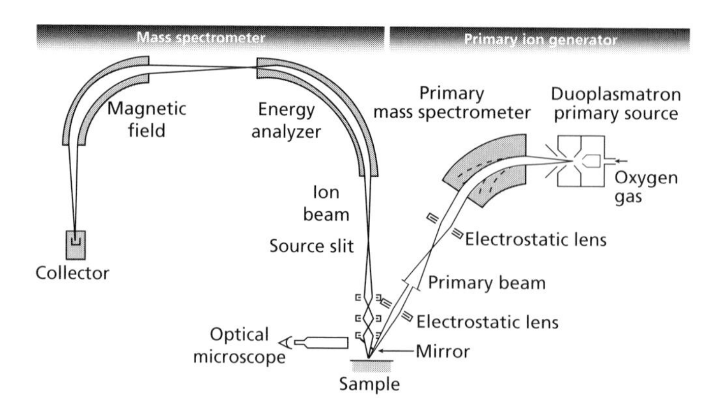

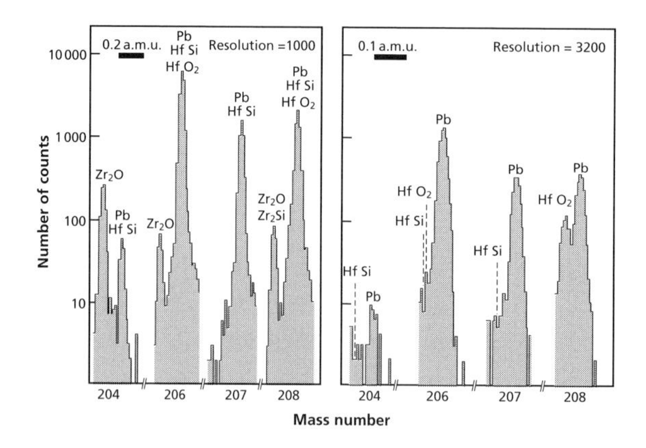

Figure 3.31 An ionic probe and mass spectra. Top: diagram of an ionic probe. Bottom: mass spectra obtained at resolutions of 1000 and then 3200. The solution to the problem of isobaric interference can be seen. Plate 3 (top) also shows an ion probe.

then sorted by mass with a magnetic analyzer and finally collected by one of various models of collector (Figure 3.31).

Drawbacks with this technique are that:

- the energies of the ions emitted are variable and therefore the *eV* term of the fundamental equation of the mass spectrometer (see Chapter 1) is not constant;
- numerous ions are ionized including complex ions formed by ionized molecular assemblages;
- of course, as there has been no chemical separation, there are isobaric interferences such as between ^{87}Rb and ^{87}Sr, to cite just one example.

In practice the problem has been solved with large appliances (large radii of curvature, large magnets) for U–Pb methods on zircon. Ancient zircon grains have been analyzed in this way, some allegedly 4.2–4.3 Ga old! Credit for this goes to **Bill Compston** at the Australian National University at Canberra who developed this method and to his team who exploited it (Compston *et al.*, 1984).

It has been shown generally that zircons are assemblages with complex geological histories, each crystal having an ancient core and younger growth zones, something like a restored cathedral where the apse is ancient and the transepts "modern." This method opens up extraordinary perspectives. First, it is fast and many zircons can be analyzed in a short time. Next, it allows the mineral to be dissected isotopically, allowing us to go beyond the standard questions of whether the system is open or closed, whether it is mixed or not. Another spectacular result has been obtained in dating meteorites by the ^{26}Al–^{26}Mg method on meteorites (Zinner, 1996). It has been possible to show the existence of several generations of fusion within a single meteorite in an interval of time of less than 1 million years, illustrating the complex early history of the Solar System.

Generalization of this method is an open question today but it has a bright future.

3.5.2 Laser ionization

Instead of a beam of ions, the surface of the rock or mineral is bombarded by a laser beam. This beam pulverizes and ionizes the elements and the isotopes extracted. They are then analyzed with a mass spectrometer. This technique is used for the ^{39}Ar–^{40}Ar method (York *et al.*, 1981). A zone in the mineral is chosen and then progressively heated and the ^{40}Ar–^{39}Ar ratios are recorded stepwise. The core of the minerals and their rims can thus be analyzed with a degassing spectrum in each case. This is an extraordinary method for deciphering the thermal history of a region and for determining true ages. The technique is now extended to other elements by coupling laser extraction and ICPMS ionization (high-temperature plasma). In the current state of knowledge this method is under development, but U–Pb analysis on zircon is already operational.

And the future? It undoubtedly lies in selective ionization using the multifrequency laser for selectively ionizing ^{87}Sr and not ^{87}Rb and vice versa; this will replace chemistry. This method is already used at the University of Manchester by **Grenville Turner** and his team for analyzing xenon isotopes. I shall say no more as it is difficult to foretell the future and science is forging ahead! But the discussions about open or closed, mixed or not will be asked in a very different context tomorrow! It is important that the theoretical models – probably statistical models (as they allow many measurements to be made) – should keep pace with analytical advances! This is not the case today as the current trend is for too-careless workers to believe that the accumulation of enormous amounts of often imprecise data by automatic machines can replace geochemical thinking and quantitative modeling!

Problems

1 Let us consider the ^{40}K–^{40}Ar system in a biotite which formed 1 Ga ago. Over the last 1 Ga the biotite has been buried at depth in the Earth's crust where it constantly lost argon by the law:

$$\frac{d^{40}Ar}{dt} = -G^{40}Ar,$$

where $G = 0.11 \cdot 10^{-9}$. Then faulting brings the material to the surface. During friction related to tectonism, the biotite loses 75% of its Ar. Supposing there was no initial ^{40}Ar, calculate the apparent age of the biotite.

2 Argon losses during a sudden event can be written:

$$\frac{d\,^{40}Ar}{dt} = -K\,^{40}Ar,$$

where $K = aT + b$, T being the temperature. What is the law for the loss of Ar with temperature?

3 Canada's Abitibi Belt contains associations of basic and ultrabasic lavas known as komatiites. The lead isotope composition of these rocks has been measured in three locations. The results are shown in Table 3.7.

(i) Calculate the ages.
(ii) The sulfides have been measured (Table 3.8). What do you conclude?

Table 3.7 Lead isotope composition in komatiites of Canada

	$^{206}Pb/^{204}Pb$	$^{207}Pb/^{204}Pb$	$^{208}Pb/^{204}Pb$	Pb (ppm)
Pyke Hill				
MT1	30.716 ± 22	17.778 ± 17	47.285 ± 63	0.02
MT2 a	16.584 ± 11	15.059 ± 15	35.651 ± 46	0.08
MT2 b	15.953 ± 15	14.972 ± 17	35.455 ± 51	
MT3 a	16.128 ± 10	14.981 ± 15	35.595 ± 45	0.11
MT3 b	15.894 ± 12	14.951 ± 16	34.400 + 47	0.13
MT3 c	16.050 ± 17	14.970 ± 15	35.541 ± 48	
MT4 a	15.613 ± 11	14.897 ± 17	34.933 ± 52	0.14
MT4 b	16.474 ± 10	15.044 ± 14	35.690 ± 45	
Fred's Flow				
F1	18.466 ± 16	15.452 ± 18	37.939 ± 58	
F2	18.083 ± 13	15.388 ± 15	37.679 ± 52	0.36
F3 a	16.184 ± 11	15.074 ± 15	35.945 ± 49	0.35
F3 b	15.842 ± 11	14.999 ± 14	35.587 ± 45	0.45
F5	21.001 ± 14	15.943 ± 16	41.554 ± 55	0.15
F6 a	24.928 ± 24	16.568 ± 20	42.688 ± 65	
F6 b	23.845 ± 16	16.406 ± 16	42.741 ± 54	0.17
F6 c	27.141 ± 31	16.928 ± 23	44.556 ± 72	
Theo's Flow				
T2 a	19.479 ± 12	15.722 ± 15	38.816 ± 49	0.69
T2 b	23.605 ± 15	16.277 ± 15	41.840 ± 53	
T3 a	17.567 ± 12	15.273 ± 15	36.978 ± 47	1.1
T3 b	20.152 ± 16	15.794 ± 16	39.319 ± 52	
T6 a	17.378 ± 11	15.343 ± 15	37.105 ± 47	0.41
T6 b	17.540 ± 12	15.288 ± 16	36.971 ± 52	

Source: Brévart *et al.* (1986).

Table 3.8 Lead isotope composition of sulfides (chalcopyrite)

	$^{206}Pb/^{204}Pb$	$^{207}Pb/^{204}Pb$	$^{208}Pb/^{204}Pb$	Pb (ppm)
F5 a	13.352	14.461	33.153	
F5 b	13.268	14.444	3.082	

4 Here Table 3.9 shows the results of an analysis of two rocks (garnet pyroxenite) from the Beni-Bossera massif of Morocco.

(i) Plot the isochrons.
(ii) Calculate the Lu–Hf and Sm–Nd ages.
(iii) What do you conclude?

Table 3.9 Analysis of garnet pyroxenite from Morocco

Samples	Lu* (ppm)	Hf (ppm)	$^{176}Lu/^{177}Hf$	$^{176}Hf/^{177}Hf$	Sm (ppm)	Nd (ppm)	$^{147}Sm/^{144}Nd$	$^{143}Nd/^{144}Nd$
M 214 rt	0.469	1.69	0.0393	0.283 19 ± 4	2.58	5.73	0.2722	0.513 043 ± 14
M 214 gt	0.974	0.604	0.2289	0.283 257 ± 11	2.22	1.42	0.9458	
				0.513 023 ± 157				
M 214 cpx	0.0631	2.42	0.0037	0.283 124 ± 10	3.05	8.29	0.2221	0.513 051 ± 10
M 101	0.498	1.04	0.0681	0.283 12 ± 10	1.25	3.05	0.2467	0.513 029 ± 9
M 101	1.61	0.268	0.8492	0.283 505 ± 14	0.34	0.12	1.7371	0.513 255 ± 37
M 101	0.0947	0.875	0.0154	0.283 107 ± 13	0.62	1.02	0.3686	0.513 044 ± 12

Source: Blichert-Toft *et al.* (1999).

5 Table 3.10 gives the results of U–Th analyses on zircon in apparent ages.

(i) Construct the (^{238}U–^{206}Pb, ^{235}U–^{207}Pb) and (^{235}U–^{207}Pb, ^{232}Th–^{208}Pb) concordia.
(ii) Calculate the ages of these three populations of zircon.

Table 3.10 Apparent ages of zircon by U–Th analysis

	Apparent ages		
Locality	$^{206}Pb/^{238}U$	$^{207}Pb/^{235}U$	$^{208}Pb/^{232}Th$
Little Belt Mountains (Montana)	560	830	320
	910	1210	700
	810	1130	950
	1140	1400	1080
	1290	1540	1550
	1860	1890	1790
	830	1190	1260
	1170	1670	1500
	1570	1980	1790
Finland	2520	2660	2750
	1890	2270	1790
	1820	2240	1815
	1820	1870	1960
	1610	1720	1860
Maryland	497	511	486
	357	370	308
	404	422	350

6 If one of the criteria for identifying a mixture is the $(R, 1/C)$ plot, can it be generalized to three dimensions?
Supposing a straight line observed in the $(^{87}Sr/^{86}Sr, ^{87}Rb/^{86}Sr)$ isotope plot is a straight line of mixing, and so satisfies the relation $(^{87}Sr/^{86}Sr, 1/C)$, what is the geometrical locus of the data points in a three-dimensional $(^{87}Sr/^{86}Sr, 1/C, ^{87}Rb/^{86}Sr)$ plot? Demonstrate this by algebra or geometry.

CHAPTER FOUR

Cosmogenic isotopes

Nuclear reactions occur in nature. As well as the reactions which take place in the stars and supernovae and which manufacture all of the chemical elements, fluxes of charged particles of cosmic or solar origin also produce nuclear reactions on rocks, both meteoritic and terrestrial. These reactions generate radioactive or stable isotopes which can be used for geological or cosmological dating. Before examining the main types, let us recall some of the basic principles of nuclear reactions.

4.1 Nuclear reactions

4.1.1 General principles

Flows of particles, whether they occur naturally or are artificial, interact with matter. Depending on their energy levels, they may cause ionization by stripping electrons from atoms, create crystal defects by displacing atoms, or trigger nuclear reactions if they have enough energy. Here we shall concentrate on this final case. Nuclear reactions lead to one nucleus being transformed into another. They are noted using a formal system similar to that used for chemical reactions. Here are a few examples:

$$^{23}_{11}\mathrm{Na} + {}^{1}_{0}\mathrm{n} \rightarrow {}^{24}_{11}\mathrm{Na} + \gamma$$
$$^{14}_{7}\mathrm{N} + {}^{1}_{0}\mathrm{n} \rightarrow {}^{14}_{6}\mathrm{C} + {}^{1}_{1}\mathrm{p}$$
$$^{12}_{6}\mathrm{C} + {}^{1}_{1}\mathrm{p} \rightarrow {}^{13}_{7}\mathrm{N} + \gamma$$

where n stands for a neutron, p for a proton, and γ for gamma radiation.

The first of these equations means that a neutron acts on a ^{23}Na nuclide to give ^{24}Na by emitting a gamma ray. The others are interpreted by following the same notations. In fact, they are symbolized in a compact form where we write in succession:

- the initial nucleus (called the target because, in experiments, it is a nucleus submitted to particle flux, which is similar to bombarding a target);
- the initial particle (here a neutron or proton);
- the particle emitted (here a gamma ray, a neutron, a proton, or a neutrino);
- the nucleus produced.

Thus the reactions above are written:

$$^{23}\text{Na (n}, \gamma)\ ^{24}\text{Na}$$
$$^{14}\text{N (n, p)}\ ^{14}\text{C}$$
$$^{12}\text{C (p}, \gamma)\ ^{13}\text{N}.$$

All of the reactions obey laws of conservation similar to the laws of conservation of chemical equations except as concerns mass and energy. In any nuclear reaction there is conservation of:

- mass–energy, and in this domain there is an equivalence between the transformation of mass and energy in accordance with Einstein's equation $\Delta E = \Delta mc^2$;
- the number of nucleons (protons and neutrons);
- the electrical charge;
- nuclear spin.

Remark

It is the conservation of the number of nucleons that balances nuclear reactions.

Take the reaction:

$$^{16}_{8}\text{O} + ^{16}_{8}\text{O} \rightarrow ^{28}_{14}\text{Si} + ^{4}_{2}\alpha + 9.6\ \text{MeV}$$

(1 MeV $= 10^6$ electronvolts. The electron volt (eV) is the energy of an electron when submitted to a potential difference of 1 V: 1 eV $= 1.602 \cdot 10^{-19}$ joules.) Conservation of protons and neutrons means the number of protons and neutrons can be balanced: $8 + 8 \rightarrow 14 + 2$ and $16 + 16 \rightarrow 28 + 4$.

Exercise

What type of nuclear reaction (write the complete reaction and its abridged form) allows us to move from ^{12}C to ^{16}O, given that the particle emitted is a gamma ray; from ^{16}O to ^{20}Ne, where a gamma ray is also emitted; from ^{13}C to ^{14}N, where once again a gamma ray is emitted?

Answer

See the end of the chapter.

4.1.2 Effective cross-sectional area

The kinetics of a nuclear reaction

How many new nuclei are produced from old nuclei? Suppose we have a layer of material with a surface area of 1 unit and thickness dx. Suppose the material contains n target atoms per unit volume. The number of target atoms is $n\ dx$. Suppose also that for the process in question, each nucleus has a probability of interaction noted Γ, termed the **effective cross-sectional area** (Figure 4.1).

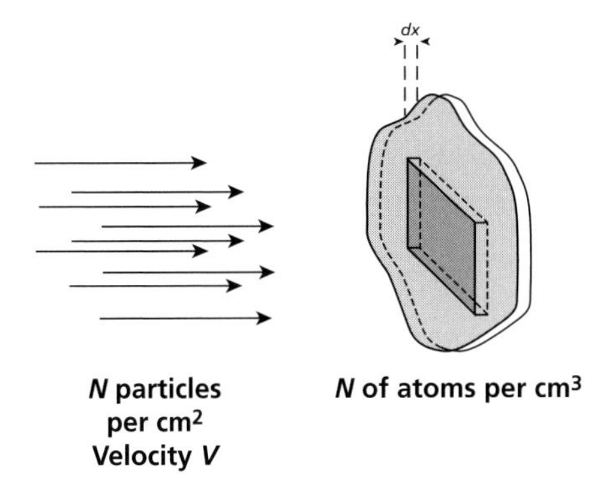

Figure 4.1 Effective cross-sectional area.

If $\dot{N}_1$ is the number of incident particles crossing a unit of surface area per unit time (flux), N_2 is the number of new nuclei produced, then the number of "actual" interactions during the time interval dt is written:

$$\mathrm{d}N_2 = \Gamma n\ \mathrm{d}x \cdot \dot{N}_1 \mathrm{d}t = -\mathrm{d}\ \dot{N}_1$$

or

$$\frac{\mathrm{d}N_2}{\mathrm{d}t} = \Gamma\ n\ \mathrm{d}x\ \dot{N}_1,$$

where $\dot{N}_1$ is the flux of incident particles which is often written $\dot{N}_1 = <N_1> V_1(E)$, in which $<N_1>$ is the mean number of incident particles per unit area and $V_1(E)$ is the speed of the particles.

The unit of effective cross-sectional area is the **barn** (1 barn $= 10^{-28}\,\mathrm{m}^2 = 100$ square femtometers). It is as though the target nuclei had an area multiplication factor (which explains its unit) relative to nuclear reactions.[1]

Exercise

A flux of 10^{12} cm^{-2} s^{-1} of slow neutrons bombards ^{23}Na to give ^{24}Na. The effective cross-sectional area of this reaction is $0.53 \cdot 10^{-24}$ cm^2 per atom. How many atoms of (radioactive) ^{24}Na are produced per gram of sodium?

Answer

For one atom of ^{23}Na, the production of ^{24}Na is $0.53 \cdot 10^{-24} \times 10^{12} = 0.53 \cdot 10^{-12}$ atoms per second per atom of ^{23}Na.

1 gram corresponds to $6.023 \cdot 10^{23}/23$ atoms, therefore $1.387 \cdot 10^{10}$ atoms of ^{24}Na are produced per second per gram of sodium.

[1] The idea of the effective cross-section dates from Rutherford's scattering experiments on calculating the size of the nucleus. He accepted that there was a halo around each nucleus. If the incident particle struck the target in this halo then an interaction occurred. The halo had width b and so the interaction surface area was πb^2.

Energy dependence of the effective cross-sectional area

The cross-sectional area Γ is specific to each nuclear reaction. The cases of charged particles, protons or α particles must be clearly distinguished from the case of neutrons.

When a **proton** penetrates matter, it interacts with electrons by electrostatic attraction. In a number of cases it captures an electron and changes into a hydrogen atom. In other instances, it strikes a target nucleus. Then, to interact with the nucleus it must overcome the electrostatic repulsion (because the two nuclei are positively charged). It must therefore have enough energy to cross the potential barrier either directly or by the tunnel effect. Then and only then does a nuclear reaction occur. An analogous phenomenon occurs with α particles, which either change into neutral helium nuclei or produce nuclear reactions.

When a **neutron** penetrates matter, no electromagnetic interaction occurs. It penetrates until it strikes a nucleus. There, either it is deflected or it causes a nuclear reaction.

Nuclear interactions depend on the energy of the incident particles. The effective cross-sectional area therefore depends also on the energy, and the quantitative relations above must be developed for each type of reaction, and for each energy domain, after which, to obtain a final result, the sum (integral) is calculated. More specifically, the energy dependence of incident particles comprises two terms. The first is a mean trend: with protons, the probability of a reaction increases with energy above a certain level; with neutrons, this probability varies with E^d, that is, the energy with an exponent ≥ 1. The second comprises specific resonances for certain energies, corresponding to frequencies of stable vibrations of nuclei, which favor the reaction for those energies.

Naturally all nuclear reactions are associated with subsequent readjustments in energy and therefore with the emission of varying numbers of gamma rays.

4.1.3 Classification of nuclear reactions

Nucleon absorption

Nucleon absorption corresponds to a (p, γ) or (n, γ) reaction. The nucleus absorbs an incident particle, vibrates, and, to return to equilibrium, emits gamma radiation. If the incident particle is a proton, the nucleus formed is chemically different. If it is a neutron, the nucleus formed is a new isotope of the same element. We have already come across such reactions, for example when studying the iodine–xenon method:

$^{127}\mathrm{I}\ (\mathrm{p}, \gamma)\ ^{128}\mathrm{Xe}$.

Another example is:

$^{23}\mathrm{Na}\ (\mathrm{n}, \gamma)\ ^{24}\mathrm{Na}$.

Proton–neutron or neutron–proton exchange

The target nucleus absorbs a proton (or neutron) and gives out a neutron (or proton). In either case there is a change of chemical element during the reaction, for example:

$^{36}\mathrm{Ar}\ (\mathrm{n}, \mathrm{p})\ ^{36}\mathrm{Cl}$, $^{14}\mathrm{N}\ (\mathrm{n}, \mathrm{p})\ ^{14}\mathrm{C}$, and $^{48}\mathrm{Ti}\ (\mathrm{p}, \mathrm{n})\ ^{48}\mathrm{V}$.

(Write out the corresponding reactions in full as an exercise.)

Reactions involving α particles

These are either (n, α) or (p, α) reactions or alternatively (α, n) or (α, p) reactions. Let us cite, for instance, the reactions that occur in rocks when the α particles emitted by radioactive chains produce isotopes of, say, neon: $^{17}O\,(\alpha, n)\,^{20}Ne$, $^{18}O\,(\alpha, n)\,^{21}Ne$, $^{24}Mg\,(n, \alpha)\,^{21}Ne$, and $^{25}Mg\,(n, \alpha)\,^{22}Ne$. We will be using these reactions later on.

Spallation reactions

Spallation reactions are much more violent and the target nucleus is broken up producing a much lighter nucleus and a suite of particles. For example, the irradiation of iron by protons of cosmic rays is written:

$$^{56}\mathrm{Fe} + \mathrm{H}^+ \rightarrow {}^{36}\mathrm{Cl} + {}^3\mathrm{H} + 2\,{}^4\mathrm{He} + {}^3\mathrm{He} + {}^3\mathrm{H}^+ + 4 \text{ neutrons}.$$

It can be seen that this reaction produces a daughter nucleus and numerous neutrons, which in turn may produce other spallation reactions.

Fission reactions

We have already spoken of spontaneous fission. Fission reactions occur also under the influence of neutrons and produce in turn a greater number of neutrons, giving rise to a chain reaction (see Chapter 2, Section 2.3). The most common example is, of course, that of man-made nuclear reactors. But as described in Chapter 1, the isotopic compositions of most of the elements found in the Oklo uranium mine near Franceville in Gabon were so strange that it was concluded there was a natural nuclear reactor there 2 billion years ago.

The result of induced fission is a distribution curve of the elements similar to that for spontaneous fission but slightly offset with two symmetrical peaks of statistical abundance (see Chapter 1). Such nuclear reactions give rise to stable or radioactive isotopes depending on whether or not they move the nucleus produced away from the **valley of stability**.

4.1.4 Absorption of particles by matter in the case of nuclear reactions

It is very important to know how easily each type of particle penetrates the different natural targets. These targets may be rocks (meteorites), gases (atmosphere), or, more rarely, fluids. Let us try to calculate the variation in the number of incident particles N_1 with the thickness of the target. We begin with the equation:

$$-\frac{\mathrm{d}N_1}{N_1} = n\Gamma\,\mathrm{d}x.$$

(N_1 is the integrated flux during the time of irradiation.) This integrates immediately to:

$$N = N_0 \exp(-n\Gamma x).$$

The number of surviving particles N decreases exponentially with distance. This decrease depends, of course, on the type of particle, the target, the effective cross-sectional areas, and the energy spectrum of the incident particles. But the exponential law is very general for all types of particle. Notice that x and also n, the number of target atoms per unit volume, are both involved in the exponential. Such behavior is therefore quantitatively very different in a gas and in a solid. In solids, the flux of incident particles falls off very quickly by interaction with matter, as there is a large number of target particles per unit volume. In gases absorption is weaker.

Exercise

Given that, in a rock, the secondary thermal neutrons are produced from primary protons, establish the law giving the number of neutrons depending on thickness. Assume the number of protons decreases in line with $P_0\,e^{-kx}$.

Answer

If N is the integrated flux of neutrons,

$$dN = (\text{production}) - (\text{destruction})$$

$$\frac{dN}{dx} = \phi P - AN$$

but the number of protons P is: $P = P_0\,e^{-kx}$.

A and $k = n\Gamma$ are the effective cross-sectional area for neutrons and for protons, respectively. Integrating gives:

$$N = \frac{\phi P_0}{A - k}(e^{-kx} - e^{-Ax}).$$

The curves representing the flux of protons and neutrons versus x are shown qualitatively in Figure 4.2.

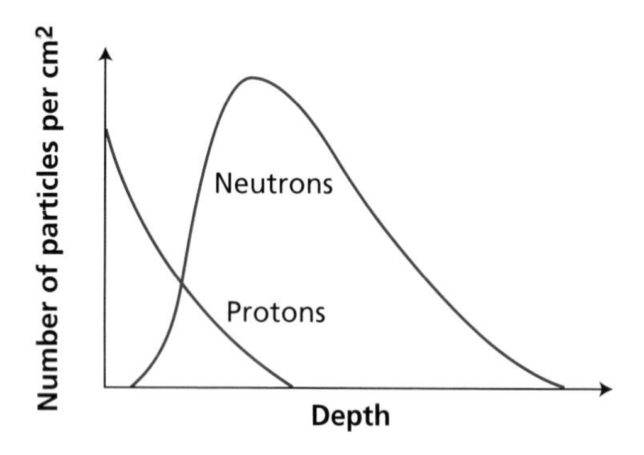

Figure 4.2 Variations in the number of protons and neutrons with depth in a rock exposed to proton radiation.

4.1.5 Galactic cosmic radiation

The Universe is traversed by a flow of charged particles reflecting its approximate chemical composition largely dominated by ionized hydrogen, that is, by protons. This particle flux traverses the Universe at kinetic energies of the order of 1 billion electron volts (GeV). This is known as **galactic cosmic radiation**. It is thought that these ions are emitted by exploding supernovae accelerated by their shock waves and by other complex phenomena of magnetic pulsation in ionized environments occurring in the magnetic fields of stars.

These charged particles interact with matter whether in gaseous state as in the atmosphere or in solid state as in meteorites or terrestrial rocks. In general, proton interaction leads above all to the production of secondary neutrons and to successive nuclear reactions as in the atmosphere (Figure 4.3). High-energy, secondary neutrons produce other neutrons in a chain whose energy decreases until it is "thermal."[2] Thermal neutrons are very efficient at causing nuclear reactions. In the atmosphere, for example, this occurs by the reaction ^{14}N (n, p) $^{14}C^*$, giving rise to ^{14}C. When thermal neutrons react with rock, they produce spallation reactions giving rise either to stable isotopes or to radioactive isotopes. It is from such nuclear reactions that we calculate what are called **exposure ages**.

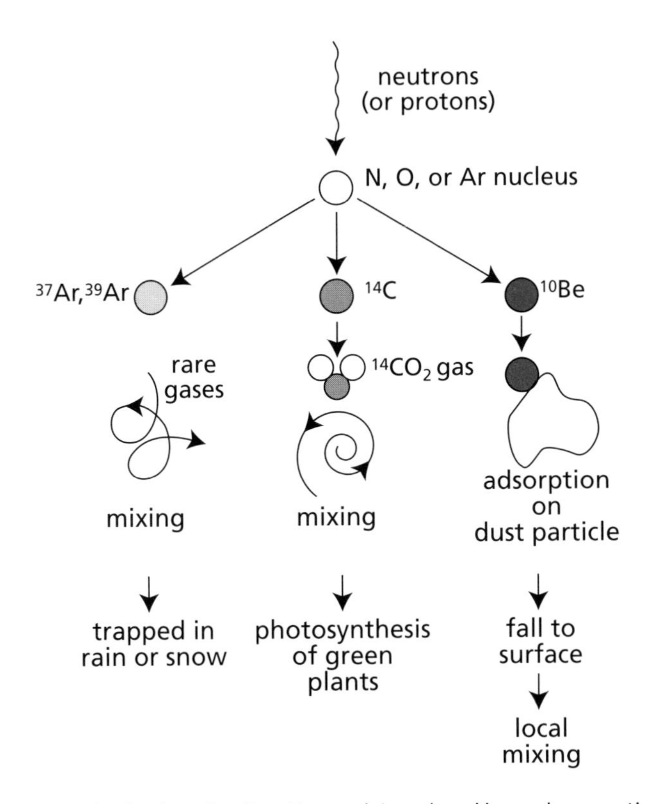

Figure 4.3 The destiny of radioactive nuclei produced by nuclear reactions in the atmosphere under the influence of cosmic rays. Gases like argon remain in the atmosphere, some like ^{14}C oxidize, while others like ^{10}Be are adsorbed onto solid particles and fall to the ground with them.

[2] Thermal neutrons are neutrons which, after colliding and interacting with matter, have energy levels of kT, where k is Boltzmann's constant and T temperature ($kT \approx 0.025$ eV). In this energy spectrum they have maximum probability of creating nuclear reactions.

All of these phenomena were brought to light and investigated little by little in particular through the pioneering work of **Devendra Lal** of the Physical Research Laboratory of Ahmedabad and the University of California at San Diego (see Lal [1988] and Lal and Peters [1967] for a review).

Each of the other three sections of this chapter deals with a type of nuclear reaction. The first concentrates on ^{14}C and concerns the radionuclides produced in the atmosphere and their use in geochronology. The second deals with exposure ages first in meteorites and then in terrestrial rocks. The third section gives an overview of stellar processes of nucleosynthesis.

4.2 Carbon-14 dating

Of all the radiometric methods, this is undoubtedly the most famous, the one that is familiar to the general public and the one that people (mistakenly) think of when speaking of the age of the Earth or of rocks. This method was developed by **Willard Libby** (1946) who eventually received the Nobel Prize for chemistry for his work.

4.2.1 The principle of ^{14}C dating

Carbon-14 is produced in the atmosphere by cosmic rays whose protons engender secondary neutrons. These neutrons react with ^{14}N:

$$^{14}N(n, p)\ ^{14}C^{*}.$$

In this reaction, ^{14}N is the target nucleus, n (neutron) is the projectile, and p (proton) is the particle ejected; $^{14}C^{*}$ is the radioactive isotope produced which disintegrates by β^{-} radioactivity to give ^{14}N. As soon as it has formed, the ^{14}C combines with oxygen to give CO_2. If we note $N(^{14}C)$ the number of ^{14}C atoms at the time of measurement t and $[N(^{14}C)]_0$ the initial number of carbon atoms, then we may write:

$$N(^{14}C) = \left[N(^{14}C)\right]_0 e^{-\lambda t}.$$

Libby showed that the proportion of ^{14}C in the atmosphere was roughly constant over time. What accounts for this constancy? Let us write out the balance for ^{14}C production:

$$\frac{d}{dt}[N(^{14}C)] = F - \lambda N(^{14}C).$$

↑ production ↑ destruction by radioactivity

If a stationary state is attained:

$$\frac{d}{dt}[N(^{14}C)] = 0 \text{ from which } N(^{14}C) = \left(\frac{F}{\lambda}\right),$$

where F is the product of the flux ϕ of neutrons by the number $N(^{14}N)$ of atoms of ^{14}N by the effective cross-section Γ.

The flux ϕ varies with **latitude** because cosmic rays, which are composed of protons, and so are positively charged, are deflected by the Earth's magnetic field. The poles receive much

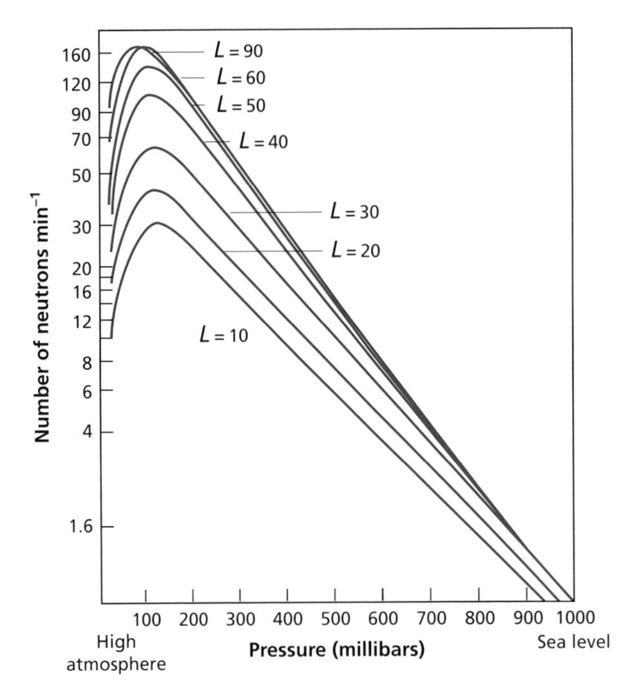

Figure 4.4 Number of neutrons with altitude and latitude. Geomagnetic latitude is given by the parameter *L*. Altitude varies with pressure, which is measured in millibars here. The value of *L* increases as we move from the equator to the poles.

more radiation than the equator. The flux also varies with **altitude**, because the Earth's atmosphere "absorbs" and transforms the incident flux (Figure 4.4).

For our purposes, the main phenomenon is that the primary protons produce secondary neutrons, which produce others in a snowball effect. It is these neutrons that produce the ^{14}C. As said, as soon as it has been produced, the ^{14}C reacts with oxygen (or ozone) to give CO_2, and this CO_2 mixes with the remainder of the atmosphere.

Since the atmosphere itself is well mixed in a few weeks, the ^{14}C is homogenized all around the planet on the timescale of interest to us. It can be taken, then, that the mean amount of ^{14}C produced in the atmosphere is a valid, uniform benchmark.

Libby and his co-workers (Libby *et al.*, 1949) determined the quantity of ^{14}C produced in the steady state. They expressed it as the number of disintegrations per minute (dpm) per gram of carbon:

$$\frac{\lambda N(^{14}C)}{N_{carbon}} = \frac{F}{N_{carbon}} = 13.5.$$

They also determined the decay constant of ^{14}C as $\lambda = 1.209 \cdot 10^{-4}\ yr^{-1}$ (the half-life is therefore 5730 years).[3]

[3] These values have been amended slightly today.

Exercise

What is the $^{14}C/^{12}C$ isotopic composition of atmospheric carbon, given that the isotopic composition of stable carbon is defined by $^{13}C/^{12}C = 0.011\,224$ or the reciprocal $^{12}C/^{13}C = 89.09$, or $^{12}C = 98.89\%$ and $^{13}C = 1.11\%$?

Answer

The atomic mass of carbon is 12.011. One gram of carbon therefore represents $1/12.011 \times 6.023\,13 \cdot 10^{23} = 5.014 \cdot 10^{22}$ atoms, including $5 \cdot 10^{22} \times 0.9889 = 4.95 \cdot 10^{22}$ atoms of ^{12}C. The basic relation of radioactivity is $\lambda N = 13.5$ dpm (where N is the number of ^{14}C atoms).

In one year there are $5.26 \cdot 10^5$ minutes, therefore there are $13.5 \times 5.26 \cdot 10^5 = 71.48 \cdot 10^5$ disintegrations per year. Since $\lambda = 1.209 \cdot 10^{-4}$ yr^{-1}, we have $5.88 \cdot 10^{10}$ atoms of ^{14}C. Therefore:

$$\frac{^{14}C}{^{12}C} = \frac{5.88 \cdot 10^{10} \text{ atoms } ^{14}C}{4.96 \cdot 10^{22} \text{ atoms } ^{12}C} = 1.1849 \cdot 10^{-12} \approx 1.18 \cdot 10^{-12}.$$

This ratio is tiny and cannot be measured by a **conventional mass spectrometer**, because the ^{12}C peak is too high compared with the ^{14}C peak. This is why it was long preferable to measure ^{14}C with a Geiger counter.

When carbon is incorporated in a living organism (plant or animal), its isotopic composition (and so its activity) is equal to that of the atmosphere and is determined by phenomena such as photosynthesis or respiration. As soon as the organism dies, such exchanges cease and radioactive decay is the only source of variation in the ^{14}C content. The time of death of an organism (or more precisely the time at which it stopped exchanging CO_2 with the atmosphere) can therefore be dated by the formula:

$$(^{14}C/C) = 13.5\, e^{-\lambda t}$$

$$t = \frac{1}{\lambda} \ln\left[\frac{13.5}{(^{14}C/C)_{\text{measured}}}\right].$$

Exercise

Let us take one of the examples that helped to make ^{14}C dating so popular: Egyptology (see Figure 4.5). A sample was taken from a wooden beam in the tomb of the vizier Hemaka at Saqqara. He was an official of the First Dynasty of Egyptian pharaohs. After measuring the ^{14}C content of the wood, Libby announced it was 4880 years old. (Archeologists reported that the royal seal-bearer had lived between 3200 and 2700 BC.) How did Libby manage this feat?

Answer

$$t = \frac{1}{\lambda} \ln\left[\frac{(^{14}C/C)_{\text{atmosphere}}}{(^{14}C/C)_{\text{sample}}}\right]$$

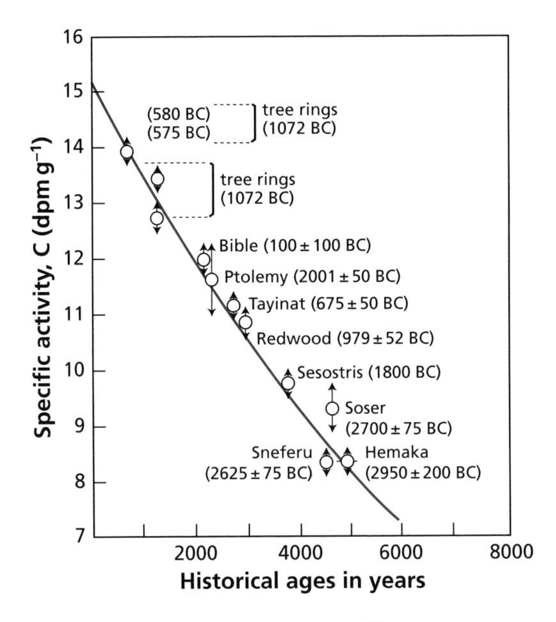

Figure 4.5 Calibration of the reliability of ^{14}C dating on historical data in Egyptology.

The intricate measurement of the $^{14}C/C$ ratio Libby made on the wooden artefact yielded 6.68 dpm g^{-1}. Applying the age measurement formula then gave $t = 4880$ years.[4]

This method was highly successful and brought about a revolution in archeology. By the same principle, a papyrus, bones, and burnt or petrified trees were dated, thereby providing an archeological chronometer that was unknown until then. This method has the drawback of destroying the object that is to be dated, which means a careful choice must be made. This is why recent advances in ^{14}C analysis made with accelerator mass spectrometers, which require only one-hundredth of the amount of material, are so important and have made the method even more incisive.

4.2.2 Measuring ^{14}C

Measuring ^{14}C is a difficult business, as we have just seen. A series of simple calculations will provide insight into this difficulty, which has already been illustrated in an earlier exercise. One gram of "young" carbon gives off 13.5 dpm, or 1 disintegration every 4.5 seconds. If we have just 10 mg of carbon, we will have one disintegration every 7.4 minutes, which is not much given a laboratory radioactive environment and also given that cosmic rays emit radiation in the same order of magnitude.

[4] In fact, Libby found a constant $\lambda = 1.244 \cdot 10^{-4}\ yr^{-1}$ and therefore $t = 5568$ years, which corresponds to 7.35 dpm g^{-1}. Oddly enough, ^{14}C specialists still use Libby's constants and then make a correction. For didactic reasons we shall **not** follow this practice.

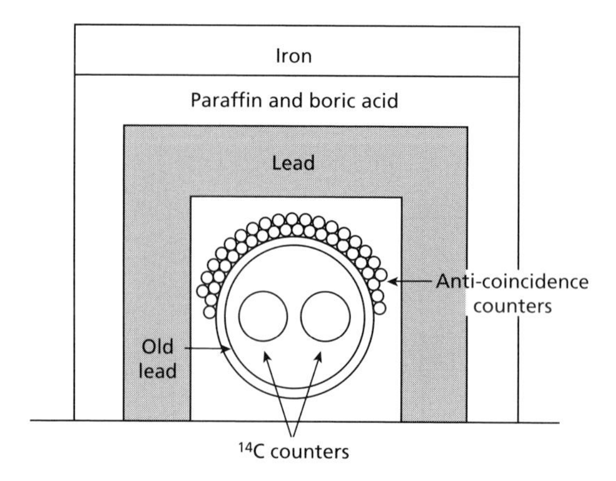

Figure 4.6 Libby's anti-coincidence counting. Geiger counters whose gas contains the ^{14}C to be measured are surrounded by a series of layers of shielding as protection from cosmic rays (only old lead is used so that the uranium chains are dead and there is no decay from them). When the small counters record a disintegration, it is due to cosmic radiation and so is subtracted from the value recorded by the central counter.

But suppose we have a sample some 55 000 years old. It now emits just $1.7 \cdot 10^{-2}$ dpm, or 1 disintegration per hour (on average, of course, since radioactivity is a statistical law). Now, over the course of the hour, other particles have been emitted in the vicinity of the gram of carbon for many reasons: the various materials surrounding the counter (brick, cement, etc.) probably contain uranium impurities of the order of 1 ppb (and therefore also their derived daughters: calculate them!), and the sky showers particle flows on the Earth from the cosmos or the Sun, etc. How can we eliminate these disturbances and make a reliable measurement?

The counting method

Libby's answer was to build a Geiger counter whose internal gas itself contained ^{14}C changed into CO_2 (Figure 4.6).

With the main counter surrounded by an array of secondary Geiger counters, any external radiation could be subtracted because it first passed through the outer counters (the anti-coincidence method). Lastly, it was all buried and surrounded by "old" lead shielding to prevent interference radiation. How much background noise did the counters measure? Without shielding 1500 disintegrations per hour were detected, with shielding the figure was just 400, and with the anti-coincidence counters it was just 8! It was therefore virtually impossible to measure an artefact 55 000 years old since it gave off just one disintegration per hour! Measurement was intricate and necessarily lasted for a long time.

The accelerator mass spectrometry method (AMS)

Since the late 1980s mass spectrometry has been adapted for ^{14}C by using small particle accelerators with energies of more than 1 MeV (see Kieser *et al.*, 1986); this is the accelerator mass spectrometry (AMS) method. The high energy imparted to the ions purifies the

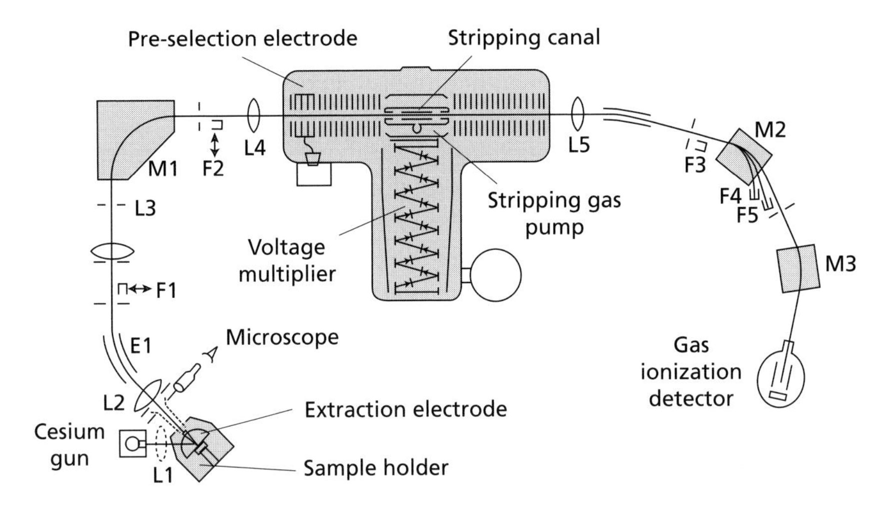

Figure 4.7 Accelerator mass spectrometer. This device is much larger and more complex than an ordinary mass spectrometer as it operates at much higher ion acceleration energies, but the principle is unchanged. The ions are extracted on the left by being bombarded with a beam of cesium ions and are then deflected first by an electrostatic field E1 and then by a magnetic field M1. In the center of the figure, a microchannel injects gas which collides with the beams and destroys any molecules (this is known as "stripping"). The purified ions are then deflected by two electromagnets M2 and M3 and collected by a detector similar to an ionization chamber used in particle physics.

organic beam of any molecules and other impurities by passing them either through thin sheets of gold or through gas streams. The carbon ions pass through these while any molecules are "deactivated" and stopped. In this way the method's speed of analysis, sensitivity, and precision have been increased compared with counting. A measurement that once took more than a week can now be made in an hour. But above all, the method requires samples of one-tenth of the size and achieves levels of precision 100 times greater (see Figure 4.7).

Exercise

If the maximum that can be measured by AMS for the $^{14}C/^{12}C$ ratio in 1 mg of carbon is $1.2 \cdot 10^{-17}$, what is the maximum age that can be measured with ^{14}C?

Answer

10 000 years.

Exercise

If the limit of detection by counting methods is 10 disintegrations per hour, what quantity of carbon is required to measure the same age?

Answer

2.213 kg!

4.2.3 Conditions for performing ^{14}C dating

Applying the formula $^{14}C = {}^{14}C(0)\, e^{-\lambda_{14} t}$ to find an age entails meeting a number of extremely stringent conditions.

The system must have remained **closed** since the time that is to be dated. This means if we wish to date a parchment, it must have neither gained nor lost any ^{14}C since the time the lamb or sheep whose hide was used to make the parchment died. We need to know $^{14}C(0)$ precisely, that is the **initial ^{14}C** content, that is the proportion of $^{14}C/C$ in the animal at the time of its death. There must remain enough ^{14}C for it to be **measurable**.

Lastly, and this is not the least of the problems, the event whose age is to be determined must be **defined**.

Can we be certain about all four conditions in practice?

The closed system

The greatest danger from an open system is contamination of the sample, whether natural or artificial. Take the case of a tree trunk we wish to date. If the trunk has been lying in humus for some time it will be impregnated, contaminated, by the carbon in the humus. Now, this carbon is "older" than the carbon of the tree that has just died and so distorts the measurement.

Let us take the example of a tree trunk whose specific activity, that is, the ^{14}C radioactivity per gram of carbon, is r_1. It lies for some time in humus whose radioactivity, in ^{14}C per gram of carbon, is r_2. What is the level of radioactivity of the mixture supposing that 10% of the carbon comes from the humus?

$$r_1 = \left(\frac{^{14}C}{C}\right)_{\text{tree}}, r_2 = \left(\frac{^{14}C}{C}\right)_{\text{humus}}$$

$$r_{\text{total}} = \frac{^{14}C_{\text{tree}} + {}^{14}C_{\text{humus}}}{C_{\text{tree}} + C_{\text{humus}}} = \left(\frac{^{14}C}{C}\right)_{\text{tree}} \frac{C_{\text{tree}}}{C_{\text{tree}} + C_{\text{humus}}} + \left(\frac{^{14}C}{C}\right)_{\text{humus}} \frac{C_{\text{humus}}}{C_{\text{tree}} + C_{\text{humus}}}$$

$$r_{\text{total}} = \left(\frac{^{14}C}{C}\right)_{\text{tree}} w + \left(\frac{^{14}C}{C}\right)_{\text{humus}} (1 - w).$$

The value of $w = C_{\text{tree}}/C_{\text{total}}$ varies between 0 and 1. (This is the same mixing formula as in the section on isotopic dilution.)

Assuming that $(^{14}C/C)_{\text{tree}} = 12.5$ dpm g^{-1} and that $(^{14}C/C)_{\text{humus}} = 4$ dpm g^{-1}, then $r_{\text{total}} = 12.5 \times 0.9 + 4 \times 0.1 = 11.65$ dpm g^{-1}, $w = 0.9$.

This corresponds to an error of 7% by default. The age measured is older than the real age. Conversely, if we suppose the tree trunk found in the sediment had been in soil which was still involved in atmospheric exchanges, then the contamination, that is, the mixture between ^{14}C in the dead tree and the ground would continually make the wood "younger."

Determining the initial ^{14}C content

Libby had assumed that the value of 13.5 dpm g^{-1} of carbon was the current value and valid as the initial value throughout time. However, several complications have arisen since then. Not all plants absorb carbon isotopes in an identical manner. If $(^{14}C/C)_{\text{present}} = r_0$, the

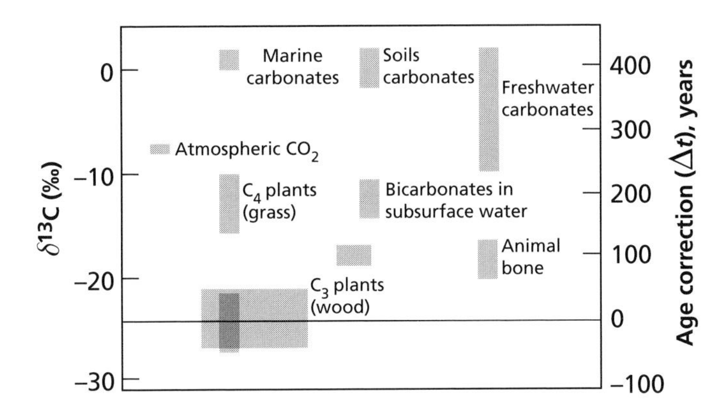

Figure 4.8 Carbon isotope fractionation relative to the atmosphere depending on the nature of the substrate.

value absorbed by living organisms is $(^{14}C/C)_{organism} = R\, r_0$, where R is a partition coefficient. This coefficient has been determined and to check it we measure the $^{13}C/^{12}C$ ratio of the object to be dated (Figure 4.8) (this is more fully explained in Chapter 7).

On the left is the $\delta^{13}C$ scale, such that:

$$\delta^{13}C = \left[\frac{\left(\frac{^{13}C}{^{12}C}\right)_{sample} - \left(\frac{^{13}C}{^{12}C}\right)_{standard}}{\left(\frac{^{13}C}{^{12}C}\right)_{standard}}\right] \cdot 10^3.$$

The standard is a carbonate. In practice, this fractionation is converted into an age difference by taking a reference standard.

We assume that fractionation is directly proportional to the difference in mass. Therefore, if $R_{3/2}$ is the partition coefficient between some plant and the atmosphere, for the $^{13}C/^{12}C$ ratio defined by:

$$\frac{\left(\frac{^{13}C}{^{12}C}\right)_{plant}}{\left(\frac{^{13}C}{^{12}C}\right)_{atmosphere}} = R_{3/2}$$

the partition coefficient for the $(^{14}C/^{12}C)$ ratio

$$\frac{\left(\frac{^{14}C}{^{12}C}\right)_{plant}}{\left(\frac{^{14}C}{^{12}C}\right)_{atmosphere}} = R_{4/2}$$

is such that $R_{4/2} = 2 \times R_{3/2}$.

It is possible, then, to **correct for natural fractionation** of ^{14}C and express this correction in age, which is done on the right of Figure 4.8.

It was realized that the ^{14}C content of the atmosphere had varied over the course of time. This effect was shown by measuring the ^{14}C of tree rings. Tree rings are evidence of annual

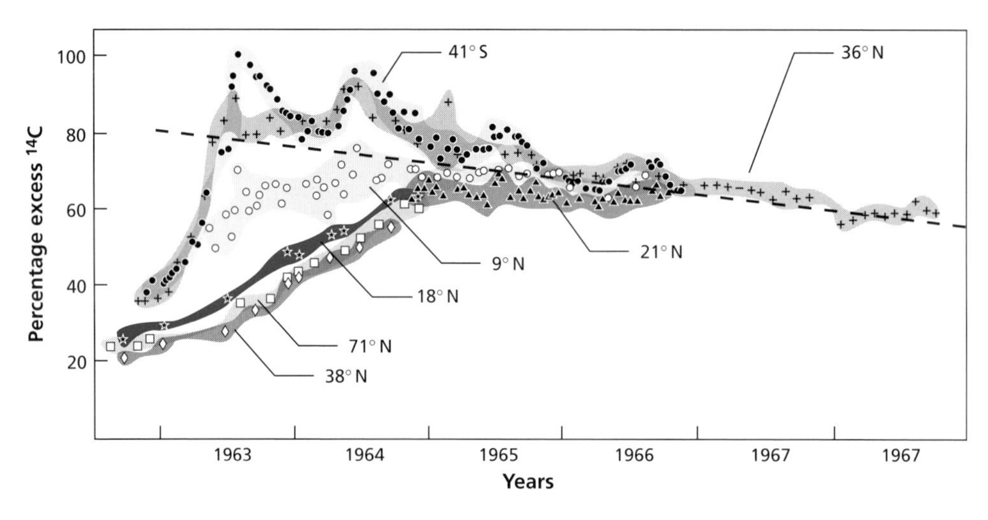

Figure 4.9 Changes in excess ^{14}C injected into the atmosphere by U.S. atom-bomb testing in the Pacific. Each symbol corresponds to a latitude. The various, ever weaker, injection dates can be seen: 1963, 1964, 1965, 1966. It took several decades to return to equilibrium. After Libby (1970).

growth, with each ring representing one year. By counting the rings, we can work back through time, year by year. This is known as **dendrochronology**. It was noticed that the ages of rings measured by ^{14}C and the ages obtained by counting the same rings failed to match.

Systematic investigation has shown that multiple phenomena are involved in this. The burning of coal and oil, that is, materials whose ^{14}C is "dead," has diluted the "natural" ^{14}C content of the atmosphere. The present-day content is too **low** compared with what it would be without this effect. Likewise, for recent periods, nuclear testing by explosions in the atmosphere has injected a large quantity of ^{14}C, which has largely disrupted the natural cycle (Figure 4.9) and, of course, increased the ^{14}C content of the atmosphere.

The flux of cosmic particles and therefore the production of ^{14}C varies over the course of time. This is related to solar activity, to fluctuations in the Earth's magnetic field, etc. We therefore need a calibration curve describing variation of the $^{14}C/C$ ratio over time. At first sight, this looks something like a chicken-and-egg situation because we wish to determine an age from the $^{14}C/C$ measurement but at the same time we wish to calculate the $^{14}C/C$ ratio by using age measurement. The curve of $^{14}C/C$ variation in the atmosphere was constructed by dating samples using a different method. For recent periods, tree rings were used as the benchmark, and for earlier periods dates were ascertained by radioactive disequilibrium methods.

To calibrate the $^{14}C/C$ ratio, we first seek to calculate $(^{14}C/C)_{\text{initial}} = (^{14}C/C)_{\text{measured}}\, e^{\lambda t}$, where t is the independently determined age. In this way a correction can be applied to the age measured by the conventional method (Figure 4.10). It is then possible to draw a curve connecting ^{14}C dates to historical calendar ages. To do this, 1950 was taken as the reference date for ^{14}C. Suppose we calculate a ^{14}C date of 3000 years before 1950 (if we made the measurement in the year 2000 we would remove 50 years!). To obtain the BC date, we must therefore subtract 1950: 3000 – 1950 = 1050 BC (Figure 4.11).

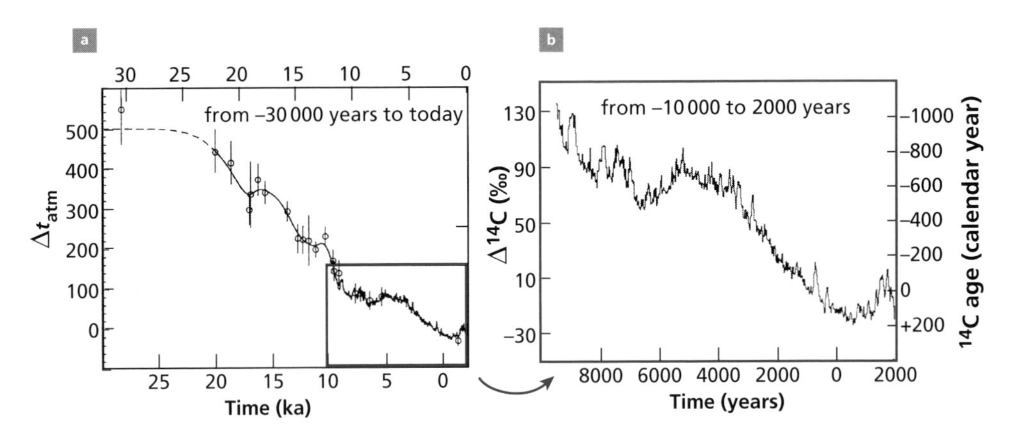

Figure 4.10 Calibration curves for ^{14}C. (a) For the older period, we use methods based on radioactive disequilibrium as did **Edouard Bard** (Bard *et al.*, 1990); (b) for the recent period we can use tree rings (Stuiver, 1965).

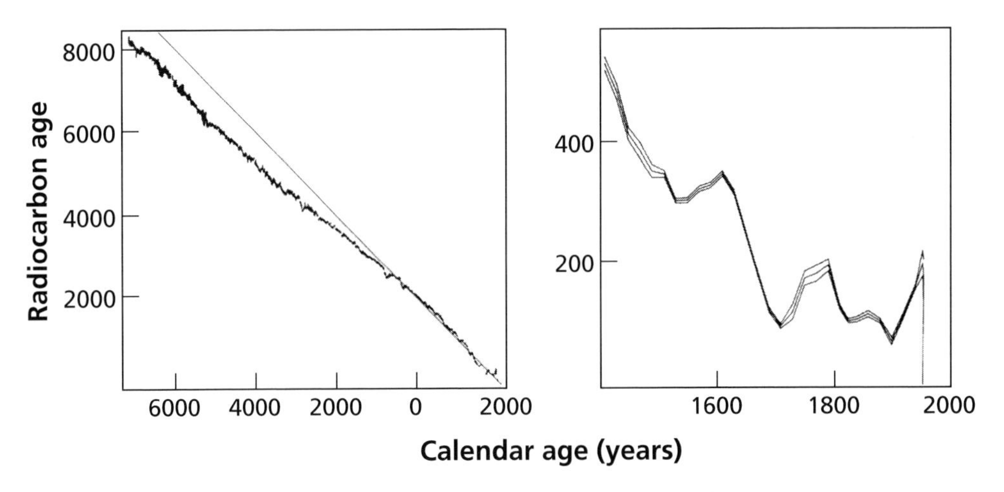

Figure 4.11 Correspondence between calendar ages and ^{14}C ages. The ^{14}C age is the time between the death of the organism and 1950. The calendar date begins at the birth of Jesus Christ. To this difference, we must add the fluctuations caused by calibrations because of historical variations of $(^{14}C/C)_{initial}$.

Exercise

We use the curve of $^{14}C/C$ variation with time (Figure 4.11). Suppose we find an age of 200 years for a piece of fabric using the standard ^{14}C formula. What date does this correspond to?

Answer

As can be seen, this "age" corresponds to two possible dates: 1630 or 1800. How can we decide between the two? The answer is obtained by using other independent information such as the nature of the fabric. This illustrates the limits of precision of ^{14}C.

Synthetic formula for age calculation

When making calculations, we can convert everything to the addition of magnitudes which themselves have age dimensions. When applying the uncorrected conventional formula we obtain:

$$t_{\text{apparent}} = \frac{1}{\lambda}\ln\left(\frac{A_c}{A_m}\right)$$

where A_c is the initial conventional activity and A_m is the activity as measured. The true age must be written:

$$t_{\text{true}} = \frac{1}{\lambda}\ln\left(\frac{A_{0,\text{real}}\,P_{2,4}}{A_m}\right)$$

where $A_{0,\text{real}}$ is the real, initial activity of the atmosphere and $P_{2,4}$ is the isotope fractionation factor. We can also introduce in the parenthesis the reference value A_c for the initial activity. Dividing the top and bottom gives:

$$t_{\text{true}} = \frac{1}{\lambda}\left[\ln\left(\frac{A_{0,\text{real}}}{A_c}\cdot\frac{A_c}{A_m}\cdot P_{2,4}\right)\right].$$

Taking advantage of the additive property of logarithms, we have:

$$t_{\text{true}} = \frac{1}{\lambda}\ln\left(\frac{A_c}{A_m}\right) + \frac{1}{\lambda}\ln\left(P_{2,4}\right) + \frac{1}{\lambda}\ln\left(\frac{A_{0,\text{real}}}{A_c}\right).$$

From which, by transforming everything into time units:

$$t_{\text{true}} = t_{\text{apparent}} + \Delta t_{\text{fractionation}} + \Delta t_{\text{atmospheric correction}}$$

where:

$$\Delta t_{\text{fractionation}} = \frac{1}{\lambda}\ln(P_{2,4}) \text{ and } \Delta t_{\text{atmospheric correction}} = \frac{1}{\lambda}\ln\left(\frac{A_{0,\text{real}}}{A_c}\right),$$

with a little extra calculation trick because $\left(A_{0,\text{real}}/A_c\right)$ can be evaluated as a function of time only.

We therefore calculate $t_1 = t_{\text{apparent}} + \Delta t_{\text{fractionation}}$, then we can use t_1 to calculate $\Delta t_{\text{atmospheric correction}}$ using the fluctuation curves of Figure 4.11, and we then obtain t_{true}.

Exercise

We measure the activity of a camel bone found in the tomb of the vizier Hemaka as 6.68 dpm g^{-1}. What is the real date of the vizier's death? (Use the existing calculations and graphs.)

Answer

In an exercise in Section 4.2.1 we dated the death of vizier Hemaka (4880 years ago) using an ordinary wooden beam.

$$t_1 = t_{\text{apparent}} + \Delta t_{\text{fractionation}}$$

Using Figure 4.8 and the previous result, we get $t_1 = 4800 + 100 = 4980$ years, or 3030 BC. By using the graph in Figure 4.10, we find for a date of 3000 BC a correction of −250 years:

$t_{real} = 4980 - 250 = 4730$ years.

4.2.4 Other forms of cosmogenic radioactivity

Carbon-14 is far more important than any other spallation product created in the atmosphere by interaction of cosmic rays, but some others exist as well. These can be divided into three groups (see Figure 4.3):

- The rare gases, which are free gases. These are incorporated into solid material by trapping part of the atmosphere. These are ^{39}Ar ($T_{1/2} = 269$ years) and ^{81}Kr ($T_{1/2} = 2.1 \cdot 10^5$ years);
- Gaseous products, which, similarly to ^{14}C, are incorporated into reactive atmospheric molecules. For example ^{3}H (tritium; $T_{1/2} = 12.43$ years) or ^{36}Cl ($T_{1/2} = 3.08 \cdot 10^5$ years);
- Products which are not gaseous but adhere immediately after their formation to dust particles and follow the history of the dust rather than the gaseous atmosphere. Examples are ^{10}Be ($T_{1/2} = 1.51 \cdot 10^6$ years) or ^{26}Al ($T_{1/2} = 7.16 \cdot 10^5$ years).

Beryllium-10

Beryllium-10 (^{10}Be) is produced by spallation of nitrogen and oxygen in the atmosphere. As soon as it forms, it is adsorbed onto particles in the atmosphere and incorporated in rain. Accordingly, it is more or less well mixed. Its lifetime in the atmosphere is short and as its production varies with latitude (as with ^{14}C), its distribution may be extremely heterogeneous and even erratic (Figure 4.12).

A further complication arises because ^{9}Be, common beryllium (which has just this one isotope), has a completely different geochemical history to ^{10}Be. It is incorporated in rocks and so in dust derived from erosion and scattered by the wind. The $^{10}Be/^{9}Be$ ratio is therefore just as erratic as the absolute ^{10}Be content. We cannot use ^{10}Be like ^{14}C, assuming a uniform

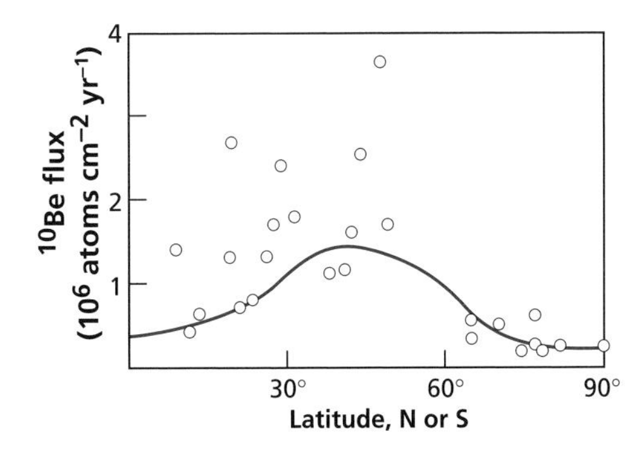

Figure 4.12 Variations in the flux of ^{10}Be with latitude display substantial dispersion.

value for the entire Earth at any given time. One method is to assume that, at a given place, fluctuations in the $^{10}Be/^{9}Be$ ratio are low and at any rate much lower than those arising from the radioactive decay of ^{10}Be. It is then possible to use ^{10}Be as a chronometer.

Exercise

We wish to measure the rate of accretion of manganese nodules found in the ocean. How can we set about this, knowing that the variation in the $^{10}Be/^{9}Be$ ratio with depth is $^{10}Be/^{9}Be = (^{10}Be/^{9}Be)_0 \, e^{-\lambda t}$?

Answer

$$t = \frac{x \text{ (thickness)}}{a \text{ (accretion rate)}},$$

from which:

$$\left(\frac{^{10}Be}{^{9}Be}\right) = \left(\frac{^{10}Be}{^{9}Be}\right)_0 e^{-\frac{\lambda x}{a}}.$$

Switching to logarithms gives:

$$\ln\left(\frac{^{10}Be}{^{9}Be}\right) = \ln\left(\frac{^{10}Be}{^{9}Be}\right)_0 - \frac{\lambda x}{a}.$$

The logarithm of the ratio $^{10}Be/^{9}Be$ plotted as a function of thickness is a straight line of slope (λ/a), which gives the rate of accretion, a.

O'Nions *et al.* (1998) measured the $^{10}Be/^{9}Be$ ratios in manganese crusts of the North Atlantic. Figure 4.13 shows the results they obtained. This method yielded an accretion rate of 2.37 ± 0.15 mm Ma^{-1}.

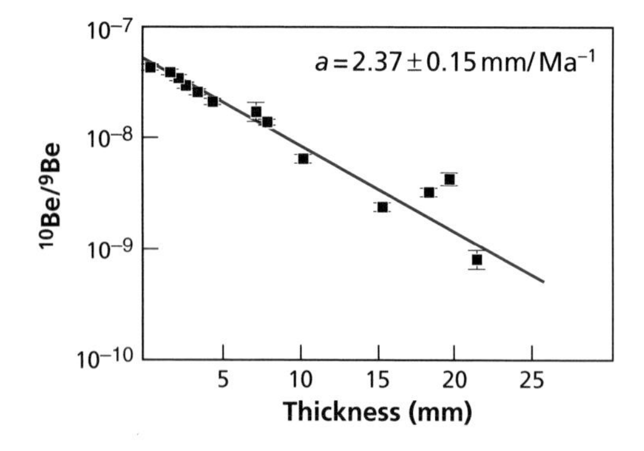

Figure 4.13 Variation in the $^{10}Be/^{9}Be$ ratios in a manganese concretion from the Atlantic Ocean. Magnitude a indicates the rate of accretion of the nodule. After O'Nions *et al.* (1998).

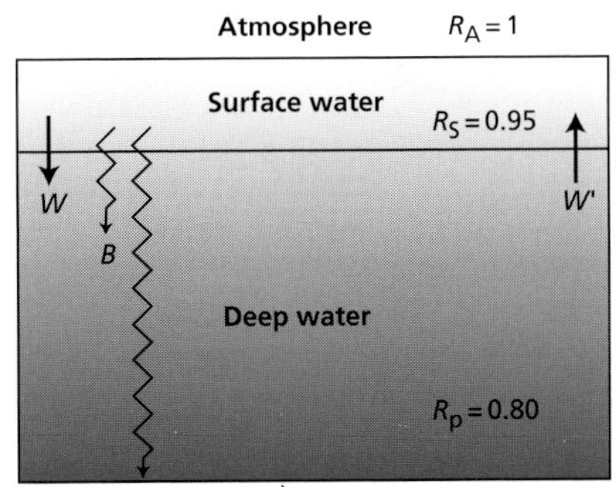

Figure 4.14 The behavior of ^{14}C in the ocean. The ocean is divided into two reservoirs: the upper layer, which is well mixed, and the deep ocean. *R* is the $^{14}C/C$ ratio. *B*, flow of solid carbonate particles; *W*, W^1, descending and ascending flows of water.

4.2.5 Dating oceanic cycles with ^{14}C

This is a rather unusual method of dating, but one that is very important in oceanography. The residence time τ of a chemical element in a reservoir is the average time the element spends in the reservoir (Figure 4.14) (see further discussion in Chapter 4). If W is the flux and V the reservoir volume, we posit that $\tau = (V/W)$.

Broecker and colleagues at the Lamont Observatory of Columbia University at New York came up with the idea of using ^{14}C to determine the residence time of water in the deep ocean (Broecker *et al.*, 1960; Broecker and Li, 1970). The ocean is divided into two layers: a surface layer that is well mixed by currents and which is constantly exchanging its CO_2 (and so its ^{14}C) with the atmosphere, and a deep layer which exchanges water and matter with the surface layer. Let B denote the flow of solid carbonate particles (moles yr^{-1}) falling from the surface and dissolving in the deep ocean; W and W' are the flows of descending and ascending water which summarize the exchange between the upper layer and the deep ocean ($m^3\ yr^{-1}$): $W = W'$; C_s and C_d are the concentrations of carbon dissolved in the surface and deep layers (moles m^{-3}); R is the $^{14}C/C$ ratio standardized relative to the atmosphere, R_s and R_d being the ratios for the surface water and deep water. The total carbon conservation equation is written:

$$(WC_s + B) = WC_d.$$

The equation for ^{14}C is:

$$(WC_s + B)R_s = WC_d R_d + V_d C_d R_d \lambda.$$

The residence time in the deep water is:

$$\tau_d = \frac{V_d\ C_d}{W\ C_d} = \frac{V_d}{W},$$

which, given the foregoing equations, is written:

$$\tau_d = \left(\frac{R_s}{R_d} - 1\right)\frac{1}{\lambda}.$$

Now, $\lambda = 1.209 \cdot 10^{-4}\,\text{yr}^{-1}$, therefore $\tau_d = 1550$ years.

4.3 Exposure ages

4.3.1 Meteorites

Meteorites are fragments of planetary bodies resulting from collisions and whose primary age is close to that of the Solar System (4.55 Ga, as an indication). They are pieces of rock "floating" in interplanetary space. Most come from the asteroid belt, between Mars and Jupiter, which is formed of rock debris, pieces of which may measure a kilometer or so across (Ceres). Some were ripped from Mars or the Moon. These rocks drift loose in space and are continually subjected to cosmic radiation whose average energy is 1 GeV. Just as happens in the atmosphere, when cosmic "primary" protons penetrate these rocks they give rise to secondary neutrons, which produce most of the nuclear reactions.[5] These particles produce nuclear reactions at depths ranging from a few tens of centimeters up to a meter at most. The surface layers of rocks exposed to cosmic rays are therefore the site of nuclear reactions, usually spallation reactions. They give rise to isotopes of lower mass than the target, which are known as **cosmogenic isotopes** (see Paneth *et al.*, 1952; Honda and Arnold, 1964). The nuclei engendered by such reactions include radioactive and stable isotopes. Both types are very numerous.

Stable isotopes

The number of cosmogenic stable isotopes N_s produced per unit time is written:

$$\frac{dN_s(t)}{dt} = \phi\Gamma_s N_{c\to s},$$

where ϕ is the particle flux, Γ_s is the effective cross-section of the reaction, that is, the probability of a nuclear reaction occurring, and $N_{c\to s}$ the number of target atoms producing the stable daughter isotope, s, by nuclear reaction. If the flux is constant in terms of intensity and spectral energy, then:

$$N_s(t) = \phi\Gamma_s N_{c\to s}\, t.$$

The number of daughter isotopes is directly proportional to the irradiation time.

It seems straightforward enough to calculate the age of irradiation, provided we know ϕ, Γ_s, and $N_{c\to s}$. It is comparatively easy to measure the parameter $N_{c\to s}$: it is the concentration of

[5] As explained, protons carry a positive charge and do not penetrate readily into matter because they are repelled electronically. Neutrons carry no charge and so penetrate much more readily.

the target product which, by nuclear reaction, yields the cosmogenic isotope. Sometimes it is single. This is so in iron meteorites composed of a metallic alloy of iron and nickel. Sometimes there are more than one, as in ordinary meteorites where krypton isotopes are produced by spallation on rubidium, strontium, yttrium, and zirconium.

The primary flux of cosmic rays ϕ is composed of protons (and some He^+ ions) (and of all the isotopes in the Universe in the ionized state but in very small abundances). It has both intensity (number of particles per unit of surface area and per unit time) and an energy spectrum, because in fact there are $N_1, N_2, \ldots, N_3$ particles, corresponding to energy levels $1, 2, \ldots, n$. As said, this primary flux of protons produces barely anything other than reactions at the meteorite's surface, since as soon as it penetrates by a few centimeters it generates a secondary flux of neutrons, which penetrate more deeply. These neutrons also have an energy spectrum which changes as they penetrate, diminishing, of course. The flux and energy spectrum of cosmic radiation may vary with time (we know neither how, nor the magnitude of such fluctuations). What we measure is the result of flux aggregated over several million or even billion years.

A further, although generally minor, complication is that in addition to cosmic radiation there is also a flux of particles from the Sun. Generally, this flux is weak and of low energy, but from time to time it may become intense and of high energy. These bursts in flux are known as solar flares. They too may engender spallation reactions.

The effective cross-sections for production of new isotopes by nuclear reactions are also dependent on energy and therefore on penetration inside the meteorite. Figure 4.15 shows the production of various isotopes versus depth.

As, in addition, some isotopes result from spallation on several target nuclei whose effects are cumulative, we can imagine the sheer complexity of the phenomenon if we wish to determine all the contributions. That would involve estimating a mean flux and its energy spectrum and localizing the sample to be measured inside the meteorite. For these reasons, a simpler way of making the calculation has been sought.

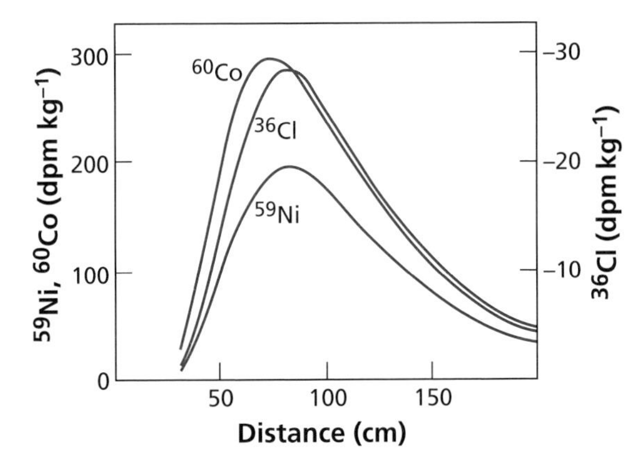

Figure 4.15 Production of ^{36}Cl, ^{59}Ni, and ^{60}Co by (n, γ) reactions in a supposedly spherical chondritic meteorite. Distance is measured from the meteorite's surface towards its center. Cl = 100 ppm, Ni = 1.34%, Co = 700 ppm. Γ ^{35}Cl = 45 barns, Γ ^{58}Ni = 4.4 barns, Γ ^{59}Co = 37 barns. Initial flux S_0 = 0.5 neutron $cm^{-3}s^{-1}$.

Radioactive isotopes

In this case, the equation in the preceding subsection must be supplemented by the decay term $\lambda_r N_r$, where N_r is the number of radioactive atoms:

$$\frac{dN_r}{dt} = \phi \Gamma_r N_{c\to r} - \lambda_r N_r.$$

Let us posit $\phi \Gamma_r N_{c\to r} = T$. It is considered to be constant to simplify things, or else it is assumed we know its mean value. The equation becomes:

$$\frac{dN_r}{dt} = T - \lambda_r N_r.$$

Integrating gives:

$$N_r = \frac{T - C e^{-\lambda_r t}}{\lambda_r}.$$

where C is the integration constant. If at $t = 0$, $N_r = 0$, $C = T$. From this:

$$N_r = \frac{T}{\lambda_r}\left(1 - e^{-\lambda_r t}\right) = \frac{\phi \Gamma_r N_{c\to r}}{\lambda_r}\left(1 - e^{-\lambda_r t}\right).$$

Now suppose we have two isotopes, one stable and one radioactive, produced by spallation under similar conditions of energy and flux. We can establish the ratio:

$$\frac{N_s(t)}{N_r(t)} = \frac{\Gamma_s N_{c\to s}}{\Gamma_r N_{c\to r}}\left(\frac{t}{1 - e^{-\lambda_r t}}\right)\lambda_r.$$

If we wait long enough for production and destruction of the radioactive product to achieve a steady state, then $e^{-\lambda t} \to 0$. This gives:

$$\frac{N_s(t)}{N_r(t)} = \frac{\Gamma_s N_{c\to s}}{\Gamma_r N_{c\to r}} \lambda_r t.$$

Notice that we have eliminated the flux factor. If, in addition, the two isotopes are products of the same element, then $N_{c\to s}/N_{c\to r} = 1$ and the equation becomes:

$$\frac{N_s(t)}{N_r(t)} = \frac{\Gamma_s}{\Gamma_r} t \lambda_r,$$

hence:

$$t = \frac{1}{\lambda} \frac{N_s}{N_r} \frac{\Gamma_r}{\Gamma_s}.$$

In this way, we determine the time elapsed since the meteorite was subjected to cosmic radiation without knowing anything more about the cosmic radiation but having just the ratio of the effective cross-sections. The exposure age is the time since the meteorite was broken into pieces, leading to its exposure to cosmic rays (Figure 4.16).

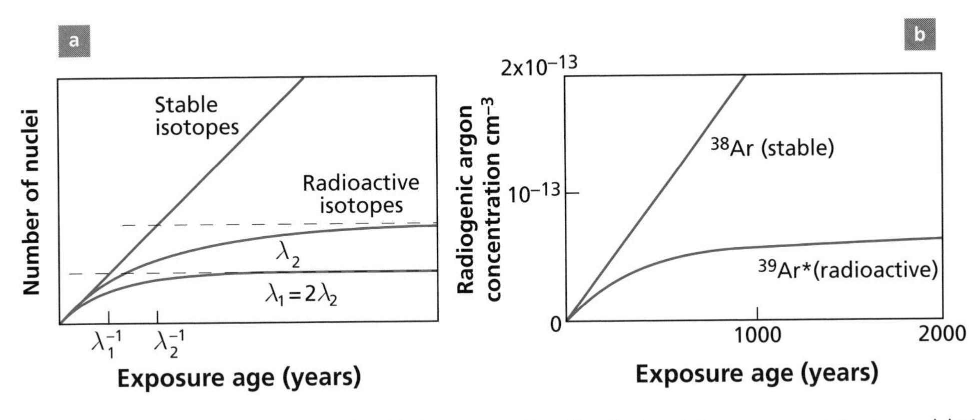

Figure 4.16 Pairwise chronometry (stable isotope, radioactive isotope) for cosmogenic isotopes. (a) The general theoretical case; (b) the example of $^{38}Ar/^{39}Ar$, which is used extensively in glaciology (see Oeschger, 1982).

One of the conditions for successfully applying this method is that the isotopes used should have effective cross-sections that are sensitive to the same ranges of energy and flux to justify the simplification of mathematically eliminating flux. Accordingly, we are interested in isotopes either of the same element or of adjacent elements in the periodic table. The pairs most used are ^{3}He–$^{3}He^*$, ^{22}Ne–$^{22}Na^*$, ^{38}Ar–$^{39}Ar^*$, ^{83}Kr–$^{81}Kr^*$, and ^{41}K–$^{40}K^*$ (* indicates that the isotope is radioactive).

The only point that remains undetermined if we are to be able to calculate the age is the ratio of effective cross-sections and their level of constancy depending on particle flux and energy. The determination of effective cross-sections has, of course, benefited from the extraordinary research activity in nuclear energy and we have a good catalog of effective cross-sections from which it has been possible to derive precise laws.

Exercise

The ^{83}Kr–^{81}Kr method is used to calculate exposure ages (**Marti**, 1982). The decay constant of ^{81}Kr is $\lambda = 0.32 \cdot 10^{-5}\ \text{yr}^{-1}$. What is the minimum age we can calculate using the steady state formula?

Answer

For this, $e^{-\lambda t}$ must be negligible compared with 1. If we accept that $e^{-\lambda t}$ must be less than 0.01, then $t = 1.4$ million years.

Exercise

Can we not use this method for ages of less than 1.4 Ma?

Answer

The evolution equation is

$$\frac{N_r}{N_s} = \frac{\Gamma_s N_{c \to r}}{\Gamma_r N_{c \to s}} \left(\frac{1 - e^{-\lambda t}}{\lambda t} \right).$$

We need to just calculate the curve $(1 - e^{-\lambda t})/t$ and use the value of the ratio N_r/N_s. We know that $N_{c\to r}$ is identical to $N_{c\to s}$ for krypton. Calculate the curve for the $^{81}Kr/^{83}Kr$ ratio, given that $\Gamma_r/\Gamma_s \approx 1.6612$.

Exercise

The ^{83}Kr–^{81}Kr method can be used for calculating exposure ages in stony meteorites by the formula:

$$T = \frac{1}{\lambda}\frac{\Gamma_{81}}{\Gamma_{83}}\left(\frac{^{83}Kr}{^{83}Kr}\right)_{\text{cosmogenic}}$$

where $1/\lambda = 0.307$ Ma. To be rid of any complications and given that many Kr isotopes are produced by spallation, the ratio of effective cross-sections is calculated from the formula:

$$\frac{\Gamma_{81}}{\Gamma_{83}} = \frac{0.95}{2}\left(\frac{^{80}Kr}{^{83}Kr} + \frac{^{82}Kr}{^{83}Kr}\right).$$

Calculate the exposure age of the Juvinas meteorite if the isotopic measurement of Kr in it is by convention:

^{84}Kr	^{78}Kr	^{80}Kr	^{81}Kr	^{82}Kr	^{83}Kr	^{86}Kr
1	0.157	0.460	0.0166	0.767	0.968	0.182

Answer

$T = 10.7 \cdot 10^6$ years.

Exercise

We use the ^{41}K–^{40}K method to date iron meteorites (Voshage and Hintenberger, 1960). The only target is therefore iron and we take it that $\lambda_{40} = 0.5543 \cdot 10^{-9}\,yr^{-1}$. As the effective cross-sections are $\Gamma_{40} = 9.4$ millibarns and $\Gamma_{41} = 14.7$ millibarns, what is the cosmogenic isotopic composition of $^{41}K/^{40}K$ for two meteorites whose exposure ages are 100 Ma and 1 Ga?

Answer

For 100 Ma $(^{41}K/^{40}K)_{\text{cosmogenic}} = 1.60$; for 1 Ga $(^{41}K/^{40}K)_{\text{cosmogenic}} = 2.038$.

What are the main conclusions to be drawn from these measurements of exposure ages?

Exposure ages are generally much younger than the ages of meteorite formation (or metamorphism). They range from a few million years to a few hundred million years. (Ages of formation are closer to 4.5 billion years, as we have seen.) This shows that for most of their lifetime meteorites are inside their parent body, where they are shielded from cosmic rays, and that the fracturing of meteorites has occurred relatively late (but long before they fall to Earth!).

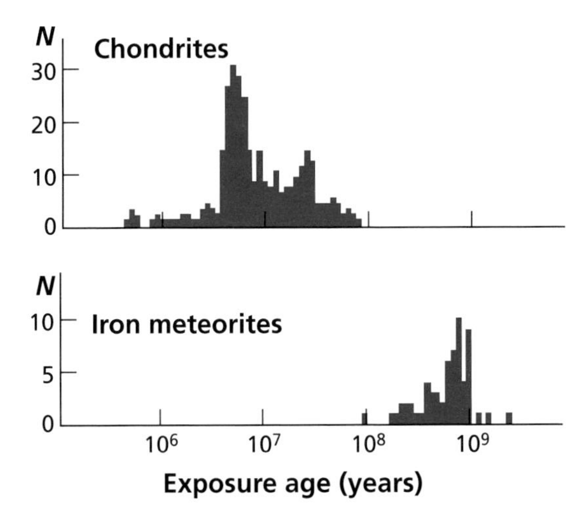

Figure 4.17 Exposure ages of various types of meteorites.

The distribution of meteorite exposure ages seems to display peaks at 5.7 and 20 million years for ordinary stony meteorites and at 700 million years for iron meteorites (Figure 4.17).

The differences in exposure ages for iron and chondritic meteorites is problematic. Why should iron meteorites have so much older exposure ages? It was first thought that they came from a different part of the Solar System. It is now thought that the exposure ages of iron meteorites are older because iron withstands impacts better than the silicate assemblages that make up chondritic meteorites, which have been subjected to impacts resulting in more recent fragmentation. In any event, the results show that meteorites have long lifetimes in interplanetary space.

4.3.2 Terrestrial rocks

The principle for terrestrial rocks (see review by Oeschger, 1982) is the same as for meteorites. **Galactic cosmic radiation** causes nuclear reactions in rocks. The nuclei produced can be measured, and if we know the flux we can calculate the duration of irradiation. The difference with meteorites is that the intensity of cosmic radiation at the Earth's surface is far lower because it has been attenuated by the absorption of protons and neutrons in the atmosphere. It was not until measurement sensitivity had been improved that these methods could be applied to terrestrial rocks. A second difficulty is that until now methods using isotope ratios employed for meteorites, such as $^{40}K/^{41}K$ and $^{83}Kr/^{81}Kr$, have not been applicable for terrestrial rocks. For the first method, no equivalent has been found for iron meteorites, which do not contain large quantities of initial K, which dilutes the effects produced by irradiation. For the second method, sensitivity is still insufficient, but that may change. Unlike the case of meteorites, where the particle flux ϕ is eliminated, here it must be calibrated carefully. Now, the flux diminishes, of course, as it penetrates into a rock. Hence there is a further difficulty, which has been solved empirically, that of calibrating the flux ϕ, or calibrating absorption.

Exploitable chronometers

In practice, we use the four radioactive isotopes $^{14}C^*$, $^{10}Be^*$, ^{26}Al, and ^{36}Cl and the two rare gases ^{3}He and ^{21}Ne, which are stable isotopes. For radioactive elements, we use the formula established for meteorite exposure ages:

$$N_r = \frac{\phi \Gamma N_{c \to r}}{\lambda} \left(1 - e^{-\lambda_r t}\right).$$

But here, equilibrium is not achieved. By noting T the production rate $\phi \Gamma N_{c \to r}$, we obtain:

$$N_r = \frac{T}{\lambda} \left(1 - e^{-\lambda_r t}\right).$$

For example, for beryllium,

$$^{10}Be = \frac{T}{\lambda_{Be}} \left(1 - e^{-\lambda_{Be} t}\right).$$

The difficulty lies in determining the flux and assuming it to be constant over time and then estimating the abundance of target nuclei.

For ^{3}He and ^{21}Ne, exposure ages are also calculated by the method described for stable isotopes in meteorites. This gives:

$$N_s = \phi \Gamma N_{c \to s}\, t.$$

The production rate is written as $P = \phi \Gamma N_{c \to s}$. The gases ^{3}He and ^{21}Ne accumulate linearly with time, therefore $^{3}He = Pt$ or $^{21}Ne = Pt$. Here too we need to know the flux (and to assume that it is constant) and the absorption laws.

For the target, ^{3}He production does not depend much on its composition whereas ^{21}Ne is produced by spallation of Mg and depends greatly on the chemical composition of the target. To make it uniform, olivine, whose Mg content is more or less constant, is separated and measurements are made on this mineral.

Calibration of production rates, erosion rates, etc.

The particle flux varies with both altitude and latitude,[6] as we said when discussing ^{14}C. The flux has therefore to be calibrated in each place where the measurement is made. There are two ways to do this. First, general laws have been established giving the value of flux by altitude and latitude. To get some idea of this, let us say that at an altitude of 5000 m, the flux is 20 times greater than at sea level. At the poles, at a constant altitude, it is 60 times greater than at the equator. This is why flux is calibrated locally, where possible, by measuring ^{10}Be, ^{26}Al, ^{3}He, or ^{21}Ne contents on samples whose age has been determined by other methods.

Let us look at two important examples that will help in understanding the thinking behind the method but which will show its "fruitful complications" and so suggest its future developments.

[6] The first measurements were made in the Antarctic where the flux is greatest and, in addition, the erosion rate is low.

Table 4.1 Usable cosmogenic nuclides

Isotope	Half-life (yr)	Measurement method	Difficulties	Production rate (atoms/yr) Sea level latitude >55°N	Time-span
^{3}He	Stable	Mass spectrometer	Diffuses easily	160 (olivine)	1 ka – 3 Ma
^{10}Be	$1.5 \cdot 10^6$	AMS	Atmospheric contamination	6 (quartz)	3 ka – 8 Ma
^{26}Al	$7.16 \cdot 10^5$	AMS	Atmospheric contamination	37 (quartz)	5 ka – 2 Ma
^{36}Cl	$3.08 \cdot 10^5$	AMS	Atmospheric contamination	8 (basalt)	5 ka – 1 Ma
^{21}Ne	Stable	Mass spectrometer	Common neon	45 (olivine)	7 ka – 10 Ma
^{14}C	5730 years	AMS	Contamination	20 (basalt)	1 ka – 40 ka
^{41}Ca	$103 \cdot 10^3$	AMS	Very difficult to measure		– 300 ka

Erosion rate measurements

Our job is to measure the exposure age of a basalt flow from the Piton de la Fournaise volcano on the island of Réunion (Staudacher and Allègre, 1993). This lava flow is at an altitude of 2300 m and there is no indication that its altitude has varied over the course of time.

We decide to use the cosmogenic age method based on ^{3}He applied to olivine. We therefore separate the olivine from a rock sampled from the surface and measure the ^{3}He content of the sample. To obtain the ^{3}He of cosmogenic origin, we must, of course, correct for ^{3}He of internal origin. To do this, we measure the $^3He/^4He$ ratio of internal origin on olivine sampled at a great distance from the surface. We can calculate the cosmogenic ^{3}He content by measuring the $(^3He/^4He)_{total}$ ratio and the total concentration in ^{3}He.

Exercise

Given $(^3He)_{total}$, $(^3He/^4He)_{internal}$, and $(^3He/^4He)_{total}$, establish the formula for calculating cosmogenic ^{3}He.

Answer

$$\left(^3He\right)_{cosmo} = \left(^3He\right)_{total}\left[\frac{\left(\frac{^3He}{^4He}\right)_{total} - \left(\frac{^3He}{^4He}\right)_{internal}}{\left(\frac{^3He}{^4He}\right)_{internal}}\right].$$

When all the calculations have been made, we find $1.3 \cdot 10^{-12}$ cm^3 g^{-1} at standard temperature and pressure of ^{3}He of cosmogenic origin. The production rate P_0 of ^{3}He in the olivine has been calculated for the latitude and altitude of the sample as $2.2 \cdot 10^{-17}$ cm^3 g^{-1} per year at standard temperature and pressure.

It is easy then to calculate the exposure age:

$$T_{\text{cosmo}} = {}^{3}\text{He}/P_0 = 59\,090 \text{ years.}$$

In addition, the "geological" age of the basalt flow has been calculated by the K–Ar method as $65\,200 \pm 2000$ years. The surface rock has been irradiated for 65 200 years and yet its exposure age is just 6000 years at the youngest. How come?

After examining the various sources of error, we accept that the age difference is due to erosion. The rock sampled at the surface today was in fact located at depth for much of its history. It only came to the surface through ablation of the material above it, by erosion.

Now, we know that the particle flux decreases exponentially with the thickness of rock it has penetrated. We need, then, to model the phenomenon.

We can write the rate of production of ^{3}He in the form:

$$P = P_0 \exp\left(\frac{-X(t)}{L}\right)$$

where $X(t)$ is the thickness counting from the surface and L is the attenuation factor of radiation with depth.

To simplify matters, we can assume that the rate of erosion is constant. We can write:

$$X = X_0 - \in t$$

where $\in$ is the erosion rate and X_0 the initial depth of the sample.

We can therefore write:

$$\frac{\mathrm{d}(^{3}\text{He})}{\mathrm{d}t} = P_0 \exp\left[\frac{-(X_0 - \in t)}{L}\right].$$

We integrate this equation between 0 and t and then replace X_0 by its value $X_0 = X + \in t$ (remembering that X is counted downwards). This gives:

$$^{3}\text{He} = \frac{P_0 L}{\in} \exp\left(\frac{-X}{L}\right)\left[1 - \exp\left(\frac{-\in t}{L}\right)\right].$$

Then X is the current depth coordinate.

Our objective is to determine the erosion rate, $\in$.

We know t (65 200 years) and P_0 ($2.2 \cdot 10^{-17}$ cm^{3} g^{-1} yr^{-1} at standard temperature and pressure), but there are still two unknowns: $\in$ and L.

To measure L, we bore a small core and so take samples at various depths. We isolate the olivine and measure the cosmogenic ^{3}He on each sample of it. We can now draw the plot:

$$\ln\left[(^{3}\text{He})_{\text{cosmo}}\right] = f(X).$$

If our assumptions about the flux and erosion rate are valid, the relation is a straight line of the form:

$$\ln\left[(3\text{He})_{\text{cosmo}}\right] = \frac{-X}{L} + A$$

where A is a constant. The slope is $(-1/L)$.

Exercise

A core was taken from the lava of the island of Réunion discussed before. Olivine was extracted at each level and the ^{3}He content measured. After correcting for ^{3}He of internal origin, the following results were found as a function of depth X (Table 4.2).

Table 4.2 Relation of ^{3}He content to depth in olivine from lava on Réunion

$X\,(\mathrm{g\,cm^{-2}})$	$^3\mathrm{He}\,(10^{-13}\,\mathrm{cm^{-3}\,g^{-1}})$
10.45	13
37.86	9.12
131.95	6.9
217.42	3.58
318.41	1.74

Calculate the attenuation factor L.

Notice that depth X in this table is expressed in grams per square centimeter, which is an unusual unit of length! The reason for this is that attenuation is, of course, dependent on the quantity of matter penetrated and so depends on the density of that matter (attenuation per unit length penetrated is not the same in rock, soil, or a layer of atmosphere). If attenuation were expressed per mil length, we would have to multiply it by density, and so the effective value would be:

$$\frac{\text{length (cm) mass (g)}}{\text{volume (cm}^3\text{)}}$$

which is equivalent to mass per square centimeter. Naturally enough, L too will be expressed in g cm^{-2}.

Answer

After calculating ln(^{3}He) as a function of X (do it) we find for the slope:

$$\mathrm{L} = 165\ \mathrm{g\ cm^{-2}} \pm 5.$$

How can we now calculate the erosion rate $\in$?

Let us go back to the expression as a function of X but now start from where $X = 0$, that is, at the present-day surface. The expression becomes:

$$^3\mathrm{He} = \frac{P_0 L}{\in}\left[1 - \exp\left(\frac{-\in t}{L}\right)\right].$$

We can develop the exponential in series limiting ourselves to the first three terms. Remembering that $^3\mathrm{He} = T_{\mathrm{cosmo}} - P_0$, we can finally write:

$$\in = 2\left(\frac{L}{t}\right) \cdot \left(\frac{t - T_{\mathrm{cosm}}}{t}\right).$$

If we note the relative difference between the geological age and cosmogenic age as $\Delta T/T$:

$$\in = 2\frac{\Delta T}{T} \cdot \left(\frac{L}{t}\right).$$

In the present case we find $\in = 4.7\ \text{g cm}^{-2}\ \text{yr}^{-1}$.

If we accept a density of 2 g cm^{-3} because the lava sampled is "bubbly" and contains many cavities, then $\in = 2.1\ \mu\text{m yr}^{-1} = 2.1$ mm per 1000 yr.

Exercise

Cosmogenic ^{21}Ne contents were measured on the same samples of olivine from the island of Réunion (Table 4.3). Given that the production rate of ^{21}Ne is $6.28 \cdot 10^{-18}\ \text{cm}^{-3}\ \text{g}^{-1}$ at the latitude and altitude in question, calculate the attenuation factor L and the erosion rate $\in$.

How do you account for the result for L compared with that found for ^{3}He? Which seems to you to be the more accurate rate of erosion, that determined with ^{3}He or that with ^{21}Ne? To what do you attribute the difference?

Table 4.3 Relation of ^{21}Ne content to depth in olivine from lava on Réunion

X (g cm^{-2})	^{21}Ne (10^{-13} cm^{-3} g^{-1})
10.45	3.29
37.86	1.73
131.95	1.59
217.42	0.96
318.41	0.52

Answer

$L = 160\ \text{gcm}^{-2}$.

It is identical to that for ^{3}He because the particle flux at the origin of ^{3}He and ^{21}Ne is the same for both. The cosmogenic age of the surface sample is $T_{\text{cosmo}} = 52\,388$ years.

$\in = 9.6\text{g cm}^{-2}\ \text{yr}^{-1} = 4.8$ mm per 1000 yr.

The ^{21}Ne age is greater than the ^{3}He age probably because helium diffuses more readily than neon and the sample has probably lost a lot of helium. This invites caution when interpreting helium ages (Figure 4.18).

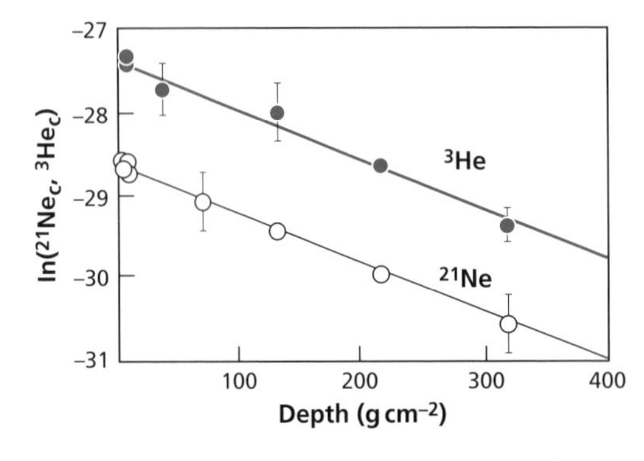

Figure 4.18 Attenuation curves for the creation of ^{3}He and ^{21}Ne in rocks of Réunion given by way of example.

Measuring the rate of uplift

As said, the particle flux at 500 meters altitude is 20 times higher than the flux at sea level. Suppose, then, we have a lava that has flowed at sea level. Subsequently the terrain it overlies has been lifted to an altitude of 500 m.

Of course, the lava would be irradiated much more over the course of uplift. This type of model can be calculated in the same way as before:

$$\frac{\mathrm{d}^3\mathrm{He}}{\mathrm{d}t} = P_0 \exp\left(\frac{X(t)}{L}\right)$$

where $X(t)$ is the altitude for which zero is taken at 100 m and L is the attenuation factor in the atmosphere. If we call the rate of uplift U, we get:

$$U = 2\left(\frac{T_{\mathrm{cosmo}} - t}{t}\right)\frac{L}{t}.$$

As can be seen, T_{cosmo} = the age of exposure and is greater than the "geological" age.

Naturally, erosion has not been allowed for in this example. If we add the erosion rate, we will need other constraints and other chronometers to solve the problem because we will then have two unknowns, $\in$ and U.

Here, we are at the frontier of current research.

Interest and limits of these methods

The methods are of interest for a number of reasons:

- erosion rates can be determined (provided they are not too high because sensitivity is lost, of course);
- we can date the time a surface comes to be exposed to the air, for example a fault plane or the wall of a collapsed caldera;
- the rate of uplift of mountains can be measured.

There are many limitations, the main one for terrestrial rocks being that flux is weak (except at high altitudes near the poles). Very sensitive methods are therefore required.

Of course, the "boxes" must have remained closed as for dating with natural radioactivity. For reliable results, measurements must be made by various methods such as ^{21}Ne or ^{10}Be or by combinations of methods (^{26}Al/^{10}Be) as proposed by **Lal** (Lal *et al.*, 1958; Lal and Peters, 1967). The various methods are dependent on the chemical composition of rocks or minerals, first because the target must not contain too many non-cosmogenic natural isotopes, and second because reaction yields depend on the chemical composition of the target.

The production of ^{21}Ne, for instance, can be written:

$$(^{21}\mathrm{Ne})_{\mathrm{cosmo}} = C\left\{\begin{array}{c} 1\,\mathrm{Mg} + 0.36\,\mathrm{Al} + 0.19\,\mathrm{Si} \\ +0.04\,\mathrm{Ca} + 0.01\,(\mathrm{Fe} + \mathrm{Ni}) \end{array}\right\}$$

in which C is a constant.

Lastly, of course, the fluxes must be calibrated in terms of altitude, latitude, and their variations over time and averaged out (see the review by Lal, 1988).

4.4 Cosmic irradiation: from nucleosynthesis to stellar and galactic radiation

We speak of cosmogenic isotopes as something special. But, in fact, all the isotopes in nature are cosmogenic. They were born of nuclear reactions in the cosmos by what is called **nucleosynthesis**. We have mentioned this phenomenon several times. It answers the question of the alchemists of old: how were the chemical elements created? For our purposes, we need to complete this question: how were the various isotopes formed and in what proportions? In the manufacture of an atom, what is important, as we have said, is the nucleus, because it is in the nucleus that all the mass is concentrated, the electrons being captured subsequently to populate the surrounding orbitals. Here, then, is a first part of the answer: the chemical elements are the outcome of nuclear reactions. Nuclear reactions produce both isotopes and chemical elements. Such reactions are not merely the invention of nuclear physicists, they occur naturally throughout the Universe, where the same causes produce the same effects, governed by the laws of nuclear reactions. Upon examination, the table of chronometers based on cosmogenic isotopes looks very similar to the table of extinct radioactivity. Why should the two converge like this? To answer this, we need to broaden our field of view and raise the more general issue of nucleosynthesis.

Nucleosynthesis and the theory explaining it are the foundation of modern astrophysics. It is also the starting point of what is called chemistry of the cosmos or **cosmochemistry**. This is hardly the place to develop this theory in full as it would take us to the heart of astronomy and very far from our present subject matter. None the less, it is worth expounding briefly a few important concepts,[7] particularly as astronomy and the earth sciences have moved closer on these topics over the last 20 years (reread Chapter 1 on this).

The chemical elements were made in the stars by nuclear reactions. The levels of energy involved (MeV or GeV) are so great that only the stars can be the sites of such synthesis on so great a scale. These are the only environments in the Universe where the "ambient energies" are intense enough and extensive enough for nuclear reactions to be generated creating new chemical species in such large masses. The alchemists of old were out by a factor of a million. They wanted to transform matter with burning coals, that is, with energies of the order of the electronvolt (eV) whereas it takes energies of the order of MeV at least to change nuclei and so atoms. With their athanors[8] they could

[7] It is worth reading the few well-documented, introductory books on this, particularly D. D. Clayton (1983), *Principles of Stellar Evolution and Nucleosynthesis*, Chicago University Press or C. Cowley (1995), *Introduction to Cosmochemistry*, Cambridge University Press.

[8] Athanors are receptacles used by alchemists to do their experiments.

alter the atoms' outer electron shells, whereas making an atom involves altering the nucleus.

The idea that the nucleosynthesis of chemical elements occurred mainly in stars, first hit upon by **Atkinson** and **Houtermans** (1929) and then by **Gamow** (1946), was confirmed only in 1952 by the astronomer **Merrill**, who observed the presence of technetium-98 around a star. Now, all the isotopes of this element are radioactive, with a period of less than 1 million years. If this element is found near an isolated star in the Universe, it must have been created recently by the star otherwise it would be "dead," destroyed by radioactive decay.

The theory of nucleosynthesis was developed synthetically in 1957 in a pioneering paper known as **B2FH** by **Margaret Burbidge**, **Jeff Burbidge**, **Willy Fowler**, and **Fred Hoyle** (Burbidge *et al.*, 1957).[9] The major stages are as follows. It all begins with the proton, that is, the hydrogen atom, synthesized at the time of the Big Bang. After that, it all happens by nuclear reactions in the stars, but not just any stars. Let us look at the successive stages in nuclear terms.

4.4.1 The transition from the proton to helium

This is no straightforward transition. It first involves the intermediate products D and ^{3}He. The entire process is written as a series of nuclear reactions in an avalanche:

$$^1H + {}^1H \rightarrow {}^2D + \beta^+ + \nu$$

$$^2D + {}^1H \rightarrow {}^3He + \gamma$$

$$^3He + {}^3He \rightarrow {}^4He + 2\,{}^1H$$

where ν is a neutrino and γ is gamma radiation.

These nuclear reactions occur in **ordinary stars** like the Sun. It is the most widespread activity in stars. It requires temperatures of several million degrees.[10]

4.4.2 The synthesis of α elements

Helium-4 (α radiation) is an exceptionally stable building block in nuclear terms (two protons and two neutrons). The nuclear reactions involving this nucleus are written:

$$^4He + {}^4He \rightarrow {}^8Be^*.$$

[9] Al Cameron of Harvard developed a similar theory at the same time but failed to publish any summary in a major scientific journal but just gave conferences and lectures on the subject.

[10] It is these reactions we cannot manage to "calm" to produce "domestic" energy, we can only produce them artificially in hydrogen bombs.

Beryllium-8* is an unstable isotope which reacts quickly to give, in its turn, ^{12}C by the equation:

$$^{8}Be^{*} + {}^{4}He \rightarrow {}^{12}C.$$

Therefore, all told:

$$3\,{}^{4}He \rightarrow {}^{12}C.$$

This is what is called **helium fusion**, that is, the formation of ^{12}C. This type of nuclear reaction continues. We have:

$$^{12}C + {}^{4}He \rightarrow {}^{16}O + \gamma$$

$$^{16}O + {}^{4}He \rightarrow {}^{20}Ne + \gamma$$

$$^{20}Ne + {}^{4}He \rightarrow {}^{24}Mg + \gamma.$$

Alongside the addition of successive "blocks" of ^{4}He, fusion of carbon and oxygen nuclei occurs, these nuclei themselves being formed by the addition of three and four ^{4}He:

$$^{12}C + {}^{12}C \rightarrow {}^{24}Mg + \gamma$$

$$^{12}C + {}^{12}C \rightarrow {}^{23}Na + p$$

$$^{12}C + {}^{12}C \rightarrow {}^{20}Ne + {}^{4}He$$

$$^{12}C + {}^{12}C \rightarrow {}^{23}Mg + n$$

$$^{12}C + {}^{12}C \rightarrow {}^{16}O + 2\,{}^{4}He$$

or

$$^{16}O + {}^{16}O \rightarrow {}^{32}S + \gamma$$

$$^{16}O + {}^{16}O \rightarrow {}^{31}P + p({}^{1}H)$$

$$^{16}O + {}^{16}O \rightarrow {}^{31}S + n$$

$$^{16}O + {}^{16}O \rightarrow {}^{28}Si + {}^{4}He$$

$$^{16}O + {}^{16}O \rightarrow {}^{24}Mg + 2\,{}^{4}He.$$

Triggering these nuclear reactions requires considerable energies to overcome the natural repulsion between positively charged nuclei. The first stages begin at 100 million degrees and end in stages at close to 1 billion degrees.

4.4.3 The iron peak

This phase of fusion (addition) of ever-heavier nuclei has its limits because with successive synthesis we arrive at minimum energy per nucleon. And the element corresponding to this minimum is iron. In other words, when there is fusion of nuclei whose mass is greater than that of iron, the resulting nuclei are less stable than the initial ones. This stability of iron underlies a statistical process of nuclear reactions. Fusion reactions occur in an anarchic fashion and are offset by destruction reactions, which may even go as far as releasing ^{4}He nuclei. We are dealing here with temperatures of 6 billion degrees. But all of these reactions have a natural limit: we cannot "exceed" the atomic number of nuclei close to iron, nickel, and cobalt, which are very stable too in nuclear terms. Accordingly, such nuclei accumulate and are very abundant. Hence we get what is known as the iron peak, corresponding to an exceptional abundance of iron and of the elements on either side of it (Figure 4.19). But how can we go "further" and make elements with atomic numbers greater than that of iron, heavier than iron? How can we cross this seemingly impassable stability barrier and make very heavy elements? After all, these elements do occur, be they strontium, rubidium, neodymium, or uranium. This is possible through the relationship of neutron addition.

4.4.4 Neutron addition

Neutrons are electrically neutral particles. They react in nuclear terms by adding to the nucleus without doing too much "damage." On the (Z, N) plot the addition of neutrons creates a new isotope with more neutrons and shifts the nucleus towards the right of the plot. When neutron addition accumulates, ever-heavier isotopes of the element are created. On the (Z, N) plot, the new nuclei move horizontally ever further to the right. There comes a

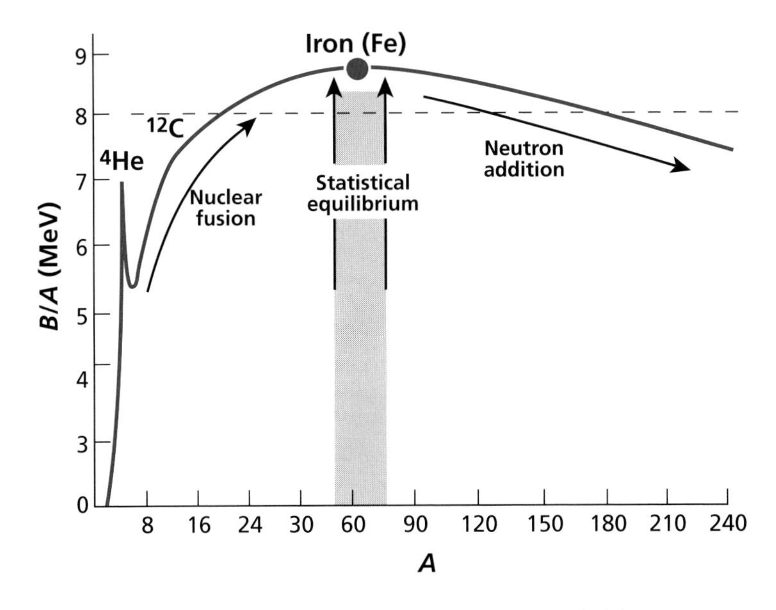

Figure 4.19 The curve of binding energies of nuclei by nucleon (B/A) versus mass number (A).

point, then, when the new nucleus lies outside the valley of nuclear stability. It is therefore unstable. To return to the valley of stability it disintegrates by β^- radioactivity. The neutron changes into a proton. By the same token, Z increases. A new chemical element has therefore been made. New isotopes and new chemical elements can be made by neutron addition. This is a general outline. In actual fact, the neutron addition process has two variants depending on the relative number of neutrons and the equilibrium established between the addition of neutrons and radioactivity.

If the neutron flux is weak, the radioactivity of isotopes to the right of the valley of stability is a barrier. Such isotopes disintegrate as soon as they form and we zigzag across the (Z, N) plot (Figure 4.20).

If the neutron flux is very high, radioactivity has no time to disintegrate fully the radioactive nuclei which are "loaded" with extra neutrons and in turn give rise to other radioactive isotopes which move further to the right of the (Z, N) plot (Figure 4.21). Of course, decay also occurs and a flux of new nuclei is formed, with larger Z numbers. New chemical elements have been synthesized.

The two figures (Figures 4.20 and 4.21) clearly show the two processes which are termed **s-process** (slow) and **r-process** (rapid), respectively.

A few simple rules can be laid down to check whether an isotope of a heavy element has been synthesized by an s-process or an r-process or by both. A stable isotope to the right of a radioactive isotope (short period) cannot have been formed by an s-process since the radioactive isotope is a barrier to any further horizontal rightwards shift. If it occurs in nature, then it must have been formed by an r-process. However, any adjacent stable isotope may be the outcome of an s-process. Conversely, any stable isotope located on the same negative diagonal as an isotope of another element of lower mass cannot have been created by an r-process as it is "shielded," protected by the other isotope (Figure 4.21).

A very simple and convenient equation which the s-process obeys is:

$$\frac{N_{\mathrm{A}-1}}{N_{\mathrm{A}}} = \frac{\sigma_{\mathrm{A}}}{\sigma_{\mathrm{A}-1}}$$

where N is the abundance, σ the effective cross-section of neutron absorption, and A and A–1 are two isotopes with decreasing mass numbers.

The abundance ratios of the two isotopes are in inverse proportion to their effective cross-sections of neutron absorption. Let us show this simply. The kinetic equation of production of isotope A is written:

$$\frac{\mathrm{d}(N_{\mathrm{A}})}{\mathrm{d}t} = \phi(\sigma_{\mathrm{A}-1}\, N_{\mathrm{A}-1} - \sigma_{\mathrm{A}}\, N_{\mathrm{A}})$$

where ϕ_{n} is the neutron flux, $N_{\mathrm{A}-1}$ and N_{A} are the numbers of atoms of atomic numbers A and A–1, and $\sigma_{\mathrm{A}-1}$ and σ_{A} are the effective cross-sections.

This is a classic destruction–production equation like the one written for ^{14}C. If equilibrium is attained:

$$\frac{\mathrm{d}(N_{\mathrm{A}})}{\mathrm{d}t} = 0.$$

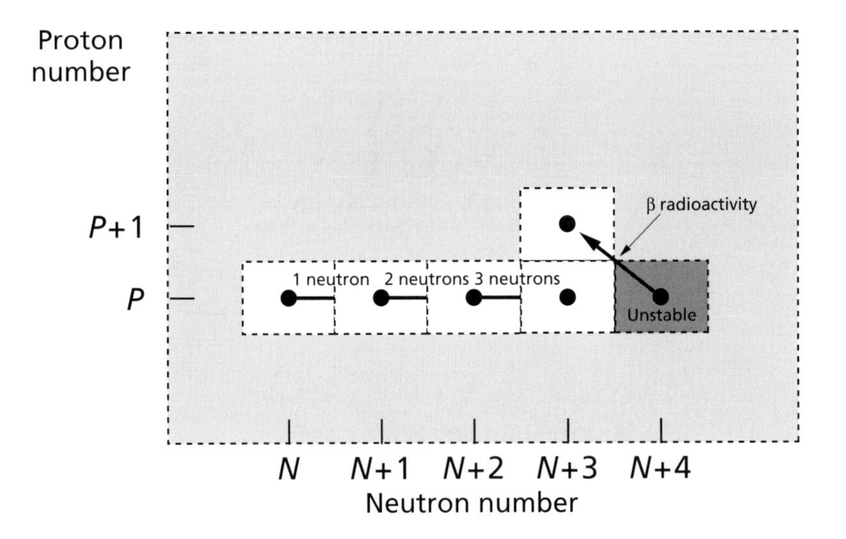

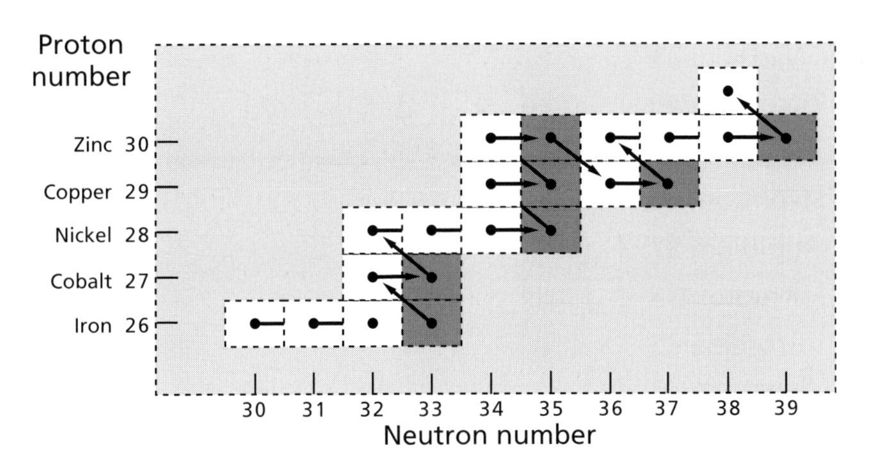

Figure 4.20 The s-process on the proton–neutron plot. Top: theoretical pathway of neutron reactions. In blue, the first radioactive isotope. There is a change of elements up to the left by β^- decay. Bottom: example of s-process pathways on the proton–neutron plot allowing the creation of elements heavier than iron. After Broecker (1986).

Therefore:

$$\sigma_{A-1} N_{A-1} = \sigma_A N_A.$$

Our assertion has been demonstrated.

Now consider the transition by β^- decay:

$$\frac{d\left({}_Z^A N^*\right)}{dt} = \phi N \sigma_{A-1} {}_Z^{A-1} N - \lambda {}_Z^A N^*.$$

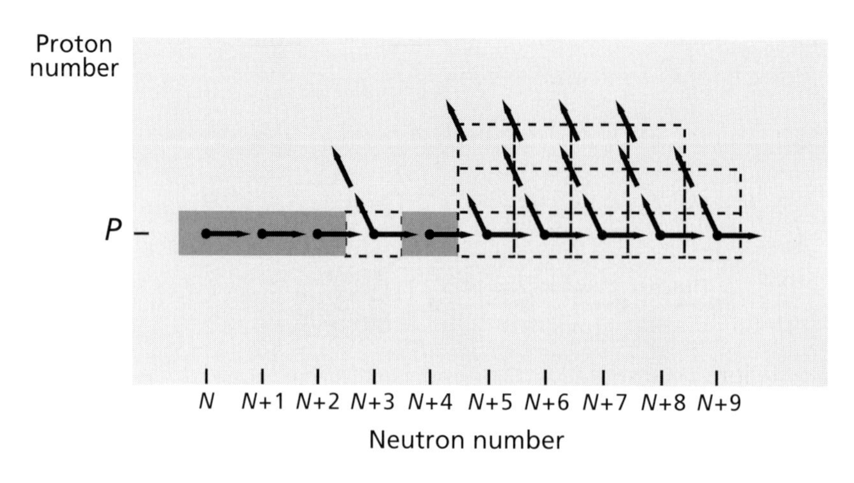

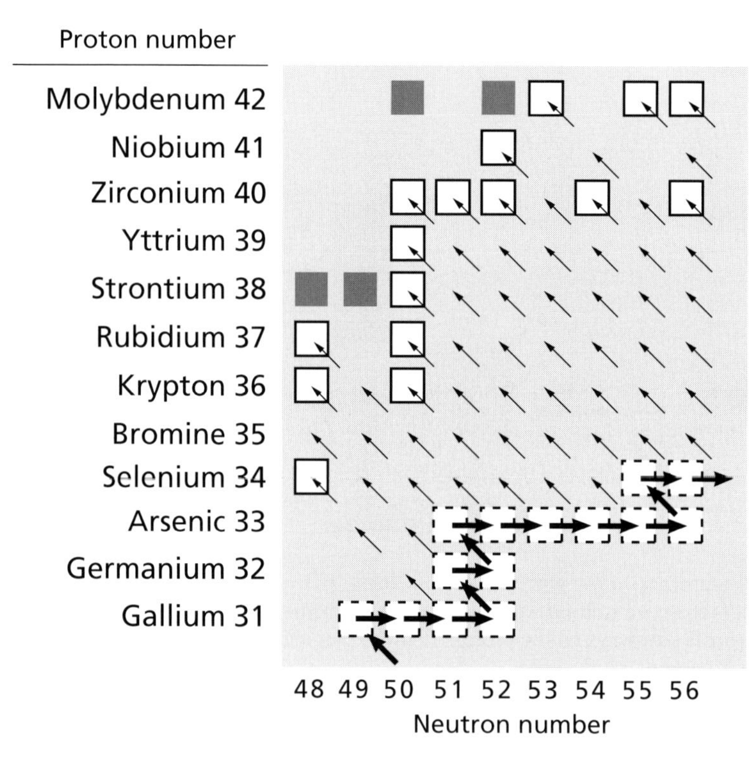

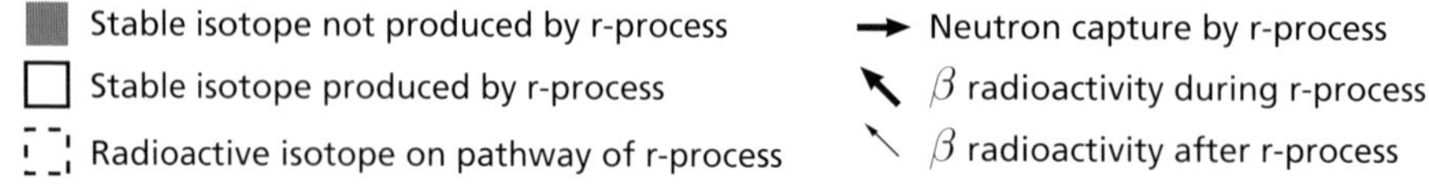

Figure 4.21 The r-process on the proton–neutron plot. The neutron flux is intense enough for considerable horizontal shift to the right before radioactivity becomes a barrier. But, of course, all radioactivity plays a part before the barrier effect kicks in. After Broecker (1986).

Production of the barrier radioactive isotope is noted with an asterisk (*), and λ is the decay constant.

$$\frac{\mathrm{d}\left({}_{Z+1}^{\ \ A}N\right)}{\mathrm{d}t} = \lambda\,{}_{Z}^{A}N^{*} - \phi_{N}\,\sigma_{\mathrm{A}}\,{}_{Z+1}^{\ \ A}N^{*}.$$

If equilibrium occurs, the radioactive term between the two equations is eliminated, giving:

$$\sigma_{\mathrm{A}-1}\ {}^{\mathrm{A}-1}_{\ \ Z}N = \sigma_{\mathrm{A}}\ {}_{Z+1}^{\ \ A}N.$$

This is the same relation as before.

The abundance of isotopes of type s (i.e., formed by the s-process) heavy elements is therefore determined by the effective **cross-sections of absorption** of neutrons. These cross-sections can be measured in the laboratory (they are required for constructing atomic reactors).

Exercise

Strontium has three isotopes, of masses 88, 87, and 86. Their effective cross-sections of neutron absorption are 4.8, 60, and 48 millibarns, respectively. The "primitive" isotope ratios of strontium are: $^{88}Sr/^{86}Sr = 8.3754$, $^{87}Sr/^{86}Sr = 0.698$, and $^{88}Sr/^{87}Sr = 11.99$.

Can strontium isotopes be synthesized by the s-process?

Answer

Let us first apply the rule $N_1/N_2 = \sigma_2/\sigma_1$. We find the ratios $^{88}Sr/^{86}Sr = 10$, $^{87}Sr/^{86}Sr = 0.8$, and $^{88}Sr/^{87}Sr = 12.5$.

Let us compare this with observations. In view of the uncertainties on effective cross-sections, the agreement is quite good. This is odd because ^{88}Sr derives from both the s- and the r-processes, which proves that the r-process is not important here.

Exercise

Table 4.4 gives the abundance N, the effective cross-section σ, and the product $N\sigma$ for samarium isotopes.

Table 4.4 Data for samarium isotopes

Samarium isotope	$N_{\mathrm{A}}\%$	Class[a]	σ_{c} (millibarn)	$N\sigma$
144	2.87	p	119 ± 55	342
147	14.94	rs	$1\,173 \pm 192$	$17\,600 \pm 2900$
148	11.24	s	258 ± 48	$2\,930 \pm 540$
149	13.85	rs	$1\,622 \pm 279$	$22\,500 \pm 4000$
150	7.36	s	370 ± 72	$2\,770 \pm 535$
152	26.90	r or rs	411 ± 71	$11\,100 \pm 1900$
154	22.84	r	325 ± 61	$7\,430 \pm 1400$

[a] For details of the p-process, see next subsection.

Calculate the proportion of the r-process involved in the formation of 147 and 149 isotopes.

Answer

If we take the 148 isotope, which is a pure s-process isotope, the (s) isotopes give a σN value $\approx$ 2930.

We can therefore write:

$$\text{percentage (r) of 147} = \left(\frac{17\,600 - 2930}{17\,600}\right) \times 100 = 83.35\%$$

$$\text{percentage (r) of 149} = \left(\frac{22\,500 - 2930}{22\,500}\right) \times 100 = 86.97\%.$$

4.4.5 The p-process

Some isotopes are depleted in neutrons (and so have "surplus" protons). In the chart of the nuclides they lie to the left of the valley of stability, off the s-process pathway, and shielded from the r-process. They include for example ^{84}Sr, ^{92}Mo, ^{124}Xe, and ^{144}Sm. They are called p-process nuclides (or p-isotopes) and are said to be formed by the **p-process**. The production process is fairly obscure but we do know that the abundance of p-isotopes plotted against (Z) forms a curve which roughly follows that of the s- and r-isotopes except that the abundance levels are much lower. To get some idea, the abundance ratios for silicon are: s-process = 1, r-process = 0.5, and p-process = 0.02.

It can be inferred, then, that the p-process is secondary and tracks the s- and r-processes, thus supplementing them. Nowadays it is thought that the p-process is mainly caused by (γ, η) reactions in supernovae.

4.4.6 The light elements lithium, beryllium, and boron

When the abundance of chemical elements is plotted against their atomic number (Figure 4.22), it can be seen that three light elements, lithium, beryllium, and boron, are underabundant compared with their atomic numbers. The He burning nucleosynthetic process seems to have leapfrogged them, with three ^{4}He nuclei eventually combining to give ^{12}C. And yet these elements do exist! How did they come to be?

Their formation is attributed to two causes:

- the Big Bang for lithium in part;
- spallation reactions in interstellar space caused by galactic cosmic radiation and acting on interstellar material.

This explains the abundance curve of the chemical elements and isotopes (Figure 4.22). But what structures produce these reactions? We now need to speak about the stars.

4.4.7 The stellar adventure: the life and death of stars

Let us now look at what nucleosynthesis is as the astrophysicist sees it.

The Universe is populated by stars clustered into galaxies. These stars are of different sizes and brightnesses but they are all gigantic nuclear reactors. Nuclear reactions take place

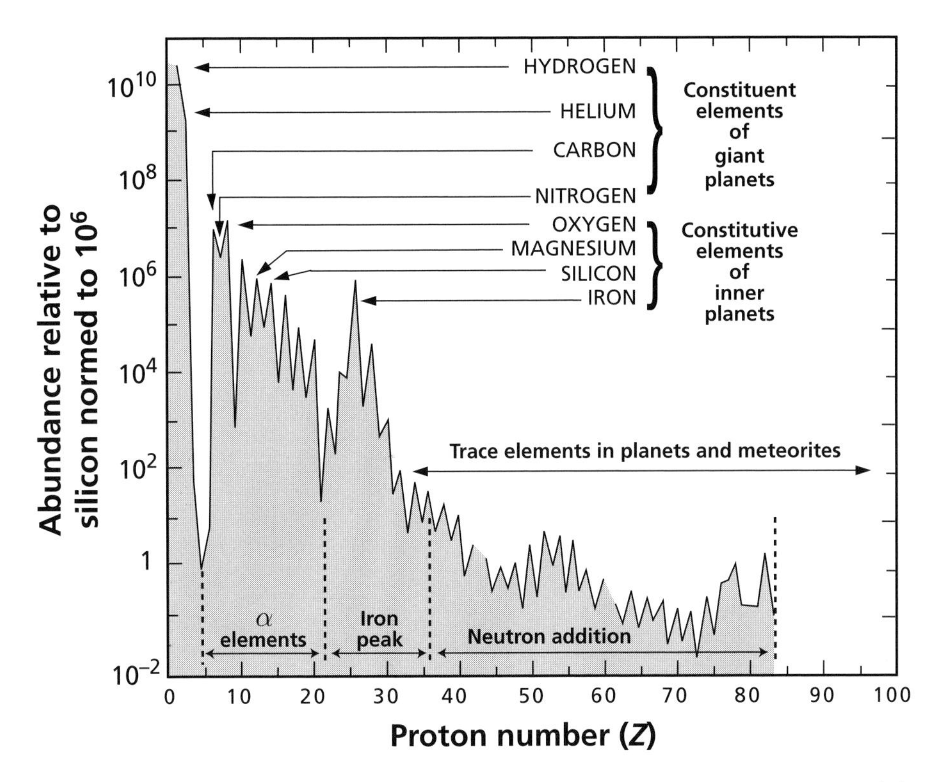

Figure 4.22 Abundance charts of elements in the cosmos. The scale is logarithmic. Silicon (Si) is taken as the reference. After Broecker (1986).

inside the stars which give off energy. Electromagnetic radiation, and light radiation in particular, is given off from the star's surface. A balance is established between the two processes and it is how long this balance is maintained that determines the star's lifetime. When the equilibrium is disrupted, that is, when the fuel is exhausted, the star changes nature.

The characteristic feature of nuclear reactions is that they need a certain amount of energy to "ignite." Once triggered, most of them produce considerable quantities of energy in their turn, because they convert matter into energy, in accordance with Einstein's famous formula $\Delta E = \Delta mc^2$.

What is the source of energy triggering these nuclear reactions? **Gravitational energy**. A star is born when an interstellar cloud of gas and dust contract under the influence of gravity, that is the mutual attraction exerted by all components of matter. The contraction of the cloud generates energy by the impacts produced between "portions" of matter. When the temperature reaches about 10 million degrees, hydrogen fusion occurs and a star is born. This is what happened for our Sun some 4.57 Ga ago. The birth of ordinary stars is the most everyday, the most commonplace phenomenon in the Universe. Most stars are of this type. Like the Sun they can make only helium. The stars are classified by the **Hertzsprung–Russell** (H–R) diagram on which we plot brightness against color, that is, the star's temperature (Figure 4.23). White is very hot and red is cooler. Most stars lie on the leading diagonal of the diagram, which is therefore called the main sequence.

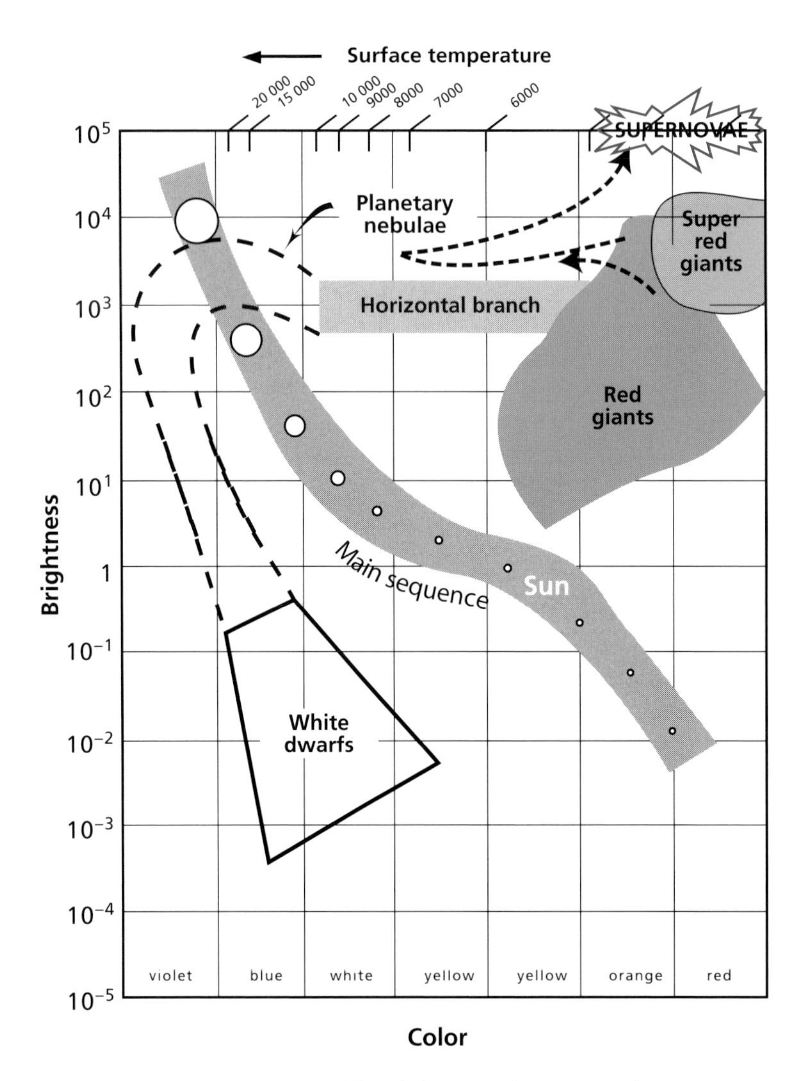

Figure 4.23 Hertzsprung–Russell (H–R) diagram of stars. Note that the temperature scale on the *x*-axis increases from right to left.

Remark

The Sun was born from gravitational contraction of a cloud of gas and dust already containing all of the chemical elements we now find in the Solar System (in the planets and in the Sun itself). The origin of these elements therefore pre-dates that of the Sun itself.

For stars with low masses (like that of our Sun or lower), when the nuclear fuel is exhausted the star burns out, expands slightly and then contracts, it shrivels up to give birth to a tiny star known as a white dwarf (bottom of Figure 4.23). This is what will become of the Sun in 5 Ga!

For more massive "ordinary" stars, the stellar adventure will continue more gloriously. The core of the star, made of helium, will contract again and its temperature rise while at the same time the outer envelope of the star will expand (and so its temperature fall). In the

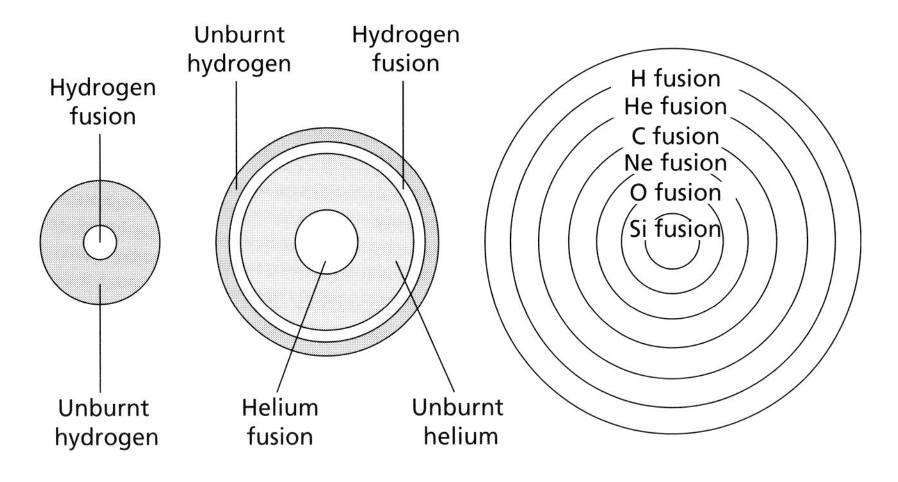

Process	Fuel	Product	Temperature
Burning hydrogen	H	He	1–60 x 10^6 K
Burning helium	He	C, O	200 x 10^6 K
Burning carbon	C	O, Ne, Na, Mg	800 x 10^6 K
Burning neon	Ne	O, Mg	1500 x 10^6 K
Burning oxygen	O	Mg to S	2000 x 10^6 K
Silicon fusion	Mg to S	Elements close to iron	3000 x 10^6 K

Figure 4.24 Stages in the evolution of a massive star.

core, fusion reactions will be triggered producing considerable energy. Little by little the star will come to be made up of a series of layers like an onion, with each layer corresponding to a temperature and therefore to a type of nuclear fusion reaction (Figure 4.24). These stars are already red giants (like Betelgeuse). Some have three or four layers, other six or seven.

At the center of these red giants reign temperatures of 1 to 6 billion degrees, while the outer envelopes are at only 100 million degrees. It is in these red giants, through the course of their evolution, that the heavy elements are manufactured and that the fusion reactions of carbon, oxygen, etc. occur and the statistical equilibrium of the iron peak is reached.

Some of these giant stars will evolve further and reach an extremely rare stage where they explode in a fraction of a second. These are **supernovae**. In our galaxy there is one supernova explosion per century. Such events generate the enormous neutron fluxes (themselves the outcome of nuclear reactions) required by the r-process. The s-process seems to occur as red giants evolve towards the explosive zone, along what astronomers know as the asymptotic giant branch (**AGB star**).

The main lines of this theory of nucleosynthesis are well understood. It unites in a single coherent whole astronomic observations, the abundance of chemical elements and isotopes, and the parameters measured in nuclear physics in reactors or accelerators.

For our own purposes, in isotope geology, we can draw a few practical conclusions from it. The Sun burns hydrogen only. The heavy elements in it therefore pre-date the Sun. They were formed in red giants or ancient supernovae. This simple remark allows us to understand that the history of the Universe is a history of stellar cycles. Stars are born, synthesize elements, and die, and their material is scattered into interstellar space. There the various clouds mix and mingle the material synthesized previously. They wander around the cosmos until the day new gravitational contraction occurs. This gives rise to a new star, which in turn makes matter and disperses it in the Universe and so on. One supernova per century is not a lot, but that still makes 10 thousand every million years, or 40 million since the Earth first formed (and in our galaxy alone).

These pre-solar heavy elements were therefore made in red giants and supernovae dating from solar prehistory and then extensively mixed in interstellar space.

For us earthlings, as parts of the Solar System, there are two major dates in this cosmic history: the **Big Bang** and the formation of the Solar System. The Big Bang was the event with which our universe began. Astronomers now date it to 13 billion years ago. For them, it is the instant at which matter as we know it came into being in a phenomenal process of expansion and heating. In the course of this process hydrogen, helium, and a little lithium were formed. Then everything arranged itself into **galaxies**, **stars**, and **planetary systems**. It was in the course of this stellar history that all the other elements heavier than lithium were synthesized.

The **formation of the Solar System** occurred 4.6–4.5 billion years ago. We shall go into this date more closely but at this point an order of magnitude is enough. The Solar System gathered up matter made in previous stellar processes including explosions of supernovae or red giants. All this matter was mixed and arranged itself into a central star with orbiting planetary bodies. But to account for the existence at the time the Solar System formed of what are now extinct forms of radioactivity we must accept that red giants and supernovae existed and exploded just before its formation and that they are the source of these forms of radioactivity.

This scenario of **late explosive nucleosynthesis** (red giant and/or supernovae) to synthesize heavy elements in the Solar System has recently been challenged (or supplemented) by a second scenario, that of **primitive irradiation**. In this model it is accepted that as it formed the young Sun emitted extremely high-energy particles which in turn produced spallation reactions on the solar material and that these reactions in particular produced the extinct forms of radioactivity. This scenario is thought necessary to produce the ^{10}Be discovered by a team from Nancy (France) (**McKeegan** *et al.*, 2000) and the ^{36}Cl discovered more recently by a Chinese team (**Lin** *et al.*, 2005). However, it does not seem to be able to make ^{53}Mn. The two processes probably occurred in succession, but in what proportions? This is a very hot subject at present.

All these events have left traces in meteorites. These objects, which date from the time the Earth, planets, and Sun were formed, contain extraordinary information. Near their surfaces are cosmogenic isotopes which inform us about cosmic radiation, but also about the time they have spent in the Universe as small pieces of rock. However, some of them contain grains of interstellar dust whose isotopic composition in certain elements informs us about the r- and s-processes of nucleosynthesis. These processes, as described, occurred well before the beginnings of our Solar System. This is true for most, save one category: extinct radioactivity, which we have already come across. Meteorites are an invaluable link between cosmic and terrestrial processes. For physics, there is a continuum between stellar nucleosynthesis and nuclear reactions produced by particle fluxes from the Sun or from cosmic radiation.

This scenario is the one that has occurred tirelessly in the Universe over 10 or 14 billion years. But in the beginning, what was there? What matter? The beginning was the Big Bang, which synthesized the elementary particles, hydrogen, helium, and lithium and enacted an essential event . . . which lies outside the scope of this book, although it impregnates every part of it!

Problems

1 Carbon-14 is produced by nuclear reactions on ^{14}N, but it also decays to ^{14}N. We propose to use this decay scheme for dating purposes as is done with ^{87}Rb–^{87}Sr or ^{40}K–^{40}Ar, for example. We consider that to make a measurement, $^{14}N/^{15}N$ must differ from the normal value (≈ 276) by $1 \cdot 10^{-5}$. What should the C/N ratio be to apply this method to objects that are, say, 15 000 years old? Do you think this would be feasible?

2 The Holy Shroud of Turin long posed a problem for believers. Was it really the shroud in which Christ had been wrapped? It first appeared in 1350 and the monks asserted it was a genuine relic. It could not be dated by the traditional ^{14}C method as there was not enough material, but with AMS it could be dated with just 150 mg of cloth. The ^{14}C age was between 600 and 800 years old. Given the correspondence curve in Figure 4.25, what is the age of the Holy Shroud?

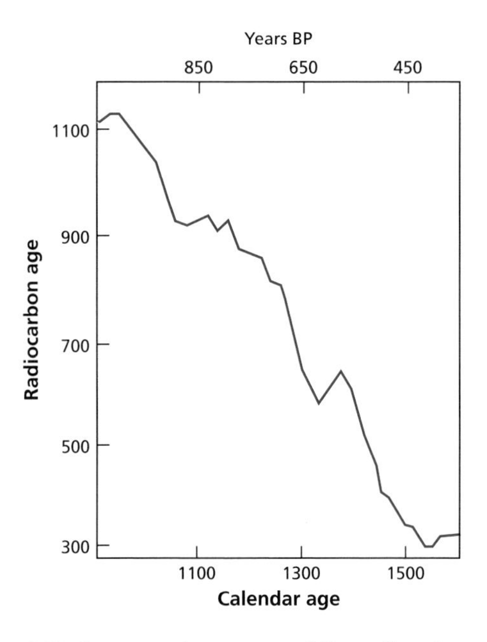

Figure 4.25 Correspondence curve of the radiocarbon age and calendar age of the Turin Shroud.

3 We measure the isotopic composition of K in an iron meteorite having undergone cosmic irradiation for 500 Ma. The mass spectrum of K measured is a mixture between the cosmogenic mass spectrum and the normal mass spectrum. Estimating that $(^{40}K)_{normal}$ is negligible, establish the equation for determining $(^{41}K/^{40}K)_{cosmogenic}$. (Suppose the normal K isotope ratios are known.)

4 Chlorine-36 is created in the atmosphere by spallation reactions on ^{40}Ar, and $T_{1/2} = 3 \cdot 10^5$ years. From the time of its creation, the chlorine is incorporated in water masses and in the hydrological cycle. The $^{36}Cl/Cl_{total}$ ratio is $100 \pm 20 \cdot 10^{-5}$ in rainwater at the Earth's surface. We analyze the water sampled from a deep aquifer and find $5 \cdot 10^{-15}$ for the $^{36}Cl/Cl_{total}$ ratio. How old is the groundwater?

5 Here is a series of $^{10}Be/^{9}Be$ measurements on a manganese crust.

Depth (mm)	$^{10}Be/^{9}Be \pm 2\,\sigma$
Surface	$3.94 \pm 0.32 \cdot 10^{-8}$
0–1.0	$3.23 \pm 0.12 \cdot 10^{-8}$
1.5–2.0	$2.19 \pm 0.25 \cdot 10^{-8}$
2.5–3.0	$1.79 \pm 0.08 \cdot 10^{-8}$
3.5–4.0	$1.41 \pm 0.06 \cdot 10^{-8}$
4.0–5.0	$1.04 \pm 0.07 \cdot 10^{-8}$
5.0–5.7	$9.00 \pm 0.52 \cdot 10^{-9}$
5.7–6.0	$5.98 \pm 0.88 \cdot 10^{-9}$
6.0–7.0	$5.20 \pm 0.44 \cdot 10^{-9}$
8.0–8.5	$3.94 \pm 0.26 \cdot 10^{-9}$
8.5–9.0	$2.69 \pm 0.62 \cdot 10^{-9}$
9.0–10.0	$2.39 \pm 0.40 \cdot 10^{-9}$
10.7–11.7	$1.85 \pm 0.44 \cdot 10^{-9}$

(i) What is the rate of accretion?
(ii) What is the error?

CHAPTER FIVE

Uncertainties and results of radiometric dating

5.1 Introduction

The purpose of this chapter is to review radiometric dating methods as applied both to terrestrial and to cosmic problems. We shall speak of the major results that now fix the chronological framework of the Earth's history, not forgetting that what they mean is inseparable from how reliable they are. The central question we shall be dealing with in this chapter is: how reliable are the geological ages we determine? In other words, what is the level of uncertainty affecting an age determination?[1] What guarantee have we that this age is geologically meaningful?

Ascertaining the uncertainty Δ, when we obtain a value T, allowing us to write that the age is $T \pm \Delta$, is a problem that breaks down into various sub-problems, some of which are uncommon even in books on uncertainty. Let us get our ideas straight with a simple example.

Suppose we measure a series of rocks by the $^{87}Rb-^{87}Sr$ method and the analytical results are just about aligned on the ($^{87}Sr/^{86}Sr$, $^{87}Rb/^{86}Sr$) diagram. How can we measure the slope of the straight line isochron to calculate the age of the rocks? Any experimental physicist will tell you that you first need to know the uncertainty affecting each individual physical measurement: the strontium isotope ratios and the absolute values in ^{87}Rb and ^{86}Sr. And then that you need to estimate the slope of the straight line by using the least-squares method where the "weight" of each data point is measured by its individual uncertainty. We then obtain an age and an uncertainty value reflecting the dispersion of data points relative to the straight line we calculated. This approach, which has been tried and tested by years of statistical practice, seems unassailable on the face of it. And yet, as soon as it is put into practice, it raises questions that are not to do with mathematical statistics but with what we might call **geological understanding**.

The first question is: should we attribute the analytical uncertainty measured in the laboratory to all the data points? Despite the reassuring look of this question, which invites us to answer "yes," we need to realize that such "automatic" attribution is not as safe as we might think. In fact, the rocks with the highest $^{87}Rb/^{86}Sr$ values and therefore $^{87}Sr^*$ values, are probably those that, in the course of geological history, have been the most likely to

[1] To avoid any confusion with the word "error" used in ordinary language it is now recommended to use the term **uncertainty** in scientific work. The word **error** is used when we know the **difference between the true value and the measured value**; uncertainty is an estimate of this error. I shall try to abide by this logical rule, although it goes against decades of habit. Indeed, it has often been used in the first three chapters.

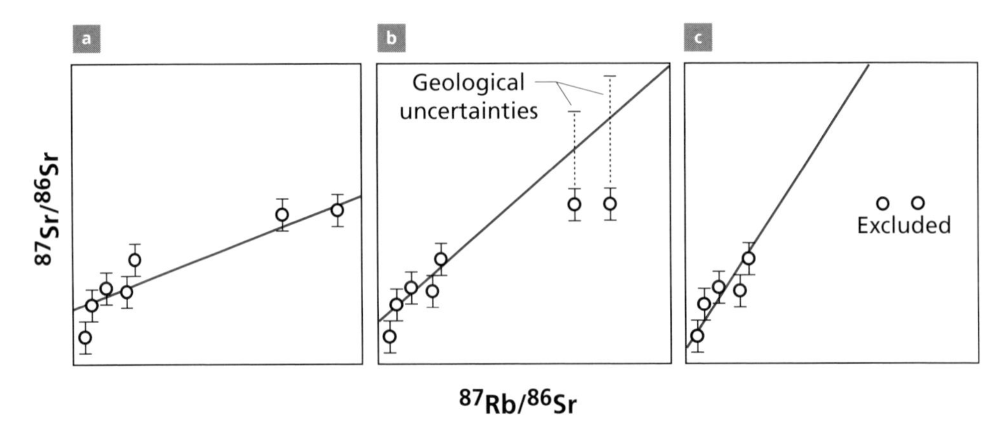

Figure 5.1 A series of Rb–Sr measurements plotted on the isochron diagram. After examining the dispersion of data points, there seem to be three options for obtaining the "best age." (a) We can calculate the best straight-line fit using all of the data points to which analytical uncertainty has been attributed and obtain an age determination. The uncertainty of the pattern suggests that this age is very much influenced by samples 6 and 7. (b) We add a geological uncertainty to samples 6 and 7 because they are the most radiogenic and so liable to have lost ^{87}Sr. We then calculate the best straight-line fit. (c) We eliminate samples 6 and 7 and use samples 1–5 alone to calculate the best straight-line fit.

lose $^{87}Sr^*$ through exchanges with adjacent rocks. Logically, we should therefore attribute greater uncertainty to their ages than to those of other rocks and measure the deficit in **geological robustness**. Moreover, such uncertainty would be asymmetrical because the loss of $^{87}Sr^*$ leads to a reduction in the $^{87}Sr/^{86}Sr$ ratio (Figure 5.1). Should we therefore increase the analytical uncertainty of a geological uncertainty?

Another question arises. Given that we are working with a natural system and that the conditions of the basic model (perfect initial isotopic homogenization, box closed since the origin) have probably not been stringently observed, the data points are distributed "statistically" (that is, more or less accurately) around the best straight-line fit. How much dispersion is acceptable for it to be considered that we are indeed working within the framework of a theoretical model and therefore that the age determination is legitimate? This is a difficult question and even more so if you think that, as we go back in time, the vagaries of disruptive geological phenomena become ever more probable and that the alignment must be less good for Precambrian rocks than for rocks of the Secondary era, say. Therefore **geological tolerance** for non-alignment must logically increase with age. By how much? To what extent can we accept non-alignment, even for very old rocks? We must remember that, in addition to measurement uncertainty, analytical uncertainty too is a function of age. There is, then, a series of prior questions to our central question: is age determination reliable?

There is a second aspect to this issue of the reliability of measurements, that of the **geological significance** of the age obtained. Imagine we have an age in thousands, millions, or billions of years, together with its uncertainty. What does the number obtained correspond to geologically? Let us return to our series of rocks whose ages have been determined by the $^{87}Rb–^{87}Sr$ method. Is the $^{87}Rb–^{87}Sr$ straight line a "mixing line" or an isochron? Chapter 3 (Section 3.4) gave us a test for deciding between these two hypotheses. Suppose we think it plausible we have an isochron. What does this mean? Our basic model is of a closed box in which we enclose radioactive and radiogenic isotopes at a well-defined "instant" T_0. But to what does this formal theoretical model correspond in the world of nature? Suppose that, to get our ideas straight, the series of rocks in question is a series of granite rocks from a single massif. Is the age obtained the age when the granite was emplaced as a magma or the age of the metamorphic basement whose partial melting gave rise to the granite mass? Assuming it is the emplacement of the granite that is dated, what is the exact time of that emplacement? When the magma first intruded? When it crystallized? Or is it when hydrothermal circulation occurred, which seems to end the emplacement of granite massifs as attested by quartz veins sometimes containing metallic minerals? Once again, this question is not unconnected to the age measured. And if we measure the minerals, have they not been subjected to some subsequent event? Did they all crystallize at the same instant?

The duration of emplacement of granite being estimated at 1 Ma in all (perhaps 2 Ma), when we date a granite that is 2 billion years old this question is irrelevant as our power of resolution is insufficient to distinguish the various episodes in its emplacement. But when we date the granites of the island of Elba (Italy) as 7 or 8 Ma old, the question becomes a crucial one. We are left to "date" something without really knowing what it is we are dating! The age will be an "average age" of the emplacement of granite on the island of Elba, but we must not read more into it than that! That is, unless we have other independent information to hand.

This is the sort of question we are going to try to answer while exploring the major results obtained by radiometric dating methods. Each will be placed in the context of how reliable it is. We shall see that the preceding chapters have given some pointers for some of the problems. For others, we shall need to give substance to our intuitions and attempt to rationalize them ... or qualify them. But the ultimate test of any age determination by radiometric methods is how coherent it is with what we already know.

This knowledge forms the chronological setting within which the new determination is to be fitted. The overall framework is formed by the geological and cosmic timescale.

The age of the Earth or of the Sun is a fixed point and our ultimate reference point in any "cosmic" chronology. At the same time, these general references are only known with a degree of uncertainty because of all the sources we have referred to and are going to study. There is, then, constant feedback between advances on uncertainties and the major results. The major results are to be looked on as numbers that are liable to vary somewhat or to see their underlying meaning change (we shall see this is the case for the age of the Earth). We shall see there is a broad range of uncertainty as regards the chronology of formation of the elements. It is to emphasize the fundamental connection between the uncertainties inherent to the methods of radiochronology and its main findings that they have been brought together here in a single chapter.

5.2 Some statistical reminders relative to the calculation of uncertainties

When we make N measurements of x, they are represented in a (N, x) plot by grouping the values of x into classes. This is a histogram. When ***N* is large** and the results are distributed

randomly, we consider that the best estimation of the true value sought is the **mean** of the measured values (noted $\bar{x}$).

$$\bar{x} = \frac{x_1 + x_2 + \cdots + x_N}{N} = \frac{\sum_i^N x_i}{N}.$$

To estimate the uncertainty associated with this, we consider the dispersion of the measurements obtained by calculating the deviations from the mean for each measurement and taking their average. In practice, we calculate the sum of the squared deviations:[2]

$$V_x = \sum_i^N (x_i - \bar{x})^2$$

and we divide by the number of measurements N minus 1 to get the **mean variance**:[3]

$$V_x = \frac{\sum_i (x_i - \bar{x})^2}{(N-1)}.$$

The **standard deviation** is the square root of the variance:

$$\sigma_x = \sqrt{V_x} = \sqrt{\frac{\sum_i (x_i - \bar{x})^2}{(N-1)}}.$$

The drawback of using σ_x is that it is expressed in units of x. So we cannot compare, say, the dispersion for the isotopic measurement of lead and that of strontium. To make such comparisons we define the **reduced dispersion** $\bar{\sigma}$ by dividing σ_x by $\bar{x}$:

$$\bar{\sigma} = \frac{\sigma_x}{\bar{x}}.$$

This dispersion or **reduced deviation** is expressed as per cent, per mil, etc.

When uncertainty is considered to be a random fluctuation due to factors that are poorly determined but which affect each measurement and make it deviate around its true value, then results of theoretical statistics can be applied. It can be shown that if there were an infinite number of random measurements, the values measured would be distributed around a mean value in accordance with what is known as the Laplace–Gauss or normal distribution (Figure 5.2). The curve of the normal distribution is written:

$$y = \frac{1}{\sigma_x \sqrt{2\pi}} \exp\left[-\frac{1}{2}\left(\frac{x - \bar{x}}{\sigma_x}\right)^2\right]$$

where y is the measurement frequency and is a function of the variable x.

[2] We use the square so that positive and negative deviations are added and do not cancel out by subtraction.

[3] Suppose we have just one measurement. If we divide by N, the uncertainty would be zero whereas if we divide by $N - 1$, that is by zero, it is indeterminate, which is indeed the case!

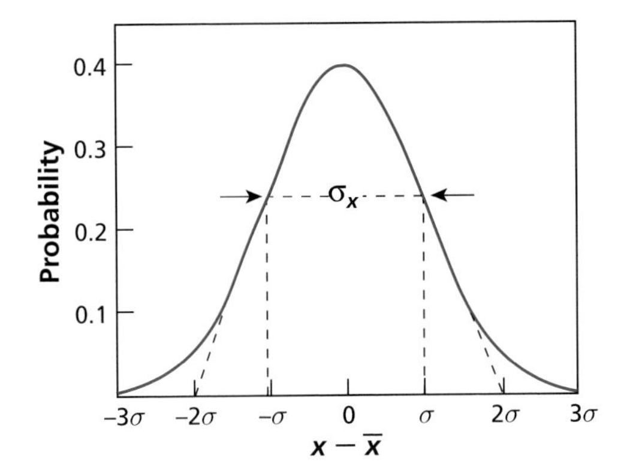

Figure 5.2 The normal curve. The probability of a measurement occurring, or its frequency, is plotted against the value of the measurement on a scale expressed in x–$\bar{x}$ (value corrected by the mean). The distance between the points of inflection is the dispersion σ. Note the geometrical signification of 2σ and 3σ.

The curve $y = f(x)$ has the famous bell shape. It is symmetrical about $\bar{x}$ (mean). The standard deviation σ_x is the distance between the point of inflection and the mean. The maximum deviation is equal to about $\pm 3\sigma_x$ or 99.7%.

Remark

The converse of this last observation is important: if a distribution is random and we know its maximum dispersion D, we can then estimate $\sigma_x = D/6$. This dispersion is known as the maximum range.

In practice, the distributions and parameters calculated are increasingly significant as we approach the theoretical distribution, that is, as the number of measurements N increases. Unfortunately, N is not always high.

When the shape of the histogram can be represented by a bell-shaped curve, we apply the results obtained for the ideal distribution with a very large N to such histograms (Figure 5.3). Let us say that in making this assimilation we are being optimistic about the uncertainty. In statistics it is shown that this uncertainty is estimated by multiplying the standard deviation of the histogram measured (σ) by $1/\sqrt{N}$. The more measurements we make, the lower the risk of uncertainty, of course, but this decrease obeys the **square-root law**. Uncertainty on x, written Δ_x, is therefore defined as:

$$\Delta_x = \left(\frac{\sigma_x}{\sqrt{N}}\right).$$

As, in the absence of other information, the standard deviation is the quadratic mean of deviations, the total uncertainty is therefore:

$$\Delta_x = \left[\sqrt{\frac{\sum_i (x_i - \bar{x})^2}{(N-1)}}\right] \cdot \frac{1}{\sqrt{N}}.$$

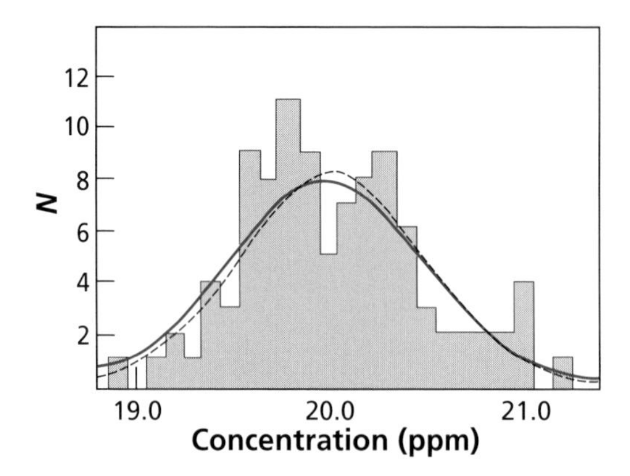

Figure 5.3 Histogram of 100 measurements of strontium concentration in a series of rocks from a single site. The histogram and the normal curve approximating it are shown. The mean is 20 ppm, the dispersion $\sigma = 0.5$ ppm. We can therefore write concentration $C = 20 \pm 0.5$ ppm. Relative uncertainty is written $0.5/20 = 2.5\%$.

Naturally, relative uncertainty may also be defined here:

$$\bar{\Delta} = \frac{\Delta_x}{\bar{x}}.$$

This uncertainty has no unit and is expressed as per cent, per mil, etc. This formula by which uncertainty varies as $1/\sqrt{N}$ is **absolutely fundamental**. It indicates that after estimating dispersion based on the histogram of experimental values, we estimate the reliability of this uncertainty by attributing to it a weight of $1/\sqrt{N}$. If we make 10, 100, 1000, and 10 000 measurements, precision improves to 30%, 10%, 3%, and 1%.

Remark

It can be shown in statistics that if, for a normal distribution, we consider an uncertainty $\pm\Delta_x$, we have a 63% chance of the real value lying within this interval. If we take $\pm 2\Delta_x$, that is a greater uncertainty, this probability reaches 95%. In the first case we speak of high-risk uncertainty, and in the second of low-risk uncertainty.

In practice, then, we measure uncertainties to the nearest sigma or two sigmas, by the expressions:

$$\pm\frac{\sigma_x}{\sqrt{N}} \text{ or } \pm\frac{2\sigma_x}{\sqrt{N}}$$

where N is the number of measurements.

A further procedure must be added to these: that of the **elimination of aberrant values (outliers)**. Some values measured are completely out of line with the histogram. These "aberrant" values are thought to arise from some random accident during measurement (typically a sudden fluctuation in electric current when making a measurement on the mass spectrometer, or the sampling of an apparently sound rock which proves to be weathered when analyzed more closely). The following criterion is used to eliminate these values: all

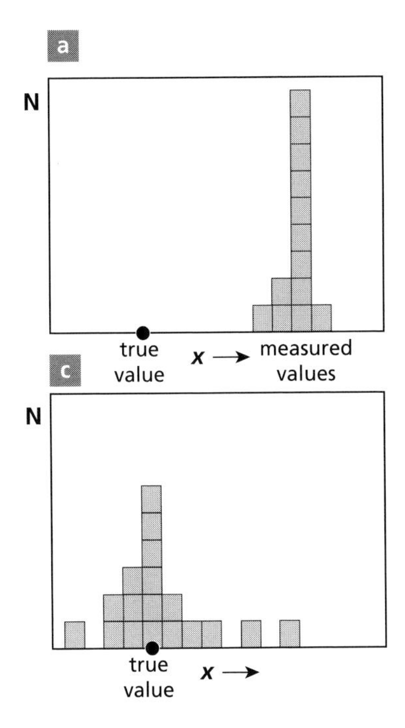

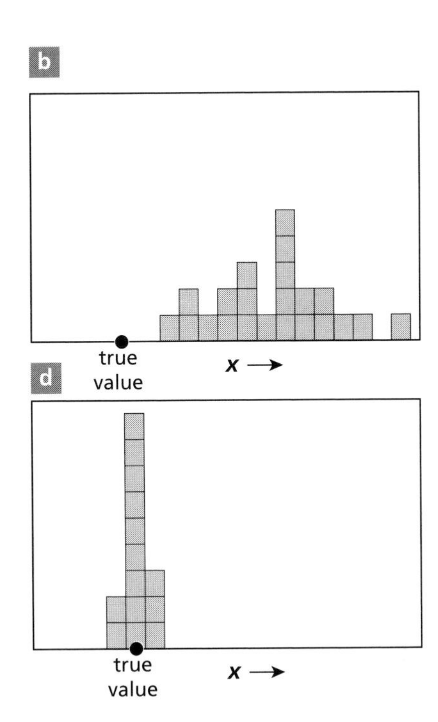

Figure 5.4 The difference between reproducibility and accuracy. (a) Good reproducibility but poor accuracy; (b) poor reproducibility and poor accuracy; (c) poor reproducibility but good accuracy; (d) good reproducibility and good accuracy.

the values beyond $\pm 3\sigma$ are eliminated, and then the entire calculation is repeated. The resulting uncertainty that has been "cleaned" of extreme outliers is denoted Δx^* or $\bar{\Delta}*$ depending on whether it is an absolute or relative uncertainty.

We now need to introduce a few useful distinctions (Figure 5.4). The **accuracy** of a measurement is estimated from the deviation between the measured value and the true value sought. **Precision** or **reproducibility** is the dispersion of a measurement repeated N times: it is this dispersion divided by the mean value. Reproducibility can be estimated from a histogram of measurements. Accuracy can only be estimated if we have independent knowledge (or an estimate) of the true values. The **power of resolution** in geochronology is the smallest age difference we can measure with any guarantee of reliability. It is defined as the quantity significantly different from zero that can be estimated between two events: $R = t_1 - t_2$. To estimate R, we assume the deviation must be greater than the sum of the standard deviations: $R \geq \sigma_{t_1} + \sigma_{t_2}$.

Exercise

The $^{206}Pb/^{204}Pb$ ratio of a rock is measured six times giving: 18.35, 18.38, 18.39, 18.32, 18.33, and 18.35. What value will we give for the isotopic composition of this rock? What is the precision achieved? Is it worth making another six measurements given that the reproducibility of each measurement is 3‰?

Answer

Supposing the uncertainty values of each measurement are equal, we calculate:

$\bar{x} = 18.353; \sigma = 0.025; \Delta_x = \sigma_x/\sqrt{N} = 0.010; \bar{\Delta} = \Delta_x/x = 5.4 \cdot 10^{-4}$.

We can therefore write $\bar{x} = 18.353 \pm 0.010$ with 63% confidence and $\bar{x} = 18.353 \pm 0.02$ with 95% confidence.

If we were to make another six measurements, precision would move from 0.01 to 0.0077, which is a gain of $4 \cdot 10^{-4}$. But the precision of a single measurement is $3 \cdot 10^{-3}$, therefore it is not worthwhile (except in special cases).

Remark

It is worth pondering that the numerical calculation gives more figures after the decimal point for a much larger number of measurements (to be exact $\sigma = 0.024\,94$). Now, we have written $\sigma = 0.025$ because the figures 94 are not significant. This approach is in line with the answer to the famous question: can we measure the length of a table to the nearest tenth of a millimeter using a measuring rod graduated in centimeters? Common sense is as good a guide to the answer as mathematics.

Exercise

Here are two histograms of measurements of $^{87}Sr/^{86}Sr$ ratios for a single sample (Figure 5.5). The first consists of 20 measurements with a mean value of 0.709 166, with $\sigma_x = 2.31 \cdot 10^{-5}$. The second consists of 400 measurements with a mean value of 0.709 184 with $\sigma_x = 4.34 \cdot 10^{-6}$.

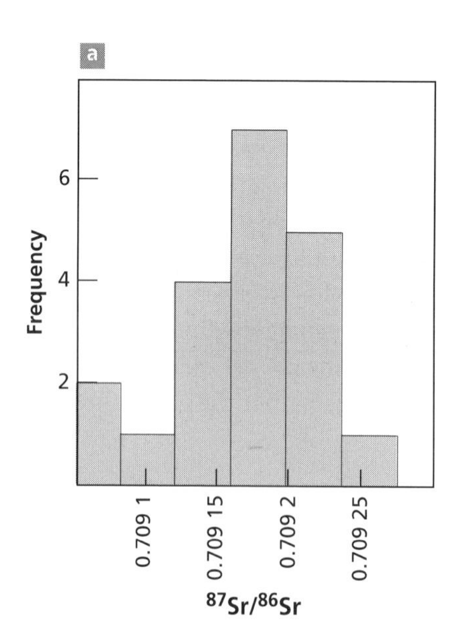

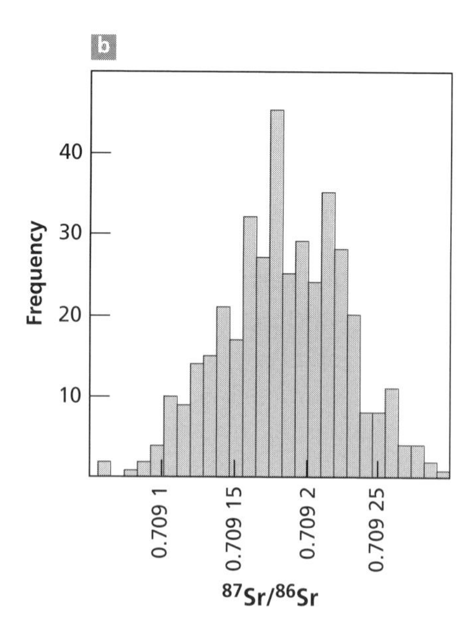

Figure 5.5 Histogram of $^{87}Sr/^{86}Sr$ ratios measured on a single sample. (a) 20 measurements, (b) 400 measurements. Notice that the classes become narrower as the number of measurements rises.

(1) What is the uncertainty affecting the measurement?
(2) How many measurements need to be made to get a standard deviation of 1 ppm?
(3) In this last case, do you think the uncertainty should be taken as $\pm\ \sigma_x/\sqrt{N}$ or $\pm\ 2\sigma_x/\sqrt{N}$?

Answer

(1) Intrinsic uncertainty is determined from the $(\ln \Delta, \ln \sqrt{N})$ curve as 111 ppm.
(2) It would take 10 000 measurements.
(3) With $\pm\sigma_x/\sqrt{N}$, the 400 ratio measurement lies outside the uncertainty limits on the 20 measurement expression. However, with $\pm 2\sigma_x/\sqrt{N}$ it is within limits. The latter expression is therefore required.

Exercise

Two ^{14}C ages are measured: 3230 ± 70 years and 3260 ± 60 years. Are these ages significantly distinct? What is the power of resolution of ^{14}C for 3000 years?

Answer

The two ages are not significantly different as their standard deviations overlap. The power of resolution of ^{14}C for 3000 years is 2%, or 60 years, which is a somewhat optimistic estimate!

5.2.1 Systematic uncertainties

Random uncertainties are deviations of measured values from the true value caused by vagaries obeying the laws of chance. "Chance is a word that hides our ignorance," as the mathematician and great scholar of probability Émile Borel used to say. But some uncertainties are systematic, that is, they affect the outcome of measurements by the same factor, although that factor is not necessarily a known one. These really are random "errors." Here's an example.

As stated, radioactive constants are affected by measurement uncertainty and so are periodically "updated," that is, improved. When we calculate an age with one of the dating formulae we have developed, with the given value of a constant, we invariably introduce the same uncertainty (but the amplitude of the uncertainty is not always the same). Such uncertainty is not really troublesome when the same method is used systematically. We then draw up a dating scale which can always be adjusted as required. But whenever we wish to use ages obtained by methods based on different decay rates and compare them, uncertainties about decay constants become very troublesome indeed.

Another systematic uncertainty may arise from the system of physical measurements. As described, international standards are used for calibrating any systematic uncertainties there may be among laboratories. A value is set for these standards although one cannot be sure it is accurate. Here again, this approach, while necessary, is not fully satisfactory whenever several methods are used. And what if some of the standards were wrong? After all, even the U.S. National Bureau of Standards, the final arbiter, makes "errors" too and gives its results with margins of uncertainty!

5.2.2 Pseudo-random uncertainties

In the problems we deal with, uncertainty is often a combination of systematic and random phenomena. When we spoke in Chapter 3 of the histogram of $^{40}K-^{40}Ar$ ages, which is asymmetrical, we said it was better to take not the mean as the most likely age value but the mode (the value most frequently measured) as the asymmetry was probably caused by diffusion of the radiogenic isotope ^{40}Ar, which tends to "lower" the age. Superimposed on a random distribution due, say, to measurement uncertainties, we have a systematic trend of argon diffusion. Even when a large number of measurements are made, pseudo-random distributions are generally asymmetrical relative to the normal distribution (Figure 5.6). Any of three types of parameter may represent the true value sought.

The **mean** is defined as in the normal distribution by:

$$\bar{x} = \sum_i \frac{x_i}{N}.$$

The **mode** is the value occurring most frequently in the distribution. It is, mathematically, the most likely value. The **median** is the value dividing the sample measured into two equal halves. It is what we might call the halfway house.

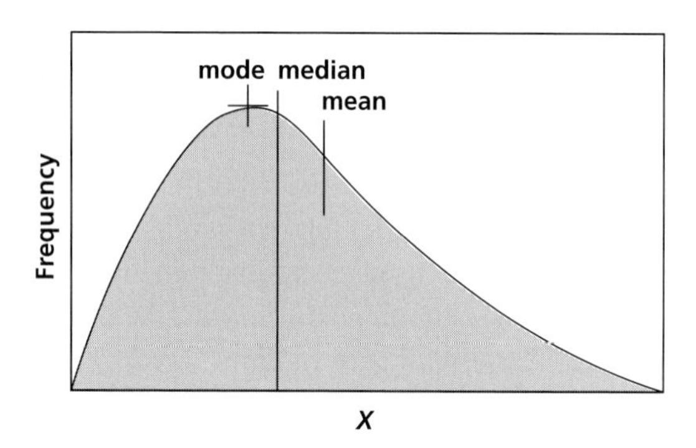

Figure 5.6 The difference between mode, median, and mean illustrated on an asymmetrical distribution.

Exercise

The U–He age of a series of magnetites is measured. Table 5.1a shows the distribution of apparent ages and Table 5.1b the distribution of uranium contents.

(1) Draw the corresponding histograms.
(2) Calculate the mean age and mean U content. Calculate the modes and medians.
(3) What is your geological interpretation of these results?

Answer

Mean age = 26.3 Ma; median age = 27 Ma; modal age = 30 Ma.
Mean U content ≈ 25 ppb; median U content ≈ 25 ppb; modal U content ≈ 25 ppb.

The most likely geological age is 30 Ma since the distribution of uranium is virtually normal, which is evidence that it has not been disrupted subsequently. The asymmetry seems to result solely from the diffusion of helium (Figure 5.7), which tends to lower the ages. Here, then, the mode is the preferred value, although there is nothing to show that the maximum value of 32 Ma is not closest to the truth.

Table 5.1a Distribution of apparent ages

	T(Ma)							
	32	30	28	26	24	22	20	18
Number of samples	3	10	7	6	5	4	3	2

Table 5.1b Distribution of uranium content

	U (ppb)							
	42.5	37.5	32.5	27.5	22.5	17.5	12.5	7.5
Number of samples	1	2	7	10	10	6	3	1

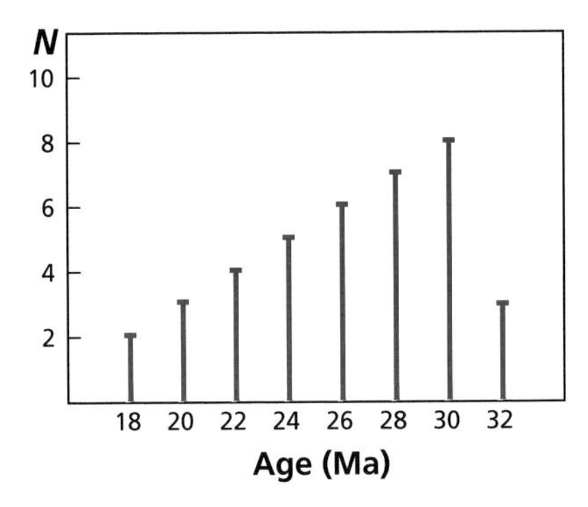

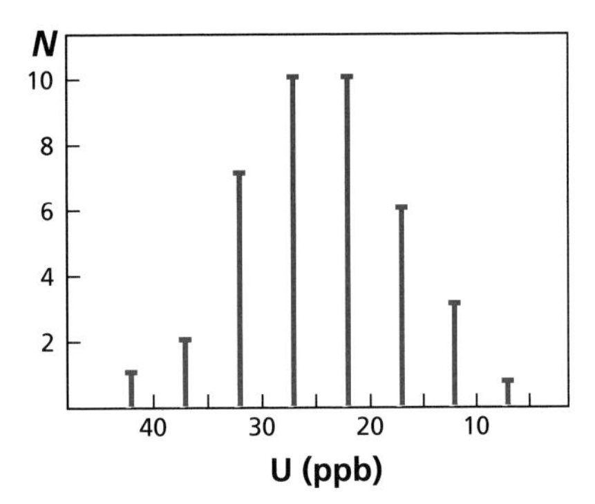

Figure 5.7 Histogram of age and uranium content measurements (see Tables 5.1a and 5.1b).

There is nothing automatic, then, about the choice of parameter (mean, mode, or median) that must be chosen to get closest to the truth. This must be decided in each individual case by a qualitative analysis (here geochemical or geological). This is a characteristic feature of **pseudo-statistics**. Let us lay down a rule of procedure: **we shall use the parameters employed in statistics but their meaning will be discussed in terms of geology and in particular the random or non-random character of the phenomena considered**.

Uncertainty is often expressed by the variance V_x, the standard deviation σ_x, and the uncertainty $\sigma_x/\sqrt{N} = \Delta x$. But here too uncertainties may be asymmetrical, as mentioned in the introductory example. We might also write 300 Ma^{+15}_{-5}. That means the age lies between 315 and 295 Ma and depends on what ultimately causes the uncertainty.

5.2.3 Composite uncertainty

We define the possible **estimated deviation** between the measured value and the true value by the absolute uncertainty Δx. This uncertainty is therefore expressed in the same units as x. We also define **relative uncertainty** $\Delta x/x$. This has no units and is expressed in per cent, per mil, etc. Both types of uncertainty are very important in radiometric dating. Absolute uncertainty determines what time interval we can measure, which is physically essential, of course, and is reflected in the expression of the **power of resolution**. Relative uncertainty represents the **measurement quality**, which is a very useful pointer too.

In what follows, we take the standard deviation of the measurement σ as the estimate of uncertainty.

When a process leading to a measurement consists in a series of operations, the calculation of the final uncertainty involves some quite strict rules.

Addition of operations

If the process is an **addition of operations** $x = au + bv$ where a and b may be positive or negative, then we add variances:

$$V_x^2 = a^2\sigma_u^2 + b^2\sigma_v^2 + 2ab\sigma_{u,v}^2$$

where $V_{u,v}$ is the correlation between variances u and v, that is, the **covariance**:

$$V_{u,v} = \sum_i \frac{(x_i - \bar{x})(y_i - \bar{y})}{N-1} = \sigma_{u,v}^2.$$

If b is negative and a positive, the covariance is subtracted. But be careful, the covariance itself may be either positive or negative. The covariance can be written:

$$\sigma_{u,v}^2 = \rho_{u,v} \cdot \sigma_u \cdot \sigma_v$$

and the linear correlative coefficent is:

$$\rho_{u,v} = \frac{\sigma_{u,v}^2}{\sigma_u \cdot \sigma_v}.$$

The value of $\rho_{u,v}$ lies between 0 and ± 1 (Figure 5.8).

If $x = u + v$, we can write:

$$\sigma_x = \sqrt{\sigma_u^2 + \sigma_v^2 + 2\rho_{u,v}\,\sigma_u\,\sigma_v}.$$

Notice that when covariance is high and is subtracted σ_x may be very low. This low variance is the result of a compensation effect and is misleading.

Thus, for example, when we wish to calculate the error on the radiogenic $(^{207}Pb/^{206}Pb)^*$ slope after making an isotopic measurement of lead, this error is very slight because of the close correlation between the errors on $^{206}Pb/^{204}Pb$ and $^{207}Pb/^{204}Pb$ created by the high uncertainty on ^{204}Pb.

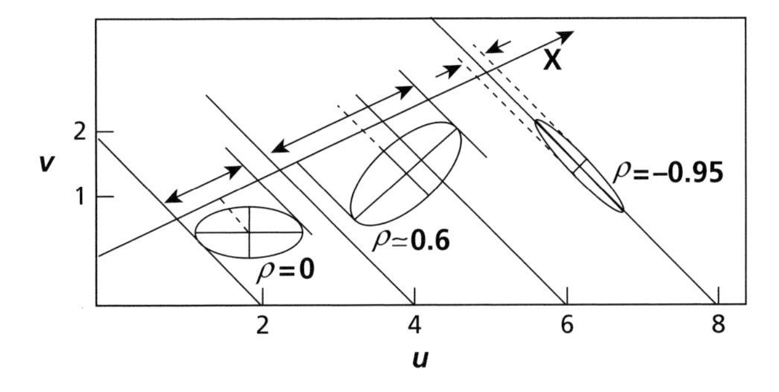

Figure 5.8 Diagram showing how errors on u and v in a product $u \cdot v$ may be correlated and how the factor $\rho_{u,v}$ varies (see text for definition of symbols).

Multiplication or division operations

If the process involves **multiplication** or **division** operations:

$$x = a \cdot u \cdot v$$

or similarly

$$x = \frac{au}{v}.$$

Once again variances are required, but this time they must be weighted:

$$\frac{V_x^2}{x^2} = \frac{\sigma_u^2}{u^2} + \frac{\sigma_v^2}{v^2} + \frac{2\sigma_{u,v}^2}{uv} \text{ for multiplication}$$

$$\frac{V_x^2}{x^2} = \frac{\sigma_u^2}{u^2} + \frac{\sigma_v^2}{v^2} - \frac{2\sigma_{u,v}^2}{uv} \text{ for division.}$$

(To obtain this expression, just shift to logs and we come back to additions.)

$$\frac{\sigma_X}{x} = x\sqrt{\frac{\sigma_u^2}{u^2} + \frac{\sigma_v^2}{v^2} \pm \frac{2\sigma_{u,v}^2}{uv}} = \bar{\sigma}_x,$$

that is, reduced dispersion. In this case, covariance is added for multiplication and subtracted for division. In fact, this **use of variance is essential when uncertainties are correlated**.

If there is no correlation, we can deal directly with standard deviations, as with **differential deviations** (obeying the rules of differentiation).

For addition

$$x = au \pm bv$$

we write:

$$\sigma_x = a\sigma_u + b\sigma_v.$$

For multiplication

$$x = a \cdot uv$$

we write:

$$\frac{\sigma_x}{x} \approx \frac{\sigma_u}{u} + \frac{\sigma_v}{v}.$$

Let us examine the scope of approximation on a simple example where $x = u + v$. If $\sigma_{u,v} = 0$ it can be seen that σ_x corresponds to the length of the diagonal of the rectangle of dimensions σ_u and σ_v:

$$\sigma_x = \sqrt{\sigma_u^2 + \sigma_v^2}.$$

If we make the approximation: $\sigma_x \approx \sigma_u + \sigma_v$ it holds good if σ_u and σ_v are quite different. If they are of the same order of magnitude, we can make a maximum error of $2/\sqrt{2} = 1.4$, which is not bad for an uncertainty given that we gain in ease of estimation compared with the calculation involving variance.

Notice that both expressions consist in differentiating and then replacing the differences by finite increases which are taken to be equal to σ. This practice is generalized when the uncertainty of a value stems from a mathematical formula.

We differentiate and then replace the differentials by the σ values.

If $x = uv$:

$$\sigma_x = \sigma(u, v) = u\sigma_v + v\sigma_u \text{ QED.}$$

Likewise if $x = u^n$:

$$\sigma_x = |n|u^{n-1}\sigma_x$$

and so on.

This is the approach we shall adopt in what follows, unless otherwise stated. If we have several measurements in each case, then we must systematically replace σ by $\Delta = \sigma/\sqrt{N}$.

Notice that to be able to add uncertainties when they are of different types, they must all be expressed in the same units. A convenient way to do this in geochronology is to express them all as ages (as we have already done for ^{14}C). This makes it easier to compare different geochronometers.

5.3 Sources of uncertainty in radiometric dating

Let us recall the stages we go through in determining a geological age:

- we collect the samples to be analyzed: rocks, minerals, wood, etc.;
- we analyze these samples in the laboratory for their isotopic and chemical ratios;
- we use these measurements to calculate an age;
- we situate this age within a geological scenario which we construct.

Each stage is affected by potential uncertainties (Figure 5.9).

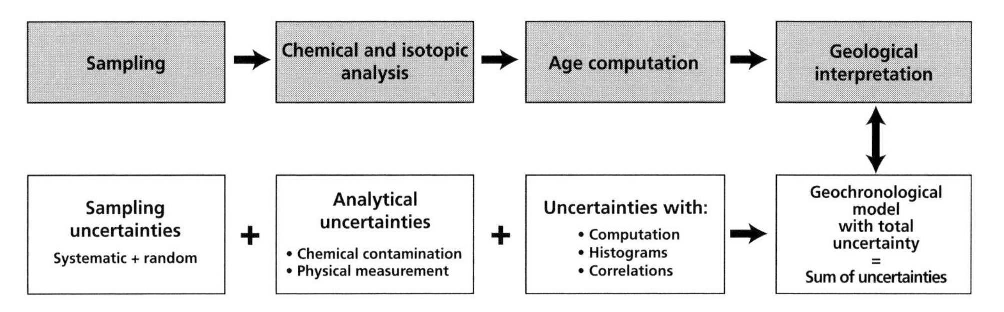

Figure 5.9 Determining a geological age while allowing for uncertainties affecting each stage.

5.3.1 Uncertainties introduced when collecting samples

A question arises before taking samples. Should the sample be taken at random? Such an approach looks rigorous as it rules out any subjectivity. Unfortunately, if we set about this "blindly" we would "knowingly" introduce systematic uncertainties. For example, if we sample rocks from a massif of granite, a random sample may include weathered rocks, rocks in contact with neighboring ones of different origins and ages, and which are liable to be contaminated by such contact at tectonic faults, etc. In practice, we try to define a homogeneous and representative geological whole (various units of the granite massif) and then try to take a random sample of that. It is difficult to evaluate the uncertainty of a sample if only one sample is taken. If several are taken, the results must be treated as for an ordinary measurement, assuming a random uncertainty of $1/\sqrt{N}$ where N is the number of samples collected. Each investigator must think about this sampling question and solve it (not forgetting any hidden structures) by combining statistics and . . . geological common sense!

5.3.2 Physical uncertainties on an individual measurement

Uncertainties on ages obtained from parent decay methods (^{14}C, ^{10}Be, etc.)

We begin with the dating equation, noting $\mu = {}^{14}C/C_{total}$ or $\mu = {}^{10}Be/{}^{9}Be$. $\mu = \mu_0 e^{-\lambda t}$ from which $t = 1/\lambda \ln(\mu_0/\mu)$.

Following the rules given earlier, the uncertainties are written:

$$\Delta t = \Delta\left(\frac{1}{\lambda}\right)\ln(\mu_0/\mu) + \frac{1}{\lambda}\frac{\Delta(\mu_0/\mu)}{(\mu_0/\mu)}$$

$$\Delta t = \frac{\Delta\lambda}{\lambda^2}\cdot\lambda t + \frac{1}{\lambda}\left(\frac{\Delta\mu}{\mu} + \frac{\Delta\mu_0}{\mu_0}\right)$$

$$\Delta t = \left(\frac{\Delta\lambda}{\lambda}\right)t + \frac{\Delta\mu}{\lambda\mu} + \frac{\Delta\mu_0}{\lambda\mu_0}$$

$$\Delta t = \left(\frac{\Delta\lambda}{\lambda}\right)t + \left(\frac{\Delta\mu}{e^{-\lambda t}} + \Delta\mu_0\right)\frac{1}{\lambda\mu_0}.$$

We suppose that $\Delta\lambda$ and $\Delta\mu_0$ are negligible. Thus:

$$\Delta t \approx \frac{\Delta\mu}{\mu_0\ \lambda e^{-\lambda t}}.$$

Exercise

We wish to calculate the uncertainty associated with the determination of a ^{14}C age. We accept that the uncertainty on the $^{14}C/C$ measurement is 2%. We ignore the uncertainty affecting the estimate of $(^{14}C/C)_0$ (which is far from negligible in practice) and consider that the uncertainty on the radioactive constant is 1/1000.

Calculate the absolute and relative uncertainties committed for ages of 500, 5000, 10 000, and 30 000 years. Plot the relative uncertainty curve against age.

Answer

We get:

Age (yr)	500	5000	10 000	30 000
Δt (yr)	176	302	554	6219
$\Delta t/t$ (%)	35	6	5.5	20

The curve of $(\Delta t/t)$ diverges towards the low and high values of t (Figure 5.10).

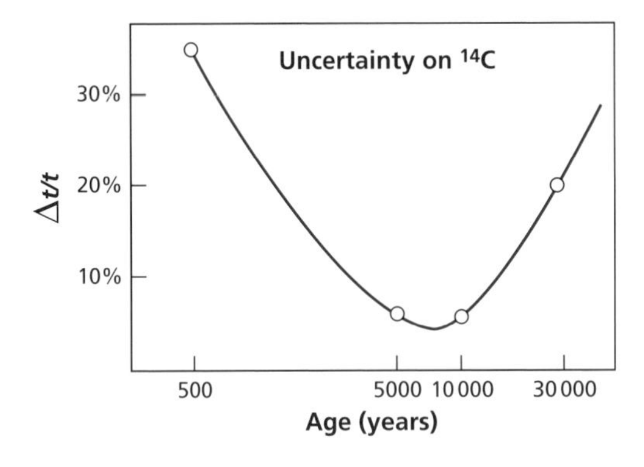

Figure 5.10 Relative uncertainty on a ^{14}C age determination as a function of that age. Notice the scale is logarithmic.

Overall estimation of uncertainty on long half-life parent–daughter methods

The chronometric equation with the usual notational conventions is written in the linear approximation, with α being the radiogenic isotope ratio and μ the parent–daughter isotope ratio:

$$\alpha = \alpha_0 + \lambda\mu t$$

from which

$$t = \frac{1}{\lambda}\left(\frac{\alpha - \alpha_0}{\mu}\right).$$

The uncertainties are then written:

$$\frac{\Delta t}{t} = \frac{\Delta\lambda}{\lambda} + \frac{\Delta(\alpha - \alpha_0)}{(\alpha - \alpha_0)} + \frac{\Delta\mu}{\mu}$$

but $(\alpha - \alpha_0) = \mu\lambda t$, hence:

$$\frac{\Delta t}{t} = \frac{\Delta\lambda}{\lambda} + \frac{1}{\mu}\left[\frac{\Delta * \alpha}{\lambda t} + \Delta\mu\right]$$

where $\Delta\alpha$ is the uncertainty on the measurement of α, and $\Delta\mu$ that on the measurement of μ. We note $\Delta^*\alpha = \alpha - \alpha_0$. Let us leave aside the uncertainty affecting the decay constant. For a given radiometric dating method:

$$\frac{\Delta t}{t} = \frac{\Delta\alpha}{\lambda\mu t} + \frac{\Delta\mu}{\mu}.$$

We notice immediately that uncertainty rises very quickly when $\lambda t \rightarrow 0$. Therefore each chronometer based on the parent–daughter pair has a lower limit of application T_{min}. In all of the long-period methods we have considered, μ never tends towards zero. So there is no upper limit. All of these chronometers can date events from T_{min} until the age of the Universe.[4]

Measurement uncertainty is related to the measurement of α and μ, and again this was considered in Chapter 1. There are two kinds of measurement uncertainty: **chemical uncertainties** in the preparation stage because of possible contamination or poor practice in isotopic dilution and **physical uncertainties** related to the precision of the mass spectrometer.

In practice, to minimize the first of these, the reagents are purified so as to make contamination negligible. For the mass spectrometer, we increase the number of measurements to improve the statistics, as already described.

Exercise

Typically, precision on the measurement of the $^{87}Sr/^{86}Sr$ ratio is 10^{-4} and does not vary much with the ratio, which is about 0.7. The uncertainty on the decay constant is $2 \cdot 10^{-2}$. Precision on measurement of the $^{87}Rb/^{86}Sr$ ratio is not as good because concentrations of elements must be measured by isotopic dilution (we shall estimate it at $3.5 \cdot 10^{-3}$). By taking the example of granite where $^{87}Rb/^{86}Sr = 3.5$, calculate Δt and $\Delta t/t$ as a function of age.

Answer

$$\frac{\Delta t}{t} = 2 \cdot 10^{-2} + \frac{1}{3.5}\left(\frac{10^{-4}}{1.4 \cdot 10^{-11} \cdot t} + 3.5 \cdot 10^{-3}\right).$$

Figure 5.11 illustrates the variation of $\Delta t/t$ and of Δt over time.

[4] This remark applies neither to radioactive chains nor to extinct radioactivity, because in both these cases there is a T_{min} and a T_{max} as in dating methods based on the parent isotope.

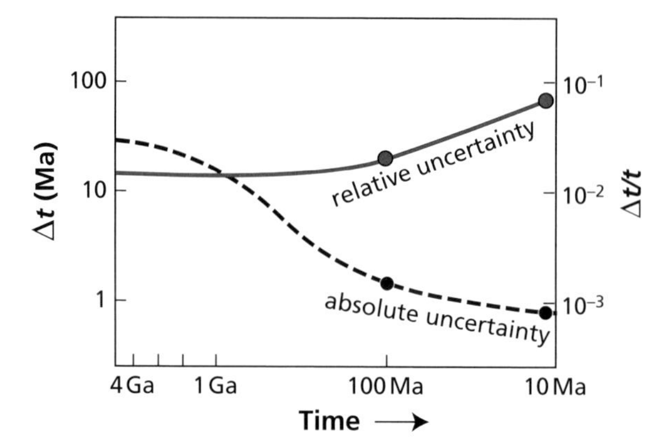

Figure 5.11 Variation of absolute and relative uncertainty on the ^{87}Rb–^{87}Sr age of a granite as a function of its age. For the granite $^{87}Rb/^{86}Sr = 3.5$. Notice that the relative uncertainty declines with age. Notice too that the age scale is logarithmic.

Exercise

Lanthanum-138 decays by electron capture to ^{138}Ba and by β^- emission to ^{138}Ce. The decay constants are $\lambda_{cap} = 4.4 \cdot 10^{-12}\,yr^{-1}$ and $\lambda_\beta = 2.255 \cdot 10^{-12}\,yr^{-1}$. (This is a similar situation to the branched decay of ^{40}K into ^{40}Ar and ^{40}Ca.)

(1) Establish the general chronometric equation and its linear approximation.
(2) Lanthanum-138 represents 0.089% of normal lanthanum. Barium-138 makes up 88% of naturally occurring barium. The uncertainty on the measurement of the $^{138}Ba/^{137}Ba$ ratio is $\pm 10^{-4}$ for a ratio greater than 6.3997, which is the normal value. This value is the value of the initial ratio for all rocks and terrestrial minerals.

What must the $^{138}La/^{137}Ba$ and therefore the La/Ba ratio be for us to use this method to date rocks $2 \cdot 10^9$ years old (neglecting the uncertainty on λ) with an uncertainty of less than $\pm 2\%$?

(Notice that if we measure a ratio with a relative precision of $\pm 2 \cdot 10^{-4}$, the absolute precision on the ratio of 6.4 is about $\pm 1.3 \cdot 10^{-3}$.)

Answer

(1) $$\frac{^{138}Ba}{^{137}Ba} = \left(\frac{^{138}Ba}{^{137}Ba}\right)_0 + \left(\frac{^{138}La}{^{137}Ba}\right)\frac{\lambda_{cap}}{\lambda_{cap} + \lambda_\beta}\left(e^{(\lambda_{cap}+\lambda_\beta)} - 1\right).$$

The linear approximation of this formula is:

$$\frac{^{138}Ba}{^{137}Ba} = \frac{^{138}Ba}{^{137}Ba} + \left(\frac{^{138}La}{^{137}Ba}\right)\lambda_{cap}t.$$

(2) To solve the problem, $^{138}La/^{137}Ba$ must be roughly equal to 3.6, that is the La/Ba ratio must be about 557. This is a substantial La/Ba ratio seldom found in nature.[5] Here are a few ratios for common rocks, to give us an idea: peridotite = 0.1, granite = 0.75, basalt = 0.5. This is why this method is little used in practice.

[5] High ratios are found only in a few rather rare minerals such as allanite. This is the mineral used to determine this date.

5.3.3 Uncertainties on age calculations

After the measurement phase we have either a single measurement or a series of measurements. In the first instance, we apply the dating formula which yields what we have called an apparent age and we move on to the next stage which is the geological interpretation. If we have a series of measurements, which is usually the case nowadays with the faster methods of analysis available, any of several situations may arise.

Case 1. The series of measurements relates to various chronometers on various cogenetic rich systems. This is the semi-quantitative problem we have already addressed. The age is limited at the lower bound by the age of the most retentive mineral, say, the $^{206}Pb-^{207}Pb$ age on zircon or hornblende if it is K–Ar alone. This situation is nowadays of historical and didactic interest only, since modern studies are made on a set of measurements.

Case 2. We have many measurements made by one method on one type of mineral. This case is becoming ever more general with the development of *in-situ* methods of analysis: series of $^{206}Pb-^{207}Pb$ measurements on zircon using an ion probe, series of $^{39}Ar-^{40}Ar$ or $^{40}K-^{40}Ar$ on mica or basalt glass using laser extraction techniques.

Case 3. A series of paired measurements may be used to define a straight-line isochron or a discordia.

In the latter two instances, which are now the most common, there is a twin problem to overcome. First, can a valid age be calculated from the values measured? In other words, do the conditions for applying the theoretical reference models prevail? Is the alignment acceptable? This is the issue of **acceptability**. After answering this first question, how do we calculate the most reliable age mathematically, and with what uncertainty? This is the age calculation proper. Obviously, the answer may involve both geological and experimental considerations. We have adopted the following rules. **We shall introduce the geological criteria at the same time as the acceptability criterion, but not when calculating the age proper.** This means we shall not attribute geological uncertainties to each sample measured because we have no rational means of fixing them quantitatively.

We shall therefore calculate ages and their uncertainties from standard statistical methods on measurements that are geologically accepted. We then have an age and an uncertainty. We shall then introduce geological uncertainties when making the geological interpretation.

To return to the example given in the introduction, if we have a series of measurements plotted in the ($^{87}Rb/^{86}Sr$, $^{87}Sr/^{86}Sr$) diagram we shall introduce a geological appraisal to decide whether the alignment of experimental measurements is acceptable or not and whether some peculiar measurements must be eliminated (we shall see how to do this with an objective statistical criterion). If the answer is positive, we shall calculate the age using a weighted least-squares method.[6] We then introduce geological uncertainties when interpreting the geological age obtained (Figure 5.12).

[6] Thus in the theoretical example in the introduction we use option (c) in Figure 5.1.

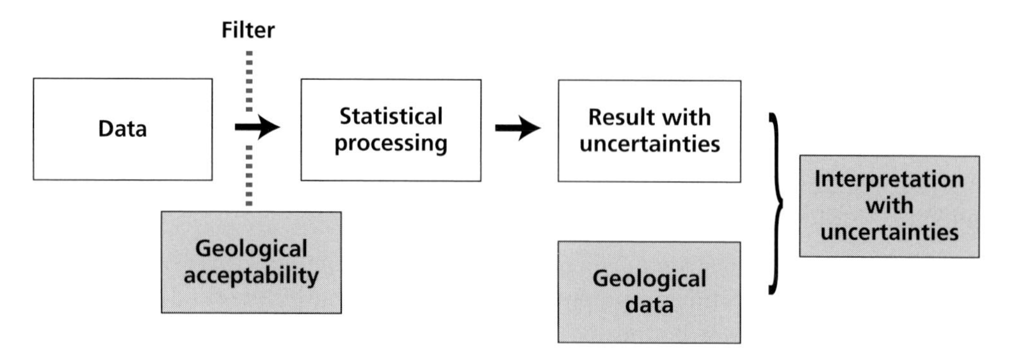

Figure 5.12 The points in the procedure at which geological knowledge is introduced.

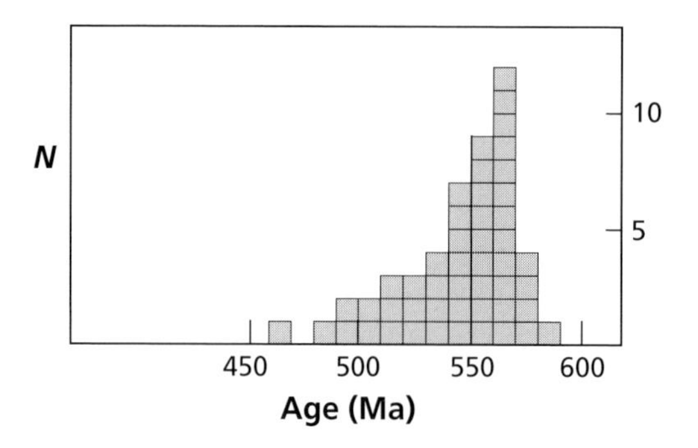

Figure 5.13 A series of apparent ^{207}Pb–^{206}Pb ages determined on zircon using ion probe spot analysis. The proposed age is 560 ± 20 Ma.

Individual age statistics

These individual apparent age statistics relate to measurements made on rich systems. The problem is analogous to that of physical measurements. For example, if we measure the apparent ^{207}Pb–^{206}Pb age on what we think is a cogenetic series of zircon samples, we construct the histogram of data. This distribution will be generally asymmetrical, with the asymmetry being established towards the younger ages because of the diffusion of the daughter isotope which tends to lower the age. From that point, we choose the modal value as the age and the standard deviation of the distribution as the uncertainty. By way of example, here is a ^{207}Pb–^{206}Pb asymmetrical age distribution obtained by using an ion probe on zircon (Figure 5.13). But when we analyze detrital zircons in sedimentary rocks, not only grain by grain but even by several measurements on a single grain, the histogram is extremely complex. It can only be interpreted after geological and quantitative statistical analyses.

Uncertainties in measuring straight lines of isotope evolution

Presence or absence of alignment

We must also know how to estimate the uncertainty on a series of measurements that do or do not define an isochron or a straight line on the concordia diagram. This is what we may call the alignment uncertainty, because for the method of the straight line of isotope evolution as for the concordia method, data points are rarely perfectly aligned and all the less so as the precision of the experimental method increases. What causes this non-alignment? A shortcoming in the initial natural isotopic homogenization, that is, at the time the rock crystallized? A slight opening of the system? The choice of samples? Admittedly, the mathematical uncertainty can be calculated with a least-squares program, which is ultimately what we do (because we know of no alternative), but is this a reliable method? As long as we ignore what caused the uncertainty, its meaning remains ambiguous.

The first estimate of the alignment or otherwise of data points must be made visually on the diagrams, making allowance, of course, for the experimental uncertainties attributed to each measurement. Next we turn to statistical methods as they allow the approach to be rationalized.

All these words of warning are intended to emphasize the fact that there is no automatic process or mathematical formula to be blindly applied. Instead, each step requires all of the physical, chemical, and geological knowledge available in each instance.

The correlation coefficient

Let there be two variables (y, x) and a series of paired values $(y_1, x_1), (y_2, x_2), \ldots, (y_N, x_N)$, that is, for us, a couple of analyses of two isotopic or chemical ratios. We can calculate the mean for both variables:

$$\bar{y} = \frac{1}{N}\sum_{1}^{N} y_i \text{ and } \bar{x} = \frac{1}{N}\sum_{1}^{N} x_i.$$

We can calculate the variances relative to the mean values of x and y:

$$V_x = \frac{1}{N}\sum (x_i - \bar{x})^2 \text{ and } V_y = \frac{1}{N}\sum (y_i - \bar{y})^2.$$

The standard deviations, which are the square roots of the variances, are written

$$\sigma_x = \sqrt{\frac{\sum (x_i - \bar{x})^2}{N}} \text{ and } \sigma_y = \sqrt{\frac{\sum (y_i - \bar{y})^2}{N}}.$$

We have analogously defined what is termed **covariance** (already encountered):

$$V_{xy} = V_{yx} = \frac{1}{N}\sum (y_i - \bar{y})(x_i - \bar{x}).$$

This covariance is zero if the values of y and x vary independently of one another. If the variations are correlated, the magnitude will be either positive or negative depending on the sign of the correlation. This covariance could reflect the value of the correlation. However, if it were left so, it would depend on the actual values of both x_i and $\bar{x}$ and of y_i and $\bar{y}$. For example, it would be different for a pair of variables measured in ppm and another measured in percent where the correlation looked the same. To make the correlation measurement independent of the units chosen for measuring y and x, we define the correlation coefficient r:

$$r = \frac{\sum(y_i - \bar{y})(x_i - \bar{x})}{\sqrt{\sum(y_i - \bar{y})^2}\sqrt{\sum(x_i - \bar{x})^2}} = \frac{V_{xy}}{\sqrt{V_{yy}}\sqrt{V_{xx}}}.$$

When x and y are uncorrelated, $r = 0$; when they are fully correlated $|r| = 1$ ($+1$ for a positive correlation and -1 for a negative correlation). This was the parameter $\rho_{u,\,v}$ used previously in combining statistics.

Exercise

We have a series of ^{87}Rb–^{86}Sr analyses given in Table 5.2. These analyses are plotted on the diagram in Figure 5.14 as stars. (The white dots are for the next exercise.)

Table 5.2 Series of $^{87}Rb/^{86}Sr$ analyses

	1	2	3	4	5	6	7
$^{87}Sr/^{86}Sr = y$	0.800	0.795	0.790	0.775	0.770	0.735	0.720
$^{87}Rb/^{86}Sr = x$	7	6	5	4	3	2	1

Calculate the coefficient of correlation.

Answer

We posit $^{87}Sr/^{86}Sr = y$ and $^{87}Rb/^{86}Sr = x$. The mean values are calculated as: $\bar{x} = 4$ and $\bar{y} = 0.769\,28$.

Sample	1	2	3	4	5	6	7
$x_i - \bar{x}$	3	2	1	0	−1	−2	−3
$y_i - \bar{y}$	0.0308	0.0258	0.0208	0.0058	0.0008	−0.0342	−0.0492
$(x_i - \bar{x})(y_i - \bar{y})$	0.0924	0.0516	0.021	0.008	−0.0684	0.1476	0.3728

Hence $\sigma_x = 2$, $\sigma_y = 0.0284$, $V_{xy} = 0.0568$, and $r = 0.9375$.

Now, as can be seen (Figure 5.14), the points are not perfectly aligned. If dealing with a very old age (>2 Ga) we would be tempted to accept this procedure for calculating ages. For younger ages, we must bear in mind that the result is affected by substantial uncertainty.

Exercise

We posit $^{87}Sr/^{86}Sr = y$ and $^{87}Rb/^{86}Sr = x$. Consider the table of measurements below (Table 5.3). Calculate the means, standard deviations, covariance, and coefficient of correlation (see Figure 5.14).

Table 5.3 Further measurements from $^{87}Rb/^{86}Sr$ analyses

	1	2	3	4	5	6	7
x	7.1	6.53	6	5.5	3	2.5	1
y	0.801	0.791	0.785	0.775	0.746	0.735	0.720

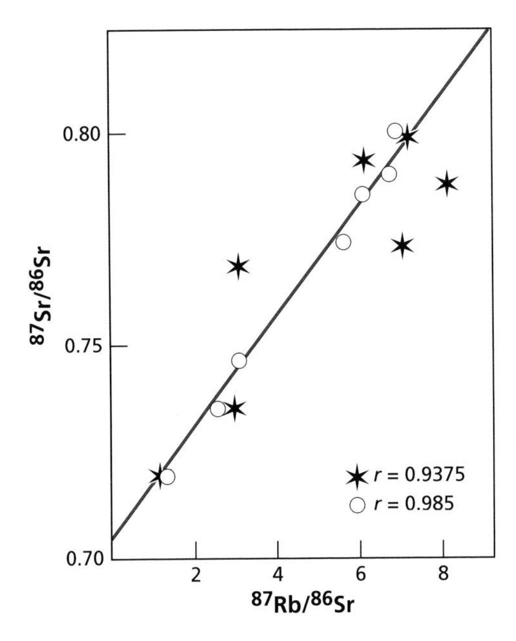

Figure 5.14 Analyses of ^{87}Rb–^{86}Sr plotted on the ($^{87}Sr/^{86}Sr$, $^{87}Rb/^{86}Sr$) diagram.

Answer

$\bar{x} = 4.5185, \bar{y} = 0.7647$.

Sample	1	2	3	4	5	6	7
$x_i - \bar{x}$	2.5815	2.015	1.4815	0.9815	−1.518	−2.0185	−3.5185
$y_i - \bar{y}$	0.0363	0.0263	0.0203	0.0103	−0.0187	−0.0297	−0.0447
$(x_i - \bar{x})(y_i - \bar{y})$	0.0937	0.0529	0.02398	0.010 10	0.002839	0.05994	0.1572

Hence $\sigma_x = 2.5158$, $\sigma_y = 0.028\,633\,6$, $V_{xy} = 0.060\,88$, and $r = 0.985$.

It can be seen that r is closer to 1 than in the previous exercise. This shows us also the sensitivity of the correlation coefficient to dispersion.

Figure 5.15 shows some examples of positive correlations with their correlation coefficients r.

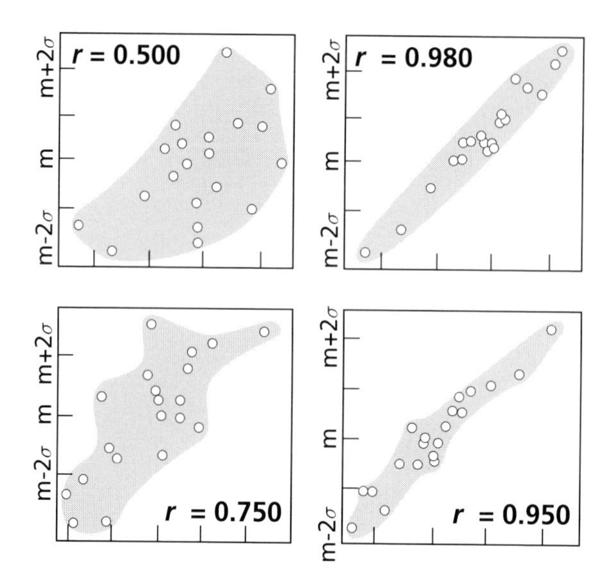

Figure 5.15 Measurements that are more or less closely correlated with various values for the corresponding correlation coefficient.

Remark

In practice, when the coefficient of correlation is less than 0.90 we consider the data points are poorly aligned. We do not attempt to identify a straight line. We are outside the conditions accepted for the various models: no initial isotopic homogeneity, complex opening of the system, etc. If $r > 0.90$ we calculate the parameters of the straight line of best fit by the least-squares method.

Inclusion of experimental uncertainties

In these calculations we have ignored the individual experimental uncertainties. Implicitly we have considered they were all equal. In practice, when these uncertainties vary between measurements, we must weight them and make allowance for them. Let us take the mean value $\bar{x}$ as an example.

Instead of simply calculating the mean by $\bar{x} = \sum x_i/N$, we weight it allowing for variance:

$$\bar{x}_\sigma = \frac{\sum_i \left(\frac{x_i}{\sigma_i^2}\right)}{\sum_i \left(\frac{1}{\sigma_i^2}\right)}.$$

Let us take a simple example. We have three very different measurements of uncertainty: 3 ± 0.1; 4 ± 2, and 2.5 ± 0.1. If we make the usual calculation, $\bar{x} = 3.16$. If we make the weighted calculation $\bar{x}_\mathrm{T} = 551/200.25 = 2.75$. There is a 14% difference between the results.

The variance of uncertainty (here it is a "true error") committed is $\sigma^2 = \left(\sum \sigma_i^2\right)^{-1}$, or $\sigma = 0.07$ and $\bar{x}_\mathrm{T} = 2.75 \pm 0.07$.

If all the uncertainties are identical, the variance is $\sigma_{\bar{x}}^2 = \sigma_i^2/N$. We find $\Delta = \sigma_i/\sqrt{N}$ again.

An analogous procedure of weighting by uncertainties is followed to calculate variance and covariance:

$$V_x = \frac{\frac{1}{N-1}\sum_i\left[\frac{1}{\sigma_i^2}(x_i-\bar{x})^2\right]}{\frac{1}{N}\sum_i\frac{1}{\sigma_i^2}} \quad \text{and} \quad V_{xy} = \frac{\frac{1}{N-1}\sum_i\frac{1}{\sigma_i^2}(x_i-\bar{x})(y_i-\bar{y})}{\frac{1}{N}\sum_i\frac{1}{\sigma_i^2}}.$$

In each case, the means are calculated by the weighted method:

$$\bar{x} = \frac{\sum_i\left(\frac{1}{\sigma_i^2}x_i\right)}{\sum_i\left(\frac{1}{\sigma_i^2}\right)}.$$

As can be seen, all of this gives expressions which are a little heavy to manipulate but which are no trouble for computer programs!

5.3.4 Calculating the equation of the straight line of correlation

We shall set out the principle of the **least-squares method** in a simple case. Let us look for the equation of the straight line that statistically describes the cloud of points $(\bar{x}_i, \bar{y}_i)$. We accept without proving it (it seems acceptable intuitively) that the straight line will pass through the mean point of coordinates $(\bar{x}, \bar{y})$. We are therefore looking for an equation of the type $y = ax + b$. The idea in the least-squares method is to look for the best (least bad) straight line passing through a cloud of points by using a precise criterion. The usual criterion is to minimize the distance between the points observed and the straight line we are looking for, which is done by minimizing the sum of the squares of the distance between the data points and their respective vertical projections. This is the simple case we shall work through mathematically to make the reasoning clear. If x and y are connected by a straight line, for the various values $(x_1, y_1), (x_2, y_2), \ldots, (x_N, y_N)$ we have $y_1 = ax_1 + b, y_2 = ax_2 + b, \ldots, y_N = ax_N + b$. But this is not quite right because the straight line does not pass exactly through each point. The square of the distances, then, is written:

$$(y_1 - b - ax_1)^2 + (y_2 - b - ax_2)^2 + \cdots + (y_N - b - ax_N)^2 = D^2.$$

We try to minimize the distances from our two unknown parameters a and b. We therefore write: $\partial(D^2)/\partial a = 0$ and $\partial(D^2)/\partial b = 0$, which gives two equations with two unknowns:

$$\frac{\partial(D^2)}{\partial a} = \sum_i -2x_i(y_i - b - ax_i) = 0$$

$$\frac{\partial(D^2)}{\partial b} = \sum_i -2(y_i - b - ax_i) = 0.$$

It can be seen then that the problem is soluble: we have to solve a system of equations with two unknowns a and b. The solution to this problem can be written very simply:

$$a = \bar{y} - r\frac{\sigma_y}{\sigma_x}\bar{x}$$

$$b = r\frac{\sigma_y}{\sigma_x}$$

where r is the correlation coefficient.
The equation of the best straight-line fit passes through $\bar{x}$ and $\bar{y}$:

$$y = \bar{y} + r\frac{\sigma_y}{\sigma_x}(x - \bar{x}).$$

The uncertainty on the slope is written:

$$\{p\} = \frac{\sigma_y}{\sigma_x}\frac{\sqrt{1 - r^2}}{\sqrt{N}}.$$

The uncertainty on the ordinate at the origin is:

$$\{I\} \approx \bar{x}\{p\}, \text{ or } \{I\} = \frac{\sigma_x}{\sigma_y}\frac{\sqrt{1 - r^2}}{\sqrt{N}}.$$

Let us turn back to Table 5.2 and Figure 5.14. We have calculated the coefficient of regression $r = 0.9375$; it is legitimate, then, to try to calculate the equation of the best straight-line fit whose slope will give us the age, and whose intercept b will give us the initial isotope ratio. The calculation gives:

$$a = r\frac{\sigma_y}{\sigma_x} = 0.013\,25 \text{ and } b = 0.704\,84.$$

By eye, taking a little care, we read $a = 0.0128$ and $b = 0.705$, which is not so bad in the end! The uncertainties are: on the slope, 0.000 347 6, therefore $p = 0.013\,25 \pm 0.000\,347\,6$, which corresponds in age to $t = 939 \pm 24$ Ma.

The ordinate at the origin is $I = \pm\,0.001\,567$. Hence

$$(^{87}\text{Sr}/^{86}\text{Sr})_0 = 0.704\,84 \pm 0.001\,56.$$

In practice, to be rigorous, we must take account of the experimental uncertainties on x and y, which are not necessarily related to the distance from the best straight-line fit. They measure the importance attributed to each point in calculating the least squares. The smaller the uncertainty, the more "important" the point is, of course. Introducing these uncertainties leads to complications in the mathematics, especially where there are many data.

All of this is done nowadays by computer programs associated with spreadsheets. But vigilance is called for since these least-squares programs available on various computers or in books are of very variable types and quality. Generally, they are programs that do not provide for weighting by experimental uncertainties. Where experimental uncertainties are taken into account, they are usually uncertainties relative to a single axis because they often

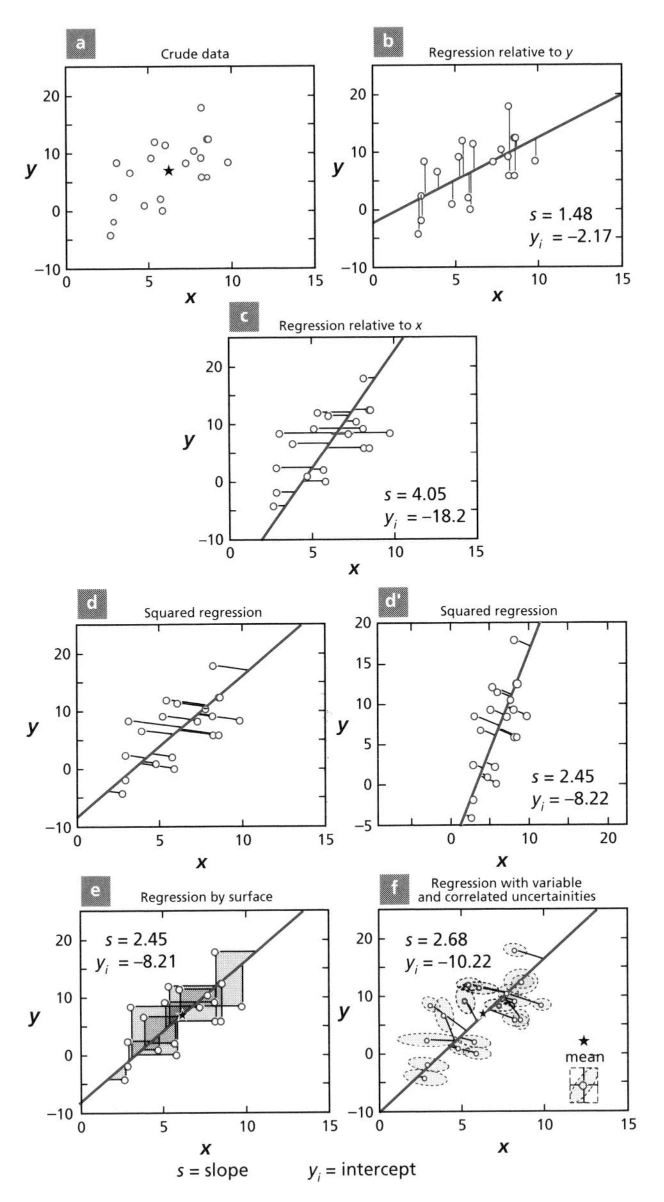

Figure 5.16 The principle of least squares with experimental uncertainties. (a) We consider a cloud of data points on x and y. (b) We have calculated least squares relative to y, which comes down to accepting that uncertainty on x is negligible compared with that on y. (This is the program on good little pocket calculators, the mathematics of which we have set out.) (c) Here the least squares are calculated relative to x, the opposite case from that in (b): an enormous difference can be seen in the outcome of these two techniques. (d) By a more elaborate procedure, the least squares are calculated taking account of uncertainties on x and y and taking as the criterion the orthogonal projection whose length is minimized. (d′) Right: the true orthogonal projection; left: the representation is distorted to give a plot with the same look as those before and after. (e) The idea here is to take account of uncertainties on x and y by minimizing the areas of the projection triangles. (This is what we did with our simplified method of calculation.) Notice this method gives acceptable results. (f) We consider each point to be affected by uncertainties of variable amplitudes on x and y, which are correlated. So each point is surrounded by an ellipse of uncertainty. This is the most elaborate program and is the benchmark. Notice that if we average out the results of (b) and (c), we get values close to (f) $p = 2.76$ and $y_i = -10.18$.

dominate the others. More often than not, such uncertainties are considered to be uniform for all data points.

Programs that do take account of individual uncertainties on x and y are much rarer and must be searched out in program libraries. Lastly, there are some even more complete ones that take account not only of individual uncertainties but also of any correlations between uncertainties (covariance again!).

To gain some idea of these different programs we have illustrated their outcomes for one example (Figure 5.16).

Again, we must remain clear-sighted. We can only make rough estimates of geological uncertainty. It is not certain, then, that the "blind" least-squares method, which admits of random uncertainties on measurements and which accordingly measures geological uncertainty implicitly by point dispersion, is totally reliable. However, for want of anything better, we choose this method as a first approximation, while bearing its limits in mind. (Never be blinded by the mathematics used in the natural sciences!) As general references for this section, see Crumpler and Yeo (1940), York (1969), Wendt (1991), Bevington and Robinson (2003), and Ludwig (2003); see also the program package Ludwig (1999).

5.4 Geological interpretations

5.4.1 General remarks

The principle consists in integrating the age (or ages) obtained into an overall geological scenario in which age becomes fully meaningful. This final step is not without its uncertainties (and risks!).

This may be an **uncertainty in identification**, if we make an age correspond to the wrong phenomenon. Thus we may attribute an age to the emplacement of a granite massif whereas the age is that of late or even subsequent isotopic re-equilibration or conversely the age of the parent rock remobilized in the materials to be dated, etc.

It may be an **uncertainty of indetermination**. An age is determined which, on the face of it, does not correspond to any phenomenon known at the time. Thus, say, we find an age of 420 Ma for a granite from southern France. Is it a mixed age between the age of Hercynian mountain-building 320 Ma ago and of the Pan-African orogeny 550 Ma ago? Or is it evidence of some previously unreported Caledonian event in the region and so a new discovery? In this event, the age determined itself poses a problem for geologists, who must identify the event by relative structural dating.

Uncertainties, then, may well arise. Remember that **an age is only the translation of an isotope ratio.** Whenever we are faced with an interpretation that is unclear, we have to return to the isotopes, their history, to the basic model of the closed box, and so on. Of course, ages must be determined by using various methods to confirm the results. Here again, the common expression for making comparisons between different methods is age, but whenever we wish to interpret a particular difference we must come back to the physical systems and to geochemical behavior. For example, argon diffuses more readily than neodymium, the Rb–Sr method is more sensitive to hydrothermal phenomena than the U–Pb system on zircon, which in turn may be aged by adjunction of inherited whole or fragmented zircon, and so on. We shall examine a few cases to illustrate these points with examples.

5.4.2 Case studies

The aim here is not to achieve systematic coverage but to give a few examples to illustrate a method, its difficulties, its limits, and its level of reliability. This approach will allow us to test the various methods of quantitative discussion of geochronology, to compare them, and to see how they can be integrated into the geological context. After noting how we shift from the qualitative use of apparent ages to quantitative models (isochron, concordia, stepwise degassing), we have to measure how reliable each approach is.

Contact metamorphism of the Eldora stock

We have made a qualitative study of discordances in apparent ages from this example. It is interesting, then, to see what the result was of the various more elaborate methods in this geologically well-defined case. As these studies, conducted by **Stanley Hart** when a student at the Massachusetts Institute of Technology (see Hart *et al.*, 1968), are rather old, we only have results for the U–Pb, Rb–Sr, and K–Ar systems (see Figure 5.17 and Chapter 3).

The U–Pb system on zircon

We obtain a discordia cutting the concordia at 1600 ± 50 Ma and 60 ± 5 Ma, giving apparently the ages of the Precambrian schist and the Eldora intrusion, respectively. The points on the discordia are placed regularly with distance from the contact. The most discordant ones are those closest to the contact.

The ^{87}Rb–^{87}Sr system in isochron construction

The measurements were made mainly on feldspars because they might prove not very significant on the whole rock of the schists. They are aligned on an approximate straight line isochron corresponding to an age of 1400 ± 100 Ma, except for the apatites and epidotes close to the intrusion (^{87}Sr acceptors).

The (Rb–Sr, K–Ar) measurements on micas discussed with the generalized concordia diagram

The discordia is a curve and not a straight line as in the U–Pb systems. We obtain the two intersections corresponding to the age of the intrusion of 60 Ma and the "schist age." However, we cannot say whether the latter is 1480 or 1600 Ma. It is 1480^{+100}_{-20} Ma.

The ^{40}Ar–^{39}Ar measurements

These were made later on hornblende, biotite, and feldspar (Berger, 1975). We can define plateaus giving increasing ages with distance from the contact. The hornblende and large biotite minerals give an age of 1450 ± 20 Ma, that is, about the same as the ^{87}Rb–^{87}Sr age.

We can envisage the age of the schists as 1450 Ma and the age measured by the U–Pb concordia of 1600 Ma corresponds to a few inherited pieces of detrital zircon. But what does the schist age mean? Is it the age of metamorphism or of earlier sedimentation? The ^{39}Ar–^{40}Ar ages on biotite and the (K–Ar, Rb–Sr) concordia age suggest the age of metamorphism, because it was then that the minerals crystallized. The age of 1600 Ma is

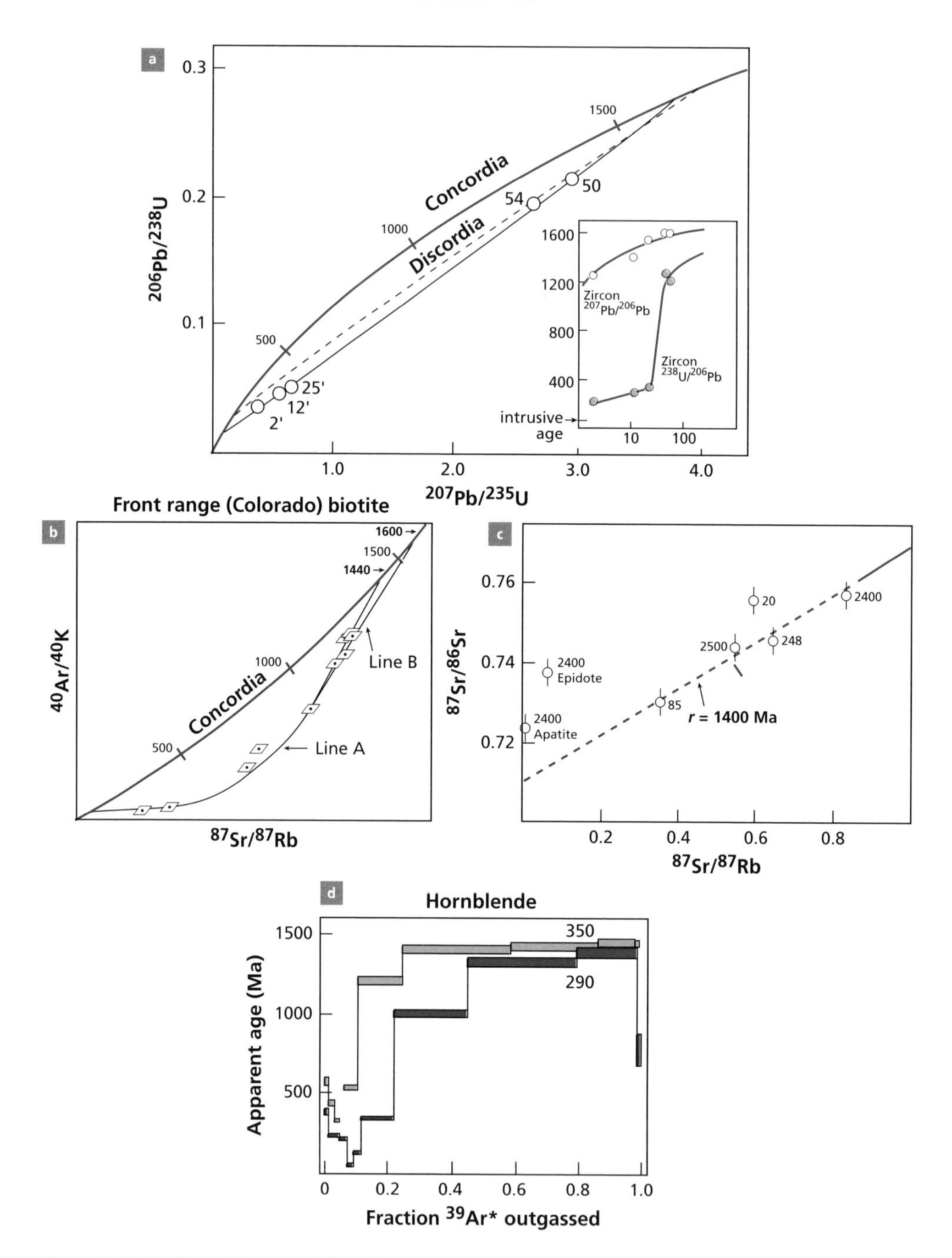

Figure 5.17 Radiometric ages of the Eldora stock examined in Chapter 3 in terms of apparent ages. (a) The U–Pb system on zircon (Hart *et al.*, 1968). (b) The two ^{87}Rb–^{87}Sr methods on biotite in the generalized Rb–Sr, K–Ar concordia. (c) The ^{87}Rb–^{87}Sr method on feldspar. (d) Stepwise ^{40}Ar–^{39}Ar ages (after Berger, 1975). The numbers indicate distances from the granite intrusion (Chapter 3).

probably that of the earlier granites whose zircon was eroded and redeposited in the clays from which the schists were formed. But the questioning and the answers are themselves a measure of uncertainty of the result. What does the age of formation of a schist mean?

The very ancient rocks of Greenland

The rocks of Greenland are the oldest in the world.[7] They have been studied by various methods: U–Pb on zircon, ^{206}Pb–^{207}Pb on whole-rock isochrons, ^{86}Rb–^{87}Sr, and ^{147}Sm–^{143}Nd. **Steve Moorbath** of the University of Oxford was both the pioneer and the principal investigator (Moorbath and Taylor, 1981; Moorbath *et al.*, 1986). These studies, combined with precise mapping of the terrain, led to the identification of two separate geological formations (Figure 5.18), the **Amitsoq** gneiss and the **Isua** Group (the mountain of

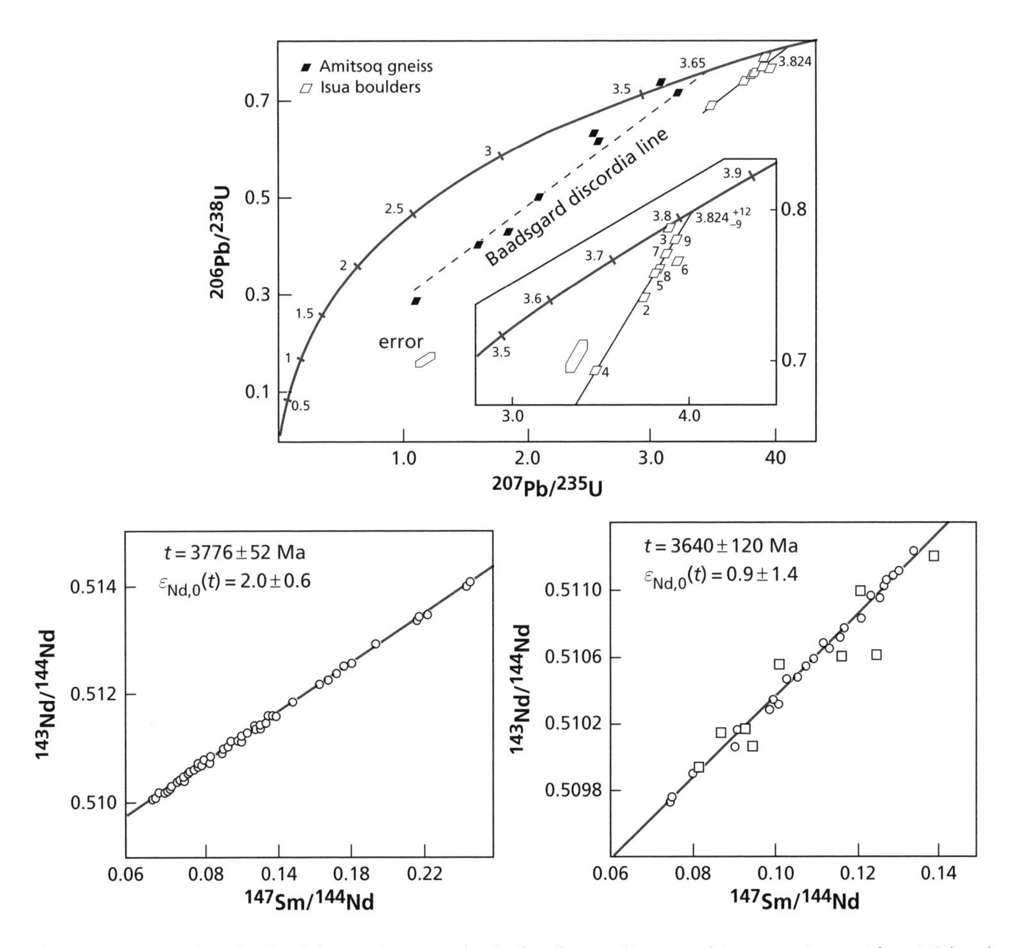

Figure 5.18 Results obtained by various methods for the Amitsoq and Isua gneisses. After Michard-Vitrac *et al.* (1977); Moorbath *et al.* (1997).

[7] There are older minerals (zircon) but they are detrital and separated from their parent rock in which they crystallized.

Table 5.4 Results obtained by various methods for the Amitsoq gneiss and the Isua Group

Dating methods	Amitsoq gneiss (Ga)	Isua Group (Ga)
U–Pb concordia (zircon)	3.45 ± 0.05	3.824 ± 0.05 (conglomerate) 3.77 ± 0.01 (massive rock)
^{87}Rb–^{87}Sr (whole rock)	3.64 ± 0.06	3.66 ± 0.06
^{147}Sm–^{143}Nd (whole rock)	3.640 ± 0.12	3.776 ± 0.05
^{207}Pb–^{206}Pb (whole rock)	3.56 ± 0.10	3.74 ± 0.12

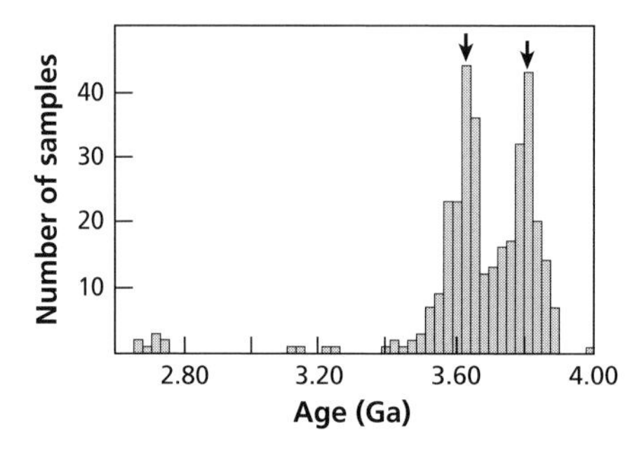

Figure 5.19 Histogram of ^{207}Pb–^{206}Pb ages obtained by ion microprobe methods on Amitsoq zircon. After Nutman *et al.* (1996).

references includes: Michard-Vitrac *et al.*, 1977; Moorbath *et al.*, 1986, 1997; Nutman *et al.*, 1996; Kamber and Moorbath, 1998).

The first is dated 3.65 ± 0.03 Ga and the second 3.74 and 3.82 Ga. These are the oldest groups of rock in the world. As can be seen by examining Table 5.4 and Figure 5.18, summarizing the results obtained by the various methods, the two groups are quite distinct. However, small differences can be noticed in the absolute values obtained by the various methods and whose precise meaning is unclear. We have no explanations other than that the rocks were subjected to subsequent metamorphism and the chronological systems were probably disrupted somewhat.

The Australian group from Canberra has used ion probe point analysis methods to measure zircon grain by grain and determine the U–Pb ages on the Amitsoq gneiss. A histogram can be made (Figure 5.19). It shows the two peaks at 3.60 Ga and 3.80 Ga, providing we take the **mode** of the two asymmetrical distributions as the value. The two statistical ages seem to be in agreement with the ages measured by conventional methods.

Let us discuss and interpret these results. The distribution of the Amitsoq and Isua formations is clear enough. The age of the Amitsoq gneiss is 3.64 ± 0.05 Ga, with $\Delta t/t = 1.3\%$. The Isua formation is a little older but not as well defined. The age of 3.66 Ga given by ^{87}Rb–^{86}Sr on whole rocks from Isua must be considered to be re-homogenization at the time the Amitsoq gneiss formed.

There seem to be rocks dating from 3.82 Ga, particularly in the form of conglomerates whose zircon “survived” the geological perturbations.

But we may surmise that the Isua surface formation is rather aged 3.77 ± 0.08 Ga ($\Delta t/t = 2\%$). The age of 3.82 Ga is the result of inherited processes. Along the same lines, we can infer that some old zircons have been incorporated into the Amitsoq formation, either through the erosion–sedimentation cycle or by processes of magmatic assimilation or metamorphism yielding the old values in the statistical distribution. As can be seen not everything is clear. But it did all happen nearly 4 billion years ago!

Archean komatiites

Komatiites are associations of basic and ultrabasic lavas found in Archean rocks alone (see, e.g., Hamilton *et al.*, 1979; Zindler, 1982; Brévart *et al.*, 1986; Dupré and Arndt, 1986). They are the only evidence of what the mantle was like at that time. These associations of rock have been dated mainly by Sm–Nd and Pb–Pb systems since the other geochronometers, particularly Rb–Sr, Ar–Ar, and U–Pb, are generally very disturbed systems. In addition, the U–Pb, concordia method is difficult to use as uranium-rich minerals are very rare in these rocks. The oldest well-identified komatiite belt is the **Barberton** Greenstone Belt in South Africa. It is dated 3.4 ± 0.12 Ga (almost as old as Amitsoq!). Several datings have been obtained on these rocks by various methods, all of them more or less concordant. Here are the results.

The whole-rock isochron method gives: ^{87}Rb–^{87}Sr = 3.35 ± 0.2 Ga; ^{147}Sm–^{143}Nd = 3.54 ± 0.07 Ga; ^{40}Ar–^{39}Ar = 3.49 ± 0.01 Ga; ^{206}Pb–^{207}Pb = 3.46 ± 0.07 Ga. This poses the question of the exact age of emplacement of the komatiites. Is it 3.35 Ga or 3.53 Ga? There is a gap of some 200 Ma between the two dates, which is as long as the time separating us from the Jurassic. Given the data we currently have, we have no criterion for deciding one way or another, and so choose the value of 3.45 ± 0.10 Ga as the most likely age. The resolution of such problems will answer the question of the duration of the emplacement episode of komatiites.

An entirely different situation is found at **Kambalda** in Western Australia. Both the ^{147}Sm–^{143}Nd and ^{206}Pb–^{207}Pb methods give very handsome alignments on the isochron diagrams. Unfortunately, these alignments do not yield the same age. The ^{147}Sm–^{143}Nd age is 3.26 Ga while the ^{206}Pb–^{207}Pb is 2.72 Ga. Both methods are reputed to be robust. Which should we choose when they fail to agree?

Dupré and Arndt (1987), then working together at the Max-Planck Institute in Mainz, showed that the ^{147}Sm–^{143}Nd straight lines were in fact straight lines of mixing, as shown in the (ε_{Nd}, 1/Nd) plot (Figure 5.20). The most likely age is therefore 2.72 Ga, which is consistent with the local geological context and datings of other associated terrains. Dupré and Arndt (1987) generalized the discussion of comparative Sm–Nd and Pb–Pb ages on komatiites and made a systematic compilation (see Table 5.5).

There are three cases where the age is fixed to within ± 20 Ma: **Barberton** in South Africa, the **Abitibi** komatiite belt of Canada, and **Zimbabwe**. This assertion is based on the concordance of ages determined by both methods and on the geological context and dating of neighboring granitic rocks. Notice that $\Delta t/t = 0.7\%$.

Cape Smith is a special case because the ^{147}Sm–^{143}Nd and ^{207}Pb–^{206}Pb ages are not very different and have overlapping margins of uncertainty. For want of any other information, we must put down an age of 1.73 ± 0.1 Ga with $\Delta t/t = 5\%$, which is not bad compared with the others.

The case of **West Pilbara** in Australia is rather similar. The two ^{207}Pb–^{206}Pb measurements seem weaker and incorrect because the geological context argues rather for an age of

Table 5.5 Comparative Sm–Nd and Pb–Pb ages on komatiites (these are solely ages for which alignments are statistically reliable)

	^{147}Sm–^{143}Nd ages	^{206}Pb–^{207}Pb ages	Most likely age
West Pilbara ≈ 3.5 Ga	3.560 Ga ± 30 Ma	≈ 3.15 Ga	3.5 Ga?
Barberton ≈ 3.5 Ga	3.540 Ga ± 30 Ma	3.460 Ga ± 70 Ma	3.45 Ga
Abitibi Munroe	2.662 Ga ± 120 Ma	2.724 Ga ± 20 Ma	2.7 Ga
Alexo (2.7 Ga)	2.750 Ga ± 90 Ma	2.690 Ga ± 15 Ma	
Newton	2.826 Ga ± 64 Ma	≈ 2.65 Ga ± 15 Ma	
Kambalda ≈ 2.7 Ga	3.260 Ga ± 40 Ma	2.720 Ga ± 105 Ma	2.7 Ga
	3.230 Ga ± 120 Ma	2.730 Ga ± 30 Ma	
Zimbabwe ≈ 2.7 Ga	2.640 Ga ± 140 Ma	2.690 Ga ± 1 Ma	2.7 Ga
	2.933 Ga ± 147 Ma		
Cape Smith ≈ 1.8 Ga	1.871 Ga ± 100 Ma	1600 Ga ± 150 Ma	1.73 Ga

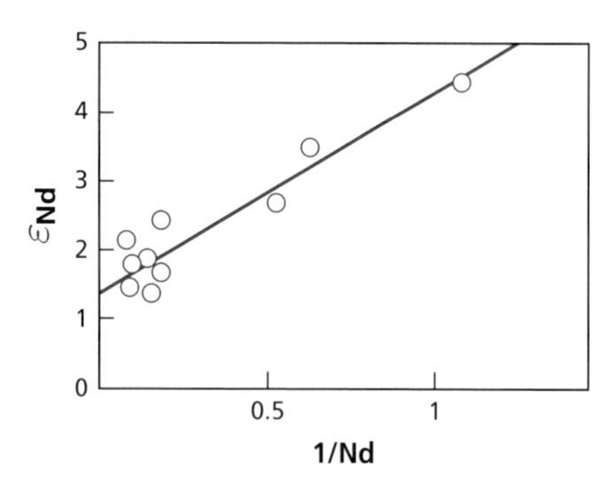

Figure 5.20 Diagram of (ϵ_{Nd}, 1/Nd) plot on komatiites of Kambalda (Australia) showing a correlation in agreement with the idea that the rocks are a mixture. After Dupré and Arndt (1987).

3.56 Ga. It is the context alone that allows any conclusion, but the ^{207}Pb–^{206}Pb results indicate there was a secondary, disruptive phenomenon.

As said, **Kambalda** is the opposite case. The geological context indicates that it is the ^{207}Pb–^{206}Pb system that yields the more reliable age.

The results of these last two cases show that there is nothing automatic about the process and other datings probably need to be made by a third method or constrained by other geological data.

The rhyolite of Long Valley, California

It is probably on this very amenable rock that the University of California – Los Angeles team around **Mary Reid** (Reid *et al.*, 1997) made the best use of the various methods for determining the ages of geological units of very recent magmatic origin. The age of eruption and of emplacement are estimated, allowing for erosion, as 140 ka. The results obtained by the different methods are: 257 ka (^{87}Rb–^{87}Sr) and 200–235 ka (isochronous ^{230}Th–^{238}U), and the ^{39}Ar–^{40}Ar ages are much younger, ranging from 100 to 150 ka (Reid *et al.*, 1997).

The discordances here (several thousand years) are construed as not arising from subsequent disruptive phenomena but rather from complex magmatic sequences (Figure 5.21).

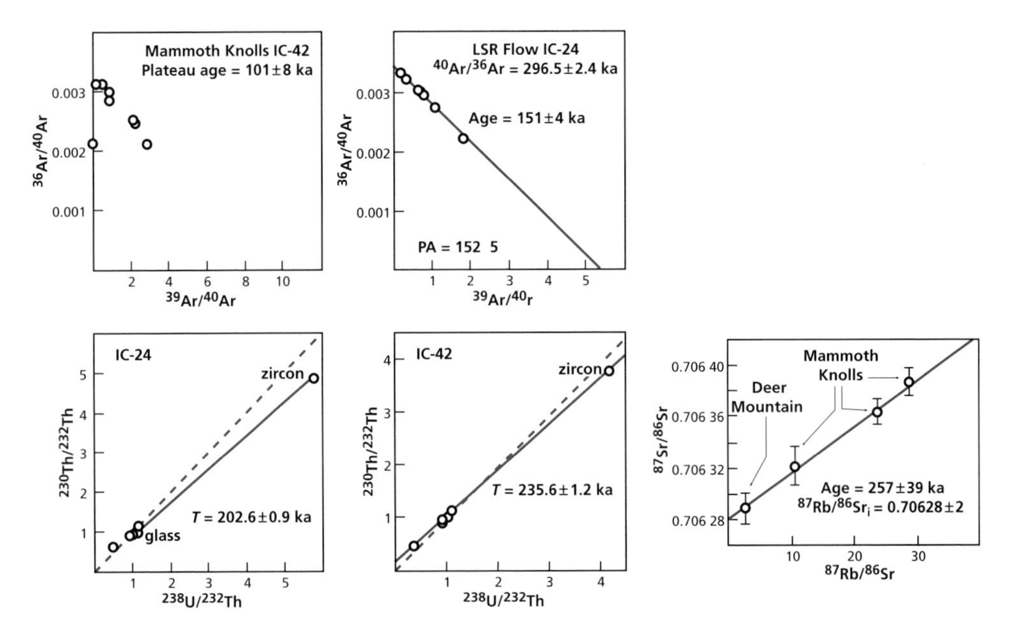

Figure 5.21 Results obtained by various methods for the rhyolite of Long Valley, California. After Reid *et al.* (1997).

The ^{87}Rb–^{87}Sr age is thought to mark the intrusion of the magma beneath the volcano and the creation of a magma chamber. The ^{230}Th–^{238}U ages supposedly date the crystallization of zircon in the same magma chamber. The ^{39}Ar–^{40}Ar ages date the eruptions. This example raises not the question of **uncertainty** but of **identification**. What age is the radiometric phenomenon associated with? What exactly is it that we date?

In the previous examples we barely posed the question just raised for Long Valley (except for the Eldora gneiss). For the komatiites the answer is clear enough: we want to date the magma intrusion. For Greenland, it is a little fuzzier. Is it the emplacement of granite before metamorphism? Is it the metamorphic episode? It is not absolutely clear.

Here, with the Long Valley rhyolite, there is no getting round the question. The ages can only be interpreted in the context of a **geological scenario**, a general model. At the same time, it can be seen that the different chronometers are not used for dating the same phenomena. The very idea of concordance of ages obtained by the various methods, so far considered as the criterion of absolute reliability, must be refined. Perhaps the uncertainty that can be estimated by the various methods reflects merely a problem of identification. Let us try to extend this type of interpretation.

Thus, when we observe that ^{87}Rb–^{87}Sr ages on whole rock measured in granite are generally a little younger than the ages obtained on zircon by the concordia method, we question what this difference means. It may arise because zircon is the first mineral to crystallize whereas ^{87}Rb–^{87}Sr is re-homogenized by the circulation of hydrothermal fluid which is driven off towards the end of granite crystallization and gives rise to the aplite and pegmatite seams. This is not the only interpretation. We might also consider that the zircon is aged because it has inherited older zircon. Ion microprobe examination of zircon minerals has now provided an answer but before we were in a position of indeterminacy!

Conclusions

There are two conclusions to be drawn from these studies (which we could have multiplied indefinitely). First, it has been seen that the quantitative methods of interpreting isotope measurements actually gave very coherent ages (despite the slight uncertainties with decay constants). Second, we have just observed that better **understanding** of these measurements **requires them to be included in geological schemas of interpretation**. The geological interpretation benefits on the one hand from the results of radiometric dating but in exchange it alone can make the figures yielded by the various methods fully meaningful.

Let us repeat, **there is nothing automatic about interpreting radiometric ages. Age is meaningful only in a given geological context. In addition, in return, inclusion in a geological context enlightens the geochemical behavior of the various chronometers. Each radiochronometer may bear original information, but for it to be credible it must be matched against other sources of information**. It must be consistent with the other things we know, with the usual behavior of the chronometer in question compared with other chronometers, and with the specific geological history under study. **The scientific approach is more like detective work than the application of automatic programming**.

5.5 The geological timescale

5.5.1 History

The making of the geological timescale, which is the greatest revolution geology has seen, paradoxically has a long history. It took longer for it to become accepted than for its main lines to be established. As stated, the idea of using radioactive decay to measure geological ages seems to have occurred independently but at about the same time to **Pierre Curie** in France (see Barbo, 1999) and to **Ernest Rutherford** in Canada (see Eve, 1939) in the early twentieth century.

It was **Rutherford** who first attempted to translate this insight into an actual measurement. He had just identified α particles as helium atoms. His colleague **Ramsay** had just shown how the proportions and abundances of the rare gases could be measured. Rutherford suggested to him measuring the amount of helium contained in a uranium ore. He did so. Rutherford found an age of more than 1 billion years. This was in 1906, about a century ago. Rutherford's discovery was a bombshell: geological terrains were more than 1 billion years old! The second stage took place in 1907 at Yale University, where Rutherford went to deliver the prestigious Silliman Lecture. Now, at Yale there was a chemist named **Boltwood**, who was interested in chemical transformations related to radioactivity. He suspected that the end product of radioactive chains, which he was analyzing by chemical methods, was lead. Rutherford suggested he should test out his idea geologically by taking what geologists reported to be young and old uranium ores and analyzing their lead content. The older ones should contain more lead than the younger ones. Boltwood (1907) made the corresponding series of measurements and confirmed there was a connection between lead and uranium chains. In turn, he then decided to use the method to date the uranium ores by analyzing the uranium and lead content of each, and he calculated the ages by applying the exponential law for decay. These were purely chemical measurements, not isotope measurements.

After Boltwood's discovery, inspired by Rutherford, the chemical dating of lead was improved by the discovery that thorium too was the parent of a radioactive chain ending in lead. The dating formula was modified slightly to read:

$$\text{Age} = \frac{(\text{Pb})_{\text{total}}}{(\text{U}) + 0.38\ (\text{Th})} \times 7400\ \text{Ma}$$

where (Pb), (U), and (Th) are the lead, uranium, and thorium contents in grams, respectively.

However, the chemical U–Pb method soon ran into what seemed at the time an insoluble technical problem. Uranium ores could be dated, but what of ordinary rocks? Chemical methods of the day were not sensitive enough and could not be used to measure either uranium or lead contents in ordinary rocks. These contents are measured in parts per million and at the time such elements could be measured only if their abundances were of the order of 1%.

That is why in 1908 **R. J. Strutt**, who later became the famous **Lord Rayleigh**, undertook to measure the age of ordinary rocks by using the helium method (Strutt, 1908). To do this, he measured the radiation produced by the rock, which gave him the instantaneous quantity of ^{4}He produced. He next measured the quantity of ^{4}He contained in the rock. It was then possible to make a simple age calculation. Although Strutt was aware from the outset that helium diffuses readily from minerals and rocks and so affects the age measured, the method developed quickly in various parts of the world: in Britain with Strutt himself, in the United States with **Piggot**, and in Germany with **Paneth**, who applied it to meteorites above all. In this race to develop a reliable geochronological method, one of Strutt's students, the Scot **Arthur Holmes**, made a name for himself and was to play an essential part in geological age determinations for the next 40 years (see Holmes, 1946). He is one of the founders of isotope geology.

Although geochronological methods based on lead and helium are subject to limitations and uncertainties, geologists embarked upon the adventure of constructing an absolute geological timescale to attribute durations to the geological eras and stages of the stratigraphers. The first successfully to build a coherent whole was **Barrell** in 1917. As Holmes was to do later, Barrell used radioactive datings and stratigraphic and geological observations to correlate them. He judged some ages unrealistic, others acceptable, and by a series of approximations and trials and errors he came up with a genuine quantitative stratigraphic scale in millions of years. Table 5.6 shows the figures of his scale alongside the "modern" figures. With hindsight, Barrell's work was outstanding – and yet no one believed it! When I began studying geology in Paris in 1958–9 we were told there was an absolute stratigraphic scale but that it was highly approximate and largely wrong and so should not be learned. If truth will out, it sometimes takes quite a while to do so!

This absolute scale, and even more so the reasoning that underpinned it, that is, the age attributions to geological phenomena in millions or billions of years, and the awareness of

Table 5.6 The absolute scale of geological times in millions of years[a]

		Barrell (1917)	Holmes (1947)	Holmes (1960)	Lambert (1971)	Modern ages
Quaternary (Noozoic)	Pleistocene	1–1.5	1	1		4
Tertiary (Cenozoic)	Pliocene	7–9	12–15	11	7	12
	Miocene	19–23	26–32	25	26	26
	Oligocene	35–39	37–47	40	38	37
	Paleocene	55–65	58–68	70	65	65
	Eocene					
Secondary (Mesozoic)	Cretaceous	120–150	127–140	135	135	141
	Jurassic	155–195	152–167	180	200	195
	Triassic	190–240	182–196	225	240	235
Primary (Paleozoic)	Permian	215–280	203–220	270	280	280
	Carboniferous	300–370	255–275	350	370	345
	Devonian	350–420	313–318	400	415	395
	Silurian	390–460	350	440	445	435
	Ordovician	480–590	430	500	515	500
	Cambrian	550–700	510	600	590	570

[a] Single dates indicate the beginning of the period.
Source: After Hallam (1983), in which the references cited may be found.

the extraordinary **geological myopia** of classical geology based on stratigraphic stages, did not come to be truly accepted until the late 1960s.

This scale must be subdivided into two separate periods because the conditions for age determination are not the same (see Plate 6).

- The **Precambrian period** extends from the formation of the Earth to the appearance of the first "unquestioned" fossils 550 million years ago. This period began with the earliest rocks, the first terrestrial rocks of Greenland, which, as we have seen, are aged 3.82 Ga.
- The **"classical" period** is that of traditional geology where fossil-based geology applies. Within this period a special place must be made for the interval ranging from 0 to 200 Ma during which we have samples of a real oceanic crust overlain by sediments that are well identified paleontologically thanks to microfossils. The undertaking was to match the terrains identified on a **worldwide scale** and to classify them, through their fossils, in time intervals measured in millions of years. The first job, then, was to put numbers on the stratigraphic scale developed by stratigraphic paleontology which divides the periods for which there is fossil evidence into eras and stages. This scale was then extended to very ancient ages, and after that, its resolution was improved for recent times.

5.5.2 The absolute scale of fossiliferous (Phanerozoic) times

The great difficulty in this undertaking is that the usual methods for dating long durations such as $^{87}Rb-^{87}Sr$, $^{238}U-^{206}Pb$, $^{147}Sm-^{143}Ne$, $^{187}Re-^{187}Os$, or $^{40}K-^{40}Ar$ cannot readily be used for dating sedimentary rocks. Now, the stratigraphic scale is defined by **index sedimentary strata** containing fossil associations chosen as the boundary beds. The reasons it is difficult to date sedimentary strata are that either sedimentary rocks are made up of inherited material (sandstone and schists) of varied provenance and age and, in such cases, the criteria for initial isotopic homogeneity are not satisfied, or they are newly formed material (limestone), in which case the rock is sensitive to secondary phenomena (particularly water circulation) and are not closed systems. Despite painstaking studies, which have not been entirely unsuccessful (which are always cited, of course) no general method has emerged.

The absolute chronology of sedimentary strata is generally established indirectly by dating igneous rocks whose geometric and chronological relations with the strata are known. Interstratified volcanism is a particularly favorable case, of course. Another case directly derived from the process of oceanic expansion is that of the ancient ocean ridges, that is, sea-floor spreading where a still-fresh layer of basalt (that can be radiometrically and magnetically dated) is overlain by a layer of limestone that is very rich in microfossils. Ocean drilling has allowed such dating (although the basalt must not be overly affected by hydrothermal circulation or by exchange with sea water at low temperature). The most commonly used method for dating basalt is the ^{39}Ar–^{40}Ar method. But such dating is increasingly difficult as we go back in time because of weathering, since the older they are, the more the basalts have been altered by reaction with sea water, with which the sediment is impregnated.

In addition, direct methods yield dates for a few reference strata only. Other results must be interpolated from them. The principle behind the method is simple enough. Suppose that in what is considered a **continuous** sedimentary series we know from one of the earlier methods the ages t_1 and t_2 of two points in the series, separated by a thickness $D(X) = X_1 - X_2$. As a first approximation, all of the strata in the interval may be dated by interpolation.

The sedimentation rate of the series is

$$V = \left(\frac{X_2 - X_1}{t_1 - t_2}\right),$$

so the age t_3 of a point located at X_3 is

$$t_3 = t_1 - \left(\frac{X_3 - X_1}{V}\right).$$

To apply the method, first the series must be continuous, in other words, there must be no **sedimentary gap or hiatus**, which stratigraphers are experienced at detecting. Next, sedimentation must be of the same type if we are to assume the sedimentation rate has remained constant. All of this is done by comparing, correlating, obtaining new dates, by modifying old data, or confirming other data, little by little and case by case. As can be seen, it is a long, hard slog and there is scope for constant improvement. It is understandable, then, that the age boundaries between stages or eras are affected by uncertainty which varies with the date of publication.

The **stratigraphic scale** is shown in Plate 6. A few brief remarks are called for. The eras, which our forebears thought were about equal in length, are in fact highly unequal:

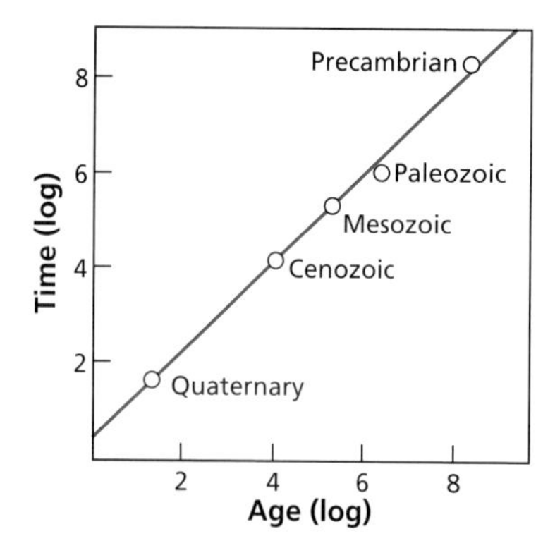

Figure 5.22 Relationship between duration and ages of geological eras. "Classical geologists" thought these durations were about equal. We are indeed dealing with a case of geological myopia.

Paleozoic (Primary) 340 Ma, Mesozoic (Secondary) 145 Ma, Cenozoic (Tertiary) 65 Ma, and Quaternary 5 Ma. If we plot the geological eras and the traditional subdivisions, it can be seen that the loss of information is about exponential with time (Figure 5.22).

Remark

With scientific advances, the names of the reference terrains do not change, but the ages of these terrains are constantly changing. Uncertainties on the limits are not symmetrical because geological constraints are not symmetrical either.

The case of the Quaternary

In recent periods, ranging from 1 to 5 Ma, which is the time in which humankind has appeared and developed, the situation is very different for two reasons. The first is that many dating methods can be applied directly to sediments. This is the case with ^{14}C, ^{10}Be, and methods based on radioactive disequilibrium. Direct dating can therefore be practiced. The second is that there are many indirect methods: fluctuations in $^{18}O/^{16}O$ isotope ratios (to which we shall return in Chapter 8), fission tracks, paleomagnetism, micropaleontology, etc. All of these methods can be used for intercalibration and refined interpolation.

In short, we can now draw up a very precise chronological scale for the Quaternary but one that is based on climatic phenomena. We shall speak of this in Chapter 6.

5.5.3 The Precambrian timescale

The Cambrian is the age at which the first common fossil faunas appeared. The two types of characteristic fossils are **trilobites** and **Archeocyatha**. Its lower boundary is dated 540^{+40}_{-10} Ma.

As Precambrian terrains cannot be subdivided on the basis of fossils, we work the other way around: we set numerical limits and then map the corresponding terrains at defined time intervals. The essential boundary has been set at 2.5 Ga. It divides the period before, known as the **Archean**, and the period ranging from 2.6 Ga to 0.4 Ga, known as the

Proterozoic because we suppose that it was during this period that life developed. The question, of course, is how to determine the lower limit to this scale.

We assume that the continental crust as it is today has not existed since the first age of the Earth but that it appeared after a time T_0 or that it appeared at an early date but was destroyed by multiple phenomena (meteorite bombardment?) and subsisted from T_0 only. This age T_0 is therefore the upper limit of the Archean. Between the formation of the Earth and T_0 we have defined a period for which there is little and often only "indirect" evidence (meaning there are no whole rocks) and that is known as the **Hadean** (see Chapter 7).

But what is the value of T_0? How old are the oldest terrains, in the sense of uninterrupted geological formations that can be mapped? The Isua and Amitsoq formations of Greenland already mentioned and dated 3.77 Ga and 3.65 Ga, respectively, are the oldest terrestrial rocks known as yet. But sedimentary rocks in Australia have been found to contain zircon whose U–Pb ages concord at 4.3 Ga (and a few even at 4.37 Ga). These are indisputably the oldest terrestrial minerals. But what is the meaning of that? Zircon is a typical mineral of granitic rocks, and so the remains seem to testify that there were pieces of continent at that time. Were those continents preserved, buried somewhere, or were they destroyed by erosion, meteorite bombardment, and Hadean tectonism? This is a question geochronology raises. We therefore provisionally fix the **Hadean–Archean boundary** at 3.8 Ga (see Plate 6 for the azoic (Precambrian) timescale). Here we have an uncertainty due to lack of information. We must therefore be prepared to change these figures as new discoveries are made.

5.6 The age of the Earth

The age of the Earth was touched upon in Chapter 2 and could in itself be the subject of an entire book, such is its historical and philosophical significance. We shall give a brief summary of it, emphasizing the approaches taken by those who attempted to answer the question.

5.6.1 History

The Bible indicates that the Earth is 4000 years old. Around 1650 Archbishop Ussher established by bibliographical studies that the Earth had been created on 26 October in the year 4004 BC at 9.00 in the morning!

By contrast with this, in the late seventeenth century, the founder of geology, Scotsman **James Hutton**, had dismissed the very idea of the Earth having an age we can estimate. The Earth had always been here and was the seat of the same phenomena described by geological cycles and would be until the end of time. "No vestige of a beginning – no prospect of an end," as he put it.

In the mid nineteenth century, while classical geology was thriving, the British physicist **William Thomson**, later to become **Lord Kelvin**, began to criticize the idea of an eternal Earth and repetitive geology. His idea was simple. Geological processes consume energy, if only to make reliefs, power volcanoes, or generate earthquakes. Now, the energy the Earth contains is not inexhaustible. He identified the source of terrestrial energy as being of thermal origin. The inside of the Earth is hot and the Earth is cooling because it loses heat from its surface by radiation. He thus came up with the figure of 100 Ma, confirmed in two ways. First, he had used the same theory to calculate an age for cooling of the Sun (this time by calculating the energy dissipated by radiation) of 100 Ma. Second, the Irishman **Joly** had also calculated the age of the ocean and found 100 Ma. The age of 100 Ma was a must!

However, geologists, and foremost among them **Charles Lyell** and **Charles Darwin**, rejected this age which they thought too young.[8] In 1910 Rutherford and then Boltwood determined the first radiometric ages exceeding 1 billion years. **Pierre Curie** and **Laborde** showed that radioactivity gave off heat, and so there was an energy source inside the Earth, which implied Lord Kelvin's calculation was false. And yet it was only after the Second World War that Kelvin's figure was seriously revised (see Burchfield, 1975; Dalrymple, 1991).

EXAMPLE

Kelvin's problem

Kelvin considered the cooling of the Earth, ignoring its roundness. He assumed the surface was plane and supposed it was maintained at constant temperature as it lost its heat by radiation into space.

For a semi-infinite solid separated by a plane surface, Fourier's theory states that if we take depth (z) with depth at the surface $z=0$ and assume the solid is at a temperature T_0 at the initial time and we allow it to cool by conduction, the temperature $T(z)$ is a function of T_0 and of $z/2(kt)^{1/2}$, where t is time and k is thermal conductivity. The thermal gradient (or geothermal gradient as we say for the Earth) is written:

$$\frac{T}{z} = \frac{T_0}{(kt)^{1/2}} \exp\left(\frac{-z^2}{4kt}\right).$$

Near the surface $z=0$ and the exponential equals 1. The geothermal gradient G is:

$$G = T_0(\pi kt)^{-1/2}.$$

If we take $T_0 = 7000\,^\circ$F, then $G = 1/1480$. With $k = 0.0118$ we get $t = 0.93$ Ma.

EXAMPLE

The age of the oceans

The Irishman Joly (who was one of the first to use radioactivity in geology) decided to calculate the age of the ocean (but attributed to it the same age as the Earth).

His reasoning ran like this: in the beginning the sea was fresh water. It became salty through the input of rivers and evaporation of water. So if we divide the mass of sodium in the ocean by the mass added annually by rivers, we will get the age of the ocean. To do this,

[8] Kelvin changed his calculation somewhat and at different times his age for the Earth varied between 25 Ma and 400 Ma, but that did not change the order of magnitude at stake!

he used the values of his day. That of sodium in rivers was $2 \cdot 10^{-4}$ moles kg^{-1} while the sodium content of sea water was 0.5 moles kg^{-1}. And he asked: how long has it taken to make the sea salty? Given that the mass of the ocean is $1.5 \cdot 10^{24}$ g and the mass of inflow from rivers is $4 \cdot 10^{19}$ g, he found an age of

$$\frac{0.50}{2 \cdot 10^{-4}} \times \frac{\text{mass of ocean}}{\text{mass of rivers}} = 0.93 \cdot 10^{7}\text{years.}$$[9]

5.6.2 The isotopic approach

In 1938 **Alfred Nier**, then doing postdoctoral work at Harvard University, measured the isotopic composition of lead in galena (PbS) from various geological settings and to which various "geological" ages had been attributed. Nier undertook to calculate the ^{207}Pb–^{206}Pb ages of the galena and found ages of 2.5 billion years. Now, at that time the astronomer **Edwin Hubble**, working at the California Institute of Technology, had used the gradual recession of galaxies to determine the age of the Universe and had come up with a figure of 2 billion years!

Rocks that were older than the Universe! Wasn't that impossible? Three scientists set about using Nier's results to recalculate the age of the Earth: **Gerling** (1942), **Holmes** (1946), and **Houtermans** (1946). Their common starting point was Nier's isotope analyses on galena in 1938 at Harvard. Some of the samples were from geological formations of known age. As galena does not contain uranium its isotopic composition is frozen from the time it becomes isolated. If we write the isotopic compositions $\alpha = {}^{206}Pb/{}^{204}Pb$ and $\beta = {}^{207}Pb/{}^{204}Pb$ and assume that the isotopic composition of the galena was "constituted" in a closed system (mantle or crust) which has been isolated since the origins of the Earth, we can write:

$$\alpha = \alpha_0 + \frac{^{238}U}{^{204}Pb}\left(e^{\lambda T_0} - e^{\lambda T_1}\right)$$

$$\beta = \beta_0 + \frac{^{235}U}{^{204}Pb}\left(e^{\lambda' T_0} - e^{\lambda' T_1}\right)$$

where T_0 is the age of the Earth, T_1 the age of the galena, α_0 and β_0 the initial isotopic compositions of the Earth, and where λ corresponds to ^{238}U and λ' to ^{235}U. As we have already done, we can rewrite these equations as:

$$\frac{\alpha - \alpha_0}{\beta - \beta_0} = 137.8\left(\frac{e^{\lambda T_0} - e^{\lambda T_1}}{e^{\lambda' T_0} - e^{\lambda' T_1}}\right).$$

Unfortunately, to calculate the age of the Earth T_0 in this equation, two unknowns are missing, the initial isotopic compositions of terrestrial lead (α_0 and β_0).

Gerling, at the Institute of Precambrian Geology at Leningrad (now St. Petersburg), simplified the problem and adopted a new assumption (Gerling, 1942) (Figure 5.23). He chose

[9] This age is now referred to as the residence time of sodium in the oceans. See Chapter 8.

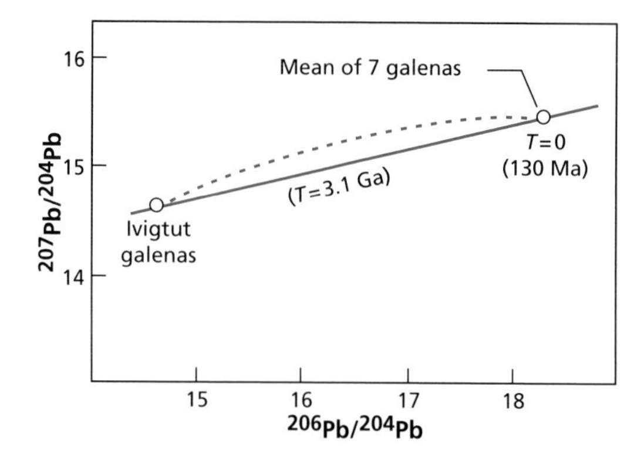

Figure 5.23 Principle of Gerling's method of calculating the age of the Earth.

the galena from Ivigtut in Greenland from among those Nier had measured, whose isotopic ratios were the lowest, and assumed those ratios corresponded almost to α_0 and β_0. To remove the complication of the geological age of the galena, he considered recent galena whose age he assumed to be 0. The equation became:

$$\frac{\alpha - \alpha_{\text{Ivig}}}{\beta - \beta_{\text{Ivig}}} = 137.8\left(\frac{e^{\lambda_8 T_0} - 1}{e^{\lambda_5 T_0} - 1}\right).$$

He thus found an age for the Earth of 3.1 Ga.

Exercise

We take two samples of galena of recent ages, one from Joplin (United States) and the other from Clausthal (Austria).

For the first, $^{206}Pb/^{204}Pb = 22.28$ and $^{207}Pb/^{204}Pb = 16.10$. For the second, $^{206}Pb/^{204}Pb = 18.46$ and $^{207}Pb/^{204}Pb = 15.6$. (For the Ivigtut galena $^{206}Pb/^{204}Pb = 14.54$ and $^{207}Pb/^{204}Pb = 14.6$.)

(1) Determine the age of the Earth by Gerling's method.
(2) What would the ages be if we assumed $\alpha_0 = 0$ and $\beta_0 = 0$?

Answer

(1) $T_{\text{Joplin}} = 2.83$ Ga and $T_{\text{Clausthal}} = 3.2$ Ga.
(2) $T_{\text{Joplin}} = 4.73$ Ga and $T_{\text{Clausthal}} = 5$ Ga.

Holmes used a more sophisticated method (Figure 5.24) but still based on the same principle of a closed system for isotopic evolution as provided by the galena measured by Nier. We take two galena specimens of "known" ages T_1 and T_2. We have two equations. For galena 1:

$$\frac{\alpha_1 - \alpha_0}{\beta_1 - \beta_0} = \frac{1}{137.8}\left(\frac{e^{\lambda T_0} - e^{\lambda T_1}}{e^{\lambda' T_0} - e^{\lambda' T_1}}\right)$$

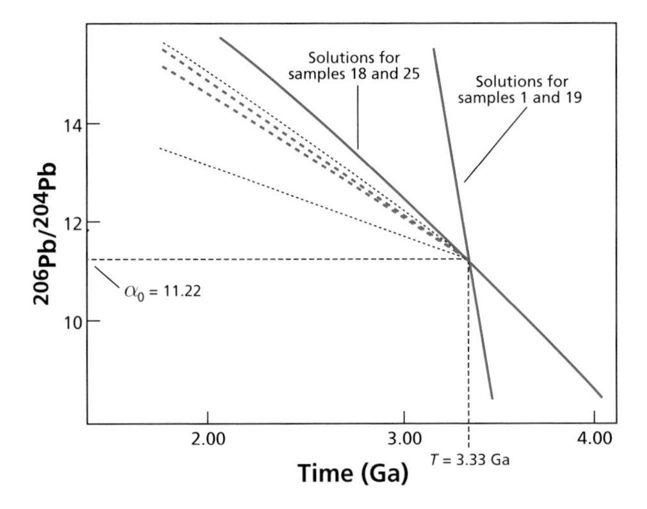

Figure 5.24 Measurements giving the age of galena by Holmes's method.

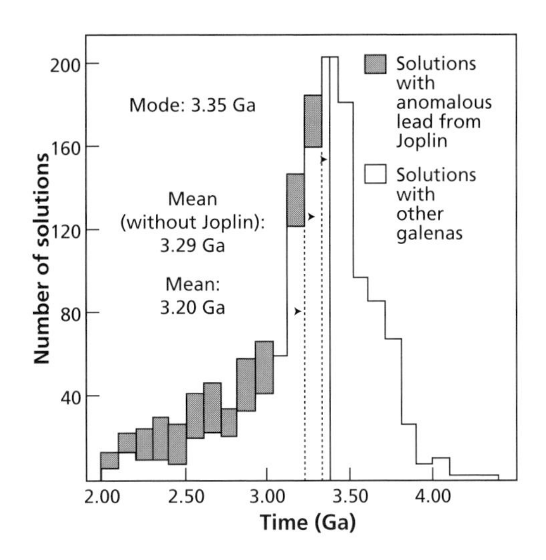

Figure 5.25 Age statistics determined by Holmes with his two-by-two galena method.

and for galena 2:

$$\frac{\alpha_2 - \alpha_0}{\beta_2 - \beta_0} = \frac{1}{137.8}\left(\frac{e^{\lambda T_0} - e^{\lambda T_2}}{e^{\lambda' T_0} - e^{\lambda' T_2}}\right),$$

and three unknown quantities α_0, β_0, and T_0.

Holmes (1946) then hit upon the idea of a sort of statistical method and reasoned like this: suppose I give myself ages of the Earth $T_0 = 2$ Ga, 3 Ga, and 4 Ga. If we replace T_0 by these values in the equations, we can calculate the two unknowns α_0 and β_0 on a case-by-case basis (Figure 5.25). So we can plot an (α_0, T_0) diagram. This operation may be repeated with

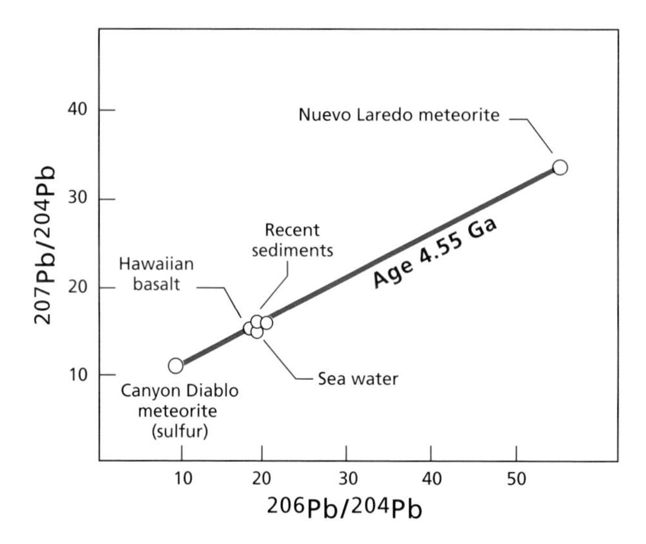

Figure 5.26 The age of the Earth as determined by Clair Patterson, 1953–6.

a second pair of galena specimens, giving us a second curve, with a third pair, and so on. We just have to take the mean of the intersections between the curves to get the real pair (α_0, T_0) and T_0. Holmes came up with 3.4 Ga, an astonishing convergence with the date calculated by Gerling. **Houtermans** (1946) adopted much the same method, leading him to a result close to 3 Ga.

In the 1950s, then, we had an estimate for the age of the Earth of about 3.5 Ga. (Meanwhile astronomers had recalculated Hubble's equation and found an age of 4 Ga for the Universe.)

In 1950, advances in mass spectrometry meant the isotopic composition of lead of ordinary rocks could be measured although this element is only found in traces. In his Ph.D. thesis at Chicago University, **Clair Patterson** (1953) first measured two meteorites, an iron meteorite (Canyon Diablo) and a basaltic meteorite (Nuevo Laredo) by the ^{206}Pb–^{207}Pb method (Figure 5.26). By joining up the two points in the ($^{206}Pb/^{204}Pb$, $^{207}Pb/^{204}Pb$) diagram, he calculated an age of 4.55 Ga, which he claimed to be the age of planetary objects (Patterson, 1953, 1956).

He then measured a present-day volcanic rock from Hawaii, which supposedly represented what was assumed to be the homogeneous interior of the Earth, fine sediments from the Pacific, and a manganese nodule supposed to represent averages of rocks from the Earth's surface. These three rocks lay roughly on the straight line at 4.55 Ga. He attributed to the Earth the same age as the meteorites and asserted this was the age at which the Solar System formed. This is the age which now features in all the textbooks. It also determines the initial values α_0 and β_0.

These values have since been determined more precisely as $(^{206}Pb/^{204}Pb)_0 = 9.307$, $(^{207}Pb/^{204}Pb)_0 = 10.294$, and $(^{208}Pb/^{204}Pb)_0 = 29.476$.

Apart from the value of the age of the Earth and the benchmark it provides for the history of geological times, Patterson's work has an important consequence. It can be used to "situate" the isotopic compositions of galena measured in geological formations relative to the

Table 5.7 Samples of galena used by Nier in two-by-two determinations

	$^{206}Pb/^{204}Pb$	$^{207}Pb/^{204}Pb$	Age (Ma)	Number[a]
Casapalca (Peru)	18.83	15.6	25	1
Ivigtut (Greenland)	14.54	14.70	600	19
Yancey (North Carolina)	18.43	15.61	220	18
Great Bear Lake (Canada)	15.93	15.30	1330	25

[a] See Figure 5.24.

initial reference point (α_0, β_0) and by the same token to allow a series of calculations for the age of the Earth from each sample of galena. We shall see what this work leads to in Chapter 6.

Exercise

Calculate the age of the Earth from Casapalca and Yancey galena samples given by Nier in Figure 5.24 and Table 5.7, assuming α_0 and β_0 to be the values known today and assuming their age is about zero.

Answer

$T_{Casapalca} = 4.39$ Ga and $T_{Yancey} = 4.45$ Ga.

5.6.3 The modern approach

A first, probably somewhat disarming, observation is that with modern analytical methods (see **Allègre** *et al.*, 1995) we cannot repeat Patterson's measurements and find the same result! The points that are supposed to represent the Earth do not lie on the straight line for meteorites but form a cloud. Moreover, the decay constants for uranium have been changed since, and so we would no longer find 4.55 Ga with modern values. And yet, Patterson's result is about right (**Tatsumoto** *et al.*, 1973)! Unfortunately for him, the very concept of the age of the Earth has changed and has become much less well defined than in his day.

First, it has been realized that the Earth's history has been very complicated and that the models of evolution of lead involving just a single episode in a closed system do not apply to any terrestrial reservoir. All of the systems – crust–mantle–ocean – are open systems that exchange material with each other: there is absolutely no ground for applying the closed-box model to reservoirs like the mantle or continental crust as Patterson did, even as a first approximation (as we shall see in Chapter 6). Then, with the advances in planetary exploration and planetological models, it was realized that the Earth had not been formed in an instant but by a slow process (named **accretion**) which probably lasted 100–150 Ma. We come back to the question we raised at the beginning of this chapter: for dating we need a box that closes at a definite given time. If we date the Earth by any particular method, what is the meaning of that age? Is it the time the core formed, or the end of degassing of the atmosphere? Is it the age of the solid material that is its main component?

We shall return to this problem in Chapter 6. Let us be content to say that the Earth formed over a period of 100 to 120 Ma and that the process began 4.567 Ga ago! The concept of

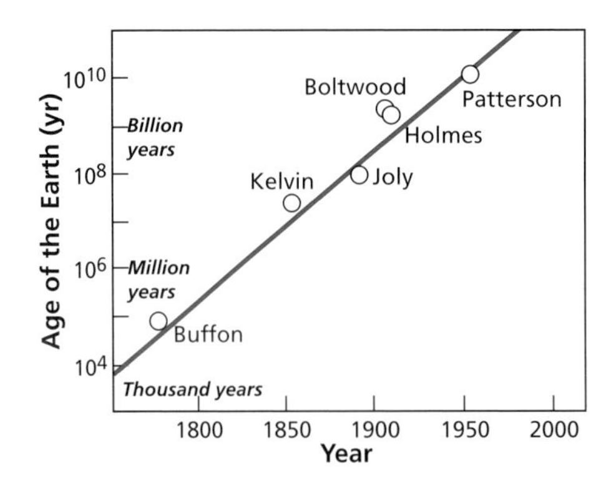

Figure 5.27 Changes in the age attributed to the Earth as geological sciences have developed.

dating can be applied in several ways. Either we date the beginning and end of accretion of the Earth, or we determine an "average age" of formation of the Earth with no precise meaning. We can see that analytical uncertainty is no longer an issue, but that it is "geological uncertainty" or even conceptual uncertainty that challenges the validity of the reference model outside of which any age is meaningless. This, then, is a good example of the different meanings we can attribute to uncertainty of a geological age measurement! For fun, we can end by noticing that if we represent the age of the earth in semi-logarithmic coordinates against the date of research into the subject (Figure 5.27), we obtain a straight line: this is food for thought about the meaning of correlations with no theoretical support. (Perhaps a sociologist or a psychologist could draw conclusions from this?)

5.7 The cosmic timescale

One of the great success stories of radiometric dating is that it has been able to extend its investigative power beyond the terrestrial domain and fix chronological markers for cosmic evolution.

5.7.1 Meteorite ages

Meteorites are pieces of embryonic planets dating from early times, which were sent hurtling off into space after collisions and spent millions of years in interplanetary space before falling to Earth.

We have said that, as a first approximation, meteorites are as old as the formation of the Solar System and the planets (around 4.55 Ga), but we have just seen that this age is poorly defined.

Modern U–Pb methods applied to uranium-rich phases developed in Paris by **Gérard Manhès**, **Christa Göpel**, and the present author have yielded meteorite ages with a precision of about 300 000 years, while the age is close to 4.55 Ga, giving a relative uncertainty of $3 \cdot 10^5 / 4.5 \cdot 10^9$ (60 ppm), which corresponds to a precision of 60 years for a date of 1 million years! Using these U–Pb methods as our reference, we have been able to establish a precise chronology (Göpel *et al.*, 1994) (Figure 5.28).

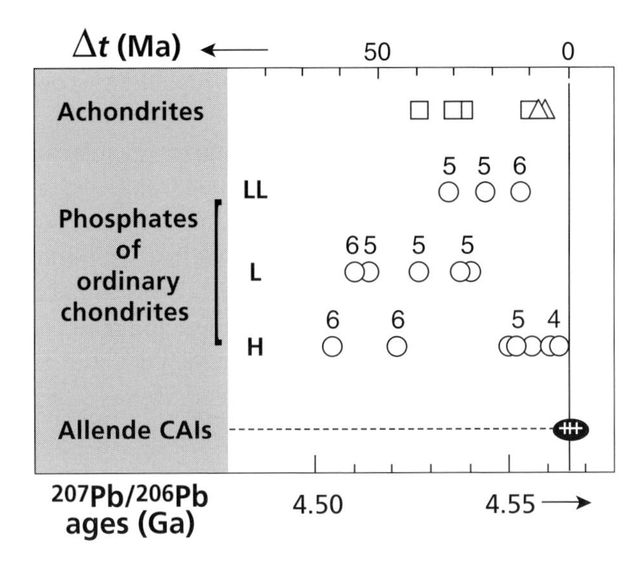

Figure 5.28 The chronological scale of meteorites after Göpel *et al.* (1994) and Allègre *et al.* (1995). CAIs, calcium–aluminum inclusions.

The results obtained with the other chronometers based on ^{87}Rb–^{87}Sr, ^{146}Sm–^{143}Nd, and ^{187}Re–^{187}Os long-period radioactivity are in general agreement with U–Pb ages but they are far less accurate. Results obtained using extinct radioactivity (^{129}I–^{129}Xe, ^{26}Al–^{26}Mg, ^{182}Hf–^{182}W, ^{53}Mn–^{53}Cr) all indicate ages close to those of the reference meteorite, confirming the great age of all these objects. However, whenever we try to use the extraordinary temporal power of resolution of these extinct forms of radioactivity, their relative ages and ages relative to the U–Pb clock, substantial inconsistency is observed. The age difference between two objects (meteorites, minerals) is positive with U–Pb, negative with Mn–Cr and Al–Mg, and positive but different with I–Xe, or the contrary. Between two other objects, the differences obey some other rationale, and so on. This raises two questions. Where does this anarchy come from? What do the "apparent ages" obtained mean?

Observation of meteorite structures and textures using the techniques of petrology has revealed that most meteorites are brecciated and impacted. Chondrites, which are the most "primitive" meteorites, are composed of spherical chondrules, evidence of transition through a molten state, surrounded by a matrix containing many low-temperature minerals. The overall impression is one of very complex relations among minerals.

Dating using ^{40}Ar–^{39}Ar methods by **Grenville Turner** since 1968 has revealed highly complicated outgassing patterns and ages of generally less than 4.56 Ga. ^{87}Rb–^{87}Sr ages obtained on the minerals also show the system did not remain closed. When alignments can be obtained from the isochron diagram they yield apparent ages of less than 4.56 Ga.

All this evidence suggests that meteorites were subjected to many complex disruptive phenomena (metamorphism, impacts, reactions with fluids, etc.) during the evolution of the solar nebula. It was long thought that differentiated meteorites (basaltic or iron meteorites) had formed much later. In fact, the use of the ^{182}Hf–^{182}W method has shown that their differentiation, that is their fusion in small bodies, was also very old, dating from 3 to 4 million years after the Allende meteorite (Lee and Halliday, 2000). It must therefore be concluded that most of the ages obtained on meteorites are disrupted by secondary phenomena that occurred right at the beginning of the Solar System.

In the current state of knowledge, we can say that the oldest objects are the calcium- and aluminum-rich, white inclusions in Allende. They are aged 4.567 $\pm$ 0.0003 Ga (Allègre *et al.*, 1995). Chondrites and differentiated iron meteorites formed in the 2 million years (probably for less) that followed. Basaltic achondrites formed in the first 4 (maybe first 2) million years. These objects were then subjected over 50 to 100 million years to impacts, heating, and chemical reactions with fluids. All these phenomena brought about chemical and mineralogical changes in these objects making the primitiveness of their dating extremely doubtful, not just in terms of the figures obtained but also in terms of what the date means. Here is an example of uncertainty that is not analytical and is mathematically difficult to quantify.

EXAMPLE

Meteorites

Meteorites are **rocks that fall from the sky** (see **Wood**, 1968; **Wasson**, 1984; **Hutchinson**, 2004). Although they have been known since ancient times, their scientific value came to be understood less than a century ago. They are pieces of small planetary bodies gravitating for the most part in the asteroid belt between Mars and Jupiter. They are produced by impacts between these bodies and then move through interplanetary space before falling to Earth.

During their travels, they are irradiated by galactic cosmic rays, as we have seen. This means they can be identified positively because they contain radioactive isotopes that are specific to such irradiation. They are of huge interest because the material dates from the time the Solar System and the planets formed. All meteorites are aged about 4.56 Ga. The chemical composition of some of them, the carbonaceous chondrites, is similar to that of the solar photosphere.

Meteorites are classified into two main categories: chondrites and differentiated meteorites (Table 5.8).

Table 5.8 Summary classification of meteorites

Chondrites They are formed by the agglomeration of small spheres of silicates (chondrules) in a material composed of small minerals	• Carbonaceous	They contain no metal. Their chemical composition is similar to that of the solar photosphere.
	• Ordinary	These are the most numerous. They are formed from a mixture of metallic iron minerals and silicate minerals. Their composition is close to a mixture of peridotites and of metallic iron.
	• Enstatite	All the iron is in the form of FeS or metallic Fe.
Differentiated meteorites	• Basaltic achondrites (eucrites)	These are basalts analogous to terrestrial basalts.
	• Mesosiderite and pallasite (iron and rock)	Large pieces of metallic iron associated with pieces of silicates.
	• Iron meteorites	Alloy of metallic iron and nickel.

The **chondrites** are the more common. They contain small spheres known as chondrules and their chemical composition is roughly similar to that of the whole Earth. They contain silicates but also scattered metallic iron minerals. The distribution of iron forms a subdivision into **enstatite chondrites, ordinary chondrites**, and **carbonaceous chondrites**. In **enstatite chondrites** (E) all of the iron is in the form of metal or iron sulfide FeS. In **ordinary chondrites**, the iron is divided between metal sulfide and silicates. Some have high (H) and others low (L) iron contents. **Carbonaceous chondrites** (C) have no metallic or sulfidic iron and all their iron is in the silicates and oxides.

A further classification is superimposed on the mineralogical–chemical one, based on the extent of metamorphism of the chondrites. Chondrites were subjected to reheating after the formation of the bodies from which they derived. This heating wrought more or less intense mineralogical changes in all of the chondrites, except for the carbonaceous ones, thus causing varying degrees of metamorphism. This metamorphism is rated on a scale from 1 to 6. So we speak of H(3), L(5), or E(3) chondrites, etc.

Differentiated meteorites are the relicts of liquid–solid separation phenomena that occurred within small planetary bodies upon their formation. Some of them are basalts (eucrites), others are enormous blocks of metallic iron whose textures show they were initially molten (iron meteorites). Yet others are made up of an assemblage of massive metallic iron and of silicates such as olivine and pyroxene (pallasites and mesosiderites). These structures are indicative of the differentiation of the metal cores of the small protoplanets.

As can be seen, the distribution of iron between metal and silicate is fundamental in meteorite classification. This must be related to the case of the Earth, where iron is mostly present in the core in metallic form and less present in the mantle where it is scattered among the silicates olivine and pyroxene. Most meteorites come from the asteroid belt between Mars and Jupiter, but some have quite different origins. They are pieces of planets broken off by impacts. Some are known to come from the Moon and it is thought others come from Mars (Figure 5.29).

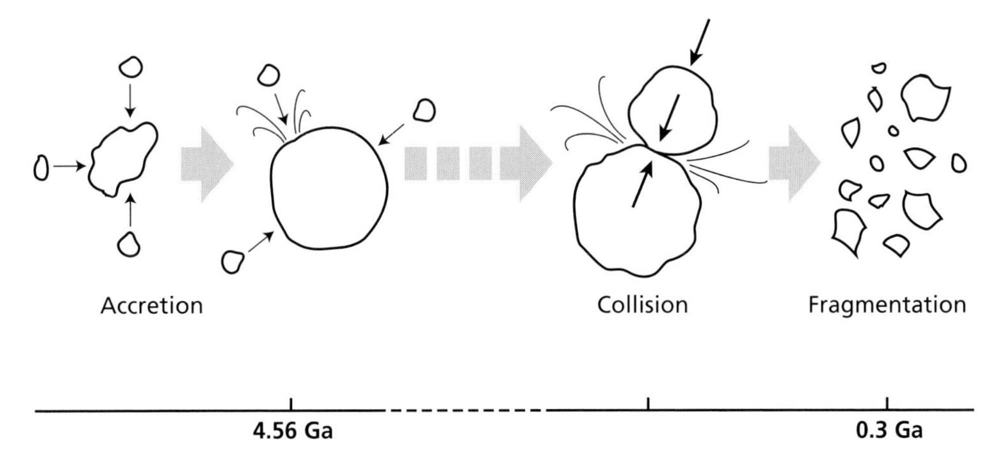

Figure 5.29 History of a meteorite.

Meteorites can be studied by using radiochronometers with long half-lives, by cosmogenic isotopes, and by extinct radioactivity. They are prize items for radiometric dating.

5.7.2 The history of the Moon

Moon rocks (see Taylor, 1982) brought back by the American Apollo and Soviet Luna missions have also been dated by various geochronological methods. Here very briefly are the main findings.

There is an ultrabasic lunar rock aged 4.50 Ga (the oldest evidence of the Moon) but that apart, the Moon is composed of two geological entities:

(1) the highlands where the dominant rock is anorthosite (calcium plagioclase) with an age range of 4.3–4.0 Ga;
(2) the maria, which are large depressions formed by gigantic impacts and are made up of basalts ranging in age from 3.7 Ga (Sea of Fecundity) to 3.2 Ga (Sea of Tranquillity).[10]

Since that date it seems there has been no internal geological activity on the Moon. The most important information derived from these studies is probably that about the number of **impact craters**, which varies with time, decreasing markedly between 4.3 and 3.2 Ga. As these craters result from the impact of celestial "thunderbolts," which after impact are captured and accreted by the Moon, it is inferred that the phenomenon of lunar accretion (and by extension accretion of the other planets) declined very rapidly in the early days of the Solar System (Figure 5.30). This provides an indirect indication about the process of

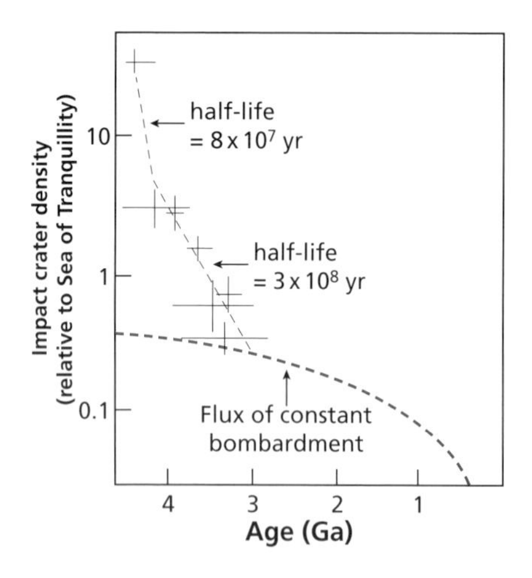

Figure 5.30 Density of impact craters on the Moon versus age as measured by the Apollo missions. The reference curve indicates what the density would be if the meteorite flux were constant. The half-lives measure the speed of decline of these meteorite straight lines.

[10] A lunar rock dated to 2.8 Ga was found in November 2004, but the context of its formation is not well understood.

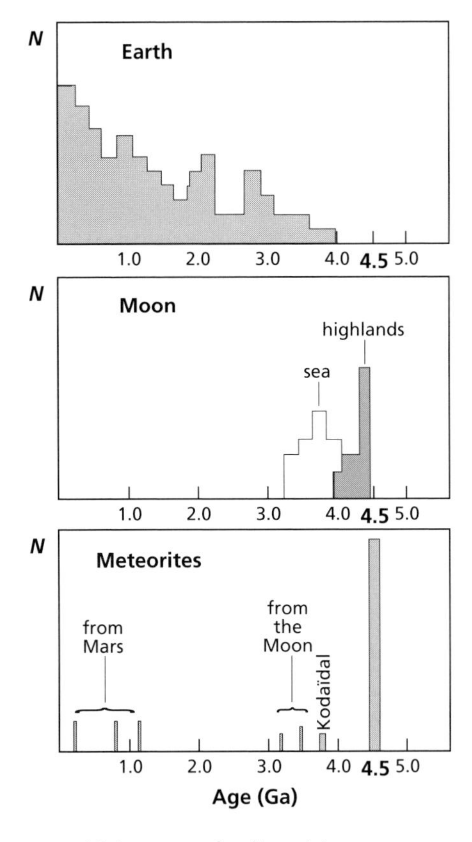

Figure 5.31 Histograms of radiometric ages measured on three types of planetary object: Earth, Moon, and meteorites. This illustrates how the dating of rock can be used for comparative planetology.

formation of the Earth by the successive, progressive build-up of ever larger planetary bodies.

To illustrate how the Moon is situated in the development of planetary objects, Figure 5.31 shows the abundance of rocks with their ages in three different planetary environments. No long speeches are need to explain the value of comparative planetology for constructing a history of planetary formation.

5.7.3 Pre-solar cosmochronology

The chemical elements were synthesized in the stars (see **Hohenberg**, 1969; **Wasserburg** *et al.*, 2006). This is the process of nucleosynthesis already discussed. The heaviest chemical elements form in massive stars only and in particular during spectacular explosions known as supernovae. The idea arose of dating the time of formation of the chemical elements by explosive nucleosynthesis. This dating system[11] is based on the abundance of heavy radioactive elements and even more so on the abundance ratio of two radioactive isotopes with different decay periods. Let us take the classic example of the

[11] Two important references are Hohenberg (1969) and Wasserburg *et al.* (2006). See also Clayton (1968).

isotopic ratio of ^{238}U and ^{235}U, both of which have long decay periods (one being much longer than the other, though). The history of these two isotopes can be broken down into two episodes:

(1) the time between the Big Bang and the formation of the Solar System during which these isotopes were synthesized in supernova type stars;
(2) the period since the formation of the Solar System during which these isotopes have been incorporated into rocky objects (meteorites or planets), after their nuclear synthesis of stellar origin had ended, and they have then decayed over these 4.55 Ga by natural radioactivity.

If we are to build a quantitative scenario of explosive nucleosynthesis, we must first go back 4.55 Ga, just before the Solar System formed, and calculate the $^{235}U/^{238}U$ isotope ratio at that time.

Exercise

Given that the $^{235}U/^{238}U$ ratio is now 1/137.8, calculate the ratio $4.55 \cdot 10^9$ years ago.

Answer

Uranium-255 decays by the law: $^{235}U = (^{235}U)_{4.55Ga}\, e^{-\lambda_5 t}$. Uranium-238 decays by the law: $^{238}U = (^{238}U)_{4.55Ga}\, e^{-\lambda_8 t}$. Therefore:

$$\left(\frac{^{235}U}{^{238}U}\right)_{present} = \left(\frac{^{235}U}{^{238}U}\right)_{4.55\,Ga} \frac{e^{-\lambda_5 t}}{e^{-\lambda_8 t}}$$

with $\lambda_5 = 0.984\,85 \cdot 10^9\,yr^{-1}$ and $\lambda_8 = 0.155\,129 \cdot 10^9\,yr^{-1}$. Therefore: $(^{235}U/^{238}U)_{4.55\,Ga} = 0.31$.

This is what we shall have to reproduce in the chronological scenario of nucleosynthesis. We shall examine two extreme scenarios, which will serve as references and on that basis we shall discuss the construction of a more plausible scenario.

The different scenarios of uranium nucleosynthesis[12]

In the **sudden nucleosynthesis model**, we assume that all the uranium in the Universe formed shortly after the Big Bang in a very short period of time when a considerable number of supernovae existed, synthesizing the heavy chemical elements in their explosions. In this scenario, the remainder of cosmic time after the Big Bang witnessed only the decay of uranium by radioactivity. We can write:

$$\left(\frac{^{235}U}{^{238}U}\right)_{4.55\,Ga} = \left(\frac{^{235}U}{^{238}U}\right) \frac{e^{-\lambda_5 \Delta t}}{e^{-\lambda_8 \Delta t}}$$

where Δt is the time between the Big Bang and the formation of the Solar System.

The question is, of course, how do we find the $^{235}U/^{238}U$ ratio at the time of nucleosynthesis, just after the supernova explosion? This production ratio is not well known.

[12] From D. D. Clayton (1968).

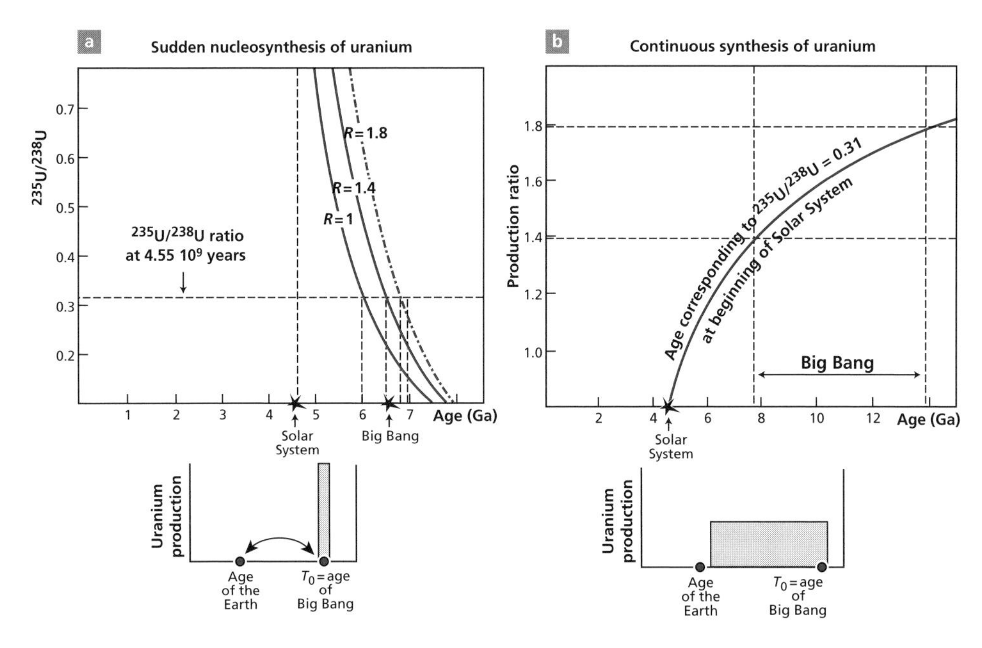

Figure 5.32 The two scenarios for uranium synthesis. (a) Total initial formation; (b) continuous uniform formation. *R*, ratio of ^{235}U versus ^{238}U production in nucleosynthetic processes.

Astrophysicists calculate it as $R = 1.4$. We shall take a figure between 1 and 2 for maximum uncertainty.

Calculations give the age of the Big Bang as 6.5 Ga, with boundaries of 6 Ga and 7.2 Ga (Figure 5.32). Comparison with the age determined by astronomers studying the recession of the galaxies seems to indicate that this age is too young (astronomers date the Big Bang to 10–15 Ga).

In the **continuous nucleosynthesis scenario**, we assume that uranium formed through stellar history at uniform rates of production. This scenario corresponds to continuous and constant stellar activity with the frequency of supernovae always identical. The equations for evolution of the two uranium isotopes can be written:

$$\frac{\mathrm{d}\,^{235}\mathrm{U}}{\mathrm{d}t} = A_{235} - \lambda_5\,^{235}\mathrm{U} \text{ and } \frac{\mathrm{d}\,^{238}\mathrm{U}}{\mathrm{d}t} = A_{238} - \lambda_8\,^{238}\mathrm{U}.$$

By accepting that there was no uranium at $t = 0$, integrating the system gives:

$$\frac{^{235}\mathrm{U}}{^{238}\mathrm{U}} = \frac{A_{235}}{A_{238}} \cdot \frac{\lambda_{238}}{\lambda_{235}} \left(\frac{1 - \mathrm{e}^{-\lambda_5 \Delta t}}{1 - \mathrm{e}^{-\lambda_8 \Delta t}} \right)$$

where A_{235} and A_{238} are the rates of production of ^{235}U and ^{238}U. Here again the big unknown quantity is the production ratio $R = A_{235} / A_{238}$. We shall take it as varying from 1 to 2 with a probable range of 1.4–1.8.

Given that $(^{235}U/^{238}U)_{4.55\mathrm{Ga}} = 0.31$, we can plot the curve of the age against R (Figure 5.32b). The age of the beginning of this process is taken as the age of the Big Bang and varies between 8 and 14 Ga. Notice that as $\lambda_{238}/\lambda_{235} = 0.157$, it would only need a production ratio R of 1.96 for the **age to be infinite**. We would then have a uniform Universe of infinite age, which shows the extreme sensitivity of the entire scenario to the uranium production ratio. This is a fine example of the relative reliability of radiochronometric methods. As with the age of the Earth, the age of the Universe is related to the definition of a scenario (here a cosmic scenario).

The contribution of extinct radioactivity

Let us consider the case of ^{129}I. We know we can write:

$$\left(\frac{^{129}I}{^{127}I}\right)_{\text{meteorites}} = \left(\frac{^{129}I}{^{127}I}\right)_{\text{initial}} e^{-\lambda t}.$$

If we know the isotopic ratio $^{129}I/^{127}I$ at the end of nucleosynthesis, we can calculate the time between nucleosynthesis and the formation of the meteorite. The various scenarios astrophysicists propose for nucleosynthesis indicate that $(^{129}I/^{127}I)_{\text{initial}} \cong 1$.

As the ratios measured for meteorites are about $10 \cdot 10^{-4}$, the age calculation $\Delta T = 1/\lambda \ln(10^4)$ gives 210 Ma. So there was a nucleosynthetic event less than 230 Ma before the formation of the Solar System during which ^{129}I was born. But the discovery of other extinct forms of radioactivity such as ^{26}Al and then ^{41}Ca and ^{36}Cl shows that nucleosynthetic events also occurred 5 Ma and 1 Ma before the formation of the Solar System for ^{26}Al and ^{41}Ca, respectively.[13] Thus all the evidence points to the occurrence of explosive nucleosynthetic events just before the Solar System formed. Some scientists claim these supernovae brought about the collapse of the proto-Sun's nebula by the mechanical action of the shock waves. Nucleosynthesis, that is, the synthesis of the chemical elements, is therefore a process that took place throughout cosmic time and is probably still going on in the stars.

Discussion and conclusions

It is immediately obvious that any combination of the two scenarios, that is, initial production followed by continuous production, will yield Big Bang ages of 6.5 to 14 Ga. Since astrophysicists think that there were supernova explosions well after the Big Bang and that extinct forms of radioactivity prove that they occurred also 4.5 Ga ago (or shortly before that), we must conclude that the sudden synthesis model of all uranium at the beginning of the Universe is unacceptable. On the other hand, all the astrophysical models seem to show that the Universe was a much more active place shortly after the Big Bang than it is today. It is therefore legitimate to give greater weight to the older epochs when it comes to nucleosynthesis (Figure 5.33).

All of that suggests a mixed model involving initial synthesis and continuous formation. If we consider that $R = 1.5$ is the most likely production ratio, the age of the beginning of the Universe can be estimated at 10 ± 2 Ga.

Naturally, one might ask why we do not try to confirm this calculation with another chronometer. In fact, we have used ^{232}Th and the $^{232}Th/^{235}U$ ratio. This ratio is chemically fractionated and we do not really know its value in the cosmos. However, taking a ratio of

[13] We have taken the criterion of 6 half-lives for the element to still be "alive."

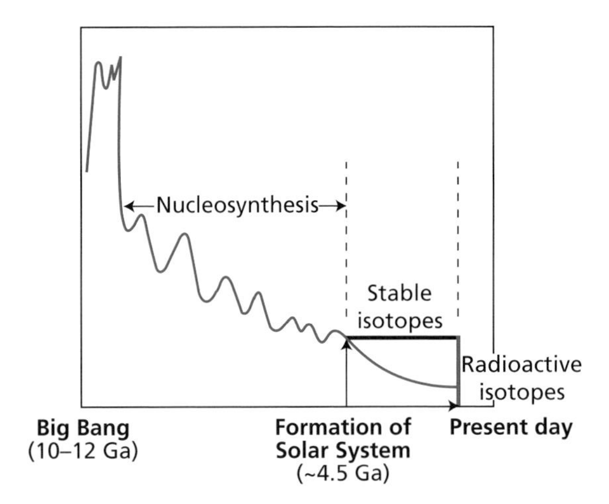

Figure 5.33 Nucleosynthesis of elements forming the Solar System before 4.5 Ga. Isotopes are synthesized (and destroyed) in stars. From 4.5 Ga, stable isotopes maintain the same abundance ratios and radioactive isotopes decay.

$^{232}Th/^{235}U = (Th/^{238}U)_{present} = 4$, we can repeat the calculations for the same scenarios and we obtain comparable results. Same thing if we use $^{187}Re-^{187}Os$ (Luck *et al.*, 1980).

By using the recession of the galaxies, astronomers date the Big Bang to 13–14 Ga. We can conclude that radiochronology "confirms" the age of the Big Bang as determined by astronomers and the validity of the scenario of continuous nucleosynthetic activity in the Universe. Here we have an example of very large uncertainties because of the concepts employed and which far outweigh the analytical uncertainties.

5.8 General remarks on geological and cosmic timescales

5.8.1 Overall geological myopia

As we have said, the deeper we go into the past, the more our absolute uncertainties increase. In recent times we can separate events 1000 years apart (and less) whereas this is impossible, of course, in the Precambrian. As spectroscopists would say, for recent periods we have high resolution and for earlier periods low resolution. Thus, we can make out clearly the alternating pattern of glacial and interglacial periods in the Quaternary. In the Secondary period we can no longer "see" these alternations clearly, and yet, if they are connected with Milankovitch cycles, as we shall see, they should also have occurred in the more remote past too. The closer we get to the present, the more clearly we can make out events separated by shorter time-spans. This is compounded by the fact that, in the course of their history, terrestrial rocks are subjected to the destructive events of erosion or metamorphism. The more time passes, the more destruction occurs and the fewer traces we find and the more the probability of sampling a rock of age *t*

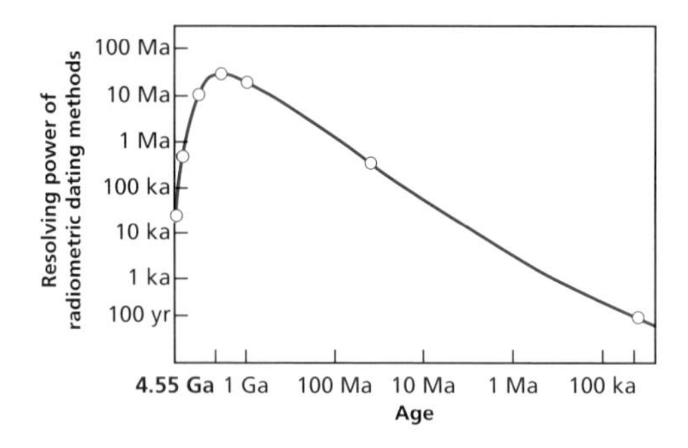

Figure 5.34 Resolving power of all dating methods versus age.

diminishes with t. There is a dual effect: the further we go back into the past, the fewer rocks we have of any given age and the less certain their age is. This is the phenomenon of **geological myopia**. Statistically, the further back into the past we travel, the lower the power of resolution. This is an intrinsic difficulty of any historical science. And yet, where radiometric dating is concerned, there is an exception!

The extinct forms of radioactivity ^{129}I, ^{182}Hf, ^{109}Pd, ^{146}Sm, etc. can be used to study the earliest times of the Earth's history with extraordinary precision, as if the time the Sun and the planetary bodies formed were suddenly "lit up." Resolution for these primordial times of the Earth's history is almost as good as for the Quaternary. This quite incredible circumstance arises because one or more supernova or AGB star explosions made radioactive isotopes 1 or 2 million years before the first rocks of the Solar System formed. Thus the curve of the power of resolution against time is not monotonic, contrary to what might have been expected (Figure 5.34)!

5.8.2 The use of radiochronometers

Advances in analytical techniques mean that we now have an impressive array of potential radiochronometers at hand: there are currently more than 40 of them and more are being developed. Thus anyone wanting to use radiometric dating is confronted with a problem of choice: which chronometer should be used for dating a particular phenomenon at such and such a period?

Without any need for long calculations, we have understood that ^{14}C can no more be used for determining the age of the Earth than ^{87}Rb–^{87}Sr can be used for dating limestones of less than 1 Ma. But beyond these obvious examples lies the more general question of our power of observation of the past: which chronometers can be used, for which periods, and for which rocks?

Domains for using geochronometers based on analytical criteria

If the time-span to be measured is far longer than the half-life of the radioactive isotope, then the radioactive product will have died prematurely and so indicate nothing more. If the time-span is very much shorter than the half-life, then the radioactivity will be insignificant and its result undetectable. We are thus brought back to the calculations of uncertainty already covered (see Plate 7).

Exercise

Can we measure an age close to 2 Ga for a rock with $^{87}Rb/^{86}Sr$ ratios whose maximum deviations from the origin are 10, 1, 0.1, 0.01? We accept that the precision of the measurement on the $^{87}Sr/^{86}Sr$ ratio is $\pm 10^{-4}$ and on $^{87}Rb/^{86}Sr$ $\pm 10^{-3}$.

Answer

We can determine an age in each of these cases. For $\mu = 10$, $T = 2 \pm 0.001$ Ga, for $\mu = 0.01$, $T = 2 \pm 0.5$ Ga.

Exercise

Why is the use of ^{14}C limited with counting to 6 half-lives whereas with mass spectrometry it can reach 10 half-lives?

Can we measure the same age with the ^{234}U–^{238}U method, given that we have a rock containing 10% uranium (uranium ore)? Why?

Answer

The answers are deliberately withheld here. Readers must find them for themselves to understand what comes next.

For methods based on decay of the parent, the dating equation is written:

$$\mu = \mu_0 \, e^{-\lambda t}$$

where μ is $^{14}C/C$, $^{10}Be/Be$, etc. Analysis of uncertainty (Section 3.2.1) has shown that this method applies only if T is neither too large nor too small. In practice, we consider T must comply with the double inequality:

$$10 \text{ half-lives} < T < 10^{-3} \text{ half-lives.}$$

Of course, the domain of application of a method is extended as improvements in analytical techniques are made, and progress is continuous with no prospect of an end.

Still, let us give an answer to the previous exercise. For modern equipment, $^{14}C/C$ activity is 13.4 dpm g^{-1} of carbon. Six half-lives represents 1 disintegration per hour, whereas the background noise is at least 5–10 disintegrations per hour. The uncertainty is intolerable. However, with the mass spectrometer we measure a ratio which at the present time is 10^{-12} and is advancing to 10^{-15}, which is not easy but can be done, especially if we increase the amount of matter analyzed by a factor of 10 or 100!

For long-period parent–daughter systems, with the usual conventions, the dating equation is written:

$$\alpha = \alpha_0 + \mu(e^{\lambda t} - 1).$$

There is no physical upper limit to the use of these methods. The lower limit of applicability arises because the quantity $(\alpha - \alpha_0)$ must be measurable and therefore depends on the precision of the isotopic measurement of the daughter product:

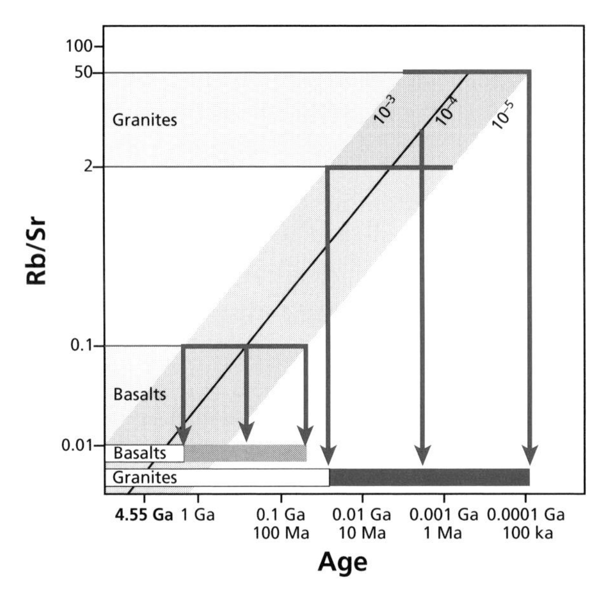

Figure 5.35 Domain of use of $^{87}Rb–^{87}Sr$ applied to whole rocks. The *x*-axis shows the minimum ages that can be measured. The *y*-axis shows the values of the chemical ratio Rb/Sr. The diagonals indicate three precisions on the measurement of the $^{87}Sr/^{86}Sr$ ratio: 10^{-3}, 10^{-4}, and 10^{-6}.

$$\mu\,(e^{\lambda t} - 1) > \Delta\alpha.$$

We know that, depending on the methods and techniques used, $\Delta\alpha$ may be 10^{-3}, 10^{-4}, or even of the order of 10^{-6} with a wealth of precautions. There is therefore a connection between μ and the minimum age $T_{\min}$ that can be measured:

$$\mu = \left(\frac{\Delta\alpha}{e^{\lambda t} - 1}\right).$$

This calculation has been done for the main long-period chronometers and in each case for the main rocks and most "useful" minerals. It is illustrated for $^{87}Rb–^{87}Sr$ in Figure 5.35.

Exercise

Calculate the range of use of $^{176}Lu–^{176}Hf$. What are the common minerals that give the best results? (Look for the necessary parameters in the scientific literature!)

Answer

Up to 1 or 2 Ma. Garnet is the most effective.

In the case of methods related to radioactive chains, the calculation is a little more complex. Allowance must be made for both the radioactive parent and the radioactive daughter

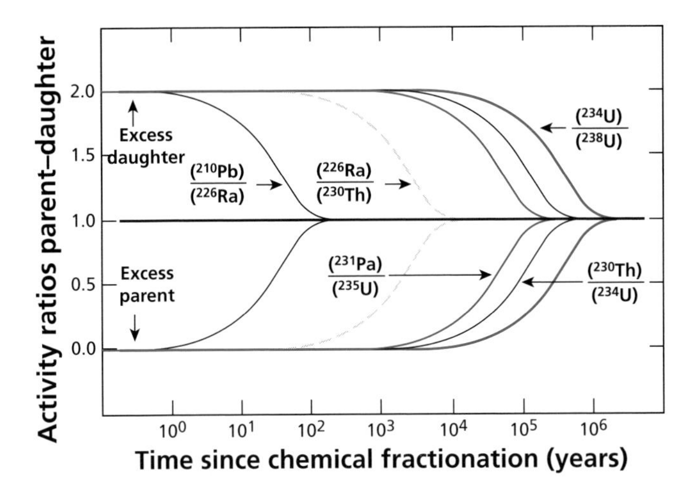

Figure 5.36 Intervals for use of the various chronometers based on radioactive disequilibrium.

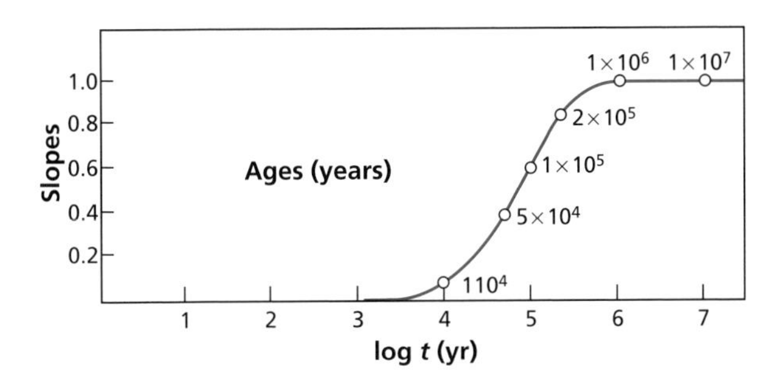

Figure 5.37 Variation of the slope of the ^{238}U–^{230}Th isochron with age. (Notice the logarithm is to base 10.)

and for the cause of disequilibrium (isolation of the parent or daughter). Without getting into the calculations, where the principle is identical to the previous ones, we can say that the ages determinable are such that: 5 half-lives $> T > 10^{-2}$ half-lives of the daughter isotope (Figure 5.36).

Exercise

We consider the isochron method applied to the ^{238}U–^{230}Th pair. Empirically, we calculate the value of the slope of the isochron against age and plot it (Figure 5.37). What do you conclude about its domain of use?

Answer

We obtain $5 \cdot 10^3 < T < 5 \cdot 10^5$.

0.06 half-lives $< T <$ 6 half-lives.

We are within the orders of magnitude defined.

Geological modulation of the time domains for the use of radiochronometers

The question here is how do the geological constraints alter the conclusion about the range of application of the various methods established based on analytical errors. As we have repeated incessantly, a radiochronometer is merely **an isotope ratio expressed as an age** by the formula reflecting radioactive decay. The concept of age in itself is "virtual" and a purely arithmetic one. We must therefore think about the physical system, its history, the chemical properties of the elements, and the initial conditions, and connect these considerations with geological history before we can attribute a significant age to a measurement (or a series of measurements).

Examination of the criteria to ensure the boxes remain closed has been one of the major concerns of geochronology. How can any "untimely" opening be detected? A form of systematic approach has arisen empirically from these studies, although it cannot be said that we have any absolute criteria because geological situations are so varied and complex. As we have already said, the reasons systems open are straightforward enough: radiogenic isotopes are intruder atoms in the crystalline structures where they are found and so tend to escape. Can this phenomenon be quantified generally? There is no mathematical method for estimating geological vagaries "for certain." We can make a few calculations on diffusion to get our ideas straight and give us orders of magnitude, but little more than that.

Let us take argon as an example. At ordinary temperatures, diffusion coefficients for ^{40}Ar are of the order of 10^{-20} cm^2 s^{-1}. If we recall the distance covered is $x \approx \sqrt{Dt}$, then, in 1 billion years $x \approx (3 \cdot 10^7 \times 10^9 \times 10^{-20})^{1/2} = 0.005$ cm. Therefore, at ordinary temperatures, even the smallest minerals are closed systems.

However, at 250 °C, which is a low-grade metamorphic situation, the diffusion coefficient of argon is of the order of 10^{-12} cm^2 s^{-1}. This time, for 100 Ma, the distance covered is 54 cm, therefore all of the minerals are open. Now, such conditions (100 Ma at 250 °C) correspond, say, to a rock located at a depth of 10 km in the Earth's crust. This may happen by chance to a granite either when caught up in tectonic folding or when covered by a thick layer of sediment during the formation of a sedimentary basin. But each region and each rock massif has its own geological history.

We have therefore given precedence to experience amassed through hundreds (or now perhaps thousands) of examples like those we have mentioned from which we have derived empirical rules, the variety of which does not lend itself to quantitative systematization. This was first done for isolated apparent ages, then for the more elaborate methods (concordia, isochrons, etc.). Here are a few examples of these results.

The most sensitive methods to diffusion phenomena are those using the rare gases $^{40}K-^{40}Ar$ or U–He. The reason for this extreme susceptibility is simple. When a rare gas has left the crystallographic site where it was created and finds itself on the rim of the crystal or in a crystal defect, as it is not electrically charged, it is gaseous and lighter than the surrounding rocks, it migrates upwards. This is why, save in exceptional circumstances (meteorites), $^{40}K-^{40}Ar$ does not provide any reliable ages for rocks older than 1 Ga. The $^{39}Ar-^{40}Ar$ method is far more effective.

For older ages, we adopt the opposite approach. Knowing the true age by other methods (Rb–Sr, U–Pb, or Sm–Nd), we determine the apparent K–Ar age to estimate the more or less complex geological history of the rock. A K–Ar age close to the true age is indicative of a quiet geological history. This is a valuable clue. If the K–Ar age is much lower than the "more robust" ages, the rock has been subjected to periods of heating. If the K–Ar age is older (which it seldom is), aqueous weathering has dissolved the potassium.

A despairing case is ^{187}Re–^{187}Os. In principle this method should be almost ideal for dating granites and basalts of all ages because the Re–Os ratios are very high (ratios of 1000 to 5000 are commonplace). We could hope to pin ages down to a few thousand years. Unfortunately, whole rocks do not generally behave like closed systems and so dating by this method is restricted and age determinations are difficult.

Exercise

We have measured the rhenium and osmium contents of olivine in oceanic basalt as 4112 pg g^{-1} and 3361 pg g^{-1}, respectively.

Supposing we know the initial ratio of the basalt $^{187}Os/^{188}Os = 0.1265$ and that the $^{187}Os/^{188}Os$ ratio measured in the olivine is 0.1859, what is the age of the olivine?

If we can measure the $^{187}Os/^{188}Os$ ratio with a precision of 1% at these very low levels, what is the age of the olivine assuming it has remained in a closed system?

Answer

565 ka.

12 300 ka.

To summarize this research which is above all empirical, the **geochronologist's toolbox** is shown in Plate 7. Beware, though, this is no automatic gearbox!

Let us repeat once again, no one radiochronometer alone can resolve a problem and guarantee a result. **Geochronological redundancy**, in other words the confirmation of a result by various other methods, is an essential concept. Now there are very many chronometers that can be used, but each series of measurements is difficult, time-consuming, and expensive. So we cannot apply every chronometer to every problem. We must choose the most suitable ones . . . and do a little thinking.

5.9 Conclusion

This chapter has allowed us to validate the various geochronometers through the analysis of uncertainties. Overall they are reliable and yield consistent results. However, we must never forget the concept of **essential uncertainty** which must always be taken into account.

Each geochronological result is affected by uncertainty. If we do not know (estimate) this uncertainty, a result is scientifically meaningless. We have analyzed these uncertainties to show they are of different types: they may be due to **sampling, laboratory measurements, adaptation of methods** to the problem in question, or to the **geological or cosmic scenario**.

In this context we must emphasize ad nauseam that age is simply an isotopic or chemical ratio that is converted into time by a mathematical formula. Such conversion is practical because we can directly compare the various methods and, of course, connect the methods with the geological history for which they are one of the reference markers. But as soon as a difficulty arises, we must come back to the isotopic ratios and the chemical properties of the elements. It is geochemical history that controls the way the chronometers behave.

This is the general context in which we have reviewed the main findings of geochronology which themselves are associated with uncertainty or even indetermination.

Problems

1 An expedition to Mars to bring back samples is under preparation and you are the Rb–Sr specialist for dating the rocks. Preliminary missions with soil analyses by *in situ* techniques and analogies with shergottite-type meteorites indicate that the rocks found will be basalts with average Rb/Sr ratios of 0.1 and 4, respectively, and average Sr contents of 300 ppm and 30 ppm, respectively.

(i) How much rock needs be collected to make analyses with a precision of 10^{-4}? (We know we will need to have 100 ng of strontium and 50 ng of rubidium.)
(ii) Given that the probable ages are between 2 Ga and 500 Ma, do you think the determination will be reliable? What sort of isotope ratios do you expect?
(iii) What will be the error you make on the age if it is 500 Ma? If it is 2 Ga?
(iv) Only 1 kg of rock can be brought back. Would you prefer to have a single compact rock or five fragments of different rocks of 200 g each? Why?

2 Here are the results obtained by the ^{238}U–^{230}Th method on a volcanic rock of Costa Rica.

	^{238}U–^{230}Th	^{230}Th–^{232}Th
Whole rock	1.41 ± 0.05	1.21 ± 0.04
Magnetite	2.41 ± 0.02	1.92 ± 0.12
Plagioclase	1.37 ± 0.03	1.18 ± 0.07
Augite	0.62 ± 0.01	0.78 ± 0.04
Hypersthene	1.02 ± 0.08	1.01 ± 0.05

After calculating the age of the rock, estimate the error on this age in various ways: graphical, least-squares method. Then state your estimate.

3 Here are results for a granite of the Montagne Noire (France).

	^{87}Sr–^{86}Sr	^{87}Rb–^{86}Sr
Whole rock	0.728 ± 0.002	3.62 ± 0.05
Biotite	2.32 ± 0.01	408 ± 4
Muscovite	0.735 ± 0.002	8.07 ± 0.05
Apatite	0.733 ± 0.002	0.03 ± 0.003
Plagioclase	0.7127 ± 0.001	1.44 ± 0.02

Can you estimate the age and the error you make on the age in this situation?

4 (i) The measurement of ^{14}C by accelerator mass spectrometry gives for modern (present-day) carbon the ratio $^{14}C/^{12}C = 1.2 \cdot 10^{-12}$. For a current of $2 \cdot 10^{-6}$ A of ^{12}C obtained by bombarding 1 mg of carbon with a stream of cesium ions, the flow of $^{14}C^{+}$ ions is 15 ions per second.

(a) For the same intensity of current of ^{12}C, how many ^{14}C ions are obtained per unit time for 1 mg of carbon aged 55 ka?

(b) If the measurement lasts 20 minutes, what is the uncertainty on the age determination? Compare it with the error on the measurement of the present-day $^{14}C/^{12}C$ ratio.

(c) What is the minimum age we can hope to measure in the same conditions if we decide the difference to be measured with the present day must be at least five times greater than the uncertainty?

(d) If we wanted to make the same measurement by counting β radioactivity and the maximum counting time for a sample was 8 days, how much carbon would need to be purified?

(ii) We wish to use the ^{238}U–^{230}Th method to test the ^{14}C age of corals assumed to be about 55 000 years old. To simplify, we ignore the $^{234}U/^{238}U$ fractionation and assume that $(^{230}Th)_0 = 0$. The dating equation is written:

$$\frac{^{230}Th}{^{232}Th} = \left(\frac{^{238}U}{^{232}Th}\right)(1 - e^{\lambda_{230} t}).$$

Given that the uncertainty on the $^{230}Th/^{232}Th$ ratio measurement is 1% and that the uncertainty on the ^{232}Th and ^{238}U measurements is about 5‰, what is the uncertainty on the age? Is this a good method for checking the results obtained by ^{14}C in this age range?

5 We are confronted with the geological situation shown in the section in Figure 5.38. A geological formation of foliated and folded gneiss is cross-cut by a granodiorite which is itself foliated.

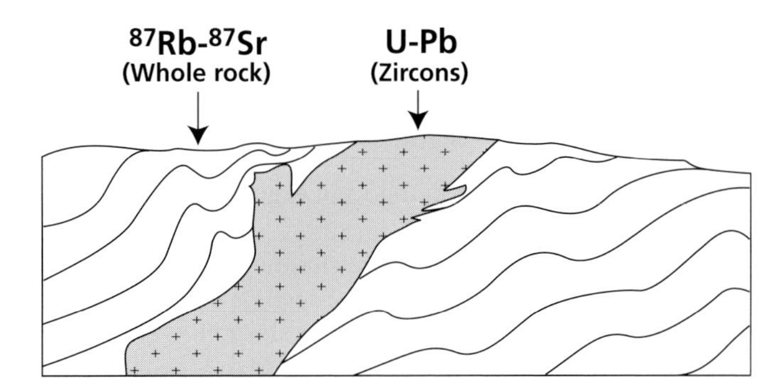

Figure 5.38 Geological formation of gneiss cross-cut by granodiorite.

(i) We wish to measure the age of the gneiss and to do so we make a series of $^{87}Rb/^{86}Sr$ measurements on the whole-rock. The results are given in Table 5.9. The Sr concentrations of the various facies are very similar. Calculate the $^{87}Rb/^{87}Sr$ age using a simple least-squares method (without experimental uncertainties).

(ii) We determine the age of the granite by U–Pb methods applied to zircons from the granite. Table 5.10 shows the results. Calculate the U–Pb age again by the simple least-squares method and the concordia method. How would you interpret the result in geological terms? Can you give a brief geological history of the region?

Table 5.9 Samples from the gneiss in Figure 5.38

Sample	$^{87}Sr/^{86}Sr$	$^{87}Rb/^{86}Sr$
1	0.713 ± 0.0005	0.1 ± 0.0005
2	0.718 ± 0.0005	0.35 ± 0.0005
3	0.719 ± 0.0005	0.45 ± 0.0005
4	0.722 ± 0.0005	0.6 ± 0.0005
5	0.724 ± 0.0005	0.8 ± 0.0005
6	0.723 ± 0.0005	0.9 ± 0.0005

Table 5.10 Samples from the granite in Figure 5.38

Sample	$^{206}U^*/^{238}U$	$^{207}Pb^*/^{235}U$
A	0.3 ± 0.01	5 ± 0.3
B	0.26 ± 0.01	4 ± 0.3
C	0.24 ± 0.01	3.5 ± 0.3
D	0.2 ± 0.01	2.5 ± 0.3

6 A granite from south-eastern Canada is dated by various methods. The apparent age results are:

Method	U–Pb concordia	Rb–Sr whole rock	Sm–Nd whole rock	Rb–Sr on biotite	^{39}Ar–^{40}Ar	Rb–Sr on muscovite
Apparent age (Ga)	2.8 ± 0.1	2.7 ± 0.1	2.7 ± 0.1	1.7 ± 0.2	1 ± 0.1	1.7 ± 0.2

Can you reconstruct the geological history of this granite and explain the ages measured?

7 We can determine the Permian–Triassic boundary at a given location. A granite cross-cutting the Permian formation but with pebbles at the base of the Triassic is dated 243 ± 5 Ma. The Triassic is dated by two basalt flows located in a section 100 m above the Permian–Triassic boundary at 229 ± 8 Ma and another 200 m above the Permian–Triassic boundary at 215 ± 3 Ma. What is the approximate age of the Permian–Triassic boundary? What is the error on the age determination at 218 ± 3 Ma?

8 You are asked to explore the possibility of using the U–He chronometer.
(i) What time-span do you think best corresponds to the potential of the method?
(ii) What types of rock or mineral do you think are interesting a priori?

9 Half a century ago the question of the origin of granite was a subject of much debate. The American Bowen showed by laboratory experiment around 1930 that rock of granitic composition could be made from basaltic magma (which we know is itself the product of partial melting of the mantle). Bowen and Winkler in Germany showed, again by experiment, that the melting of detrital sediments (shales and sandstones) to which sodium has been added also gives rise to a granitic magma. Subsequently, other experiments showed that the partial melting of "wet" basalts produced granitic fluids.

These experimental data were supplemented by field observations and then more recently by $^{87}Sr/^{86}Sr$ measurements supporting the idea that almost all continental granites result from the remelting of continental crust. The model of granites derived directly from the mantle was abandoned. Yet, in the middle of the mid-Atlantic ridge, in Iceland, the famous Hekla volcano has spewed out lava of granitic composition (rhyolites). The $^{87}Sr/^{86}Sr$ ratios of about 0.7032 give no clear answer to their origin as they are similar to those of many basaltic lavas of Iceland.

The Icelander Sigmarsson studied the ($^{230}Th/^{232}Th$, $^{238}U/^{232}Th$) ratios of these rhyolites with the ^{230}Th–^{238}U isochron diagram. The mantle beneath Iceland is in secular equilibrium, corresponding to $^{238}U/^{232}Th$ ratios of 1.1 to 1.3 (in activity). The oldest basalt crust of Iceland has (active) $^{238}U/^{232}Th$ ratios ranging from 0.9 to 1.1. It too is in secular equilibrium. The rhyolites have a $^{230}Th/^{232}Th$ ratio of 0.95 and $^{238}U/^{232}Th$ of about 0.8. What do you think is the origin of this rhyolite?

10 Calculate the range of application of the ^{238}U–^{206}Pb dating method for rocks and minerals with the chemical ratios shown in the table below, given that the isotopic measurement is affected by several relative uncertainties Δ_x of 10^{-3}, 10^{-4}, and 10^{-5}.

Type of rock or mineral	Granite	Basalt	Ultramafic	Zircon	Sphene
U/Pb	0.2	0.1	0.05	100–10	2

Draw the graphs corresponding to Figure 5.35. Calculate for the Rb/Sr ratio, and for U–Pb applied to whole rocks and the main minerals that can be dated by this method.

CHAPTER SIX

Radiogenic isotope geochemistry

Our main concern in the previous chapters has been to calculate geological (or cosmological) ages and to clarify their geological meanings. In the next chapters we are going to address **isotope geochemistry**, which has a markedly different viewpoint although it is based on the same physical laws and on the same equations. We are going to look not at age but at the isotope ratios measured and the ways in which their natural variations may be interpreted.

This chapter, then, focuses on isotope ratios whose variations arise mainly from radioactive decay. Chapter 7 will be about isotope variations caused by physical and chemical processes. Examination of the variations in isotopic ratios will lead us to develop a methodology in which these ratios are used as **tracers** for major geological and geodynamic phenomena. The purpose of these studies is therefore to determine the major structures and exchanges of matter occurring (or having occurred) among the major terrestrial reservoirs (crust, mantle, core, and atmosphere). The notion of age will crop up again, but in a much more general context, although in complete continuity with the earlier chapters.

6.1 Strontium isotope geochemistry

6.1.1 Continents and oceans: granite and basalt

The natural isotopic composition of strontium varies with ^{87}Rb decay. Systematic collation of ^{87}Sr/^{86}Sr isotope compositions of terrestrial basalt and granite (basalt being an isotopic "messenger" from the mantle while granite is a sample of continental crust) reveals very different distributions. Strontium isotope ratios for granites from the various continents are highly variable, ranging from 0.705 to 0.850 and more. Strontium isotope ratios for basalts are much more uniform. They range from 0.7020 to 0.7070 and for oceanic basalts from just 0.7022 to 0.7045.

Let us get these orders of magnitude clear. "Ordinary" analytical precision on the ^{87}Sr/^{86}Sr isotope ratio is at least $1.5 \cdot 10^{-4}$ (relative) and therefore, on ratios close to 0.7, the uncertainty is better than 10^{-4} (absolute). This means that 0.7022 can be distinguished from 0.7021 for certain. For granite, variations range from 0.850 to 0.705, corresponding to 1500 measurement units. For oceanic basalts, the range is 25 measurement units, or a dispersion ratio of 60 times less than for granites.

This difference in the distributions of strontium isotope ratios between continents and mantle can be explained by the geological history of the two reservoirs from which granite

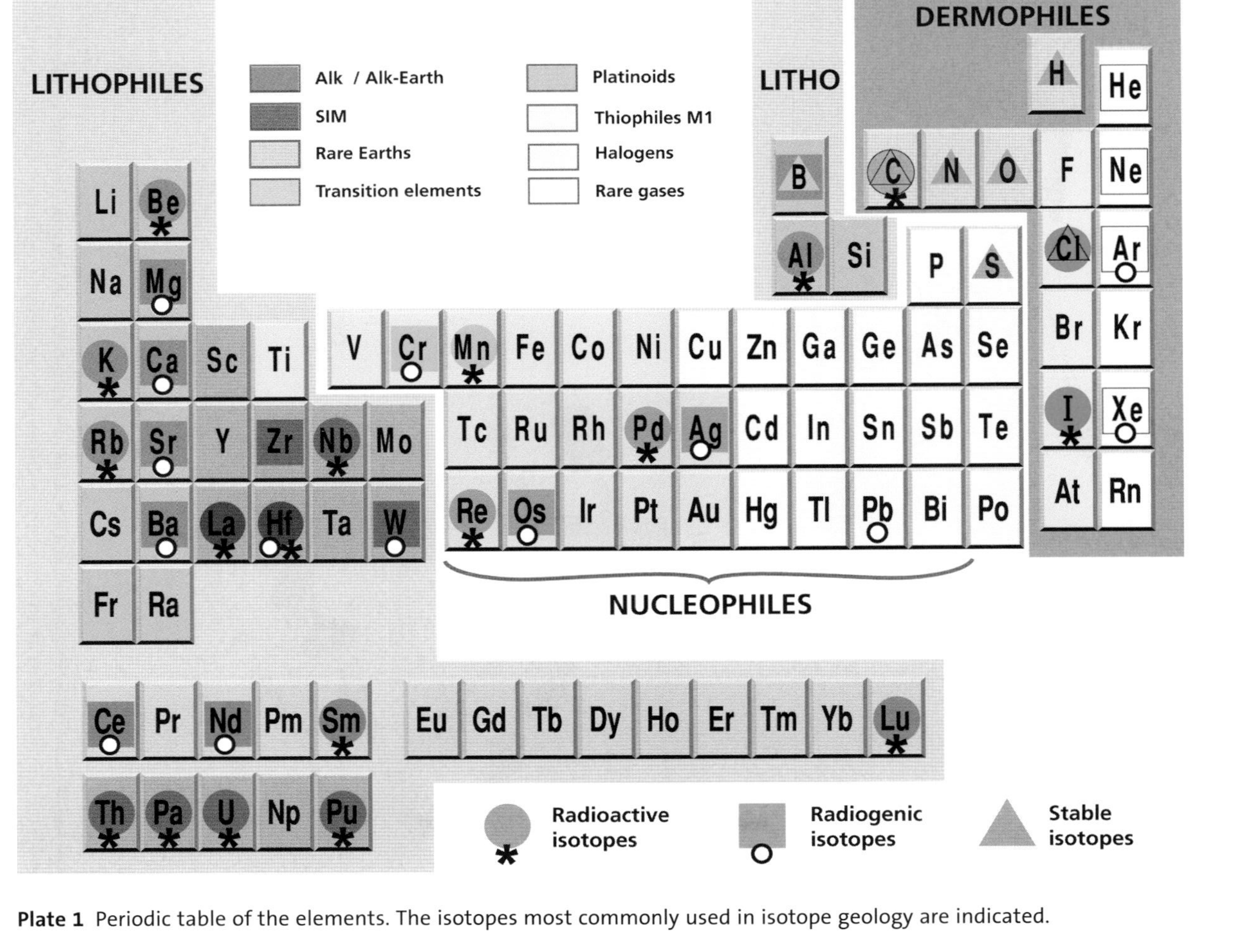

Plate 1 Periodic table of the elements. The isotopes most commonly used in isotope geology are indicated.

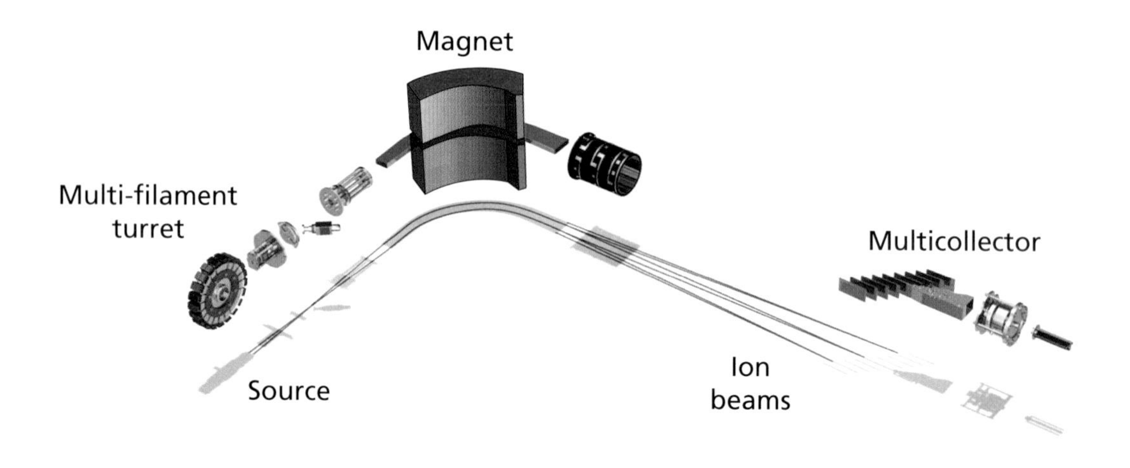

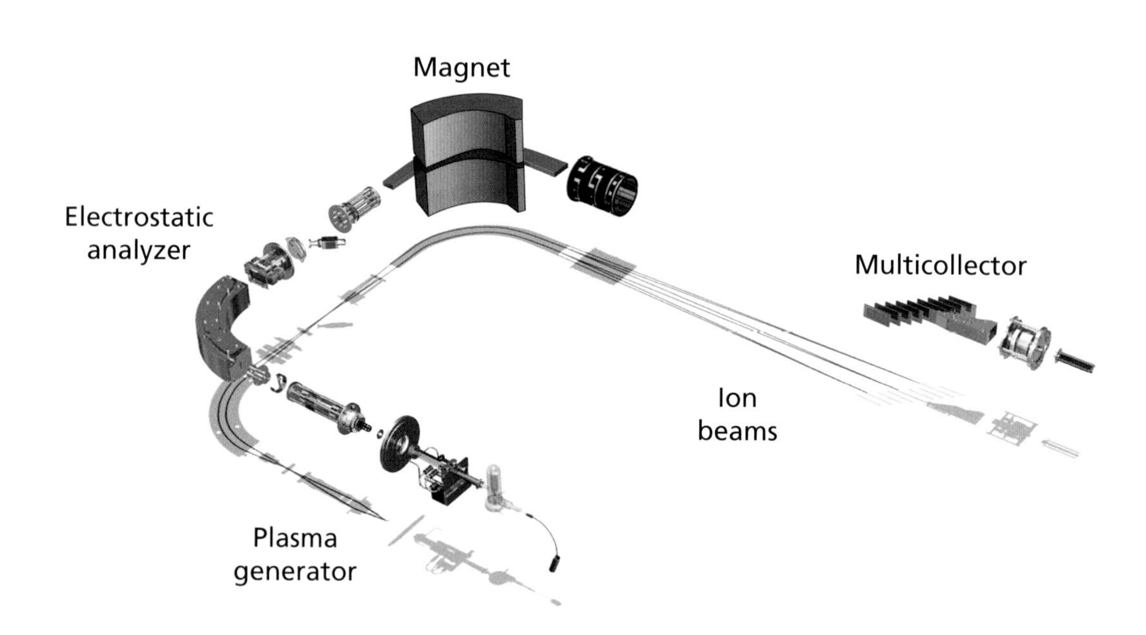

Plate 2 Modern mass spectrometers. In each case, the mechanical components are shown at the top and the ion beam pathways below. Top: TIMS (thermal-ionization mass spectrometry). At the source, a turret allows several samples to be tested in succession; at the collector, multicollection allows the various beams from different isotopes to be analyzed without scanning. Bottom: ICPMS (inductively coupled plasma mass spectrometry). In addition to the magnet, an electrostatic analysis sorts the ions by energy levels. The source is a plasma generator. Multicollection is used as in TIMS.

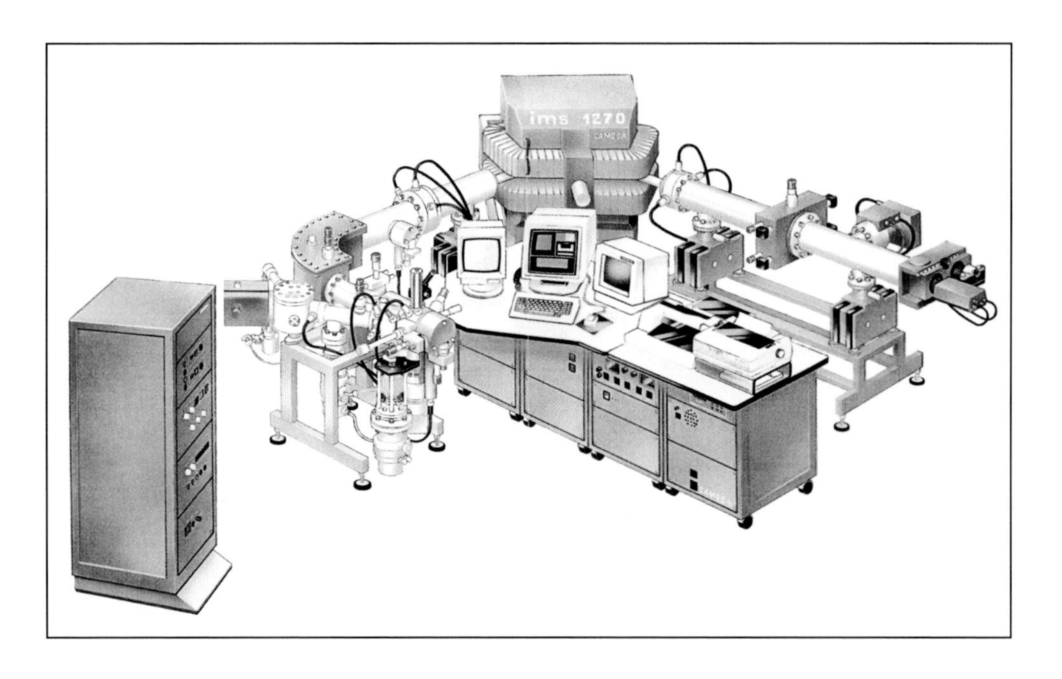

Plate 3 Actual laboratory equipment. Top: A Cameca ion probe. Bottom: A clean room where samples are prepared.

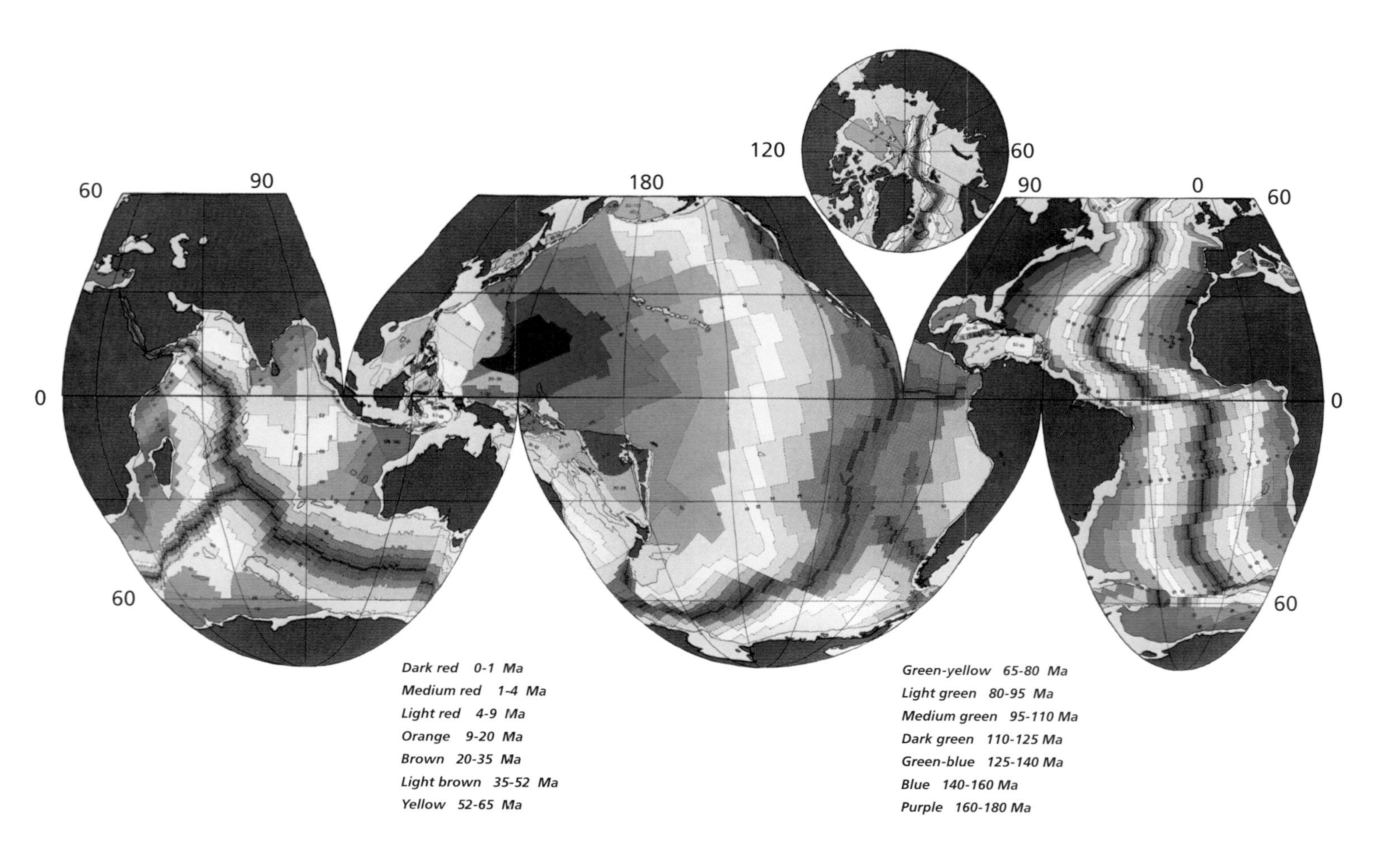

Plate 4 Map of ocean floor ages. After Sclater, Jaupart, and Garlson (1980).

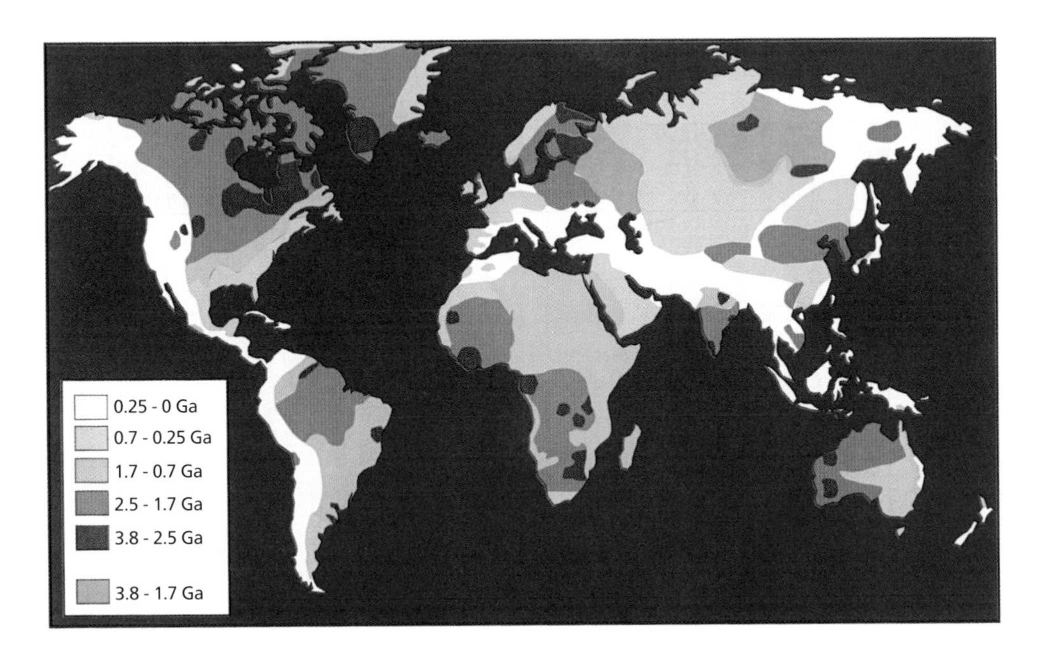

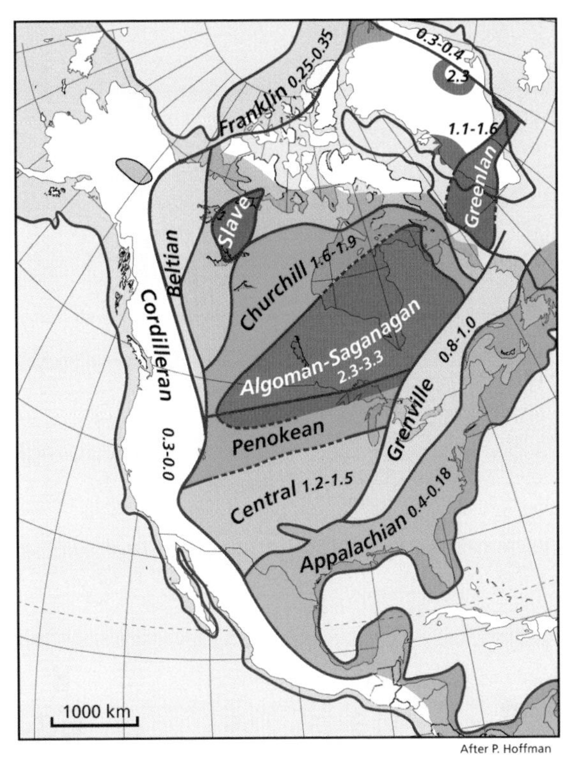

Plate 5 Map of age provinces of continents. Top: world map. Bottom: more detailed map of North America. After Hofmann (personal communication, modified).

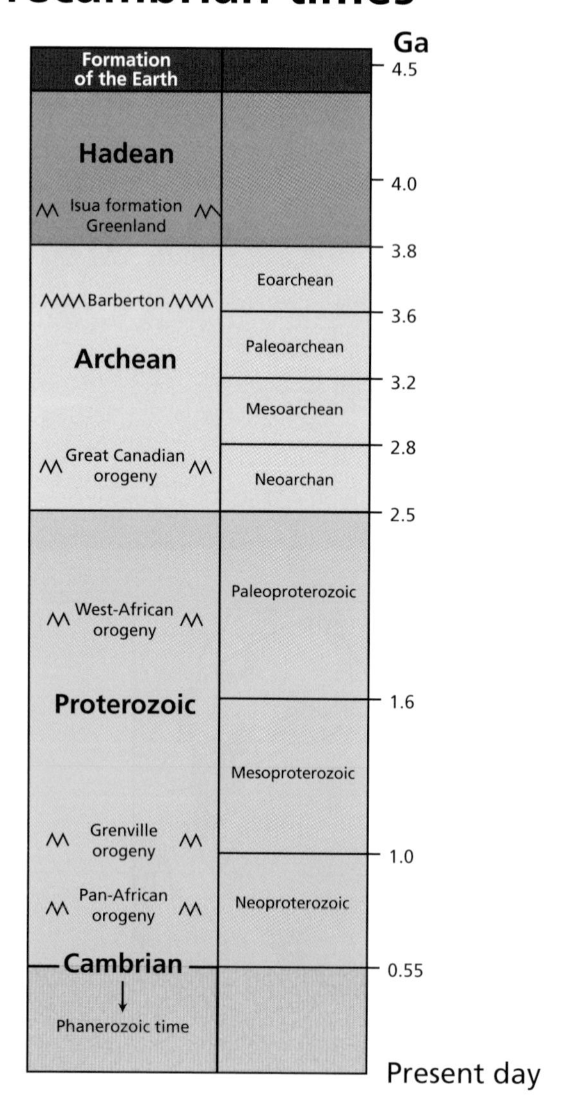

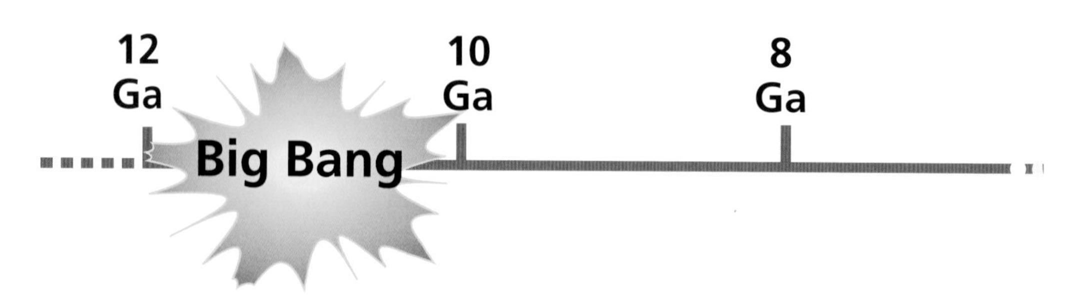

Plate 6 Geological timescales. Left: Precambrian times. Right: Phanerozoic times.

Phanerozoic times

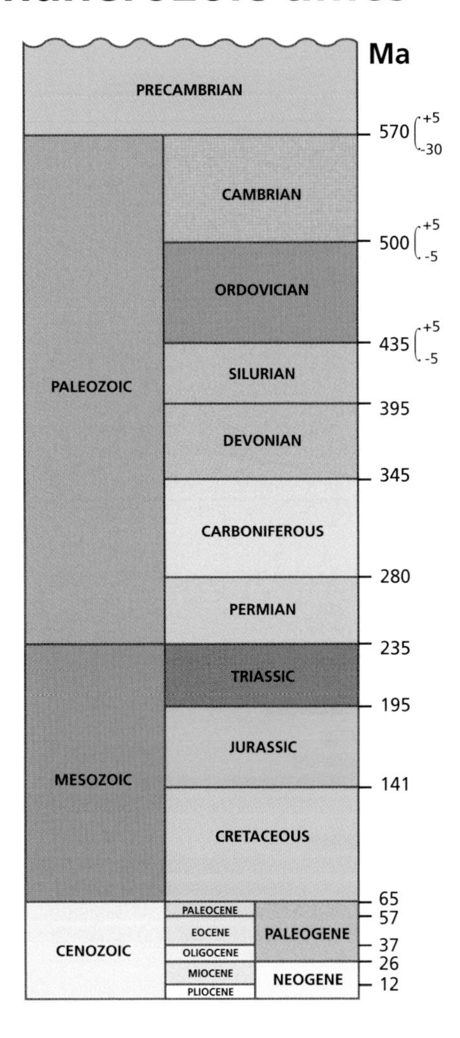

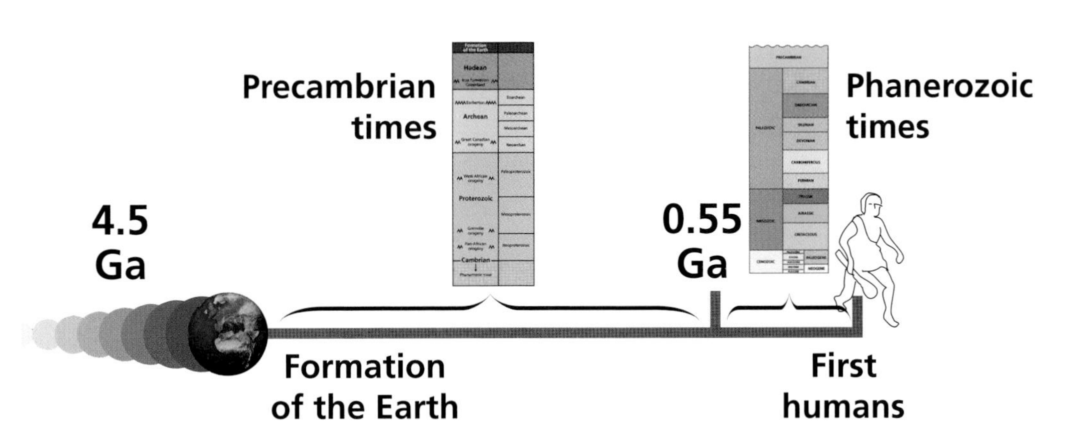

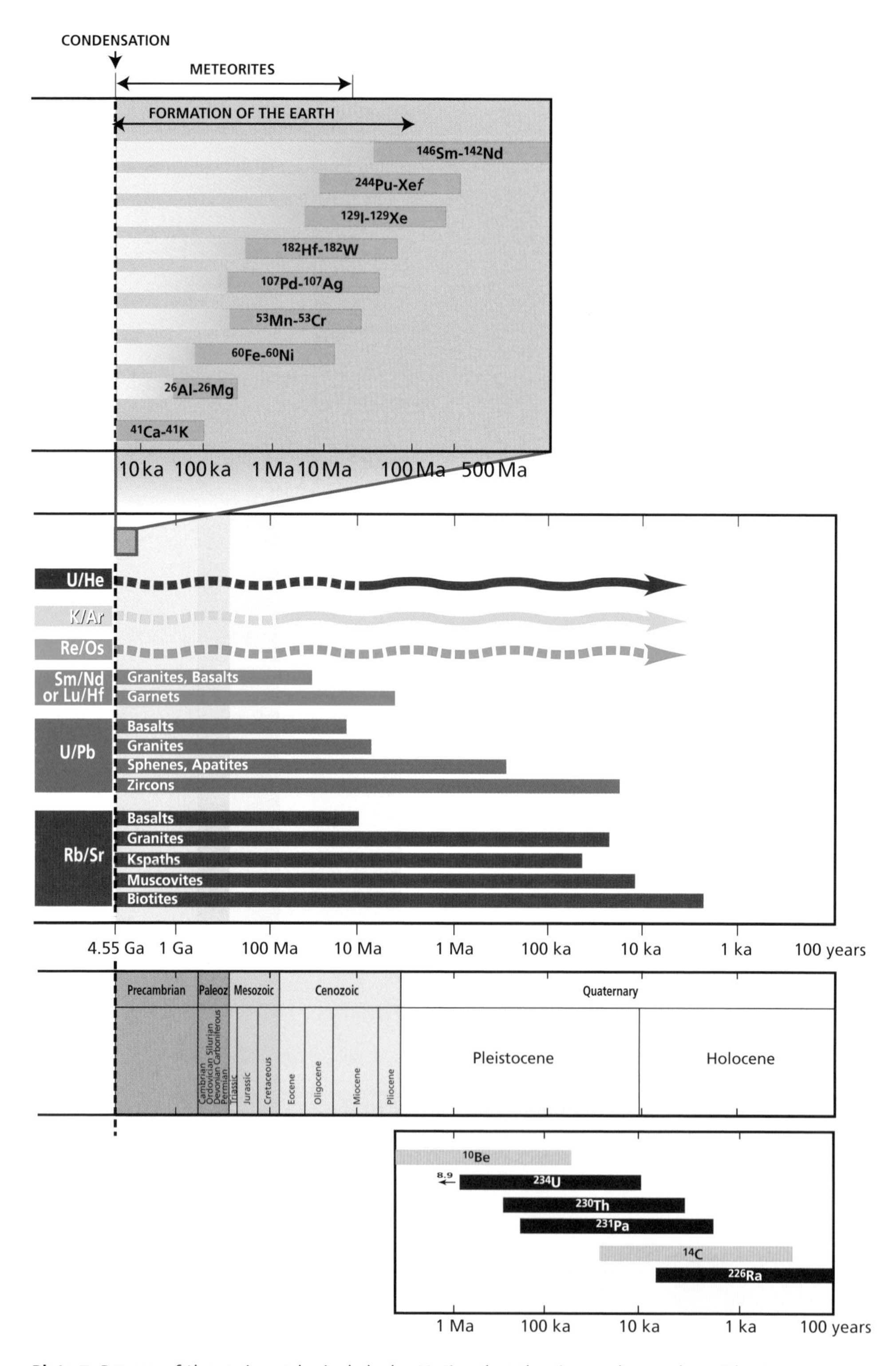

Plate 7 Ranges of the main geological clocks. Notice that the timescales are logarithmic.

and basalt arise. **Granite** is the essential constituent of **continental crust**. The continental crust is a permanent feature of the Earth's surface and is very, very old, at least 4 Ga and possibly older. When granites are systematically dated on a continent and the datings mapped, geological age provinces are defined. The classical example is that of North America, as we have seen, and the same is true of other continents. There are granites aged 3.8 Ga (Greenland), 2.7 Ga (Canada, Scandinavia), and 2 Ga (Ivory Coast, Sahara), but also granites aged 50 Ma (Himalayas) or even 5 Ma (Himalayas, Andes, Alps) (see Plate 5).

As geological studies have shown, these provinces correspond to well-defined episodes in geological history during which orogeny, that is, mountain building, occurred. Rock was folded and metamorphosed and granite was formed by melting of the lower crust and was injected into the upper part of the crust during these orogenic episodes. In accordance with one of the basic principles of plate tectonics, these pieces of continent float on the surface of the mantle, they are broken up and drift about but are not swallowed up as such within the mantle by subduction. They therefore form a mosaic of blocks of different geological ages which have accumulated, broken up, and drifted together at the Earth's surface throughout geological time.

In the course of geological cycles, with the succession of erosion, sedimentation, metamorphism, and folding and formation of granites, the isotopic geological clocks we have studied in the previous chapters are more or less reset and so date these episodes more or less accurately. But a great part of the ancient continental material is preserved and recycled, whether it is rejuvenated or not.

Basalt is the primary constituent of oceanic crust. Unlike continental crust, oceanic crust is very young. Oceanic crust arises from the mid-ocean ridges, spreads, and then plunges at subduction zones to be recycled in the mantle. The oldest oceanic crust is 200 Ma, the average age being 80 Ma. By contrast, the rocks of the present-day ocean ridges are virtually of zero age. Oceanic basalts are the product of mantle-melting and as that melting occurs at high temperatures **isotopic equilibrium** is achieved. This is why oceanic basalt reflects the isotopic compositions of the present-day mantle. Now, the mantle, being subjected to vigorous convective motion, is probably well mixed. It is only to be expected, then, that the isotopic compositions of such a medium should be far less heterogeneous than those of a medium that is segmented like a mosaic of old pieces of various ages, which is what the continental crust is (Figure 6.1).

Let us now try to explain the difference in the strontium isotope structure of the continents and of the oceanic basalts in more quantitative terms. By the same token, we shall bring together the reasoning already employed and which helped us in developing geochronology.

6.1.2 Isotope evolution diagrams and multi-episode evolution models

One simple way of representing changes in $^{87}Sr/^{86}Sr$ ratios is to plot a diagram of $^{87}Sr/^{86}Sr$ versus time. If we write as our reference the evolution of an isotope ratio in a closed environment, by using the linear approximation, we obtain an equation we have already come across:

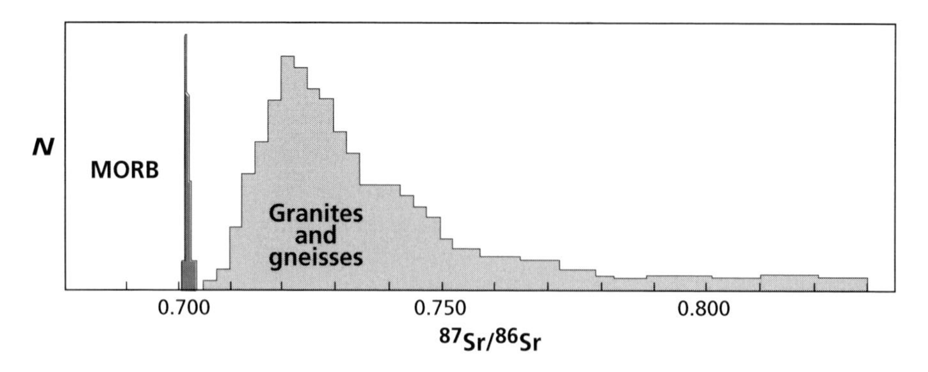

Figure 6.1 Histogram of $^{87}Sr/^{86}Sr$ ratios. Ratios are measured in the mid-ocean-ridge basalts (MORBs) and in the granulites of the lower continental crust and the granitoids of the upper continental crust. Notice the wide dispersion of values for granite compared with MORB values.

$$\left(\frac{^{87}\mathrm{Sr}}{^{86}\mathrm{Sr}}\right)_t = \left(\frac{^{87}\mathrm{Sr}}{^{86}\mathrm{Sr}}\right)_0 + \left(\frac{^{87}\mathrm{Rb}}{^{86}\mathrm{Sr}}\right)\lambda t.$$

This is the equation of a straight line in the $\left(\frac{^{87}\mathrm{Sr}}{^{86}\mathrm{Sr}}, t\right)$ diagram whose slope is $\left(\lambda\frac{^{87}\mathrm{Rb}}{^{87}\mathrm{Sr}}\right)$ and so is proportional to the Rb/Sr chemical ratio. The higher $\mu = {}^{87}\mathrm{Rb}/{}^{86}\mathrm{Sr}$ is, the faster the $^{87}Sr/^{86}Sr$ isotope ratio grows, and the steeper the slope of the straight line of evolution (Figure 6.2).

Let us now consider a system comprising two episodes. We begin with a $(^{87}\mathrm{Sr}/^{86}\mathrm{Sr})_0$ isotope ratio $= \alpha_0$. During the first episode from T_0 to T_1 (the ages are counted from the present day) the system has a $^{87}Rb/^{86}Sr$ ratio $= \mu_0$. Then, at T_1, the chemical ratio changes because the initial system is separated into several sub-systems, A, B, and C, whose $^{87}Rb/^{86}Sr$ ratios are noted μ_A, μ_B, and μ_C, respectively. The three sub-systems evolve as closed systems until the present day. The evolution equations for each medium are written:

$$\text{Sub-system A} = \alpha^{\mathrm{A}}_{\text{present day}} = \alpha_0 + \lambda\mu_0(T_0 - T_1) + \lambda\mu_{\mathrm{A}} T_1 = \alpha_{T_1} + \lambda\mu_{\mathrm{A}} T_1$$

$$\text{Sub-system B} = \alpha^{\mathrm{B}}_{\text{present day}} = \alpha_0 + \lambda\mu_0(T_0 - T_1) + \lambda\mu_{\mathrm{B}} T_1 = \alpha_{T_1} + \lambda\mu_{\mathrm{B}} T_1$$

$$\text{Sub-system C} = \alpha^{\mathrm{C}}_{\text{present day}} = \alpha_0 + \lambda\mu_0(T_0 - T_1) + \lambda\mu_{\mathrm{C}} T_1 = \alpha_{T_1} + \lambda\mu_{\mathrm{C}} T_1$$

where α_{T_1} is the common isotope ratio at $\alpha_{T_1} = \alpha_0 + \lambda\mu_0(T_0 - T_1)$.

Graphically, in the (α, T) plane, these three equations are represented first by a single straight line joining the origin to the point of convergence (T_1, α_1) and then by three straight lines which diverge in proportion to their different μ values (Figure 6.2).

Let us return now to the question of the difference in **dispersion** between **granites** and **basalts**, that is, between continental crust and mantle materials.

The mantle has very low $^{87}Rb/^{86}Sr$ ratios, ranging from 0.01 to 0.1. The evolution of its $^{87}Sr/^{86}Sr$ on the isotope evolution diagram is almost a horizontal line if we take dispersion

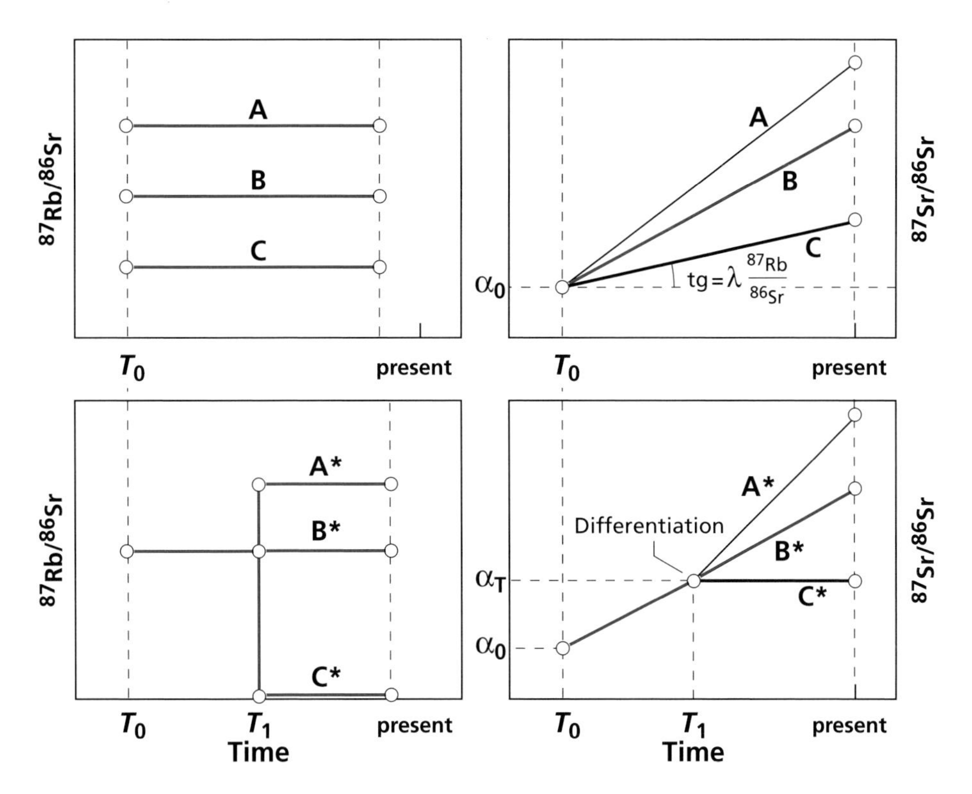

Figure 6.2 Diagrams of the evolution of isotope systems with time. Left: the $^{87}Rb/^{86}Sr$ parent–daughter ratio versus time. Right: the $^{87}Sr/^{86}Sr$ ratios for the same systems under the same conditions. Top: evolution of several closed systems where the $^{87}Rb/^{86}Sr$ ratios are constant. Bottom: a two-stage evolution. The first episode has a constant Rb/Sr ratio and evolves in a closed system. At T_1 a sudden chemical differentiation event occurs giving rise to two systems (A) and (B) with two different Rb/Sr ratios. Afterwards, the systems evolve separately as closed systems. Right: the **chemical** scenario as it translates on the **isotope** evolution diagram.

measured on granite as our scale. The ratios representing the continental crust are much higher: they range from 0.3 to 5. Let us get the order of magnitude clear in our minds.

Exercise

Suppose that granites from the continental crust derive from the mantle, which has a $(^{87}Sr/^{86}Sr)$ isotopic ratio $\alpha_0 = 0.701$. Calculate the present-day $^{87}Sr/^{86}Sr$ ratios of the granites assuming that their ages are 3, 2, and 1 Ga, respectively, with $^{87}Rb/^{86}Sr$ ratios of 1, 2, and 3, corresponding to the time when they became granites, therefore when they were extracted from the mantle. Plot the isotope evolution on an (α, T) diagram.

Answer

Here is the matrix of the results for the $^{87}Sr/^{86}Sr$ ratios of these granites (Table 6.1). Assuming the mantle has a $^{87}Rb/^{86}Sr$ ratio of 0.01, the evolution of the granites is shown in the (α, T) diagram in Figure 6.3. It shall be seen that the Rb/Sr ratios and age dispersions assumed in the model are enough to explain the wide dispersion of continental granites observed.

Table 6.1 Present-day ($^{87}Sr/^{86}Sr$) ratios

	Age (Ga)		
μ	3	2	1
3	0.8288	0.7862	0.7436
2	0.7862	0.7578	0.7290
1	0.7436	0.7294	0.7152

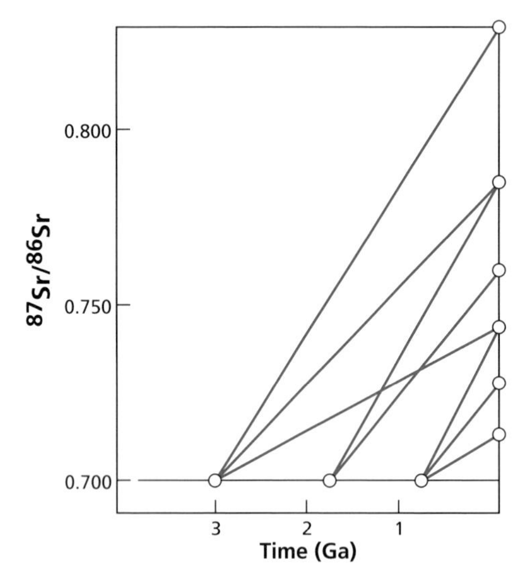

Figure 6.3 Theoretical ($^{87}Sr/^{86}Sr$, T) isotope evolution diagrams.

EXAMPLE

The history of the granites of western Greenland

A real case is represented in the strontium isotopic evolution diagram (Figure 6.4), that of the granites of western Greenland studied by **Steve Moorbath** of the University of Oxford and his various co-workers. Notice that this real diagram (almost) matches the diagram for the theoretical model (see Moorbath and Taylor, 1981).

Let us tie this in with what we learned in Chapter 3 about the isochron diagram. In the theoretical two-stage model of evolution with differentiation at T_1, we represented the system in an (α, T) diagram. The equations for the three straight lines were:

$$\alpha^{\mathrm{A}} = \alpha_{T_1} + \lambda \mu_{\mathrm{A}} T_1$$

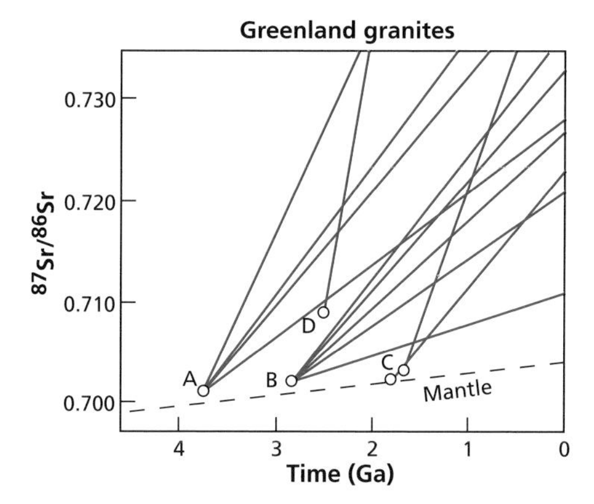

Figure 6.4 Isotope evolution diagrams for rocks of western Greenland. A, Amitsoq gneiss; B, Nuk gneiss; C, Ketelidian gneiss; D, Quorguq granites. After Moorbath and Taylor (1981).

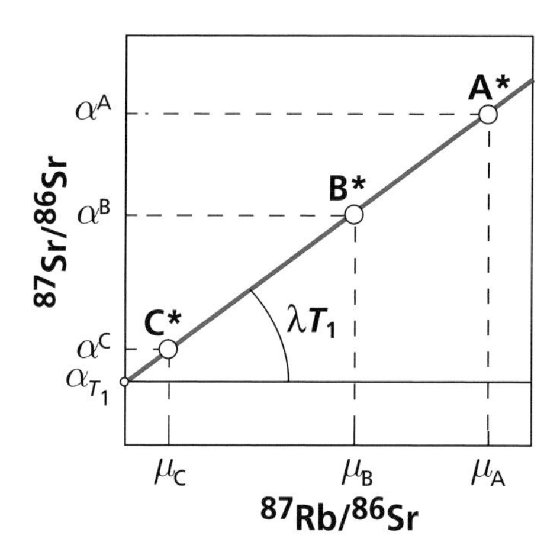

Figure 6.5 Equivalent to the diagram at the bottom of Figure 6.2, but plotted differently.

$$\alpha^{B} = \alpha_{T_1} + \lambda \mu_{B} T_1$$

$$\alpha^{C} = \alpha_{T_1} + \lambda \mu_{C} T_1.$$

In these equations the common isotope composition at time T_1 is noted α_{T_1}. The three systems could have been represented in an (α, μ) diagram (Figure 6.5). This is the isochron diagram, which is one of the foundations of geochronology. Why use (α, μ) in geochronology and (α, T) in isotope geochemistry? The answer is important as it defines a method.

In geochronology, what we measure directly are α and μ. It is therefore normal to plot the two types of measurement that are to define a correlation diagram whose slope, that is, the age, we shall calculate. In isotope geochemistry, we determine ages T and initial

ratios α_0 by isochron construction diagrams on the various sets of rock, of varied ages. Our data will therefore be α_0 and T. It is normal, once again, to plot the data for one against the other. The unknown (provided by the slope) is the $^{87}Rb/^{86}Sr$ ratio, which is therefore a **chemical parameter**.

6.1.3 Granites and granites

In the theoretical model of evolution of granites in the exercise, we assumed that the granites derived from differentiation of the mantle. Their initial isotope ratios were therefore those of the mantle. This idea of granite deriving from the mantle by chemical differentiation from basaltic magma was first proposed by **Norman Bowen**. In the 1920s and 1930s he conducted the first petrology experiments in the course of which he made a piece of granite artificially by crystallizing a melt of basalt by stages (fractional crystallization) (Bowen, 1928).

After the war, a second idea arose. Having first been evoked by **Bowen** himself it was actually developed by the German school and particularly **Helmut Winkler**. He too made granite in the laboratory, but this time in a quite different way, by melting clastic sediments (Winkler, 1974). In petrology, this process is known as **anatexis**. Granite is thought to be the end product of extreme metamorphism where the increase in temperature causes melting.

These two theories on the origin of granite have repercussions for isotope geochemistry. If a granite derives from basalt, its initial $^{87}Sr/^{86}Sr$ ratio is close to that of basalt and so very low (close to 0.703). On the contrary, if it derives from ancient sediment, which itself derived from the erosion of ancient granite, its $^{87}Sr/^{86}Sr$ ratio will be much higher (more like 0.720 to 0.780). The corollary of this observation is that, by measuring the initial isotope ratios (that is, by correcting for age by the isochron construction), the origin of granites can be systematically studied.

EXAMPLE

Himalayan granites

If we measure the initial $^{87}Sr/^{86}Sr$ isotope ratio of Lhasa granite (Tibet) dated 50 Ma, we find 0.705 whereas the corresponding ratio for granite of the geographically very close High Himalayas near Mount Everest, dated 30 Ma, is 0.780. Comparing these ratios with the histogram in Figure 6.1 reveals that the two granites have very different sources, origins and histories. The Lhasa granite Sr initial isotope ratio is similar to $^{87}Sr/^{86}Sr$ ratios for the mantle. It may originate directly from the mantle or indirectly from the intrusion of magma from the mantle which has crystallized and remelted. By contrast, the High Himalayas granite collected from the slopes of Mount Everest has a Sr isotope ratio situating it clearly among values for the continental crust. This granite derives from remelting of older continental crust. Himalayan granites are therefore of two separate origins (author's unpublished results).

There, then, is a first example of an isotopic tracing technique. A straightforward measurement of strontium isotope ratios has provided a solution to a problem over which generations of petrologists have argued, that of the origin of granites, and to make a quick and simple inventory of them. We shall return to this issue later on as it is central to the origin of the continental crust.

6.1.4 Basalt and oceanic basalt

Given the great difference in $^{87}Sr/^{86}Sr$ isotopic composition between continental crust and mantle, it is better, if we wish to determine the chemical and isotopic composition of the mantle, to avoid taking samples from continental volcanoes (such as those of the Massif Central in France, for example). In this way, we avoid contamination by continental material, which is always possible and would skew results. We shall therefore look first at oceanic volcanism. Since the oceanic crust is basaltic, the $^{87}Sr/^{86}Sr$ ratios of magmas and the oceanic crust are similar, which reduces the risk of any disastrous contamination.

Exercise

A basaltic magma with a $^{87}Sr/^{86}Sr$ ratio of 0.7025 and a Sr concentration of 300 ppm passes through granitic crust with a $^{87}Sr/^{86}Sr$ ratio of 0.750 and a Sr concentration of 150 ppm. The basalt assimilates granite in the proportions of 1%, 2%, and 5%. What are the isotope ratios of the contaminated magmas? What precision is required on the $^{87}Sr/^{86}Sr$ ratio measurement to detect the phenomenon?

Answer

1%	2%	5%
0.702 72	0.702 96	0.703 65

Notice that if we measure the isotope ratio to the nearest 1%, that is, ±7 on the third figure, we cannot detect these finer points. At 1% we cannot detect anything for sure either. It can only be seen with precision of at least 1%.

There are two types of oceanic basalt: **mid-ocean-ridge basalt (MORB)**, which is emplaced on the sea floor at oceanic ridge crests and has very characteristic forms of flow known as pillow lavas, and **ocean-island basalt (OIB)** produced very largely by subaerial volcanism and which forms island chains (Hawaii, Tahiti) or archipelagos (Azores, Canaries, Galápagos). These two types of basalt are **statistically** different in terms of their petrology and chemistry.

The MORBs are said to be **tholeiitic**. They contain lower concentrations of alkali elements (sodium, potassium) and are slightly richer in silicon. Conversely, OIBs tend to have **higher alkali contents**. (But again this is a statistical observation, because tholeiitic basalt is found on ocean islands and some alkali basalts are found in the vicinity of mid-ocean ridges.) Isotope analysis of both types of basalt (removing the altered parts for submarine basalts) indicates that MORBs have $^{87}Sr/^{86}Sr$ ratios ranging from 0.7022 to 0.7034 while the corresponding range for OIBs is 0.7030–0.7050.

The $^{87}Sr/^{86}Sr$ isotope ratios of MORBs are statistically lower than the ratios for OIBs. To understand this, we need to turn to geological observation, that is, the geological context in which samples are collected. Samples of MORBs are collected from zones where the ocean floor is spreading, as if the mantle were almost cropping out at the surface, whereas OIB forms island chains "independently" of sea-floor spreading. These are plumes of magma that cross the tectonic plates and so come from reservoirs that lie

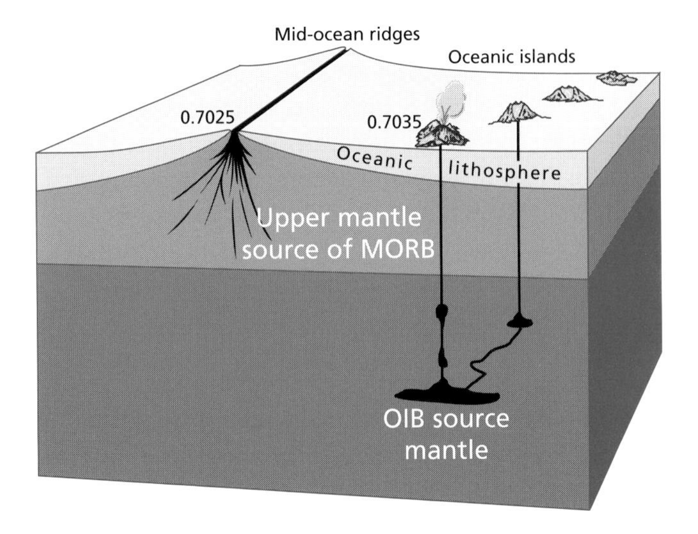

Figure 6.6 The origins of mid-ocean-ridge basalt (MORB) and ocean-island basalt (OIB). The diagram shows how the two types of oceanic basalt may have arisen on the basis of isotope analyses and elementary geodynamics. The values are $^{87}Sr/^{86}Sr$ ratios.

deeper than the ocean-ridge reservoirs. This is the **hot spot** hypothesis proposed by **Jason Morgan** of Princeton University (Morgan, 1971). If we accept that Sr isotope ratios are preserved during partial melting of the mantle and transport, this suggests that the mantle is isotopically stratified, with an upper layer as the source of MORB, with a very low Rb content (compared with Sr) and a lower layer as the source of OIB, which is a little richer in Rb (compared with Sr) (Figure 6.6). Let us introduce some quantitative constraints to make this clearer.

6.1.5 Comparative "prehistoric" evolution of MORB and OIB: the concept of a model age

Suppose we have two oceanic volcanic rocks, one collected by dredging along the center of an oceanic ridge and the other collected from a Recent volcano of the Canary Islands. Their $^{87}Sr/^{86}Sr$ ratios are 0.7025 and 0.7045, respectively. What can we say about the prehistory of the two rocks? Using just the Sr isotope ratios, we can assert that the rocks are from different sources with different $^{87}Sr/^{86}Sr$ signatures and that those differences are ancient. But how ancient? Suppose now that we measure the $^{87}Rb/^{86}Sr$ ratios of these rocks and find 0.01 and 0.1, respectively. Of course, these ratios are not those of the sources in the mantle. To make a volcanic rock from the solid mantle, a liquid magma has had to be created, which implies partial melting of the mantle and associated chemical fractionation, and so variations in the chemical Rb/Sr ratio. This is followed by the transfer of the magma to the surface with possible contamination on its way up. But laboratory experiments, observations, and measurements on volcanic series have taught us that rubidium enters the magma much more readily than strontium. Consequently, we can infer that:

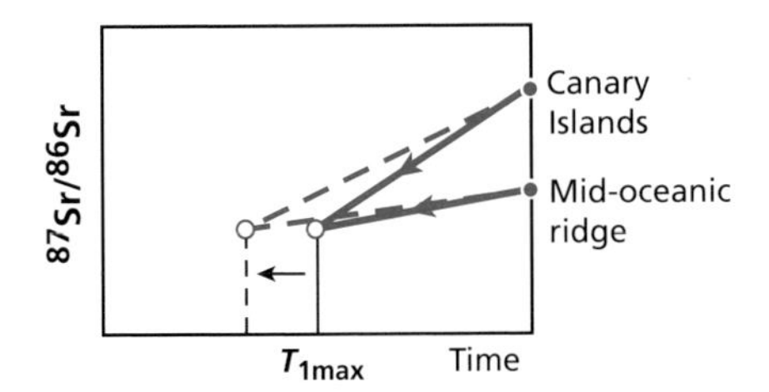

Figure 6.7 Calculation of a model age of differentiation of two basalts, one from the Canary Islands and one from a mid-ocean ridge. The solid lines show the vectors of isotope evolution for (Rb/Sr) measured on the rocks and the dashed lines are the more plausible vectors.

$$\left(\frac{^{87}\text{Rb}}{^{86}\text{Sr}}\right)_{\text{volcanic rock}} \geq \left(\frac{^{87}\text{Rb}}{^{86}\text{Sr}}\right)_{\text{mantle source}}.$$

The chemical Rb/Sr ratios measured on the basalts are therefore the upper bounds of the source ratios of these basalts. Let us make a further assumption: as these two volcanic rocks come from the mantle (we shall suppose that at one time they both belonged to the Earth's primitive mantle which was homogeneous, identical throughout) and then that their source domains separated (or became individualized) through geodynamic and geochemical processes of varying complexity.

Can this scenario be put into quantitative terms? Let us put it graphically first. We plot a ($^{87}Sr/^{86}Sr$, T) graph and construct **vectors of evolution** for the sources of the two rocks (Figure 6.7). They pass through the present-day measurement points, the two isotope compositions. Their maximum slopes are $\lambda\left(^{87}\text{Rb}/^{86}\text{Sr}\right)$. The two vectors intersect at about 1.4 billion years. (Plot the graph to check this as an exercise.)

If we address the question using algebra, we can write the evolution equations for the two sources by noting the $^{87}Sr/^{86}Sr$ isotope ratios α and the $^{87}Rb/^{86}Sr$ ratios μ. Using a linear approximation, we get:

$$\begin{cases} \alpha^t_{\text{ridge}} = \alpha_0 + \lambda\mu_{\text{ridge}}\, T_1 \\ \alpha^t_{\text{Canaries}} = \alpha'_0 + \lambda\mu_{\text{Canaries}}\, T'_1. \end{cases}$$

If the two volcanic rocks are from identical media which separated at time T_1, we can calculate T_1 and their composition by positing $\alpha_0 = \alpha'_0$ and $T_1 = T_1'$.

This then gives:

$$T_1^{\text{Sr}} = \frac{1}{\lambda}\left(\frac{\alpha_{\text{ridge}} - \alpha_{\text{Canaries}}}{\mu_{\text{ridge}} - \mu_{\text{Canaries}}}\right).$$

With $\lambda = 1.42 \cdot 10^{-11}\,\text{yr}^{-1}$, we get $T_1^{\text{Sr}} = 1.40$ billion years.

But, one might object, we have used the $^{87}Rb/^{86}Sr$ ratios of the volcanic rocks and not those of the mantle. Now, as we said, these are maximum values and not real values. This means that the 1.4 billion is a minimum value. The two real vectors may well intersect at an earlier time. So the true age of differentiation T_D is greater than (or equal to) the age calculated with strontium $= T_1 \geq T_1^{Sr}$.

To interpret the $^{87}Sr/^{86}Sr$ ratios of the two basalts, one of MORB and one of OIB, we presume the two domains of the mantle from which the basalts were extracted were separated more than 1.4 billion years ago and evolved independently thereafter. This is a strong constraint as to the present-day structure of the mantle and as to the way it has evolved historically. However, one point has to be clarified. This reasoning assumes, as said, that when the magma forms at depth, it is isotopically re-equilibrated with the source domain, that is, it has the same isotopic composition as the source domain (this is the **isotopic equilibration** assumption) and that subsequently, during its time at the surface, the basalt's isotopic composition is not altered either by mixing with the crust (which is why we chose two oceanic basalts), or by interaction with surface fluids (the closed system assumption), or by weathering. With a few precautions, these assumptions are generally borne out.

As can be seen, the **principles** behind this model are **very simple** and yet they lead to **very important** results.

Remark

Until now, with dating, we were concerned about the history of rocks. In studying initial isotope ratios, we are interested in their **pre-history** and in that of the chemical elements of which they are composed.

Exercise

Exercise

The $^{143}Nd/^{144}Nd$ isotope ratios were measured on the same basalts as before from the mid-oceanic ridge (0.513 50) and the Canaries (0.512 95). The $^{147}Sm/^{144}Nd$ ratios are 0.25 for the MORB and 0.20 for the OIB. Supposing the two reservoirs of these rocks first had a common history in the same mantle reservoir before separating and differentiating at time T_1, write the evolution equations for the $^{147}Sm/^{143}Nd$ system. What is the approximate age T_1 at which the two media of origin of the two basalts separated?

Given that during genesis of basalts from the mantle, the magma is more enriched in neodymium than in samarium, is the age calculated a minimum or a maximum? By comparing it with the model age for strontium, can you improve your knowledge of T_1? What do you conclude from this?

Answer

$T_1^{Nd} = 1.69$ Ga and $T_1^{Nd} \geq T_1 \geq T_1^{Sr}$, therefore $T_1 = 1.55 \pm 0.13$ Ga.

6.1.6 The problem of mixtures

In the phenomena we have considered so far, the process by which magma is generated is straightforward enough: partial melting with isotope equilibration followed by transfer towards the surface. In fact, actual geological phenomena are more complex because mixing phenomena occur frequently in nature. We have already come across these when dealing with problems of geochronology and then in the exercise in Section 6.1.4.

To get our ideas straight, let us begin with an example. Let us consider the case of two continental basalts located in, say, La Chaîne des Puys in the Massif Central (France). One of the basalts contains **xenoliths**[1] of mantle rock (peridotite), showing that the magma did not remain in a magma chamber as it was transferred to the surface, otherwise the heavier xenoliths would have settled out. Its $^{87}Sr/^{86}Sr$ ratio is similar to that of peridotitic xenoliths at 0.7035. The second basalt contains slightly more silica and does not contain any xenoliths of deep origin, but, on the contrary, xenoliths of granite which were probably scoured from the continental crust. The second basalt has a $^{87}Sr/^{86}Sr$ ratio of 0.706. The granite xenoliths have $^{87}Sr/^{86}Sr$ ratios of 0.740.

We interpret such an occurrence by considering that the second basalt has been contaminated by continental crust. What is the extent of this contamination? From the mixing formula already studied at the end of Chapter 3:

$$\left(\frac{^{87}\mathrm{Sr}}{^{86}\mathrm{Sr}}\right)_{\text{mixture}} = \left(\frac{^{87}\mathrm{Sr}}{^{86}\mathrm{Sr}}\right)_{\text{mantle magma}} \cdot x + \left(\frac{^{87}\mathrm{Sr}}{^{86}\mathrm{Sr}}\right)_{\text{continental crust}} \cdot (1 - x)$$

where $x = (m_{\mathrm{m}}\, C_{\mathrm{m}}^{\mathrm{Sr}})/(m_{\mathrm{m}}\, C_{\mathrm{m}}^{\mathrm{Sr}} + m_{\mathrm{cc}}\, C_{\mathrm{cc}}^{\mathrm{Sr}})$ is the mass of uncontaminated magma, m_{cc} the mass of continental crust dissolved in the magma, $C_{\mathrm{m}}^{\mathrm{Sr}}$ the strontium concentration of the uncontaminated magma = 300 ppm, and $C_{\mathrm{cc}}^{\mathrm{Sr}}$ the strontium concentration of continental contaminants = 100 ppm.

With $(^{87}Sr/^{86}Sr)_{\text{mixture}} = 0.705$, $(^{87}Sr/^{86}Sr)_{\text{magma}} = 0.7035$, and $(^{87}Sr/^{86}Sr)_{\text{cc}} = 0.740$, we get:

$$x = \frac{0.740 - 0.706}{0.740 - 0.7035} = 0.931$$

for the proportion of strontium of deep origin (mantle).

Let us now determine the contaminated quantity. If we posit $u = m_{\mathrm{c}}/(m_{\mathrm{c}} + m_{\mathrm{m}})$, which is the proportion of the continental crust assimilated, we find $u = 0.18$ or 18%. The magma is contaminated by about 20% in mass.

Where we have several samples, mass contamination phenomena may be shown up by the linear relation $\left(^{87}\mathrm{Sr}/^{86}\mathrm{Sr},\ 1/C^{\mathrm{Sr}}\right)$ where C^{Sr} is the strontium concentration. (Look back at Chapter 3.)

Exercise

Analysis of a series of rocks related to a carbonatite[2] volcano yields the results shown in Table 6.2.

Given that the granite country rock contains 100 ppm of strontium and has a $^{87}Sr/^{86}Sr$ ratio of 0.715, can this suite of rocks be interpreted as contamination of a deep magma in a granitic environment? How can this be demonstrated? Can the isotope composition of the deep magma be ascertained and what is its value?

Answer

Yes, it is a mixture. This can be shown by applying the (1/*C*) test of Chapter 3, Section 3.4.2. The composition of the uncontaminated magma is 0.7030.

[1] Xenoliths are pieces of rock that are "foreign" to the magma, having been picked up during its transfer towards the surface.

[2] Look up what this and the words in the table mean in a petrology textbook.

Table 6.2 Analysis of volcanic rocks

Rock	$^{87}Sr/^{86}Sr$	C^{Sr} (ppm)
Carbonatite	0.7040	1000
Ijolite	0.7062	350
Syenite	0.7070	280
Nephelinite	0.7107	150

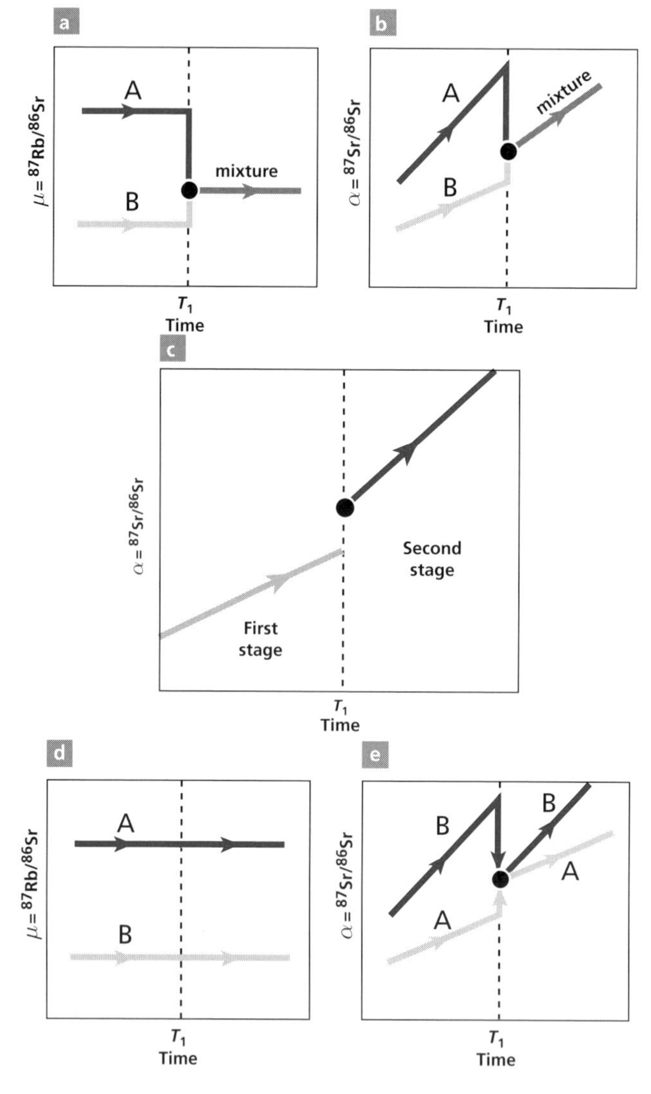

Figure 6.8 Evolution over time of two systems involved in mixing at T_1. (a) The $^{87}Rb/^{86}Sr$ ratios do not change until T_1, when mixing occurs. (b) The corresponding evolution of the $^{87}Sr/^{86}Sr$ ratio. (c) Reconstructed evolution with a shift in the $^{87}Sr/^{86}Sr$ ratio at T_1. The shift could also be negative! **This is the only case where the radiogenic $^{87}Sr/^{86}Sr$ ratio can decrease. Otherwise, it invariably increases as a result of radioactivity**. (d, e) Evolution of two systems which underwent isotopic exchange at T_1 without changing their chemistry.

These mixing phenomena may occur at depth or near the surface during the system's isotopic evolution, for example, during tectonic phenomena where rocks are mixed by thinning and folding or by breaking and mixing, or during mantle convection processes associated with stirring. They also occur, of course, in the processes of erosion, sedimentation, transport, and the formation of oceans and lakes. They have substantial repercussions, as we shall see on many occasions in what follows. **Each time, they introduce a hiatus in the ($^{87}Sr/^{86}Sr,t$) isotope evolution diagram.** This isotope ratio may **rise** or **fall** suddenly. The chemical $^{87}Sr/^{86}Sr$ ratio may also change abruptly. Thus, the slope of evolution after mixing changes suddenly (Figure 6.8).

6.1.7 Isotopic exchange

We spoke of isotope exchange at the beginning of this book when discussing the isotopic dilution method. But the phenomenon extends well beyond the confines of the laboratory. In nature, when two systems (rocks, minerals, water/rock mixtures) are in chemical equilibrium but have different isotope compositions, both systems exchange their atoms to tend towards isotopic homogeneity (while being, it is worth repeating, chemically heterogeneous). The final isotopic composition for heavy elements is equal, of course, to the weighted mean of the systems involved.

As we have said, it is the diffusion processes that control the rate at which isotopic exchanges occur. They are faster in liquid phases than in solid phases and are activated by temperature.

We have made use of this phenomenon in the case of the isochron diagram. Here, we wish to remind readers that it plays a role in all chemical processes, but naturally this isotopic exchange may be partial or total depending on circumstances.

6.1.8 The Schilling effect

Jean-Guy Schilling of Rhode Island University discovered an intermediate structure between MORB and OIB, which appears to complicate matters but in fact confirms the dual origins of the basalts. Hot-spot injections occur at some points on the mid-oceanic ridges, as reflected by the ridges being subjected to topographic bombardment (see review in Schilling, 1992) (Figure 6.9).

The first reported and most spectacular case of such hot-spot injection beneath mid-oceanic ridges is that of Iceland. An immense volcanic structure of deeper origin lies astride the mid-Atlantic ridge creating a huge oceanic island. A less spectacular case is found on the Atlantic ridge near the Azores. There the bombment remains submarine, because the emerged islands are shifted to the east.

Now, analysis of the isotope compositions of these mixed structures reveals $^{87}Sr/^{86}Sr$ values that are intermediate between MORB and OIB values. The distribution of $^{87}Sr/^{86}Sr$ isotope ratios along the mid-ocean ridge follows the profile of the topography (Figure 6.9). It is only at some considerable distance from the hot spot that isotope values typical of MORB recur.

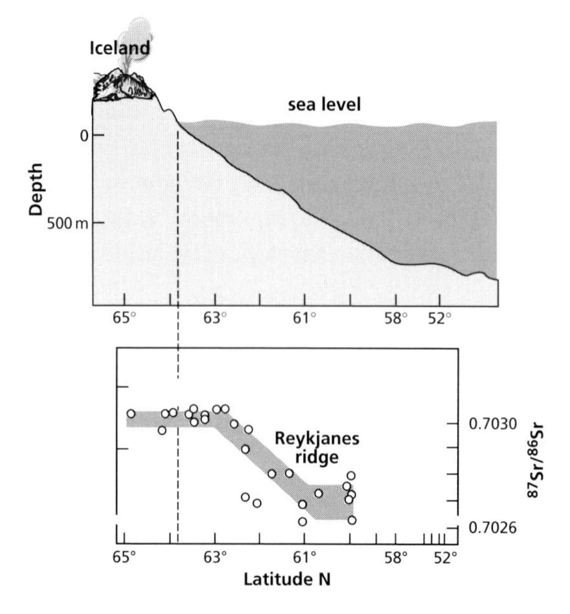

Figure 6.9 Study of samples collected south of Iceland. Variation in the $^{87}Sr/^{86}Sr$ ratio along the mid-oceanic ridge southwards from Iceland along with variation in submarine topography. The parallel is astonishing. After Hart *et al.* (1973).

When the areas of the world where the Schilling effect occurs are removed from the statistics, the dispersion of MORB values declines. The range of variation is then just 0.7022–0.7027, and so is far narrower than the variation observed for OIB, which is 0.7032–0.7050. This observation, which has been confirmed for many ocean ridges, shows there is a clear statistical distinction between MORB sources with a narrow range of very low $^{87}Sr/^{86}Sr$ ratios and the higher and more scattered $^{87}Sr/^{86}Sr$ ratios of the hot spots. This shows that samples must be collected from ocean ridges using knowledge of the geological and geodynamic context before interpreting the isotope measurements and before distinguishing the MORB interacting with hot spots from the others (**Hart** *et al.*, 1973) (Figure 6.10).

6.2 Strontium–neodymium isotopic coupling

The development of neodymium isotopic geochemistry was a key moment in isotope geology. It was brought about by three teams: **the present author's** team at the Institut de Physique du Globe in Paris (Richard *et al.*, 1976), **Jerry Wasserburg**'s team at the California Institute of Technology (DePaolo and Wasserburg, 1976a, b), and the team of **Keith O'Nions** then at Columbia University, New York (O'Nions *et al.*, 1977). This was a decisive stage in the use of isotope tracers for understanding geodynamic processes and the historical evolution of terrestrial reservoirs because, for the first time, two chemically different isotopic tracers yielded coherent, complementary information.

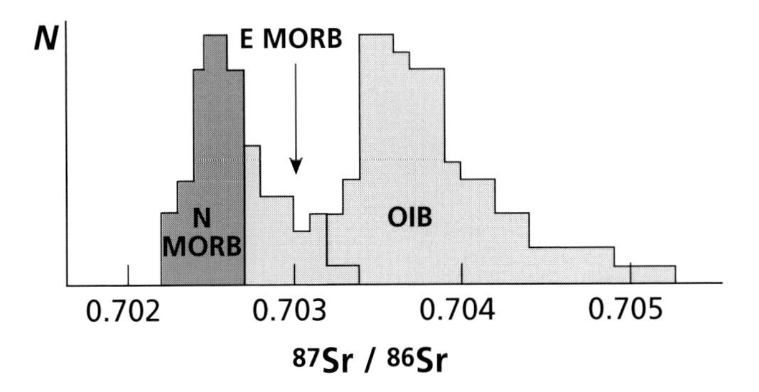

Figure 6.10 Histogram of isotope ratios measured on various types of oceanic basalts. Hoffmann has called the MORB subjected to the Schilling effect E-MORB. Normal N-MORB is located outside of these peculiar areas.

As said in the previous chapters, ^{147}Sm decays by α radioactivity to ^{143}Nd with a half-life of $1.5576 \cdot 10^{11}$ years and a decay constant λ of $6.42 \cdot 10^{-12}$ yr^{-1}. It is possible, then, in principle to study variations in ^{143}Nd/^{144}Nd isotope ratios in basalts and granites as with ^{87}Sr/^{86}Sr ratios. The difficulty is that these Nd isotope variations are very small, of the order of 1 per 10 000. It was therefore not until very precise mass spectrometry techniques were developed for the Apollo missions that this program could be carried out.

6.2.1 Neodymium isotope variations

It was therefore not until 1975 in Paris, and then 6 months later in Pasadena, that variations in the isotope composition of neodymium of terrestrial origin were shown to exist by drawing on the giant leap in mass spectrometry:

- MORBs have ^{143}Nd/^{144}Nd isotope ratios of about 0.513 20;
- OIBs have ^{143}Nd/^{144}Nd isotope ratios of about 0.5128;
- the isotope ratios of granitic rocks range widely from 0.508 to 0.511 depending on the age of the rock.

Qualitatively, the phenomenon is the same as for ^{87}Sr/^{86}Sr ratios, except that the variations are in the opposite direction. The most radiogenic strontium ratios are for granites, while the most radiogenic neodymium ratios are found in MORBs. This inverse variation called for closer study and examination of the isotope correlation between Sr and Nd measured on the same samples.

6.2.2 The ε_{Nd} notation

To make the variations easier to read, we shall express the Nd variations as variations relative to a reference sample. This reference sample is chosen from among meteorites whose present-day isotope ratio is constant (whichever meteorite it may be) at 0.512 638.[3] The ε_{Nd} unit is defined (see **DePaolo** and **Wasserburg**, 1976a) by:

$$\varepsilon_{Nd} = \left[\frac{\left(\frac{^{143}Nd}{^{144}Nd}\right)_{sample} - \left(\frac{^{143}Nd}{^{144}Nd}\right)_{chondrite}}{\left(\frac{^{143}Nd}{^{144}Nd}\right)_{chondrite}} \right] \times 10^4.$$

In short, ε_{Nd} is a relative variation expressed per 10 mil, which is more convenient for expressing variations and can be used to manipulate numbers like 2, 10, or 20 whether positive or negative.

Exercise

Calculate the ε_{Nd} ratios for two samples: one is a MORB whose $^{143}Nd/^{144}Nd$ ratio is 0.513 10 and the other a billion year-old granite whose $^{143}Nd/^{144}Nd$ ratio is 0.509 00.

Answer

By using the isotope ratio of chondrites as a reference, we find $\varepsilon = +9.36$ for the MORB and $\varepsilon = -70.6$ for the granite.

Exercise

The isotopic composition of a mixture (noted R_M) of two components A and B of isotopic composition R_A and R_B is

$$R_M = R_A x_A + R_B(1 - x_A),$$

x_A being the mass fraction of component A in the mixture. Show that if we use the ε notation, the mixture ε_M between two components (A) and (B) with ε_A and ε_B can be written:

$$\varepsilon_M = \varepsilon_A x_A + \varepsilon_B(1 - x_A).$$

Answer

Begin with $R_M = R_A x_A + R_B (1 - x_A)$. Subtracting R_S (standard) from each side gives:

$$R_M - R_S = (R_A - R_S)\, x_A + (R_B - R_S)(1 - x_A).$$

Dividing both sides by R_S and multiplying by 10^4 gives:

$$\varepsilon_M = \varepsilon_A\, x_A + \varepsilon_B(1 - x_A)$$

6.2.3 The Sr–Nd isotope correlation of basalts

As might be expected after the initial measurements, when the Sr and Nd isotope compositions are measured on a series of oceanic basalts, they are found to be anti-correlated. The

[3] Sometimes another value is given in the literature because of the way stable Nd isotopes are normalized. This difference is removed with the ε notation.

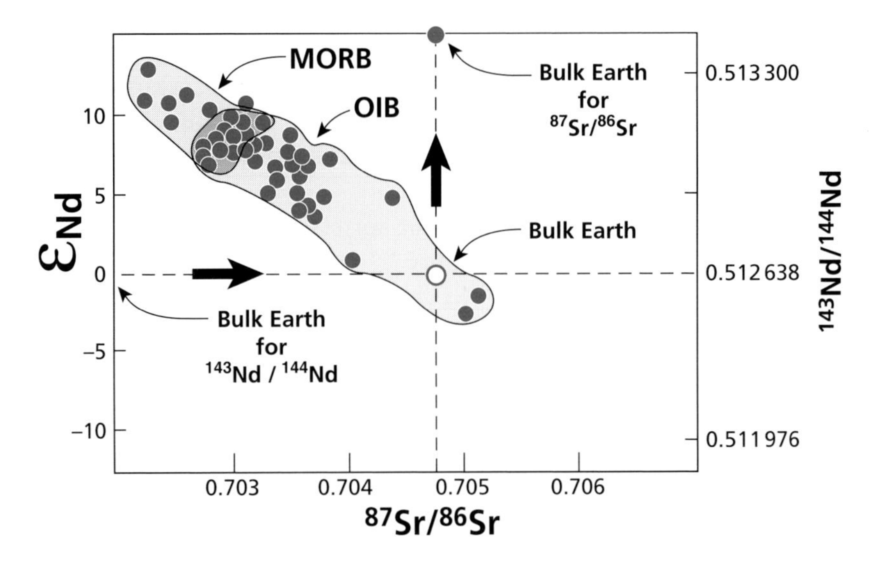

Figure 6.11 Strontium–neodymium isotope correlation established by the Paris, Pasadena, and Columbia teams (1976–7) on oceanic basalts. Thousands of measurements have been added since, and we shall see later how the initial correlation has been altered. The correlation is used for determining the $^{87}Sr/^{86}Sr$ ratio of the Earth from the $^{143}Nd/^{144}Nd$ value of the Earth as determined from meteorites (and then taken as a reference for calculating ε).

correlation observed is good (Figure 6.11) and is an inverse correlation (**DePaolo** and **Wasserburg**, 1976b; **Richard** *et al.*, 1976; **O'Nions** *et al.*, 1977).

Oceanic basalts define a straight line of correlation with low dispersion. This correlation is very important as it shows coherence between Nd and Sr isotope variations. These variations, then, are not caused by some minor chemical processes specific to Sr or Nd elements but by general grand-scale geological processes. It can be considered, at least as a working hypothesis, that the Sr and Nd isotope ratios together act as **tracers** for the major geological phenomena, because together they keep a record of these phenomena.

First of all, two essential reference parameters can be extracted from this correlation: **the Nd and Sr isotope values for the Bulk Earth and the Sm/Nd and Rb/Sr ratios for the Bulk Earth.** The Bulk Earth is a complex system comprising various entities: the continental crust, oceanic crust, mantle, and core. It is not easy to obtain a single mean value for the whole. But, conversely, when we do obtain such a value, it acts as a "universal" reference for all the others.

Let us come back briefly to the meteorites we took as the reference standard for calculating the ϵ parameters of neodymium (as already specified in Chapters 4 and 5). Meteorites have extremely varied chemical compositions. Some have chemical compositions quite close to that of the Sun. These "primitive" meteorites are **carbonaceous chondrites**. Others, by contrast, have chemical compositions analogous to that of basalt. They are the products of volcanism and are known as **basaltic achondrites**. Yet others are made up of iron–nickel alloys: these are the **iron meteorites**. Other better known because more abundant ones are chemically intermediate: these are the **ordinary chondrites** as seen in the previous chapters (see **Jacobsen** and **Wasserburg**, 1980). All of these meteorites are of the same age to within a few million years: around 4.55 ± 0.05 Ga. Now, all of them have the same present-day $^{143}Nd/^{144}Nd$ ratio of 0.512 638, but not the same $^{87}Sr/^{86}Sr$ ratio![4] It is therefore reasonable to suppose that if these meteorites, which represent the primordial planetary material, have the same Nd isotope composition and a near-constant $^{147}Sm/^{144}Nd$ ratio, the primitive material of the Earth must also have the same isotopic and chemical composition. These ratios are therefore probably those of the Bulk Earth.

On the strength of this and of the (Sr–Nd) isotope correlation, we can determine the strontium isotope composition of the Earth (Figure 6.11) and, assuming an age of 4.55 Ga for the Earth, we can determine its $^{87}Rb/^{86}Sr$ ratio. Note that we know the initial $(^{87}Sr/^{86}Sr)_0$ ratio which is the initial ratio of meteorites of 0.698 98 and is assumed to be the same for the early material of the Earth.

Exercise

Calculate the Rb/Sr ratio of the Earth given that the present-day $^{87}Sr/^{86}Sr$ ratio is 0.7050.

Answer

We write the equation for evolution in a closed system, which is legitimate for the Bulk Earth since neither Rb nor Sr escapes from it. If the age of the Earth is T_0, we have:

$$\left(\frac{^{87}\mathrm{Sr}}{^{86}\mathrm{Sr}}\right)_t = \left(\frac{^{87}\mathrm{Sr}}{^{86}\mathrm{Sr}_0}\right)_0 + \left(\frac{^{87}\mathrm{Rb}}{^{86}\mathrm{Sr}}\right)(e^{\lambda T_0} - 1).$$

Hence:

$$\frac{^{87}\mathrm{Rb}}{^{86}\mathrm{Sr}} = \frac{\left(\frac{^{87}\mathrm{Sr}}{^{86}\mathrm{Sr}}\right)_t - \left(\frac{^{87}\mathrm{Sr}}{^{86}\mathrm{Sr}}\right)_0}{(e^{\lambda T_0} - 1)} = \frac{0.705 - 0.6989}{0.006\,674\,5}$$

$$\left(\frac{^{87}\mathrm{Rb}}{^{86}\mathrm{Sr}}\right)_{\mathrm{Earth}} = 0.091 \pm 0.004.$$

The second stage is more difficult than it looks. We must shift from an **atomic** isotope ratio to a **gravimetric** chemical ratio:

$$\mathrm{Rb_{total}} = (^{87}\mathrm{Rb} + ^{85}\mathrm{Rb}) \times \text{atomic mass of Rb}$$

$$\mathrm{Rb_{total}} = \left[^{87}\mathrm{Rb}\left(1 + \frac{^{85}\mathrm{Rb}}{^{87}\mathrm{Rb}}\right)\right] \times \text{atomic mass of Rb}$$

$$\mathrm{Sr_{total}} = (^{88}\mathrm{Sr} + ^{87}\mathrm{Sr} + ^{86}\mathrm{Sr} + ^{84}\mathrm{Sr}) \times \text{atomic mass of Sr}$$

[4] The explanation for this is that Sm and Nd are "refractory" elements, that is, they do not vaporize easily. Strontium is refractory too, but Rb is volatile. When solid bodies were created in the primordial solar nebula, fractionation between refractory and volatile elements was an essential phenomenon. However, refractory elements behave quite coherently. Thus Sm/Nd ratios of meteorites are almost constant whereas Rb/Sr ratios vary greatly.

$$Sr_{total} = {}^{86}Sr\left(1 + \frac{^{88}Sr}{^{86}Sr} + \frac{^{87}Sr}{^{86}Sr} + \frac{^{84}Sr}{^{86}Sr}\right) \times \text{atomic mass of Sr.}$$

If we wish to make a precise calculation, we must recalculate the atomic mass of strontium because the $^{87}Sr/^{86}Sr$ ratio is not exactly that of the standard samples, which is 0.709. This is not an easy operation because we must calculate the atomic mass of each isotope (or look it up in tables).[5] We shall ignore this difficulty. The standard isotope ratios are:

$^{85}Rb/^{87}Rb = 2.5926$; $^{88}Sr/^{86}Sr = 8.3752$; $^{84}Sr^{86}Sr = 0.056\,584$.

The atomic mass of Sr is 87.613 and of Rb is 85.35.

We therefore find:

$$\left(\frac{Rb}{Sr}\right)_{mass} = \frac{^{87}Rb}{^{86}Sr} \times 0.341,$$

which gives:

$$\left(\frac{Rb}{Sr}\right)_{Earth} = 0.031$$

Exercise

Calculate the relation between the $^{147}Sm/^{144}Nd$ ratio and the Sm/Nd gravimetric chemical ratio.

Answer

Read on!

6.2.4 Position of granites in the (Sr, Nd) diagram

Analysis of granites of various ages initially by teams in Paris and Cambridge (1978–80) (see **Allègre** and **Ben Othman**, 1980; **O'Nions** *et al.*, 1983; **Ben Othman** *et al.*, 1984) showed, as expected, much wider dispersion of isotope ratios than for the mantle. Instead of having a simple almost linear correlation as for oceanic basalts, continental granites, gneisses, and granulites define a fairly broad domain, but in a "symmetrical" position to the basalts relative to the reference point of the primitive mantle (Figure 6.12). This pattern of distribution of points representing the continental crust is the outcome of two combined effects: for one thing, the continental crust is made up of tectonic segments of different ages fitted together like a mosaic; for another, in the complex processes occurring in the continents (erosion, sedimentation, metamorphism, hydrothermalism, anatexis, folding), fractionation between Rb and Sr and between Sm and Nd are different and so the isotopic results are varied.

[5] The precise atomic mass of an isotope is not just the number of neutrons × neutron mass + number of protons × proton mass, expressed as a unit of mass (1/12 of ^{12}C), because the bonding energy $\Delta E = \Delta mc^2$, related to the energy of formation of the atomic nucleus, must be subtracted.

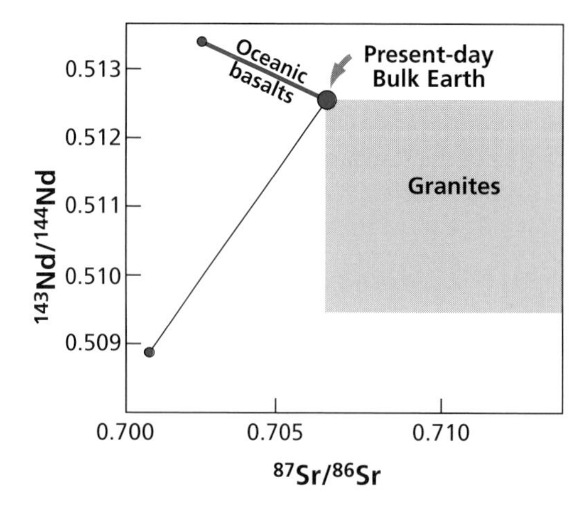

Figure 6.12 Position of granites in the (Sr, Nd) isotope correlation diagram. Notice the domain is much broader than that of basalt. The straight lines of evolution of oceanic basalts and of the Earth's primitive mantle are shown.

6.2.5 Back to the isotope evolution diagrams

We can now plot the isotope evolution diagram for the $^{87}Sr/^{86}Sr$ ratio versus time for basalt and granite. The novelty is that we can now plot the straight line of isotope evolution of the Earth, that is, of the primitive mantle as the reference since none of Rb, Sr, Nd, or Sm is incorporated in the Earth's core.[6] Figure 6.13 shows that granites, which constitute most of the continental crust, are situated above the primitive mantle, and oceanic basalts below, with the MORB being the most extreme.

By measuring the Rb and Sr contents of the continental crust, we can estimate the slope of the straight line of evolution for the continents as equal to $\lambda\ \left(^{87}Rb/^{87}Sr\right)$. The $^{87}Rb/^{86}Sr$ ratios of continental rocks range from 0.5 to 3. The value 1 is a good average. Conversely, the $^{87}Rb/^{86}Sr$ ratios of MORBs are very low ($\sim$0.001). The straight line of evolution of the MORB source medium is virtually horizontal.

As said, the continents are made up of segments of tectonic provinces that are pieced together and formed at different times. It is therefore a series of vectors of evolution that must be considered and which lie behind the wide dispersion of the present-day $^{87}Sr/^{86}Sr$ isotope ratios (Figure 6.1).

Let us try to construct an analogous diagram for the $^{143}Nd/^{144}Nd$ ratio. To do this, we must measure the composition of granites which turn out to have very negative ε parameters (the Bulk Earth being taken as the reference with $\varepsilon_{\mathrm{Earth}} = 0$). When we plot the results on the Nd isotope evolution diagram, we observe a similar distribution as for Sr but in reverse. The values of the granites are below those of the primitive mantle and the basalt values above, with the MORB lying furthest out, with the highest ε parameters (Figure 6.13).

[6] We shall use the terms "Earth," "Bulk Earth," and "primitive mantle" interchangeably when speaking of Rb/Sr or Sm/Nd ratios, since the corresponding ratios are all identical.

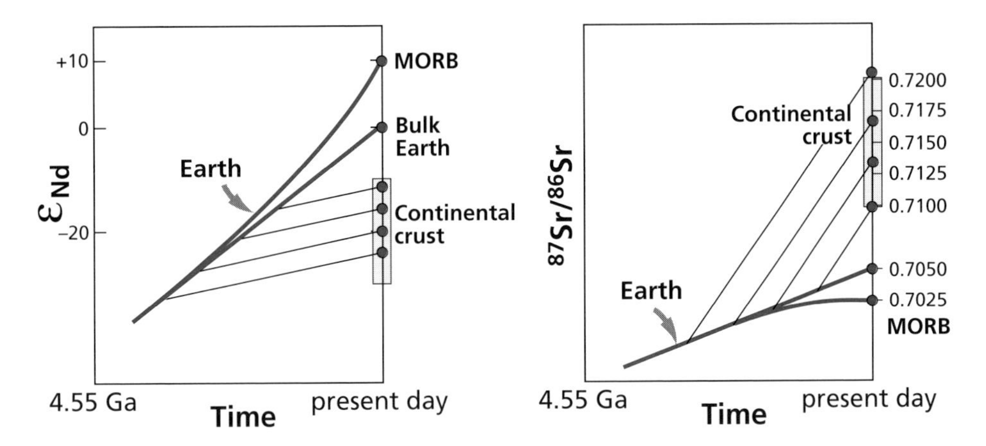

Figure 6.13 Isotope evolution diagrams inferred directly from the (Sr, Nd) isotope correlation. These diagrams show how we can imagine the surface part of the Earth evolved, with continental crust having preferentially extracted some elements and a depleted oceanic mantle (MORB source) as a result of that extraction. Positions in the isotope evolution diagrams are "symmetrical" relative to the Bulk Earth.

To pin down the quantitative evolutions of $^{143}Nd/^{144}Nd$ ratios, we have sought to evaluate the $^{147}Sm/^{144}Nd$ ratios of the two "symmetrical" reservoirs: the continental crust and the mantle, which is the MORB source. For the crust, the $^{147}Sm/^{144}Nd$ ratios are close (0.10–0.12). For the MORB source, the $^{147}Sm/^{144}Nd$ ratios range from 0.15 to 0.3. Unlike with $^{87}Sr/^{86}Sr$, the evolution of the MORB sources in terms of $^{147}Sm/^{144}Nd$ is not a horizontal line.

On this basis, a **simple geodynamic model** can be developed. The continental crust has differentiated at the expense of the mantle probably through volcanism or magmatic processes related to subduction. The continental crust is enriched (compared with the mantle) in elements such as potassium, rubidium, cesium, rare earths, thorium, and uranium. According to the plate tectonics paradigm, the continents "float" on the mantle and are never swallowed up again by it: the enriched elements of the continental crust are therefore stored at the Earth's surface. Thus they are removed from the upper mantle, which is consequently depleted in these elements.

Measurements of concentrations reveal that the continental crust is greatly enriched in Rb and less so in Sr. Therefore, the Rb/Sr ratio of the continental crust is far higher than that remaining in the mantle after extraction (residual mantle). Likewise, the continental crust is much more enriched in Nd than in Sm, and so the Sm/Nd ratio of the continental crust is lower than in the residual mantle. This explains the origin of the inverse isotope correlation. The differences in chemical behavior between the Rb–Sr and Sm–Nd systems are important in dispersion. As Figure 6.14 shows, the Sm/Nd ratio is just as widely dispersed in MORBs as it is in granite, whereas the dispersion of Rb/Sr is very different in the two media (MORB values showing little dispersion). This is because of the properties of rare earths, which are very similar to each other.

Extraction of elements by the continental crust at the expense of the mantle is a complex geological process that must be evaluated but which probably involves magmatic processes.

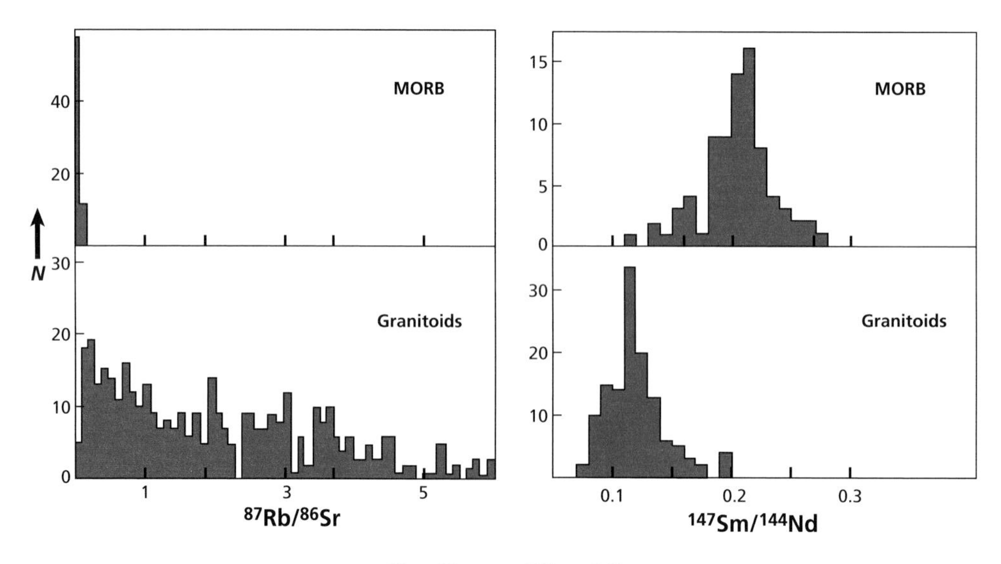

Figure 6.14 Statistical distribution of $^{87}Rb/^{86}Sr$ and $^{147}Sm/^{144}Nd$ ratios in MORB and granitoids.

Geological mapping (see Plate 5) and the isotopic dispersion of continental rocks suggest that this differentiation of the continents has been going on throughout geological time in successive episodes. The upper mantle from which the continental crust has been extracted and which is termed "depleted" mantle (that is, depleted in certain elements such as potassium, uranium, thorium, rubidium, etc.) is therefore the residue of this extraction process. But the upper mantle is also involved in the gigantic process of plate tectonics. It is therefore a medium in which convection processes occur, with transport of matter over great distances but also with vigorous **mixing**. This medium is therefore **relatively homogeneous**, which explains the narrow isotopic spread of MORBs in Figure 6.1.

It seems that MORBs are samples from the part of the mantle that is most affected by extraction of the continental crust. As MORBs crop out at the ocean floor along oceanic ridges, it is assumed that this reservoir is the upper part of the mantle (as confirmed now by tomographic imagery obtained by seismology).

In contrast, OIB comes from the more primitive, deeper mantle; it is contaminated as it crosses the upper mantle and so its isotopic compositions lie midway between those of MORB and those of the more primitive mantle (Figure 6.15).

6.2.6 Evolution of time in the (Nd, Sr) isotope diagram and the question: geochemical differentiation or mixing?

In the foregoing reasoning, to illustrate changes in the Sr and Nd isotope ratios, we returned to the diagrams of isotope ratios versus time, adding an essential component, the reference constituted by the straight lines of evolution of the Bulk Earth (**Allègre** *et al.*, 1979). Can these evolutions not be shown directly on the (Sr, Nd) isotope diagram? Of course they can, and this presentation has the advantage of clearly situating the isotope domains observed.

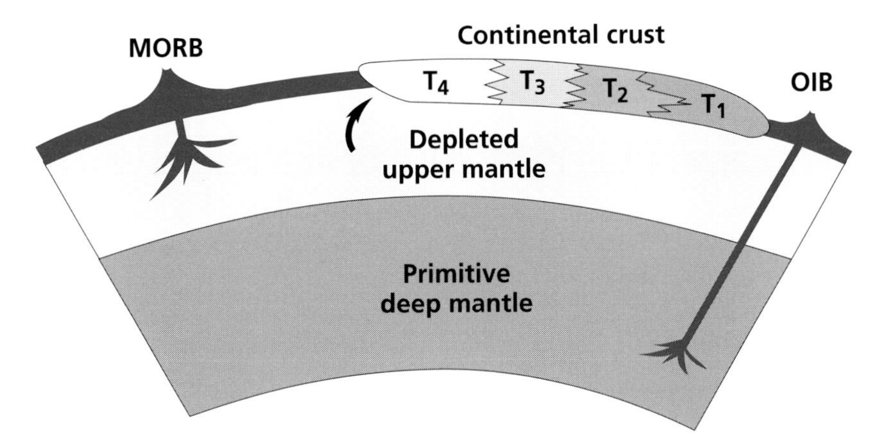

Figure 6.15 The standard model of the mantle. The mantle is separated into two layers: the MORB derived from the upper layer and the OIB from the deeper layer. The upper layer is depleted in some elements by extraction of the continental crust. The deeper layer is more primitive (that is, closer to the value of the Bulk Earth). The continents are made up of a series of provinces of varied ages T_1, T_2, T_3, T_4.

The straight line of evolution of the Earth system can be drawn. The two evolution equations are:

$$\begin{cases} \dfrac{^{87}\mathrm{Sr}}{^{86}\mathrm{Sr}} = \left(\dfrac{^{87}\mathrm{Sr}}{^{86}\mathrm{Sr}}\right)_0 + \left(\dfrac{^{87}\mathrm{Rb}}{^{86}\mathrm{Sr}}\right)_{\mathrm{Earth}} \lambda_{\mathrm{Rb}} t \\ \dfrac{^{143}\mathrm{Nd}}{^{144}\mathrm{Nd}} = \left(\dfrac{^{143}\mathrm{Nd}}{^{144}\mathrm{Nd}}\right)_0 + \left(\dfrac{^{147}\mathrm{Sm}}{^{144}\mathrm{Nd}}\right)_{\mathrm{Earth}} \lambda_{\mathrm{Sm}} t \end{cases}$$

Eliminating t gives:

$$\frac{\dfrac{^{143}\mathrm{Nd}}{^{144}\mathrm{Nd}} - \left(\dfrac{^{143}\mathrm{Nd}}{^{144}\mathrm{Nd}}\right)_0}{\dfrac{^{87}\mathrm{Sr}}{^{86}\mathrm{Sr}} - \left(\dfrac{^{87}\mathrm{Sr}}{^{86}\mathrm{Sr}}\right)_0} = \frac{\left(\dfrac{^{147}\mathrm{Sm}}{^{144}\mathrm{Nd}}\right)_{\mathrm{Earth}}}{\left(\dfrac{^{87}\mathrm{Rb}}{^{86}\mathrm{Sr}}\right)_{\mathrm{Earth}}} \frac{\lambda_{\mathrm{Sm}}}{\lambda_{\mathrm{Rb}}} = \frac{0.196 \times 6.54 \cdot 10^{-12}}{0.091 \times 1.42 \cdot 10^{-11}} = 0.9919.$$

Therefore in the $\left(^{143}\mathrm{Nd}/^{144}\mathrm{Nd},\ ^{87}\mathrm{Sr}/^{86}\mathrm{Sr}\right)$ diagram, the evolution is a straight line with a slope very close to unity. This straight line is graduated in time, of course. The two straight lines of evolution leading to the average of the continental crust and of the residual mantle, the MORB source, can be drawn. Their slopes can be easily calculated and are equal to:

$$\frac{\left(\dfrac{^{147}\mathrm{Sm}}{^{144}\mathrm{Nd}}\right)}{\left(\dfrac{^{87}\mathrm{Rb}}{^{86}\mathrm{Sr}}\right)} \frac{\lambda_{\mathrm{Sm}}}{\lambda_{\mathrm{Rb}}}.$$

With the approximate values for the continental crust and the primitive mantle, we find the slope of the straight line of evolution of the depleted mantle is very, very steep (more than 100) and that of the continental crust is close to 0.05.

Exercise

Given that the slope of the straight line of evolution of the depleted mantle is very high and the extreme value of MORB is $^{143}Nd/^{144}Nd = 0.513\,50$ and $^{87}Sr/^{86}Sr = 0.7021$, what is the minimum age of differentiation of the crust–mantle?

Answer

The age is greater than 2 Ga.

On the basis of this reasoning, we can trace the curves of evolution of the depleted mantle and the continental crust (Figure 6.16), assuming that the continental crust has been extracted in a series of episodes. When the theoretical models are compared with experimental data, a question arises. How can we explain the dispersion of values for the continental crust and the lesser dispersion of basalts?

A priori, there are two hypotheses: **differentiation** or **mixing**? (or **a combination of the two**?). Were there multiple crust–mantle differentiations giving rise to different pieces of crust and leaving different pieces of depleted mantle? Or, on the contrary, are all the intermediate points the result of mixing between two extreme components? Before discussing these questions, we need to know how a mixture is represented in the ($^{143}Nd/^{144}Nd$,

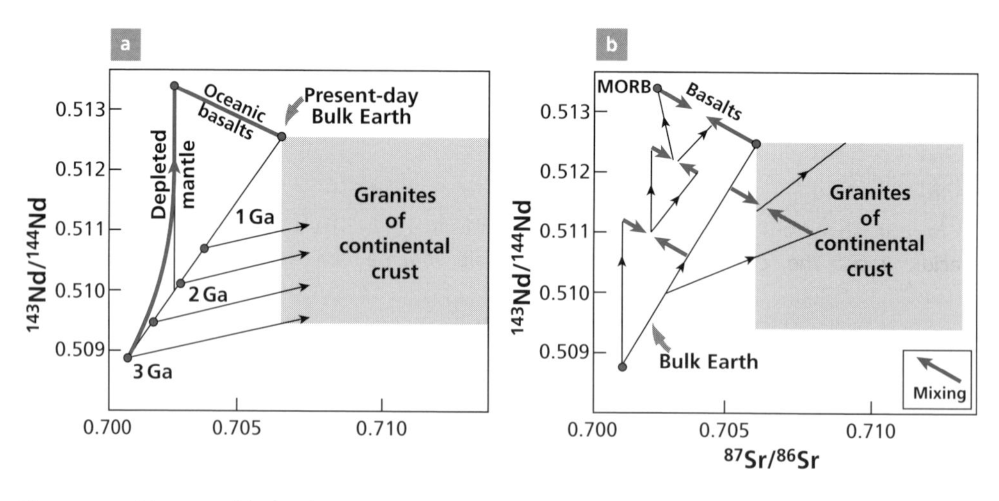

Figure 6.16 Diagram of (Nd–Sr) isotope evolution over geological time. (a) We have supposed that the continental crust was extracted in a series of episodes and that the residual mantle reflected these extractions in a continuous, gradual manner. (b) We have supposed the extraction episodes of pieces of continents were followed by mixing phenomena both in the residual mantle to the left of the straight line of evolution of the Earth and in the continental pieces (one case is shown, to the right of the straight line of evolution of the Earth). After Allègre *et al.* (1979).

$^{87}Sr/^{86}Sr$) diagram. We have spoken of mixing in the isochron diagrams but not yet in the case of paired isotope ratios.

6.2.7 Mixing in isotope ratio correlation diagrams

Imagine we have two reservoirs characterized by their isotope ratios noted here R and ρ (e.g., $^{87}Sr/^{86}Sr$ and $^{143}Nd/^{144}Nd$). How is a mixture represented in the (R, ρ) diagram? If we write the equations of the mixture M between two extreme reservoirs (1) and (2), where A and B are the two chemical elements whose ratios are under consideration, we get:

$$R_M = R_A x_1 + R_B(1 - x_1)$$

$$\rho_M = \rho_A y_1 + \rho_B(1 - y_1)$$

where $x_1 = m_1 C_A^1 / m_1 C_A^1 + m_2 C_A^2$ and $y_1 = (m_1 C_B^1)/(m_1 C_B^1 + m_2 C_B^2)$, C_A^1 and C_A^2 being the concentrations of element A in poles 1 and 2, and C_B^1 and C_B^2 being the concentrations for element B. Let us calculate the mixing equation in the (R, ρ) diagram. Combining the two equations for the two ratios gives:

$$\left(\frac{R_1 - R_M}{R_M - R_2}\right) = K\left(\frac{\rho_1 - \rho_M}{\rho_M - \rho_2}\right),$$

with $K = (C_A^1/C_B^1)/(C_A^2/C_B^2)$. If the denominator of the two ratios is identical (as is the case, for example, with lead isotopes whose denominator is common lead, ^{204}Pb), the mixing equation is a **straight line**. The same is true if the denominators are different, but $K = 1$, that is, if the chemical ratios of the two elements are identical for 1 and 2 (for example if Nd/Sr is constant in both systems).

Generally, when K is different from 1, the mixing curve is a **hyperbola** whose shape varies with K (Figure 6.17). However, if K is not too different from 1, the mixing curve is almost a **straight line**. (We shall make much use of this property in what follows.) This is generally the case for mixtures between mantle rocks or between continental rocks for the Nd–Sr system.

Exercise

We suppose a granite has been formed from a mixture of a basalt intrusion and pre-existing acidic rock. The characteristics of the basalt magma are $\varepsilon_{Nd} = +10$, $C_{Nd} = 7$ ppm, $^{87}Sr/^{86}Sr = 0.7025$, and $C_{Sr} = 300$ ppm; those of the pre-existing rock are: $\varepsilon_{Nd} = -20$, $C_{Nd} = 20$ ppm, $^{87}Sr/^{86}Sr = 0.725$, and $C_{Sr} = 100$ ppm. Calculate the mixing curve for the various granite facies, supposing the m_{crust}/m_{magma} ratio varies from 0.1 to 1. (See Figure 6.18.)

Answer

Table 6.3 Mixing curve

Number	m_{crust}/m_{mantle}	ε_{Nd}	$^{87}Sr/^{86}Sr$
1	0.1	−18.98	0.719 00
2	0.25	−17.58	0.715 200
3	0.5	−15.53	0.711 50
4	0.75	−13.75	0.709 33
5	1	−12.2	0.708 12

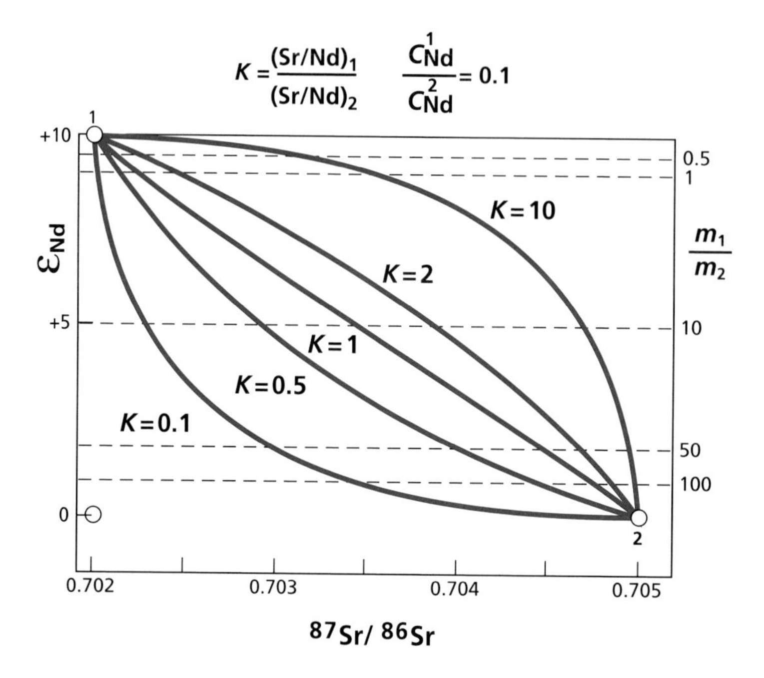

Figure 6.17 Theoretical mixing diagram between two poles 1 and 2 in the (Nd, Sr) isotope diagram for various values of relative concentrations of Nd–Sr of 1 and 2. Parameters are shown in the figure: K is the ratio of Sr/Nd chemical ratios, the ratio of Nd concentrations is taken as 0.1, and the result is graduated in mass ratio m_1/m_2.

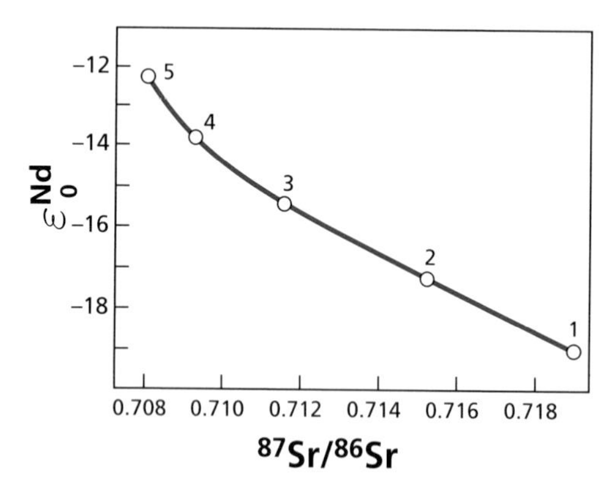

Figure 6.18 Representation of the various granite facies resulting from the crust–mantle mixture in the exercise above.

Differentiation results in straight lines, while the mixtures in the Nd–Sr isotope diagrams are almost straight lines. By virtue of a powerful mathematical theorem by which, in a linear (or near-linear) system, when there are two solutions, any combination of the two solutions are also solutions, there is no means of distinguishing by analyzing the experimental data which of differentiation or mixing predominate in explaining the Sr–Nd isotope correlation diagram as a whole. Geological considerations suggest that both phenomena are involved in the dispersion of the points observed. In the mantle, magma arises from differentiation which separates a magma composition from that of a residue. Conversely, convection in the mantle results in mechanical mixing. In the continental crust, the formation of granites as well as the erosion cycle, chemical sedimentation, and metamorphism also correspond to chemical differentiation. Sedimentation and tectonic folding are mixing processes. These two phenomena probably occurred in the past and in recent periods (genesis and transfer of magma). We asked the question: differentiation *or* mixing? The answer seems to be **differentiation *and* mixing**. There's a simple example of a non-unique solution, or, if you like, of quantitative uncertainty.

6.2.8 The Sr–Nd–Hf system

We have just explored the coherence within the Nd–Sr isotope systems. We are going to add to this the ^{176}Lu–^{176}Hf system, which will strengthen the cohesion without contributing any fundamental new information.

Lutetium is a heavy rare earth while hafnium has geochemical properties close to those of light rare earths. Lutetium-176 disintegrates into ^{176}Hf at a decay rate $\lambda = 1.94 \cdot 10^{-11}$ yr^{-1}. It was only natural, then, to try to connect the variations in $^{176}Hf/^{177}Hf$ isotope composition with the variations observed in $^{143}Nd/^{144}Nd$. But this endeavor was long unsuccessful because of difficulties in correctly analyzing the isotope composition of Hf in the low amounts in which it occurs in rocks. This difficulty was overcome in 1980 by **Patchett** and **Tatsumoto** (1980), working for the U.S. Geological Survey in Denver. Since then, a substantial amount of rock has been measured by teams at the University of Arizona at Tucson and the *École Normale Supérieure* in Lyon and the result is an extraordinarily coherent correlation between the isotope compositions of Nd and Hf (Figure 6.19). This correlation is important because it strengthens cohesion (**Blichert-Toft** and **Albarède**, 1997; see also **Patchett**, 1983; **Vervoort** and **Blichert-Toft**, 1999). It shows that the isotopic variations observed for Nd are not the outcome of some property peculiar to Nd but that they obey more general rules. As we did with Nd, we define:

$$\varepsilon_{\mathrm{Hf}}(0) = \left[\frac{\dfrac{^{176}\mathrm{Hf}}{^{177}\mathrm{Hf}} - \left(\dfrac{^{176}\mathrm{Hf}}{^{177}\mathrm{Hf}}\right)_{\mathrm{p}}}{\left(\dfrac{^{176}\mathrm{Hf}}{^{177}\mathrm{Hf}}\right)_{\mathrm{p}}} \right] \times 10^4.$$

The constants required for these calculations are:

Ratio at 4.55 Ga $\quad (^{176}\mathrm{Hf}/^{177}\mathrm{Hf})_0 = 0.279\,78$

Present-day Bulk Earth $\quad (^{176}\mathrm{Hf}/^{177}\mathrm{Hf})_p = 0.282\,95.$

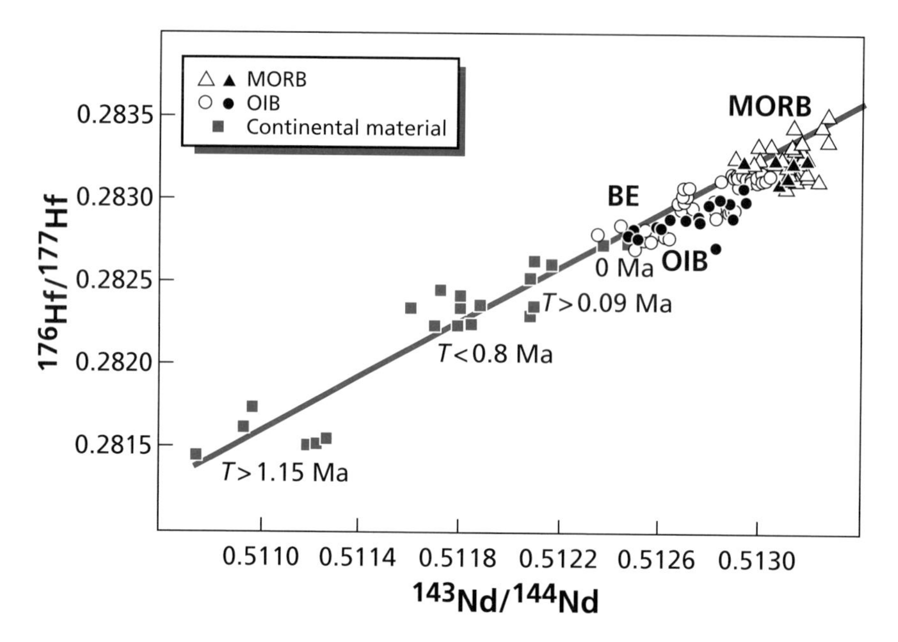

Figure 6.19 Modern (Nd, Hf) isotope correlations after data from Patchett and Tatsumoto (1980) and Vervoort and Blichert-Toft (1999). The correlation is excellent and extends linearly to rocks of the continental crust for which only a few points are shown.

The value of $(^{176}\text{Lu}/^{177}\text{Hf})_\text{p}$ can be calculated from the growth of the $(^{176}\text{Hf}/^{177}\text{Hf})_\text{p}$ ratios, giving $\mu_\text{p}^{\text{Lu}} = 0.036$.

Exercise

Three rocks (1, 2, and 3) are measured with $^{176}\text{Hf}/^{177}\text{Hf}$ isotope compositions of 0.2835, 0.2832, and 0.281, respectively. Calculate $\varepsilon_{\text{Hf}}(0)$. Can you identify the rocks from the Nd–Hf correlation?

Answer

$\varepsilon_1 = +19$, $\varepsilon_2 = +8.8$, and $\varepsilon_3 = -68.9$.
1 = MORB, 2 = OIB, and 3 = very old granitoid rock.

6.3 The continental crust–mantle system

Examination of the regularities and correlations in Sr–Nd–Hf systems sparked the idea that the fundamental process behind chemical differentiation of the upper part of the planet was the extraction of continental crust and its repercussions on the mantle. We shall try to look more closely at this process in quantitative terms using the model developed by **Allègre, Hart**, and **Minster** (1983a).

6.3.1 The chronology of extraction of continental crust

We saw when examining the results of geochronology that the continents are made up of provinces of variable but well-defined ages (see Plate 5). This geological distribution seems to show that continental crust has been progressively extracted from the mantle throughout geological time. But if this is so, what is the exact rate of extraction of the material making it up? Is it a uniform rate? Is it modulated?

In opposition to this idea of continuous extraction, it can be assumed that continental crust was extracted from the mantle right at the beginning, in early times, when differentiation of the Earth occurred, and that since then it has been recycled continuously throughout geological time (erosion, sedimentation, orogeny) but without any new material being created from the mantle, just "old material reworked into new." Age provinces would then be evidence of these recycling processes or internal rearrangements of the continental crust (reworking).

One way of addressing this issue is to calculate what is called the **mean age of continental crust materials**. Let us suppose the continental crust is made up of segments of ages $t_1, t_2, \ldots, t_n$ of respective masses $m_1, m_2, \ldots, m_n$. The fractions by mass of these segments are therefore $x_1 = m_1/M, x_2 = m_2/M, \ldots, x_n = m_n/M$, where the magnitude $M = m_1 + m_2 + \cdots + m_n$ is the mass of the continental crust.

The **mean age** of the materials is equal to

$$\langle T \rangle = x_1 t_1 + x_2 t_2 + \cdots + x_n t_n,$$

in other words it is an age weighted by the fraction by mass:

$$\langle T \rangle = \sum_{i=1}^{n} x_i t_i$$

(see Jacobsen and Wasserburg, 1979).

We can define an isotopic age of the materials assuming that the differentiation of the continental crust occurred all at one time at T_{diff}. Let us note:

- α_{cc} is the $^{87}Sr/^{86}Sr$ or $^{143}Nd/^{144}Nd$ isotope ratio of the continental crust;
- α_p is the $^{87}Sr/^{86}Sr$ or $^{143}Nd/^{144}Nd$ isotope ratio of the primitive mantle;
- μ_{cc} is the $^{87}Rb/^{86}Sr$ or $^{147}Sm/^{144}Nd$ parent–daughter isotope ratio of the continental crust;
- μ_p is the $^{87}Rb/^{86}Sr$ or $^{147}Sm/^{144}Nd$ parent–daughter isotope ratio of the primitive mantle.

Then, $\alpha_{cc}(\text{present-day}) = \alpha_p(T) + \mu_{cc}\lambda T_{\text{diff}}$, but since the primitive mantle follows the evolution $\alpha_p(\text{present-day}) = \alpha_p(T) + \mu_p \lambda T_{\text{diff}}$, from which, eliminating $\alpha_p(T)$ gives:

$$\alpha_{cc} - \alpha_p \text{ (present-day)} = \lambda(\mu_{cc} - \mu_p)T.$$

The model age is therefore written:

$$\overline{T}_{\text{diff}} = \frac{1}{\lambda}\left(\frac{\alpha_{cc} - \alpha_p}{\mu_{cc} - \mu_p}\right).$$

Moreover, the model age of the continental material is equal to the model age of mantle depletion, which is written (subscript d denotes the depleted mantle):

$$\overline{T} = \frac{1}{\lambda}\left(\frac{\alpha_{d} - \alpha_{p}}{\mu_{d} - \mu_{p}}\right).$$

We shall now demonstrate that this age is indeed the mean age defined above. To do this, we are going to exploit mathematically a geochemical property of the continental crust, which is that its average chemical composition is almost constant whatever its age. Let us calculate, then, the mean age of the continental crust using the constancy of μ_{cc} (and of course of μ_{p}):

$$\langle T \rangle = \frac{1}{\lambda(\mu_{cc} - \mu_{p})}\{[\alpha_{cc}(t_1) - \alpha_{p}]x_1 + [\alpha_{cc}(t_2) - \alpha_{p}]x_2 + \cdots + [\alpha_{cc}(t_n) - \alpha_{p}]x_n\}$$

where $\alpha_{cc}(t_n)$ is the present-day isotope ratio of a segment of continental crust of age t_n and α_{p} is the present-day Bulk Earth value.

$$\alpha_{cc}(t_1)x_1 + \alpha_{cc}(t_2)x_2 + \cdots + \alpha_{cc}(t_n)x_n = \alpha_{cc} \text{ present day.}$$

Since $x_1 + x_2 + \cdots + x_n = 1$,

$$\overline{T} = \frac{1}{\lambda}\left(\frac{\alpha_{cc} - \alpha_{p}}{\mu_{cc} - \mu_{p}}\right).$$

This is the model age. Therefore we do have the fundamental relationship of:

mean age = model age

$$\langle T \rangle = \overline{T}.$$

As seen, we know α_{p} and μ_{p} for Rb–Sm and Sm–Nd isotope systematics. We can estimate α_{d} quite well because the upper mantle corresponding to the MORB source is fairly homogeneous. Estimating μ_{d} is more difficult since, when basalt is formed, chemical fractionation occurs and the Rb/Sr or Sm/Nd ratios measured in MORB are not those of the mantle. It is difficult to evaluate α_{cc} (average continental crust) since the continents are made up of provinces of different ages and these must be averaged. This is a difficult operation because different segments have different means, different isotopic compositions, and so on. It is much easier to estimate μ_{cc} as the chemical composition of continental crust is almost constant and can be measured directly by sampling and analyzing surface rocks. To overcome these difficulties, we look to use a few tricks of the trade (Figure 6.20).

For the Rb–Sr system, it can be observed that $\mu_{d} = 0$ since the Rb/Sr ratio of MORB is very low ($\approx$0.001) and the ratio measured in basalts is a maximum value, as we have said, because the Rb/Sr ratio increases during basalt formation. We can therefore calculate $\overline{T}$ with α_{d}, α_{p}, and μ_{p}. Using the ratios $\alpha_{d}^{Sr} = 0.7022$, $\alpha_{p}^{Sr} = 0.705$, and $\mu_{p} = 0.0091$, we get $\overline{T} = 2.16$ Ga. Graphically, this amounts to drawing a horizontal line from the value 0.7022 and observing the age of intersection with the straight line of evolution of the primitive mantle (Figure 6.20).

The second trick concerns the Sm–Nd system, in which the average isotope ratio of the continental crust can be evaluated because rare earths are insoluble elements and fine oceanic

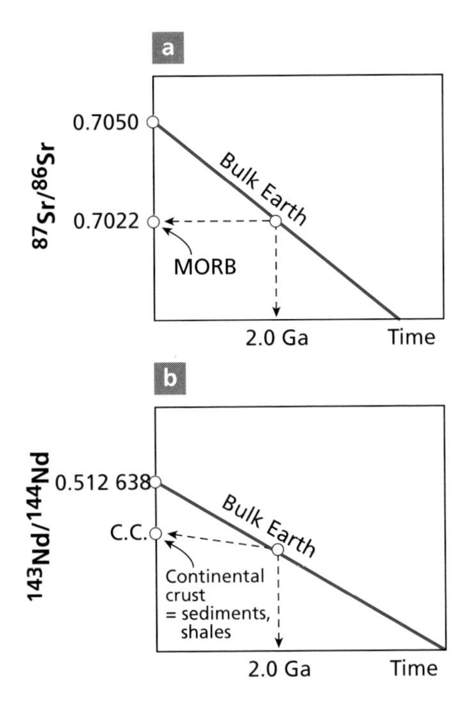

Figure 6.20 Calculating the model age (average age) of crust–mantle differentiation. (a) The $^{87}Rb–^{87}Sr$ system and determination of the model age of MORB using direct analysis of basalt. (b) The $^{147}Sm–^{143}Nd$ system and use of shales of mostly continental origin.

sediments provide a natural average of the continental crust subject to erosion. With the values $\alpha_{cc}^{Nd} = 0.5116$, $\alpha_{p}^{Nd} = 0.512\,638$, $\mu_{cc}^{Sm} = 0.11$, and $\mu_{p}^{Sm} = 0.196$, we find $\overline{T} \approx 2$ Ga.

Both methods therefore yield an age close to 2 Ga. One concerns the mantle and the Rb–Sr system, the other the continental crust and the Sm–Nd system. We can therefore consider this age of 2 Ga as a good approximation of the **average age of crust–mantle differentiation**.[7] Continental crust did not therefore form by primordial differentiation, otherwise $\overline{T}$ would be somewhere around 4.5 or 4 Ga, but probably formed throughout geological time. However, the idea of an average age allows great simplification because it reduces the problem to a two-stage story, which simplifies all quantitative reasoning but, on the other hand, conceals the real processes (Figure 6.21).

Exercise

The MORB isotope composition used in the previous calculation is $^{87}Sr/^{86}Sr = 0.7022$. This value implies that the lowest value 0.7022 measured on the mid-ocean ridges is chosen. Suppose the most representative value is 0.7025 or even, less probably, 0.7028. What would be the model age of differentiation in this case? How can you express the uncertainty on T_{diff}?

[7] A more refined calculation yields an age of 2.3 Ga, which is not fundamentally different.

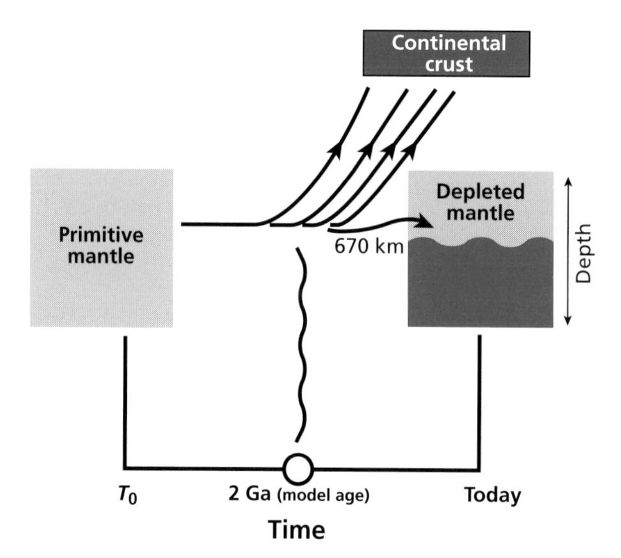

Figure 6.21 Diagram summarizing the approach and results obtained with the simple model of evolution of the crust–mantle system, assumed to be a two-stage process separated by a sudden period of differentiation.

Answer

$(^{87}Sr/^{86}Sr)_{MORB} = 0.7025$ and $T = 1.93$ Ga.
We find $(^{87}Sr/^{86}Sr)_{MORB} = 0.7028$ and $T = 1.70$ Ga.
As can be seen, the first result barely changes the reasoning. Taking the first value, the result, allowing for uncertainty, may be written $T = 2^{+0.1}_{0.3}$ Ga.

6.3.2 The mass of the depleted mantle and the mantle structure

Let us consider the system made up of three reservoirs: mantle, continental crust, and primitive mantle. Extraction of the continental crust and the corresponding depletion of a fraction of the mantle is written like a mass balance relation:

primitive mantle = continental crust + depleted mantle.

This is the opposite of a mixture, although arithmetically it is the same thing. Returning to the formula for isotope dilution or mixing, we can write:

$$\alpha_p = \alpha_{cc} W + \alpha_d (1 - W)$$

with $W = (m_{cc} C_{cc})/(m_p C)$.

Here, W is the mass fraction of continental crust, m_{cc} is the mass of the continental crust, which we know from seismology: $m_{cc} = 2.2 \cdot 10^{22}$ kg, m_p is the mass of the primitive mantle that has been depleted and which we are trying to estimate, C_{cc} is the concentration of the element in the continental crust, while C_p is the initial concentration of the element in the primitive mantle.

The idea is to calculate W from the first equation, and then to calculate m_p, the mass of the mantle which has been depleted. Is this the total mass of the mantle? Is it merely a fraction? To bring this out, we can write:

$$W = \frac{m_{cc}}{M_p}\left(\frac{M_p}{m_p}\right)\cdot\frac{C_{cc}}{C_p}$$

where M_p is the total mass of the mantle: $M_p = 4.10 \cdot 10^{24}$ kg. If we note $m_p/M_p = f$ the fraction of the depleted mantle, we get:

$$f = \frac{m_{cc}}{M_p}\cdot\frac{C_{cc}}{C_p}\cdot\frac{1}{W}$$

with $m_{cc}/m_p = \dfrac{2.2 \cdot 10^{22}}{4.10 \cdot 10^{24}} = 0.0054.$

The difficulty lies in estimating the α parameters. We know α_p and we can estimate α_d quite well from MORB measurements. Here again the difficulty is in estimating α_{cc} because of the segmentation of the continents. How do you take the average value of a mosaic? For Sr, it is difficult to do. A statistical estimate of 0.720 by **Patrick Hurley** *et al.* (1962), then at the Massachusetts Institute of Technology, is a likely value but subject to considerable error. For Nd, we have seen that fine marine sediments seem to be a good approximation with $\varepsilon_d = +10$. Then $W^{Nd} = 0.32$. If we take for the absolute concentrations $C_{cc}^{Nd} = 28$ ppm, $C_p^{Nd} = 1.1$ ppm we calculate $f = 0.43$. The depleted mantle therefore makes up about 43% of the total mantle.

Seismology has shown there is a marked seismic discontinuity at a depth of 670 km, which is interpreted as the transition from a silicate with tetrahedral structures (SiO_4) to perovskite-type octahedral structures (SiO_6) and that this discontinuity is a strong barrier against movement of matter. This boundary is taken as the lower limit of the upper mantle in a convective mantle.

The upper mantle above 670 km represents 25% of the mantle. So the mass of the depleted mantle is greater than the upper mantle alone. While the continental crust has differentiated from the upper mantle, the upper mantle has exchanged matter with the lower mantle.

Exercise

Write the equation giving W in the ε notation. If the extreme values of ε_{cc} are −16 and −25, and those of ε_d are +15 and +10, what are the extreme values of the estimate of W? What extreme values of f do they correspond to?

Answer

Results are given by the matrix. The first number is W and the second in brackets is f.

		ε_d	
		+10	+15
ε_{cc}	−25	0.285 (0.4822)	0.375 (0.366)
	−16	0.3846 (0.3617)	0.483 (0.286)

The extreme values of W are 0.28 and 0.48, which correspond to f-values of 0.482 and 0.286. As can be seen, these results differ with the extreme values but the reasoning does not change qualitatively, namely a part only of the mantle is depleted and that part is greater than the upper mantle alone. This implies that exchanges have occurred between the upper and lower mantle.

Exercise

The estimation of the Nd concentration of the primitive mantle is highly constrained: its error is estimated at ± 0.1 ppm at most; that of the continental crust is estimated at ± 2 ppm. Do possible errors with the absolute Nd concentrations in the continental crust and in the primitive mantle greatly increase the estimated limits of f? What are the two extreme values we obtain for f by combining the preceding uncertainties and uncertainties on concentrations?

Answer

The extreme values for f are 0.567 and 0.231. Once again, quantitatively the difference is great, but qualitatively the result is not overturned in geochemical terms and so yields quantitative limits.

How can we calculate W^{Sr} and the parameters for the Rb–Sr system for differentiation of the crust and mantle?

The idea is to come back to the Nd system and to exploit the reliable cross-referenced data we have between the Sm–Nd and Rb–Sr systems. The way to evaluate W^{Sr} is to return to neodymium.

We write:

$$\frac{W^{Nd}}{W^{Sr}} = \frac{(Nd/NdSrSr)_{cc}}{(Nd/NdSrSr)_{p}},$$

with $Nd_{cc} = 28$ ppm, $Sr_{cc} = 230$ ppm, $Nd_p = 1.1$, and $Sr_p = 20$. This gives $W^{Sr} = 0.144$. Then $W^{Nd}/W^{Sr} = 2.212$.

Exercise

Roughly speaking, the continental crust is made up of a deep layer (lower crust) and a surface layer (upper crust). Studies of heat flow have shown that the deep crust is depleted in certain soluble elements (uranium, potassium, and probably rubidium too) which have been transferred to the upper crust. Accordingly, the Rb/Sr ratio measured in the upper parts of the crust is not representative of the whole crust. Given that W^{Sr} is 0.144 and that the $^{87}Rb/^{86}Sr$ ratio of surface continental rocks is 0.8, what is the maximum thickness of the deep crust? The Sr content is assumed to be identical for the whole of the continental crust, which is certainly an oversimplification.

Answer

The value of W^{Sr} is the same for budget equations of α and μ values. Since $\mu_d^{Rb} \approx 0$, we have:

$$\mu_{cc}^{Rb} = \frac{\mu_p^{Rb}}{W} = \frac{0.09}{0.144} = 0.625.$$

Let us write the equation giving the average composition of the continental crust.

$$\mu_{cc}^{Rb} = \mu_u^{Rb} Q + \mu_1^{Rb}(1 - Q).$$

where μ_u^{Rb} and μ_1^{Rb} are the μ parameters for the upper and lower crust.

If we note

Q = mass of upper crust/mass of whole crust

thickness is maximum when $\mu_1^{Rb} = 0$, corresponding to all Rb being transferred to the upper continental crust. In this case $Q = 0.625/0.8 = 0.78$.

Maximum thickness of the upper crust is about 78% of that of the whole crust. Here is a new structural constraint derived from isotope data. If we consider the upper crust is enriched in Rb by a factor of 2 compared with the lower crust, then the lower crust represents 56% of the continental crust.

Remark

In practice, all of the unknowns such as ages, W, α, μ, etc. are calculated globally by what is termed the generalized inverse model in mathematics, by which all the equations can be solved simultaneously. All the equations, including the μ balances, α balances, and the model ages as well as the various relations among the elements, for example on the Sr/Nd ratio measured in the crust or the mantle samples, are calculated with a margin of uncertainty for each, which is essential. We then calculate all the values in one go. To do this, we introduce values estimated on an a-priori basis, optimizing the result with their estimated uncertainties and we solve a gigantic least-squares problem. It is difficult to solve as it is non-linear, with some unknowns multiplying each other. We shall see the spirit of it in one of the problems proposed at the end of the chapter (see **Allègre** *et al.*, 1983b).

6.3.3 Changes in the Rb–Sr and Sm–Nd systems over geological time

In the foregoing, we have used present-day results and have projected into the past. But this technique has a limited power of resolution.

We know that the average age of the continental crust is about 2 Ga. But we would also like to know exactly how this growth came about: was the rate of extraction from the mantle uniform, modulated, etc.?

Exercise

We consider the following eight models of continental growth (A–H). Each is represented by a distribution of continental orogenic segments for which the table shows the contribution from continents and the average age of each segment.

	Segments			
Model A	(1)	(2)	(3)	(4)
	3.5–2.5 Ga	2.5–1.5 Ga	1.5–0.5 Ga	0.5–0 Ga
Average ages	3 Ga	2 Ga	1 Ga	0.25 Ga
Proportions	39%	32%	18%	11%

	Segments			
Model B	(1)	(2)	(3)	(4)
	3.5–2.5 Ga	2.5–1.5 Ga	1.5–0.5 Ga	0.5–0 Ga
Average ages	3 Ga	2 Ga	1 Ga	0.25 Ga
Proportions	21%	59%	19%	1%

	Segments			
Model C	(1)	(2)	(3)	(4)
	3.5–2.5 Ga	2.5–1.5 Ga	1.5–0.5 Ga	0.5–0 Ga
Average ages	3 Ga	2 Ga	1 Ga	0.25 Ga
Proportions	40%	10%	10%	40%

	Segments				
Model D	(1)	(2)	(3)	(4)	(5)
	4.7–3.8 Ga	3.8–2.5 Ga	2.5–1.5 Ga	1.5–0.5 Ga	0.5–0 Ga
Average ages	4 Ga	3 Ga	2 Ga	1 Ga	0.25 Ga
Proportions	20%	20%	20%	20%	20%

	Segments			
Model E	(1)	(2)	(3)	(4)
	3.5–2.5 Ga	2.5–1.5 Ga	1.5–0.5 Ga	0.5–0 Ga
Average ages	3 Ga	2 Ga	1 Ga	0.25 Ga
Proportions	10%	20%	30%	40%

	Segments				
Model F	(1)	(2)	(3)	(4)	(5)
	4.2–3.8 Ga	3.8–2.5 Ga	2.5–1.8 Ga	1.8–0.8 Ga	0.8–0 Ga
Average ages	4 Ga	3 Ga	2 Ga	1 Ga	0.5 Ga
Proportions	40%	15%	15%	15%	15%

	Segments				
Model G	(1)	(2)	(3)	(4)	(5)
	4.2–3.8 Ga	3.8–2.8 Ga	2.5–1.8 Ga	1.8–0.8 Ga	0.8–0 Ga
Average ages	4 Ga	3 Ga	2 Ga	1 Ga	0.5 Ga
Proportions	3%	3%	85%	5%	4%

	Segments				
Model H	(1)	(2)	(3)	(4)	(5)
	4.2–3.8 Ga	3.8–2.5 Ga	2.5–1.8 Ga	1.8–0.8 Ga	0.8–0 Ga
Average ages	4 Ga	3 Ga	2 Ga	1 Ga	0.5 Ga
Proportions	10%	50%	30%	10%	0%

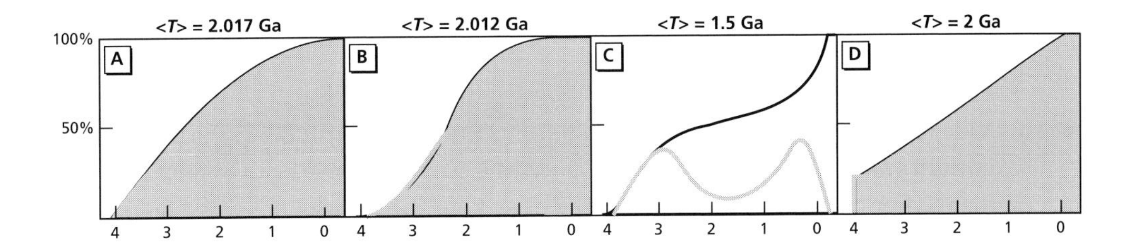

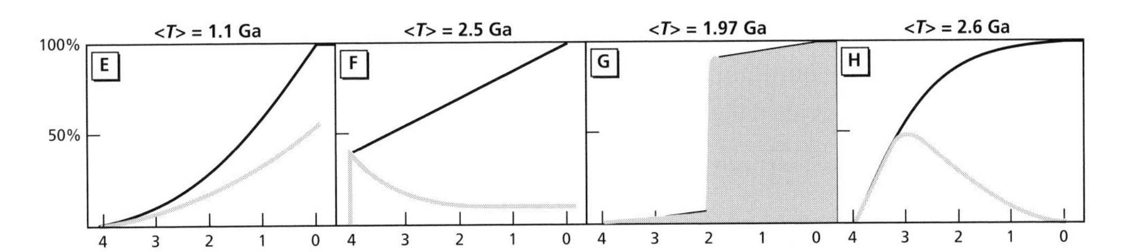

Figure 6.22 The results for the various models calculated in the exercise. On each graph, the x-axis shows time in Ga and the y-axis the percentage of continental crust formed. The integral curve is in black, the derivative curve, that is the growth rate, in blue. At the top of each graph $\langle T \rangle$ indicates the average age calculated. The models shaded in blue are those that are compatible with an average age of about 2 Ga.

In each case, calculate the average age of the continent and draw the cumulative growth curve of the continents and variations in the rate of growth (see Figure 6.22). Which models do you think are compatible with the results for the crust–mantle system with current data?

Answer

Four models are acceptable with an age of 2 ± 1 Ga: A, B, D, and G.

Comments

As the exercise has just demonstrated, the average age found for the continental crust using the calculation of the previous paragraph can indeed be used to eliminate some scenarios, but still leaves scope for a range of possible scenarios. For example, one might equally well envisage either steady continuous growth (model D) or sudden growth (model G).

To get a real picture of the kinetics of continental extraction (examination of the shapes of the curves of growth rates is extremely demonstrative) outside the mantle, we had the idea of trying to obtain isotopic information for Sr and Nd for different geological epochs so as to calculate W as a function of time $W(t)$.

These studies, which looked as if they should be straightforward, have proved very difficult. The first problem was the relevance of the evidence of ancient massifs. For recent rocks, we know that basalt rocks collected from ocean ridges are evidence of depleted mantle. But in the past, when we no longer have mid-ocean ridges, how can we recognize which basalts can be used as evidence of ancient depleted mantle like the present-day MORB?

For continents, how could we establish an average sample of the continents at a given period in the past? Are the current mappings of various age provinces representative in relative abundance of what they were in the past?

- **For evidence about the upper mantle**, we adopt a pragmatic attitude. Up to 600 Ma, we use **ophiolite massifs**, which are pieces of oceanic crust that have been obducted on the continents. For terrains older than 2 Ga, we use **komatiites**, which are associations of rock put in place under the seas and which associate ultrabasic and basic lava (which is peculiar to Archean times). But these are working hypotheses that have not really been demonstrated (other than that they do not seem absurd).
- **For the continental crust**, the only evidence of what it was like at a given time are the shales of the time in question. They are a sample of everything that had been eroded and then mixed in the ocean at the time. They are an average of the upper continental crust.

In any event, the following program can be set.

(1) Choose a group of rocks whose origin has been identified for a given time-span.
(2) Determine its age and initial isotope ratio precisely (which takes us back to the considerations on isochrons in Chapter 3).

We then just need to repeat the operation for several geological epochs and so obtain Sr and Nd isotope evolution curves for our two reservoirs (Figure 6.23).

This program has involved some 20 or so research teams worldwide over 10 years. The results of the quest have been very disappointing. Why? Not because the teams were bad, but because, in the past, isotope systems seldom remained strictly closed and therefore uncertainties on initial ratios increase statistically with time. Without going into details that extend beyond the scope of this book, we can remember two related ideas: considerable difficulties and poor and inconsistent results. However, let us summarize the all too scarce positive findings.

Strontium

Whole-rock isochrons of ancient basic and ultrabasic rocks are very rarely reliable. To resolve the problem, we bring in analysis of pure clinopyroxene extracted from rocks containing almost no Rb and providing a fairly accurate reflection of initial $^{87}Sr/^{86}Sr$ ratios. We determine the age of the basic massifs in question independently by other chronological methods.

This method has yielded few reliable results. Even so, the overall shape of the curve of evolution (Figure 6.24) is clear enough. It is virtually aligned on the straight line of isotope evolution of the primitive (closed) medium up to 3 Ga and then is practically horizontal from 1 Ga onwards.

It is impossible to obtain the curve of isotope evolution of the average continental crust for Sr from the analysis of ancient shales. There are two reasons for this: the first is that a part of the erosion of Sr is soluble and was deposited in limestone; the second is that shales themselves are open systems for Sr.

Neodymium

The situation is far more favorable because both Sm and Nd are much more chemically inert (less soluble) elements than Rb and Sr. But the variations are, of course, very small: to

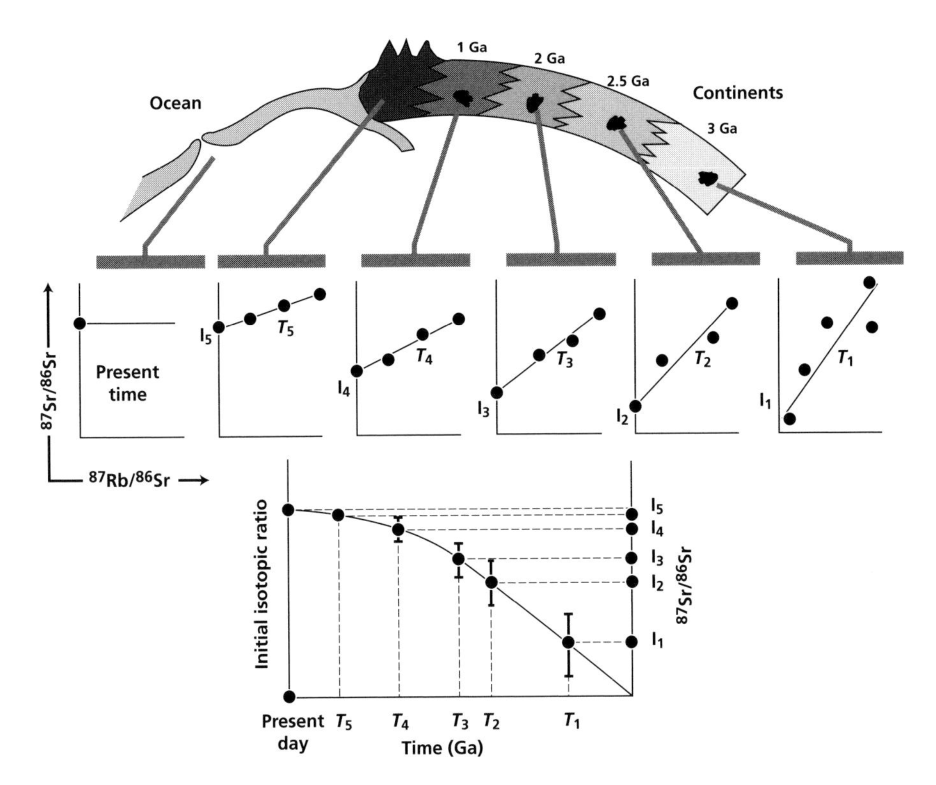

Figure 6.23 Theoretical diagram of the approach we would like to take to determine the isotope evolution of strontium in the mantle over geological time. Top: we sample rocks from the mantle whose age is determined by an isochron diagram. Middle: each isochron corresponds to an initial $^{87}Sr/^{86}Sr$ isotope ratio. Bottom: the initial ratios are plotted as a function of age.

express variations in the $^{143}Nd/^{144}Nd$ ratio, we adopt the ε notation. Instead of referring to the present-day primitive mantle, we refer to the primitive mantle at the time for which the $^{143}Nd/^{144}Nd$ ratio is being estimated.

We have:

$$\varepsilon_{\text{Nd}}(T) = \left[\frac{\alpha^{\text{Nd}}(t) - \alpha_{\text{p}}^{\text{Nd}}(T)}{\alpha_{\text{p}}^{\text{Nd}}(t)}\right] \times 10^4.$$

On a plot like this, the primitive mantle evolves along a horizontal line, the initial ratios of the various massifs considered as evidence of the upper mantle are located relative to this straight line, the equation for which is:

$$\alpha_{\text{p}}^{\text{Nd}}(T) = \alpha_{\text{p}}^{\text{Nd}}(\text{present day}) - \lambda\mu_{\text{p}}T,$$

T being the age, with $\alpha_{\text{p}}^{\text{Nd}}(\text{present day}) = 0.512\,638$ and $\mu_{\text{p}} = 0.196$.

For the depleted mantle, **Don DePaolo** of the University of California at Berkeley managed to establish a pretty acceptable curve of $\varepsilon(t)$ (see DePaolo, 1988). This curve follows the horizontal line of the reference primitive mantle up to 2.7 Ga and then curves around to

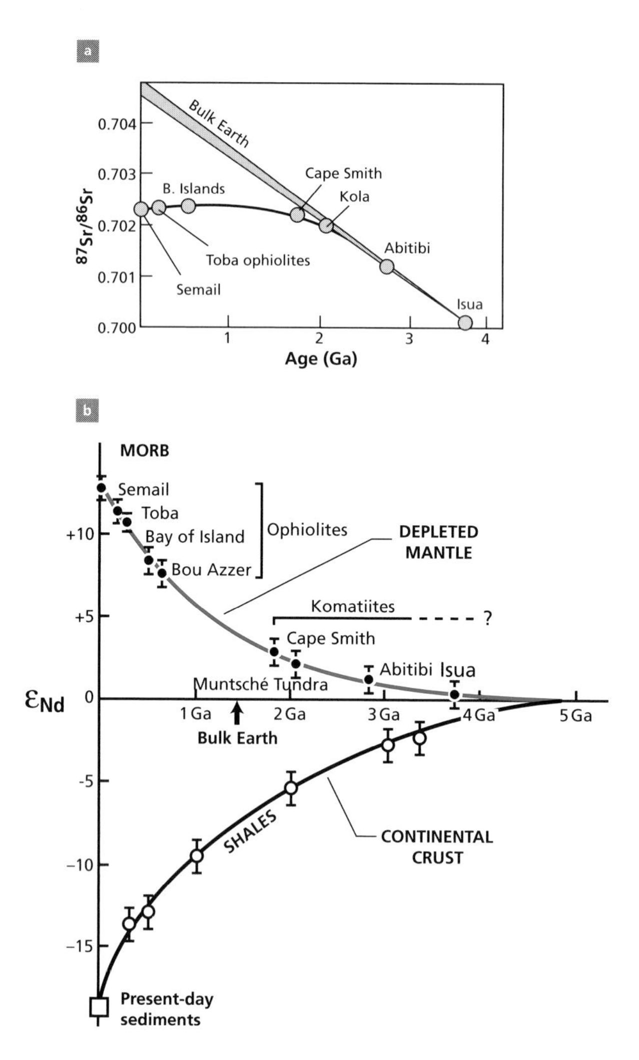

Figure 6.24 (a) Isotope growth curve of the mantle for strontium based on analysis of purified clinopyroxenes. (b) Isotope growth curve of the mantle and continental crust (shales) for neodymium.

reach +12, the value of present-day MORB. Notice, though, that in the 4–3 Ga time-span, uncertainties with low absolute values $\pm 0.2\,\varepsilon$ have considerable repercussion on the deductions that can be drawn, as we shall see.

For continental crust, shales provide a good record of the average of the landmasses for a given period. As their $^{147}Sm/^{144}Nd$ ratio of about 0.11 is roughly constant, it is enough to correct for the ^{143}Nd produced *in situ* from sedimentation in order to obtain the $^{143}Nd/^{144}Nd$ ratio at the time of sedimentation. This work on the isotope composition of shales was conducted by teams from Caltech, Cambridge, and Paris in a complementary manner (**McCulloch** and **Wasserburg**, 1978; **O'Nions** *et al.*, 1983; **Allègre** and **Rousseau**, 1984; **Jacobsen**, 1988).

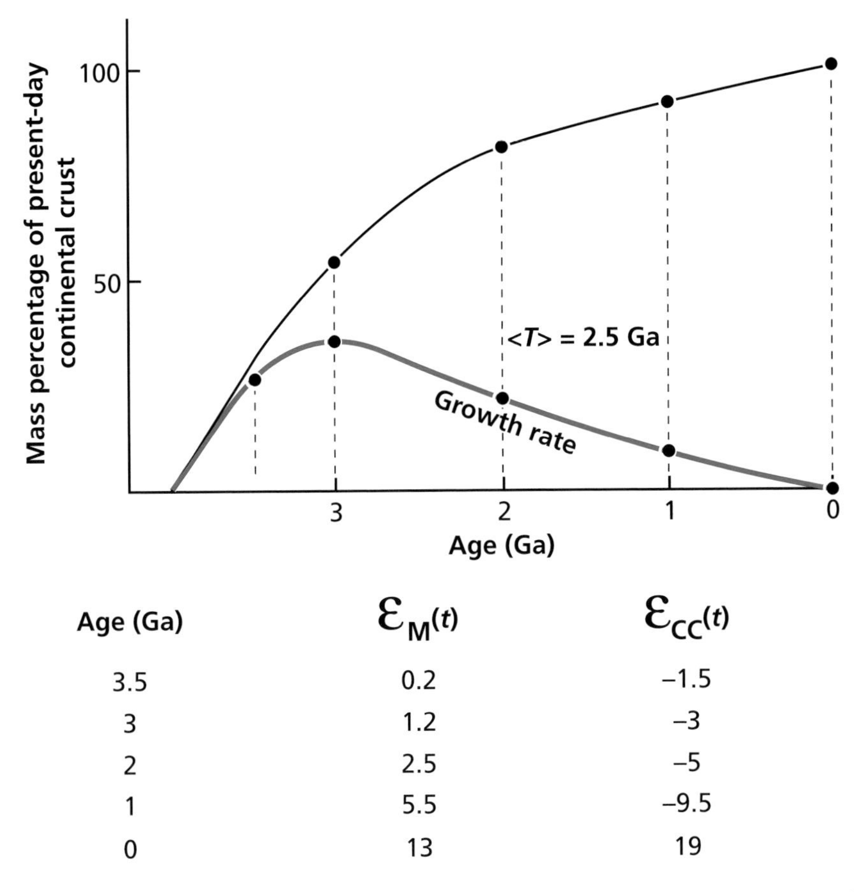

Age (Ga)	$\varepsilon_{M}(t)$	$\varepsilon_{CC}(t)$
3.5	0.2	−1.5
3	1.2	−3
2	2.5	−5
1	5.5	−9.5
0	13	19

Figure 6.25 Growth curve of continents deduced from the curves of strontium and neodymium isotope evolution. The numerical values used are shown below.

We therefore have two curves of $\varepsilon(t)$ evolution of Nd, one for the depleted mantle and one for the continental crust. We can therefore write epoch by epoch the balance equation:

$$0 = \varepsilon_{\rm cc}(t) \cdot W(t) + \varepsilon_{\rm M}(t)[1 - W(t)]$$

and therefore calculate $W(t)$ step by step.

Accepting that the volume of the upper mantle has remained constant and corresponds to that determined on the present-day balance (which is not proved at all), the evolution of the continental mass extracted from the mantle can be calculated. The outcome is shown in Figure 6.25. Naturally, this curve involves considerable uncertainty for very ancient periods.

Exercise

To get this point across, let us consider the evolution between 2.8 and 2.7 Ga, between the Isua time and the great Canadian orogeny. Now, the literature contains extremely varied values for ancient rocks, whose reliability is unknown, leaving considerable scope for interpretation.

Let us assume the values from 2 Ga are well constrained and identical to the overall calculation. We consider two extreme scenarios corresponding to the ε_{Nd} values below.

	Average age (Ga)	$\varepsilon_M(t)$	$\varepsilon_{cc}(t)$
Scenario A	3.8	+1.9	−0.5
	3	+0.5	−3
	2.7	+1	−3.5
Scenario B	3.8	0	0
	3	+1	−3
	2.7	+2	−4

Given that the current $W_{Nd} = 0.4$ and assuming the same values of the upper mantle mass entering into the process and the Nd concentration of continental crust, calculate the growth curves of the continents and the average age in each case.

Answer

	Age (Ga)	Percentage of present-day continents	$\langle T \rangle$
Scenario A	3.8	200	
	3	24	1.82 Ga
	2.7	57	
Scenario B	3.8	0	
	3	65	2.08 Ga
	2.7	82	

This result of the exercise (see Figure 6.26) shows us the possibility of an a-priori surprising phenomenon in scenario A. Some 3.8 billion years ago, the volume of continental crust might have been twice what it is today.

The conclusion of this is that this primitive crust, which may be the result of major chemical differentiation of the primordial Earth, disappeared, was destroyed, and was swallowed up in the mantle, and that only later did the crust as we know it now begin to form. This is just one extreme scenario, but it raises such fundamental issues that it must be explored further.

Let us leave this subject for the end of the chapter when we shall deal with the primordial Earth. These examples show how important it is to take account of measurement uncertainties which sometimes lead to considerable uncertainties in the scenarios proposed. So how can we improve our knowledge? By getting away from the narrow scenario and introducing other constraints from other sources of information.

For example, we can test the models using the results of evaluation of the crust–mantle system obtained with present-day data. The average age, it will be remembered, is $2\,\text{Ga} \pm 0.1$. In this way, we can calculate the average ages of the various scenarios (see Figure 6.26). It shall be seen that the model extracted from the two growth curves gives $\langle T \rangle = 2.5$ Ga, which is a little high. Scenario A, where the initial peak disappears, of course,

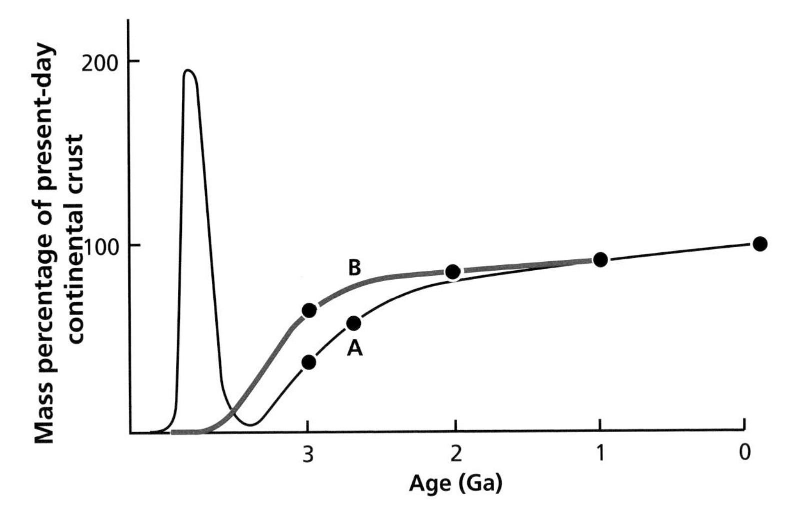

Figure 6.26 Growth curve of continents in scenarios A and B of the exercise.

gives $\langle T \rangle = 1.82$ Ga, which is a little low. Scenario B gives $\langle T \rangle = 2.08$ which is quite acceptable. In fact, it is the most acceptable.

A second constraint is provided by the strontium isotope evolution curve, which shows that not much continental crust has been extracted from the mantle over the last 1 billion years. Yet another way of constraining the most probable model is the direct simulation method. A scenario is suggested and from that stage on it is calculated. The next exercise provides an example of this.

Notice that, even if it matches the information, direct simulation does not guarantee it is the one and only right model. There may be others. We have illustrated this matter of the uniqueness in exploring the scenarios explaining the current model age of the continental crust. This is a question we must always ask. We have found a solution, fair enough. Is it the only solution? We have no guarantee at all!

Exercise

Suppose continental crust has been extracted from the mantle continuously over 4 billion years, with the growth rate A such that $m = At$. We assume the volume of the depleted mantle has remained the same and has been confined to the upper mantle above the seismic discontinuity at a depth of 670 km. Given that the ratio of the current masses of the continental crust and upper mantle is 0.011, that $C_{cc}^{Sr} = 150$ ppm, $\mu_{cc}^{Rb} = 0.35$, $C_p^{Sr} = 20$ ppm, and $\mu_p^{Rb} = 0.1$, calculate the curves of α_{cc}^{Sr} and α_d^{Sr} over the course of geological time.

Answer

We begin by estimating variation in W^{Sr} over time:

$$W^{Sr} = \frac{m_{cc}}{m_p}\frac{C_{cc}}{C_p}$$

Table 6.4 Isotope evolution of the depleted mantle and of the continental crust

Time (Ga)	μ^{A}	W^{Sr}	α_{p}^{Sr}	α_{d}^{Sr}
1	0.0817	0.068	0.701	0.70074
2	0.06064	0.136	0.7024	0.70158
3	0.03592	0.204	0.7038	0.70208
4	0.00659	0.27	0.705	0.702172

therefore

$$W^{Sr}(t) = \frac{At}{m_p} \cdot \frac{C_{cc}}{C} = A^* t.$$

The value of $A^* = 0.0205$ if t is measured in Ma.

We then estimate the variation in μ_d^{Rb} from the mass balance equation:

$$\mu_d = \frac{(\mu_T - \mu_c A^* t)}{(1 - A^* t)}.$$

Table 6.4 shows the calculation for 1, 2, 3, and 4 billion years. It is possible then to calculate the isotope evolution of the depleted mantle starting from α_pSr (4 Ga) = 0.6996 and then calculating by stages with $\mu^{Rb}(t)$ being determined. To calculate the evolution of the continental crust, we use the mass balance equation again. We obtain α_{cc}^{Sr} from the $W(t)$ values. A comparison is made with the curve for the total differentiation of the continental crust at 4 Ga (Figure 6.27).

Remark

Comparison of the $\alpha_d^{Sr}(t)$ curve of evolution of the depleted mantle from the synthetic model with the actual curve (which requires care!) from measurements on rock shows that the theoretical evolution of the exercise is more progressive and less sudden than the real curve. This reflects the fact that production of continental crust is probably shifted more towards the past than the uniform growth model indicates. Past growth was therefore greater than recent growth.

Exercise

Calculate the average age from the model in the previous problem.

Answer

The average age written as an integral is defined by

$$\langle T \rangle = \frac{1}{M} \int_0^T t\, m\, dt$$

where m is the fraction produced per unit time (here $m = A$). We therefore have:

$$\langle T \rangle = \frac{1}{M} \int_0^T t\, A dt = \frac{AT^2}{2M} = \frac{AT^2}{2AT} = \frac{T}{2}.$$

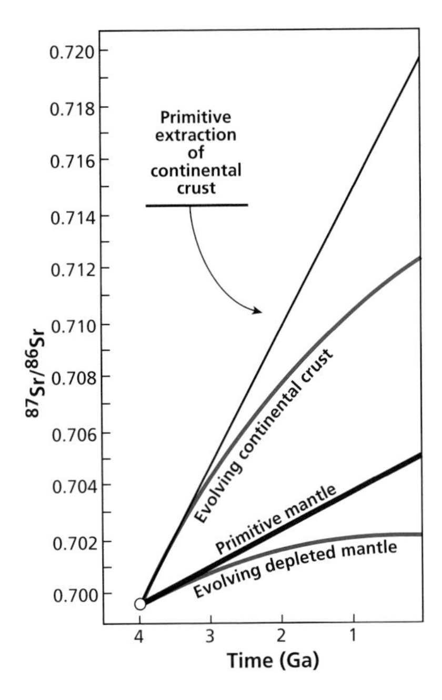

Figure 6.27 The ($^{87}Sr/^{86}Sr$, t) isotope evolution curves in the model of the previous exercise. Blue curves are for continuous evolution. Black curves are for initial differentiation of continental crust at 4 Ga. Clearly, present-day ratios do not result from simple initial differentiation.

As $T = 4$ Ga, $\langle T \rangle = 2$ Ga. We find the age calculated before. In a population where the rate of accumulation is constant, the average age is half the maximum age. In the human population, the maximum age is 80 years, the average age 40 years! (We shall be discussing this idea again.)

6.3.4 The process of development of continental crust over geological time

Continental reworking

We have just established a growth curve for continental crust using strontium and neodymium (above all the latter!) isotope balances. Now, we have seen (see Plate 5) that the world has been mapped in terms of age provinces by systematic geochronology. **Hurley** and **Rand** (1969) of the Massachusetts Institute of Technology have transformed this map into percentages by age (Figure 6.28).

Exercise

Translate the graph (Figure 6.28) into percentages and calculate an average age using the proportions of age provinces.

Answer

$\langle T \rangle = 0.7$ Ga.

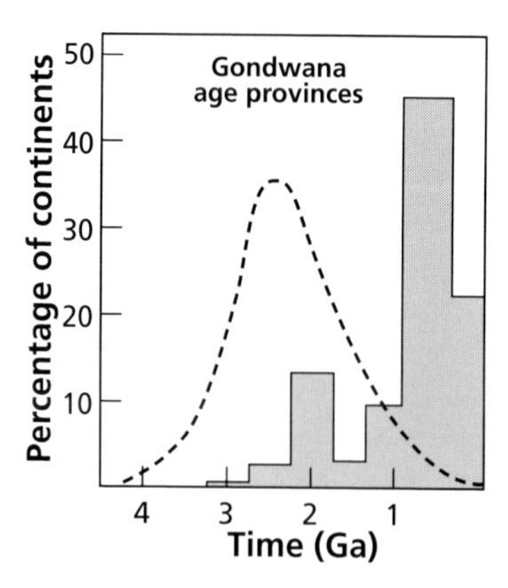

Figure 6.28 Continent growth rates: established by geochronology and mapping by Hurley and Rand (1969) (blue), and also showing isotope ratio evolution models (dashed curve).

As shown by the exercise and by comparing the two growth curves, there is a contradiction between the isotope geochemistry approach and the geochronological approach in the plates. How can this difference be accounted for? By reworking of the continental materials (see Figure 6.29). Provinces of recent age are the result of reworking in the geological cycle (erosion, sedimentation, metamorphism, anatexis) of ancient pieces of continental crust. The ancient provinces have been **cannibalized**, that is eroded and incorporated into younger provinces. The older they are, the more likely it is that they will have been given a new lease of life in this way. Notice that this interpretation is consistent with the idea of constancy of geological phenomena over time. The processes by which continental crust is formed have operated throughout geological time in much the same way but the type of material involved in the processes has changed. In ancient times these materials must have been mostly mantle. In more recent times, on the contrary, it was above all recycled material of the continental crust, making new out of old!

Exercise

The initial Nd isotope composition of a granite dated 1 Ga is measured as $^{143}Nd/^{144}Nd = 0.51062$. The local gneiss basement has a present-day isotope composition of $\varepsilon_{Nd}(0) = -28$ and a $^{147}Sm/^{144}Nd$ ratio of 0.11. Assuming the mantle contribution to the granite mixture is from the primitive mantle, calculate the percentage of this contribution if the Nd concentrations of the mantle and recycled components are identical.

Answer

The isotope compositions of the gneiss and mantle 1 billion years ago must be calculated first and then the proportions of the mixture. We find a 13% contribution from the mantle.

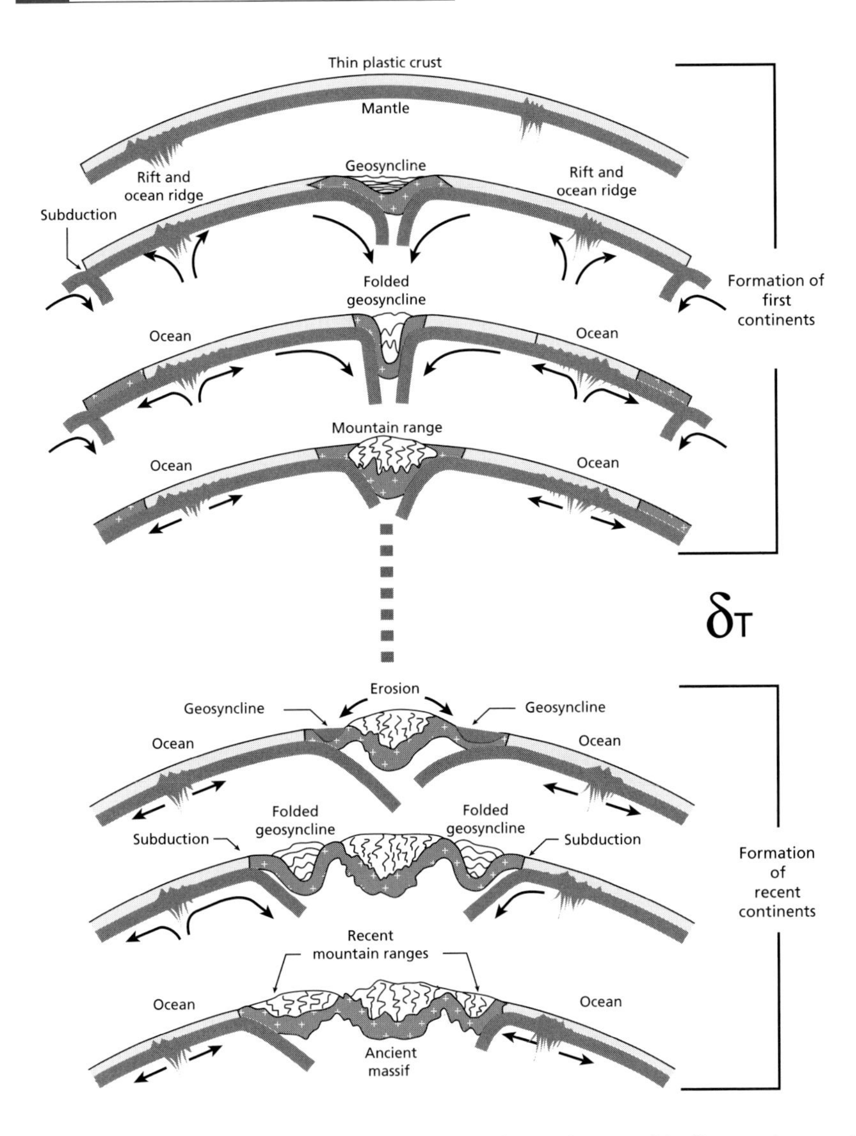

Figure 6.29 The formation and growth of the early continents. This diagram is reworked from Dietz (1963) and shows how the early continents might have formed and grown over geological time, with the proportion of recycled material increasing constantly.

Exercise

Geological history is divided into four periods: 4–3 Ga, 3–2 Ga, 2–1 Ga, and 1–0 Ga. Mapping formations dated by radiometric methods converted into percentage of materials gives:

Period (Ga)	4–3	3–2	2–1	1–0
Percentage	5	15	20	60

Conversely, the study of growth of continents by initial Nd ratios gives as proportions of continental crust made from the mantle:

Period (Ga)	4–3	3–2	2–1	1–0
Percentage	20	50	30	0

It is assumed that at each new period, part of the continent is made of a neoformed (from the mantle) fraction and a fraction of material reworked from earlier periods.

(1) For each segment, calculate the proportion of material from the period in question and from the preceding periods.

(2) The ratio (R) of reworked mass to total mass of a segment is known as the reworking coefficient. Calculate this coefficient for the four periods.

(3) By taking the median age for each segment (3.5 Ga, 2.5 Ga, 1.5 Ga, and 0.5 Ga, respectively), calculate the **mean age** for each segment. Plot the mean age (or model age) on a graph as a function of geological age.

Answer

The 20% created in the 4–3 Ga interval is divided as:

Period (Ga)	4–3	3–2	2–1	1–0
Share of 20%	5	3.5	2.87	8.6
Percentage	25	17.5	14.35	43

The 50% created in the 3–2 Ga interval is divided as:

Period (Ga)	4–3	3–2	2–1	1–0
Share of 50%		11.5	9.62	28.9
Percentage		23	19.24	57.9

The 30% created in the 2–1 Ga interval is divided as:

Period (Ga)	4–3	3–2	2–1	1–0
Share of 30%			7.5	22.5
Percentage			25	75

The reworking coefficient is given in the table below.

Period (Ga)	4–3	3–2	2–1	1–0
R	0	0.23	0.37	1

The mean ages are given below.

Period (Ga)	4–3	3–2	2–1	1–0
Mean age (Ga)	3.5	2.7	2.25	2.25

See Figure 6.30 for a summary of data and results.

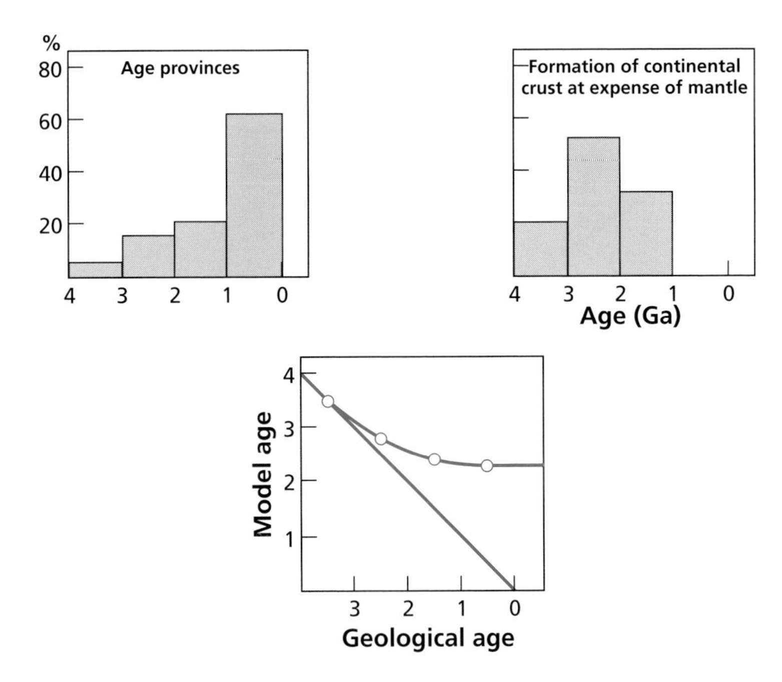

Figure 6.30 Data and results of the exercise. Top: starting data. Left: proportions of continents determined by mapping age provinces. Right: proportions of continents as a function of extraction from mantle. (Data corresponding to the two tables.) Bottom: what becomes of the various materials extracted from the mantle, then the model ages as a function of geological age.

The neodymium model age and its geological applications

The fact that Sm/Nd fractionation is almost zero in geological surface processes after extraction from the mantle means a model age can be attributed to any neodymium isotope measurement of a material belonging to the crust. In an ideal scenario, this is the age at which the material (or its precursors) was extracted from the Earth's mantle to begin its "continental geological life." This is the idea hit upon by **Malcolm McCulloch** and **Jerry Wasserburg** of the California Institute of Technology in 1978. No such model age can be calculated for Rb–Sr as the Rb/Sr ratio varies considerably during the surface history (although we shall see in Chapter 7 that it can still be used for limestone). As will be seen, it is possible to use Lu–Hf in a similar manner because lutetium and hafnium are also almost insoluble in water.

To calculate this model age (which we have already come across in calculating present-day balances of the crust–mantle system), let us consider the straight line of isotope evolution of the closed mantle (Figure 6.31). Imagine that a piece of continental crust separated from the mantle at time T_{M}, and has evolved since with constant $\mu_{\mathrm{cc}}^{\mathrm{Sm/Nd}}$ growth. This straight line of evolution intersects the time axis at $\alpha_{\mathrm{cc}}^{\mathrm{Nd}}$. How can we calculate time T_{M} in return?

Let us call the two parameters of evolution of the primitive mantle $\alpha_{\mathrm{pm}}^{\mathrm{Nd}}$ and $\mu_{\mathrm{pm}}^{\mathrm{Nd}}$, the two parameters for the segment of continental crust $\alpha_{\mathrm{cc}}^{\mathrm{Nd}}$ and $\mu_{\mathrm{cc}}^{\mathrm{Sm/Nd}}$, and the common isotope ratio corresponding to T_{M} we call I.

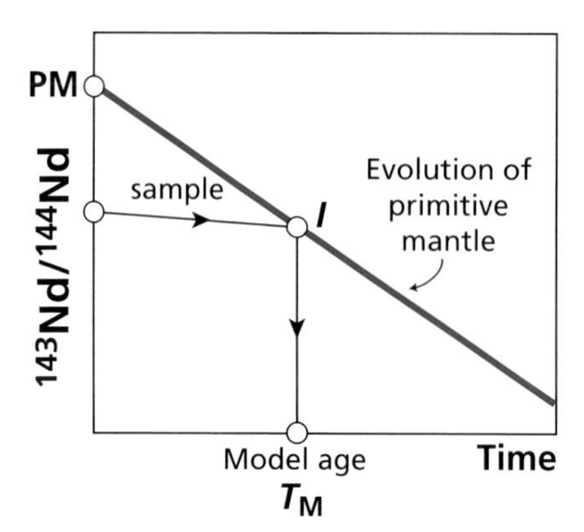

Figure 6.31 Principle of model age calculation. Start from the sample. Draw a straight line of slope 0.11 cutting the straight line of mantle evolution at *I*, corresponding to the model age T_M.

We then have the two equations:

$$\alpha_{pm}^{Nd} = I + \lambda \mu_{pm}^{Sm/Nd} T_M$$

$$\alpha_{cc}^{Nd} = I + \lambda \mu_{cc}^{Sm/Nd} T_M.$$

By eliminating I, we get:

$$T_M = \frac{1}{\lambda}\left(\frac{\alpha_{cc}^{Nd} - \alpha_{pm}^{Nd}}{\mu_{cc}^{Sm/Nd} - \mu_{pm}^{Sm/Nd}}\right).$$

For the continental crust, $\mu_{cc}^{Sm/Nd} = 0.11$ and for the primitive mantle $\mu_{pm}^{Sm/Nd} = 0.196$.

But $\mu_{cc}^{Sm/Nd} - \mu_{pm}^{Sm/Nd} = 0.11 - 0.196 = 0.086$ is constant. Recalling the definition of $\varepsilon(0)$:

$$\varepsilon_{Nd}(0) = \left(\frac{\alpha_{cc}^{Nd} - \alpha_{pm}^{Nd}}{\alpha_{pm}^{Nd}}\right) \times 10^4$$

we obtain:

$$T_M = \frac{10^{-4}\,\varepsilon(0)\,\alpha_{pm}}{6.54 \cdot 10^{-12} \times 0.086} = -0.091\,\varepsilon(0).$$

If we want an approximation of T_M in Ga, we have about $T_M = 0.09\,\varepsilon(0)$, which is handy for switching from $\varepsilon(0)$ to T_M. (It is almost $\varepsilon(0)$ divided by 10 with a change of sign!)

Naturally, if the crustal material contains a little of the original mantle material, the Sm/Nd ratio is slightly higher than 0.11 and α_{cc}^{Nd} is also a little larger and the previous formula is not strictly applicable. But it provides a first approximation.

Exercise

We have a sediment of mixed origin: 90% is from continental crust with an average $\varepsilon(0) = -25$; 10% is from the ancient mantle with $\varepsilon(0) = +5$. The Sm/Nd ratio of the continental crust is 0.11 and that of the mantle is 0.21. It is taken that the two components have the same Nd concentration.

(1) Calculate $\epsilon(0)$ of the sediment and its conventional model age.
(2) Calculate the model age using the Sm/Nd ratio measured on the sediment itself.
(3) Compare these ages with the model age of continental crust.

Answer

(1) 2 Ga.
(2) 2.26 Ga.
(3) 2.27 Ga.

By using the effective Sm/Nd age of the sediment, the dilution effect is "corrected" and we get the age of differentiation of the continental fraction.

Exercise

If a shale is found with $\varepsilon(0) = -20$, what is its model age?

Answer

$T_M = 1.82$ Ga.

Exercise

Supposing we apply the same formalism to Lu–Hf, given that $(^{176}Lu/^{177}Hf)_{cc} = 0.02$ and $(^{176}Lu/^{177}Hf)_{pm} = 0.036$, calculate the model age of a schist whose $\varepsilon_{Hf}(0) = -25$.

Answer

If $\varepsilon^{Hf}_{Sch}(0) = -25 \quad \alpha^{Hf}_{Sch} = (\varepsilon(0)10^{-4} + 1)\,\alpha^{Hf}_{pm} = 0.282\,24$

$$\bar{T} = \frac{1}{\lambda}\left(\frac{\alpha^{Hf}_{Sch} - \alpha^{Hf}_{pm}}{\mu^{Hf}_{Sch} - \mu^{Hf}_{pm}}\right) = 2.3 \text{ Ga}.$$

Neodymium model ages of granites over geological time

Dalila Ben Othman and the present author measured the initial isotope ratios of granites (major constituents of continental crust of varied ages) for the first time in 1979 (Allègre and Ben Othman, 1980; Ben Othman *et al.*, 1984). The Nd model age is plotted against geological age (Figure 6.32). If all the granites were derived from the mantle, the points would lie on a straight line of slope 1. Now, this is generally so for ancient granites (Mont d'Or granite of Zimbabwe is a spectacular exception) while for more recent granites, their model age is much older than their "geological age." This confirms the idea of recycling which increases with time. Let us try to be even more specific.

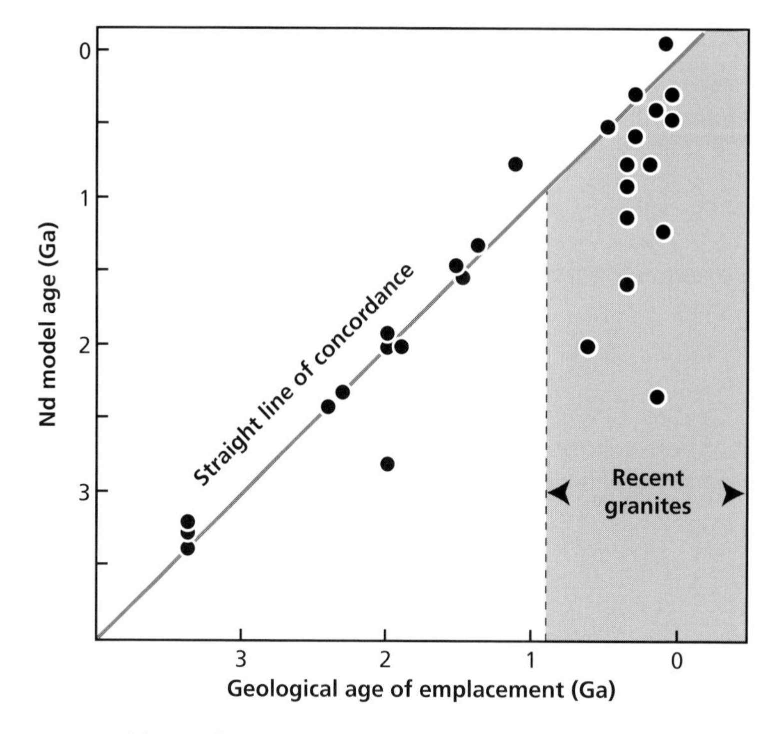

Figure 6.32 Model ages of granites of varied geological ages. After Allègre and Ben Othman (1980); Ben Othman *et al.* (1984).

Genetic cartography of continents

We have spoken of geological maps showing the different tectonic provinces which are assembled to form the continents and we have given an overall explanation of these maps. **Don DePaolo** (1981a, 1981b) and his group at the University of California at Berkeley undertook a similar approach (Farmer and DePaolo, 1983; Bennett and DePaolo, 1987), considering the model ages of Nd only, that is, by trying to eliminate continental recycling. He and his team studied two cases from the western United States: that of Colorado and the neighboring states where the age of emplacement of granites is 1.8 Ga and the west of the region (Rocky Mountains) where the granite intrusions are of Tertiary age (Figure 6.33).

In the Colorado province, there is a central part formed 1.8 Ga ago and which differentiated from the mantle at that time, and then bordering parts to the north and west whose model age is 2–2.3 Ga. It can be seen that the 1.8 Ga materials also extend into the province to the south in New Mexico and Texas dated 1.2–1.5 Ga. Heading west, towards the Rocky Mountains, the Berkeley team traced isoclines with the same $\varepsilon_{Nd}(0)$ value showing that when moving from the continent toward the oceanic margins, $\varepsilon_{Nd}(0)$ becomes increasingly positive, that is ever closer to mantle values (Figure 6.34). The percentages of mantle material in the make-up of the granite can be calculated: they increase from east to west.

These two examples show how continental tectonic segments are built up by addition to ancient sediment. Probably through subduction processes, as shown in the Rocky

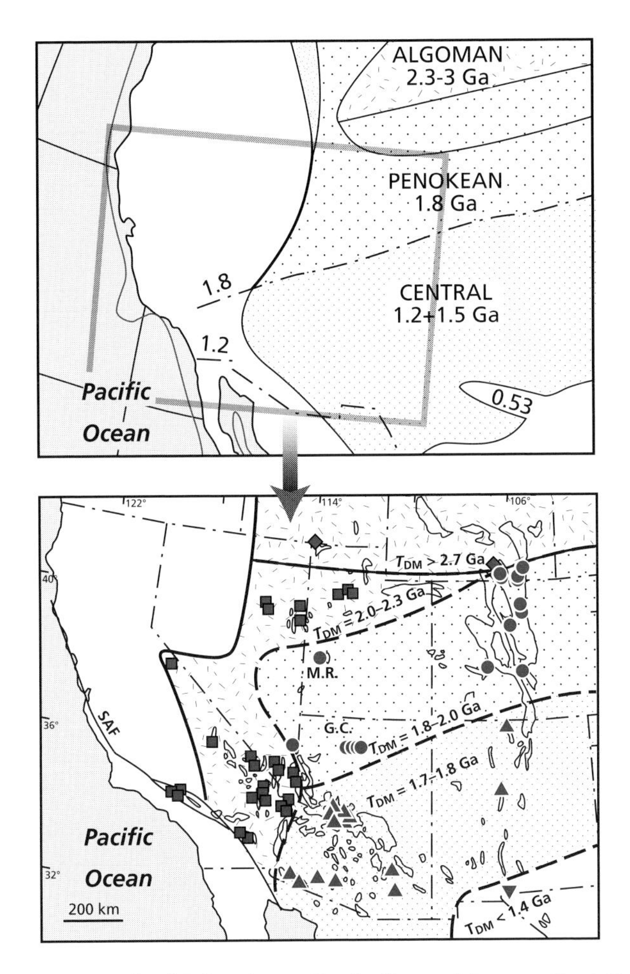

Figure 6.33 Study of Colorado granites by the model age method. Top: tectonic provinces (see Plate 5). Bottom: provinces mapped by neodymium model ages. The two distributions do not coincide. The difference can be explained by reworking. After Bennett and DePaolo (1987).

Mountains, but with reworking and reuse of older sediment. If a similar study is made in the Himalayas, that is, in a collision range, the geographical distribution is markedly different but the division between newly formed crust and recycled crust remains. The same goes for all of Europe where the Caledonian, Hercynian, and Alpine orogenies essentially reworked ancient pieces of continental crust, some of which are very old, as reflected by the Nd model ages calculated on granites or sediments.

We therefore have two very different situations:

- continents that have grown through new segments, which are very clearly mapped as in North America or Scandinavia;
- continents where it is difficult to identify large age provinces because everything has been mixed and recycled, as in Europe and the middle part of Asia.

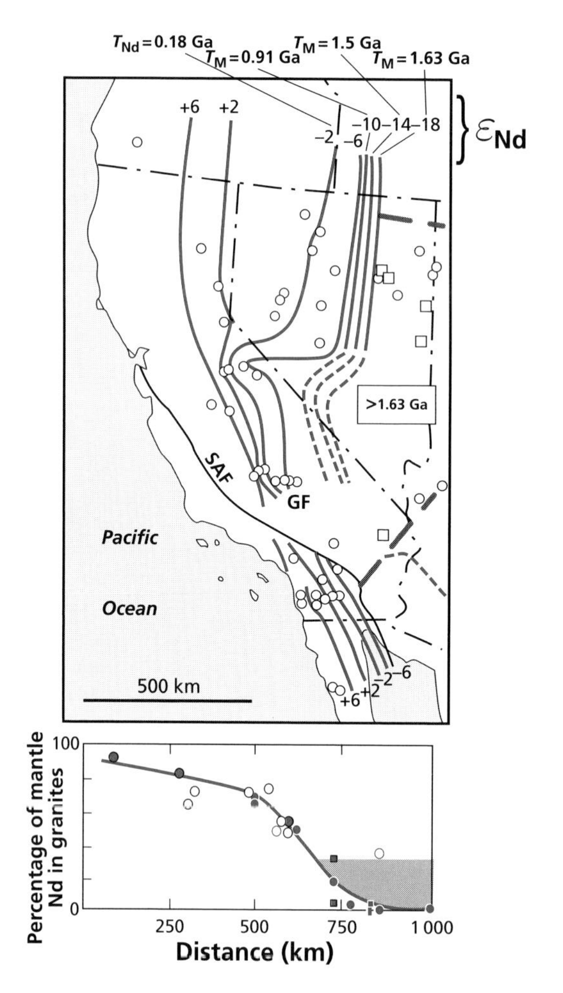

Figure 6.34 Neodymium isotope study of the Rocky Mountains in California. The curves of the ε_{Nd} isotope ratios are plotted. An E–W curve is shown below. The proportion of mantle decreases very rapidly eastwards. After Farmer and DePaolo (1983).

Remark

The main mechanisms of continental growth seem to be well understood. One question remains. We have drawn a general map of the tectonic provinces. The Colorado study gives details of one specific case. There is no doubt that continental crust in a given continent is extracted during mountain-building episodes of well-defined ages. But is this true at the scale of all the continents? Does not the combination of all orogenies lead to continuous extraction?

Australian detrital zircons

These ideas about dual mechanisms of formation of continental crust – reworking of old material and formation of new segments of continent material – are wonderfully illustrated by work on Australian detrital zircons by **Chris Hawkesworth**'s Bristol team using the latest *in situ* isotope analysis technology (Hawkesworth and Kemp, 2006).

This work began with the analysis of detrital zircons. Zircons ($ZrSiO_4$), which are prime minerals for U–Pb radiometric dating, are extremely resistant to erosion. They are engendered by the formation of granitic rocks. They withstand weathering very well and are transported mechanically and mix with clastic sediments (sandstones and, to a lesser extent, shales). They may undergo a second phase of erosion and be re-sedimented. At that time they can mix with zircons born in a new generation of granites. Thus, a sandstone may contain zircons of various ages. This pseudo-immortality of zircons was revealed in studies by **Ledent** *et al.* (1984), who were attempting to determine a mean age for continental crust by analyzing a zircon population. Using grain-by-grain zircon analysis, **Gaudette** *et al.* (1981) showed there were many episodes of granitization recorded in a single population of detrital zircons. But such studies using conventional methods were time-consuming and tedious. This approach was revolutionized by the team from the Australian National University when **Bill Compston** and **Ian Williams** (1984) developed the SHRIMP ion probe for U–Pb isotope analysis of zircons. Advances in automation mean that 500 zircons can now be mounted on a plate and their U–Pb analyses completed in a matter of days. It was with this method that the Australian team discovered the existence of zircons aged 4.3 Ga and even a few grains aged 4.4 Ga in Precambrian clastic sediments. They also made a further discovery that had been suspected for some time. Zircon grains have complex individual histories. Around an ancient core, which is often rounded by erosion, new growth zones have developed giving the zircon crystals the appearance of a double pyramid. The zircon crystals contain a record of the different periods of their individual histories.

This is proof, if any were needed, that some granites were formed from the remelting of earlier sediments, which themselves contained detrital zircons. These ancient zircons acted as seed crystals for new additions of zircon around them. A single zircon may tell the complex geological history of a region!

The Bristol team working in conjunction with the Australian National University team took a sandstone from the Primary period (400 Ma) as their starting point. After mechanically separating the zircons, they analyzed the U–Pb ages of several hundred zircons. They also analyzed the cores of zircons from a granite dated 430 Ma. They obtained ages between 3.2 Ga and the age of the granite. A whole series of ages, with maxima and minima, is shown in Figure 6.35.

They then used the fact that zircons are rich in hafnium (Hf is the element just below Zr in the periodic table) and very poor in lutetium. They thus managed to analyze the Hf isotope composition and calculate their model age for each zircon, in the same way as is done for Nd. This yielded a model age at which the material from which the zircon derives became separated from the mantle. The high Hf content of zircon meant this analysis could be performed by laser ablation followed by ICPMS analysis. But they added a further criterion. They analyzed the $^{18}O/^{16}O$ isotope composition of zircons with an ion probe. As we shall see in the next chapter, basic magmatic rocks have very constant $^{18}O/^{16}O$ compositions with δ varying from 5.5 to 6.5. They therefore selected zircons grains with $\delta^{18}O < 6.5$. In this way zircons derived from materials extracted from the mantle could be selected. This double-sorting process yielded an extraordinary result. The model ages were clustered around two values of 2 Ga and 3.2 Ga (Figure 6.36).

The conclusion is that new continental crust was only formed from the mantle at these two periods. However, continental crust (granites) has formed throughout geological

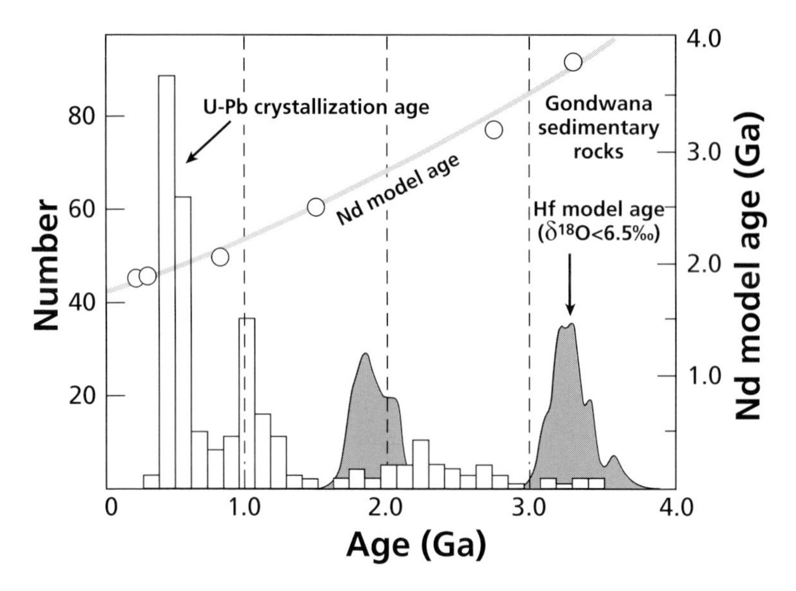

Figure 6.35 Composite zircon age diagram. The histogram in black shows the crystallization ages determined on different zircons using an ion probe. The histogram in blue is the model age of hafnium computed for individual zircons with $\delta^{18}O < 6.5$. The blue curve is the neodymium model age obtained on Australian sediments. After Hawkesworth and Kemp (2006).

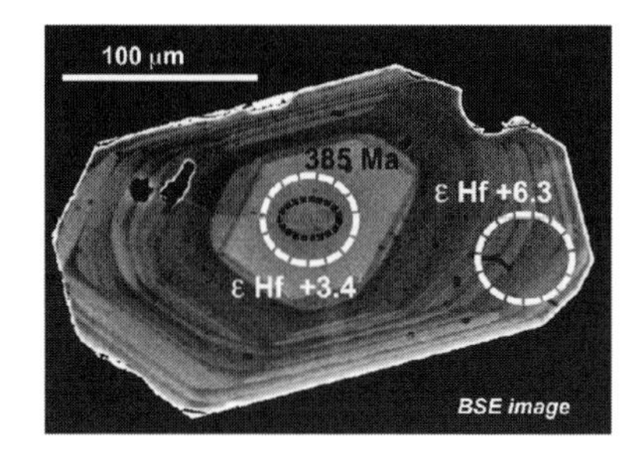

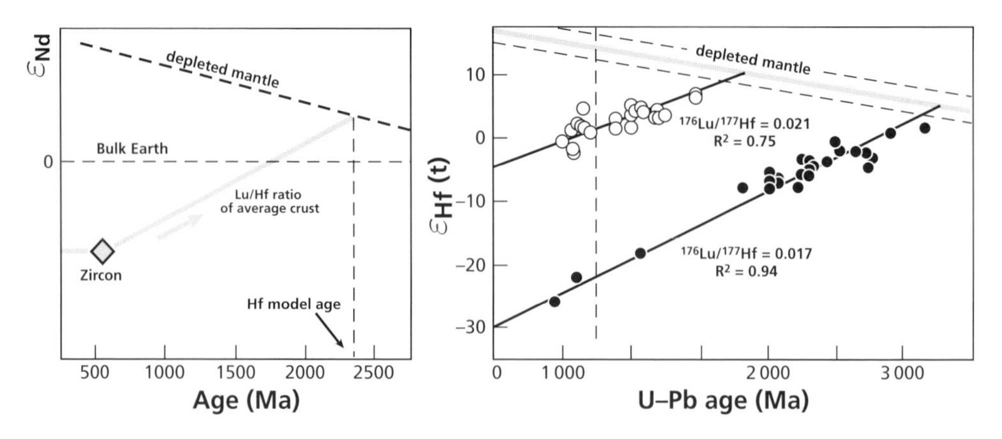

Figure 6.36 How hafnium model ages are obtained by laser ablation ICPMS. Zircons on spot size on the left. The principle of hafnium model age competition. Results for Australian zircons.

history, as shown by U–Pb dating, but these episodes were merely the reworking of ancient crust. New things were made out of old!

This new method, once extended to various regions of the planet, will indicate exactly how continental crust formed. Did it form continuously throughout geological time, more or less in relation with the activity of subduction zones, but in different geographical locations? Or did it form worldwide during specific episodes of intense activity?

When a histogram is drawn of geological ages measured on the various continents, peaks are found at 600 Ma, 1100 Ma, 1600 Ma, 2100 Ma, 2700 Ma, and 3200 Ma. Their dispersion is ±100 Ma on average. What do these peaks mean? Are they the heartbeat of the planet or the reflection that some regions like China or India are still under-explored? Studies of the western United States by the team from Berkeley seem to show that the formation of new continental crust is also associated geographically with reworking processes. But is this a general result? We can entertain high expectations of the results of this research because the method developed at Bristol and at the Australian National University suggests that we now have the means to solve the problem.

6.4 Isotope geochemistry of rare gases

There are five rare gases. They form the final column of the periodic table of the elements: helium (He), neon (Ne), argon (Ar), krypton (Kr), and xenon (Xe). Some isotopes of rare gases are produced by long-period radioactive processes and so cause variations in isotope abundance (Tables 6.5 and 6.6). We have already referred to ^{4}He and ^{40}Ar when dealing with geochronology (Chapter 5). The decay schemes leading to rare gas isotopes are recalled in Table 6.7.

All rare gas atoms share the common property of having their outer electron shell saturated and so being chemically inert. They are transported by physical processes only and tend to migrate upwards, towards the atmosphere where they accumulate. In the atmosphere, their behavior varies depending on their atomic mass.

Helium, a very light element, is not retained by the Earth's atmosphere as its mass is too low. It escapes continuously into space like (and with) hydrogen. Neon is retained by the present-day atmosphere, but it is thought that some neon was lost in the Earth's early history. Argon, krypton, and xenon are retained by the atmosphere. How can the escape of these elements be interpreted? As a first approximation, it can be understood quite simply. For an object to escape from the Earth, its velocity must exceed the Earth's gravitational attraction. Since $1/2\, mV_e^2 = mgR$, the velocity V_e must be greater than $\sqrt{2gR}$. Now, acceleration due to gravity is $g \approx 10$ m s^{-2} and the Earth's radius R is 6400 km, therefore the escape velocity is $V_e = 11.28$ km s^{-1}.

Let us suppose that escape is by thermal means only.

$$\frac{1}{2}mV_e^2 = \frac{3}{2}kT \text{ and } V_e = \sqrt{\frac{3kT}{m}}.$$

Hence:

$$\frac{3kT}{m} > (11.2 \text{ km s}^{-1})^2$$

where k is Boltzmann's constant, T is absolute temperature, and m is mass.

Table 6.5 Composition of the atmosphere

Isotope	Atomic abundance (%)
Helium	
3	0.000140
4	≈ 100
Neon	
20	90.50
21	0.268
22	9.23
Argon	
36	0.3364
38	0.0632
40	99.60
Krypton	
78	0.3469
80	2.2571
82	11.523
83	11.477
84	57.00
86	17.398
Xenon	
124	0.951
126	0.0887
128	1.919
129	26.44
130	4.070
131	21.22
132	26.89
134	10.430
136	8.857

			Total balance	
Major gases	Molecular mass	Fraction by volume	By mass (kg)	By volume at standard temperature and pressure
N_2	28.0134	0.78014	$3.866 \cdot 10^{18}$	$3.093 \cdot 10^{24}$
O_2	31.9988	0.20948	$1.185 \cdot 10^{18}$	$8.298 \cdot 10^{23}$
CO_2	44.0099	$(3.40) \cdot 10^{-4}$	$2.450 \cdot 10^{15}$	$1.248 \cdot 10^{21}$
He	4.0026	$(5.24 \pm 0.05) \cdot 10^{-6}$	$3.707 \cdot 10^{12}$	$2.076 \cdot 10^{19}$
Ne	20.179	$(1.818 \pm 0.004) \cdot 10^{-5}$	$6.484 \cdot 10^{13}$	$7.202 \cdot 10^{19}$
Ar	39.948	$(9.34 \pm 0.01) \cdot 10^{-3}$	$6.594 \cdot 10^{16}$	$3.700 \cdot 10^{22}$
Kr	83.80	$(1.14 \pm 0.01) \cdot 10^{-6}$	$1.688 \cdot 10^{13}$	$4.516 \cdot 10^{18}$
Xe	131.30	$(8.7 \pm 0.1) \cdot 10^{-8}$	$2.019 \cdot 10^{12}$	$3.446 \cdot 10^{17}$

Table 6.6 Rare gases in sea water

	Concentration (at standard temperature and pressure)			
	He (10^{-8})	Ne (10^{-7})	Ar (10^{-4})	Xe (10^{-8})
Surface sea water (25 °C)	3.7	1.47	2.26	0.65
Deep water (4 °C)	4.22	1.85	3.51	1.1

Table 6.7 Decay systems leading to rare gas isotopes

Radioactivity	Radioactive product	Isotope ratios studied
$^{238}U, ^{235}U, ^{237}Th$ radioactive chains	$\rightarrow {}^{4}He$	$\rightarrow \left(\frac{^{4}He}{^{3}He}\right)$
$^{18}O(\alpha, n)\, ^{21}Ne$	$\rightarrow {}^{21}Ne$	$\rightarrow \left(\frac{^{21}Ne}{^{20}Ne}\right)$
$^{40}K\, e^{-}$ cap	$\rightarrow {}^{40}Ar$	$\rightarrow \left(\frac{^{40}Ar}{^{36}Ar}\right)$
^{129}I (extinct) β^{-}	$\rightarrow {}^{129}Xe$	$\rightarrow \left(\frac{^{129}Xe}{^{130}Xe}\right)$
^{238}U spontaneous fission	$^{131}Xe, ^{132}Xe, ^{134}Xe, ^{136}Xe$	$\rightarrow \frac{^{131,132,134,136}Xe}{^{130}Xe}$
^{244}Pu extinct spontaneous fission	$^{131}Xe, ^{132}Xe, ^{134}Xe, ^{136}Xe$	$\rightarrow \frac{^{131,132,134,136}Xe}{^{130}Xe}$

Exercise

The temperature of the very high atmosphere is typically 700–900 K. Calculate the escape velocity of atomic hydrogen and then compare it with that of the molecule H_2.

Answer

The mass of an atom or molecule is the molar mass divided by Avogadro's number, $6.023 \cdot 10^{23}$. The mass of a hydrogen atom is $0.001/6.023 \cdot 10^{23}$, which is $1.65 \cdot 10^{-27}$ kg.

For H,

$$\sqrt{\frac{3kT}{m}} = \left(\frac{3 \times 1.381 \cdot 10^{-23} \times 800}{1.65 \times 10^{-27}}\right)^{1/2}$$

therefore $V_e = 4.48$ km s^{-1}.

For H_2, $V_e = 3.16$ km s^{-1}.

The two velocities calculated in the exercise above are less than 11.2 km s^{-1}. They seem to indicate that hydrogen does not escape from the Earth. Now, this is not so, as observations and measurements prove! Where is the mistake? In fact, the velocities of the various atoms or molecules do not have constant values: they obey **Boltzmann's statistical distribution**. And for a velocity of 4.48 km s^{-1}, some 10% of ^{2}H particles escape every second.

For ^{4}He, $V_e = 2.24$ km s^{-1}. Some 10^{-15} atoms have the necessary velocity to escape per second. That represents a substantial quantity of atoms in geological terms, when the time measured is multiplied by millions of years. However, for the other rare gases, the number of particles attaining escape velocity is too low to be effective: they remain trapped by the gravitational field.

The upshot for us is that, as the atmosphere derives from degassing of the mantle and not (as for the major planets) from retention of a primary gas envelope, the atmosphere may be considered as the complement of the mantle (as continental crust was for strontium and neodymium) for argon, krypton, and xenon, but not for helium and probably not for neon either, if we consider all of geological time. For argon, krypton, and xenon, we can write a balance equation:

atmosphere + mantle = whole Earth.

The isotope composition of the rare gases of the atmosphere has been known for about 50 years. However, it was for a long time difficult to obtain a measurement of the isotope compositions of the rare gases of the mantle. This was because it was too easy for samples to be contaminated by the atmosphere, which skewed the results. Rare gases occur in low concentrations in rocks from the mantle (basalts) and any contact with the atmosphere contaminates them. Where molten lava is in contact with the air, contamination is catastrophic. It is much the same for submarine contact, as rare gases are soluble in sea water, which contaminates the gas itself (hence the importance of knowing the solubility of rare gases in water). The geochemistry of rare gases began with the discovery of pillow lavas, whose rims turn to glass at the contact of sea water, preventing the sea water containing dissolved rare gases from contaminating the lava. Moreover, when pillow lavas are emplaced, the rare gases migrate and concentrate in gaseous inclusions concentrating the rare gases 1000 times compared with magmas. More recently it has been possible to analyze He and Ne in gaseous inclusions in olivine phenocrysts.

The second factor making this analysis difficult is the low abundance of rare gases, which decreases with their mass. The atmosphere does not retain He and Ne quantitatively, as said, so He and Ne concentrations are relatively low in the atmosphere (and in sea water). As the concentrations of rare gases are higher in magmas, magmas are the less difficult to analyze because they are less likely to be contaminated by the atmosphere.

Measuring rare gases with a mass spectrometer is a difficult but very sensitive business. Special equipment is required to extract gases without them being contaminated by the atmosphere or by previous sampling (see Figure 6.37).

Exercise

The rare gas composition of the atmosphere is expressed in cubic centimeters at standard temperature and pressure in Table 6.8 below.

(1) What is the composition of the atmosphere in rare gases expressed in moles?
(2) What is the composition of the atmosphere in rare gases expressed in grams?
(3) What is the composition in ^{3}He and ^{40}Ar in moles and grams, given that $^{40}Ar/^{36}Ar = 296.8$ and that $^{3}He/^{4}He = 1.4 \cdot 10^{-6}$?
(4) What is the concentration of these gases if related to the mass of the Earth?

Answer

Under standard conditions, 1 mole of an ideal gas occupies 22.4 liters. The table below shows the answers to questions (1) and (2).

	^{4}He	^{20}Ne	^{36}Ar	^{84}Kr	^{130}Xe
Composition (mole)	$0.0926 \cdot 10^{16}$	$0.29098 \cdot 10^{16}$	$0.0555 \cdot 10^{17}$	$0.1149 \cdot 10^{15}$	$0.06263 \cdot 10^{13}$
Composition (g)	$0.3704 \cdot 10^{16}$	$5.8196 \cdot 10^{16}$	$1.998 \cdot 10^{17}$	$9.6516 \cdot 10^{15}$	$8.1449 \cdot 10^{13}$

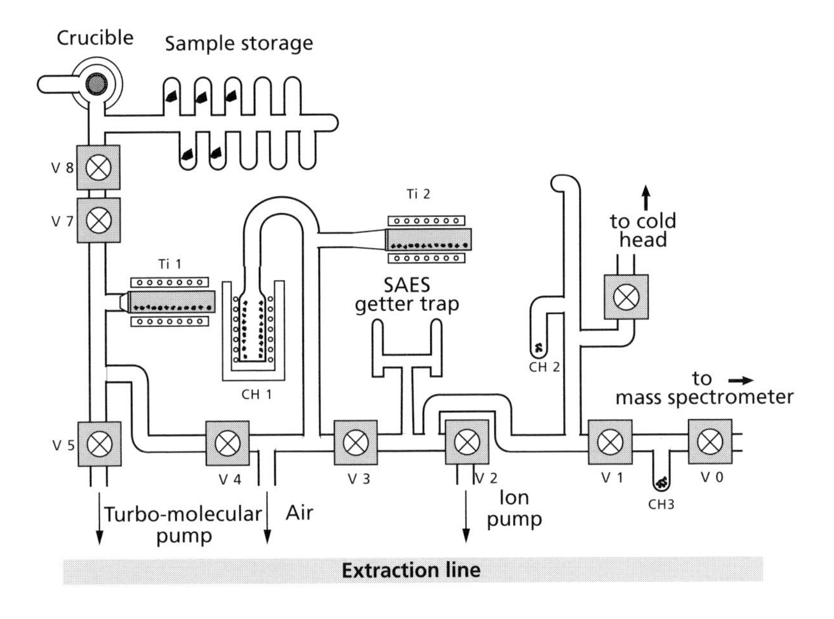

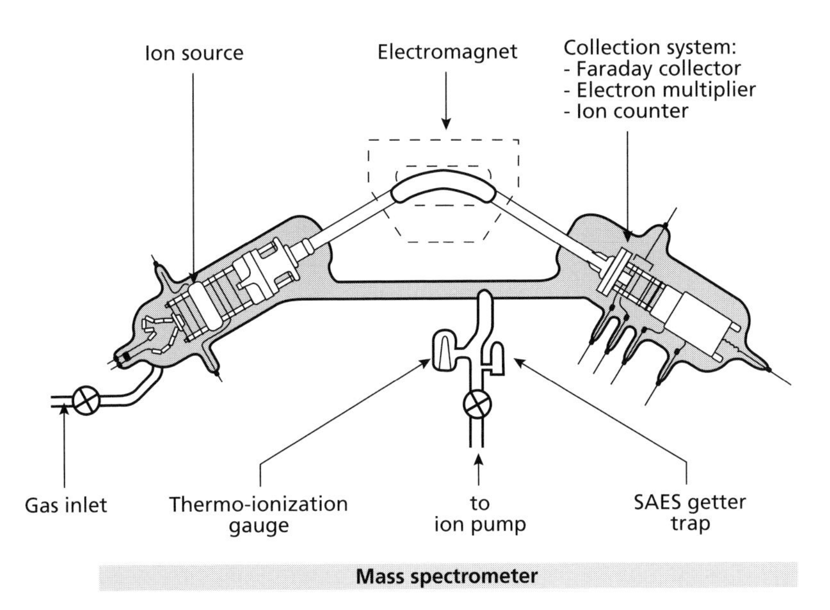

Figure 6.37 Laboratory equipment for measuring the isotope composition of rare gases. The fundamental point is that the measurement enclosure is at very low pressure of 10^{-9} mm Hg. It comprises two parts: (i) the extraction line made of special glass (or metal). This is a circuit where the much more abundant gases (H_2O, O_2, N_2, CO_2, CH_4) are captured because their presence in the mass spectrometer would lower the partial pressure of the rare gases too much for them to be measured. (ii) The mass spectrometer. The purified gases are fed one at a time into the mass spectrometer with its electron bombardment source and where the gases are enclosed (we speak of static measurement): helium, neon, argon, krypton, and xenon are measured in turn. Concentrations are very low. The signal is measured either with a Faraday cup or by an ion counter.

Table 6.8 Rare gas composition of the atmosphere

	^{4}He	^{20}Ne	^{36}Ar	^{84}Kr	^{130}Xe
Composition (cm^3)	$2.076 \cdot 10^{19}$	$6.518 \cdot 10^{19}$	$1.245 \cdot 10^{20}$	$2.245 \cdot 10^{18}$	$1.403 \cdot 10^{16}$

(3) The results are shown in the table below.

	^{40}Ar	^{3}He
Concentration	$65.8\ 10^{18}$	$1.647\,24 \cdot 10^{18}$
Composition (mole)	$3.8892 \cdot 10^{9}$	$1.296 \cdot 10^{8}$

(4) The mass of the Earth is $6.057 \cdot 10^{27}$ kg. The results are shown in the table below.

	^{4}He	^{36}Ar
Concentration	$3.47 \cdot 10^{-8}$	$2.083 \cdot 10^{-8}$

6.4.1 Isotope geochemistry of helium

Helium-3 was first discovered by **Alvarez** and **Cornog** (1939). After the Second World War, **Aldrich** and **Nier** (1948) discovered the variations in terrestrial abundances, but helium isotope geochemistry was initiated by the Soviet team under the impetus of **Igor Tolstikhin** (Mamyrin *et al.*, 1969; Tolstikhin *et al.*, 1974) and by **Brian Clarke** and **Harmon Craig** (Clarke *et al.*, 1969) at the Scripps Institution of Oceanography and their students and later by **Mark Kurz** and **Bill Jenkins** (1981) at the Woods Hole Oceanographic Institution.

The principles are as follows. Helium-4 is thc product of collatcral disintegration of radioactive chains (α particles): 8α for the ^{238}U chain, 7α for the ^{235}U chain, and 6α for the ^{232}Th chain. Helium-3 is a stable isotope. The $^{4}He/^{3}He$ ratio in a closed system varies with the equation:

$$\frac{^{4}\mathrm{He}}{^{3}\mathrm{He}} = \left(\frac{^{4}\mathrm{He}}{^{3}\mathrm{He}}\right)_0 + \left(\frac{^{238}\mathrm{U}}{^{3}\mathrm{He}}\right) \cdot f(t).$$

The function $f(t)$ is defined as below:

$$\text{If } t \text{ is small, } f(t) = \left[8\lambda_8 + \frac{7\lambda_5}{137.8} + 6\left(\frac{\mathrm{Th}}{\mathrm{U}}\right)\lambda_2\right] t \approx 2.47t, \text{ if } t \text{ is in Ga.}$$

$$\text{If } t \text{ is large, } f(t) = \left[8\left(e^{\lambda_8 t} - 1\right) + \frac{7}{137.8}\left(e^{\lambda_5 t} - 1\right) + 6\frac{\mathrm{Th}}{\mathrm{U}}\left(e^{\lambda_2 t} - 1\right)\right].$$

In keeping with an odd practice introduced by **Harmon Craig**, the helium isotope ratio in basalt rocks is often expressed "upside down":

$$\frac{^{3}\mathrm{He}}{^{4}\mathrm{He}} = R \cdot \left(\frac{^{3}\mathrm{He}}{^{4}\mathrm{He}}\right)_{\text{atmosphere}} \quad \text{with} \left(\frac{^{3}\mathrm{He}}{^{4}\mathrm{He}}\right)_{\text{atmosphere}} = 1.4 \cdot 10^{-6}.$$

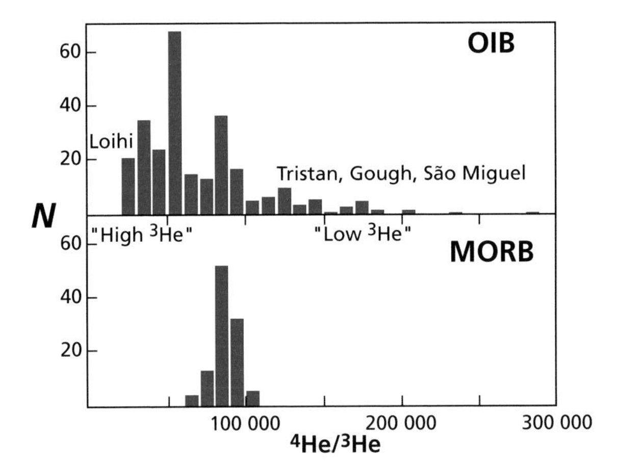

Figure 6.38 Distribution of $^4He/^3He$ ratios in MORB and OIB. The difference in dispersion can be observed.

When $R = 8$, the value is 8 times that of the atmosphere. If $R = 30$, it is 30 times that of the atmosphere, and so on. In what follows, we shall give the results in $^4He/^3He$ ratios but we shall add the $R(^3He)$ notation for comparison with papers and books using this notation.

Measurements of $^4He/^3He$ on ocean basalts display a very different distribution for MORB and OIB (Figure 6.38).

The MORB distribution is tightly clustered around $^4He/^3He = 90\,000$ whereas the OIB distribution is very dispersed with ratios ranging from 13 000 to 130 000. However, looking more closely, large islands like Hawaii or Iceland, with large volumes of basalt, have values between 13 000 and 36 000 (Figure 6.38).

The simplest interpretation resulting from the standard model is to say that MORB derives from the upper-mantle reservoir with a high $^{238}U/^3He$ ratio and so by radioactive decay a high $^4He/^3He$ ratio; OIB derives from a more primitive reservoir where the ^3He content is higher, and so the $^{238}U/^3He$ ratio is smaller as is the $^4He/^3He$ ratio. This is because the upper reservoir, which is directly involved in the tectonic plate mechanism, is highly degassed, whereas the deep reservoir is much less degassed (Figures 6.39 and 6.40). This dual origin is confirmed by the existence of the Schilling effect, that is a mixing of OIB and MORB along some mid-ocean ridges. Thus, the mid-Atlantic ridge southwards from Iceland displays a variation in $^4He/^3He$ ratios, which increase from Iceland (14 000) southwards where they reach 120 000. This is analogous to what we saw for Sr isotope composition (**Kurz** and **Jenkins**, 1981).

Exercise

Accepting that the lower mantle is a closed system for U and He, calculate the $^{238}U/^3He$ ratio $= \mu_1^{He}$ given that $^4He/^3He = 15\,000$ and that $(^4He/^3He)_{initial} = 2500$. If the $^4He/^3He$

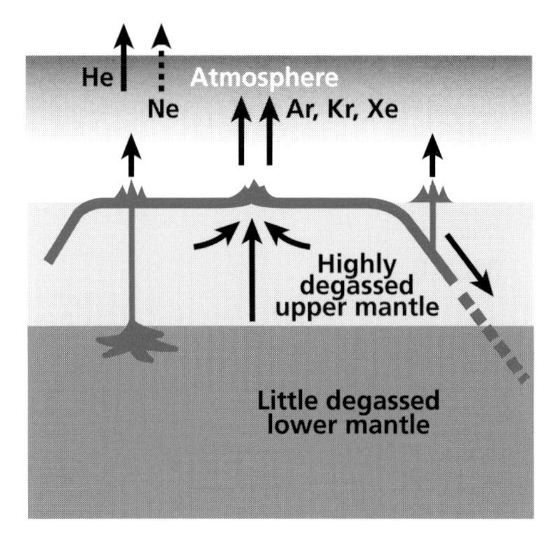

Figure 6.39 The standard model developed from the results for helium isotope analysis and extended to all rare gases. With an atmosphere that retains argon, krypton and xenon but lets helium (and neon in the past) diffuse, the upper mantle is highly degassed but not so the lower mantle.

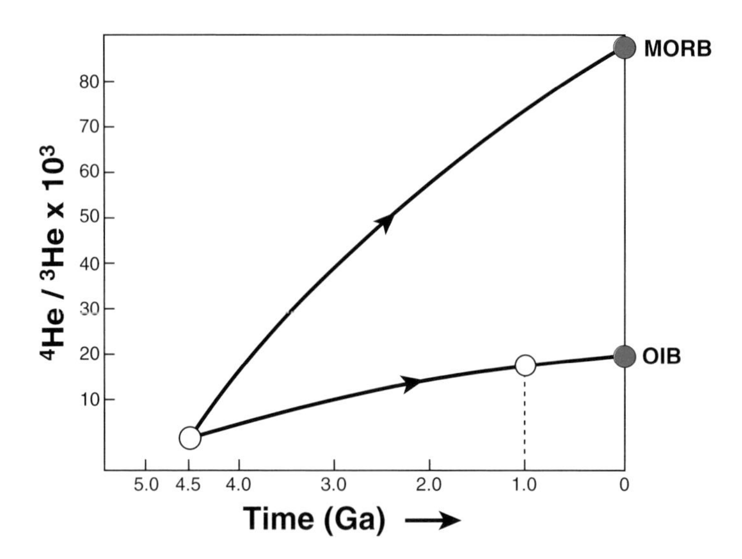

Figure 6.40 Possible changes in $^4He/^3He$ ratios in the MORB (upper mantle) and OIB source reservoirs. It is assumed they evolved in a closed system over $4.5 \cdot 10^9$ years in both cases, which is undoubtedly an extreme oversimplification, but gives an order of ideas. The values of $^{238}U/^3He = m_{U,\,He}$ considered are 4900 for the upper mantle and 800 for the lower mantle.

ratio = 100 000 (still the hypothesis of a closed system), calculate μ_2^{He}. (We take Th/U ≈ 4.)

Answer

We use the dating formulae established in Chapter 2 and recalled above. We find: $\mu_1^{He} = 684$ and $\mu_2^{He} = 5300$.

Exercise

For four basalts, we give the helium isotope composition with Craig's $^{3}He/^{4}He = NR_A$ notation in the table below.

	Basalt			
	B_1	B_2	B_3	B_4
N	40	5	10	20

(1) Calculate the $^{4}He/^{3}He$ composition of the four samples and plot the curve $NR_A = f(^{4}He/^{3}He)$.
(2) Suppose we have measured the Sr isotope ratios in the same samples as set out below.

	Basalt			
	B_1	B_2	B_3	B_4
N	0.7035	0.7022	0.7029	0.7033

Draw the Sr–He isotope correlation using both notations (see Figure 6.41). What do you conclude?

Answer
(1)

	Basalt			
	B_1	B_2	B_3	B_4
$(^{4}He/^{3}He)$	17 800	142 000	71 400	35 710

(2) The three curves for the abundances of the two radiogenic isotopes ^{4}He and ^{87}Sr are shown below. Craig's R_A notation **destroys** perfect linear correlation.

6.4.2 Isotope geochemistry of neon

Neon has three isotopes: ^{20}Ne, ^{21}Ne, and ^{22}Ne. The abundance of ^{21}Ne varies with nuclear reactions caused by α particles emitted by uranium and thorium chains $^{18}O(\alpha, n)\,^{21}Ne$ or, for 17%, $^{24}Mg(n, \alpha)\,^{21}Ne$. These variations are said to be **nucleogenic**, although in fact they follow the mathematical laws of radiogenic production (**Wetherill**, 1954). We show the isotope variations observed in nature in a plot of ($^{20}Ne/^{22}Ne$, $^{21}Ne/^{22}Ne$) where the two ratios vary in the basalts but for two different reasons. The $^{21}Ne/^{22}Ne$ ratios vary for nucleogenic reasons. The $^{20}Ne/^{22}Ne$ ratios vary in nature because the Earth's atmosphere has a different value from that of the Earth's mantle (which is closer to that of the Sun) (**Craig** and **Lupton**, 1976).

$$\left(\frac{^{20}Ne}{^{22}Ne}\right)_{\text{atmosphere}} = 9.5 \qquad \left(\frac{^{20}Ne}{^{22}Ne}\right)_{\text{Sun}} = 13.5.$$ [8]

[8] Terrestrial values can vary by only $^{20}Ne/^{22}Ne$ from the Sun's values but may be very different from the atmospheric value.

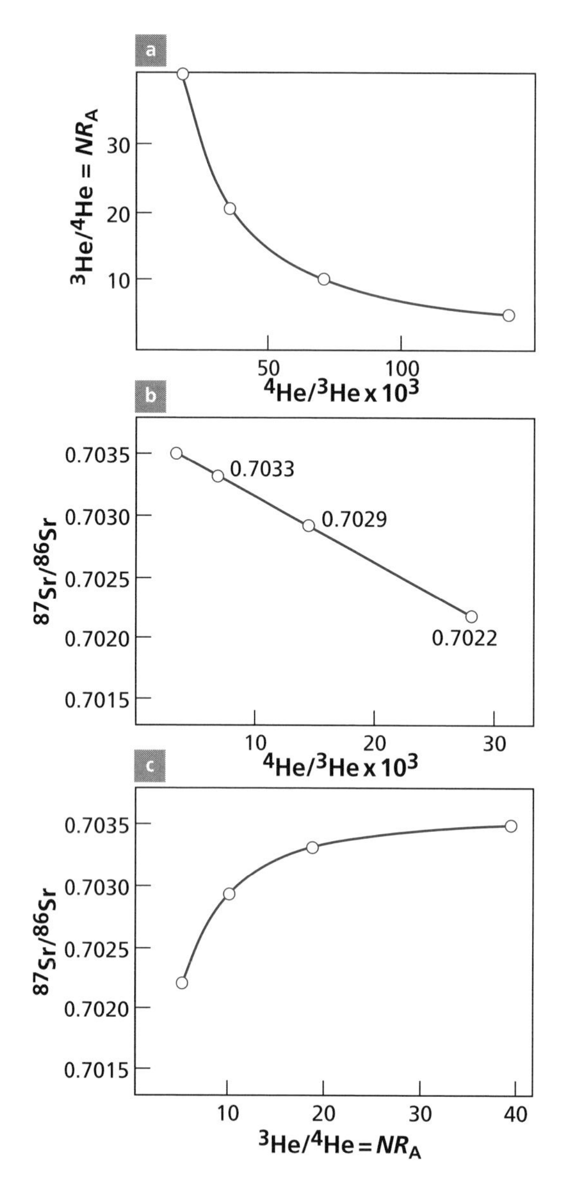

Figure 6.41 Results. (a) ^{4}He/^{3}He–^{3}He/^{4}He in NR_A. (b) ^{87}Sr/^{86}Sr–^{4}He/^{3}He linear correlation. (c) Same relations but with ^{3}He/^{4}He = NR_A notation.

This circumstance is not fully elucidated[9] but is a godsend because, for each isotope ratio measurement, the proportion contaminated by the Earth's atmosphere can be calculated and **uncontaminated values** obtained. Measurements on MORB and OIB by the Paris

[9] It is thought that, early in the Earth's history, the atmosphere was very hot and neon probably escaped by isotope fractionation by a process known as hydrodynamic escape.

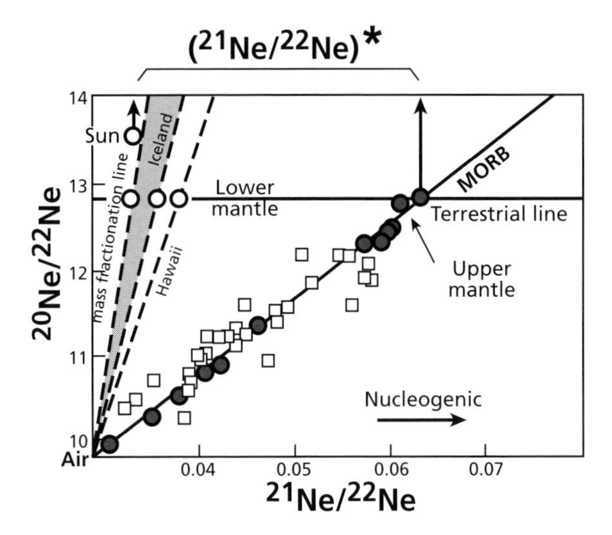

Figure 6.42 Correlation diagrams for ($^{20}Ne/^{22}Ne$, $^{21}Ne/^{22}Ne$). Results for oceanic basalts are used to define what is called [$^{21}Ne/^{22}Ne$]*.

group (**Sarda** *et al.*, 1988; **Moreira** and **Allègre**, 1998; **Moreira** *et al.*, 1998) can be used to define two clearly distinct distributions (Figure 6.42). The experimental measurements form straight lines, connecting the pure composition to be measured with the atmospheric value. The MORB defines a "low-angle" straight line passing through the atmospheric value. The OIB defines steeper straight lines, close to the atmosphere–Sun straight-line segment.

We define ($^{21}Ne/^{22}Ne$)* ratios, which are uncontaminated by the atmosphere at the intersect of the horizontal line of the solar ratio (or slightly lower) with the straight lines (sample–atmosphere). These ($^{21}Ne/^{22}Ne$)* ratios are automatically corrected for atmospheric contamination.

These values are close to $7.5 \cdot 10^{-2}$ for MORB and to $3.8 \cdot 10^{-2}$ for OIB. (By taking 10^{-2} as the unit, we deal with figures like 7.5 or 3.8, which is handier.) Moreover, it can be seen that intermediate values occur where hot spots underlie mid-oceanic ridges (Schilling effect) (Figure 6.43).

Neon isotopes therefore confirm the idea of the mantle being two reservoirs (an upper highly degassed and lower more primitive mantle) described from helium isotopes.

Exercise

Imagine we have a measurement of the raw isotope composition of neon of an oceanic basalt: $^{20}Ne/^{21}Ne = 11.5$ and $^{21}Ne/^{22}Ne = 5 \cdot 10^{-2}$. Calculate ($^{21}Ne/^{22}Ne$)*.

Answer

($^{21}Ne/^{22}Ne$)* $= 7 \cdot 10^{-2}$.

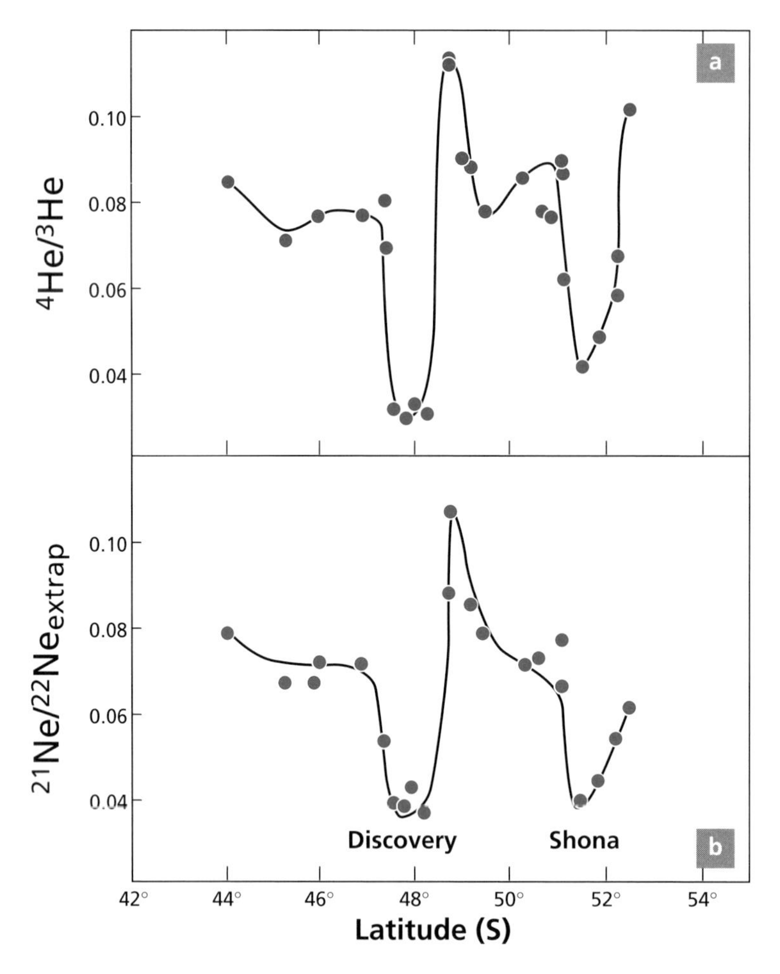

Figure 6.43 Variation of $^4He/^3He$ and $[^{21}Ne/^{22}Ne]^*$ ratios on the mid-Atlantic ridge between 42° and 54° S. There are two hot spots beneath the ridge in this area, with topographic effects known as the Discovery and Shona seamounts. They are illustrations of the Schilling effect. In both cases the two hot spots are clearly detected by helium and neon systematics. After Sarda *et al.* (2000).

6.4.3 Isotope geochemistry of argon

This is undoubtedly the oldest form of rare gas geochemistry. It was first introduced by **Paul Damon** and **Larry Kulp** (1958) and by **Karl Turekian** (1959) when they were working at Columbia University but suffered greatly from the difficulty in correcting the values measured for atmospheric contamination. And yet, argon has considerable advantages compared with other rare gases. It is retained by the atmosphere, meaning that a balance equation can be written. There is no initial ^{40}Ar, as ^{40}Ar is not made in stellar nucleosynthesis and is entirely produced by ^{40}K decay. However, the big disadvantage is its extreme sensitivity to atmospheric pollution. Therefore aerial basalts, for example, are unsuitable for measurement, as are many submarine basalts. All of the $^{40}Ar/^{36}Ar$ isotope ratios published

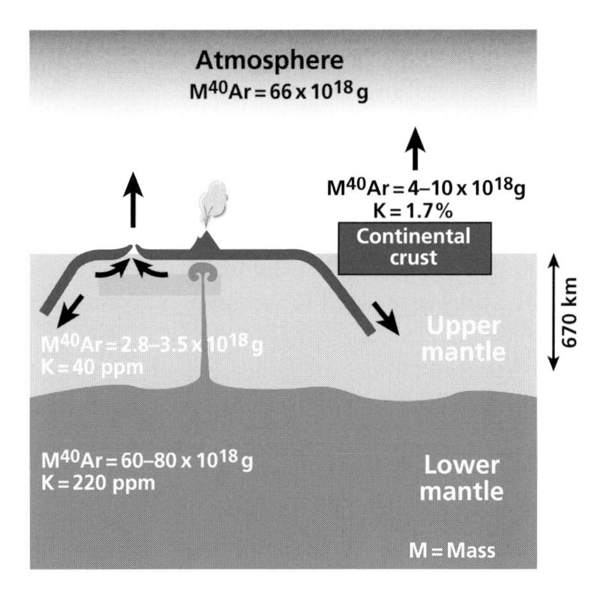

Figure 6.44 The ^{40}Ar mass balance of the Earth. After Allègre *et al.* (1996).

in the literature over a long period were heavily contaminated and so too low. In 1985, the author's Paris team finally managed to measure reliable $^{40}Ar/^{36}Ar$ ratios for MORB by using pillow lava glass, yielding $^{40}Ar/^{36}Ar = 30\ 000$ (compared with the $^{40}Ar/^{36}Ar = 296$ ratio of the atmosphere) (**Sarda** *et al.*, 1985). From this, a mass balance for the Earth's ^{40}Ar could be estimated (**Allègre** *et al.*, 1996) (Figure 6.44). The K content of the silicate Earth is quite well known (250–280 ppm).[10] Knowing that ^{40}K is $1.16 \cdot 10^{-4}$ of total K, we can calculate the total ^{40}Ar produced in $4.55 \cdot 10^9$ years: from $140 \cdot 10^{18}$ g to $156 \cdot 10^{18}$ g, depending on the value used for K. Now, the total quantity of argon in the atmosphere is $66 \cdot 10^{18}$ g. If we evaluate ^{40}Ar in the continental crust at a maximum of $4–10 \cdot 10^{18}$ g, there remains $60–86 \cdot 10^{18}$ g of ^{40}Ar, which is the missing argon inside the Earth. This means that for ^{40}Ar, the Earth is only half degassed.[11] There is little ^{40}Ar in the upper mantle. The flow of ^3He from the ocean ridges is estimated at $1.1 \cdot 10^3$ moles yr^{-1}. Using the $^4He/^3He$ and $^4He/^{40}Ar$ ratios measured in MORB, we find an ^{40}Ar flux of $2 \cdot 10^7$ moles yr^{-1}. If we accept that the oceanic lithosphere is entirely degassed when it passes through the ocean ridges by the fusion processes occurring there, we can calculate the quantity of argon in the mantle if the mantle were homogeneous. We find $2.4 - 1.8 \cdot 10^{18}$ g, which does not match the amount of missing argon.

Where is the missing argon then? Experimental measurements have shown that neither argon nor potassium is soluble in iron even at high pressures. Therefore the argon is not stocked in the Earth's core. That leaves just the lower mantle. Supposing, in the standard model, an upper layer above 670 km convecting separately from the lower

[10] We know the terrestrial uranium value quite well as it is analogous to that of carbonaceous chondrites (as is Nd which is also a refractory chemical element). Now, the K/U ratio, which is about constant on Earth, is between 10 000 and 12 000. We can therefore deduce K.

[11] This is not the case for ^{36}Ar, but ^{40}Ar was produced later by K decay. We shall see the explanation at the end of this chapter.

mantle: by the same calculation as before we find that the upper mantle contains $2 \cdot 10^{18}$ g of argon. Therefore $60–86 \cdot 10^{18}$ g of ^{40}Ar is stored in the lower mantle. This model therefore also supports the standard model with two layers which are degassed to different extents.

In all, then, the three rare gases confirm the two-layer **standard model**: a degassed upper mantle and a less-degassed (more primitive) lower mantle. This is an extremely important confirmation because degassing of rare gases has nothing to do with the geological phenomena associated with differentiation of the continental crust and which causes Rb/Sr, Sm/Nd, and Lu/Hf fractionation. Yet both approaches lead to the same outcome: a mantle divided into two layers.

Two important consequences follow from this pattern of argon distribution in the mantle. First, given the present-day distribution of roughly half of the total ^{40}Ar being found in the atmosphere and half in the lower mantle but none in the upper mantle, and assuming the initial distribution was uniform, the atmospheric ^{40}Ar is derived from complete outgassing of the upper mantle and one-third outgassing of the lower mantle. Since this calculation concerns ^{40}Ar, which has no primitive component, the observation relates to "geological outgassing" from 4.4 Ga to the present day.

Second, the (primitive) ^{36}Ar concentration is about 200 times higher in the lower mantle than in the upper mantle. Therefore the outgassing of the lower mantle has to be such that it travels through the upper mantle as a transient. Otherwise it would contaminate the upper mantle.

Exercise

Just as we calculated the age of the continental crust we can calculate the mean age of the atmosphere using argon isotopes. Given that the mass of the mantle is about $4 \cdot 10^{27}$ g, the potassium content is 250 ppm with $^{40}K = 1.16 \cdot 10^{-4}\ K_{total}$; the mass of ^{40}Ar in the atmosphere is $66 \cdot 10^{18}$ g and we can estimate the $^{40}K/^{36}Ar$ ratio of the Earth from $(^{40}Ar/^{36}Ar)_{atmosphere} = 296$. In addition we assume ^{36}Ar is degassed by 90–100%. Calculate the age of the atmosphere.

Answer

We write the equation

$$\left(\frac{^{40}Ar}{^{36}Ar}\right)_{atmosphere} = \left(\frac{^{40}Ar}{^{36}Ar}\right)_{Earth} \left(e^{\lambda T_0} - e^{\lambda T_{atm}}\right).$$

We obtain $T = 3.3$ Ga. So the indication from argon is that the atmosphere formed much earlier than the mean age of the continents. **But this is not the end of the story!**

6.4.4 Isotope geochemistry of xenon

Xenon has nine isotopes, of masses 124, 126, 128, 129, 130, 131, 132, 134, and 136. On Earth, variations in abundance related to radioactive decay are of two types: extinct forms of radioactivity and spontaneous fusion of ^{238}U. To these must be added the exceptional fission process induced by ^{235}U in the Oklo nuclear reactor. Xenon-130 is taken as the reference isotope because it does not derive from any of these processes. Hence we can speak of $^{129}Xe/^{130}Xe$, $^{131}Xe/^{130}Xe$, $^{132}Xe/^{130}Xe$, and $^{136}Xe/^{130}Xe$ ratios.

The first of these ratios varies because ^{129}Xe is the decay product of ^{129}I, which is an extinct form of radioactivity (Reynolds, 1960). We saw in Chapter 3 how this form of radioactivity was exploited for determining age differences among meteorites. Now, this ratio varies on Earth and is an essential datum.

The $^{131-136}Xe/^{130}Xe$ ratios vary too, but the situation here is more complex as their variations may be attributed to extinct spontaneous fission from ^{244}Pu or to very-long-period spontaneous fission of ^{238}U. These isotope ratios were measured in primitive carbonaceous meteorites, in the atmosphere, and also in rocks from the Earth's mantle. That is, like helium, neon, and argon, xenon too is found in the glassy margins of MORBs.

It has long been known that xenon isotope compositions in the Earth's atmosphere are different from those in carbonaceous chondrites, proving that the Earth's atmosphere formed after the carbonaceous chondrites did, because the isotopes in the denominators of the isotope ratios are more abundant in the atmosphere than in carbonaceous chondrites. As they are of radiogenic origin, they are demonstrably younger.

The second important discovery, made by **Thomas Staudacher** and the present author (1982), is that the $^{129}Xe/^{130}Xe$ isotope ratios and the $^{132}Xe/^{130}Xe$ isotope ratios (to choose just one fissiogenic ratio) of MORBs are greater than those of the atmosphere. They vary in a correlated manner (Figure 6.45).

By contrast, the isotope ratios of what corresponds to OIBs, whether in Hawaii or in Iceland, are only marginally greater than those of the atmosphere. For the $^{129}Xe/^{130}Xe$ ratio, this is unambiguous. It is proof of the intense ^{129}I activity in the upper mantle early in the Earth's history.

The situation is more complex for $^{131-136}Xe/^{130}Xe$ isotope ratios because their variations stem either from extinct fission of ^{244}Pu or from long-period ^{238}U fission (Kuroda, 1980). The very difficult work of distinguishing between the effect of each has led to acceptance of the following approximation:

- in MORBs, the excess of $^{131-136}Xe$ compared with the atmosphere (normed to ^{130}Xe) is mostly due to ^{244}Pu;
- in granites, on the contrary, it is ^{238}U fission that is responsible for the essential variations.

Analysis of MORBs plotted on the ($^{129}Xe/^{130}Xe$, $^{132}Xe/^{130}Xe$) diagram (Figure 6.45) shows there is an excellent correlation passing through the atmospheric value.

To derive the most information possible from this observation we concentrate here on ^{129}Xe from the mantle. The value observed in MORBs is $^{129}Xe/^{130}Xe = 7.65$. The value found in OIBs (Hawaii, Iceland, the Galápagos Islands) is 7 at most. The atmospheric value is 6.5. These values confirm the difference between the MORB and OIB reservoirs. As with neon, helium, and argon, the MORB reservoir (upper mantle) is more radiogenic than the OIB reservoir.

There is coherence then. Except that the difference here can only have been established in the early history of the Earth, since ^{129}Xe came from the decay of ^{129}I, which is extinct radioactivity. The MORB and OIB reservoirs must have separated in the first 150 million years of the Earth's history and not have been merged since. Exchanges of the MORB

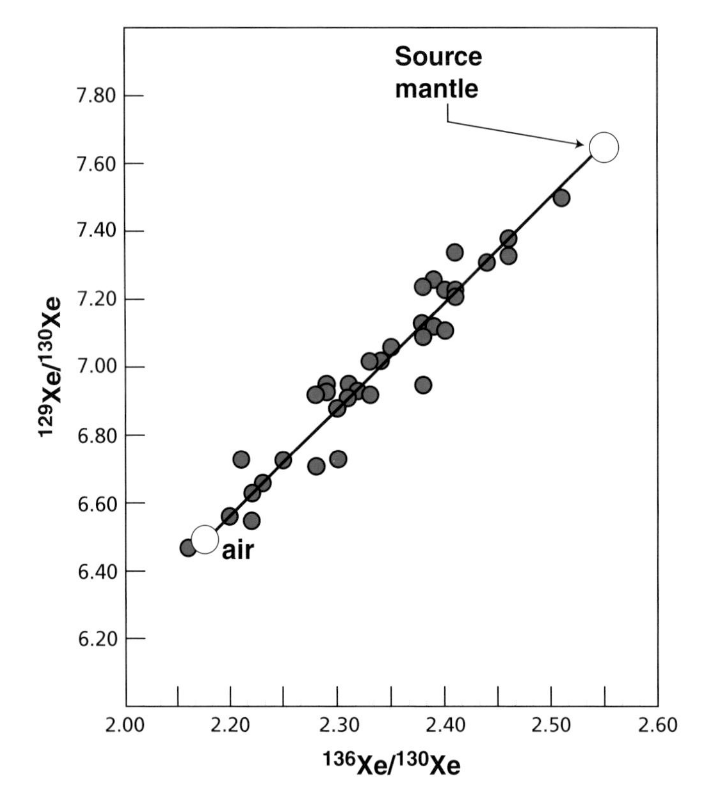

Figure 6.45 Correlation diagram of xenon isotope ratios in MORB. The $^{129}Xe/^{130}Xe$ ratios are evidence of extinct ^{129}I radioactivity. The $^{136}Xe/^{130}Xe$ ratios are mostly from extinct ^{244}Pu fission and are coupled with spontaneous long-term ^{238}U fission. The variations are due to mixing with atmospheric air either in nature or in the laboratory during measurement.

reservoir both with the atmosphere and with the OIB reservoir have been limited for 4.3 billion years.

We shall return to other repercussions for the age of the early Earth, but it is already apparent that the ages we are dealing with here are much older than the "age of the atmosphere" calculated from a simple mass balance equation for the $^{40}K-^{40}Ar$ system.

6.4.5 Coherence in rare gas geochemistry

We have seen that the use of four rare gases yielded coherent and complementary results when taken independently (Allègre *et al.*, 1983c). It is natural and essential to see whether analyses on the same samples yield coherent results too. Two graphs give the correlations obtained between all the rare gases in MORBs and in OIBs (Figure 6.46) obtained by **Moreira** *et al.* (1998).

For all rare gases (helium is not shown here because it is not contaminated by the atmosphere) we can distinguish two separate reservoirs with distinctive isotopic signatures. Notice, though, that the essential thing about these correlations is the mixing that occurs

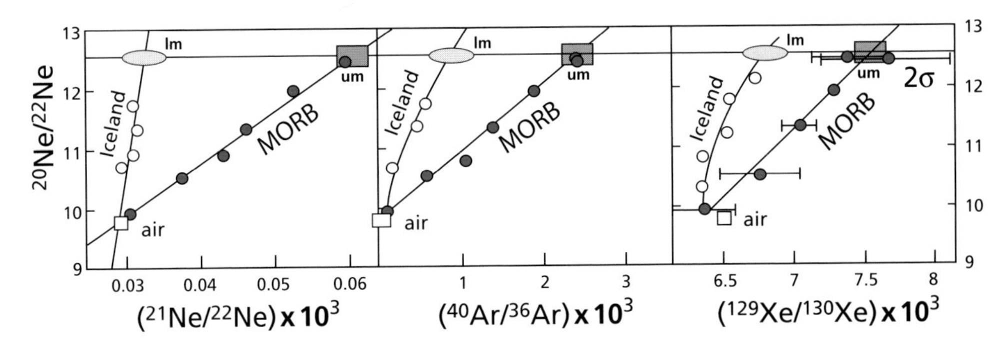

Figure 6.46 Correlation diagram of various rare gas isotope ratios related to radioactivity with the $^{20}Ne/^{21}Ne$ ratio. The diagrams can be used to correct for atmospheric contamination. On one side is the slope for MORBs and the other the result of the study on Iceland by Trieloff *et al.* (1998), supposedly representing the lower mantle. lm, lower mantle; um, upper mantle.

between the gas in the original rock and the air, which invariably contaminates the measurement to some degree and makes rare gas geochemistry so very difficult. In all cases MORBs are more radiogenic than OIBs.

Exercise

Isotope ratios measured in the same glassy margin of a MORB are $^{21}Ne/^{22}Ne = 0.055$, $^{20}Ne/^{22}Ne = 11.8$, $^{40}Ar/^{36}Ar = 18\,000$, and $^{129}Xe/^{130}Xe = 7.2$. Can you estimate the isotopic compositions of the source, correcting for atmospheric contamination?

Answer

Yes, assuming contamination is identical for all rare gases. We take the value of neon as the reference: 12.5 (or 13.5) and given the values of the isotope ratio of the atmosphere, the other ratios can be computed. The answers are dependent on the chosen reference for neon.

Exercise

Why do we not adopt the same reasoning for helium and neon as we used for argon by computing the budget between the atmosphere, upper mantle, and lower mantle?

Answer

Because the atmosphere is not closed for helium and so the extent of outgassing is unknown. The situation is even worse for neon. The atmosphere has a very different $^{20}Ne/^{22}Ne$ isotopic composition from the mantle. In other words, the present-day atmospheric content cannot be considered as the product of mantle outgassing.

Despite these intrinsic difficulties with the light rare gases, we can estimate the differences in concentrations between the upper and lower mantle for non-radiogenic isotopes by using measured isotope ratios:

$$({}^4\mathrm{He}/{}^3\mathrm{He})_{\mathrm{um}} = ({}^4\mathrm{He}/{}^3\mathrm{He})_{\mathrm{initial}} + (\mathrm{U}_0/{}^3\mathrm{He})_{\mathrm{um}} f_{\mathrm{um}}(t)$$

$$({}^4\mathrm{He}/{}^3\mathrm{He})_{\mathrm{lm}} = ({}^4\mathrm{He}/{}^3\mathrm{He})_{\mathrm{initial}} + (\mathrm{U}/{}^3\mathrm{He})_{\mathrm{lm}} f_{\mathrm{lm}}(t)$$

where um and lm are the upper and lower mantle, $f_{\mathrm{um}}(t)$ and $f_{\mathrm{lm}}(t)$ the expressions of radioactive decay, and U the uranium concentration.

By taking the ratios between the two expressions:

$$\frac{{}^3\mathrm{He}_{\mathrm{lm}}}{{}^3\mathrm{He}_{\mathrm{um}}} = \frac{\left(\frac{{}^4\mathrm{He}}{{}^3\mathrm{He}}\right)_{\mathrm{um}} - \left(\frac{{}^4\mathrm{He}}{{}^3\mathrm{He}}\right)_{\mathrm{initial}}}{\left(\frac{{}^4\mathrm{He}}{{}^3\mathrm{He}}\right)_{\mathrm{lm}} - \left(\frac{{}^4\mathrm{He}}{{}^3\mathrm{He}}\right)_{\mathrm{initial}}} \cdot \frac{\mathrm{U}_{\mathrm{lm}}}{\mathrm{U}_{\mathrm{um}}} \cdot \frac{f_{\mathrm{lm}}(t)}{f_{\mathrm{um}}(t)}$$

we obtain $({}^4\mathrm{He}/{}^3\mathrm{He})_{\mathrm{initial}} \approx 6000$ and $\mathrm{U}_{\mathrm{lm}}/\mathrm{U}_{\mathrm{um}} \approx 4$.

If we take the mean age of the lower mantle to be $4.5 \cdot 10^9$ years and less than $1 \cdot 10^9$ years for the upper mantle,

$$f_{\mathrm{lm}}(t) \approx 18.58 \text{ and } f_{\mathrm{um}}(t) \approx 2.12.$$

$${}^3\mathrm{He}_{\mathrm{lm}}/{}^3\mathrm{He}_{\mathrm{um}} \approx 360.$$

This yields similar ratios to argon and neon.

The difference in concentration between the lower and the upper mantle is such that a piece of contaminated lower mantle can only display upper-mantle values if mixed in minute proportions. Therefore, if the OIBs were from the lower mantle exclusively, the isotopic composition of the rare gases they contain would be uniform (except in the unlikely case of the lower mantle being highly heterogeneous). Since OIB composition is not uniform (see Figure 6.38 for He), it must be concluded that OIBs are a mixture of lower and upper mantle, but with only small proportions of material from the lower mantle.

6.5 Isotope geology of lead

As seen when discussing geochronology, three of the isotopes of lead are produced by the final decay of radioactive chains: ^{206}Pb by ^{238}U, ^{207}Pb by ^{235}U, and ^{208}Pb by ^{232}Th. Assuming the chains are in secular equilibrium (given the length of geological time), we can suppose that U and Th decay directly to the Pb isotopes (see Chapters 2 and 3). Just as with Sr and Nd, we can try to find out whether the lead isotope ratios vary in basalts and granites and so seek confirmation of the models developed with the threesome Rb–Sr, Sm–Nd, and Lu–Hf and with the rare gases.

In fact, the advantage with lead isotopes is that the results are naturally correlated since the two decay schemes are chemically similar and it is only the decay constant that differs.

It should also be remembered that as the Pb isotope ratios have the same denominator (^{204}Pb), mixtures are always shown by straight lines on the various plots.

6.5.1 The (Pb–Pb) isotope diagram or Holmes–Houtermans diagram[12]

One of the unique advantages with uranium – lead systems is that it is a priori easy to calculate theoretical models to act as references for the experimental data involving only isotope compositions that are more robust than the parent – daughter ratios (see **Russell** and **Farquhar**, 1960). Let us see how a model Earth would behave if it differentiated into two envelopes, say, from the beginning of the Earth's history if the two envelopes (continental crust and mantle) evolved ever since as closed systems. As usual $(^{238}\text{U}/^{204}\text{Pb})_{\text{today}}$ is written μ^{U}. Then $\mu^{\text{U}}_{\text{cc}} = 14$ and $\mu^{\text{U}}_{\text{M}_1} = 8$. It is assumed that the initial value of lead isotope composition is given by the analysis of sulfides from iron meteorites which, as they do not contain any uranium, record the isotope composition of the first instants of the Solar System.

By noting:

$$^{6}\alpha = \frac{^{206}\text{Pb}}{^{204}\text{Pb}}, \quad ^{7}\beta = \frac{^{207}\text{Pb}}{^{204}\text{Pb}}, \lambda_{238} = \lambda_8, \lambda_{235} = \lambda_5, \mu = \frac{^{238}\text{U}}{^{204}\text{Pb}},$$

and given that $\left(^{238}\text{U}/^{235}\text{U}\right)_{\text{today}} = 137.8$, the equations for the evolution of the four supposedly closed systems up to the present can be written:

$$^{6}\alpha_{\text{cc}} = {}^{6}\alpha_0 + \mu_{\text{cc}}\left(e^{\lambda_8 T_0} - 1\right)$$
$$^{7}\beta_{\text{cc}} = {}^{7}\beta_0 + \frac{\mu_{\text{cc}}}{137.8}\left(e^{\lambda_5 T_0} - 1\right),$$

where the subscript (cc) stands for continental crust. Similarly, for the mantle (m) we can write:

$$^{6}\alpha_{\text{m}} = {}^{6}\alpha_0 + \mu_{\text{m}}\left(e^{\lambda_8 T_0} - 1\right)$$
$$^{7}\beta_{\text{m}} = {}^{7}\beta_0 + \frac{\mu_{\text{m}}}{137.8}\left(e^{\lambda_5 T_0} - 1\right).$$

where T_0 is the age of the Earth. The initial ratios $\alpha_0 = 9.307$ and $\beta_0 = 10.294$ have been determined exactly by various analyses of iron meteorites and carbonaceous chondrites (Tatsumoto *et al.*, 1973).

Remark

Look back at Chapter 2 for how to reach these equations from the laws of radioactivity. Reread the exercise on the Rb/Sr ratio at the beginning of this chapter to calculate μ from chemical compositions of uranium and lead.

[12] We have already mentioned the Pb–Pb method when discussing the age of the Earth. We shall review it more fully here. It might then be useful to reread the part of Chapter 5 about the age of the Earth.

These equations are the parametric equations of two curves, the leading parameter being μ, that is the ratio $(^{238}U/^{204}Pb)_{today}$. Notice that these are the same equations as for the Rb–Sr or Sm–Nd systems, except that, in view of the values of the decay constants, the linear approximation is no longer valid. Let us represent them as a function of time (calculate them from the formula $\alpha = \alpha_0 + \mu(e^{\lambda T_0} - e^{\lambda T})$. It will be observed that the form of variation with time is very different for the $^{206}Pb/^{204}Pb$ ratio and the $^{207}Pb/^{204}Pb$ ratio. The latter increases very rapidly at first and then much more slowly. It is almost an extinct form of radioactivity. The $^{206}Pb/^{204}Pb$ ratio increases much more steadily, but not linearly. Once again, this is because of the difference between the decay constants. The isotope evolution referred to in correlation diagrams can also be represented: ($^{206}Pb/^{204}Pb$, $^{207}Pb/^{204}Pb$) (Figure 6.47c).

Notice first that the three curves are concave downwards but to different degrees. This is because of the different decay constants. Let us try to define the isochrons (geometric locus of points of the same age) in Figure 6.47c. We can do this graphically by taking, for example, points for 3 billion, 2 billion, and 1 billion years and for the present time in the three systems defined above (let us take the α_0 and β_0 values used for calculating the age of the Earth in Chapter 5). We observe that **each** defines a straight line, with the whole determining an array of straight lines converging towards (α_0^6, β_0^7) (dashed in Figure 6.47).

Let us try to demonstrate this mathematically. For a given age T, what is the equation of the geometric locus of representative points independently of μ? From the equations giving α^6 and β^7, we can therefore eliminate μ^8 by writing:

$$\frac{\beta^7(T) - \beta_0^7}{\alpha^6(T) - \alpha_0^6} = \frac{1}{137.8}\left(\frac{e^{\lambda_5 T_0} - e^{\lambda_5 T}}{e^{\lambda_8 T_0} - e^{\lambda_8 T}}\right).$$

Here T_0 is common at $4.55 \cdot 10^9$ years (or $4.50 \cdot 10^9$ years), and T is fixed. The equation therefore takes the form:

$$\frac{y - y_0}{x - x_0} = C.$$

This is the equation of a straight line through the point (x_0, y_0), that is the original isotope compositions. We find the result of our numerical construction. This array of straight lines therefore calibrates the diagram in ages. These straight lines are the geometric locus of points of the same age and are known as **isochrons**.

We therefore have an (α, β) plot formed by a series of curves cut by an array of converging straight lines. The point of convergence is (α_0, β_0), the initial composition of the Earth (Figure 6.47c). This is known as the **Holmes–Houtermans** (H–H) diagram.[13]

If we assume the Earth is made up of concentric envelopes isolated from each other and having evolved as closed systems since the "beginning," then the initial ratios of any geological object (mineral, rock, massif) (α_i^6, β_i^6) plotted on the diagram show the age of the object and the μ of the system in which it has been since the "beginning."

Much use was made of this diagram in the early days of isotope geochemistry for extracting information from Pb isotope measurements on galena. Galena crystals with the formula

[13] After the two scientists, the Scotsman **Arthur Holmes** and the Russo-Swiss **Fritz Houtermans**, who developed it independently in 1946 (see Holmes, 1946; Houtermans, 1946).

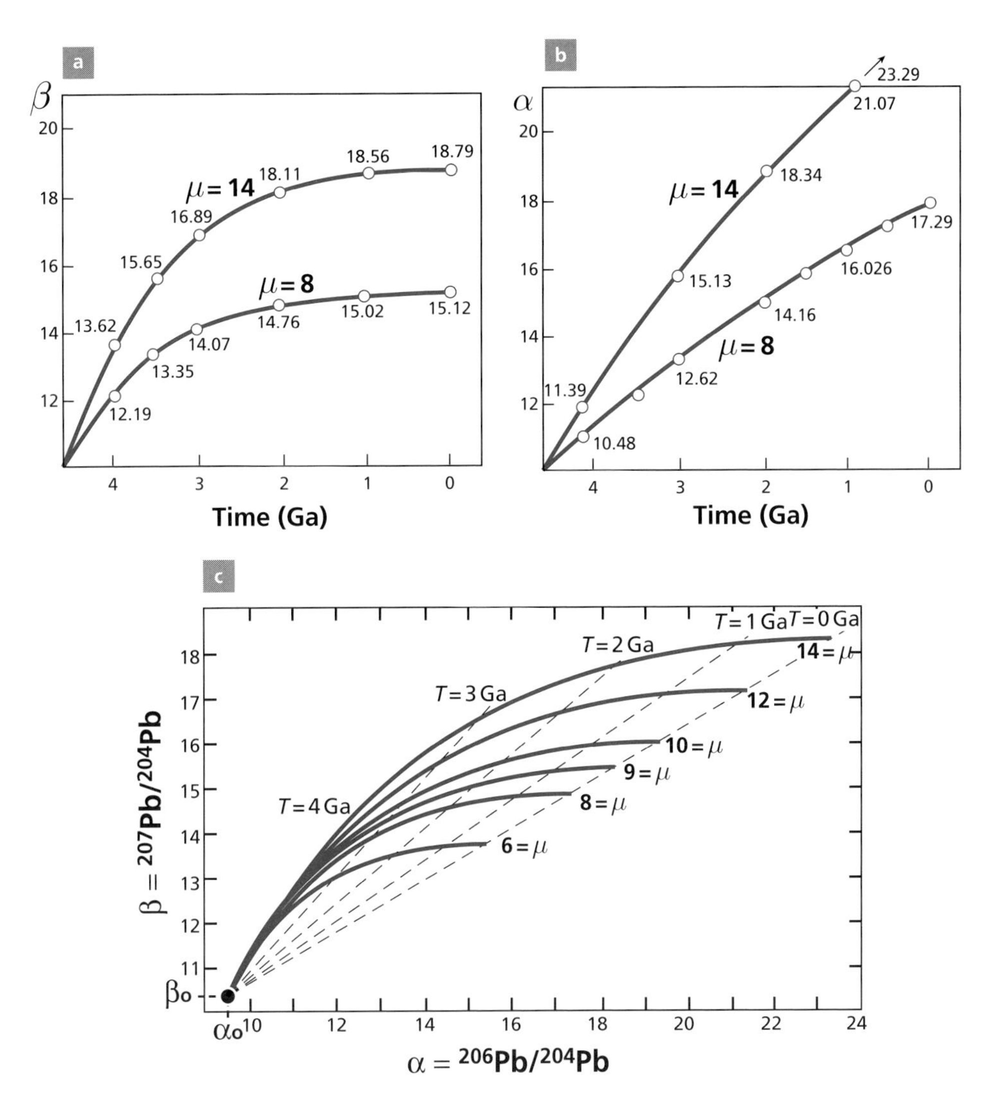

Figure 6.47 Evolution of $^{206}Pb/^{204}Pb$ and $^{207}Pb/^{204}Pb$ ratios as a function of time for two μ values ($\mu = 14$ and $\mu = 8$). (a, b) Notice that the $^{206}Pb/^{204}Pb$ variation is very progressive while the $^{207}Pb/^{204}Pb$ ratio varies very little after 2 Ga. (c) Plot of α versus β: $^{206}Pb/^{204}Pb = \alpha$, $^{207}Pb/^{204}Pb = \beta$ for the previous two μ values (and some other values of μ). Time is present here in parametric form. We distinguish the geometric locus of points evolving over time with the same μ (growth curves) from the geometric locus of points of the same age having evolved with different μ values (isochrons) (dashed lines).

PbS contain neither uranium nor thorium and so the lead isotope measurement corresponds to initial isotope ratios at the time they formed. Unfortunately it is often difficult to determine the absolute age of galena directly with precision and very often we must settle for an age of the geological setting. When working with galena, it can be seen that very few galena crystals lie on the isochrons corresponding to their supposed ages, meaning that generally the idea of evolution in the closed system of envelopes that have been isolated since

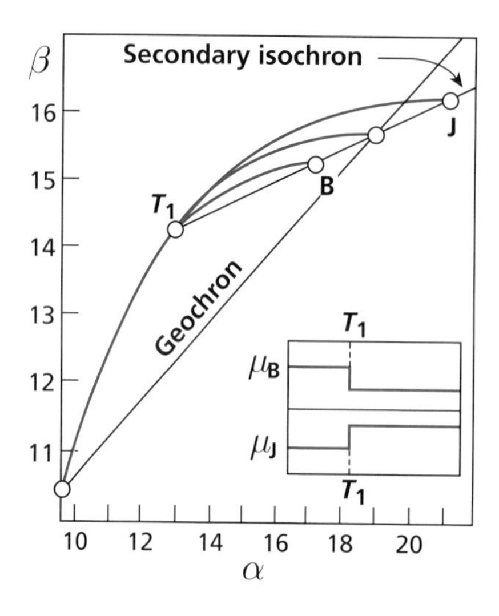

Figure 6.48 Diagram of (α, β) explaining how J-type or B-type lead is produced with a two-stage history and change of μ. The first stage is evolution from T_0 to T_1 in the same reservoir with μ_1. Two new reservoirs are formed at T_1. One has a μ value greater than μ_1 and another a μ value less than μ_1. The first yields J-type lead, the second B-type lead.

the beginnings of the Earth does not correspond to geological reality. Nowadays, that does not surprise us after all that has been said about the evolution of the crust–mantle system and its complexity. These results simply confirm this complexity, but in 1950 this was a real discovery.[14]

In practice, galena crystals have H–H model ages that are either older or younger than their geological ages. Some even have "model ages" in the future. Lead with H–H model ages that are older than the actual age are known in geochemical jargon as B-type lead (after Bleiberg mine in Austria) while those with model ages younger than the real age are known as J-type lead (after Joplin mine in Missouri). These ages were first interpreted by **Johannes Geiss** of Berne in 1954, by assuming a two-stage geological history with a change in $^{238}U/^{204}Pb = {}^{8}\mu$ ratios (Figure 6.48).

The equations of a two-stage model are written:

$$\alpha = \alpha_0 + \mu_1\left(e^{\lambda_8 T_0} - e^{\lambda_8 T_1}\right) + \mu_2\left(e^{\lambda_8 T_1} \quad 1\right)$$

$$\beta = \beta_0 + \frac{\mu_1}{137.8}\left(e^{\lambda_5 T_0} - e^{\lambda_5 T_1}\right) + \frac{\mu_2}{137.8}\left(e^{\lambda_5 T_1} - 1\right).$$

When $\mu_2 > \mu_1$ the model age is younger (J-type). When $\mu_2 < \mu_1$ the model age is older (B-type). Everything then comes down to constructing a grid analogous to the H–H diagram with a growth curve and isochron, but making the system start from a certain time T_1 with initial ratios $\alpha(T_1)$ and $\beta(T_1)$ (Figure 6.48).

[14] Nier was already alert to this, which is probably why he did not want to calculate an "age of the Earth."

Exercise

We have lead with a two-stage geological history. Stage 1: evolution in a closed system from $T_0 = 4.55 \cdot 10^9$ years to $T_1 = 2 \cdot 10^9$ years with $\mu = {}^{238}U/{}^{204}Pb = 8$. Stage 2: from T_1 to the present day in system A with $\mu = 10$ and in system B with $\mu = 5$.

Calculate the isotope compositions of lead given that $\alpha_0 = 0.30$ and $\beta_0 = 10.29$. Show by calculation and then analytically that A, B, and point P representing the closed system are aligned and that the straight line cuts the primordial evolution curve at a point whose age shall be determined.

Answer

System A: $\alpha = 16.28$ and $\beta = 14.97$. System B: $\alpha = 18.1$ and $\beta = 15.41$.

For the mathematical demonstration, just notice that in the equations

$$\alpha = \alpha_0 + \{\mu_1(e^{\lambda_8 T_0} - e^{\lambda_8 T_1}) + \mu_2(e^{\lambda_8 T_1} - 1)$$

$$\beta = \beta_0 + \frac{\mu_1}{137.8}(e^{\lambda_5 T_0} - e^{\lambda_5 T_1}) + \frac{\mu_2}{137.8}(e^{\lambda_5 T_1} - 1)$$

the episode from T_0 to T_1 is the same in both scenarios. We can therefore posit:

$$\alpha_0(T_1) = \alpha_0 + \mu(e^{\lambda_8 T_0} - e^{\lambda_8 T_1})$$
$$\beta_0(T_1) = \beta_0 + \frac{\mu}{137.8}(e^{\lambda_5 T_0} - e^{\lambda_5 T_1}).$$

The equations are then written:

$$\alpha = \alpha_0(T_1) + \mu_2(e^{\lambda_8 T_1} - 1)$$

$$\beta = \beta_0(T_1) + \frac{\mu_2}{137.8}(e^{\lambda_5 T_1} - 1).$$

Hence:

$$\frac{\alpha - \alpha_0(T_1)}{\beta - \beta_0(T_1)} = \left(\frac{e^{\lambda_8 T_1} - 1}{e^{\lambda_5 T_1} - 1}\right) \times 137.8.$$

This is the equation of a straight line cutting the curve of primordial evolution at $\alpha_0(T_1)$ and $\beta_0(T_1)$, that is at $T = 2$ Ga.

This is the same demonstration as for the isochrons in the H–H diagram.

6.5.2 The geochron and the age of the Earth's core

Let us now look at the isochron corresponding to $T = 0$, that is the present day. It is now common practice to call this particular isochron, corresponding to the present time, the **geochron**. All points having evolved in a closed system since the origin of the Earth must lie on this straight line for which the equation is:

$$\frac{\beta^7 - \beta_0^7}{\alpha^6 - \alpha_0^6} = \frac{1}{137.8}\left(\frac{e^{\lambda_5 T_0} - 1}{e^{\lambda_8 T_0} - 1}\right).$$

The slope of this straight line depends on T_0 alone, which is the age of the Earth. **The older the Earth, the steeper the geochron.** The younger the Earth, the lower the angle of the geochron. As they all go through (α_0^6, β_0^7), the lower the angle of the slope, the further to the right it is shifted (Figure 6.49).

Exercise

Calculate and draw the geochrons for ages of the Earth of 4.65, 4.55, 4.50, and 4.40 Ga. Use values $\alpha_0 = 9.307$ and $\beta_0 = 10.294$.

Answer

The answer is given in Figure 6.49.

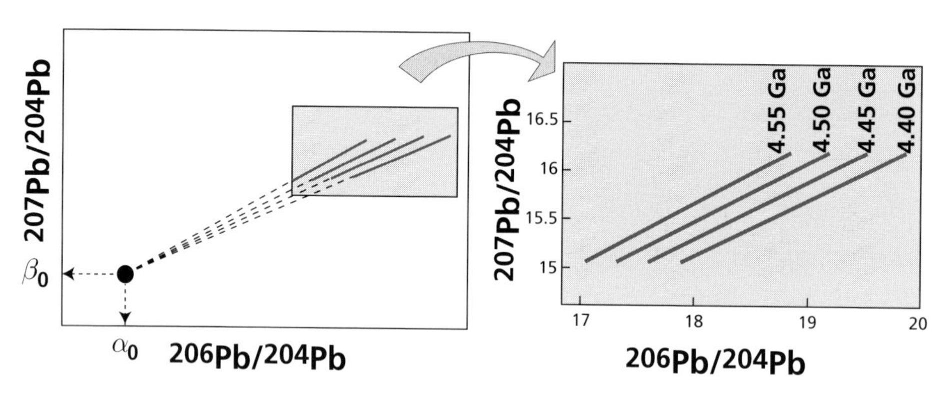

Figure 6.49 Geochrons calculated for various values of the "age of the Earth." Notice that when age **declines**, the geochrons shift to the right.

It can therefore be supposed that the Bulk Earth, as with the references considered for the Sr and Nd isotopes, must lie on the geochron. By simple analogy with what we noticed for Sr–Nd, it can be supposed that the points representing the continental crust and the depleted mantle, which is the MORB source, will lie on either side of the geochron (or if the extraction of continental crust is a very, very ancient phenomenon, on the geochron itself). Now, we observe that the MORBs like granites lie to the right of what we shall call the "classical" geochron with an age of the Earth of 4.55 Ga. How can we explain this anomaly contrary to expectations? Do we need to revise the age of the Earth? It seemed to have been well established by meteorites, though!

To account for this anomaly, it has been proposed to shift the geochron, invoking the formation of the core. The Earth's core probably formed very early in the Earth's history, at the time the Earth was first differentiating. It is essentially an iron and nickel alloy. But to form, the iron was probably associated with sulfur, because the Fe–FeS eutectic has a melting point below that of silicates, allowing the molten metal to percolate through the porous silicate lattice. The core therefore contains sulfur. (This scenario has been largely confirmed by other means.) But if the core contains sulfur, it also contains lead because PbS is a very stable compound and forms easily. The $\mu = {}^{238}U/{}^{204}Pb$ of the mantle is therefore affected

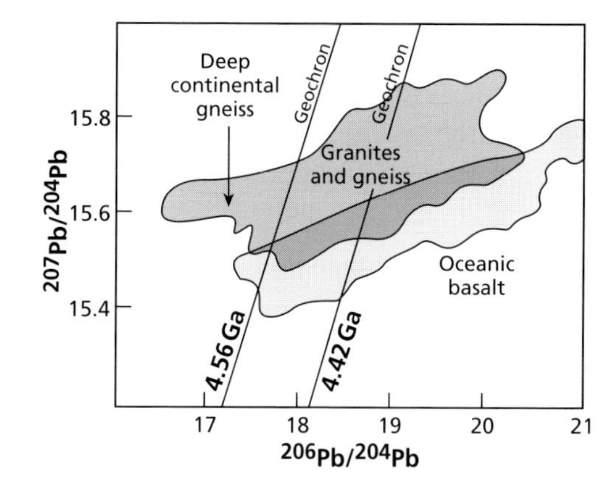

Figure 6.50 Position of domains representing rocks from the continental crust, rocks from the mantle, and the two reference geochrons in the (α, β) lead diagram. If the geochron is at 4.55 Ga (or older), most of the data points are in the J domain. Notice also that for the $^{206}Pb/^{204}Pb$ ratio, the extent of the domain is not very different for continental crust and ocean basalt.

by extraction by the core. Imagine the core took several hundred million years to form: then, the relevant geochron for our reasoning corresponds **not to the age of the Earth** (in Patterson's sense) but to the **mean age of differentiation of the core**. Let us explore this scenario, remembering that the smaller the age T_0, the further to the right the geochron is shifted.

By going back to the graph and shifting the geochron to have a continental crust-depleted mantle arrangement comparable with Sr–Nd, we obtain an average age of the core of 4.45–4.42 Ga (Gangarz and Wasserburg, 1977; Doe and Zartman, 1979; Allègre *et al.*, 1999) (Figure 6.50).

If the age of the Earth, or rather that of the meteorites, is 4.55 Ga, we can conclude that it took the core 100 million years to segregate, relative to the age of formation of the meteorites. We saw when examining the Sr–Nd isotopes that the average age of differentiation of the continental crust was about 2 billion years, with an S-shaped extraction curve. Examination of the (α, β) Pb isotope diagram shows that the main difference between the distribution of points for the continental crust and for the mantle source of MORB is in the $^{207}Pb/^{204}Pb$ ratios. The $^{207}Pb/^{204}Pb$ ratios of the continental crust (that is, granites) are higher than those of basalts. Now, that could only have come about in the past, when ^{235}U was abundant enough to vary the $^{207}Pb/^{204}Pb$ ratios. This confirms that a good part of the continental crust became differentiated very early in geological history and the more recent continental crust has been formed by recycling of ancient crust.

Can we go beyond such qualitative reasoning? Notice first that the difference is not great for the $^{206}Pb/^{204}Pb$ ratio. This suggests that for most of geological history μ_{crust} and μ_{mantle} of the MORB reservoir have not been very different. Now, the μ values of the oceanic crust are higher than those of the mantle: therefore, in the formation of continental crust there is a process which compensates and enriches the crust more in Pb than in U, so that the outcome is a lack of fractionation. (Is it island-arc volcanism and magmatism that is rich in H_2O and that fractionates Pb more than U?)

This state of affairs has an unfortunate and a fortunate consequence. The unfortunate consequence is that the U–Pb systems are not very effective for confirming the Rb–Sr and Sm–Nd systems in the crust–mantle differentiation process. The fortunate consequence is that we can clearly distinguish differentiation of the oceanic crust from differentiation of the continental crust which, further to fractionation, is often confused for both Sm–Nd and Rb–Sr. We shall take advantage of this.

6.5.3 The (Pb, Pb) isotope diagram and the OIB source

We have not spoken much of the origin of OIB since the standard model was developed and we have continued to accept that the OIB source was the lower mantle, which is similar if not identical to the primitive mantle. Let us plot the results of OIB lead isotope analyses on the (α, β) diagram. It is observed that a large category of OIBs lies well to the right of the geochron at 4.55 Ga and even at 4.42 Ga (Figure 6.51). No modification of the age of the Earth can account for this. It is particularly true of islands like St. Helena in the Atlantic or Mangaia in the Pacific Ocean.

Now, by definition, the geochron is the geometric locus of systems that have evolved in closed systems since the origin of the Earth. **The OIB source is therefore not the primordial mantle**. Moreover, the OIBs do not lie between the MORBs and the geochron, contrary to the arrangement in the Nd–Sr isotope diagrams. We must either abandon the idea of a primitive, closed lower mantle, which is one of the components of the standard model, or challenge the idea that the OIBs come from the lower mantle. Lead isotopes therefore invite us to question some of the ideas accepted so far. But which ones? **The standard model is no longer tenable!**

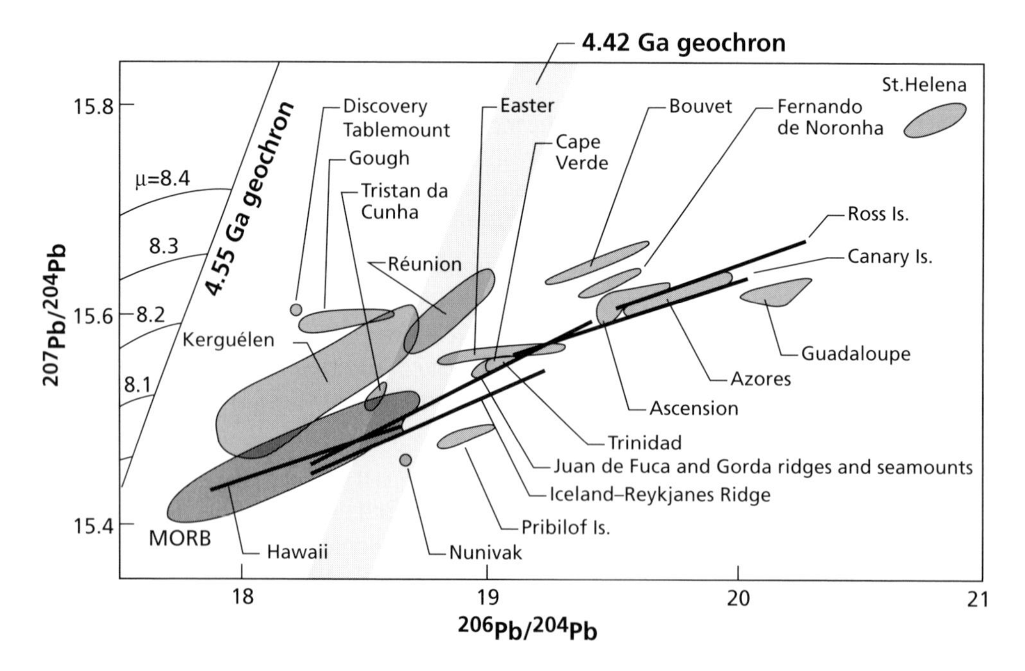

Figure 6.51 Diagram of (α, β) for OIBs. We shall see they are all in the J domain compared with the 4.55 Ga and 4.42 Ga geochrons. The first comprehensive modern synthesis was by Sun (1980).

Exercise

Take a two-stage geological history of a basalt. Stage 1: from $T_0 = 4.45$ Ga to $T_1 = 2$ Ga, its "ancestor" evolved in a system with $\mu^{U} = 7.5$. Stage 2: from T_1 to the present day, it evolved in a system with $\mu^{U} = 14$. Calculate the α^6, β^7 values of the basalt. Do the same calculation if $T_1 = 3.5$ Ga, 3 Ga, and 0.5 Ga. What do you conclude?

Answer

The equations for the evolution of a two-stage model are written:

$$\alpha^6 = \alpha_0^6 + \mu_1\left(e^{\lambda_8 T_0} - e^{\lambda_8 T_1}\right) + \mu_2\left(e^{\lambda_8 T_1} - 1\right).$$

$$\beta^7 - \beta_0^7 + \frac{\mu_1}{137.8}\left(e^{\lambda_5 T_0} - e^{\lambda_5 T_1}\right) + \frac{\mu_2}{137.8}\left(e^{\lambda_5 T_1} - 1\right).$$

Therefore $\alpha_0^6 = 9.30$ and $\beta_0^7 = 10.29$, $\lambda_8 = 0.155\,512\,5 \cdot 10^9$, and $\lambda_5 = 0.984\,85 \cdot 10^9$.

T_1 (Ga)	α^6	β^7
3.5	21.66	16.28
3	20.72	15.66
2	19.23	15.10
0.5	17.39	14.84

Two points on the geochron can be calculated for $\mu = 7.5$ and $\mu = 14$. It is observed that the curve joining up the points for the different ages of differentiation is neither a straight line nor a "common" function. (Drawing is believing!)

In addition, if we suppose evolution in a closed system represents the evolution of the primitive mantle and that the various two-stage models represent a theoretical continental crust, the relative size of $\Delta(207/204)$ to $\Delta(206/204)$ increases from $T_1 = 1$ to $T_1 = 4$.

6.5.4 The (Sr, Pb) isotope diagram

Let us now try to connect up the (Hf, Nd, Sr) systems with their own internal coherence and the (Pb^6, Pb^7, Pb^8) systems which also have their own, but apparently different, internal coherence. To do this we shall use the ($^{87}Sr/^{86}Sr$, $^{206}Pb/^{204}Pb$) diagram. Let us plot the data points measured on basalts in this diagram as in Sun and Hanson (1975) (Figure 6.52).

The MORBs define a restricted domain. The OIBs cover much of the diagram, with no general correlation. It was first thought that this dispersed pattern showed the Sr and Pb isotope tracers were incoherent. However, **Bernard Dupré**, **Bruno Hamelin**, and the present author were able to provide new data for deciphering this diagram. We first showed that most MORBs were aligned, with a positive slope, seeming to indicate covariation between $^{87}Sr/^{86}Sr$ and $^{206}Pb/^{204}Pb$. Such variation seems to link many of the samples from the North Atlantic and Pacific (Dupré and Allègre, 1980).

However, examination of oceanic ridges in the South Atlantic and Indian Ocean showed that while there were alignments, their slopes were different from those of the North Atlantic and the Pacific. Close study by the team of **Jean Guy Schilling** of Rhode Island University showed that, in each case, the alignments ended with a nearby OIB (Figure 6.53). We are dealing, basically, with a series of **Schilling effects**, which look to be general but are

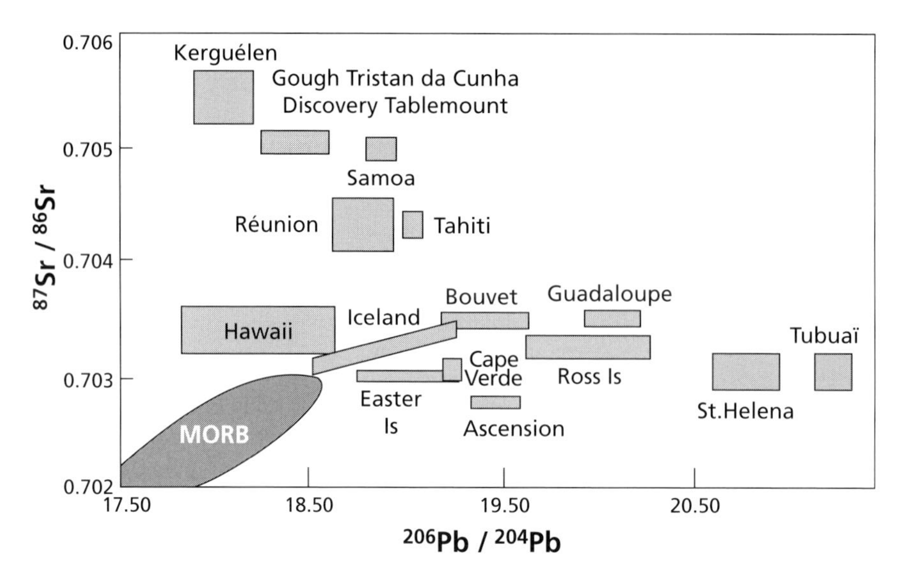

Figure 6.52 Correlation diagram for ^{87}Sr–^{207}Pb showing the somewhat elongated MORB domain and the much larger and dispersed OIB domain. Modified from Sun and Hanson (1975).

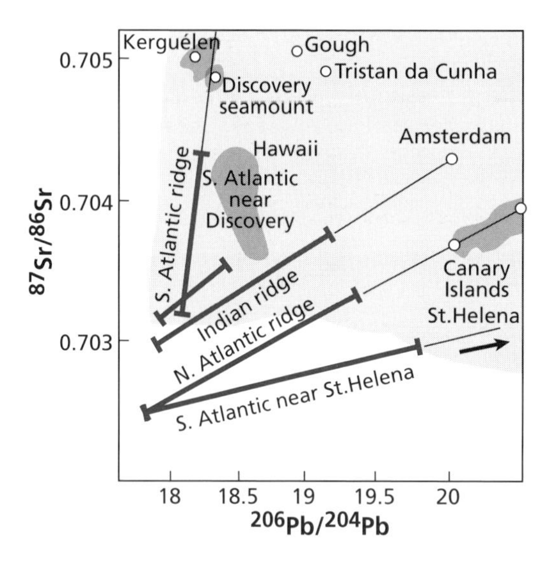

Figure 6.53 Various ($^{87}Sr/^{86}Sr$, $^{206}Pb/^{204}Pb$) alignments of MORBs of various regions, one of whose components is invariably an OIB of the same region. These alignments support the idea that (Sr, Pb) isotope alignments are brought about by generalization of the Schilling effect.

sometimes very difficult to detect depending on topographical criteria, as these are not readily visible (see Schilling, 1992).

Counter-proof that these correlations are not of the same kind as the (Sr, Nd) or (Nd, Hf) isotope correlations is that, if we determine the $^{206}Pb/^{204}Pb$ ratio corresponding to

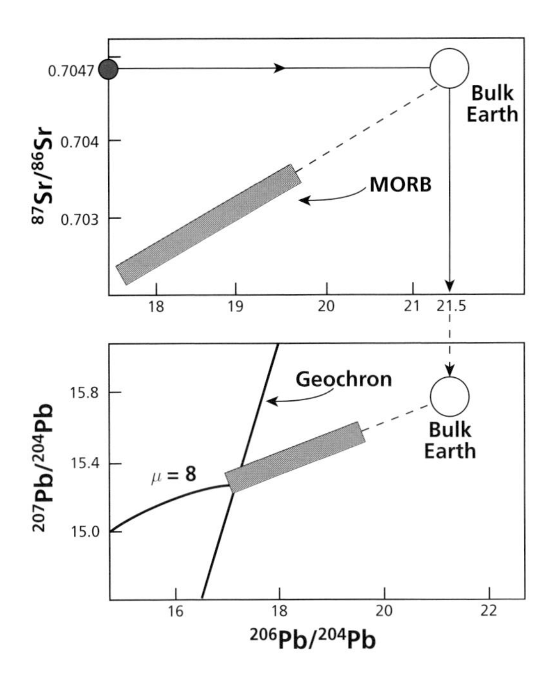

Figure 6.54 Isotope correlations for ($^{87}Sr/^{86}Sr$, $^{206}Pb/^{204}Pb$) for the North Atlantic. The (Sr, Pb) isotope correlation for the North Atlantic is in blue. Determining the $^{206}Pb/^{204}Pb$ ratio for the value of the Earth ($^{87}Sr/^{86}Sr$) = 0.7047 gives $^{206}Pb/^{204}Pb = 21.5$. The same correlation in the ^{207}Pb–^{206}Pb diagram shows that the point is in the J domain.

the $^{87}Sr/^{86}Sr$ of about 0.7047 of the Bulk Earth, for example, the (Sr, Pb) correlation of the North Atlantic (Figure 6.54), and if we plot this $(^{206}Pb/^{204}Pb)^*$ value on the equivalent ($^{206}Pb/^{204}Pb$, $^{207}Pb/^{204}Pb$) correlation obtained from the same samples, we get a point lying well away from the geochron. Now, lets us repeat, the geochron is the geometric locus of all systems having evolved as closed systems since the beginning of the Earth's history.

The same observation applies for the other straight lines of (Sr, Pb) correlation for MORB of other regions. This clearly shows that (Sr, Pb) correlations are different from the general (Sr, Nd) or (Nd, Hf) correlations.

The OIB domain in the (^{87}Sr, ^{207}Pb) isotope diagram is highly dispersed, but it begins to make sense if we try to examine how it ties in with the geographical distribution. **Bernard Dupré**, **Bruno Hamelin** (now at the universities of Toulouse and Marseille, respectively), and the present author identified a very characteristic province now known by the name of **Dupal** (Hart, 1984).[15] This province comprises the Indian Ocean and South Atlantic. Its isotopic signature is very clear: a high $^{87}Sr/^{86}Sr$ ratio for the same $^{206}Pb/^{204}Pb$ ratio. The important point is that the Indian Ocean MORBs also seem to be enriched in ^{87}Sr compared with those of the Pacific and North Atlantic (Dupré and Allègre, 1983; Hamelin *et al.*, 1984; Hamelin and Allègre, 1985; Hamelin *et al.*, 1986) (Figures 6.53 and 6.55).

[15] Dupal is formed from the contraction of the names Dupré and Allègre.

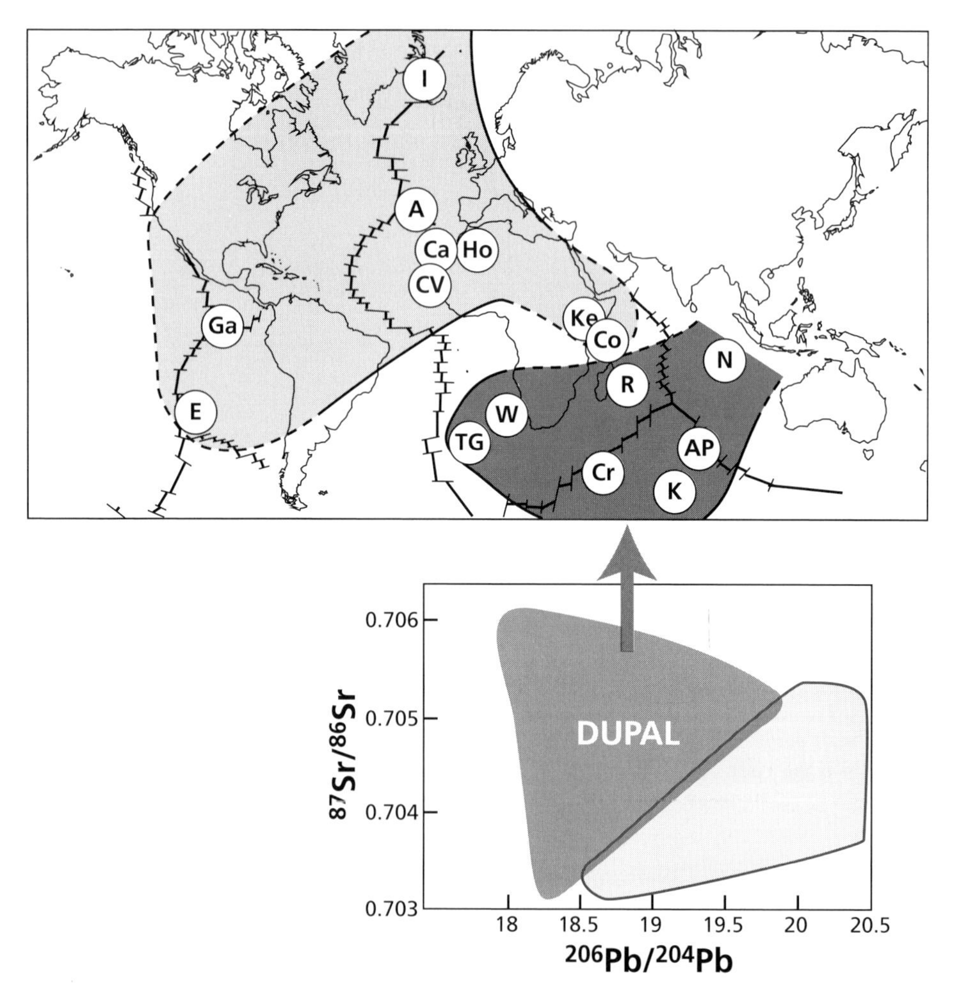

Figure 6.55 Geographical correspondence of the Dupal domain identified in the ($^{87}Sr/^{86}Sr$, $^{206}Pb/^{207}Pb$) diagram as plotted originally by Dupré and Allègre (1983). I, Iceland; A, Azores; Ca, Canaries; CV, Cape Verde; Ho, Hoggar; Ga, Galápagos; E, Easter Island; TG, Tristan da Cunha and Gough; W, Walvis Ridge; Cr, Crozet; K, Kerguélen; R, Réunion; AP Amsterdam and St. Paul; N, Ninety East Ridge; Ke, Kenya; Co, Comoros.

What would be a coherent interpretation of this? The initial idea of the standard model 1, that OIB came from a closed primordial mantle, has already been undermined by lead isotope observations. The (Sr, Pb) isotope diagram confirms that OIBs do not come from a closed system, otherwise they would all lie on a curve joining the MORBs and the homogeneous source reservoir in question.

At this point, we need to introduce the **Hofmann–White (H–W) hypothesis**. After many quite technical geochemical comparisons, **Al Hofmann** and **Bill White** of the Max-Planck Institute in Mainz came up in 1982 with the hypothesis that OIBs come from remelting of old oceanic crust reinjected into the mantle at a depth which they situated at the core–mantle boundary while subsequent workers (including the present author) preferred to situate

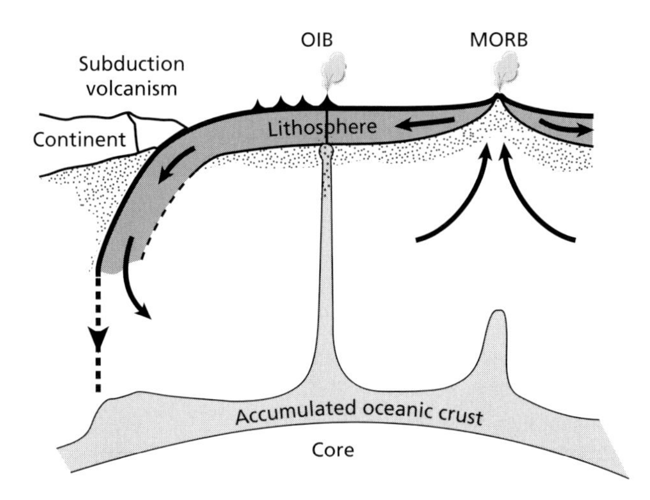

Figure 6.56 Hofmann and White (1982) diagram explaining the origin of OIBs by recycling of oceanic crust.

this source at 670 km. These two positions are chosen because, in a convecting reservoir, the plumes are generated by an unstable boundary layer. If the mantle is divided into two layers, there is a boundary layer at 670 km, where there is the great seismic discontinuity transition (tetrahedral silicon – octahedral silicon) or the core – mantle interface, which is the iron – silicate boundary and where seismologists have detected what they call the D″ layer.

Little by little the **H–W** hypothesis has become more complex and it is accepted that recycled oceanic crust might include not just basaltic ocean crust altered by reaction with sea water but also the sediments lying on it on the ocean floor (Figure 6.56). This supplement makes the H – W idea very appealing. It provides a very good explanation for such OIBs as those of St. Helena and the Canaries, or Hawaii. Old oceanic crust weathered by sea water is rich in U/Pb and poor in Rb/Sr. Marine sediments produced by weathering of the continents are rich in Rb/Sr and in U/Pb. After a transit time, these chemical values are translated into isotope values. However, the H – W hypothesis is not nearly as straightforward when explaining the OIBs of the Indian Ocean (Dupal). Additional components have to be considered.

In conclusion, OIBs do not come from some primitive reservoir. So how can we explain the Nd, Sr, and Hf isotope ratios which that hypothesis seemed to account for?

Exercise

Altered oceanic crust that has been reinjected into the mantle has the following characteristics: $^{87}Sr/^{86}Sr = 0.7025$, $^{206}Pb/^{204}Pb = 18.5$, $^{87}Rb/^{87}Sr = 0.1$, and $^{238}U/^{204}Pb = 15$. The Sr content is 150 ppm, that of Pb is 1.3 ppm. The characteristics of the mid–upper mantle are: $^{87}Sr/^{86}Sr = 0.7022$, $^{206}Pb/^{204}Pb = 17.5$, $^{87}Rb/^{87}Sr = 0.001$, and $^{238}U/^{204}Pb = 5$. The Sr content is 15 ppm, that of Pb is 0.15 ppm.

Suppose the weathered crust remained in the mantle for 1 Ga and was then mixed with the upper mantle to give OIBs. Calculate the trajectory of the mixture in the (Sr, Pb) isotope diagram.

Answer

We first calculate the isotopic characteristics of the altered oceanic crust after 1 Ga: $^{87}Sr/^{86}Sr = 0.703\,92$, $^{206}Pb/^{204}Pb = 21.01$.

We then observe that Sr/Pb is about constant. Therefore the mixtures are straight lines. All that remains to do is to draw the straight line joining the weathered oceanic crust aged 1 Ga and the mid–upper mantle.

Exercise

Suppose now that sediments are injected along with the oceanic crust. The characteristics of these sediments at the time they are reinjected are: $^{87}Sr/^{86}Sr = 0.712$ and $^{87}Rb/^{86}Sr = 0.1$ (because of the amount of limestone which is rich in Sr), $^{206}Pb/^{204}Pb = 18.5$, and $^{238}U/^{204}Pb = 10$. The Sr content is 400 ppm and the Pb content 3.46 ppm. These sediments also remain for 1 Ga with the oceanic crust. Calculate the trajectories of the upper mantle–sediment and the sediment–buried oceanic crust mixtures. If 5% of the mass of sediments is mixed, draw the mixing trajectory.

Answer

Isotopic characteristics of sediments aged 1 Ga: $^{87}Sr/^{86}Sr = 0.713\,42$ and $^{206}Pb/^{204}Pb = 20.17$. Once again the Pb/Sr ratios are about constant and so the mixing "curves" are straight lines (Figure 6.57). The mixture of 5% mass corresponds to the mass fraction of Sr at 34%; we can therefore obtain the point on a straight line between that of sediments and of the oceanic crust. It can be seen we have thus roughly reproduced the OIBs which are not Dupals.

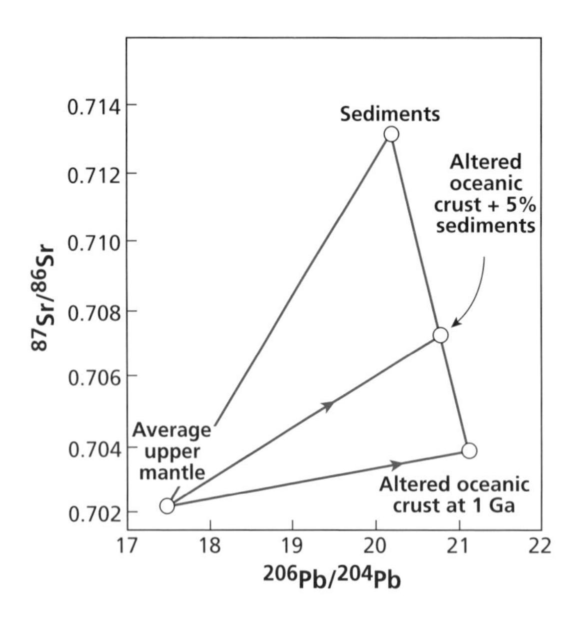

Figure 6.57 Graph of the results for the two exercises above.

In conclusion, when we observe the (Sr, Pb) isotope diagram which seeks to unify the two series of correlations (Sr–Nd–Hf) on the one hand and (^{206}Pb–^{207}Pb–^{208}Pb) on the other, it can first of all be seen there are not two completely separate domains for OIBs and MORBs. On the contrary, there is no gap, even in regions in which plumes are injected into ocean ridges (**Schilling effect**), where the two domains overlap. This topology does not suggest a model whereby OIBs come from a separate and relatively homogeneous lower mantle. The second observation, which is brought out by the (Sr, Pb) isotope diagram, is that regionalization occurs. There are genuine isotopic provinces.

The most general division, as said, is between basalts from the Indian Ocean and South Atlantic on one side and those of the Pacific and the North Atlantic on the other side. This is marked by higher ^{87}Sr/^{86}Sr ratios than for the North Atlantic–Pacific province, for a same ^{206}Pb/^{204}Pb ratio and therefore occupies the top part of the (Sr, Pb) isotope diagram. This distinction is particularly true for OIBs but is also found for MORBs.

Moreover, this "Indian" isotope signature is very old, as it is found in Tibetan ophiolites (135 Ma pieces of oceanic crust trapped in the India–Asia collision zone in Tibet) (Göpel *et al.*, 1984). How can this be accounted for? What readily distinguishes rocks from continental crust and rocks from the mantle is their much higher ^{87}Sr/^{86}Sr ratio while the ^{206}Pb/^{204}Pb ratios are similar. If a piece of continental crust is swallowed up by the mantle and mixes with it, the ^{206}Pb/^{204}Pb ratio increases little while the ^{87}Sr/^{86}Sr increases greatly. The difference with sediment recycling is that sediments contain a large amount of limestone whose isotope ratios are close to 0.707–0.709 and are very rich in Sr ($\approx$ 1000 ppm). They therefore buffer the Sr isotope ratios. When continental crust associated with its lithosphere is delaminated, there is no "limestone effect." Now, the province where the Indian signature is found is the province where continents existed (Gondwana) in the past and broke up, probably leading to delamination. The overall conclusion from examining the (Sr, Pb) isotope diagram is that OIBs come from an isotopically very heterogeneous source but having affinities with the MORB source of the same region.

Thus, we began by drawing a clear distinction between MORB and OIB and now we are saying there is no gap between them. How can this be explained? How can rare gases be worked into the schema since we have seen an extremely clear distinction in their isotopic ratios between OIBs and MORBs? This is an as-yet-unanswered question.

6.5.5 The ^{187}Re–^{187}Os system and (Os, Pb) isotope diagrams

We have already come across the ^{187}Re–^{187}Os system for which the decay constant is $\lambda = 1.64 \cdot 10^{-11}\,\text{yr}^{-1}$. How does the ^{187}Os/^{188}Os ratio evolve in the Earth's mantle? By examining osmiridiums, that is, osmium-rich minerals containing no rhenium, associated with ultrabasic rocks, Allègre and Luck (1980) were able to show the isotope evolution of mantle osmium corresponded to a chondritic mantle with ^{187}Re/^{186}Os $\approx$ 1. This is because osmium is an element that remains in the mantle when partial melting gives rise to oceanic crust. It is more "compatible" than either nickel or chromium. The Os concentration of ultrabasic rocks is close to 6 ppm and is 0.06 ppm in the continental crust. If we write the crust–mantle balance equation:

$$\alpha_{\text{T}}^{\text{Os}} = \alpha_{\text{cc}}^{\text{Os}} W^{\text{Os}} + \alpha_{\text{m}}^{\text{Os}} (1 - W^{\text{Os}})$$

the value of W^{Os} is:

$$W^{\mathrm{Os}} = \frac{m_c}{m_d} \cdot \frac{C_{cc}^{\mathrm{Os}}}{C_{\mathrm{T}}^{\mathrm{Os}}} \approx 0.15 \times \frac{0.06}{6} = 1.5 \cdot 10^{-4}.$$

Therefore:

$$\alpha_{\mathrm{m}}^{\mathrm{Os}} = \frac{\alpha_{\mathrm{T}}^{\mathrm{Os}} - \alpha_{\mathrm{c}}^{\mathrm{Os}} W}{(1 - W)} \approx \alpha_{\mathrm{T}}^{\mathrm{Os}}.$$

This is what we actually observe: extraction of continental crust has no effect on the isotope evolution of Os of the upper mantle. However, what we do notice is that Re is fractionated a little like lutetium during magmatic processes. It concentrates in the liquid, but weakly. Accordingly the Re/Os ratios of basalts are extraordinarily high.Values of 50 or 100 are commonplace for the $^{187}\mathrm{Re}/^{188}\mathrm{Os}$ ratio, whereas the mantle has ratios close to 0.1.

The continental crust has Os isotope ratios similar to basalts. This means there is a high level of fractionation at the mid-ocean ridges but that it is neutral at the subduction zones. The diagram of isotope evolution therefore looks like that in Figure 6.58.

Perhaps the most spectacular illustration of this difference in concentration between crust and mantle is the (Nd, Os) isotope diagram for the Stillwater Complex in Montana, which is an ultrabasic massif. The magma is known to have been contaminated by the crust. The mixing curve is spectacular and needs no comment (Lambert *et al.*, 1989) (Figure 6.59).

While the $^{187}\mathrm{Re}-^{187}\mathrm{Os}$ system does not provide many resources for studying the differentiation of continental crust and mantle, it is an excellent tracer for studying the reinjection

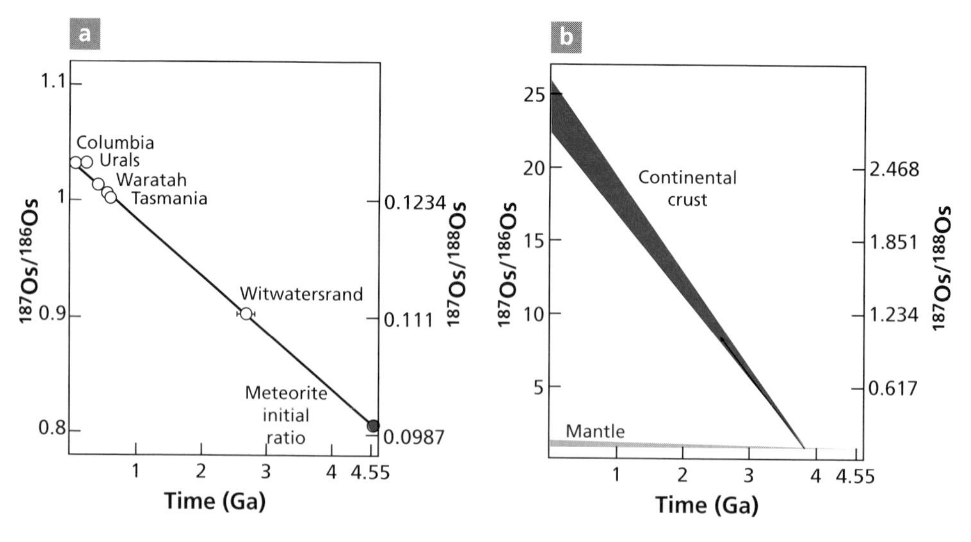

Figure 6.58 (a) Evolution of the $^{187}\mathrm{Os}/^{186}\mathrm{Os}$ ratio of the mantle from osmiridiums associated with ultrabasic rocks. (b) Isotope evolution of the continental crust and mantle. The mantle slope corresponds to a chondritic Re/Os ratio as shown by Allègre and Luck (1980).

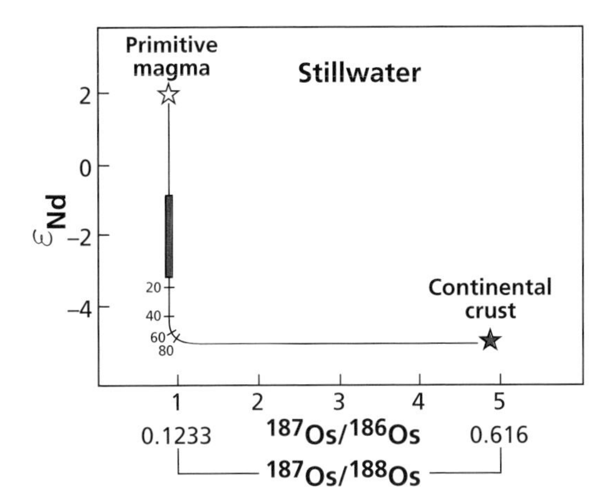

Figure 6.59 Isotope diagram for (Os, Nd) in the Stillwater Complex, a mixture of ultra-basic magma and contaminating continental crust. The Nd/Os ratio is 400 for the ultrabasic magma and 400 000 for the continental crust. After Lambert *et al.* (1989).

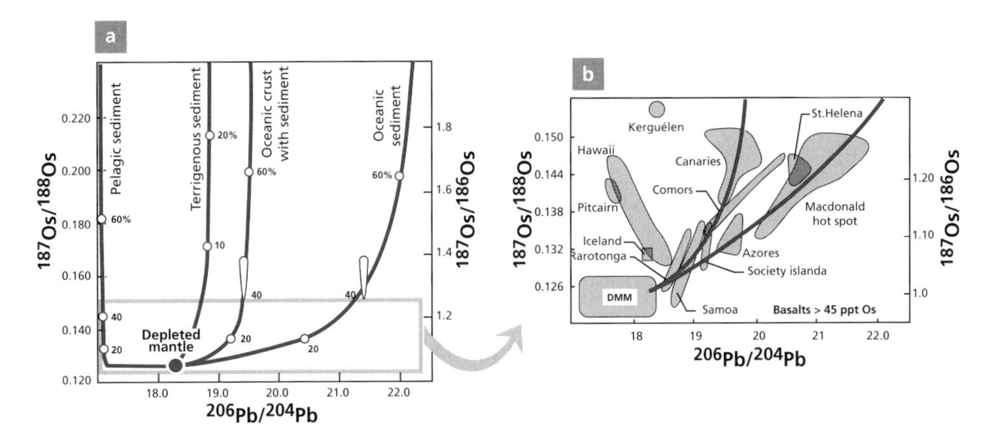

Figure 6.60 Isotope correlation diagram for (Os, Pb) and theoretical mixing models. (a) Theoretical mixing models between the mantle and the various components swallowed up by the mantle. The boxed part corresponds to (b). (b) Isotope correlation diagram for (Os, Pb) for basalts. DMM, depleted upper mantle. After Hauri and Hart (1993); Roy Barman and Allègre (1994).

of oceanic crust or continental materials into the Earth's mantle. With this objective in mind, we shall examine the (Os, Pb) isotope correlation diagrams obtained for basalts by comparing the (Pb, Sr) diagrams (Figure 6.60). Let us examine the ($^{187}Os/^{188}Os$, $^{206}Pb/^{204}Pb$) diagram for basalts. We consider only basalts whose Os content is greater than 45 ppt (the others are suspected of being either victims of analytical error or of secondary contamination during their genesis).

This diagram brings out something very different from the corresponding (Sr, Pb) diagram. There is no sample in the bottom right. This is because the mantle is very rich in Os and the mixtures are shown by very curved paths. **Mathieu Roy Barman** and the present author (1994), and also **Eric Hauri** and **Stan Hart** (1993) of the Woods Hole Oceanographic Institution, calculated a theoretical mixing model with various components. Its trajectories provide quite good explanations for the observations. The important point here is that the recycled materials must be in quite large proportions for them to be detected. Nonetheless, this model gives quite a good explanation of the Os isotope diversity in oceanic basalts in the context of the H–W model.

6.5.6 Conclusions

After these first five dense and often difficult sections, it is probably worthwhile (or essential?) to review these matters. Our method has been to use isotope correlation diagrams two by two, to extract the useful information from them and then to combine the results crosswise to confirm them, refute them, clarify them, or develop them:

- Examination of the Sr–Nd–Hf isotope systems established that extraction of continental crust from the mantle left an upper mantle that was "depleted" in some elements (and so in some isotope ratios) and was of a lower mass than the total mantle.
- This idea of a two-layer structure of the mantle was confirmed by examination of the rare-gas isotope composition, the upper mantle being both "depleted" and "degassed." The upper mantle is the MORB source. The complementary reservoirs of these transfers are the continental crust and the atmosphere where the elements expelled from the mantle have accumulated.
- However, since the mass of the depleted mantle is greater than that of the upper mantle, we have to accept mass exchanges between the upper and lower mantle (the upper mantle being defined as everything above 670 km).
- The continental crust is structured into tectonic segments, representing the episodes during which continental crust formed at the expense of the mantle. However, the continental crust is composed of "young" materials from the mantle and also reworked materials that are older continental materials which have been eroded, deposited as sediment, metamorphosed, and granitized. The whole process seems to be well understood and, of course, to fit in with the schemas developed by geologists.
- Studies of lead isotopes confirmed that a large part of the continental crust was extracted from the mantle before 3 Ga, as best evidenced by the $^{207}Pb/^{204}Pb$ isotope signature compared with ratios for the mantle.
- The atmosphere seems to have been sealed for argon, krypton, and xenon but permeable for helium, which escapes into space, and "semipermeable" for neon. The age of formation of the atmosphere is very ancient and greater than 3 Ga, and, so its seems, older than the mean age of the continents. But we have said that this question was not settled and we shall return to it.
- Difficulties arose when it came to the genesis of OIBs. Initially, it was considered that they stemmed from the lower mantle, which is both more primitive and less degassed than the upper mantle. This was what was termed the standard model. Study of the lead isotope composition overthrew this simple model. Either the OIBs do not come from the lower mantle, or the lower mantle is not primitive. This brings in the subsidiary question: why do OIBs have rare-gas isotope signatures some of which seem very primitive?

- At this point in the discussion we introduced the Hofmann–White hypothesis by which OIBs are related to recycling of oceanic crust. But where does this recycling occur? In the upper mantle? Then how can the rare-gas measurements from Hawaii or Iceland be interpreted? Or in the lower mantle? But how can the upper mantle mechanism and its formation be explained?

All told, then, two major questions remain unanswered: the origin of OIB and the nature of the lower mantle (which questions are connected). This uncertainty notwithstanding, we can develop an overview of the physical and chemical processes related to the geodynamic cycle.

6.6 Chemical geodynamics

This model, which the present author formalized first in 1980 and then in 1982 (see Allègre, 1987), is the result of a long evolution of ideas. The first attempts to relate isotope variations and major geological phenomena concerned the interpretation of variations in the isotopic composition of galenas. For example, studies by **Paul Damon** at Columbia University, New York (Damon, 1954) and **Don Russell** at the University of Toronto (Russell, 1972) sought to go beyond the age of the Earth and to look at the working of the planet as a whole. The decisive steps that followed were made by **Patterson** and **Tatsumoto** (1964) studying the formation of continental crust with lead isotopes and **Patrick Hurley**'s team at the Massachusetts Institute of Technology using strontium isotopes (1962).

The next stage was by **Paul Gast** in the 1960s with two significant papers, one on the limitation of mantle composition (1960) while the other, written with **Tilton** and **Hedge** (1964), showed for the first time that the mantle was not isotopically homogeneous (Patterson, 1963). **Wasserburg** (1964) sought to tie in the development of continental crust and the complementary evolution of the mantle using a quantitative model. Other important contributions confirming the isotopic heterogeneity of the mantle were made by **Tatsumoto** (Tatsumoto *et al.*, 1965; Tatsumoto, 1966) for lead and **Hart** *et al.* (1973) for strontium.

The chemical geodynamic approach came about after the discovery of Nd isotope variations, which began with the (Sr, Nd) correlation, and gave some coherence to the various isotopic measurements. The approach involved combining observed variations in isotopic composition into a logical and coherent scheme within the framework of plate tectonics (Allègre, 1982; Allègre *et al.*, 1982; Hart and Zindler, 1986; Zindler and Hart, 1986; Hart, 1988; Hofmann, 1988).

6.6.1 The foundations of the model

In this approach, it was proposed to construct a coherent model of the Earth based on three fundamental features:

(1) The Earth's structure, first of all, as determined by seismology: a continental crust and an oceanic crust, a marked p- and s-wave seismic discontinuity at 670 km, and a mantle–core discontinuity.

(2) Plate tectonics, next, with cycles of spreading of ocean floors based on the formation of lithosphere at the mid-ocean ridges and its recycling at subduction zones. On this cycle is superimposed the injection of magmas from deeper-lying hot spots. The continents break up, drift around, and collide but remain at the Earth's surface. The use of this model for the entire geological timescale results from the application to geochemistry of a founding principle of geology, **uniformitarianism**.[16] No call is made on "extraordinary" phenomena to explain the past. Yet, nothing shows it is true. Perhaps the workings of the Earth in the past were quite different from what we have reconstituted for the last few million years. The supposition in the chemical geodynamic model is that if it was the case, the variations remained within a context similar to the one we find nowadays.

(3) The data of isotope geochemistry, lastly, and to begin with those for Sr and Nd. In this model, it is assumed that the continents are the "scum" of the Earth, and result from differentiation of the mantle which has extracted from the mantle elements like potassium, rubidium, and the rare earths as well as uranium and thorium. Geological mapping suggests that the extraction of continental material from the mantle occurred over the course of geological time, giving rise to a new tectonic province each time (see Plate 4).

This convection process divided the Earth's mantle little by little into two reservoirs: the upper reservoir, the MORB source, depleted in some chemical elements by extraction of the continents, and the lower mantle less depleted (but not primitive!). This division also affects the rare gases which have degassed from the (consequently depleted) upper mantle but remain more concentrated in the lower mantle, the escaped heavy gases having collected in the atmosphere.

The cycle of oceanic crust which is extracted from the mantle and then returned to it after a time of spreading at the surface acts as the extractor of mantle elements. The magmaphile elements (potassium, rubidium, cesium, uranium, thorium, etc.) were enriched for the first time in liquid during the magmatic processes of formation of oceanic crust. (The partition coefficient $D = C_{\text{solid}}/C_{\text{magma}}$ of these elements is very low for solid–magma partition phenomena.) Then, they were enriched a second time during subduction processes, which transfer elements from the oceanic crust to the continents via more complex fractionation, but whose partition coefficients are somewhat different from those of the mid-ocean ridges. If we term the processes occurring along the mid-ocean ridges the **R-process** (ridge) and the processes occurring during subduction the **S-process** (subduction), the extraction of elements from the mantle to the continental crust corresponds to the sum (R + S). We can write symbolically:

$$\text{continental crust} = \text{S}(\text{R} \times \text{mantle}),$$

with R operating on the mantle and S on the product of this first operation.[17]

The oceanic crust modified by the S-process plunges back into the mantle and is **mixed mechanically** with it by mantle **convection**. In this way, an increasingly "depleted" mantle is formed. The rare gases are outgassed by the R-process but S-process volcanism also leads to outgassing, preventing the rare gases from being recycled into the mantle.

[16] This dates back to one of the founders of geology, **Charles Lyell**, and underpins the development of this discipline on a scientific basis.

[17] Not to be confused with the notation for the r- and s-processes in nucleosynthesis!

The mantle is therefore made up of an **upper mantle**, which is "depleted" in elements such as potassium, uranium, thorium, and rubidium, and degassed of most of its rare gases and of a **lower mantle**, which, while not pristine, is less depleted in potassium, uranium, thorium, and rubidium, and is richer in rare gases.

If, as some investigators assert on the basis of seismic tomography, the downgoing plates were recycled in the lower mantle throughout geological history, the lower mantle would be depleted because it is the downgoing plates that convey the depletion (the extraction of elements concentrated in the continental crust). It is essential, therefore, that the plates have been reinjected **for the most part** in the upper mantle. Likewise, if all the mantle were totally outgassed, there would not be any marked difference in the isotope ratios of the rare gases like helium and neon contained in MORB and OIB. The mantle is therefore necessarily composed of two separate (but not isolated) reservoirs.

6.6.2 The internal geochemical cycle

Two fundamental geochemical processes that are an integral part of geodynamic processes occur in the upper mantle. The oceanic lithosphere is formed at the mid-ocean ridges and separates a basaltic ocean crust enriched in certain elements (relative to the mantle) such as potassium, rubidium, uranium, the rare earths, etc. and an underlying, ultrabasic residual lithosphere depleted in these elements and whose remelting cannot create much basalt because the essential elements making up certain basalt minerals (feldspars, clinopyroxenes) such as calcium and aluminum have been extracted to manufacture oceanic crust. It is often said in petrologists' jargon that the deep ocean lithosphere is **infertile** because it can no longer produce basalt by fusion. This oceanic lithosphere (crust and residual lithosphere) is reinjected into the mantle where it mixes back with the upper mantle (McKenzie, 1979; Allègre *et al.*, 1980; White and Hofmann, 1982).

Mixing is mechanical. Pieces of the lithosphere are caught up in the convective movements of the mantle. In this process, rock is stretched, folded, and pinched. Rock from the ocean crust, generally basaltic, is transformed mineralogically by metamorphism during subduction into rocks such as eclogite (assembly of garnet and pyroxene). The upper mantle looks like a marbled cake (Allègre and Turcotte, 1986) with a basic peridotite (pyroxene and olivine) composition interspersed with strands of eclogites, drawn out and stretched to varying degrees. These strands are shreds of ancient oceanic crust which has been laminated, pinched, and folded by mantle movements. After some time, these ever-thinner strands will end up reacting mineralogically with the peridotite mantle and will disappear (Hofmann and Hart, 1978; Allègre and Turcotte, 1985), while the mantle will be **fertilized** by this **absorption**, that is, made capable of basalt extraction by melting again. But these slivers of continental crust are not uniform. Some have dragged layers of sediments down with them. These sediments carried by the downgoing plate are dehydrated, compacted, and then metamorphosed. They are also stretched, thinned, pinched, and folded (Figure 6.61).

Some of the pieces of oceanic crust reinjected into the mantle are very peculiar as they have lost the chemical elements that will give rise to volcanism in the subduction zones and through that to continental crust. These are the materials which effect the depletion of the mantle in some elements (rubidium, uranium, etc.) and which is reflected in isotope ratios,

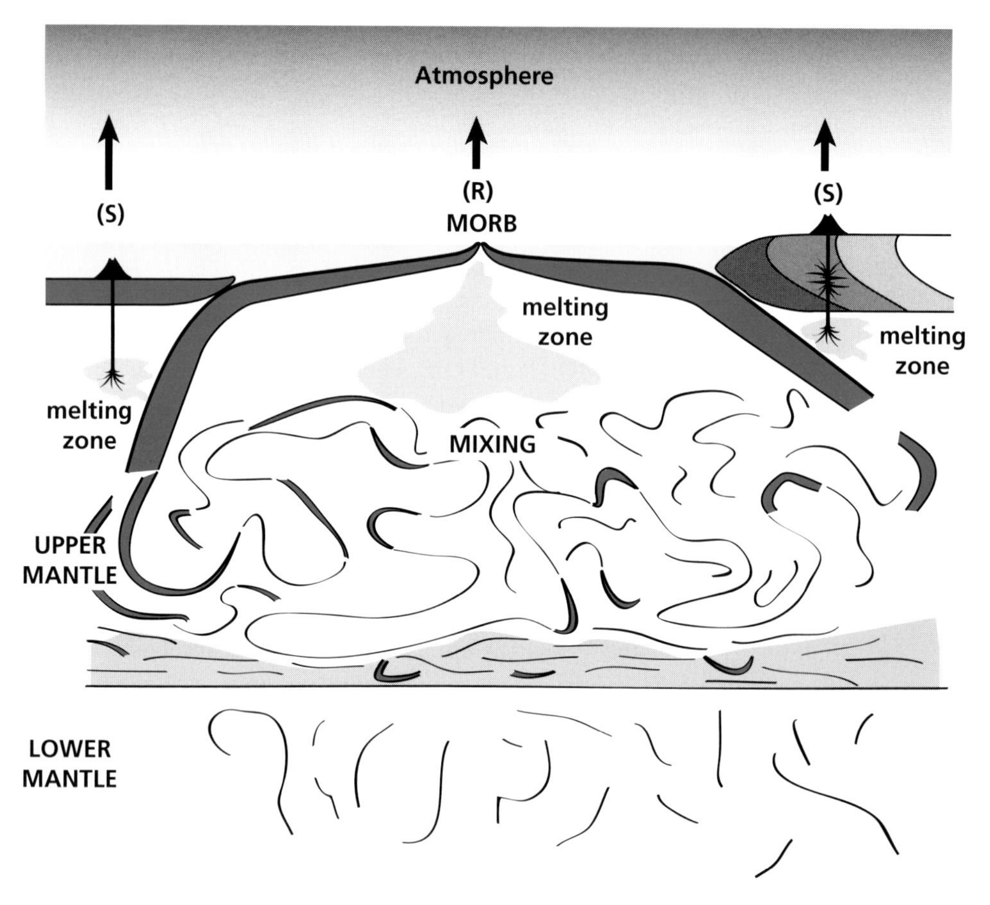

Figure 6.61 Fractionation of the geodynamic cycle. The internal geodynamic cycle exhibits mixing, differentiation, and removal phenomena, mostly in the upper mantle.

because all of the materials swallowed up by the upper mantle have varied isotope ratios and chemical ratios.

Is this marbled mantle a purely theoretical model dreamt up by geochemists and geophysicists? Not a bit of it! It is real enough. It can be seen cropping out in a few special places where tectonic processes have brought pieces of mantle to the surface. This is the case in southern Spain near the little town of Ronda (the temple of bullfighting), in northern Morocco near the village of Beni-Bousera (Figure 6.62), at Lhez (France) in the Pyrenees, and at Lanzo near Turin (Italy).

These massifs are often made up of peridotite, often depleted in aluminum and calcium, with isotope signatures of the upper mantle. They contain strands of eclogites which have been shown to come from ancient oceanic crust. They also contain ancient sediments and even diamond transformed into graphite, that is, old limestones rich in organic material that has been reduced and changed into diamonds. The upper mantle is indeed made up of a thorough mixture of peridotite and eclogite. By contrast, the lower mantle contains far fewer strands and is much more uniform.

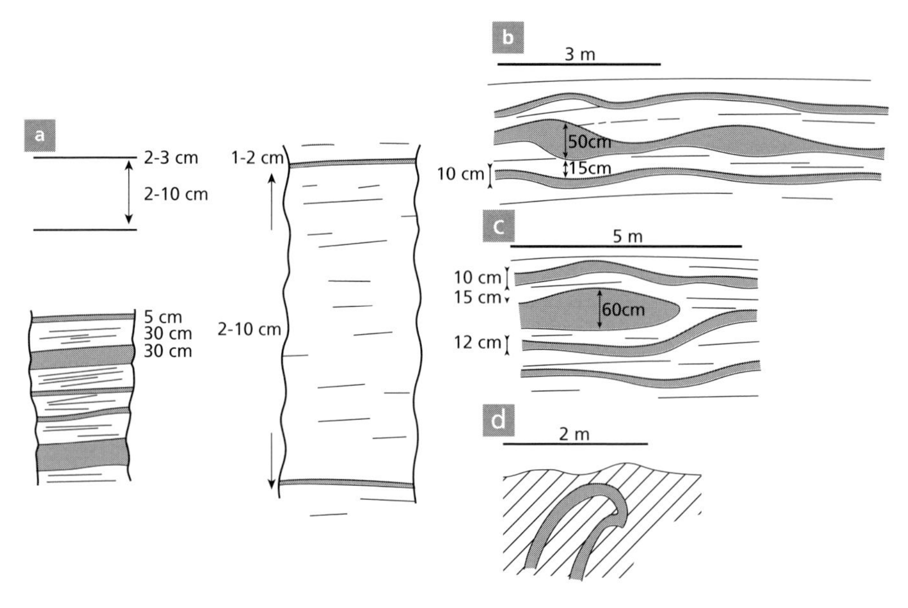

Figure 6.62 Garnet pyroxenite (eclogite) bed structures in the peridotites of Beni-Bousera massif, Morocco. Pyroxenite in blue, peridotite in white. (a) Series of pyroxenite beds of different sizes in the peridotite; (b) pinching and swelling; (c, d) folding. After Allègre and Turcotte (1986).

6.6.3 Statistical geochemistry

So far we have spoken a great deal of reservoirs exchanging material, considering that continental crust is segmented into age provinces and therefore heterogeneous but that the mantle is homogeneous (at least at regional scale). The upshot of this is that we have considered an **average value** to characterize each reservoir. Without relinquishing such modeling, which is straightforward and extremely useful, we have to supplement it with another description, which is statistical. Each reservoir is described not by a single average value but by a distribution of values, isotope ratios, or chemical concentrations.

When an oceanic plate dips into the mantle it is chemically and isotopically heterogeneous. Admittedly, when oceanic crust forms at the mid-ocean ridges it is isotopically homogeneous (by melting) but chemically heterogeneous, with the crust being enriched in rubidium compared with strontium, the lithosphere being depleted in Rb/Sr, and the sediments rich in Rb/Sr when detrital and poor when they are limestone. These chemical heterogeneities are reflected by isotopic heterogeneities during the ridge–subduction zone journey, by radioactive decay. Then, of course, the same happens in the mantle with radioactivity continuing its inexorable decay and its creation of new isotopes.

The upper mantle, then, is not isotopically or chemically homogeneous, but **heterogeneous**. It is not characterized by a single average isotope ratio, as in the "box" model we dealt with when examining the earlier crust–mantle system, but by a **spread of isotope ratios**, with an **average**, **variance** (standard deviation), and a **degree of asymmetry** for each type of ratio. This statistical distribution also has, of course, a geographical distribution.

For example, the α^{Sr} and μ^{Rb} parameters are each replaced by a distribution notated as:

$$\langle \alpha^{Sr} \rangle = \{\alpha^{Sr}(\overline{m}, \sigma, d)\} \text{and} \langle \mu^{Rb} \rangle = \{\mu^{Rb}(\overline{m}, \sigma, d)\}$$

where $\overline{m}$ is the mean of α^{Sr} or μ^{Rb}, σ the standard deviation, and d the asymmetry.

Naturally enough, the more the system is subjected to convection the more it is mixed and the more homogeneous it is, and the smaller σ is. Two antagonistic phenomena are therefore in action: chemical fractionation near the Earth's surface, which tends to increase chemical (and so isotopic) heterogeneity, and convective mixing phenomena within the mantle, which tend to reduce them. It is a contest of **chemical differentiation** versus **mixing**.

Partial melting and volcanism (at the ridges and hot spots) of the heterogeneous mantle yield samples of the mantle, which are more or less random, from which we can measure the $^{87}Sr/^{86}Sr$, $^{143}Nd/^{144}Nd$, $^{87}Rb/^{86}Sr$, and $^{147}Sm/^{144}Nd$ ratios, etc., for which we can also define distributions and with them a mean, a standard deviation, and an asymmetry. (We notate these distributions by $\{\ \}$: $\{\alpha^{Sr}\} = \{\alpha^{Sr}(\overline{m}, \sigma, d)\}$ and similarly for $\{\mu^{Rb}\} = \{\mu^{Rb}(\overline{m}, \sigma, d)\}$.)

An important step in our project is, of course, to determine the actual distributions of isotope ratios in the mantle from the distributions sampled and measured at the surface. A priori, it might be thought that the larger the sample, the closer $\{\alpha\}$ is to $\langle\alpha\rangle$. We take as a working hypothesis that $\langle\alpha\rangle = K\{\alpha\}$, that is, **the distribution measured from surface samples is proportional to the distribution of the mantle at depth.** When the distributions measured on various samples of a same type of basalt are more dispersed than those measured on another type, it is legitimate to suppose that the sources of the former are more heterogeneous than the sources of the latter or that the phenomena giving rise to the basalts did not homogenize them as much (because we are mindful, of course, that the phenomena of partial melting of the mantle, then of pooling of the liquids giving rise to magmas, tend to homogenize the isotope ratios). They therefore effect a sort of natural average of the zone where partial melting occurred (statistical sampling). As it is accepted that isotope homogenization occurs at such temperatures, basalt sampling provides a representative average of the sample space. Study of the basalt population is therefore a sample of the mantle on the scale of a few kilometers. But at the same time, these phenomena of magmatogenesis erase any smaller local heterogeneities, as the magma averages them out. Now, when we speak of heterogeneity of the mantle, we must be clear what we mean by it. There are various types of heterogeneity, as Figure 6.63 shows.

6.6.4 Geochemical fractionation

Since the beginning of this chapter, we have spoken of chemical fractionation which modifies the Rb/Sr, Sm/Nd, U/Pb chemical ratios, etc., modifications that are reflected by isotopic changes. The variations in isotope ratios are the reflection, with a time lag, of chemical fractionation. It is customary to say that radiogenic isotope ratios are **chemical fossils.** We should therefore look a little more closely at geochemical fractionation.

The essential fractionations in chemical geodynamics are those related to the formation of magma, at the mid-ocean ridges to manufacture oceanic crust, at the subduction zones to produce arc volcanism and extract continental crust, and in the deep continental crust to yield granites. All are related to the partial melting process (Gast, 1968).

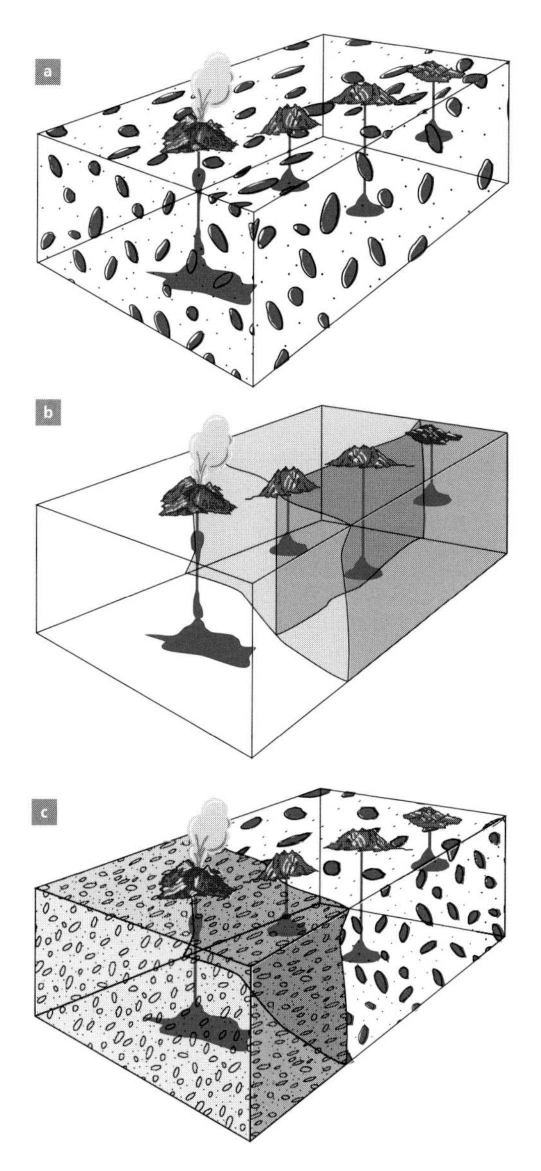

Figure 6.63 The types of heterogeneity the mantle may conceal. (a) Different domains scattered in a whole. Sampling a small part brings out the heterogeneities, while more extensive sampling makes them disappear. (b) Large-scale heterogeneities with several domains, each of which is homogeneous. Heterogeneity is observed geographically. (c) Types (a) and (b) combined. (Of course, other types may exist.)

When a polycrystalline assemblage is in conditions of temperature and pressure at which it melts, melting is **partial melting** pretty well obeying the thermodynamic rules of **eutectic equilibrium**. Melting is characterized by **a degree of partial melting** F and a chemical composition of the liquid which is different from that of the solid.

Trace elements are partitioned between liquid and solid states. This partition is characterized by the partition coefficient D^i = concentration in solid/concentration in liquid of element i.

It obeys a simple balance equation:

$$C_0^i = C_1^i F + C_s^i (1 - F)$$

where C_0^i is the initial concentration of the solid for element i, C_1^i is the concentration of element i in the magma, C_s^i the concentration of element i in the residual solid, and F the degree of partial melting. If the overall partial melting coefficient of element i is defined in the process considered by $C_s^i/C_1 = D^i$, without worrying whether the process is in equilibrium or not, whether it is complex or simple, but such that D^i is about constant, then the partial melting equation can be written:

$$C_1^i = \frac{C_0^i}{F + D^i(1 - F)}.$$

The concentration in the residual solid is therefore:

$$C_s^i = D^i \frac{C_0^i}{F + D^i(1 - F)}.$$

A fundamental result of these formulae for the geodynamic cycle is **Treuil's approximation**. **Michel Treuil** of the Institut de Physique du Globe in Paris noticed in 1973 (Treuil, 1973; Treuil and Joron, 1975) that for some elements with low D, $D^i\,(1 - F)$ becomes negligible compared with 1; therefore he named those elements "magmaphile" but some others have named them "incompatible."

$$C_1 \simeq \frac{C_0}{F}.$$

For these elements the concentration depends directly on F (the smaller F is the more concentrated the liquid) and, yet more importantly, if the ratio of the two elements is taken, F is eliminated:

$$\frac{C_1^i}{C_1^j} = \frac{C_0^i}{C_0^j},$$

that is, the ratios are conserved across partial melting, which is true for isotope ratios of heavy elements and for ratios of a number of trace elements whose partition coefficients D are very, very small and so negligible compared with 1, such as Th, U, Rb, Nd, and Pb under certain circumstances. This is fundamental because not only can we obtain the isotope ratios of the mantle such as $^{87}Sr/^{86}Sr$, $^{143}Nd/^{144}Nd$, etc. but we can also obtain the Rb/Sr and Sm/Nd chemical ratios of the mantle (or at any rate good estimates of them).

Exercise

Exercise

Is Treuil's approximation by which (for low values of D) we can write:

$$\frac{C_1^i}{C_1^j} = \frac{C_0^i}{C_0^j}$$

still valid in the two instances below?

- $D_{Th} = 0.01$ and $D_U = 0.02$. Study the Th/U ratio as a function of the degree of partial melting F, taking the values of 0.11, 0.05, and 0.01.
- $D_{Th} = 0.02$ and $D_U = 0.1$.

In both cases, plot the curve $(Th/U)_{liquid}/(Th/U)_{source} = f(F) = R$.

Answer

We write:

$$\left(\frac{C_l^1}{C_l^2}\right) \bigg/ \left(\frac{C_0^1}{C_0^2}\right) = \frac{D_2 + F(1 - D_2)}{D_1 + F(1 - D_1)} = R.$$

In the first case, we find $R = 1.08$ (0.1), 1.159 (0.05), 1.4897 (0.01). In the second case, we find $R = 1.61$ (0.1), 2.10 (0.05), 3.65 (0.01). This shows the limits of Treuil's approximation.

Exercise

At the ocean ridges, U/Pb fractionation is such that $\mu = {}^{238}U/{}^{204}Pb$ shifts from 5 in the upper mantle to 10 and more in the oceanic crust. If $D_U^{(R)} = 0.03$ and if $F = 0.08$, what is the value of $D_{Pb}^{(R)}$?

This oceanic crust is melted in the S-process. In the andesite lava $\mu = 8.5$. Supposing that $D_U^{(S)} = 0.03$ and $F = 0.10$, what is the value of $D_{Pb}^{(S)}$?

Answer

First we find $D_{Pb}^{(R)} = 0.135$ and then $D_{Pb}^{(S)} = 0.01$. Lead is less compatible than uranium during subduction.

We have the answer to an earlier question: why does lead distinguish the mantle from the continental crust so poorly? Because the U/Pb R and S chemical fractionation processes partially offset one another in this case. There are two results from these investigations. First, if the melting rates are high, the ratios between the elements that interest us (Rb/Sr, Sm/Nd, U/Th, and U/Pb) are just about preserved.

Accordingly, the mixing lines in the isotope correlation diagrams (see the beginning of this chapter) are more or less straight. This is why roughly correct correlations were found between many isotope ratios. Because everything is **almost linear**: fractionation, mixing, and radioactive decay. (For lead isotope ratios, where growth of radioactive origin is non-linear, the parent–daughter identity preserves linearity; when such linearity is lost, the isotope diagrams become more complex.)

By contrast, when dealing with elements that fractionated during different processes, linearity is not preserved. This is the case, for example, of isotopes of lithophile elements (Sr, Nd, Hf) and of the rare gases, because they obey two different rationales. Fractionation for one is related to extraction from the continental crust, and for others to mantle outgassing. The former are sensitive to processes of reinjection of oceanic or continental crust, the others are not reinjected. An example of this is the (Sr, He) non-correlation established by **Mark Kurz** of the Woods Hole Oceanographic Institution (Kurz *et al.*, 1982). Another example is the case that implies osmium, because its geochemical behavior is atypical compared with that of other elements (**Lambert** *et al.*, 1989).

For elements whose only isotope fractions are created by extraction of continental crust, a double chemical fractionation process occurs: one at the mid-ocean ridges, the other in the subduction zones. We have just seen with the example of lead that the two processes are not identical (in the case of Pb/U, fractionation is reversed). We shall look more closely at this question using other isotope tracers, those from radioactive disequilibria.

6.6.5 Consequences for the Earth's energy regime

Uranium, thorium, and potassium, which are all elements whose radioactivity is an important source of heat for the internal activity of our planet, are greatly enriched in the continental crust and therefore depleted in the upper mantle. The lower mantle is much richer in these elements. **It is therefore the lower mantle that supplies the upper mantle's convective heat**. The upper mantle is heated from below and cooled from above by the ocean. It is therefore in an almost ideal situation for convecting, by an admittedly complex process but which is akin to what is known in physics as **Rayleigh–Bénard** convection (see Turcotte and Schubert, 2002), which can be observed when a pan of water is put on the cooker to boil. The lower mantle too convects, because it is very hot and is of large dimensions, but it convects in a different manner, by internal heating particularly by the radioactive elements it contains (it might be worth looking back at the end of Chapter 1, where the calculations set out were preparation for current reasoning).

Exercise

We wish to test the two-layer convective model in respect of heat flow (see Chapter 1). The measurements on MORB give the results: U = 50 ppb, Th/U = 2.2, and K/U = 10^4. Accepting that the three elements are wholly incompatible, that is, that $C_{liquid} = C_{solid}/F$ where F is the degree of partial melting estimated at 10%, calculate the heat flow of radioactive origin produced by the upper mantle. Calculate the heat flow produced by the lower mantle if we assume that for the mantle U = 21 ppb, K/U = 10^4, and Th/U = 4.2. Calculate the heat flow of the continental crust given that K = 1.2%, K/U = 10^4, and Th/U = 5. Given that the total flow is $41 \cdot 10^{12}$ W, what proportion is produced by radioactivity? What makes up the difference?

Answer

The heat flows from radioactivity are: upper mantle: $0.95 \cdot 10^{12}$ W (terawatts), lower mantle $15.3 \cdot 10^{12}$ W, continental crust: $6.25 \cdot 10^{12}$ W, making a total of $22.5 \cdot 10^{12}$ W. The proportion created by radioactivity (Urey ratio) is 0.5. The difference comes from energy stored when the Earth formed and from differentiation of the core. It is stored in the lower mantle and core (gravitational energy transformed into heat). The lower mantle therefore transmits $36.8 \cdot 10^{12}$ W whereas the upper mantle produces just $1 \cdot 10^{12}$ W.

6.6.6 Radioactive disequilibria and the difference between R- and S-processes

We have just touched upon it with regard to U/Pb fractionation, but we shall see in detail with the example of ^{230}Th/^{238}U systems, which we have already dealt with in terms of chronology,

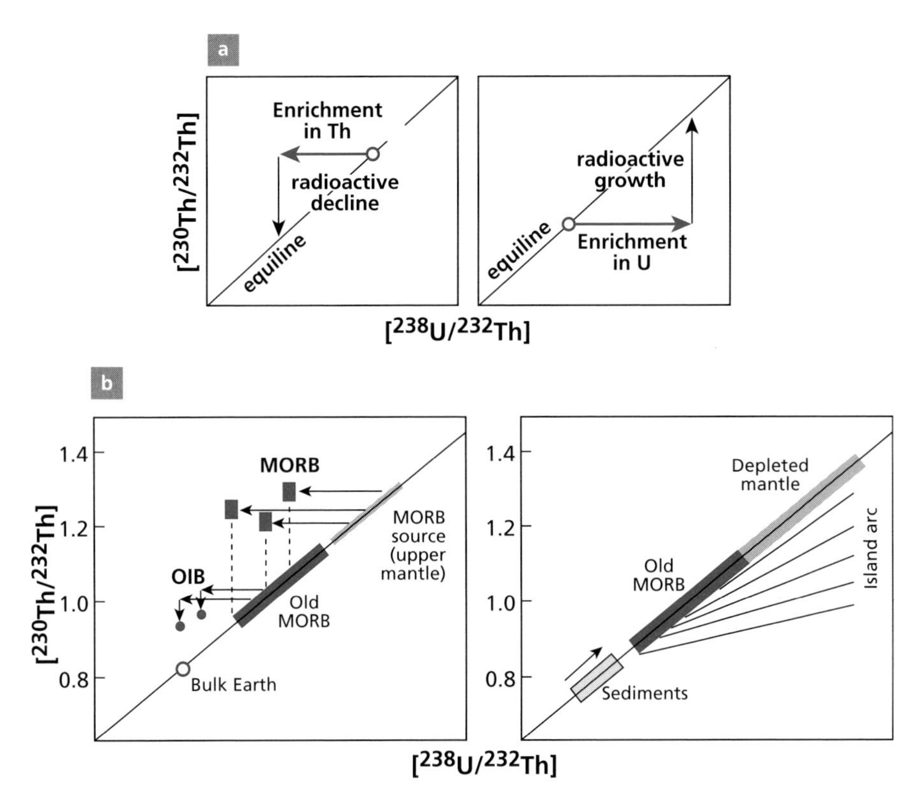

Figure 6.64 General principles of the reasoning in this section. (a) The two types of fractionation are shown in the ($^{230}Th/^{232}Th$, $^{238}U/^{232}Th$) diagram. Left: fractionation where Th is enriched in the fluid compared with U; right: enrichment of U compared with Th. (b) The two types of observations and explanations. Left: genesis of MORB and OIB; right: genesis of island arc volcanoes. After Allègre and Condomines (1982).

there is a considerable difference between the S- and R-processes, as **Michel Condomines** and the present author established in developing this method in 1982 (Allègre and Condomines, 1976; 1982). It was extremely well explored and explained later on by a large amount of outstanding work.[18]

Let us plot the results on the graph (of activity)[19] $[^{230}Th/^{232}Th,\ ^{238}U/^{232}Th]$ (Figure 6.64), in which (a) is a reminder of how it is read.

Let us pick up on what was developed in Chapter 3 about dating based on $^{238}U-^{230}Th$ radioactive disequilibrium using the isochron construction. We take the ($^{230}Th/^{232}Th$, $^{238}U/^{232}Th$) diagram in the form of activities. Any system in secular equilibrium has its data point on the equiline. This can be assumed to be the case of mantle rock.

[18] See the excellent book by B. Bourdon, G. M. Henderson, C. C. Lundstrom, and S. P. Turner (2003) *Introduction to U-Series Geochemistry*, Reviews in Mineralogy and Geochemistry Vol. 52. Washington, DC: Mineralogical Society of America.

[19] See Chapter 3. Remember that the activity of a radioactive element R* is written $[R] = \lambda_R R^*$, where R* is the number of atoms and λ_R the decay constant.

- If the partial melting process involves enrichment of the liquid in (Th) compared with (U), the point representing the magma will shift horizontally to the left. The system will return to equilibrium through decay of its excess ^{230}Th (excess relative to the ^{238}U value). The ^{230}Th/^{232}Th isotope ratio of the magma will decrease over time with the data point shifting vertically downwards. When it returns to equilibrium, it will lie on the equiline again but in a lower position (with a lower ^{230}Th/^{232}Th ratio in U/Th).
- If, on the contrary, the partial melting process involves enrichment of the melt in (U) compared with (Th), the data point will shift horizontally to the right. The system will return to equilibrium by recreating ^{230}Th through ^{238}U decay. The point will therefore rise vertically. When it reaches equilibrium, the ^{230}Th/^{232}Th ratio will have increased. The U/Th ratio will then be far greater than that of the initial solid.

The evolution equation is, it will be recalled (with ratios in square brackets in the form of activities):

$$\left[\frac{^{230}\mathrm{Th}}{^{232}\mathrm{Th}}\right] = \left[\frac{^{230}\mathrm{Th}}{^{232}\mathrm{Th}}\right]_0 \mathrm{e}^{-\lambda_{230}t} + \left[\frac{^{238}\mathrm{U}}{^{232}\mathrm{Th}}\right]\left(1 - \mathrm{e}^{-\lambda_{230}t}\right).$$

If we measure the composition of a magma, the age since it underwent partial melting is written:

$$t = \frac{1}{\lambda}\ln\left(\frac{\left[\frac{^{230}\mathrm{Th}}{^{232}\mathrm{Th}}\right]_0 - \left[\frac{^{238}\mathrm{U}}{^{232}\mathrm{Th}}\right]}{\left[\frac{^{230}\mathrm{Th}}{^{232}\mathrm{Th}}\right] - \left[\frac{^{238}\mathrm{U}}{^{232}\mathrm{Th}}\right]}\right).$$

Calculating an age like this involves measuring the isotopic composition of thorium and the ^{238}U/^{232}Th (chemical) ratio. But one datum is missing: the initial isotope ratio of the thorium. One can also reason backwards and calculate the initial isotope ratio of the thorium if the age is known. Let us leave aside this issue of calculating the age and the initial ratios for the moment and turn to the results of analyses on present-day lavas from various types of volcanic rocks.

Of course, the rocks must be from the present day so significant comparisons can be made. Otherwise radioactive decay since eruption will have shifted the position of the data point vertically in the (^{230}Th/^{232}Th, ^{238}U/^{232}Th) diagram. We must be wary of contamination processes, too. As for rare gases, sea water is a powerful contaminant for the ^{230}Th/^{232}Th ratio because it has extremely high ^{230}Th/^{232}Th ratios (values of 50 or 100 are commonplace while the values of basalt lavas are close to unity). Even slight contamination by sea water shifts the data point upwards.

Having taken the requisite precautions, what do we observe? MORB and OIB are located to the left of the equiline. They therefore correspond to cases of magma being enriched in Th relative to U since the U/Th ratio has fallen.

It is observed that statistically the OIB lies below the MORB. An enrichment coefficient is defined $k = (\mathrm{U/Th})_{\mathrm{magma}}/(\mathrm{U/Th})_{\mathrm{source}}$. From the exercises in the previous section, the lower the degree of partial melting, the smaller k is. Petrologists tell us that the extent of partial melting of OIB is much less than that of MORB (1% versus 10%). Now, the relative position on the (^{230}Th/^{232}Th, ^{238}U/^{232}Th) diagram is not so clear. This can be explained by accepting that the genesis of OIB is often complex, involving phenomena of mixing, contamination by ancient oceanic crust, and multiple transfer times (Figure 6.65). When the data points for island arc volcanoes are plotted on the same diagram, they are found, on the contrary, to lie to the right of the equiline. Their genesis involved processes which enriched the magma in U more than in Th. It is also observed that many points representing these volcanoes have much greater thorium isotope ratios than are found for MORB and OIB. This is interpreted in the framework of models which have been developed by the genesis of magmas in subduction zones. It is accepted that the downgoing plate begins to melt, but that it passes on to the overlying mantle wedge a silica-rich fluid with a very high water content. This fluid has probably also derived some of its material from the subducted sediments. The fluid induces melting of the material in the mantle wedge overlying the plunging plate. The initial material is therefore a mixture of upper mantle similar to that which gave rise to the MORB, of a subducted sediment whose ^{238}U/^{232}Th ratios may be as low as 0.4, and of a water-rich fluid which is enriched more in U than in Th as U is more soluble in water than Th.

There is a big difference, then, between magmas which result from partial melting of the mantle, like MORB, and island arc magmas whose genetic processes are more complex. The genesis of OIB belongs to the family where melting processes dominate, but in detail it is observed that the Th isotope ratios of one and the same island are variable.

Detailed studies have shown complex phenomena to be involved. First of all, there are variable types of source, as seen with lead isotopes. Then there are processes of interaction with the oceanic crust through which the OIB makes its way. So there are three types of magma-producing processes. Simple partial melting (MORB), partial melting followed by mixing (OIB), and partial melting in the presence of water and a mixture of multiple components (island arcs). But the diagrams show a continuous spread of data points among these different types, indicating the existence of certain affinities.

Exercise

A domain of the mantle is considered whose [^{230}Th/^{232}Th] ratio in terms of activity is 1.4. Suppose this domain undergoes 8% partial melting to form oceanic crust. This oceanic crust returns to radioactive equilibrium. Then it undergoes 5% partial melting in a subduction zone. The transfer time of the magma to the surface is 40 000 years. What are the [^{230}Th/^{232}Th] and [^{238}U/^{232}Th] ratios of magma in terms of activity? It is assumed the solid/melt global fractionation coefficients are $D_U = 0.015$, $D_{Th} = 0.0034$.

Answer

[^{230}Th/^{232}Th] = 1.15. [^{238}U/^{232}Th] = 1.

Let us move on now to transit times between the source and the surface. Despite complex theoretical models that predicted a quite long and complex transfer from the melting zone to the surface (McKenzie, 1989) the transfer rate for MORB is quite high. For MORB, the issue has been addressed in two ways. The first is to calculate the initial value of [^{238}U/^{232}Th] of the source mantle by estimating the degree of partial melting using the fractionation

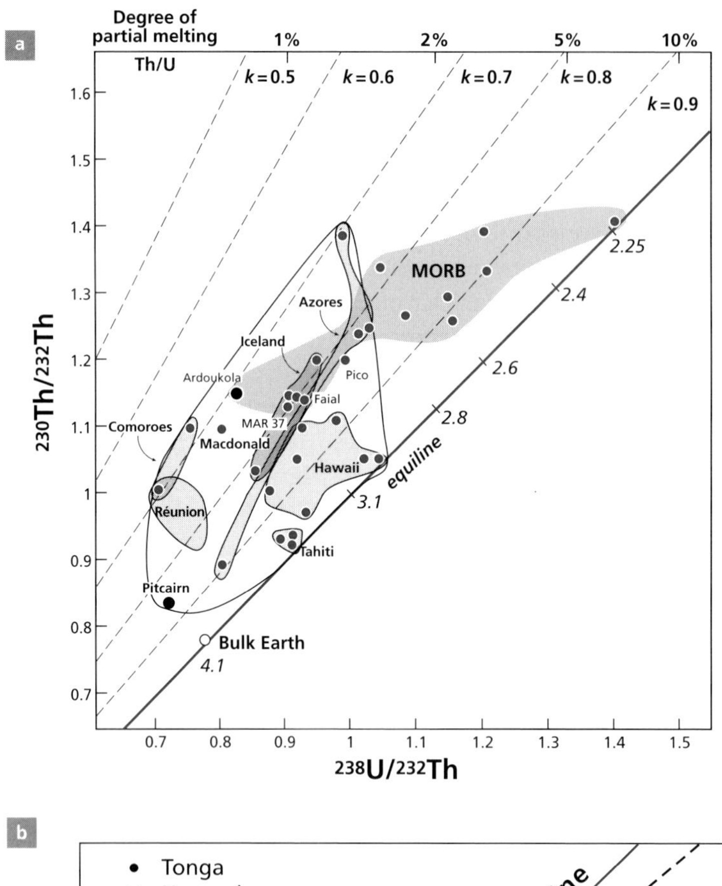

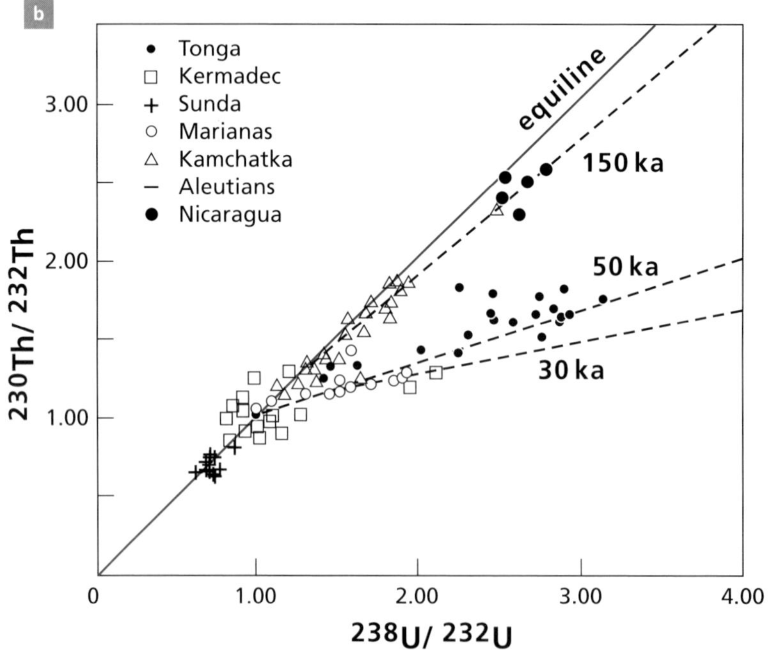

Figure 6.65 Actual cases of disequilibrium data. (a) The MORB zone in blue, the OIB zone in white with the main OIBs shown. Note the two domains are continuous. (b) Subduction zones with various isochrons showing a very different representation of MORB and OIB. After Condomines *et al.* (1988); Condomines and Sigmarsson (1993).

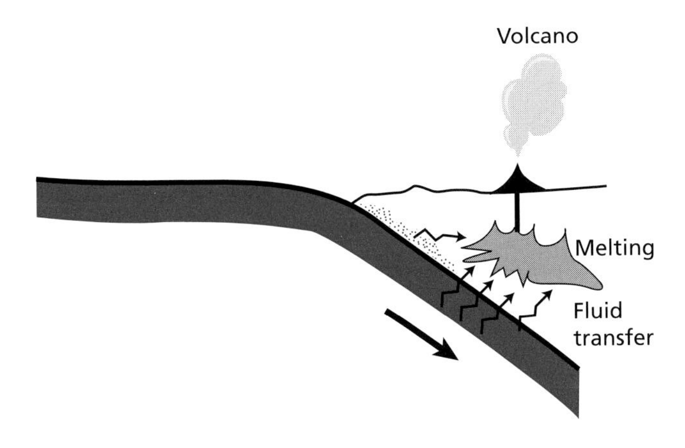

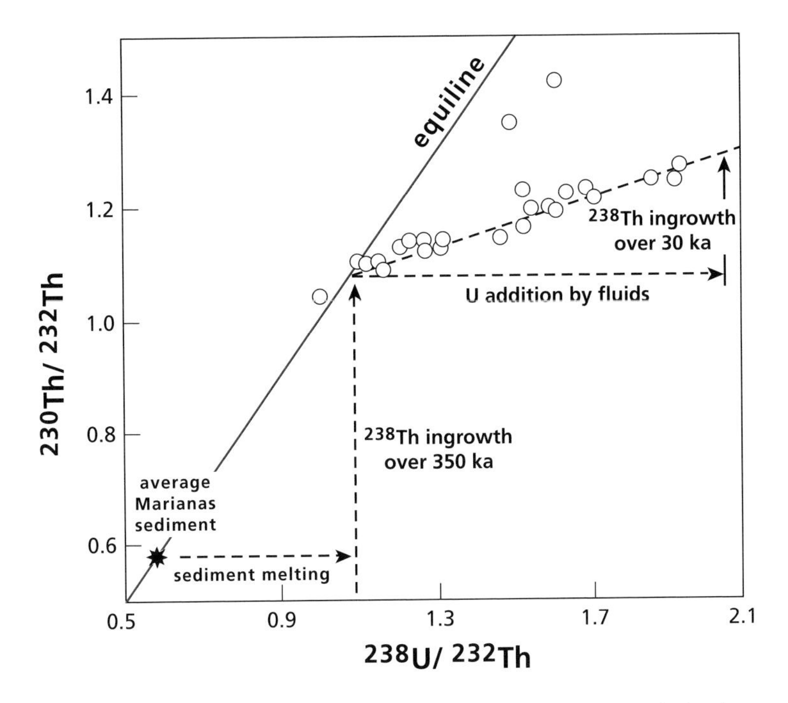

Figure 6.66 (a) Diagram of current understanding of the formation of island arc volcanoes; (b) the various stages of genesis explaining the observations. The example of the Marianas Islands, studied by Elliot *et al.* (1997).

coefficients determined in the laboratory for U and Th. The second is based on the following remark. A radioactive pair has returned to equilibrium when the time since the disequilibrium was created is six times the shorter half-life. As many $^{230}Th/^{232}Th$ ratios of MORB are far from equilibrium, it can be estimated that the transit times are very much less than 300 000 years.

But there is a more interesting observation. The $^{226}Ra/^{230}Th$ and $^{210}Pb/^{226}Ra$ ratios are not in equilibrium. The transit time is less than 6 times the half-lives of ^{226}Ra and ^{210}Pb of $1.622 \cdot 10^3$ years and 22.26 years, respectively. This corresponds to a date for ^{226}Ra of less than 10 000 years, but for ^{210}Pb of less than 135 years. If it is accepted that mid-ocean ridge magmas form at depths of 30 km, this corresponds to transit speeds of 200 m yr^{-1}. The constraint of radium corresponds to 3 m yr^{-1}. Therefore, the speed is instantaneous for MORB and it should be considered that the $^{230}Th/^{232}Th$ ratios measured on present-day rocks yield the U/Th value of their source mantle directly and better than the directly measured chemical ratio because it is fractionated.

Calculating transfer times for OIB is much more complicated because of the complex processes involved, such as magma chamber processes, lithosphere interaction, and so on (see Hawkesworth, 1979; Elliot *et al.*, 1997; Turner *et al.*, 2003).

Island arcs have very long and variable transit times. Some are more than 300 000 years because the data points are very high on the equiline, far above the MORB points. In the case of the Marianas Island arc, **Tim Elliott** of the University of Bristol and his colleagues have managed to put a figure on the final stages and to show that there were two stages of input from the downgoing plate. A first stage involving sediments having occurred 300 000 years before a second melting stage, which itself occurred 150 000 years before the eruption (see Figure 6.66a). All of this clearly shows the difference in behavior of island arc volcanoes compared to MORB and OIB.

Exercise

The radioactive disequilibrium of an OIB is measured giving $[^{230}Th/^{232}Th] = 1.08$ and $[^{238}U/^{232}Th] = 0.8$. In addition, $[^{238}U/^{230}Th]_{source} = 1.1$. Calculate the fractionation coefficients $D = (U/Th)_{source}/(U/Th)_{magma}$, the transit time between fractionation and reaching the surface, as well as the rate of ascent, if it is assumed the process began 20 km beneath the ocean ridge.

Answer

$D = 1.43$, $T = 1380$ years, $v = 15$ m yr^{-1}.

6.6.7 The major elements

The major elements are those that make up minerals and therefore rocks. Their abundance is measured in percent by mass. They are oxygen, silicon, magnesium, aluminum, calcium, iron, and sodium. They determine the physical properties of the Earth's interior and so are responsible for the properties measured in geophysics (velocity of seismic waves, density, magnetism). Now, what is important is that these major elements are practically unaffected by the major processes governing chemical geodynamics. To show this, let us take an extremely simple calculation.

Let us consider the weight analyses of peridotite (an essential constituent of the mantle) and of granite (an essential constituent of the continental crust) (Table 6.9).

The proportions by mass between continental crust and upper mantle are 0.01 (for the upper mantle above the seismic discontinuity at 670 km). Let us suppose the peridotite

Table 6.9 Composition of granite and of peridotite of the mantle

Oxide	Element	Peridotite		Granite	
		percent weight of oxide	percent weight of major element	percent weight of oxide	percent weight of major element
	O		43.36		50.38
SiO_2	Si	41.82 ± 0.80	19.51	72.30 ± 0.50	33.13
Al_2O_3	Al	3.50 ± 0.10	1.26	14.15 ± 0.10	5.094
Fe_2O_3	Fe	9.06 ± 0.15	6.33	1.91 ± 0.02	1.33
MgO	Mg	37.99 ± 0.5	22.86	0.040 ± 0.008	0.03
MnO	Mn	0.17 ± 0.01	0.14	0.370 ± 0.001	0.22
CaO	Ca	3.02 ± 0.06	2.15	1.33 ± 0.01	0.949
Na_2O	Na	0.82 ± 0.1	0.6	3.49 ± 0.03	2.589
K_2O	K	0.03 ± 0.006	0.0250	2.55 ± 0.06	2.12

Table 6.10 Composition of the primitive mantle

	O	Si	Al	Fe	Mg	Mn	Ca	Na	K
"Reconstructed" primitive mantle (%)	43.42	19.64	1.298	6.28	22.603	0.1389	2.137	0.6198	0.4595
Depleted mantle	43.36	19.51	1.26	6.33	22.86	0.14	2.15	0.6	0.025

comes from the depleted upper mantle, and let us calculate the composition of the "reconstructed" primitive mantle (Table 6.10) before extraction of the continental crust by mixing of continental crust with depleted mantle:

$$[\mathrm{A}]_{\mathrm{primitive}} = [\mathrm{peridotite}] \times 0.99 + [\mathrm{granite}] \times 0.01.$$

This composition is almost identical to that of the peridotite analyzed. The only element for which a difference can be clearly seen is potassium, which is not a major element! Those difference in composition cannot be distinguished by any geophysical, seismological, or gravimetric measurements.

Remark

Remember this warning in chemical geodynamics: minor elements mean major problems. Whereas in petrology, major elements mean key problems!

We have translated this in Figure 6.67, remembering we have calculated factors W. It will be recalled that:

$$W^i = \frac{m^i_{\mathrm{cc}}\ C^i_{\mathrm{cc}}}{m^i_{\mathrm{p}}\ C^i_{\mathrm{p}}}$$

where m^i_{cc} is the mass of continental crust, and m^i_{p} is the mass of the primary mantle before depletion; we can take it to be equal to the upper mantle, such that $m_{\mathrm{cc}}/m_{\mathrm{p}} = 0.015$;

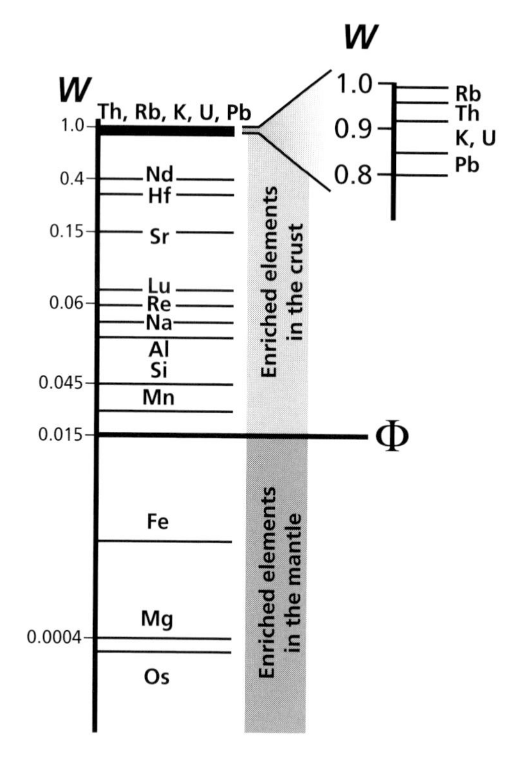

Figure 6.67 The W scale for some elements. W is the fraction of an element contained in the continental crust relative to the sum total (crust + upper mantle), and Φ is the proportion by mass of the lower mantle–upper mantle. It separates the elements enriched in the crust from those enriched in the mantle.

C_{cc}^{i} is the concentration of element i in the continental crust, C_{p}^{i} is the concentration of element i in the primitive mantle, and W is the proportion of the element concentrated in the continental crust.[20]

It is possible to calculate W for a few major and minor elements. The scale (in $\log_{10}$) is separated in two by the values $\Phi = 0.015$; an element that does not fractionate between crust and mantle has this value. Above this value the element is enriched in continental crust; below, it is depleted. It shall be seen that while silicon and manganese are not very different from it, below Φ the elements are enriched in the mantle, while above they are enriched in the crust. Table 6.11 gives the values of the various isotope ratios that can be used for studying the continental crust–mantle system.

Hofmann's two-stage model

Hofmann (1988) developed a simple model that gives a pretty good account of the chemical composition of the continental crust, the oceanic crust, and the upper mantle. Although unsophisticated, it does allow us to get our thinking straight. It assumes, as already established, that the continental crust became differentiated from the primitive mantle through

[20] We know it is an approximation because the mass of the depleted mantle is greater than that of the upper mantle. However, it helps to get our ideas straight.

Table 6.11 Average isotope ratios of the three reservoirs of the silicate Earth

	Continental crust	Depleted mantle	Primitive mantle
$^{87}Rb/^{86}Sr$	0.35	0.0014	0.087
$^{87}Sr/^{86}Sr$	0.7123	0.7021	0.70475
$^{147}Sm/^{144}Nd$	0.11	0.252	0.1975
$^{143}Nd/^{144}Nd$	0.51155	0.51330	0.51265
$^{176}Lu/^{177}Hf$	0.0073	0.0459	0.0333
$^{176}Hf/^{177}Hf$	0.28169	0.28343	0.28287
$^{187}Re/^{188}Os$	37.54	0.3622	0.3850
$^{187}Re/^{186}Os$	312	3.01	3.20
$^{187}Os/^{188}Os$	1.40	0.1251	0.1254
$^{187}Os/^{186}Os$	11.7	1.040	1.0422
$^{238}U/^{204}Pb$	9.58	5.77	8.78
$^{232}Th/^{238}U$	4.55	2.54	4.27
$^{206}Pb/^{204}Pb$	18.59	17.43	18.34
$^{207}Pb/^{204}Pb$	15.586	15.42	15.55
$^{208}Pb/^{204}Pb$	39.56	37.13	39.05

partial melting with a degree of melting F_1 and that this extraction left a depleted mantle. The depleted mantle then underwent a second stage of partial melting in much more recent times with a degree of melting F_2. For example, the differentiation of the continental crust may have a mean age of about 2 Ga. The differentiation of ocean crust is far more recent, less than 100 Ma.

The equations with which to compute such a model are straightforward. We begin with the formula for the creation of melt by partial melting at equilibrium, assuming that the composition of this melt represents the average composition of the continental crust:

$$C_{\mathrm{e}}^{i} = \frac{C_{0.\mathrm{s}}^{i}}{D_{\mathrm{s}/1}^{i} + F\left(1 - D_{\mathrm{s}/1}^{i}\right)}$$

where C_{e}^{i} is the concentration of element (i) in the melt, $C_{0.\mathrm{s}}^{i}$ is the concentration in the initial solid, $D_{\mathrm{s}/1}^{i}$ is the fractionation factor between the magmatic melt and the residual solid, and F is the degree of partial melting.

After the first melting episode producing the continental crust with a degree of melting (F_1), the concentration for an element (i) of the residual mantle is:

$$C_{\mathrm{s}}^{i} = \frac{D_{\mathrm{s}/1}^{i} \cdot C_{0.\mathrm{s}}^{i}}{D_{\mathrm{s}/1}^{i} + F_1\left(1 - D_{\mathrm{s}/1}^{i}\right)}.$$

If the residual solid is subjected to a new melting episode, the concentration of the melt (magma) for an element (i) is written:

$$C_{1_2}^{i} = \frac{D_{\mathrm{s}/1}^{i} \cdot C_{0.\mathrm{s}}^{i}}{D_{\mathrm{s}/1}^{i} + F_1\left(1 - D_{\mathrm{s}/1}^{i}\right) \cdot D_{\mathrm{s}/1}^{i} + F_2\left(1 - D_{\mathrm{s}/1}^{i}\right)}.$$

The degree of melting needed to produce the continental crust may be estimated from the concentration of the most highly magmaphile elements (rubidium, barium, thorium) in the crust. For such elements, the coefficient $D_{s/l}$ is close to zero so that the equation

$$C_l^i = \frac{C_{0/s}^i}{D_{s/l}^i + F\left(1 - D_{s/l}^i\right)}$$

reduces to

$$C_l^i \approx \frac{C_{0/s}^i}{F}.$$

Since the average concentration of these elements in the continental crust is approximately

$$\frac{C_l^i}{C_{0/s}^i} \approx 100$$

we obtain a melt fraction of

$$F_1 \approx 0.01.$$

We can then calculate empirical partition (fractionation) coefficients for all the other elements from their estimated mean concentrations in the continental crust, C_l, the value of $F_1 = 0.01$, and the initial bulk mantle composition C_0, using the equilibrium melting equation. Finally, we can calculate the composition of the second melting event producing the oceanic crust using these formulae with the same partition coefficients and letting F_2 vary around 0.1.

This is the approach **Al Hofmann** took in 1988. Hofmann assumed, and this is his big hypothesis, that fractionation coefficients are identical for the creation of both continental crust and oceanic crust. The model therefore gives an exclusive role to partial melting of the mantle. Again, this hypothesis is debatable and even simplistic, but it does have the advantage of being clear and of setting limits and providing a benchmark for other more elaborate models.

Because there are good reasons to think that the processes forming the continental crust are more complex than just simple, equilibrium partial melting, Hofmann *et al.* (1986) also used an alternative method for determining solid–liquid fractionation factors. This method is based entirely on observations from naturally occurring oceanic basalts.

For this purpose, he used a method originally proposed by Michel Treuil for another use. Suppose that the partial melting formula might be approximated by:

$$C_l^i = \frac{C_0^i}{D_{s/l}^i + F}$$

where C_l^i is the concentration of element (i) in the liquid, and C_0^i the concentration of element (i) in the initial solid.

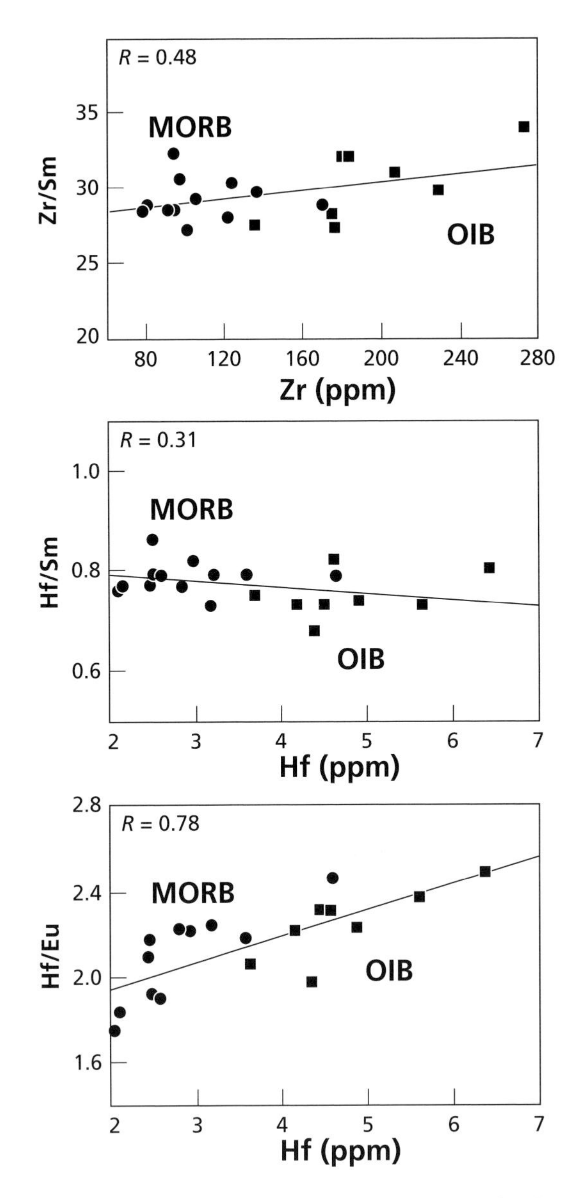

Figure 6.68 Illustration of Hofmann's second method for determining the fractionation factors of the various magmaphile elements. The slope defines the relative value of the solid/magma fractionation factors. It is seen here that $D_{Zr} > D_{Sm}$, $D_{Sm} > D_{Hf}$, and $D_{Hf} > D_{Eu}$.

This means considering that $D_{s/l}$ is small relative to the unit in the general formula. We can then write the ratio between two elements for which this approximation can be made, which we note (i) and (j):

$$\frac{C_1^{(i)}}{C_1^{(j)}} = \frac{C_0^{(j)}}{C_0^{(i)}} + \frac{D_{s/l}^{(i)} - D_{s/l}^{(j)}}{C_0^{(i)}} C_1^{(j)}.$$

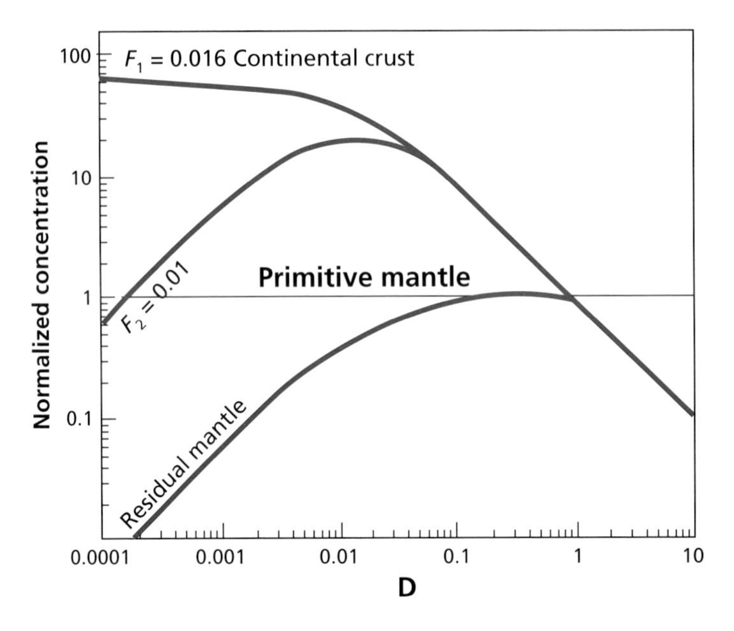

Figure 6.69 Hofmann's calculation to explain the composition of continental crust and oceanic crust. After Hofmann (1988).

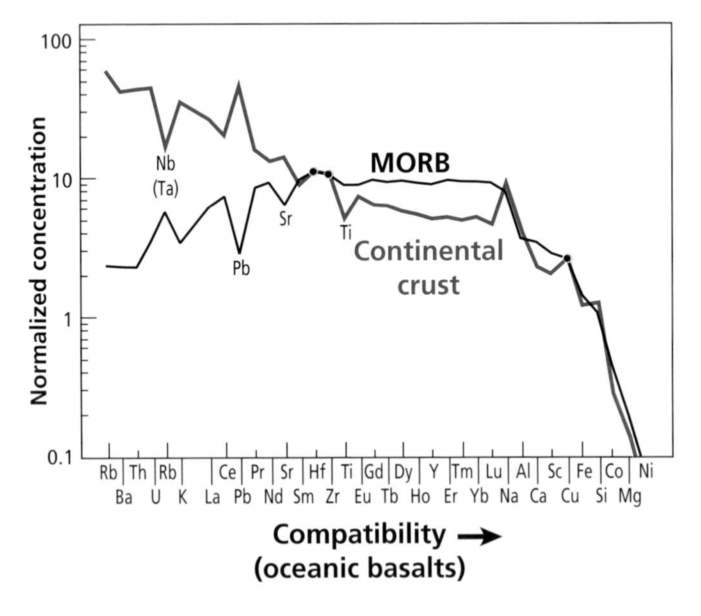

Figure 6.70 Actual distribution of the composition of continental crust and of MORBs. Notice the anomalies for niobium, (tantalum), lead, strontium, and titanium. From Hofmann (1988).

This equation, from which the melt fraction F has been eliminated, gives the concentration ratios of two elements in the melt as a function of their respective ratio in the source and of their partition coefficients.

In a $\left(C_1^i/C_1^i, C_1^j\right)$ diagram, that is a ratio of elements as a function of the concentration of one element in the liquid (magma), the process is represented by a straight line whose slope is proportional to $D^j_{s/l} - D^i_{s/l}$. If $D^j_{s/l} > D^i_{s/l}$, the slope is positive and vice versa. It can be seen then that by plotting the ratios between two elements as a function of the concentration of one of them, it can be determined whether $D^j_{s/l}$ is greater than or less than $D^i_{s/l}$ (Figure 6.68).

This approach of estimating the relative magnitudes of the partition coefficients can be used to test the above crust–mantle differentiation model: if we plot abundances on the y-axis and the sequence of chemical elements on the x-axis, ordered according to their decreasing solid–liquid fractionation factors, there should be no major discontinuities for MORBs. Hofmann *et al.* (1986) also calculated theoretical models (Figure 6.69).

When the calculations are compared with actual observations, it can be seen that the model does not reflect the observations exactly although the difference is not great (Figure 6.70).

(1) Elements with very low fractionation factors are more abundant in the real world than in the model. Hofmann *et al.* (1986) overcame this difficulty by assuming that a small part of the melt is retained in the mantle. This is an ad hoc hypothesis but there is no reason to rule it out since, through erosion and subduction, this process certainly occurred.
(2) The second difficulty concerns the elements Nb and Ta, Ce and Pb, Sr and Ti, which "cause" peaks or valleys in the abundance diagram. Hofmann *et al.* (1986) ascribed these "anomalies" to the fact that the actual fractionation factors involved in the differentiation of continental crust and those involved in the differentiation of oceanic crust are not identical. This relates back to the difference between the R- and the S-processes, one without water, the other with water.

Accordingly, Hofmann's approach is interesting for teaching purposes. It shows that with a very simple model, the general relationships between continental and oceanic crust can be reasonably well modeled, but that a precise description involves distinguishing between differentiation of continental crust and partial melting occurring at the mid-ocean ridges. It is a good example of modeling to meditate on.

6.6.8 The origin of MORB and OIB and the mantle structure

Let us turn back now to the question of the origin of oceanic basalts. Let us examine the problem of the origin of the two main types of oceanic basalt, MORB and OIB, on the basis of the statistical distribution of their isotope composition. Both types of basalt are the result of a threefold process: partial melting of the mantle, collection of magma, and transfer to the surface. These three processes apply to a mantle source made up of the mixing of thin folded layers of a basaltic component (eclogite) in an ultrabasic matrix (peridotite). We know from petrological experiments (partial melting experiments in the laboratory in varied conditions of pressure and temperature) that the MORBs correspond to a greater degree of partial melting (5–10%) than the OIBs which are closer to 0.1–1%. Therefore OIBs sample a larger volume of mantle per cubic meter of magma than MORBs do. But on the other hand, the volume given out by the ocean ridges (MORB) is so much greater than the volume from plumes (OIB) that a much greater volumetric proportion of the mantle is sampled by the MORBs than by the OIBs.

Exercise

Calculate the volume of mantle sampled in 5 million years by the MORB. What proportion does that correspond to for the upper mantle?

Answer

$4 \cdot 10^7$ m (ridge length) $\times$ $4 \cdot 10^{-2}$ (spreading) $\times$ $6 \cdot 10^4$ (thickness) $\times$ $3.2 \cdot 10^3$ (density) $\times$ $5 \cdot 10^6$ (duration) $\times$ $10 = 1.5 \cdot 10^{22}$ kg. Now, the mass of the mantle is $1 \cdot 10^{24}$ kg, therefore more than 100% is sampled in 500 Ma. If the sampling were random, all of it would be sampled! (But in practice some domains are sampled several times because of recycling.)

That said, the statistical distribution of $^{87}Sr/^{86}Sr$, $^{4}He/^{3}He$, or $^{206}Pb/^{204}Pb$ isotope ratios for MORB and OIB exhibits much greater dispersion for OIB than for MORB (White, 1985; Allègre, 1987).

This is particularly obvious if all the zones where the Schilling effect is observed are removed from the statistics.

Measurements of lead isotope compositions have shown that the signatures of the various OIBs are very varied (Figure 6.51). The accumulation of Sr and Nd isotope data have also revealed the same phenomenon. The Sr and Nd isotope correlation is not a simple straight line as early studies had suggested (Figure 6.71).

The MORBs, then, have very homogeneous isotopic compositions. This corresponds to what we know about their source (the upper mantle), which is well mixed by mantle convection processes.

In contrast, OIBs come from a much less well mixed zone of the mantle. The idea that has developed, and which combines this last observation and the physical conditions for manufacturing plumes, as in fluid mechanics experiments, is, as said, that OIBs are generated in the boundary layers at the base of the convective systems (Figure 6.72). Plumes arise from the instability of the boundary layers, like those that give rise to hot spots. These boundary layers are relatively stagnant zones located at the frontier between convecting systems. They are stagnant layers which receive surface material (oceanic crust, oceanic crust and sediment, continental lithosphere). They are therefore poorly mixed and so heterogeneous. The dispersion observed in the various isotopic diagrams is interpreted as the result of mixing between a mantle that is analogous to that which generated the MORBs (depleted mantle) and various extraneous components (e.g., sediments, recycled oceanic crust, or delaminated lithosphere). In the

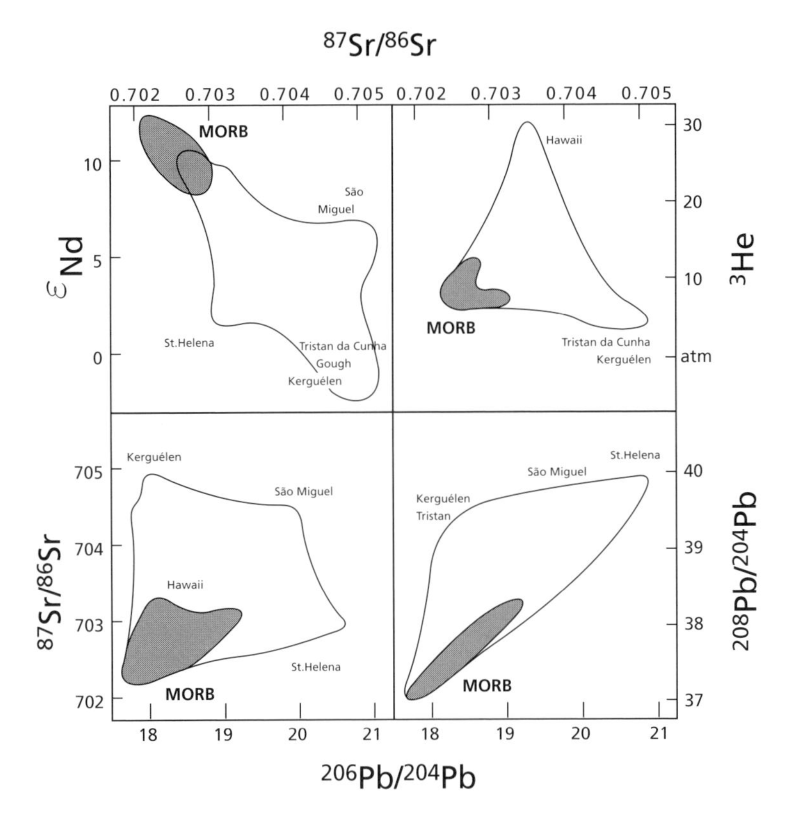

Figure 6.71 With the increased number of measurements, multiple correlations, which were simple when chemical geodynamics began, have become more complicated. This diagram (after Allègre and Turcotte, 1985) shows the comparative dispersion of MORB and OIB measurements in various correlation diagrams but also the continuity between their domains.

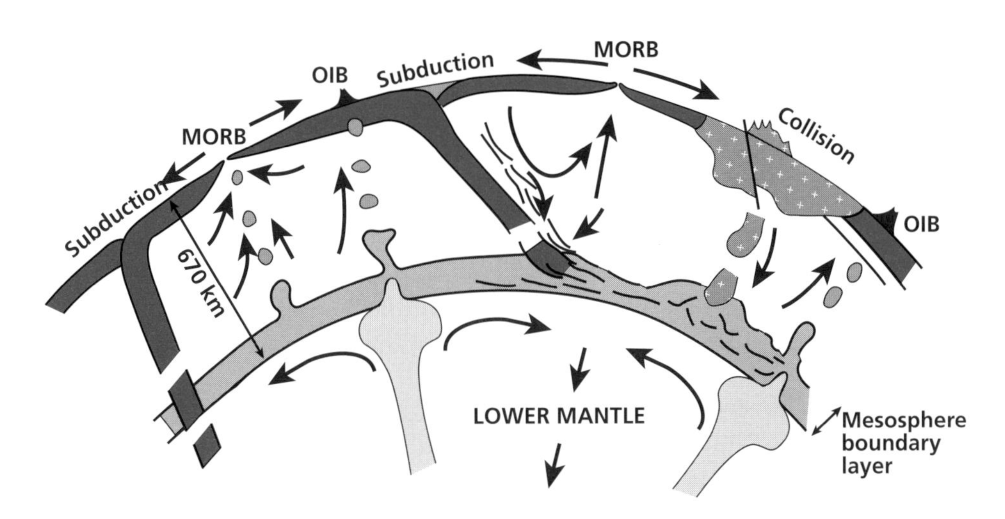

Figure 6.72 Diagram showing how one can think of the upper mantle convecting separately from the lower mantle but exchanging matter with the lower mantle. After Allègre and Turcotte (1985).

Dupal zone, these zones receive pieces of continents delaminated by continental collision phenomena. Elsewhere, they receive recycled oceanic crust with or without sediments.

The question is: where is (are) the boundary layer(s) from which the OIB take their source? Two zones of the mantle are to be considered: the 670-km boundary layer above the seismic discontinuity and the core–mantle interface boundary layer.

The present author still believes most observations can be explained by OIB originating at a depth of 670 km (mesosphere boundary layer) as proposed by Allègre and Turcotte (1985).

(1) In the context of a two-reservoir mantle as required by the isotope geochemistry of rare gases and as is consistent with the isotopic mass balances of Sr, Nd, and Hf, the idea of OIB rising from the 670-km boundary provides a natural explanation of the continuity observed between MORB and OIB in the various isotope diagrams if they come from the same reservoirs. This includes the fact that the Dupal signature exists for both OIB and MORB (Figure 6.71).

(2) The duality observed in the isotopic composition of rare gases and not in the isotopic composition of the elements Sr, Nd, Hf, or Pb can be accounted for by the entrainment phenomenon proposed by O'Nions and Oxburgh (1983). When an instability gives rise to a plume in a boundary layer, the process entrains a fraction of the lower layer of less than 2–5%.

The concentration ratio of rare gases between the lower mantle and the upper mantle is 100 to 500, as can be extracted from the mass balance of various rare gases. When plumes form, the mixture so created is dominated by the signature of the lower layer, but to variable degrees depending on the proportion of entrainment. Large plumes entrain a greater proportion of the lower layers compared with small plumes. This is indeed what is observed: the big hot spots (Hawaii, Iceland) have more "primitive" He or Ne signatures than small hot spots. However, such entrainment phenomena do not leave traces in the Sr and Nd isotope signatures because, for these elements the difference in concentration between lower and upper mantle is a factor of 2 or 3 and not 100 or 500 (Allègre, 1987; Allègre and Moreiva, 2004).

Exercise

Imagine a plume starting from the boundary layer and entraining 5% by mass of material from the lower mantle. Suppose that $\varepsilon^{\mathrm{Nd}}_{\text{plume source}} = +5$, $\varepsilon^{\mathrm{Nd}}_{\mathrm{lm}} = +2$, and the concentration ratio $C^{\mathrm{Nd}}_{\mathrm{lm}} / C^{\mathrm{Nd}}_{\mathrm{um}} = 2$. Calculate the plume's $\varepsilon^{\mathrm{Nd}}$ value.

Answer

$\varepsilon_{\text{plume}} = +4.7$, which is difficult to distinguish from values like $+5$ or $+6$, which are usual for OIB.

The outcome is that we have no direct information about Sr, Nd, Hf, or Pb for the lower mantle. It is a hidden reservoir and its properties can only be determined differentially.

The opposing model by which OIB stems from the core–mantle boundary at a depth of 2900 km is based primarily on seismic tomography. The method yields two essential results:

(1) In various locations corresponding to subduction zones, slabs can be "seen" plunging into the mantle beyond the 670-km boundary (see Van der Hilst *et al.*, 1991; Grand *et al.*, 1997).
(2) Fine-scale imagery seems to show some plumes arise from the core–mantle boundary while others seem to originate at 670 km (Montelli *et al.*, 2004).

This model is not supported by any clear and direct geochemical arguments. There are many counterarguments to the interpretations of these observations:

(1) If downgoing plates had plunged into the lower mantle completely throughout geological time, we would have a global mantle convection regime. But, as we have seen, the rare gases require a two-layer mantle. As these slabs are chemically "depleted structures" because of the extraction of continental crust, it is the lower mantle that would have an isotopic signature depleted in Sr and Nd. Conversely it is the mid-ocean ridges, whose surface origin is attested by the most reliable part of seismic tomography, that have such a signature.

    There are several ways of overcoming this **contradiction between seismic tomography and isotope geology**.

The first is to accept that the geophysical observations relate to the present period alone. The situation must have been different in the past. There must have been a two-layer mantle for most of geological time, and this layering must have been destroyed only recently. In the past, only a small fraction of the slabs might have plunged into the lower mantle. This would easily explain the isotopic mass balances of Nd and Sr.

The second is to accept that what the seismic tomography images show is not a downgoing plate but a cold current in the lower mantle convection pattern. A combination of both interpretations is also possible.

The third is to consider that only a part, about one-third, of the plate is subducted into the lower mantle while most of the slabs pile up in the upper mantle as observed by Fukao *et al.* (1992).

(2) How, then, can the seismic images of plumes be interpreted? By accepting that the regimes between upper and lower mantle are thermally and not mechanically coupled. Lower mantle plumes provide heat, which triggers certain instabilities in the 670-km boundary layer. And this is probably the general explanation. The lower mantle–upper mantle coupling is largely thermal. This in no way means there are no exchanges of mass between the upper mantle and the lower mantle. Such exchanges have been substantial and about one-third of the lower mantle is reinjected material from the upper mantle according to the Sr and about Nd isotopic mass balance calculation. This provides a picture, then, of the mantle and of exchanges which have taken place between the various reservoirs throughout geological time (Figure 6.73).

But we have to distinguish between structure and texture. The **structure** consists of two layers convecting independently but exchanging mass and energy continuously. The **texture** is the intimate mixture of pyroxenite strips which are elongated, folded, and

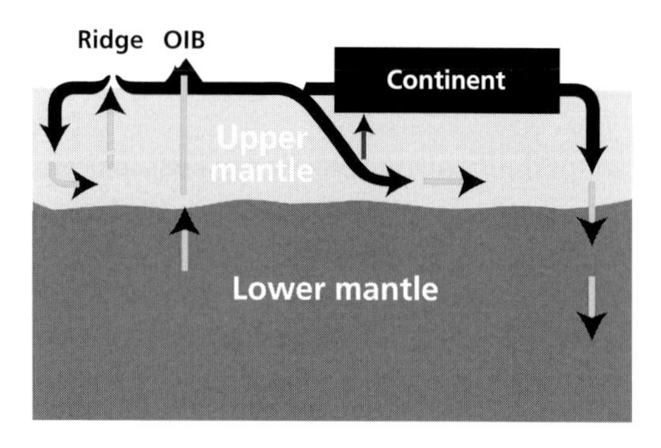

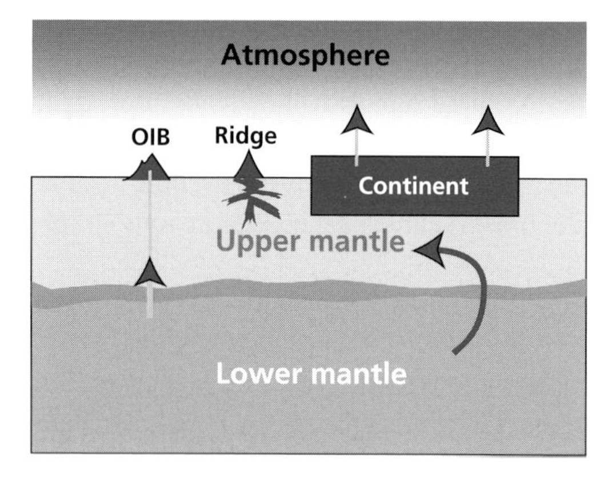

Figure 6.73 Two diagrams of geodynamic mechanisms. Top: using strontium, neodymium, and hafnium isotopic tracers. Bottom: using rare gases.

embedded into peridotite in the upper mantle or ringwoodite (garnet plus spinel) in the transition zone.

Since pyroxenite melts more readily than peridotite, the proportion of pyroxenites contributing to the basaltic melt is higher than their actual presence in the mantle. This proportion depends on the degree of partial melting and also the proportion of pyroxenite in the source. Recent work by **Sobolev** *et al.* (2007) estimates those proportions in various types of basalt at 10% in MORB and 70% in OIB. Remember that MORB corresponds to 10% of partial melting and OIB to 2%.

The MORB source is a well-stirred reservoir where pyroxenite has been partially destroyed. OIBs came from a source where pyroxenite makes up 10–15% of the mass because they are formed when subducted slabs pile up.

6.6.9 Comparison and limits of the model

We now have a coherent model of mantle structure and dynamics. While it should not be taken as an eternal truth, it should not be abandoned either, until we have something better.

Direct comparison of geochemical models and geophysical observations[21] is a delicate business, because, as pointed out, the fundamental processes do not involve the major elements much. Yet it is the major elements that, through mineralogy and the physical properties of minerals, condition geophysics observations. A second reason is that geophysical observations are made on the Earth as it is today, whereas isotopic geochemical observations cover the 4.5 billion years of its history. And as we know, the Earth has evolved during its history.

One very important point to make is that there is no guarantee that this cyclic process has been constant throughout geological times, either in terms of speed or in terms of the "quality" of the process. As said, one of the sources of the Earth's energy powering mantle convection is the radioactivity of uranium, potassium, and to a lesser extent thorium. This radioactivity has declined over geological time. It may be, then, that convection was stronger in the past, or even that its overall structure was different from what we know today.

The formation of oceanic crust and the extraction of continental crust may have been different in the past. What suggests this is that we find in the Archean submarine ultrabasic lavas (komatiites), which are unknown nowadays and the continental crust contains acid rocks that are unknown today such as the **TTG** association (tonalite, trondhjemite, granodiorite).

The model we have developed, based on the paradigm of plate tectonics, covers the 4.4 billion years of the Earth's history, provided the phenomena are qualitatively similar. However, studies of ancient rocks have not shown any major contradiction. The agreement obtained between a vision by extrapolation into the past of present-day Sr and Nd isotope data and the study of past variations of these isotopes in shales or komatiites shows they are consistent. We therefore have no hard facts that currently require us to radically change our model, but there is no guarantee this position will not change in the future if new studies so require. Meanwhile, let us work with it, without being dogmatic, but without qualms. **Science is neither an unchangeable dogma nor a domain where "anything goes."**

6.7 The early history of the Earth

Isotope geology has achieved results of considerable scope concerning the Earth's structure and evolution. Why not apply the same methods of reasoning to extinct forms of radioactivity, which would allow us a more precise vision of our planet's early history? Why deprive ourselves of a close-up of the first days of the Earth's history?

The initial idea is very straightforward, then. Before setting it out in a few examples – because here as throughout this textbook, the focus is on method – it is useful to recall a little vocabulary. It is generally considered that the solid material making up the telluric planets **condensed** from a mostly gaseous primordial nebula to form solid dust particles. This solid material then **accreted** to form ever larger bodies: grains formed beads, then balls, and then bodies of 1 km, 10 km, 100 km in diameter, and so on. This was the process of **accretion** during which a fraction of the nebula's gases were trapped inside the solids. In this context, there are two extreme scenarios for explaining the formation of the Earth (Figure 6.74).

[21] Especially the wonderful pictures produced by seismic tomography.

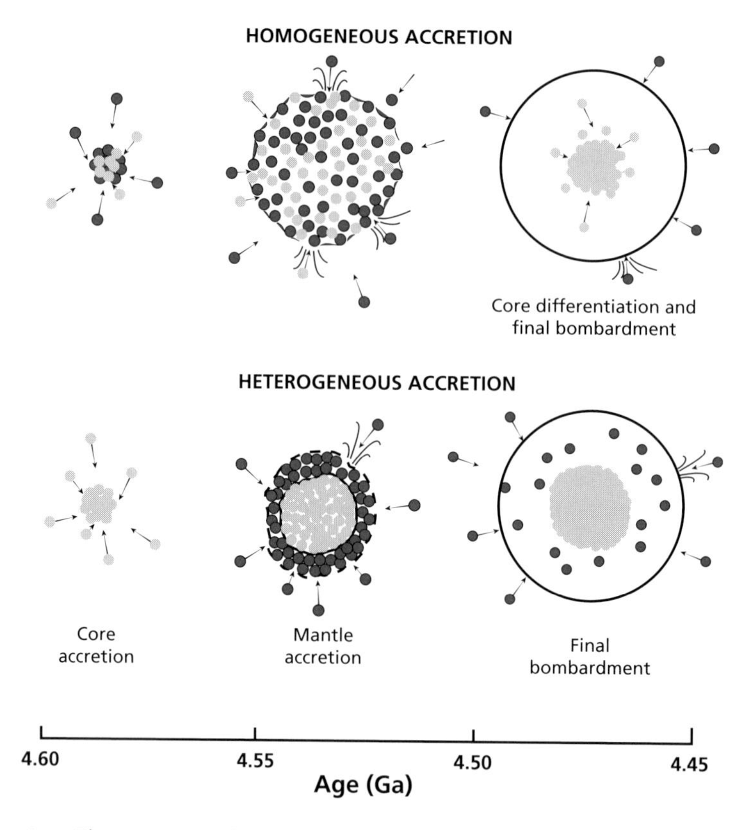

Figure 6.74 The two types of accretion.

In the **homogeneous accretion** scenario, the accreted materials were all of the same composition. The primordial Earth was therefore a large sphere of uniform composition which then **differentiated** through a large melting phenomena to make the core, the mantle, and the atmosphere and perhaps even an embryonic crust as on the Moon. The energy of this general melting episode came from accretion (transformation of potential energy into heat above all through impacts at the end of accretion), extinct radioactivity and above all ^{235}U and ^{40}K radioactivity, which were more "active" at the time (see Chapter 1).

The other scenario is that of **heterogeneous accretion**, which assumes that the Earth first accreted its iron core, then subsequently the silicate mantle, and finally the atmosphere and ocean. This raises many questions. What was the relative timing of these processes? What are the mutual relations between them? In an extreme scenario, condensation, accretion, and differentiation occurred in succession. In more complex scenarios all three operations are considered as more or less simultaneous or combined.

We have already answered three essential questions using radiogenic isotopes with long half-lives. First of all, the differentiation of the continental crust occurred throughout geological time but was more active in the past and has an average age of 2–2.3 Ga (this was derived from Sr–Nd isotope pairs). The core differentiated very early on, with a mean age of 4.4 Ga (this information was obtained from lead isotopes). The atmosphere outgassed relatively early on with an average age of 3.3 Ga (this was obtained from ^{40}Ar). We are going to test or refine these assertions using extinct radioactivities.

6.7.1 The age of the atmosphere and the way it formed

From the ^{40}Ar balance, it has been established that 50% of the ^{40}Ar was in the atmosphere and 50% in the mantle. From this, and from the $^{40}Ar/^{36}Ar$ isotope compositions measured in the atmosphere and mantle, we have calculated a "model age" for the atmosphere of 3.3 Ga. Is this model age reliable?

Let us look first at the balance of ^{36}Ar, the non-radiogenic isotope. We know that the mantle is made up of two layers convecting separately: the upper one is largely degassed, the lower one little degassed. The $^{40}Ar/^{36}Ar$ ratios of these two reservoirs have been measured. In the upper mantle, $^{40}Ar/^{36}Ar = 40\ 000$ while the ratio in the lower mantle is 4000. The ratio in the atmosphere is $^{40}Ar/^{36}Ar = 296.8$.

Exercise

What is the extent of degassing of ^{36}Ar?

Answer

From the quantities of ^{40}Ar in the atmosphere and mantle, it can be estimated that the quantity of ^{36}Ar in the atmosphere is $2.3 \cdot 10^{17}$ g and the quantity of ^{36}Ar in the lower mantle is $1.7 \cdot 10^{16}$ g (the quantity of ^{36}Ar in the upper mantle $\approx 5 \cdot 10^{13}$ g is negligible). Therefore, in total $^{36}Ar = 2.3 \cdot 10^{17} + 1.7 \cdot 10^{16} = 2.47 \cdot 10^{17}$g. Therefore the extent of degassing is 93%.

How can we explain the enormous difference highlighted in the preceding exercise between the extent of degassing of ^{40}Ar (50%) and of ^{36}Ar (93%)? It is because ^{40}Ar is purely radiogenic. There was no ^{40}Ar at the beginning of the Earth, only ^{36}Ar and ^{38}Ar. The extent of degassing of ^{40}Ar measures above all outgassing throughout geological history by volcanism and particularly along ocean ridges. The extent of degassing of ^{36}Ar measures both that related to the primordial history and that related to subsequent geological history. The Earth's history falls into two episodes:

Primordial history | Geological history

If we call the extent of primordial degassing P and the extent of subsequent degassing G, the ^{36}Ar content of the mantle is equal to $(1 - P) \times (1 - G)$. We can estimate $1 - G = 0.5$ and $1 - P = 0.07$, hence $P = 86\%$. The extent of primordial degassing of the Earth is enormous and dominant. Most of the atmosphere formed therefore at the beginning of the Earth's history (Figure 6.75). This is not the impression we gained from the ^{40}Ar balance! Remember, though, that ^{40}Ar was not present initially. It has built up progressively over time. It is not a good recording system for early time, but it is a good recording system for geological time.

We are going to re-examine the problem addressed with the argon isotopes but this time using extinct radioactive ^{129}I, which, as we know, decays to give ^{129}Xe. Xenon-130 is the

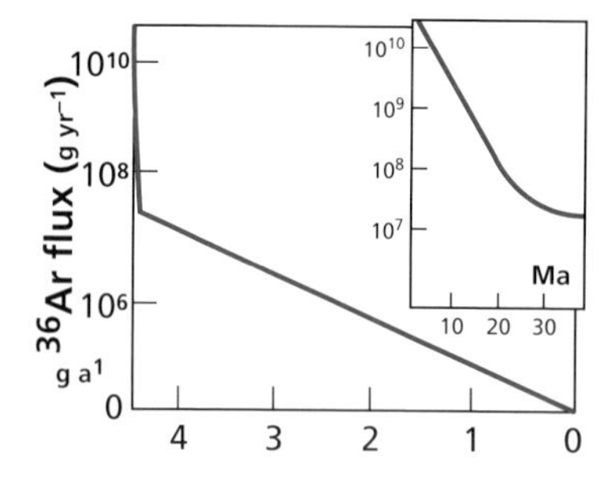

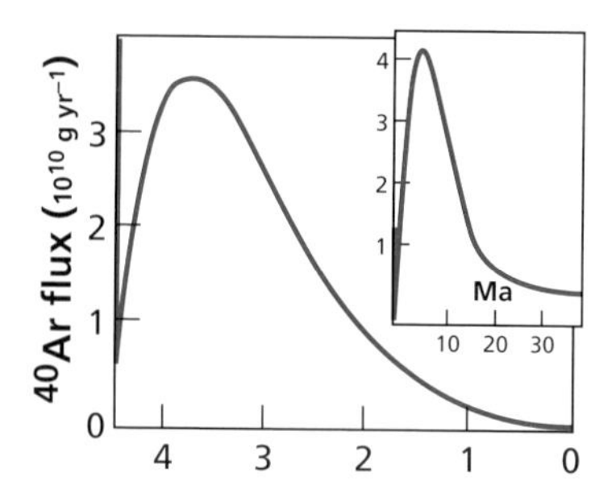

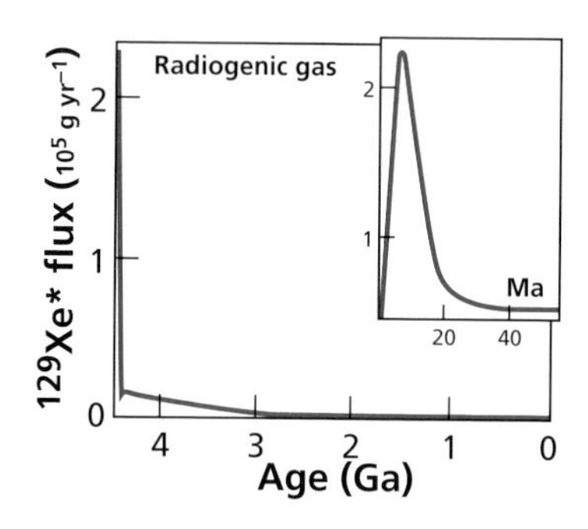

Figure 6.75 Scenario of primordial degassing of the Earth's atmosphere. After Sarda *et al.* (1985).

stable reference isotope. To do this, we take a balance of radiogenic ^{129}Xe. Suppose the Earth is degassed to x%. The quantity of ^{130}Xe in the atmosphere is $^{130}\mathrm{Xe}_{\mathrm{total}} \cdot x$. The quantity of ^{129}Xe* (^{129}Xe* is the excess of ^{129}Xe relative to the usual abundance in meteorites relative to other isotopes) of the atmosphere is estimated at $3.63 \cdot 10^{13}$ g. If x% degassing is high, the quantity of ^{129}Xe* in the atmosphere is equal to that of ^{129}I at the time the Earth began to retain its atmosphere quantitatively, that is, the time the Earth was big enough for xenon not to escape.

For primordial gas, we take a rate analogous to that of ^{36}Ar of 90%. If $^{130}\mathrm{Xe}_{\mathrm{atmosphere}} = 8 \cdot 10^{13}$ g and the extent of degassing is 90%, we therefore have a balance: $^{130}\mathrm{Xe}_{\mathrm{total}} = 8.8 \cdot 10^{13}$ g.

Therefore $^{129}\mathrm{Xe}/^{130}\mathrm{Xe} = (^{129}\mathrm{I}/^{130}\mathrm{Xe})_{\mathrm{accretion}} = 0.4125$.

But $^{129}\mathrm{Xe}/^{130}\mathrm{Xe} = (^{129}\mathrm{I}/^{127}\mathrm{I}) \times (^{127}\mathrm{I}/^{130}\mathrm{Xe})$.

This expression is intended to bring out the isotope ratio of iodine which does not fractionate at the time of accretion but varies with the radioactive decay of ^{129}I. Iodine-127 may be estimated at 25 ppb of iodine in $4 \cdot 10^{27}$ g, which is the mass of the mantle, making $1 \cdot 10^{20}$ g in all. We therefore obtain $^{127}\mathrm{I}/^{130}\mathrm{Xe} = 1.13 \cdot 10^{6}$, hence

$$^{129}\mathrm{I}/^{127}\mathrm{I} = 0.36 \cdot 10^{-6}.$$

We can then calculate a model age relative to a meteorite whose initial iodine isotope ratio has been determined. The carbonaceous meteorite taken as the reference has a $(^{129}\mathrm{I}/^{127}\mathrm{I})_{\mathrm{initial}}$ ratio $= 1.1 \cdot 10^{-4}$.

$$\Delta T = \frac{1}{\lambda} \ln \left[\frac{\left(\frac{^{129}\mathrm{I}}{^{127}\mathrm{I}} \right)_{\text{initial}}}{\left(\frac{^{129}\mathrm{I}}{^{127}\mathrm{I}} \right)_{\text{Earth}}} \right],$$

therefore $\Delta T = 148$ Ma. The Earth therefore retained its atmosphere 150 Ma after the formation of the carbonaceous chondrites, or as an absolute value,

$$T_{\text{acc}} = T_{\text{ch}} - 0.148 = 4.565 - 0.148 = 4.41 \text{ Ga},$$

an average age, which, it should be noted, corresponds almost to the mean age of the core determined using lead isotopes.

We have just seen then that most of the atmosphere was degassed very early on, but 10% or so comes from subsequent degassing, probably from the upper mantle and in conjunction with the plate tectonics cycle. We accordingly have a picture of the primordial Earth with a very high extent of degassing, intense bombardment by large objects, volcanism, and differentiation of the core.

6.7.2 The age of the Earth's core and the mode of terrestrial accretion

We have already determined an age for the Earth's core with lead isotopes using the fact that lead has a great affinity for lead sulfide (PbS). We can test this result using extinct ^{182}Hf–^{182}W radioactivity, for which $T_{1/2} = 9$ Ma. This isotope pair has a peculiar chemical property which is that tungsten is a siderophile element: it enters metallic iron whenever it can, whereas hafnium is lithophile. The iron–silicate division therefore corresponds to a very important W–Hf chemical fractionation. This is how it has been possible to date meteorites and study the evolution of this form of radioactivity in the solar nebula.

We shall examine this aspect first before concerning ourselves with the age of the Earth. It is possible, as we saw when studying chronological methods, to make precise datings of events which took place when ^{182}Hf radioactivity was not yet "dead."

To study these events, we have isochron diagrams of ancient objects (^{182}Hf/^{184}W, ^{182}W/^{184}W) (see Chapter 4). The slope of the isochron gives the ^{182}Hf/^{184}W ratio. This ^{182}Hf/^{180}Hf isotope ratio does not fractionate during the chemical fractionation processes. It decays by the law

$$^{182}\mathrm{Hf}/^{180}\mathrm{Hf} = (^{182}\mathrm{Hf}/^{180}\mathrm{Hf})_0 e^{-\lambda_{\mathrm{Hf}} t},$$

which allows us to determine a relative age as soon as a reference has been fixed for the $(^{182}\mathrm{Hf}/^{180}\mathrm{Hf})_{\text{initial}}$ ratio. In this way time can be measured in absolute terms.

We also have the $(^{182}\mathrm{W}/^{184}\mathrm{W}_{\text{initial}})$ ratios of isochrons constructed for meteorites (Figure 6.76). This ratio increases as age increases (therefore as ^{182}Hf decreases).

If it is supposed that meteorites derive from the condensation of the primordial solar nebula and supposing that this nebula evolves in a closed system, the $(^{182}\mathrm{W}/^{184}\mathrm{W})_{\text{initial}}$ ratio varies with the equation:

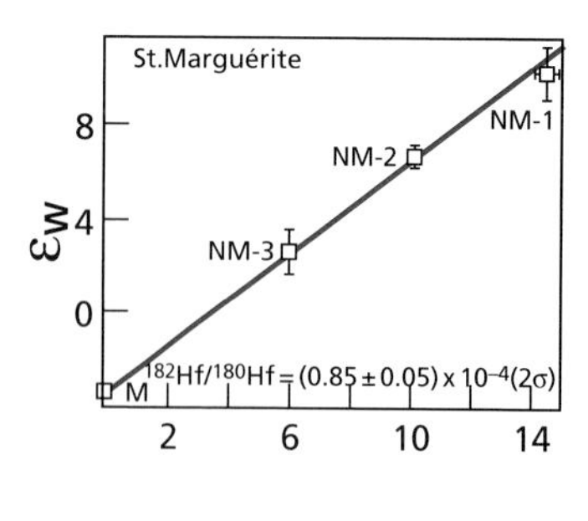

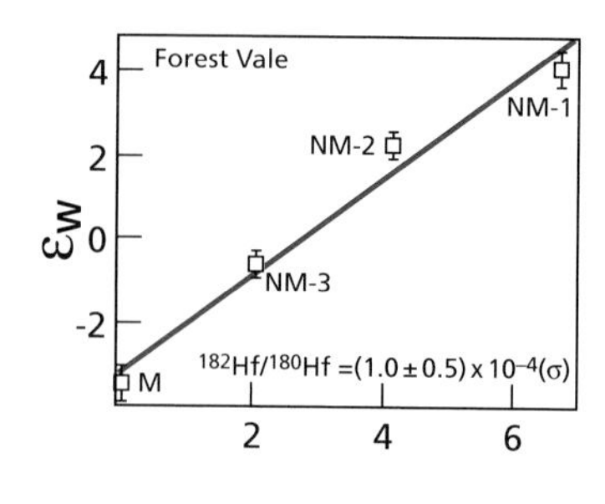

Figure 6.76 Graphs of ^{186}Hf–^{186}W isochrons of two chondrites. After Kleine *et al.* (2004).

$$\left(\frac{^{182}\mathrm{W}}{^{184}\mathrm{W}}\right)_{\mathrm{initial}}(t) = \left(\frac{^{182}\mathrm{W}}{^{184}\mathrm{W}}\right)_{\mathrm{present}} - \left(\frac{^{180}\mathrm{Hf}}{^{184}\mathrm{W}}\right)\cdot\left(\frac{^{182}\mathrm{Hf}}{^{180}\mathrm{Hf}}\right).$$

If we draw a graph, positing $Y = \left(^{182}\mathrm{W}/^{184}\mathrm{W}\right)$ and $x = \mathrm{e}^{-\lambda t} = {}^{182}\mathrm{Hf}/^{180}\mathrm{Hf}$, the evolution of the primordial nebula is therefore a straight line whose (negative) slope is $-(^{180}\mathrm{Hf}/^{184}\mathrm{W})$, and therefore proportional to the Hf/W chemical ratio. When we plot on the graph the values extracted from the chondrite isochrons and the current $^{182}\mathrm{W}/^{184}\mathrm{W}$ ratio of carbonaceous chondrites $\mathrm{C_I}$, we do indeed define a straight line whose slope corresponds to the chemical Hf/W ratio of $\mathrm{C_I}$: $^{180}\mathrm{Hf}/^{184}\mathrm{W} = 1.55$, which value corresponds to Hf/W = 1.31.

This graph is the equivalent of the ($^{87}\mathrm{Sr}/^{86}\mathrm{Sr}$, t) isotope evolution diagram, except that time is graduated exponentially by the $^{182}\mathrm{Hf}/^{180}\mathrm{Hf}$ ratio.

It is interesting to situate two types of differentiated meteorite on this diagram: iron meteorites and basaltic achondrites. For iron meteorites, the straight line of evolution in the isotope evolution diagram is horizontal since Hf = 0. Now, the $^{182}\mathrm{W}/^{184}\mathrm{W}$ ratios are very low, among the lowest measured for meteorites. These ratios correspond to very ancient, primordial ratios. This shows that iron differentiated very early on in the small parent bodies known as **planetesimals**. Basaltic achondrites have very high $^{182}\mathrm{W}/^{184}\mathrm{W}$ ratios and their initial ratios lie far above the straight line of evolution of the nebula. The Hf/W ratio of silicates of the parent body, which can be estimated, set the differentiation at 3 to 8 Ma after that of the iron meteorites, but at the same time as differentiation of meteorites composed of the iron–metal mixture.

And the Earth? The Earth has a $^{182}\mathrm{W}/^{184}\mathrm{W}$ value that stands apart from that of the carbonaceous chondrites. This shows that Hf/W differentiation, that is, segregation of the core from the mantle, occurred when ^{182}Hf was not yet extinct (Figure 6.77). But what, then, is the model age of this differentiation?

Exercise

We are given $\left(^{182}\mathrm{W}/^{184}\mathrm{W}\right)^{\mathrm{Earth}}_{\mathrm{present}} = 0.8650$. The $\left(^{182}\mathrm{W}/^{184}\mathrm{W}\right)^{\text{carbonaceous}}_{\text{chondrites present}}$ ratio is equal to that of the nebula, 0.864 83. The Hf/W ratio of the nebula is 1.31 and the Hf/W ratio of the Earth's mantle is 20. We suppose that:

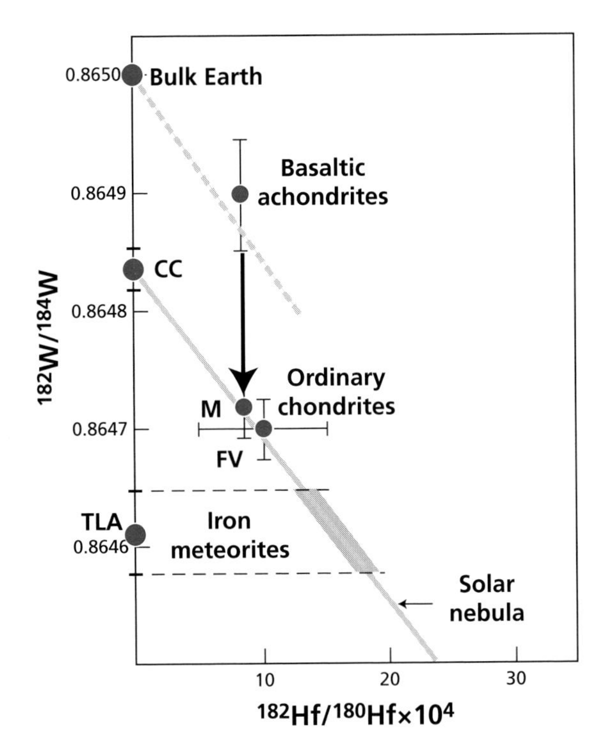

Figure 6.77 Isotope evolution diagram for W–Hf. M, St. Marguerite; CC, carbonaceous chondrites; TLA, Tlacotepec (an iron meteorite); FV, Forest Vale; carbonaceous chondrites.

$$\frac{^{180}\text{Hf}}{^{184}\text{W}} = \left(\frac{\text{Hf}}{\text{W}}\right)_{\text{chemical}} \times 0.845.$$

Calculate the model age of core–mantle differentiation estimated by ^{182}Hf–^{182}W chronology.

Answer

We wish to calculate the initial $(^{182}\text{Hf}/^{180}\text{Hf})_{\text{initial}}$ ratio of the Earth's mantle. The time at which it acquired its current chemical ratio is written T_{diff}. The equation for evolution of the solar nebula is:

$$\left(\frac{^{182}\text{W}}{^{184}\text{W}}\right)_{\text{present}} = \left(\frac{^{182}\text{W}}{^{184}\text{W}}\right)_{T_{\text{diff}}} - \left(\frac{^{180}\text{Hf}}{^{184}\text{W}}\right)_{\text{nebula}} \left(\frac{^{182}\text{Hf}}{^{180}\text{Hf}}\right)_{\text{initial}}.$$

Likewise:

$$\left(\frac{^{182}\text{W}}{^{184}\text{W}}\right)_{\text{Earth}} = \left(\frac{^{182}\text{W}}{^{184}\text{W}}\right)_{T_{\text{diff}}} - \left(\frac{^{180}\text{Hf}}{^{184}\text{W}}\right)_{\text{mantle}} \left(\frac{^{182}\text{Hf}}{^{180}\text{Hf}}\right).$$

The solution to this system of two equations with one unknown gives:

$$(^{182}\text{Hf}/^{180}\text{Hf})_{\text{initial}} = 1.1 \cdot 10^{-5}.$$

If we consider the initial reference ratio for the Solar System is $^{182}Hf/^{180}Hf = 2.4 \cdot 10^{-4}$, then:

$$\Delta T = \frac{1}{\lambda} \ln\left(\frac{2.4 \cdot 10^{-4}}{1.1 \cdot 10^{-5}}\right)$$

and $1/\lambda = 13$ Ma. Hence $\Delta T = 40$ Ma after the formation of the chondrites.

The Hf–W age of 40 Ma calculated in the previous exercise is very different from the 150 Ma estimated from Pb isotopes. How can we overcome this contradiction? By coming back to meteorites and remembering that in their small parent bodies, iron became differentiated very early on. The bodies that accreted to form the Earth were therefore probably already differentiated bodies where the iron was already separated from the silicates. The age of 40 Ma is a mixed age between these primordial "meteorite" ages and the average age of differentiation of the core indicated by lead isotopes. It is therefore an indication that accretion of the Earth involved solid bodies that were already differentiated. This is an **intermediate scenario** between **homogeneous** and **heterogeneous accretion**. It is homogeneous accretion of a heterogeneous product! The core differentiated 150 Ma after primitive condensation but its differentiation occurred on solid bodies in which silicate–iron had already differentiated. The Hf–W 40 Ma age is a mixture between the two events.

Exercise

The isotope ratios of the Earth and carbonaceous chondrites are known with a precision of 15 ppm. The initial ratio is known with precision of 20%. Uncertainties on Hf/W ratios are assumed negligible. What are the uncertainties on the estimation of ΔT of Earth–core differentiation calculated by Hf–W?

Answer

$\Delta T = 40^{+44}_{-35}$ Ma, or 10 Ma, which does not alter the reasoning.

6.7.3 The primitive crust and ^{146}Sm–^{142}Nd

Samarium-146 decays to ^{142}Nd by α decay with a decay constant $\lambda = 6.849 \cdot 10^{-9}$ (nearly a thousand times less than ^{147}Sm!). It is an extinct radioactivity. We have seen that the continental crust has a lower Sm/Nd ratio than the mantle. Accordingly $^{143}Nd/^{144}Nd$ ratios of MORB are positive when expressed in ε_{Nd}, whereas values for the continental crust are negative. The idea then was to try to determine whether a primitive crust had separated out during the early differentiation of the Earth and if such segregation had left traces in the $^{142}Nd/^{144}Nd$ ratio such as are found for the $^{129}Xe/^{130}Xe$ ratio.

This idea was put forward by **Stein Jacobsen** of Harvard University. He and his student **Chris Harper** (Harper and Jacobsen, 1992) claimed to have found a positive ^{142}Nd anomaly in very ancient rocks of Greenland. This was not ϵ_{Nd} measured in 10^4 but differences in ppm 10^{-6} $(100\,\varepsilon)^{-1}$. They reported enrichments of 30 ppm of ^{142}Nd compared with uniform and identical values of MORB, OIB, and recent and ancient rocks. The measurements were

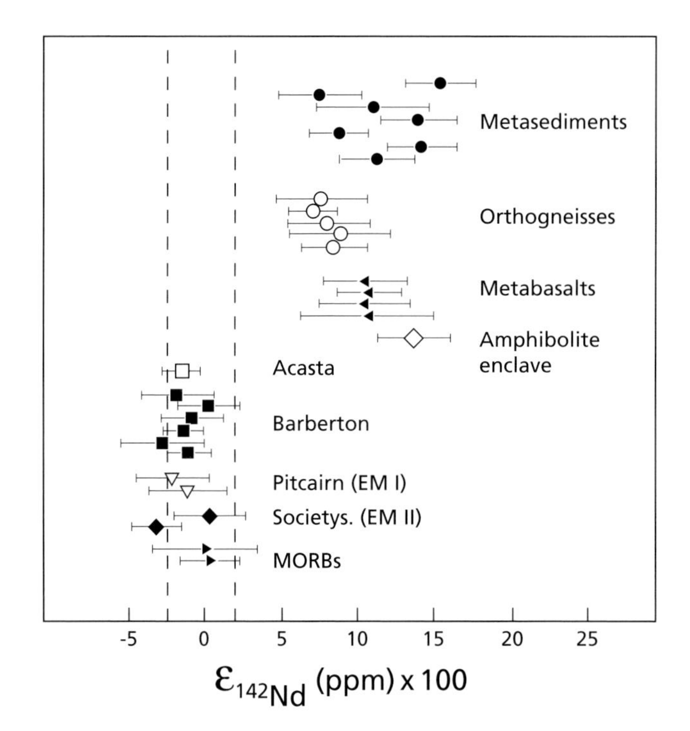

Figure 6.78 Measurements of ^{142}Nd anomaly in the Isua rocks of Greenland. After Caro *et al.* (2005).

repeated by the teams of **Bernard Bourdon** in the Paris laboratory (Caro *et al.*, 2003) yielding an excess of ^{142}Nd in the Isua rocks of Greenland dated 3.8 Ga, but of only +10 to +15 ppm (Figure 6.78). Notice that the precision reported in the measurement of isotopic compositions is the extreme limit of what current techniques can achieve.

Dating of the same rocks by the conventional ^{147}Sm–^{143}Nd method confirmed the age of 3.8 Ma and yielded the initial ^{143}Nd/^{144}Nd ratio expressed in $\varepsilon_{CHUR} = +2$. It is possible, then, to reason in terms of a two-stage model. The first stage involves meteoritic material from 4.576 Ga to T_1. Then, from T_1 to 3.8 Ga, the primitive crust (or the primitive residual mantle), T_1 being the age of terrestrial differentiation.

To do the calculation, we take 4.576 Ga as the origin, which is more convenient for extinct radioactivities. Let us call the two isotopic ratios of Nd α and β.

$$\alpha^{143}\mathrm{Nd}/^{144}\mathrm{Nd}, \beta =^{142}\mathrm{Nd}/^{144}\mathrm{Nd}.$$

We note ε_{143} in 10^{-4} and f_{142} in 10^{-6} as relative values against a standard:

$$\epsilon_{143_{\mathrm{ND}}} = \left(\frac{143/144\ \mathrm{Nd}_{\mathrm{sample}}}{143/144\ \mathrm{Nd}_{\mathrm{standard}}} - 1\right) \times 10^4$$

$$f_{143_{\mathrm{ND}}} = \left(\frac{143/144\ \mathrm{Nd}_{\mathrm{sample}}}{143/144\ \mathrm{Nd}_{\mathrm{standard}}} - 1\right) \times 10^6$$

$$\mu_0 = \frac{^{147}\mathrm{Sm}}{^{144}\mathrm{Nd}},$$

but is calculated at 4.576 Ga and not at the present time.

$$a = \frac{^{146}\mathrm{Sm}}{^{147}\mathrm{Sm}} \text{ at 4.576 Ga.}$$

The two evolution equations are written as multi-stage model with κ as the fractionation factor between the chondritic episode and the primitive Earth episode.

$$\alpha = \alpha_0 + \mu_0\left(1 - \mathrm{e}^{-\lambda_7 T_1}\right) + \kappa\mu_0\left(\mathrm{e}^{-\lambda_7 T_1} - \mathrm{e}^{-\lambda_7 T_2}\right)$$
$$\beta = \beta_0 + \mu_0\, a\left(1 - \mathrm{e}^{-\lambda_6 T_1}\right) + \kappa\mu_0^a\left(\mathrm{e}^{-\lambda_6 T_1} - \mathrm{e}^{-\lambda_6 T_2}\right),$$

α_0 and β being the initial isotopic ratios, λ_7 and λ_6 the two radioactive constants, and T_2 being the time 3.8 Ga calculated since 4.576, that is, $T_2 = 0.776$ Ga.

This non-linear system of two equations with two unknowns κ and T_1 can be solved graphically by constructing the two curves corresponding to the two equations with the values measured for α, α_0, β, β_0, a, and μ_0.

A few values are needed for the calculation.

$$\left(^{143}\mathrm{Nd}/^{144}\mathrm{Nd}\right)_{t=0} = 0.506\,677\,48$$

$$\mu_{\text{present day}} = \left(^{147}\mathrm{Sm}/^{144}\mathrm{Nd}\right)_{\text{Bulk Earth}} = 0.1966, \text{ or } \mu_0 = 0.201\,942 \text{ for 4.567 Ga}$$

$$\left(^{147}\mathrm{Sm}/^{144}\mathrm{Sm}\right)_{\text{present day}} = 4.888\,99, \text{ or } 5.037\,316\,6 \text{ for 4.567 Ga}$$

$$^{142}\mathrm{Nd}/^{144}\mathrm{Nd} = 1.141\,838\,2, \text{ or } 1.141\,478\,8 \text{ for 4.567 Ga}$$

$$\left(^{146}\mathrm{Sm}/^{144}\mathrm{Sm}\right)_0 = 0.008.$$

From this it can be deduced that $^{146}\mathrm{Sm}/^{147}\mathrm{Sm} = a = 0.001\,588\,178$ and therefore the parameter $\mu_0\, a = 0.000\,320\,719$.

The curves $\kappa = f(T_1)$ are therefore drawn for the pairs $^{147}\mathrm{Sm}-^{143}\mathrm{Nd}$ and $^{146}\mathrm{Sm}-^{142}\mathrm{Nd}$ taking two values for observations at 3.8 Ga.

$$\varepsilon_{143} = +1.5 \text{ and } +2\,\mathrm{f}_{142} = 15 \text{ and } 10 \text{ ppm}.$$

It can be seen that the two curves intersect between 4.46 and 4.35 Ga for (T_1) (Figure 6.79). This is about the age determined for the differentiation of the core and the atmosphere, with μ for the present day of 0.216.

The question is what does this age mean? The rocks analyzed by **Caro** *et al.* (2003) were mostly metamorphic and metasedimentary rocks. They interpreted the results as being the value of a primitive depleted mantle which was the complement of differentiation of the first continental crust. Given the nature of the rocks, it is not obvious that this was their origin. An alternative interpretation would be that these values are those of a lunar-type primitive crust, that is, one rich in plagioclase. The residual mantle (not found) would have a negative ε_{142} value. As things stand, both interpretations are possible. In addition Harrison *et al.* (2005) using hafnium isotopes in old zircons claim the existence of a very early crust. The precise age of this crust and its nature are still subjects of debate.

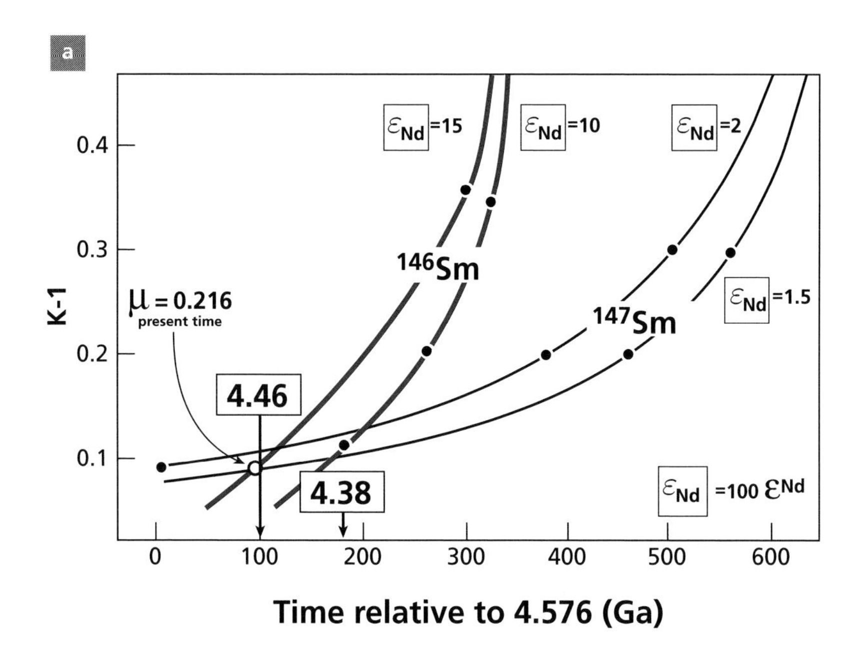

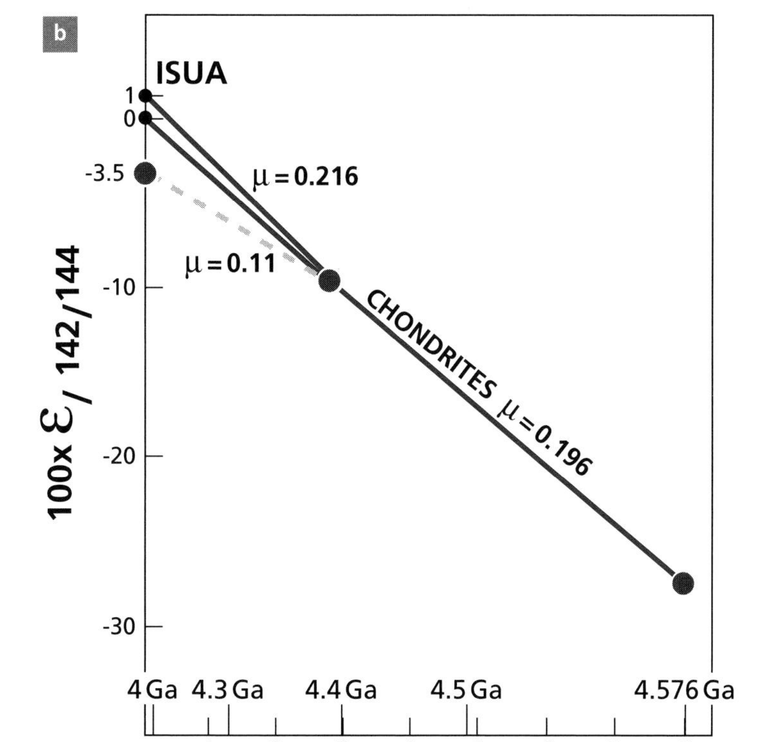

Figure 6.79 (a) Age determination of differentiation of the primitive crust using the two Sm–Nd systems. (b) Isua anomalies represented on the isochron diagram.

Remark

More recently **Boyet** and **Carlson** (2005) have asserted that ordinary chondrites have a f_{142} value of −20 ppm compared with the Earth. Unfortunately, there are few C1 and C2 carbonaceous chondrites in their measurement sets and the dispersion is quite wide. What are the values for Bulk Earth? It is best to wait a while, then, before incorporating these results into this textbook although readers need to know of their existence and should follow the development of this chapter.

6.7.4 The ^{187}Re–^{187}Os system and accretion–differentiation chronology

Examination of the evolution of the $^{187}Os/^{186}Os$ ratio of the Earth's mantle showed us that the Re/Os chemical ratio of the mantle was very close to the Re/Os ratio of the carbonaceous chondrites. Now, this is a surprising result given that both Re and Os are highly siderophile elements, that is, they enter liquid iron in preference to silicates.

Laboratory experiments for these elements give a partition coefficient $D = C_{Fe}/C_{silicate} \approx 10^4$–$10^5$. Most Re and Os therefore passes into the iron core, as they are in the iron phase of ordinary chondrites and in the metallic iron of iron meteorites. But Re and Os do not have exactly the same affinity for iron: Os is more siderophile than Re. We speak of a factor of 10 to 100 between them.

Let us look at this with a simple model. From the partition equation of trace elements:

$$C_{mantle} = C_0/F + D(1-F)$$

where F is the proportion of the mantle (between 0 and 1).

In the case of Re and Os, because D is very large, $F << D(1-F)$, therefore:

$$C_{mantle} = C_0/D(1-F).$$

Accordingly, the ratio between the two elements Re and Os is written:

$$\left(\frac{C^{Re}}{C^{Os}}\right)_{mantle} = \left(\frac{C^{Re}}{C^{Os}}\right)_0 \frac{D^{Os}}{D^{Re}}.$$

Therefore:

$$(Re/Os)_{mantle} > 10(Re/Os).$$

Now, observation shows that $(Re/Os)_{mantle} = (Re/Os)_{initial}$.

How can this apparent contradiction be accounted for? By admitting that the accretion process continued after differentiation of the core and so contributed enough Re and Os, incorporated in the mantle (probably by primary subduction processes) for them to dominate the Re and Os balance of the present-day mantle. Let us try to put this in somewhat more quantitative terms. Let us write a balance equation:

$$C^{Os}_{mantle} = \frac{C^{Os}_{meteorite}\, m}{M_{mantle}}$$

where C^{Os} is the concentration in osmium, M_{mantle} the mass of the mantle, and m the added mass. Now, $C^{Os}_{mantle} = 3$ ppb and $C^{Os}_{chondrites} = 490$ ppb. Hence $m/M = 0.6\%$. Less than 1% of the mass of the mantle was added after 4.4 Ga by the accretion processes. This indicates that accretion probably decreased very rapidly after 4 Ga, as the curve of lunar craters also indicates.

Exercise

(1) Given that $C^{Re}_{mantle} = 0.26$ ppb and $C^{Re}_{chondrites} = 37$ ppb, calculate the ratio (m/M).
(2) If we underestimate the abundance in the mantle by 100%, how is the result altered?

Answer
(1) 0.7%.
(2) The ratio changes to 1.4%, which does not alter the conclusion in qualitative terms.

If we attempt to make a (provisional) review of these studies of the primordial Earth, what can we say today? First of all, a methodological point we shall return to in quantitative terms in the final chapter: **extinct forms of radioactivity do not record primordial phenomena in the same way as long-period forms of radioactivity. The former are sensitive to short-term fluctuations, the latter on the contrary average variations out more over the long term**.

Together they give complementary views of primordial phenomena, each of which is insufficient in itself. This is the case for ^{40}Ar and ^{129}Xe for the atmosphere and for ^{206}Pb, ^{207}Pb, and ^{182}Hf for the core, and will be the case for any new form of extinct radioactivity that might be used in the future.

Accretion is a process that lasted for quite some time after 4.4 Ga. The formation of the core is a phenomenon that was prepared by differentiation of iron in planetesimals and itself occurred rather suddenly (perhaps in less than 100 or 1000 years) sometime between 4.4 and 4.35 Ga. The atmosphere was produced by outgassing of the mantle. It first formed very suddenly at 4.4 Ga and has been enriched throughout geological time through volcanic eruptions.

And the continents? The interpretation of the ^{142}Nd anomalies is still a matter of debate, but workers agree that if there was a primordial crust dating from the great differentiation of 4.4 Ga, that crust was destroyed and digested by the mantle, because we have no record of that ^{142}Nd in rocks as old as 3.4 Ga. The existence of detrital zircons dated 4.3 to 4.4 Ga seem to carry a similar message. No continental rocks. Only detrital minerals. All of this suggests that the primordial continental crust was ephemeral. Apart from this possible occurrence of a very primordial continental crust (perhaps a plagioclast crust like that on the Moon) the growth curves of continents in the geological and geochemical sense of the term is what we have established. All of this is rounded up in a summary figure. The PAT (Patterson) age is the mythical age of 4.55 Ga determined by Patterson (Figure 6.80) (see Chapter 5).

6.8 Conclusion

Like the remainder of this textbook, this chapter places more emphasis on methods of reasoning than on any would-be firm and final results. However, it exposes results that seem either to be robust or which probably form a first approximation of reality.

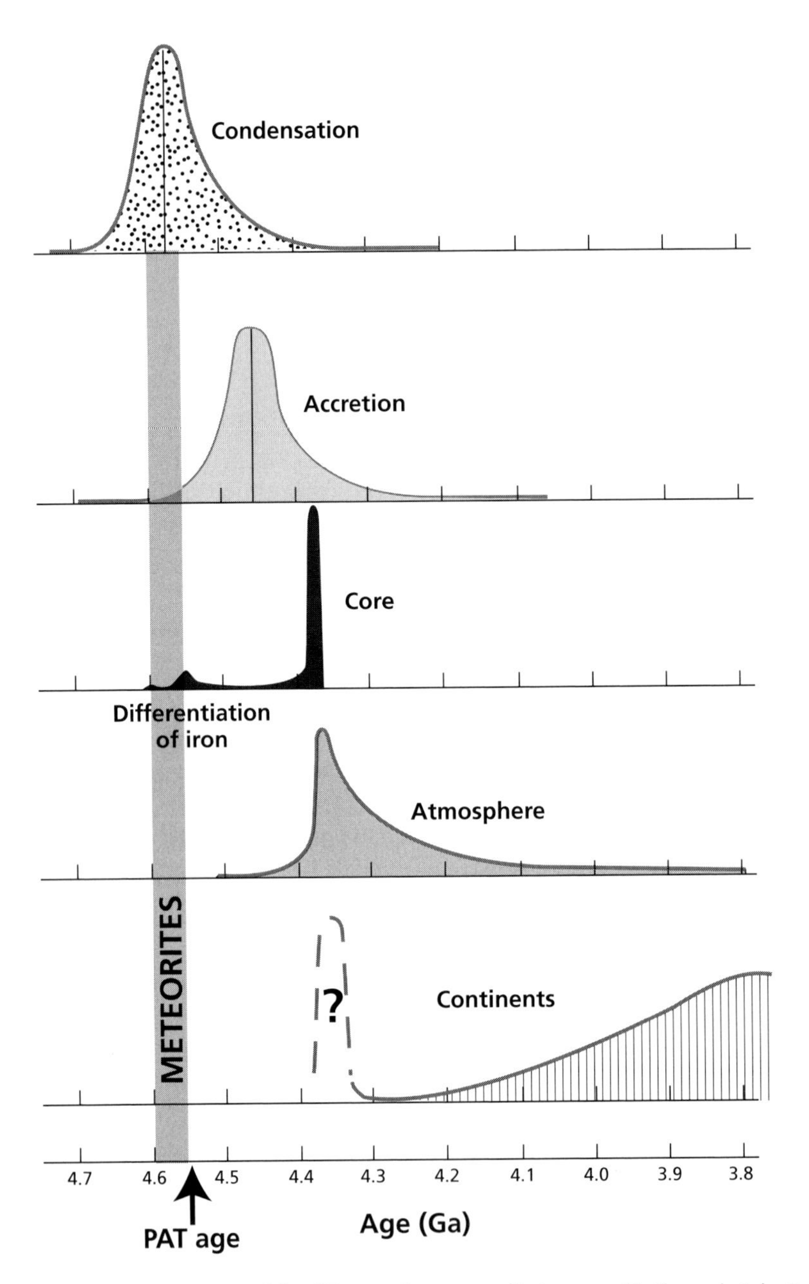

Figure 6.80 Timescale of the different phenomena that occurred in the early Solar System and the different phases of early formation and differentiation of the Earth from the isotope results. PAT age, Patterson age.

The general method developed in this chapter is based on the idea of using radiogenic isotopes as **tracers of phenomena**. Biologists inject radioactive tracers to understand the physiology of an animal body or a cell; hydrologists or chemists use coloring agents to monitor fluid movement. Like them, we too use tracers to study geological phenomena. This method has proved extraordinarily effective for deciphering the structure and physiology of the planet.

Geological tracers have an original feature compared with other tracers in that they are chronometers too. They move around with the motion of matter but keep a record of their history, which is what makes them so valuable. In addition, they can be used for exploring a wide time range: not only the 12 billion years of the history of the Universe and the 4.5 billion years of the Earth's history but also the details of planetogenesis some 4.5 billion years ago or recent phenomena of the last millennia.

A second original feature is that the isotopic variations explored here are tiny, ranging from 10^{-3} to 10^{-5}. In other words, **none of this would be feasible without the incredibly precise and sensitive technique of mass spectrometry**.

Lastly, and this is not the least advantage, we have a high number of isotopic tracers (more than 40). Thus extremely varied problems can be addressed with the same techniques and the same methods. There is no doubt that the future will see more detailed studies of the various fundamental phenomena (ocean ridges, subduction) and of the primordial Earth using the same methodology.

Problems

1 We consider the following geological history of a granite. Some 1500 Ma ago a volcanic–detrital sediment formed by mixing in the proportions of 1/3 and 2/3. The volcanic sediment has a $^{87}Sr/^{86}Sr$ ratio of 0.703, and the Sr and Rb concentrations are 300 ppm and 10 ppm, respectively. For the detrital sediment, the $^{87}Sr/^{86}Sr$ ratio $= 0.720$ and the Sr and Rb concentrations are 100 ppm and 100 ppm, respectively. The sediment is changed into rock and sinks progressively over 500 Ma. At that time, caught up in orogenic convulsions, it undergoes anatexis, which gives rise to a granite by partial melting. The Rb/Sr ratio of the melt is $(Rb/Sr)_{melt} = 3(Rb/Sr)_{sediment}$. The granite then evolves over 1 Ga.

(i) Show the isotope history of the granite on a graph.
(ii) Calculate the initial and present-day $^{87}Sr/^{86}Sr$ ratios of the granite.

2 In some cases, workers ascribe an important role to the metal core to modify classical conclusions about the mantle.

(i) We assume the core contains potassium in a concentration of about 100 ppm. How does that affect the ^{40}Ar balance as it has been envisaged? Calculate the mass of ^{40}Ar in the core and the lower mantle.
(ii) Core–mantle reactions are also evoked to explain the lead isotope compositions of OIB. Take the $^{238}U/^{204}Pb$ ratios for the current μ value of the Earth of 1; the ratio of the primordial mantle is 7 and that of the core is 0. Calculate the lead concentration in the core, given that the total lead concentration of the core–lower mantle system is $C_m^{Pb} = 0.696$ ppm and that in the lower mantle, after initial differentiation of the core, $C_m^{Pb} = 0.1611$.
(iii) Suppose that after initial differentiation, the core continues to deplete the lower mantle of lead. Schematically, let us accept that 20% of the lead disappears (increasing the μ_{mantle} value) and that the phenomenon can be modeled by a two-stage process involving an extraction episode followed by an episode with no extraction, the age of the change in regime being 3 Ga. Calculate the $^{206}Pb/^{204}Pb$ and $^{207}Pb/^{204}Pb$ ratios of the lower mantle. Locate them relative to the 4.5 Ga geochron.
(iv) Do the same calculation with $T_{diff} = 4$ Ga.
(v) Can you draw any geochemical conclusions from these calculations?

3 The Canadian **Dick Armstrong** (1981) proposed a new model of continent growth. This problem looks more closely at his idea. Armstrong supposed that all of the continental crust differentiated say 4.5 Ga ago (like the core and most of the atmosphere) but that since then, during each mountain-building episode, part of the continent is reinjected into the mantle (as sediment or as detachment from the lithosphere) and that symmetrically an identical piece of continent is formed at the expense of the mantle.
In this model, the μ of the mantle and continental crust remain constant. It is supposed there were four periods of exchange: at 3.5 Ga, 2.5 Ga, 1.5 Ga, and at the present day. It is assumed that at present a quarter of the continental crust is swallowed up and regenerated.

(i) Calculate and draw the two (α, t) curves of evolution for average continental crust and for the supposedly isolated upper mantle.
(ii) Calculate and draw the statistical age diagrams for $^{143}Nd/^{144}Nd$ represented in $\tau(t)$, $^{87}Sr/^{86}Sr$, $^{147}Sm/^{144}Nd = \mu^{Sm}$, $^{87}Rb/^{86}Sr = \mu^{Rb}$. The subscripts c, m, and 0 represent the continent, mantle, and Earth, respectively. We note $^{143}Nd/^{144}Nd = \alpha^{Nd}$, $^{87}Sr/^{86}Sr = \alpha^{Sr}$ at 4.4 Ga, the starting point of the calculation.

So:

$$\alpha_0^{Nd} = 0.506\,677;\ \alpha_0^{Sr} = 0.6989$$
$$\mu_c^{Sm} = 0.11;\ \mu_m^{Sm} = 0.25$$
$$\mu_c^{Rb} = 1;\ \mu_m^{Rb} = 0.05.$$

(iii) What do you conclude?

4 The residence time of oceanic lithosphere in the primitive upper mantle is ≈ 1 Ga. It can be assumed that when it goes through the mid-ocean ridge, the corresponding 70 km of mantle is entirely degassed. The ^4He in the upper mantle is the sum of two terms. The radiogenic part, formed *in situ* over 1 Ga, and the part from the lower mantle at the same time as ^3He during the same period. This second part will be ignored.

(i) Calculate the quantity of ^4He accumulated in 1 Ga in the upper mantle, knowing that $U = 5$ ppb and $Th/U = 2.5$.
(ii) Given that degassing of ^3He from the ocean ridges is $1 \cdot 10^3$ moles yr^{-1} and that $(^4He/^3He) = 10^5$, calculate the residence time of ^4He in the upper mantle. What do you conclude about the processes involved?

5 We consider the continental crust–depleted mantle system, as in Section 6.3.

(i) We wish to calculate the best $^{143}Nd/^{144}Nd$ and $^{147}Sm/^{144}Nd$ ratios for the continental crust $(\,)_{cc}$, primitive mantle $(\,)_{pm}$, and residual mantle $(\,)_{dm}$.

We note the isotope ratios α_{Nd} and the $^{147}Sm/^{144}Nd$ ratio $= \mu^{Sm/Nd}$.
We give:

$$\alpha_{pm}^{Nd} = 0.512\,62 \pm 0.000\,01$$
$$\alpha_{dm}^{Nd} = 0.513\,15 \pm 0.000\,01.$$

From which α_{cc}^{Nd} is between 0.511 and 0.5120.

$$\mu_{pm}^{Sm/Nd} = 0.197 \pm 0.01$$
$$\mu_{cc}^{Sm/Nd} = 0.11 \pm 0.01.$$

From which $\mu_{dm}^{Sm/Nd}$ is between 0.227 and 0.280.

Calculate the factor $W^{\mathrm{Nd}} = \dfrac{M_{\mathrm{CC}}\, C_{\mathrm{CC}}^{\mathrm{Nd}}}{m_{\mathrm{T}}\, C_{\mathrm{T}}^{\mathrm{Nd}}}$

where m_{cc} is the mass of the continental crust and m_{T} the mass of the primitive mantle, which differentiated by extraction of the continental crust.

Calculate the model age of differentiation of the continental crust.

(ii) We also give the values for the $^{87}\mathrm{Sr}/^{86}\mathrm{Sr}$ ratio noted α^{Sr} and $\mu^{\mathrm{Rb/Sr}}$ for $^{87}\mathrm{Rb}/^{86}\mathrm{Sr}$:

$$\alpha_{\mathrm{pm}}^{\mathrm{Sr}} = 0.7047, \quad \alpha_{\mathrm{dm}}^{\mathrm{Sr}} = 0.7020, \quad \alpha_{\mathrm{cc}}^{\mathrm{Sr}} = \text{variable}$$

$$\mu_{\mathrm{cc}}^{\mathrm{Sr}} = 0.3 - 0.5, \quad \mu_{\mathrm{pm}}^{\mathrm{Rb/Sr}} = 0.09, \quad \mu_{\mathrm{dm}}^{\mathrm{Rb/Sr}}\mu_{\mathrm{dm}}^{\mathrm{Rb/Sr}} = 0.$$

Given that $W^{\mathrm{Nd}}/W^{\mathrm{Sr}} = 1.5$, calculate W^{Sr}, $\alpha_{\mathrm{cc}}^{\mathrm{Sr}}$, the strontium model age, and $\mu_{\mathrm{cc}}^{\mathrm{Rb/Sr}}$.

CHAPTER SEVEN

Stable isotope geochemistry

When defining the properties of isotopes we invariably say that the isotopes of an element have the same chemical properties, because they have the same electron shell, but different physical properties, because they have different masses. However, if the behavior of isotopes of any chemical element is scrutinized very closely, small differences are noticeable: in the course of a chemical reaction as in the course of a physical process, isotope ratios vary and isotopic fractionation occurs. Such fractionation is very small, a few tenths or hundredths of 1%, and is only well marked for the light elements, let us say those whose atomic mass is less than 40. However, thanks to the extreme precision of modern measurement techniques, values can be measured for almost all of the chemical elements, even if they are extremely small for the heavy ones.

When we spoke of isotope geochemistry in the first part of this book, we voluntarily omitted such phenomena and concentrated on isotope variations related to radioactivity, which are preponderant. We now need to look into the subtle physical and chemical fractionation of stable isotopes, the use of which is extremely important in the earth sciences.

7.1 Identifying natural isotopic fractionation of light elements

The systematic study of the isotopic composition of light elements in the various naturally occurring compounds brings out variations which seem to comply with a purely naturalistic logic. These variations in isotope composition are extremely slight, and are generally expressed in a specific unit, **the δ unit**.

$$\delta = \left(\frac{\text{sample isotope ratio} - \text{standard isotope ratio}}{\text{standard isotope ratio}}\right) \times 10^3.$$

Ultimately, δ is a relative deviation from a standard, expressed as the number of parts per mil (‰). Isotope ratios are expressed with the heavier isotope in the numerator.

If δ is positive then the sample is richer in the heavy isotope than the standard. If δ is negative then the sample is poorer in the heavy isotope than the standard. The terms "rich" and "poor" are understood as relative to the isotope in the numerator of the isotope ratio in the formula above: by convention it is always the heavy isotope. Thus we speak of the $^{18}O/^{16}O$, D/H, $^{13}C/^{12}C$ ratio, etc. The standard is chosen for convenience and may be naturally abundant such as sea water for $^{18}O/^{16}O$ and D/H, a given carbonate for $^{13}C/^{12}C$, or even a commercial chemical (Craig, 1965).

Exercise

Oxygen has three stable isotopes, ^{16}O, ^{17}O, and ^{18}O, with average abundances of 99.756%, 0.039%, and 0.205%, respectively. The $^{16}O/^{18}O$ ratio in a Jurassic limestone is 472.4335. In average sea water, this same ratio is $^{16}O/^{18}O = 486.594$. If average sea water is taken as the standard, what is the δ of the limestone in question?

Answer

By convention, δ is always expressed relative to the heavy isotope. We must therefore invert the ratios stated in the question, giving 0.002 116 7 and 0.002 055 1, respectively. Applying the formula defining $\delta^{18}O$ gives $\delta^{18}O = +30$.

Exercise

The four naturally occurring, stable isotopes of sulfur are ^{32}S, ^{33}S, ^{34}S, and ^{36}S. Their average abundances are 95.02%, 0.75%, 4.21%, and 0.017%, respectively. Generally, we are interested in the ratio of the two most abundant isotopes, ^{34}S and ^{32}S. The standard for sulfur is the sulfide of the famous Canyon Diablo meteorite[1] with a $^{32}S/^{34}S$ value of 22.22. We express δ relative to the heavy isotope, therefore:

$$\delta = \left(\frac{(^{34}S/^{32}S)_{\text{sample}}}{(^{34}S/^{32}S)_{\text{standard}}} - 1 \right) \times 10^3.$$

If we have a sample of sulfur from a natural sulfide, for example, with $^{32}S/^{34}S = 23.20$, what is its $\delta^{34}S$?

Answer

Given that the standard has a $^{34}S/^{32}S$ ratio of 0.0450 and the sample a ratio of 0.0431, $\delta^{34}S = -42.22$. Notice here that the sign is negative, which is important. By definition, the standard has a value $\delta = 0$.

7.1.1 The double-collection mass spectrometer

Variations in the isotope composition of light elements are small, even very small. A precise instrument is required to detect them (and a fast one, if we want enough results to represent natural situations). We have already seen the principle of how a mass spectrometer works. Remember that in a scanning spectrometer, the magnetic field is varied and the ion beams corresponding to the different masses (or different isotopes) are picked up in turn in a collector. The collector picks up the ions and provides an electric current which is fed through a resistor to give a voltage read-out.

As we have already said, in multicollector mass spectrometers, the collectors are fixed and the beams of the various isotopes are received simultaneously. In this way we get around

[1] Canyon Diablo is the meteorite that dug Meteor Crater in the Arizona desert.

the temporary fluctuations that may occur during ionization. However, the recording circuits for the various collectors must be identical.

Since 1948, the double-collection mass spectrometer invented by **Nier** has been used for measuring slight isotopic differences for elements which can be measured in the gaseous state and which are ionized by electron bombardment (Nier, 1947; Nier *et al.*, 1947).[2] The two electrical currents, picked up by two Faraday cups, are computed using a Wheatstone bridge arrangement, which we balance (we measure the resistance values required to balance the bridge). The ratio of electrical currents $I_{a/b}$ is therefore directly related to the isotope ratio $R_{a/b}$ by the equation:

$$I_{a/b} = K R_{a/b}$$

where K is a fractionation factor and reflects bias that may occur during measurement. It is evaluated with an instantaneous calibration system using a standard. The standard sample is measured immediately after the unknown sample x. This gives:

$$I_s = K R_s.$$

Eliminating K from the two equations gives:

$$\frac{I_x}{I_s} = \frac{R_x}{R_s}.$$

The measurement of the relative deviation is then introduced quite naturally:

$$\Delta_x = \frac{R_x - R_s}{R_s} = \left(\frac{R_x}{R_s} - 1\right)\left(\frac{I_x}{I_s} - 1\right).$$

As we are handling small numbers, this number is multiplied by 1000 for the sake of convenience. This is where the definition of the $\boldsymbol{\delta}$ unit comes from, which is therefore provided directly by the mass spectrometer measurement, since $\boldsymbol{\delta} = \Delta_x \cdot 10^3$.

This gas-source, double-collection mass spectrometer automatically corrects two types of effect. First, it eliminates time fluctuations which mean that when we "scan" by varying the magnetic field (see Chapter 1), the emission at time t when isotope 1 is recorded may be different from emission at time $(t + \Delta t)$ when isotope 2 is recorded. Second, it corrects errors generated by the appliance by the sample–standard switching technique.

The measurement sequence is straightforward: sample measurement, standard measurement, sample measurement, etc. The operation is repeated several times to ensure measurement reproducibility. Fortunately, many light elements can enter gas compounds. This is the case of hydrogen in the form H_2 (or H_2O), of carbon and oxygen as CO_2, of sulfur (SO_2) or (SF_6), of nitrogen (N_2), of chlorine (Cl_2), and so on. For other elements such as boron, lithium, magnesium, calcium, and iron, it was not until advances were made in solid-source mass spectrometry or the emergence of inductively coupled plasma mass spectrometry (ICPMS), originally developed for radiogenic isotope studies, that an

[2] Multicollector mass spectrometers for thermo-ionization or plasma sources have been routinely used only since the year 2000 because of electronic calibration difficulties.

effective multicollection technique could be used. This domain is booming today and we shall touch upon it at the end of this chapter.[3]

7.1.2 Some isotope variations and identifying coherence

Oxygen

This is the most abundant chemical element on Earth, not only in the ocean but also in the silicate Earth (Figure 7.1). Its isotope composition varies clearly, which is a godsend!

Oxygen has three isotopes: ^{18}O, ^{17}O, and ^{16}O (the most abundant). We generally study variations in the $^{18}O/^{16}O$ ratio expressed, of course, in δ units, taking ordinary sea water as the benchmark (with $\delta = 0$ by definition).[4] Systematic measurement of various naturally occurring compounds (molecules, minerals, rocks, water vapor, etc.) reveals that they have characteristic isotope compositions that are peculiar to their chemical natures and their geochemical origins, whatever their geological ages or their geographical origins. For igneous or metamorphic silicate rocks δ is positive, ranging from +5 to +13. Such rocks are therefore enriched in ^{18}O (relative to sea water). Limestones are even more enriched since their δ values vary from +25 to +34. Of course, we may ask what "offsets" such enrichment in ^{18}O.

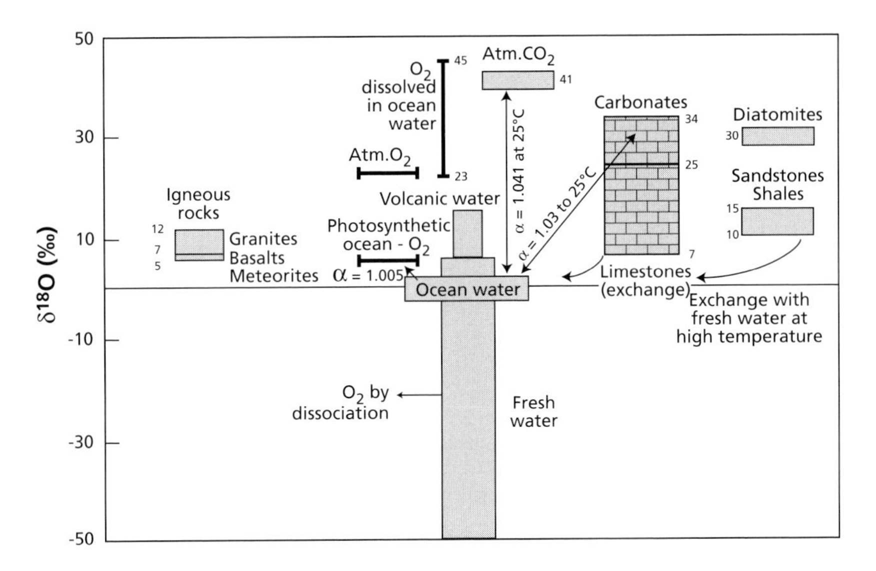

Figure 7.1 Distribution of oxygen isotope compositions in the main terrestrial reservoirs expressed in $\delta^{18}O$. The isotope fractionation factors are shown for various important reservoirs. The smaller numbers indicate extreme values. Values are of $\delta^{18}O$ expressed relative to standard mean ocean water (SMOW). After Craig and Boato (1955).

[3] The technique of alternating sample and standard used with electron bombardment of gas sources is difficult to implement whether with sources working by thermo-ionic emission or by ICPMS because of the possible memory effects or cross-contamination.

[4] It is called standard mean ocean water (SMOW).

Which compounds have negative δ values? We observe that those of fresh water are negative, ranging from −10 to −50. A few useful but merely empirical observations can be inferred from this. As we know that limestones precipitate from sea water, enrichment in ^{18}O suggests that limestone precipitates with enrichment in the heavy isotope. Conversely, we know that fresh water comes from evaporation and then condensation of a universal source, the ocean. It can therefore be deduced that there is depletion in ^{18}O during the hydrological cycle (evaporation–condensation). These observations suggest there is a connection between certain natural phenomena, their physical and chemical mechanisms, the origin of the products, and isotope fractionation.

Hydrogen

Let us now look at the natural isotopic variations of hydrogen, that is, variations in the (D/H) ratio (D is the symbol for deuterium). Taking mean ocean water as the standard, it is observed that organic products, trees, petroleum, etc. and rocks are enriched in deuterium whereas fresh water contains less of it (Figure 7.2).

We find similar behavior to that observed for oxygen, namely depletion of the heavy isotope in fresh water and enrichment in rocks and organic products. The product in which hydrogen and oxygen are associated is water (H_2O). It is important therefore to know whether the variations observed for D/H and $^{18}O/^{16}O$ in natural water are "coherent" or not. Coherence in geochemistry is first reflected by correlation. **Epstein** and **Mayeda** (1953) from Chicago and then **Harmon Craig** (1961) of the Scripps Institution of Oceanography at the University of California observed excellent correlation for rainwater between D/H and $^{18}O/^{16}O$, which shows that there is "coherence" in isotopic fractionation related to the water cycle (Figure 7.3). This invites us therefore to look more closely at any quantitative relations between isotope fractionation and the major natural phenomena.

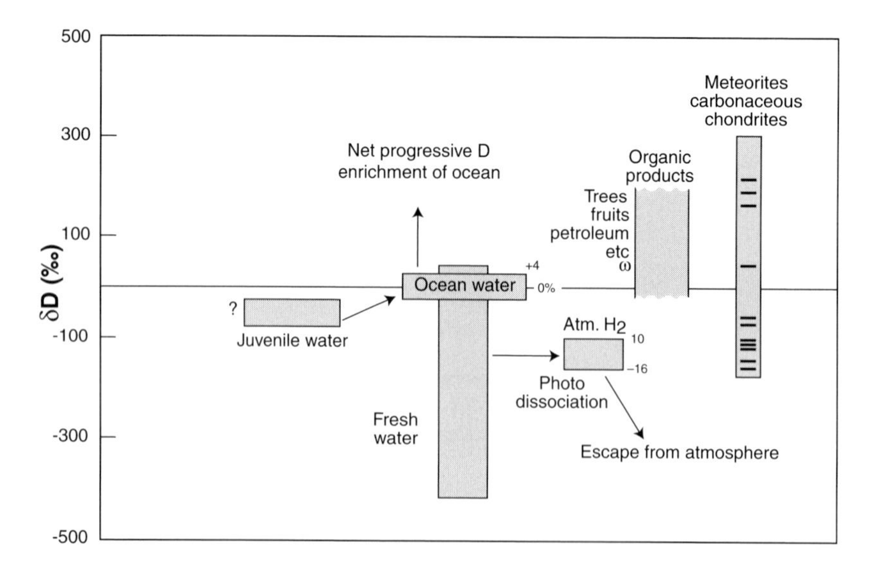

Figure 7.2 Distribution of isotope compositions of hydrogen expressed in δD in the main terrestrial reservoirs. After Craig and Boato (1955).

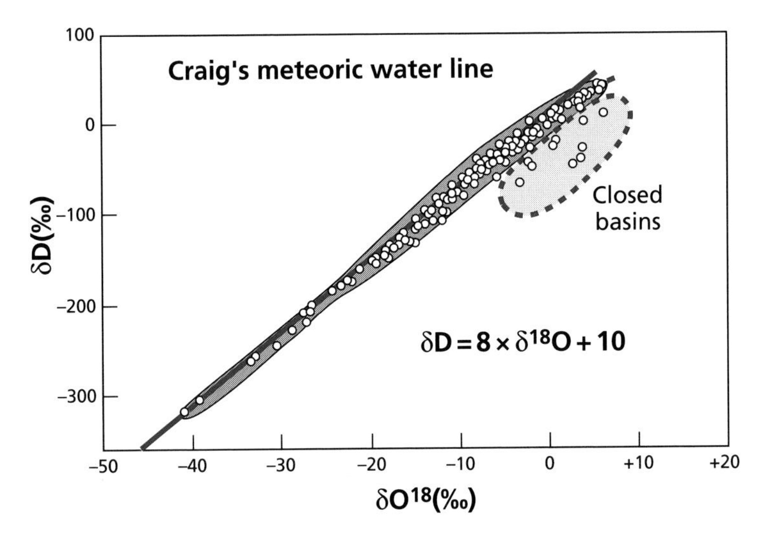

Figure 7.3 Correlation between (D/H, $^{18}O/^{16}O$) of rainwater. After Craig (1961).

7.1.3 Characterization of isotope variations

Between two geological products A and B, related by a natural process, and whose isotope ratios are notated R_A and R_B, we can write:

$$\theta_{AB} = \frac{R_A}{R_B}$$

where θ_{AB} is the overall fractionation factor between A and B. With $\boldsymbol{\delta}_A$ and $\boldsymbol{\delta}_B$ being defined as previously, we can write:

$$\theta_{AB} = \frac{1 + \frac{\delta_A}{1000}}{1 + \frac{\delta_B}{1000}} \approx 1 + \frac{(\boldsymbol{\delta}_A - \boldsymbol{\delta}_B)}{1000}$$

following the approximation $(1 + \varepsilon)/(1 + \varepsilon') \approx 1 + (\varepsilon - \varepsilon')$.

We note $\Delta_{AB} = \boldsymbol{\delta}_A - \boldsymbol{\delta}_B$. This yields a fundamental formula for all stable isotope geochemistry:

$$1000(\theta_{AB} - 1) \approx \Delta_{AB}.$$

Exercise

Given that the $\delta^{18}O$ value of a limestone is +24 and that the limestone formed by precipitation from sea water, calculate the overall limestone–sea water fractionation factor θ.

Answer

$\Delta_{Lim-H_2O} = \boldsymbol{\delta}_{Ca} - \boldsymbol{\delta}_{H_2O} = 24 - 0$. We deduce that $\theta = 1.024$.

It is possible, then, to calculate the overall fractionation factors for various geological processes: the transition from granite to clay by weathering, the evaporation of water between ocean and clouds, the exchange of CO_2 in the atmosphere with that dissolved in the ocean or with carbon of plants, and so on.

This is a descriptive approach, not an explanatory one. Various chemical reactions and physical processes have been studied in the laboratory to determine the variations in their associated isotope compositions. Thus, for instance, it has been observed that when water evaporates, the vapor is enriched in light isotopes for both hydrogen and oxygen. Fractionation factors have been defined for each process from careful measurements made in the laboratory. These elementary fractionation factors will be denoted α.

Geochemists have endeavored to synthesize these two types of information, that is, **to connect θ and α**, in other words, to break down natural phenomena into a series of elementary physical and chemical processes whose isotope fractionations are measured experimentally. This approach involves making models of natural processes. We then calculate θ from measurements of α made in the laboratory. When the agreement between θ so calculated and θ observed in nature is "good," the model proposed can be considered a "satisfactory" image of reality. Thus, while the study of the isotopic compositions of natural compounds is interesting in itself, it also provides insight into the underlying mechanisms of natural phenomena. Hence the role of tracers of physical–chemical mechanisms in geological processes that are associated with studies of light-isotope fractionation.

In attempting to expose matters logically, we shall not trace its historical development. We shall endeavor first to present isotope fractionation associated with various types of physical and chemical phenomena and then to look at some examples of natural isotope fractionation.

7.2 Modes of isotope fractionation

7.2.1 Equilibrium fractionation

As a consequence of elements having several isotopes, combinations between chemical elements, that is molecules and crystals, have many isotopic varieties. Let us take the molecule H_2O by way of illustration. There are different isotopic varieties: $H_2{}^{18}O$, $H_2{}^{17}O$, $H_2{}^{16}O$, $D_2{}^{18}O$, $D_2{}^{17}O$, $D_2{}^{16}O$, $DH^{18}O$, $DH^{17}O$, $DH^{16}O$ (omitting combinations with tritium, T). These different molecules are known as **isotopologs**. Of these, $H_2{}^{16}O$ accounts for 97%, $H_2{}^{18}O$ for 2.2%, $H_2{}^{17}O$ for about 0.5%, and $DH^{16}O$ for about 0.3%. When the molecule H_2O is involved in a chemical process, all of its varieties contribute and we should write the various equilibrium equations not just for H_2O alone but for all the corresponding isotopic molecules.

Chemical equilibria

Let us consider, for example, the reaction

$$Si^{18}O_2 + 2H_2^{16}O \rightleftarrows Si^{16}O_2 + 2H^{18}O,$$

which corresponds to a mass action law:

$$\frac{(H_2{}^{18}O)^2(Si^{16}O_2)}{(H_2{}^{16}O)^2(Si^{18}O_2)} = K(T).$$

Harold Urey (1947), and independently **Bigeleisen** and **Mayer** (1947), showed using statistical quantum mechanics that this kind of equilibrium constant, although close to 1, is different from 1.

More generally, for an isotope exchange reaction $aA_1 + bB_2 \rightleftarrows aA_2 + bB_1$, where B and A are compounds and the subscripts 1 and 2 indicate the existence of two isotopes of an element common to both compounds, we can write in statistical thermodynamics, following Urey (1947) and Bigeleisen and Mayer (1947):

$$K = \left[\frac{Q(A_2)}{Q(A_1)}\right]^a \cdot \left[\frac{Q(B_1)}{Q(B_2)}\right]^b.$$

Functions Q are termed partition functions of the molecule and are such that for a given single chemical species we can write:

$$\frac{Q_2}{Q_1} = \frac{\sigma_1}{\sigma_2}\left(\frac{M_2}{M_1}\right)^{3/2} \frac{\sum \exp\left(\frac{-E_{2i}}{kT}\right)}{\sum \exp\left(\frac{-E_{1i}}{kT}\right)} \cdot \frac{I_1}{I_2}.$$

In this equation σ_1 and σ_2 are the symmetry numbers of molecules 1 and 2, E_{2i} and E_{1i} are the different rotational or vibrational energy levels of the molecules, M_1 and M_2 are their masses, and I_1 and I_2 are their moments of inertia.

The greater the ratio M_1/M_2 the greater the fractionation between isotope species, all else being equal. It can also be shown that ln K, as for any equilibrium constant, can be put in the form $a' + b'/T + c'/T^2$, which induces the principle of the isotopic thermometer. It can be deduced from the formula that as T increases K tends towards 1. At very high temperatures, isotope fractionation tends to become zero and at low temperature it is much greater.[5] If we define the isotope fractionation factor α associated with a process by the ratio $(A_2/A_1)/(B_2/B_1) = \alpha_{AB}$, α and K are related by the equation $\alpha = K^{1/n}$, where n is the number of exchangeable atoms. Thus, in the previous example, $n = 2$ as there are two oxygen atoms to be exchanged, but usually $\alpha = K$.

Let us now write the fractionation factor α_{AB} in δ notation, noting each isotope ratio R_A and R_B:

$$\delta_A = \left(\frac{R_A}{R_S} - 1\right)10^3 \qquad \delta_B = \left(\frac{R_B}{R_S} - 1\right)10^3,$$

R_S being the standard.

$$\alpha = \left(\frac{1 + \frac{\delta_A}{1000}}{1 + \frac{\delta_B}{1000}}\right) \approx 1 + \frac{(\delta_A - \delta_B)}{1000},$$

since δ_A and δ_B are small.

[5] Remember that isotope geology studies phenomena from −80 °C (polar ice caps) to 1500 °C (magmas) and in the cosmic domain the differences are even higher.

We come back to the equation $(\alpha_{AB} - 1)\,1000 = \boldsymbol{\delta}_A - \boldsymbol{\delta}_B = \Delta_{AB}$, which we met for the factor θ.

Exercise

We measure the $\boldsymbol{\delta}\,^{18}O$ of calcite and water with which we have tried to establish equilibrium. We find $\boldsymbol{\delta}_{cal} = 18.9$ and $\boldsymbol{\delta}_{H_2O} = -5$. What is the calcite–water partition coefficient at 50 °C?

Calculate it without and with the approximation $(1 + \boldsymbol{\delta}_1)/(1 + \boldsymbol{\delta}_2) \approx 1 + (\boldsymbol{\delta}_1 - \boldsymbol{\delta}_2)$.

Answer

(1) Without approximation: $\alpha_{cal-H_2O} = 1.024\,02$.

(2) With approximation: $\alpha_{cal-H_2O} = 1.0239$.

Physical equilibria

Such equilibrium fractionation is not reserved for the sole case where chemical species are different, but also applies when a phase change is observed, for instance. The partial pressure of a gas is $Pg = P_{total} \cdot Xg$, where Xg is the molar fraction. Moreover, the gas–liquid equilibrium obeys Henry's law. Thus, when water evaporates, the vapor is enriched in the light isotope. If the mixture $H_2{}^{18}O$ and $H_2{}^{16}O$ is considered perfect, and if the water vapor is a perfect gas, we can write:

$$P(H_2{}^{16}O) = X^{e}_{H_2{}^{16}O} \cdot P^0(H_2{}^{16}O)$$

$$P(H_2{}^{18}O) = X^{e}_{H_2{}^{18}O} \cdot P^0(H_2{}^{18}O)$$

where P is the total pressure, X designates the molar fractions in the liquid, and $P^0(H_2O)$ the saturated vapor pressure. Then (prove it as an exercise):

$$\alpha(\text{vapor}-\text{liquid}) = \frac{P^0(H_2{}^{18}O)}{P^0(H_2{}^{16}O)},$$

the denser liquid being the less volatile $P^0(H_2{}^{18}O) < P^0(H_2{}^{16}O)$ and $\alpha < 1$. Like all fractionation factors, α is dependent on temperature. Using Clapeyron's equation, it can be shown that $\ln\alpha$ can be written in the form $\ln\ \alpha = (a/T) + b$. For water at 20 °C (this is the vapor–liquid coefficient, not the opposite!), $\alpha_{^{18}O} = 0.991$ and $\alpha_D = 0.918$. At 20 °C fractionation is therefore about eight times greater for deuterium than for ^{18}O. (Remember this factor of 8 for later.)

Exercise

What is the law of variation of α with temperature in a process of gas–liquid phase change? We are given that $\alpha = P^0(X_1)/P^0(X_2)$, where X_1 and X_2 are the two isotopes.

Answer

Let us begin from Clapeyron's equation:

$$\frac{dP}{dT} = \frac{L_{vapor}}{TV_{vapor}}$$

where T is the temperature, V the volume, and L_{vapor} the latent heat of vaporization.

$$\frac{1}{P}\frac{dP}{dT} = \frac{L}{TVP}.$$

Since $PV = nRT$ (Mariotte's law):

$$\frac{1}{P}\frac{dP}{dT} = \frac{L}{RT^2} \text{ hence } \frac{dP}{P} = \frac{L}{RT^2}dT.$$

Integrating both terms gives $\ln P = \frac{L}{RT} + C$.

Since $\alpha = P^0(X_1)/P^0(X_2)$, we have:

$$\ln \alpha = \ln P^0(X_1) - \ln P^0(X_2) = \frac{L_{X_1} - L_{X_2}}{RT} + C.$$

Exercise

The liquid–vapor isotope fractionation is measured for oxygen and hydrogen of water at three temperatures (see table below):

Temperature (°C)	α_D	$\alpha_{^{18}O}$
+20	1.0850	1.0098
0	1.1123	1.0117
−20	1.1492	1.0141

(1) Draw the curve of variation of α with temperature in (α, T), [ln(α), $1/T$], and [ln(α), $1/T^2$].
(2) What is the δ value of water vapor in deuterium and ^{18}O at 20 °C and at 0 °C, given that water has $\delta = 0$ for (H) and (O)?
(3) Let us imagine a simple process whereby water evaporates at 20 °C in the temperate zone and then precipitates anew at 0 °C. What is the slope of the precipitation diagram (δ D, $\delta\,^{18}$O)?

Answer

(1) The answer is left for readers to find (it will be given in the main text).
(2) At +20 °C, δ D = −85 and $\delta\,^{18}$O = −9.8, and at 0 °C, δ D = −112.3 and $\delta\,^{18}$O = −11.7.
(3) The slope is 14.3. In nature it is 8, proving that we need to refine the model somewhat (the liquids have as starting values at 20 °C, δ D = 0 and $\delta\,^{18}$O = 0 and at 0 °C, δ D = −27.3 and $\delta\,^{18}$O = −1.9).

7.2.2 Kinetic fractionation

For a general account of kinetic fractionation see **Bigeleisen** (1965).

Transport phenomena

During transport, as isotopic species have different masses, they move at different speeds. The fastest isotopes are the lightest ones. Isotopic fractionation may result from these differences in speed. Suppose we have molecules or atoms with the same kinetic energy $E=\frac{1}{2}mv^2$. For two isotopic molecules 1 and 2 of masses m_1 and m_2, we can write v_1, v_2 being the velocities:

$$\frac{v_1}{v_2} = \left(\frac{m_2}{m_1}\right)^{1/2}.$$

The ratio of the speed of two "isotopic molecules" is proportional to the square root of the inverse ratio of their mass. This law corresponds, for example, to the isotopic fractionation that occurs during gaseous diffusion for which the fractionation factor between two isotopes of ^{16}O and ^{18}O for the molecule O_2 is written:

$$\alpha = \left(\frac{32}{34}\right)^{1/2} = 1.030.$$

Note in passing that such fractionation is of the same order as the fractionation we encountered during equilibrium processes! Such fractionation is commonplace during physical transport phenomena. For example, when water evaporates, vapor is enriched in molecules containing light isotopes (H rather than D, ^{16}O rather than ^{18}O). In the temperate zone ($T = 20\ ^\circ C$), for water vapor over the ocean $\delta^{18}O = -13$, whereas for vapor in equilibrium the value is closer to $\delta^{18}O = -9$, as seen.

Chemical reactions

Isotopically different molecules react chemically at different rates. Generally, the lighter molecules react more quickly. Lighter molecules are therefore at a kinetic advantage. This is due to two combined causes. First, as we have just seen, light molecules move faster than heavy molecules. Therefore light molecules will collide more. Second, heavy molecules are more stable than light ones. During collisions, they will be dissociated less often and will be less chemically reactive. The details of the mechanisms are more complex. During a chemical reaction, there is a variation in isotopic composition between the initial product and the end product. Let us consider, for example, the reaction:

$$C + O_2 \rightarrow CO_2.$$

In terms of oxygen isotopes, there are two main reactions:

$$C +^{16} O^{18}O \rightarrow C^{16}O^{18}O$$

$$C +^{16} O^{16}O \rightarrow C^{16}O_2.$$

Remark

The other possible reactions are not important. The reaction $C + ^{18}O^{16}O \rightarrow C^{16}O^{18}O$ is identical to the first in terms of its result. The reaction $C + ^{18}O^{18}O \rightarrow C^{18}O_2$ yields a molecule of very low abundance as ^{18}O is much rarer than ^{16}O.

These two reactions occur at different speeds, with two kinetic constants, K_{18} and K_{16}. Let us note the initial concentrations of the product containing the isotopes 18 and 16 as U_{18} and

U_{16}, giving $^{16}O^{16}O$, and note as Y_{18} and Y_{16} the concentrations of $C^{18}O^{16}O$ and $C^{16}O_2$. We can write:

$$-\frac{dU_{18}}{dt} = K_{18} \qquad U_{18} = \frac{dY_{18}}{dt}$$

and

$$-\frac{dU_{16}}{dt} = K_{16} \qquad U_{16} = \frac{dY_{16}}{dt}.$$

If the concentration of initial products is kept constant

$$\frac{Y_{18}}{Y_{16}} = \frac{K_{18}\,U_{18}}{K_{16}\,U_{16}}.$$

Therefore

$$\left(\frac{^{18}O}{^{16}O}\right)_{CO_2} = \alpha\left(\frac{^{18}O}{^{16}O}\right)_{O_2},$$

or:

$$\alpha = \frac{K_{18}}{K_{16}}.$$

The isotopic fractionation factor is equal to the ratio of the kinetic constants for each isotope.

A fuller expression of this ratio may be obtained by statistical mechanics by using the fact that the kinetic process consists of two transitions, one towards the activated complex and the other towards the stable compound. Naturally, we usually have very few data on this activated complex which is very short-lived. Two reactions with two different isotopes (see Lasaga, 1997) are written:

$$A + BC \xrightarrow{K_1} AB + C \text{ and } A + BC' \xrightarrow{K_2} AB + C'.$$

It can be shown that

$$\frac{K_1}{K_2} = \frac{Q_{ABC'}\,Q_{BC}}{Q_{ABC}\,Q_{BC'}},$$

Q being partition functions corresponding to the activated complex and to the molecules. It should be possible to determine the parameters by spectrometry and so check the precision of this theory but in fact the problem is so complex that we are far from having resolved the theoretical approach and having determined the necessary spectroscopic parameters. But we do understand the general sense of the mechanisms, which is the most important thing. Experimental data are therefore used to model natural phenomena.

The temperature effect

During transport, isotopic fractionation is insensitive to temperature as it is in $(m_1/m_2)^{1/2}$. However, collisions and molecular recombinations are a function of energy and therefore of temperature and are theoretically activated. It is understandable, then, that isotopic fractionation varies with temperature during kinetic processes.

Roughly speaking, temperature should promote kinetic fractionation. Having made this simple observation, things become more complicated. Isotopic exchange, the process by which equilibrium is attained, is itself a kinetic process and is therefore activated by temperature, so much so that the increased fractionation because of kinetic effects is progressively cancelled because the equilibrium processes become dominant and therefore fractionation will diminish with the increase in temperature.

This double general process will thus lead to a law of kinetic fractionation represented by a bell-shaped curve: fractionation increasing with temperature at first, and then declining beyond a certain temperature. This rule is modulated by specific kinetic mechanisms. This is why, despite many attempts, we have never managed to give a general expression for kinetic isotopic fractionation based on statistical mechanics.

Biological effects

Many (if not all) biochemical reactions involve isotopic fractionation. A number of these fractionation phenomena have been studied in vitro and in vivo, elucidating the intimate mechanisms of certain important biochemical reactions. It is understandable, then, that some biological mechanisms, formed by the combination or the succession of biochemical reactions, produce isotopic effects some of which are particularly important in geochemistry and so deserve our attention. Let us discuss two of them: sulfate–sulfide reduction by *Desulfovibrio desulfuricans* bacteria and chlorophyll photosynthesis (Harrison and Thode, 1957, 1958).

Sulfate–sulfide reduction by *Desulfovibrio desulfuricans* bacteria The reaction for the reduction of sulfate to sulfide is written $SO_4^{2-} \Rightarrow S^{2-}$. It involves a big change in the degree of oxidation of sulfur (+6) to (−2), which is made possible at low temperature only by the intervention of the bacteria in question (conversely, the reaction $S^{2-} \rightarrow SO_4^{2-}$ is easy). This bacterial reduction goes along with isotopic fractionation favoring the light isotope of sulfur but whose amplitude is well below that of the sulfide ⇔ sulfate equilibrium process, governed by the mass action law ($\alpha = 1.025$ at 25 °C versus $\alpha = 1.075$ for the equilibrium process). This means the sulfate is enriched in the heavy isotope (^{34}S) when there is fractionation with the sulfide. This fractionation plays a role in nature and helps to fix the isotopic composition of low-temperature naturally occurring sulfides (see the end of this chapter).

Chlorophyll photosynthesis During this process atmospheric CO_2 is fixed and the reduced carbon is incorporated into organic molecules. An enrichment in ^{12}C compared with ^{13}C is observed. The $\delta^{13}C$ value of atmospheric CO_2 is −8‰. For carbonate sediments, $\delta^{13}C$ varies from +5 to −5‰. However, plants have $\delta^{13}C$ values ranging, depending on varieties, from −15 to −35‰. **Park** and **Epstein** (1960) of the California

Institute of Technology showed that an important step in ^{12}C enrichment occurred in the process of photosynthesis. They were even able to attribute partition coefficients to the different photosynthetic mechanisms (this is outside our field but is important in biochemistry).

In short, let us say that the biochemical effects are important. They are even fundamental in some instances in geochemistry for understanding a whole series of phenomena such as those related to the CO_2 cycle or the sulfur cycle. But need they be considered as specific effects of living organisms that are not bound by ordinary physical and chemical laws? Various studies have shown on the contrary that biological processes involving enzymes are in fact a series of chemical reactions. These reactions are associated with isotopic fractionation, generally of the kinetic type. There do not seem to be certain specific mechanisms (such as the spin effect) for biological reactions. These biological fractionations of isotopes have been discussed in detail by **Eric Galimov** (1985).

7.2.3 The effects of molecular symmetry: mass-independent fractionation

All the effects we have examined so far fractionate isotopes according to laws proportional to the difference in mass of the isotopes. Thus, in carbonate precipitation, $^{18}O/^{16}O$ fractionation is twice $^{17}O/^{16}O$ fractionation. In bacterial reduction of sulfate, $^{34}S/^{32}S$ fractionation is half $^{36}S/^{32}S$ fractionation. However, kinetic fractionation has been discovered where differences do not depend on the mass difference but on the symmetry of the molecule. Thus, $^{18}O/^{16}O$ and $^{17}O/^{16}O$ fractionation is the same. **Mark Thiemens** of the University of California at San Diego has referred to these phenomena explaining some fractionation observed by **Robert Clayton** in meteorites (Figure 7.4). He has proved the reality of this phenomenon in the laboratory (Thiemens and Heidenreich, 1983). These effects also occur in nature, for instance, with ozone (O_3) in the atmosphere and for sulfides in meteorites and also in Precambrian rocks. Although their theoretical explanation is complex,[6] it does seem that the decisive parameter in such fractionation is molecular symmetry.

In this sense, two molecules $^{16}O-^{18}O$ or $^{16}O-^{17}O$, both equally asymmetrical, should have similar degrees of fractionation. During the ozone-forming reaction in the high atmosphere (at an altitude of 50 km), which reaction is extremely important as ozone not only absorbs ultraviolet radiation and protects the Earth,

$$O + O_2 \rightarrow O_3^*$$

and then

$$O_3^* + M \rightarrow O_3 + M$$

in which O_3^* is the excited molecule, and M is the molecule with which O_3^* collides and becomes de-excited.

[6] This explanation was given by **Rudy Marcus**'s team at the California Institute of Technology chemistry department, but is quite complicated. See Gao and Marcus (2001) for an example.

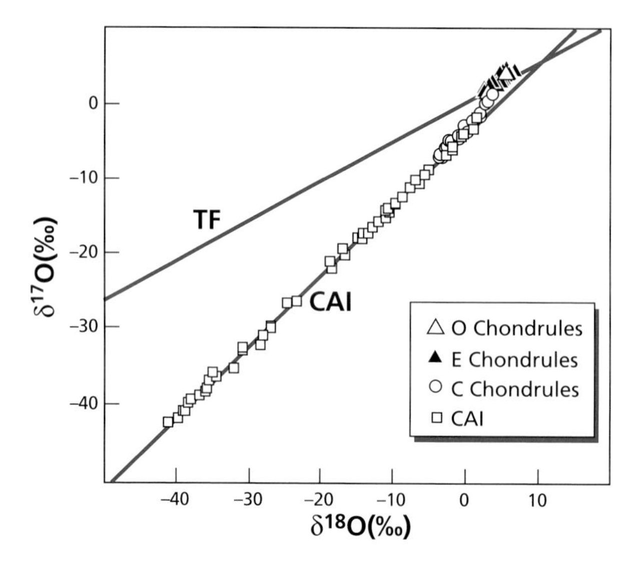

Figure 7.4 The $\delta^{17}O$, $\delta^{18}O$ relation in chondrules and refractory inclusions of various meteorites (CAI, calcium–aluminum inclusions). For these objects the correlation is of slope 1 whereas the usual terrestrial fractionation (TF) correlation observed is of slope $\frac{1}{2}$, in line with the mass difference between ^{17}O and ^{16}O and ^{18}O and ^{16}O. This discovery made by Robert Clayton *et al.* (1973) is interpreted by Thiemens (1999) as mass-independent fractionation, unlike Clayton who interpreted it as a nucleosynthetic effect, and later as a photochemical effect (Clayton, 2002).

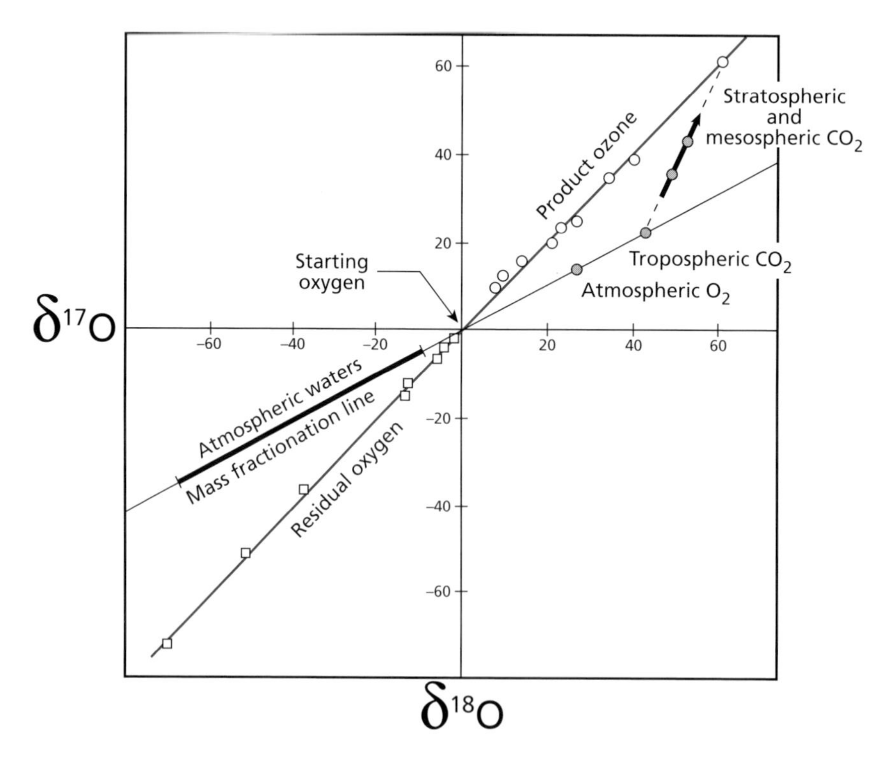

Figure 7.5 Mass-independent fractionation (MIF) for oxygen isotopes in atmospheric material compared with classical mass-dependent fractionation. The line of slope 1 is MIF; the line of slope $\frac{1}{2}$ is mass-dependent fractionation.

It has been shown that ozone of mass 54 ($^{18}O^{18}O^{18}O$) is not enriched relative to $^{16}O^{16}O^{16}O$ ozone of mass 48, whereas the asymmetrical molecule $^{16}O^{17}O^{18}O$ of mass 51 is enriched by 200%. It has also been shown that symmetrical ozone molecules $^{17}O^{17}O^{17}O$ or $^{18}O^{18}O^{18}O$ are depleted, whereas all the asymmetrical molecules $^{16}O^{17}O^{17}O$ or $^{17}O^{18}O^{18}O$, etc. are enriched. This effect, which is called **mass-independent fractionation** and might be more appropriately termed the molecular symmetry effect, seems to act with reactions such as $O + CO \rightarrow CO_2$, $O + SiO \rightarrow SiO_2$, etc.

This is an important process in the atmosphere and seems to have played a role in the presolar primitive nebula as a linear relation of slope 1 is found in carbonaceous meteorites between $\delta^{17}O$ and $\delta^{18}O$ (Figure 7.5). This is an important effect but highly specific to certain processes. It is just beginning to be exploited but already very successfully (see below).

7.3 The modalities of isotope fractionation

7.3.1 Kinetic effects or equilibrium effects? Isotopic exchange

We have already spoken of this in the earlier chapters. Let us recall a few facts here, as it is a very important but often neglected phenomenon. Let us bring into contact two chemical compounds, AO and BO, with at least one element in common, for example, both having oxygen in their formulas. One of these species has been prepared with ^{18}O exclusively, the other with ^{16}O. After a certain time in contact it can be seen that the $^{18}O/^{16}O$ composition of the two compounds is such that:

$$\frac{(^{18}O/^{16}O)_{AO}}{(^{18}O/^{16}O)_{BO}} = K(T)$$

where $K(T)$ is the equilibrium constant. In other words, the isotopes ^{18}O and ^{16}O have exchanged such that equilibrium has been attained. The rate of this isotope exchange can be measured and several phenomena observed:

(1) It is faster at higher temperatures.
(2) It is faster in gases or liquids than solids. If one of the compounds is a solid it becomes very slow (in this case the rate of diffusion in the solid limits the kinetics of the process).
(3) It depends largely on the position oxygen occupies in the steric configuration of compounds AO and BO,[7] that is, the nearer oxygen is to the outside of the molecular structure, the faster the kinetics[8] – this isotope exchange is essential in geochemistry as it provides understanding of various fundamental observations (Figure 7.6).

[7] Which relates to the spatial arrangement of the atoms composing the molecule.
[8] For example, in the complex ion SO_4, oxygen exchanges much faster than sulfur. This is why in sulfate water S retains the memory of its source but O does not.

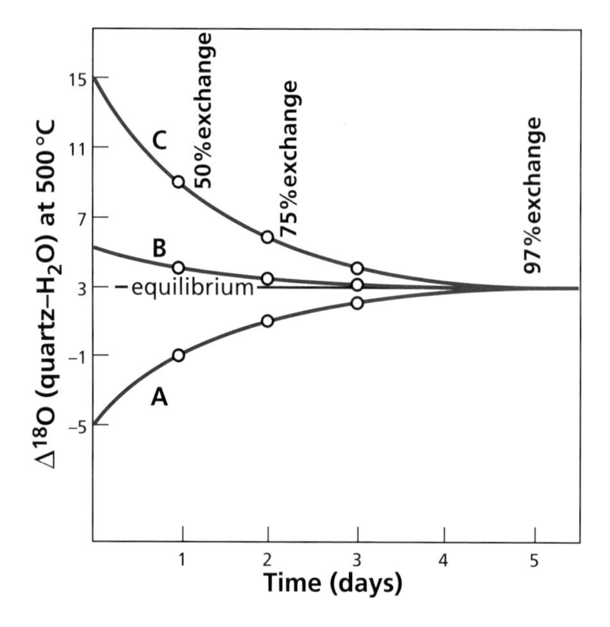

Figure 7.6 Kinetic curve showing the speed of equilibration by water–quartz exchange. The quartz has a $\delta^{18}O$ value of 10. Three types of water with different compositions are brought into contact with the quartz at 500 °C. The initial isotope compositions of the waters are expressed in δ: A (−5), B (+5), and C (+15). The equilibrium value is 3. It can be seen that the three equilibration curves converge towards the equilibrium value in a matter of days. After O'Neil (1986).

Let us suppose we have a reaction A → B together with kinetic isotope fractionation. If A and B are left in contact for long enough, the isotopes of A and B swap over, and eventually the fractionation between A and B is of the equilibrium fractionation type. To maintain kinetic fractionation, the initial product and the end product must not be left in contact. An example of this is the reduction of the sulfate ion SO_4^{2-} to the sulfide S^{2-} (by bacteria) which goes along with an out-of-equilibrium isotope effect. If, after partial reduction, the sulfate ion remains in contact with the sulfide ion, the system tends to establish sulfate–sulfide isotopic equilibrium. Conversely, if the sulfide ion S^{2-} is in the presence of a ferrous ion Fe^{2+}, the following reaction occurs: $2S^{2-} + Fe^{2+} \rightarrow FeS_2$. This iron sulfide crystallizes and "isolates" the sulfide from any further isotopic exchange which would cancel out the kinetic effect. This is why a number of naturally occurring sulfides have isotope compositions reflecting the kinetic effect (bacterial) related to sulfate reduction.

Isotope exchange is activated by temperature; therefore, at high temperatures, only swift and complete isolation of the resulting product can prevent equilibrium fractionation from taking over. In practice, except for the case of gases that escape and become isolated, such as gases from volcanoes, it is generally difficult to observe kinetic effects at high temperatures. In these circumstances, equilibrium effects are mostly preponderant.

7.3.2 A consequence: isotopic memory

As we have already said when discussing radiogenic isotopes, it is fundamental to understand that all isotope geochemistry, including that of stable light isotopes, is based on the fact that isotope exchange in the solid phase at low temperatures is very slow and the system is not constantly re-equilibrated, otherwise there would be no isotopic memory. This derives from the issues of diffusion covered previously.

Let us take the example of calcareous fossil shells. A shell records the $^{18}O/^{16}O$ isotope composition of the sea water it was formed in and also the ambient temperature. Once formed, the shell moves around with the animal that carries it and when the animal dies the shell falls to the sea floor. There it is incorporated into sediments and with them will be petrified in a certain proportion and possibly, much later, will be brought to the surface on the continents by tectonic processes. It will remain there for millions of years before a geologist comes along and collects it for analysis. During this time, the fossil shell is in contact with the groundwater that circulates in the outer layer of the Earth. How does the shell behave in contact with this new water? If it is isotopically re-equilibrated with the fresh water whose δ value is very different from zero, it loses its former isotopic composition and so its paleothermal memory. Its isotopic composition no longer reflects the conditions of the old ocean but the conditions of recent aqueous circulation. In fact, in most (but not all!) cases, the shell remains compact and no isotope exchange occurs. The low rate of diffusion of oxygen in calcite at low or moderate temperatures limits the mechanism. And all the better for geologists! They can determine the past temperature of the ocean where the animal whose shell it was lived.

An important phenomenon is cooling. Isotopic equilibrium among minerals is established at high temperature. The mineral assemblage cools and so follows a decreasing thermal trajectory. The isotope equilibrium constant is dependent on temperature, and isotope reactions should continue to take place constantly matching temperature and isotope composition. If this were so, the system would lose all memory of its past at high temperature and isotope analysis would merely reflect the low-temperature equilibrium. In fact, as isotope exchange at low temperatures occurs very slowly, if cooling is rapid, the minerals often retain the composition acquired at high temperature. But this is not always so. Cooling is not always rapid. In metamorphism especially, exchanges are sometimes accelerated by certain factors and "initial" isotope compositions are not always maintained. But as the oxygen diffusion constants of the various silicate minerals are different, the temperatures indicated by the various minerals also differ. There is a sort of disequilibrium allowing us to detect the occurrence of any secondary effect.

All of this means that when measuring a compound's isotopic composition we must question the meaning of the message it carries and the time it was encoded. Does it correspond to the period when the object formed? Is it the outcome of secondary phenomena? If so, what phenomena? Once again, everything is dominated by isotope exchange mechanisms. The importance of these effects is attested by the answer to the following general observation. Why is sulfur isotope geochemistry not used more often, since it has substantial natural variations (from +60 to −40)? Because in many compounds, and particularly in sulfides, secondary isotope exchange occurs very rapidly. Through this exchange, the compounds lose much of the isotope memory of their origins. Another reason is the fact that sulfur geochemistry is highly complex with many degrees of oxidation, etc. However, interesting results have been obtained with sulfur isotopes.

7.3.3 Open system or closed system

The open system or infinite reservoir

When one of the reservoirs present is of infinite size (or is in direct contact with a boundless reservoir) the modalities of isotope fractionation are governed by the initial fractionation conditions and by conditions related to subsequent isotope exchange. No mass balance effect disturbs the relation between θ and α:

$$\theta = \alpha_{\text{equilibrium}}, \theta = \alpha_{\text{kinetic}}, \text{or } \theta = \alpha_{\text{mixed}},$$

depending on the nature of the initial fractionation and the subsequent isotope exchange. If the isotope composition of the infinite reservoir is R_0, the "large" reservoir imposes its isotope composition through the fractionation factor:

$$R = \alpha R_0 \text{ and } \delta \approx \delta_0 + (\alpha - 1)\, 1000.$$

Exercise

Sea water has a $\delta^{18}O$ value of 0. Liquid–vapor fractionation at equilibrium at 20 °C is $\alpha = 1.0098$. What is the composition of the water vapor evaporating if it is in equilibrium with the water?

Answer

The fractionation factor $(^{18}O/^{16}O)_{\text{vapor}}/\left(^{18}O/^{16}O\right)_{\text{liquid}} = 1/\alpha = 0.990\,29$. Therefore $(\alpha - 1) = -0.0097$, or $\delta^{18}O = -9.7‰$.

The closed system

Where the system is closed, a balance effect is superimposed on the modalities described. We note the isotope composition of the initial system R_0 and assume that from there two compounds, A and B, are produced with isotopic ratios R_A and R_B. We can write an isotope fractionation law (without specifying whether it is for equilibrium or not) characterized by Δ_{AB}, and an atom conservation equation. This gives: $R_0 = R_A x + R_B (1 - x)$, where x is the molar fraction of the element. In δ notation, this gives:

$$\boldsymbol{\delta}_0 = \boldsymbol{\delta}_A x + \boldsymbol{\delta}_B (1 - x) \text{ or } \boldsymbol{\delta}_0 = (\boldsymbol{\delta}_A - \boldsymbol{\delta}_B)x + \boldsymbol{\delta}_B \text{ or } \boldsymbol{\delta}_0 = \Delta_{AB} x + \boldsymbol{\delta}_B.$$

Exercise

Let us consider bacterial reduction $SO_4^{2-} \rightarrow S^{2-}$ by *Desulfovibrio desulfuricans*. The kinetic fractionation factor $^{34}S/^{32}S$ between sulfate and sulfide at 25 °C is 1.025 (Harrison and

Thode, 1958). Let us suppose that bacterial reduction occurred in oceanic sediment that was continually supplied with sulfate ions. The sulfate stock can therefore be considered infinite. What is the composition of the S^{2-} on the ocean floor if the $\delta^{34}S$ of the sulfate is +24?

Answer

Applying the equation $\Delta_{AB} = 10^3 \ln \alpha$ gives $\Delta = +24.6$.

$\delta_{sulfate} - \delta_{sulfide} = +24.6$ hence it can be deduced that $\delta_{sulfide} = -0.6$.

Exercise

Let us suppose now that the sediment becomes isolated from the ocean and is no longer supplied with sulfate ions and that the same phenomenon occurs. The quantity of organic matter is such that the proportion of sulfur in the state of sulfate is 1/3. Suppose that, in the initial state, all of the sulfur was in the sulfate state at $\delta^{34}S = +24$. What is the isotope composition of S^{2-}? What is the isotope composition of the sulfate?

Answer

We apply the equation:

$$\delta_0 = \Delta_{AB} x + \delta_B, \text{ or } \delta_B = \delta_0 - \Delta_{AB} x.$$

From this we obtain $\delta_{S^{2-}} = 15.8, \delta_{SO_4} = 40.4$.

As seen in the previous exercise, the result is markedly different for an open system, as the δ value is then positive. The effect of the closed system has shifted the isotope values of the sulfate and sulfide, but not the fractionation factor, of course! (The limiting cases where $x = 0$ and $x = 1$ should be examined.)

However, a flaw can be found in the foregoing reasoning. If the sulfides remained in a closed system as ions long enough, it might be that there was some isotopic exchange and that the sulfate and sulfide attained thermodynamic equilibrium. In this case $\alpha = 1.075$ at 25 °C (Tudge and Thode, 1950). Repeating the calculation with this value gives $\delta^{34}{}_{sulfide} = 0.14$ and $\delta^{34}{}_{sulfate} = 72.4$.

Intermediate scenarios can be imagined and therefore, in nature, the values will probably be intermediate ones.

As just seen, then, widely different isotope values are obtained for the same phenomenon but different modalities. It is probably the diversity of modalities that accounts for the great isotopic variation in sulfides of sedimentary origin (Figure 7.7).

Distillation

Here we look at a rather special (but widely applicable!) case where the system is closed but where the product is isolated as it forms. Let X_2 and X_1 represent the number of atoms of the two isotopes. At each moment in time, we have:

$$\frac{\mathrm{d}X_2/\mathrm{d}X_1}{(X_2/X_1)_A} = \alpha$$

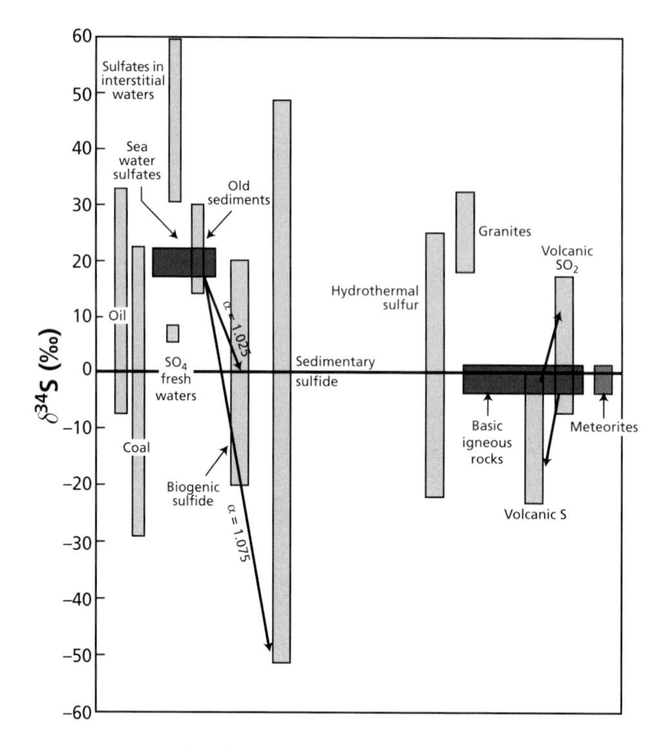

Figure 7.7 Analysis of $^{34}S/^{32}S$ isotope composition in the main terrestrial reservoirs. Notice that the domains are very extensive for all reservoirs. This corresponds to highly variable reducing conditions to which sulfur is subjected.

where α may be an equilibrium or kinetic value, dX_2 is the quantity of isotope 2 of A which is transformed into B, and dX_1 is the quantity of isotope 1 of A which is transformed into B. By separating the variables and integrating, we get:

$$X_2^{\mathrm{A}} = cX_1^{\alpha}$$

therefore:

$$(X_2/X_1)_{\mathrm{A}} = c\,X_1^{\alpha-1}.$$

At time $t = 0$ $(X_2/X_1)_{\mathrm{A}} = (X_2/X_1)_0$ and $X_1 = X_{1,0}$, therefore: $c = (X_2/X_1)\ \frac{1}{X_{1,0}^{\alpha-1}}$. Hence:

$$(X_2/X_1)^{\mathrm{A}} = (X_2/X_1)_0\left(X_1/X_{1,0}\right)^{\alpha-1}.$$

If the transformed remaining fraction of X_1 is called f, we get the famous **Rayleigh distillation law**:

$$R_{\mathrm{A}} = R_0 f^{\,\alpha-1}.$$

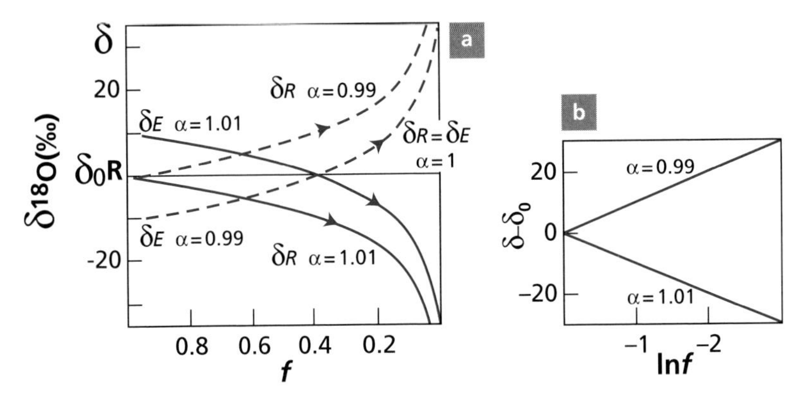

Figure 7.8 Changes in the instantaneous isotopic composition of a reservoir (*δ***R**) and an extract (*δ***E**) during a Rayleigh distillation process as a function of the partition coefficient (1.01 and 0.99 respectively). We have $\alpha_{\text{ext-res}} > 1$, $\alpha_{\text{ext-res}} = 1$, and $\alpha_{\text{ext-res}} < 1$ and an initial isotopic composition of the reservoir $\delta_{0R} = 0$; *f* is the remaining fraction of the reservoir and $(1-f)$ the extent to which the reaction has progressed. After Fourcade (1998).

Figure 7.8 shows the Rayleigh law as a function of f where $\alpha > 1$ and $\alpha < 1$. We shall see that the effects are opposite but are only extreme when f is very small. We see how A evolves, and also B, for which, of course, we have

$$R_B = \alpha R_0 f^{\alpha-1}.$$

The mean composition of A is written:

$$\overline{R}_A = R_{A,0}\left(\frac{f^\alpha - 1}{f-1}\right).$$

It can be seen that when f is small, the compositions of the two compounds seem to converge. And yet their partition coefficient remains constant! But it is clear that as small variations in f lead to large variations in δ, the optical illusion gives the impression of convergence. Notice too that when $f=0$, $R = R_{A,0}$, because of course "matter is neither created or destroyed" as Lavoisier said (except in nuclear reactions at high energy!).

Exercise

Find the Rayleigh formula expressed in δ.

Answer

$\delta = \delta_0 + 10^3\,(\alpha - 1)\ln f$. See the next exercise.

Exercise

Let us go back to our example of the formation of sedimentary sulfides. For the time being, we assume that as soon as the sulfide is formed, it reacts with iron dissolved in solution and forms FeS_2, without isotope fractionation (in fact, things are more complex than this). Being heavy in its solid state, the iron sulfide settles out and is removed from contact with the sulfates. This is a distillation effect. Given that in the end sulfates make up only one-third, what are the sulfide compositions?

Answer

The initial $\delta\,^{34}S$ is still +24. The kinetic coefficient α is 1.025. Let us first apply the Rayleigh equation, which we can use in a handier form with δ. Its mathematical form invites us to shift to logarithms. The formula becomes:

$$\ln R = \ln R_0 + (\alpha - 1) \ln f.$$

Given that $R = R_S(1 + \delta/1000)$ with the logarithmic approximations $\ln(1+\varepsilon) \approx \varepsilon$, and approximating the two terms $\ln R_S$, we get:

$$\delta = \delta_0 + 10^3 (\alpha - 1) \ln f.$$

This is the form we shall use. The final composition of the sulfates is $\delta = 24 + 25 \ln(1/3) = 24 + 27.7 = 51.7$.

The sulfides precipitating in the end have a δ value of +27.1. The average sulfide is obtained by the balance equation $\delta_{S\ \mathrm{average}} = +10.4$.

Exercise

In the first quantitative studies to estimate the degassing rate of magmas, **Françoise Pineau** and **Marc Javoy** (1983) of the Institut de Physique du Globe in Paris measured the $^{13}C/^{12}C$ partition coefficient of CO_2 in a magma at 1200 °C and found 4.5‰ (CO_2 being enriched in ^{13}C). Let us take a basalt with an initial $\delta^{13}C$ value of −7. After degassing we find $\delta^{13}C = -26$‰, with a carbon content of 100–150 ppm. If we assume a Rayleigh distillation, what is the extent of degassing of the magma? What was the initial carbon content of the magma?

Answer

We apply the Rayleigh law in δ:

$$\delta - \delta_0 = 1000 (\alpha - 1) \ln f.$$

Hence: $-20 = 4.5 \ln f$ and $f = 0.011$, therefore the magma was degassed to 98.8%. Its initial carbon content was therefore 9000–13 000 ppm.

EXAMPLE

Isotopic evolution of a cloud shedding rain

A cloud forms over the sea. It then migrates over a landmass or migrates to higher latitudes and loses rain. It is assumed that the cloud formed by the evaporation of sea water and that the fractionation factor for the oxygen isotopes remains constant at $\alpha = 1.008$. Figure 7.9 summarizes the isotope evolution of the cloud and of the rain that falls as it evolves. It is described by a simple Rayleigh distillation.

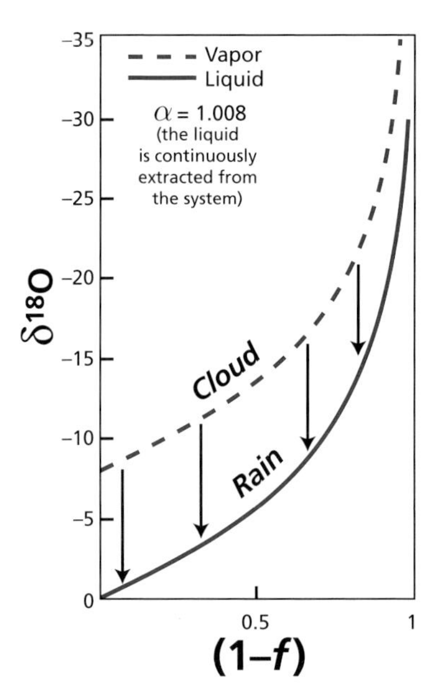

Figure 7.9 Rayleigh distillation between a cloud and rain for $\delta^{18}O$. The liquid (rain) is continuously removed. The vapor fraction is $1 - f$. After Dansgaard (1953).

7.3.4 Mixing

As we have already seen several times, mixing of two sources is an extremely important phenomenon in geochemistry. For example, sea water is a mixture of the various inputs of rivers, submarine volcanoes, rain, and atmospheric dust. We have a mixture of two components A_1 and A_2 with isotopic compositions:

$$\left(\frac{{}^{x}A}{{}^{y}A}\right)_1 \text{ and } \left(\frac{{}^{x}B}{{}^{y}B}\right)_2.$$

The isotope composition of the mixture is:

$$\left(\frac{{}^{x}A}{{}^{y}A}\right)_m = \frac{{}^{x}A_1 + {}^{x}A_2}{{}^{y}A_1 + {}^{y}A_2} = \frac{\left(\frac{{}^{x}A}{{}^{y}A}\right)_1 {}^{y}A_1 + \left(\frac{{}^{x}A}{{}^{y}A}\right)_2 {}^{y}A_2}{{}^{y}A_1 + {}^{y}A_2}.$$

If we posit:

$$\frac{{}^{y}A_1}{{}^{y}A_1 + {}^{y}A_2} = x_1 \text{ and } \frac{{}^{y}A_2}{{}^{y}A_1 + {}^{y}A_2} = 1 - x_1,$$

and if we write the ratios

$${}^{x/y}R : {}^{x/y}R_m = {}^{x/y}R_1 x_1 + {}^{x/y}R_2(1 - x_1),$$

then replacing R by the δ notation gives:

$$\delta_{\mathrm{m}} = \delta_1 x_1 + \delta_2 (1 - x_1).$$

We find a familiar old formula!

Exercise

Carbonates have a $^{13}C/^{12}C$ isotope composition expressed in $\delta^{13}C$ of 0‰. Organic products precipitating on the sea floor have a $\delta^{13}C$ value of −25‰. What is the mean value of $\delta^{13}C$ of the sediments, given that 80% of the sedimentary carbon is in the carbonates and 20% in the organic products?

Answer

The main isotopic component of carbon is ^{12}C. Therefore x and $(1-x)$ are 0.2 and 0.8, respectively. This gives $0.2 \times (-25‰) + 0.8 \times 0‰ = -5‰$. The average composition of the sediments is therefore −5‰.

Mixing in a correlation diagram of two isotope ratios obeys the equations already developed for radiogenic isotopes. Let the two elements whose isotopes are under study be A and B. Remember that if the (C_A/C_B) ratio is constant for the two components of the mixture, the mixture is represented by a straight line. If the two ratios are different, the mixture is represented by a hyperbola whose direction of concavity is determined by the concentration ratios of A and B.

7.4 The paleothermometer

In some sense, paleothermometry is to stable isotopes what geochronometry is to radiogenic isotopes, both an example and a symbol.

7.4.1 The carbonate thermometer

An example of this field of research has become a legend of sorts. In 1947, **Harold Urey** (1934 Nobel Prize winner for his discovery of deuterium, the hydrogen isotope ^2H) and **Bigeleisen** and **Mayer** published two theoretical papers in which they calculated isotope fractionation occurring in a series of chemical equilibria. In 1951, while professor at Chicago University, Urey and his co-workers used his method of calculation to determine the isotope equilibrium of carbonate ions CO_3^{2-} and water (H_2O) and calculated the isotopic fractionation that must affect the ^{18}O and ^{16}O oxygen isotopes whose common natural abundances are 0.205% and 99.756%, respectively. The $(^{18}O/^{16}O)_{\mathrm{carbonate}}/(^{18}O/^{16}O)_{\mathrm{water}}$ ratio must be a function of the temperature at which the two species are in equilibrium. The variations Urey predicted were small but could be measured, after converting the CO_3^{2-} into CO_2 gas, on the double-collection mass spectrometer already developed by **Alfred Nier** and his students at the University of Minnesota at the time. This fractionation was measured experimentally by Urey's team with the special involvement of **Samuel Epstein**, who was to become one of the big names in the speciality. Together, they developed the simple thermometric equation (in fact, the original coefficients were slightly different):

$$T_{^\circ\mathrm{C}} = 16.5 - 4.3\left(\delta^{18}_{\mathrm{CO_3}} - \delta^{18}_{\mathrm{H_2O}}\right) + 0.13\left(\delta^{18}_{\mathrm{CO_3}} - \delta^{18}_{\mathrm{H_2O}}\right)^2$$

where $T_{°C}$ is the temperature in degrees centigrade, and $\delta^{18}_{CO_3}$ the isotope composition of the CO_2 extracted from the carbonate, which is expressed by a deviation from the reference carbonate sample:[9]

$$\delta^{18}_{CO_3} = \left[\frac{\left({}^{18}O/{}^{16}O\right)_{CO_2,\,\text{carbonateX}} - \left({}^{18}O/{}^{16}O\right)_{CO_2,\,\text{standard}}}{\left({}^{18}O/{}^{16}O\right)_{CO_2,\,\text{standard}}} \right] \cdot 10^3.$$

The standard chosen is a reference limestone known as PDB. The Chicago team decided to use its carbonate thermometer to measure geological temperatures. To do this, they chose a common, robust fossil, the rostrum (the front spike on the shell) of a cephalopod known as a belemnite that lived in the Jurassic (–150 Ma) and was similar to present-day squids. Suppose that in the course of geological time, the isotopic composition of oxygen in sea water had remained constant at $\delta^{18}O = 0$. Then the $^{18}O/^{16}O$ oxygen isotopic composition of the carbonate of the fossils reflects the temperature of the sea water in which the shell formed. This isotope composition became fixed when the carbonate was incorporated as calcite crystals in the fossil shells (as solid-phase reactions at low temperature are very slow, there is little chance that the composition was altered by secondary processes). By measuring the isotope composition of fossils, it is possible to determine the temperature of the ancient seas. To confirm this idea, the Chicago team therefore measured a series of belemnite rostra from various geographic areas and of different stratigraphic ages (Figure 7.10).

The results, first announced in preliminary form at the 1950 annual meeting of the Geological Society of America were spectacular and immediately claimed the attention of the entire geological community. Let us summarize them.

At the scale of the planet, for the Jurassic, when belemnites lived, isotope temperature obtained varied from 12 to 18 °C. These are likely and coherent temperatures; likely because other paleoecological indicators are in agreement with them, coherent because variations over time in various measurements in various parts of the world concord. Thus it has been determined that the maximum temperature was in the Late Cretaceous, using samples from a single area (Sweden, Britain) or samples including fossils collected from North America and Europe.

Encouraged by these worldwide results, the Chicago scientists set about dissecting individual rostra. Each rostrum is made up of concentric layers which are evidence of belemnite annual growth. Layer-by-layer analysis revealed regularly alternating temperatures. There were therefore summers and winters at the time! They even managed to show that one particular individual was born in the fall and died in springtime!

Exercise

The standard chosen for oxygen is SMOW ($\delta^{18}O = 0$). McCrea and Epstein's simplified thermometric equation is:

$$T_{°C} = 16.5 - 4.3\,\delta^{^{18}O}_{CO_3}.$$

[9] This is an important detail. It is not the isotopic composition of the CO_3^{2-} that is measured but that of the CO_2 in equilibrium with the carbonate!

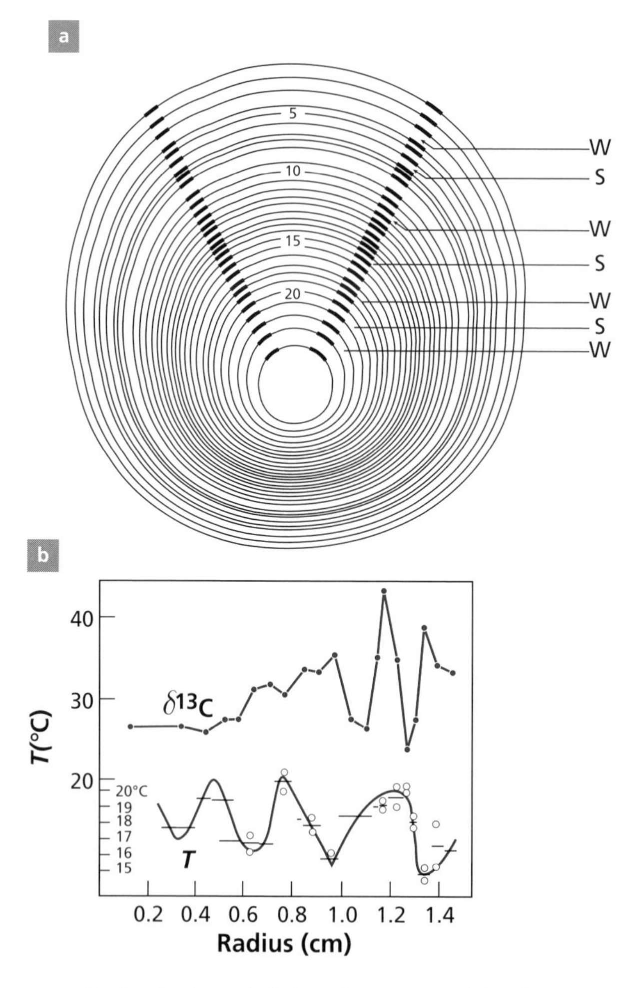

Figure 7.10 Study of a Jurassic belemnite rostrum. (a) A famous figure of a cross-section through a Jurassic belemnite rostrum. Samples were taken a different radial distances (S, summer; W, winter; numbers of rings are counted from the outside). (b) Values of $\delta^{13}C$ and below $\delta^{18}O$ converted directly into temperature. The curve shows that the belemnite was born in the fall and died in spring! After Urey *et al.* (1951).

The precision of measurement of oxygen isotope composition is 0.1 in δ units. What is the power of resolution in temperature of the isotope method defined by Urey?

Answer

Differentiating the formula above gives $\Delta T = 4.3\ \Delta\delta$. So the precision is 0.43 °C. One might envisage further increasing the precision when making measurements with the mass spectrometer to attain 0.01%, but this raises a geochemical problem: what do the tiny differences revealed signify? We shall get some inkling of an answer in what follows.

This exceptional scientific success story opened the way to a new geological discipline, **paleothermometry**, or the study of past temperatures on a precise scientific basis, which gave tremendous impetus to paleoclimatology. It also encouraged researchers to forge ahead. If stable isotopes of oxygen had yielded such significant results in their first application in geology, it could be hoped that the examination of other problems, other properties, and other elements would be equally successful. This hope gave rise to the work that founded **stable isotope geochemistry**. However, the Chicago team's paleothermometer was based on the assumption that $\delta_{\text{sea water}} = 0$ has been constant throughout geological times. As we shall see, this hypothesis probably holds over the average for millions of years but not on the scale of thousands of years which is the timescale of the Quaternary era (Epstein *et al.*, 1953; Epstein, 1959).

7.4.2 The $^{18}O/^{16}O$ isotope composition of silicates and high-temperature thermometry

It is relatively easy to measure the isotopic composition of oxygen in carbonates since CO_3^{2-} reacts with phosphoric acid to transform into CO_2, which can be measured directly in double-collector mass spectrometers. It is far more difficult to extract oxygen from silicate minerals. This means using fluorine gas or even the gas BrF_5 and then transforming the oxygen into CO_2 by burning. Of course, all such processes should be performed with no isotopic fractionation or well-controlled fractionation! These techniques were developed at the California Institute of Technology by **Hugh Taylor** and **Sam Epstein** in the late 1960s (Epstein and Taylor, 1967).

Measuring the oxygen isotope composition of silicate minerals reveals systematic variations with the type of mineral and the type of rock to which the mineral belongs. These compositions can be characterized by measuring isotope fractionation between minerals. Now, one of the great features of isotopes is that isotope fractionation is very largely independent of pressure and dependent mainly on temperature. Variations in volume associated with exchange reactions are virtually zero. Therefore isotope equilibrium reactions are very useful for determining the temperatures at which natural mineral associations formed. Indeed α varies with temperature and tends towards unity at very high temperatures. As we have said, the variation of α with T takes the form:

$$\ln \alpha = B + \frac{C}{T} + \frac{A}{T^2}.$$

The form of this equation is preserved for α and δ. Between two minerals m_1 and m_2 in equilibrium:[10]

$$\Delta_{m_1 m_2} = \delta_{m_1} - \delta_{m_2} \approx A\left(10^6\, T^{-2}\right) + B = 1000 \ln \alpha.$$

The term $1/T$ is generally negligible. Oxygen isotopes are especially useful here. Oxygen is the most abundant element in silicates and the ^{18}O and ^{16}O isotopes fractionate in nature in proportions that can be easily measured by mass spectrometry. Experimental studies conducted mostly by the Chicago University group under **Robert Clayton** and

[10] Tables usually give absolute temperatures so degrees must be converted from Celsius to Kelvin.

Table 7.1 Isotope fractionation for mineral–water pairs

Mineral	Temperature (°C)	*A*	*B*
Calcite (CO_3Ca)	0–500	2.78	− 2.89
Dolomite	300–500	3.20	− 1.5
Quartz	200–500	3.38	− 2.90
Quartz	500–800	4.10	− 3.7
Alkali feldspar	350–800	3.13	− 3.7
Plagioclase	500–800	3.13	− 3.7
Anorthite	500–800	2.09	− 3.7
Muscovite	500–800	1.9	− 3.10
Magnetite	(reversed slope) 0–500	− 1.47	− 3.70

Table 7.2 Results of ^{18}O isotope thermometry based on $^{18}O/^{16}O$ fractionation of mineral pairs

Pair	*A*	*B*
Quartz–albite	0.97	0
Quartz–anorthite	2.01	0
Quartz–diopside	2.08	0
Quartz–magnetite	5.57	0
Quartz–muscovite	2.20	− 0.6
Diopside–magnetite	5.57	0

Source: After O'Neil (1986) modified by Bottinga and Javoy (1975).

Jim O'Neil and supplemented by theoretical work of **Yan Bottinga** and **Marc Javoy** at the Institut de Physique du Globe in Paris have provided a series of reliable values for coefficients *A* and *B* (see O'Neil and Clayton, 1964; Bottinga and Javoy, 1975; Javoy, 1977).

In the experimental procedure, the isotope fractionation between minerals and water is measured first. This is a convenient method as isotope equilibration is attained quite rapidly at about 80–100 °C. The fractionation between minerals is then calculated.

Tables 7.1 and 7.2 show the values of coefficients *A* and *B* for various mineral–water equilibria (we shall see the intrinsic importance of such fractionation later) and then for fractionation between pairs of minerals.

Exercise

What is the $\delta^{18}O$ composition of a muscovite in equilibrium with water at 600 °C whose $\delta = -10$?

Answer

The Δ is written:

$$1.9\left(10^6 \times \frac{1}{(873)^2}\right) - 3.1 = -0.6$$

where $\Delta = \delta_{musc} - \delta_{water}$.

From this we obtain $\delta_{musc} = -10.6$.

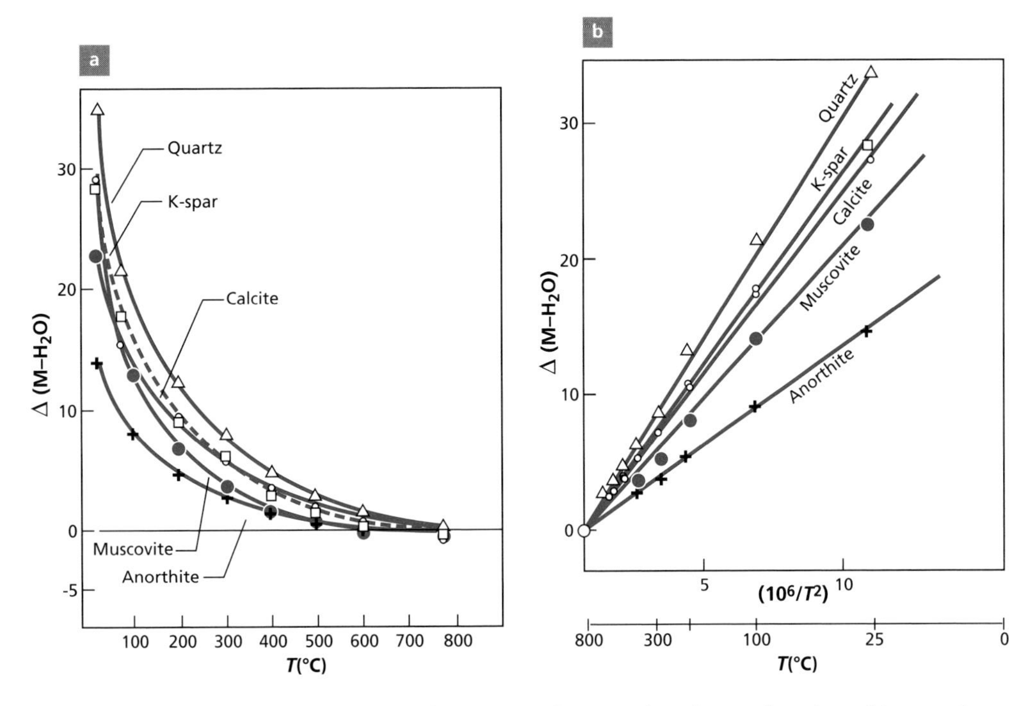

Figure 7.11 Isotope fractionation curves for water and some minerals as a function of temperature (T, or $10^6/T^2$). Notice that the curve should theoretically converge to zero. The error is the result of experimental uncertainty. After O'Neil (1986).

These are shown in Figure 7.11 in two ways: as a function of temperature (°C) and as a function of $10^6/T^2$ because the fractionations are linear. We plot $1000 \ln\alpha$, that is Δ, on the ordinates, which means we can calculate $\Delta_{water} = \delta_{mineral} - \delta_{water}$ directly. Notice that fractionation cancels itself out at high temperatures. On the experimental curves, this convergence seems to occur at less than $\Delta = 0$, but this effect is probably due to experimental errors. That would mean that minerals and water were of the same composition at high temperatures.

Exercise

Water with $\delta_{water} = -10$ and rock (composed of several minerals) with an initial δ value of $\delta_{(0)rock} = +6$ are put together. If we mix 100 g of rock and 110 g of water and heat them to high temperature (500 °C in an autoclave) for which we take a zero overall Δ value, what will be the composition of the rock and water after the experiment, given that the rock contains 50% oxygen and 90% water?

Answer

$\delta_{water} = \delta_{rock} = -4.29.$

So having the values A and B for several minerals, we can calculate fractionation between mineral pairs for each temperature:

$$\Delta_{m_1-m_2} = \Delta_{m_1-\text{water}} - \Delta_{m_2-\text{water}}.$$

Let us take the case of quartz–muscovite between 500 and 800 °C:

$$\Delta_{\text{quartz-musc}} = 2.20 \cdot 10^6/T^2 - 0.6.$$

We can set about geological thermometry using these various pairs of minerals. Having measured $\Delta_{m_1-m_2}$, we return to the established formula and calculate T.

In this way, the temperatures of various metamorphic zones have been determined. But, of course, much as with concordance of ages by various methods, we must make sure the various pairs of minerals yield the same temperature.

Marc Javoy, Serge Fourcade, and **the present author**, at the Institut de Physique du Globe in Paris, came up with a graphical discussion method: after choosing a reference mineral, we write for each mineral:

$$\Delta_{\text{quartz-mineral}} - B = A/T^2.$$

In a plot of $\Delta - B$ against A, the various minerals of a rock in isotopic equilibrium are aligned on a straight line through the origin whose slope $(1/T^2)$ gives the temperature at which they formed (Figure 7.12). If the points are not aligned, the rock is not in equilibrium and the temperature cannot be determined. It was thus possible to draw up a table of the thermal domains where the main rocks were formed (Figure 7.13). These findings are consistent with indirect evidence from mineral synthesis experiments and metamorphic zoneography.

Exercise

The $\delta^{18}O$ values of the minerals of a metamorphic rock are: quartz +14.8, magnetite +5.

(1) Calculate the equilibrium temperature of quartz–magnetite.
(2) Calculate the $\delta^{18}O$ of an aqueous fluid in equilibrium with the rock.

Answer

(1) 481 °C.
(2) +11.3.

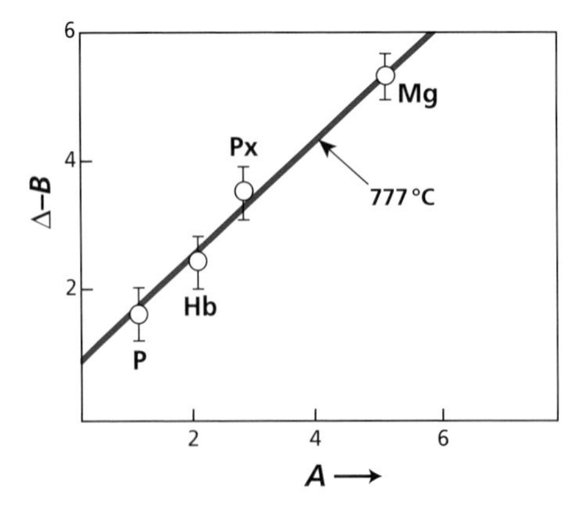

Figure 7.12 Javoy's method of determining paleotemperatures, used here for San Marcos gabbro. P, plagioclase; Hb, hornblende; Px, pyroxene; Mg, magnetite; *A* and *B* are defined in the text.

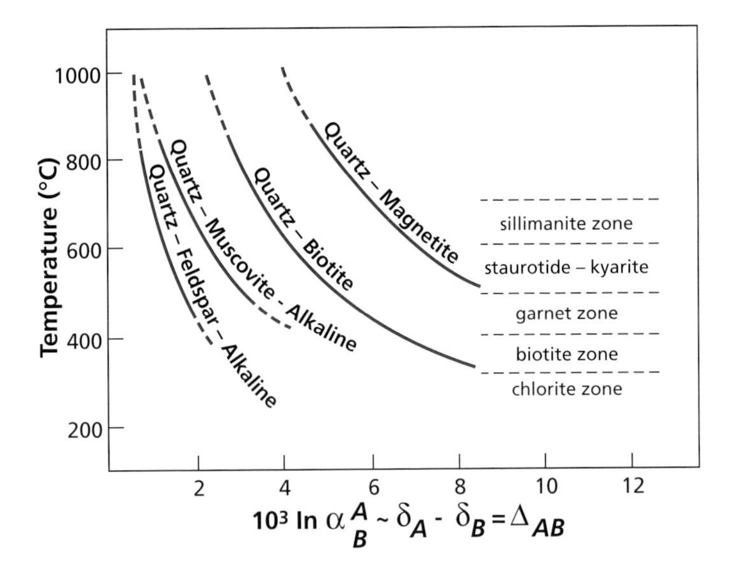

Figure 7.13 Isotope temperature of different metamorphic grades determined from pairs of minerals. After Garlick and Epstein (1967).

7.4.3 Paleothermometry of intracrystalline isotopic order/disorder

After the paleothermometry of silicate rocks, one might legitimately ask with hindsight why the same approach was not adopted for low-temperature paleothermometry and why several minerals were not used instead of calcite alone to break free of the hypothesis of a constant δ value for sea water? In fact, research was conducted along these lines and, for this, the isotopic fractionation between water and calcium phosphate and water and silica was measured since these minerals are commonplace in marine sediments and in particular in fish teeth for phosphates and diatoms for silica. Unfortunately, as Figure 7.14 shows, while the fractionations are different for the three minerals ($CaCO_3$, $CaPO_4$, and SiO_2), their variations with temperature are parallel. They may therefore not be used two-by-two to eliminate the unknown factor which is the isotopic composition of sea water!

A new method has very recently emerged to eliminate the unknown quantity of the isotopic composition of ancient water. It was developed by the new team around **John Eiler** at the California Institute of Technology. It is based on isotopic fractionations existing within a single molecular species among the different varieties of isotope (see Ghosh *et al.*, 2006b). Let us take the carbonate ion CO_3^{2-} as an example. This ion comprises numerous isotopic varieties: $^{12}C^{16}O^{16}O^{16}O$, $^{12}C^{16}O^{16}O^{18}O$, $^{12}C^{16}O^{18}O^{18}O$, . . ., $^{13}C^{16}O^{16}O^{16}O$, $^{13}C^{16}O^{16}O^{18}O$, . . ., etc. These are what are called **isotopologs** (see Section 7.2.1). Table 7.3 provides an inventory and gives their mean proportions in the "ordinary" carbonate ion. Each is characterized by a different molecular mass.

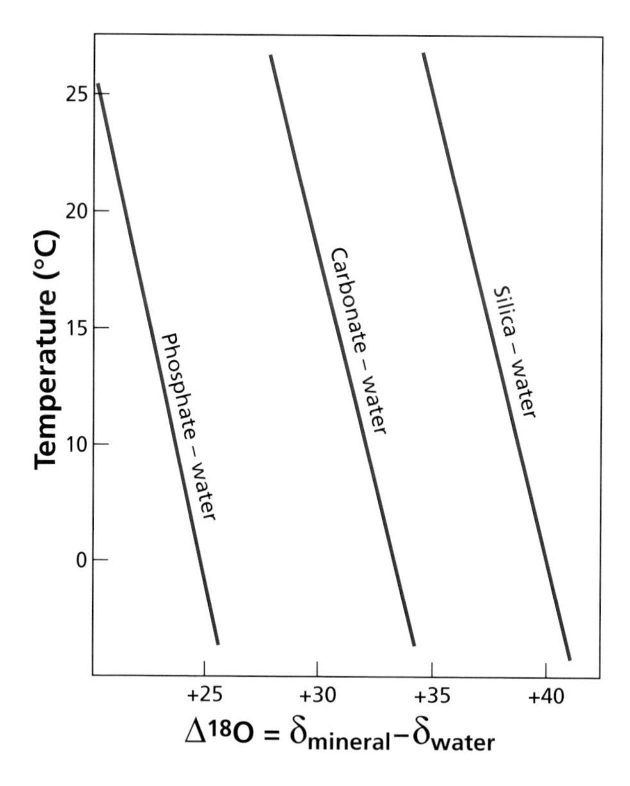

Figure 7.14 Fractionation for $^{18}O/^{16}O$ for various minerals with water. The curve shows clearly that they are parallel. After Longinelli and Nutti (1973); Labeyrie (1974).

In a calcium carbonate crystal, thermodynamic equilibria in the sense of Urey occur among the various isotopic species. Keeping to the most abundant varieties, we can write the equilibrium:

$$^{13}C^{16}O_3^{2-} + {}^{12}C^{18}O^{16}O_2^{2-} \Leftrightarrow {}^{13}C^{18}O^{16}O_2^{2-} + {}^{12}C^{16}O_2^{2-}.$$

masses: (61) (62) (63) (60).

The equilibrium constant depends on temperature. The lower the temperature, the more the reaction favors the right-hand members, that is the members with the heavy isotopes of carbon and oxygen (the most advantaged would be $^{13}C^{18}O^{18}O^{18}O$, but as its abundance is 94 ppt, it can barely be measured). In fact, this reaction may be considered an order/disorder reaction. The lower the temperature, the greater the ordering (light species with light, heavy species with heavy). The higher the temperature, the more disordered the assembly and the equilibrium constant tends towards unity.

It is a smart idea to use these equilibria within the calcite crystal, but there is a major difficulty in practice. Calcium carbonate isotopic compositions cannot be measured directly in the laboratory (they may be measurable one day with instruments for *in situ* isotope analysis, but for the time being they are not precise enough). To measure the isotopic

Table 7.3 Isotopologs

	Mass	Abundance
CO_2		
$^{16}O^{12}C^{16}O$	44	98.40%
$^{16}O^{13}C^{16}O$	45	1.10%
$^{17}O^{12}C^{16}O$	45	730 ppm
$^{18}O^{12}C^{16}O$	46	0.40%
$^{17}O^{13}C^{16}O$	46	8.19 ppm
$^{17}O^{12}C^{17}O$	46	135 ppm
$^{18}O^{13}C^{16}O$	47	45 ppm
$^{17}O^{12}C^{18}O$	47	1.5 ppm
$^{17}O^{13}C^{17}O$	47	1.5 ppm
$^{18}O^{12}C^{18}O$	48	4.1 ppm
$^{17}O^{13}C^{18}O$	48	16.7 ppm
$^{18}O^{13}C^{18}O$	49	46 ppb
CO_3		
$^{12}C^{16}O^{16}O^{16}O$	60	98.20%
$^{13}C^{16}O^{16}O^{16}O$	61	1.10%
$^{12}C^{17}O^{16}O^{16}O$	61	0.11%
$^{12}C^{18}O^{16}O^{16}O$	62	0.60%
$^{13}C^{17}O^{16}O^{16}O$	62	12 ppm
$^{12}C^{17}O^{17}O^{16}O$	62	405 ppb
$^{13}C^{18}O^{16}O^{16}O$	63	67 ppm
$^{12}C^{17}O^{18}O^{16}O$	63	4.4 ppm
$^{13}C^{17}O^{17}O^{16}O$	63	4.54 ppb
$^{12}C^{17}O^{17}O^{17}O$	63	50 ppt
$^{12}C^{18}O^{18}O^{16}O$	64	12 ppm
$^{13}C^{17}O^{18}O^{16}O$	64	50 ppb
$^{12}C^{17}O^{17}O^{18}O$	64	828 ppt
$^{13}C^{17}O^{17}O^{17}O$	64	0.5 ppt
$^{13}C^{18}O^{18}O^{16}O$	65	138 ppb
$^{12}C^{17}O^{18}O^{18}O$	65	4.5 ppb
$^{13}C^{17}O^{17}O^{18}O$	65	9 ppt
$^{12}C^{18}O^{18}O^{18}O$	66	8 ppb
$^{13}C^{17}O^{18}O^{18}O$	66	51 ppt
$^{13}C^{18}O^{18}O^{18}O$	67	94 ppt

composition of CO_3^{2-} radicals they are transformed into CO_2 molecules by a reaction with phosphoric acid.

The breakthrough by the Caltech team was to have developed a technique for extracting carbonate isotope varieties and transforming them into clearly identifiable CO_2 molecules and in particular for distinguishing $^{13}C^{18}O^{16}O$ (mass = 47), $^{12}C^{16}O^{16}O$ (mass = 44), $^{12}C^{18}O^{16}O$ (mass = 46), and $^{13}C^{16}O^{16}O$ (mass = 45) and showing they reflect the proportions of CO_3^{2-} molecules (by adding ^{16}O to each). To do this, they defined the unit Δ_{47} between the ratios measured for masses 47 and 44:

$$\Delta_{47} = \left[(47/44)_{\text{sample}} - (47/44)_{\text{reference}}\right] \times 10^3.$$

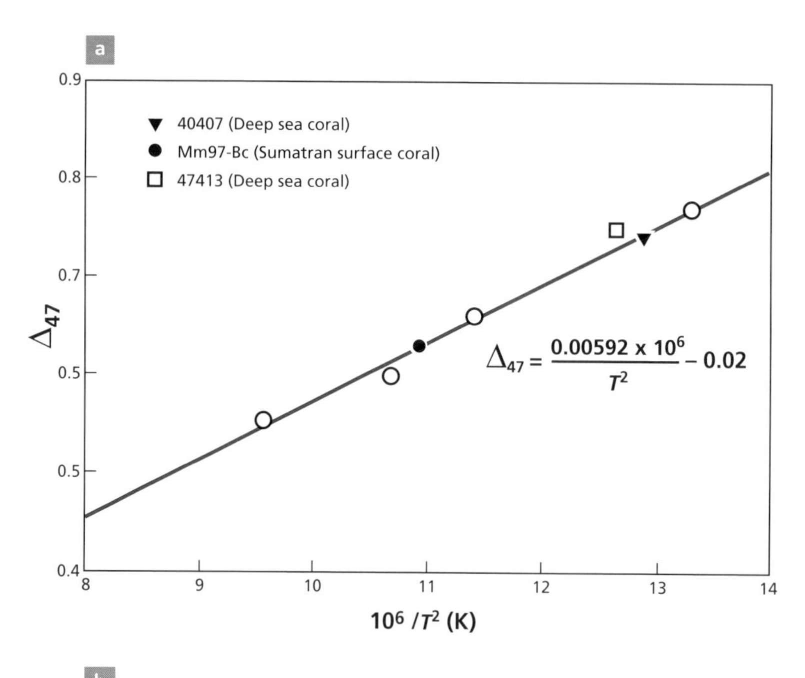

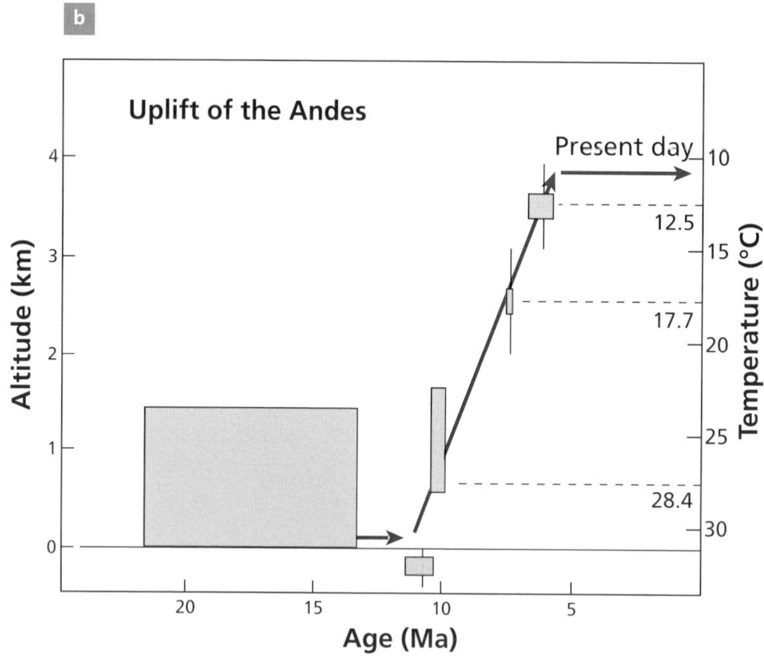

Figure 7.15 (a) Calibration of the isotopic order/disorder thermometer with the corresponding formula. (b) Uplift of the Andes reconstructed by the isotopic order/disorder chemometer. After Ghosh *et al.* (2006).

The reference (47/44) is the ratio that would pertain if the isotopic distribution among the varieties of isotopes were purely random. They established the fractionation curve (Δ_{47}, as a function of temperature).

The temperature can therefore be determined from a measurement of Δ_{47}. The exact formula (between 0 and 50 °C) is:

$$\Delta_{47} = 0.0592 \cdot 10^6 T^{-2} - 0.02.$$

Precision is estimated to be ±2 °C.

An interesting application of this method has been to determine the rate of uplift of the Bolivian Altiplano. Samples of carbonates contained in soil were taken from the plateau but of different ages and dated by other methods. The temperature at which these carbonates formed was then calculated. As the curve of temperature variation with altitude in the Andes is known, the curve of altitude versus time could be determined (Figure 7.15).

Exercise

Do you think this isotopic order/disorder method could apply to SiO_2 at low temperature (diatoms)? Write the equivalent equation to that written for carbonate. What would the isotopic parameter be? Do you see any practical difficulty in this?

Answer

Yes, in principle. The order/disorder equilibrium equation would be:

$$^{30}Si^{16}O_2 + ^{28}Si^{18}O^{16}O \Leftrightarrow ^{30}Si^{18}O^{16}O + ^{28}Si^{16}O_2$$

mass : (62) (62) (64) (60)

$$\Delta_{64} = \left[(64/60)_{sample} - (64/60)_{reference}\right] \times 10^3 \text{ (or } 10^4 \text{ as necessary).}$$

The difficulty is that with the present-day method, Si is measured in the form of SiF_4 on the one hand, oxygen being extracted on the other hand. To apply the method, direct measurement by an *in situ* method in the form of SiO_2 would be required. This will probably be feasible in the future with ion probes or laser beam ionization.

7.5 The isotope cycle of water

Let us return to the water cycle mentioned at the beginning of this chapter. On Earth, it is dominated by the following factors.

(1) *The existence of four reservoirs.* A series of exchanges among the ocean, the ice caps, fresh water, and the atmosphere make up the water cycle. It is another dynamic system. The reservoirs are of very different dimensions: the ocean (1370 million km^3), the ice caps (29 million km^3), river water and lakes (0.002 12 million km^3). The transit time of water in each reservoir varies roughly inversely with its size, each reservoir playing an important geochemical role. Thus the quantity of water that evaporates and precipitates

is 500 million km^3 per thousand years, or more than one-quarter of the volume of the oceans.[11]

(2) *The ocean–atmosphere hydrological cycle.* Water evaporates from the ocean and atmospheric water vapor forms clouds that migrate and may occasionally produce rain. Thus salt water is changed into fresh water and transferred from tropical to polar regions and from the ocean to the landmasses. The hydrological cycle has a double effect. Clouds move from low to high latitudes and also from the ocean to the continents. The fresh water that falls as rain over the landmasses re-evaporates in part, runs off or seeps in, thus forming the freshwater reservoir which ultimately flows back to the ocean.

(3) *The polar regions.* When precipitation from clouds occurs in polar regions, we no longer have rain but snow. The snow accumulates and changes into ice forming the polar ice caps. These ice caps flow (like mountain glaciers, but more slowly) and eventually break up in the ocean as icebergs and mix with the ocean.

The whole of water circulation on the planet and the various stages of the cycle have been studied in terms of isotopes. We have seen, when examining theoretical aspects, that when water and water vapor are in equilibrium, oxygen and hydrogen isotope fractionation are associated. This double pair of isotopes has allowed us to construct quantitative models of water circulation. However, the problems raised by these studies are not as simple as the theoretical study suggested.

7.5.1 Isotope fractionation of clouds and precipitation

A cloud is composed of water droplets in equilibrium with water vapor. Water vapor and droplets are in isotopic equilibrium. All of this comes, of course, from water which initially evaporated.

Let us take a cloud near the equator and follow it as it moves to higher latitudes. The cloud is enriched as a whole in ^{16}O relative to sea water, as we have seen, and so has a negative δ value. As it moves it discharges some of its water as rainfall. The rainwater is enriched in the heavy isotope, and so the cloud becomes increasingly enriched in the light isotope. The precipitation is increasingly rich in light isotopes, which effect is offset in part by the fact that the fractionation factor varies with $1/T$. As we move away from the equator, it can be seen statistically that the precipitation has increasingly negative $\delta^{18}O$ values (Figure 7.16).

As clouds undergo genuine distillation, by progressively losing their substance, their isotope composition obeys a Rayleigh law, but a "super law" because as they move polewards, the temperature falls, the fractionation factor also increases and distillation becomes increasingly effective (Figure 7.17), so much so that at the poles the $\delta^{18}O$ values are extremely negative.

We observe geographical zoning for which the $\delta^{18}O$ value and mean air temperature can be related (Epstein *et al.*, 1965; Dansgaard and Tauber, 1969) (Figure 7.18).

The general cycle of clouds is repeated at local scale, when clouds move over landmasses and progressively shed their water. Thus, fresh water has negative δ values. This phenomenon has been studied using the paired tracers $^{18}O/^{16}O$ and D/H. **Harmon Craig** of the Scripps Institution of the University of California showed that rain and snow precipitation and the

[11] $1\ km^3 \approx 10^{12}$ kg.

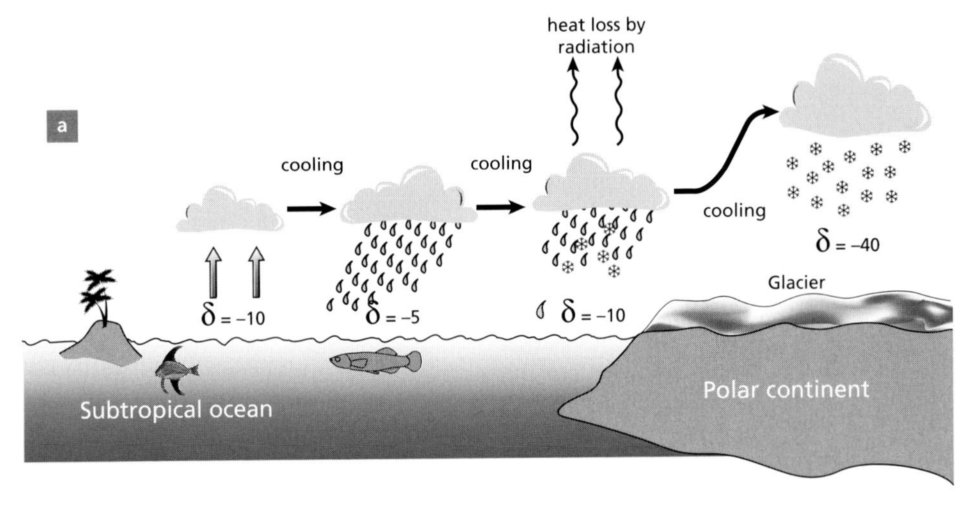

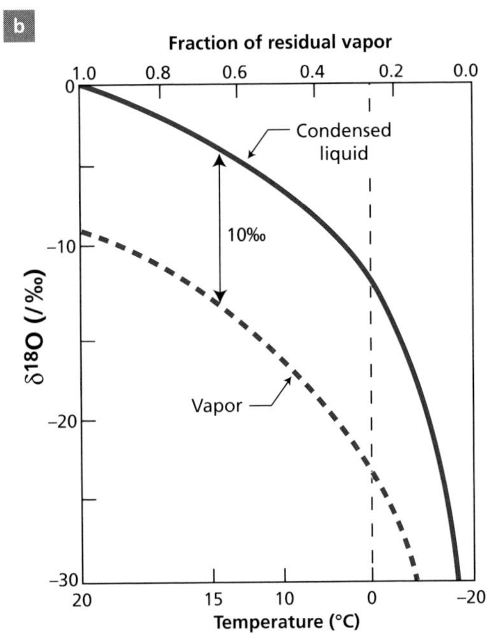

Figure 7.16 Fractionation of $\delta^{18}O$ in a cloud as a function of Rayleigh distillation. The cloud forms at the equator and moves to higher latitudes, losing water. The fractionation factor varies with temperature. Modified after Dansgaard (1964).

composition of glaciers lie on what is known as the meteoric water line: $\boldsymbol{\delta}D = 8 \times \boldsymbol{\delta}^{18}O + 10$ in the ($\boldsymbol{\delta}D$, $\boldsymbol{\delta}^{18}O$) diagram (Figure 7.3). The slope of value 8 corresponds to an equilibrium fractionation between the water and its vapor at around 20 °C. We have good grounds to think, then, that precipitation occurs in conditions of equilibrium. It was thought in early studies of the water cycle that evaporation was also statistically an equilibrium phenomenon. In fact, this is not so. Evaporation, which is a **kinetic phenomenon** in isotopic terms, leads to

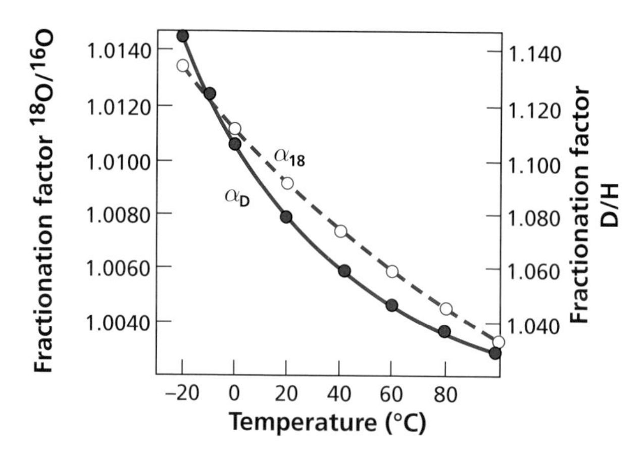

Figure 7.17 Study of $^{18}O/^{16}O$ fractionation. Liquid–vapor fractionation of H_2O for $^{18}O/^{16}O$ and D/H as a function of temperature. Notice that the scales are different. After Jouzel (1986).

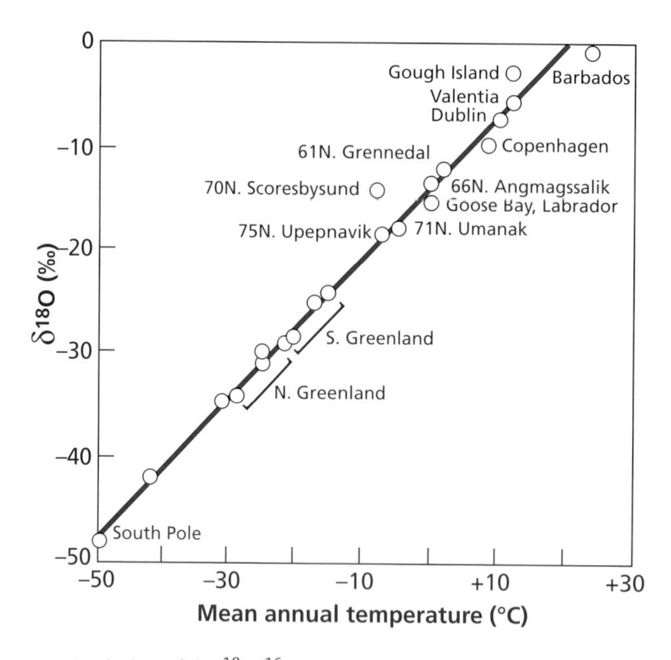

Figure 7.18 Variation of the $^{18}O/^{16}O$ ratio in rainwater and snow with latitude and so with temperature. After Dansgaard (1964).

^{18}O contents of vapor that are much lower than they would be in equilibrium. But depending on the climate, kinetic evaporation may or may not be followed by partial isotope re-equilibration which means the vapor composition does not lie on the straight line of precipitation. The same is true, of course, of surface sea water, which forms the residue of evaporation. Its ^{18}O composition is variable and depends on the relative extent of

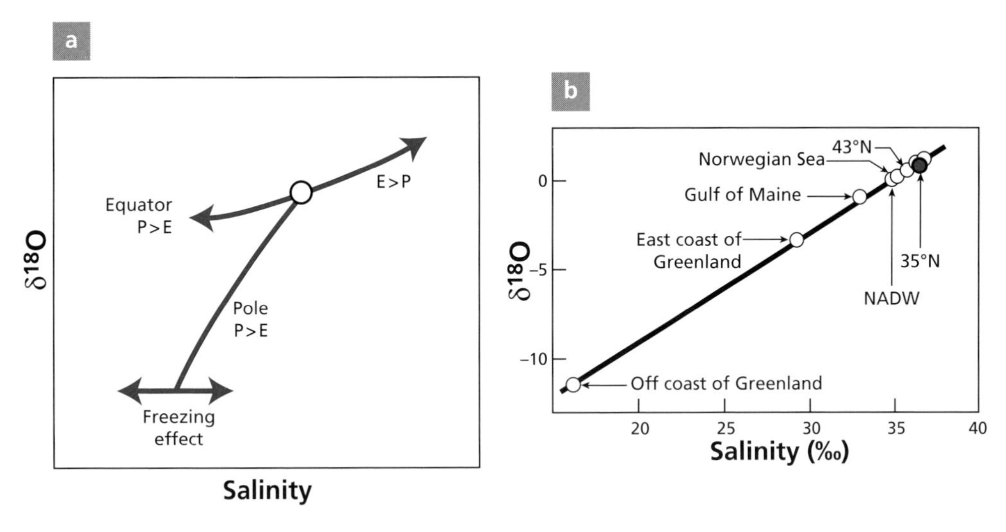

Figure 7.19 Relations between δ^{18}O and salinity. (a) Theoretical relation. P, precipitation; E, evaporation. (b) Various measurements for the North Atlantic. NADW, North Atlantic Deep Water. After Craig (1965).

evaporation and of precipitation (which are substantial over the ocean) and of the input of fresh water. These variations are particularly sensitive in the North Atlantic (Epstein and Mayeda, 1953). We visualize the variations and the influence of the various phenomena that causes them in a (δ^{18}O, S‰) plot, where S‰ is the salinity of sea water (Figure 7.19). As can be seen, there is a very close correlation between the two. All of this shows that this is a well-understood field of research.

EXAMPLE

Precipitation in North America

This is a map of δD and δ^{18}O for precipitation in North America (Figure 7.20). From what has just been said about the effect of isotope distillation of clouds, the pattern of rainfall over North America is described. The main source of rainfall comes from the Gulf of Mexico with clouds moving northwards and becoming distilled. This distribution is modified by several factors. First, the relief, which means the clouds penetrate further up the Mississippi valley but discharge sooner over the Appalachian Mountains in the east and the Rocky Mountains in the west. Other rain comes in from the Atlantic, of course, so the distribution is asymmetrical. Conversely rain from the Pacific is confined to the coast and moves inland little, so the lines are more tightly packed to the west.

Exercise

From the information given since the beginning of this chapter, use theoretical considerations to establish Craig's equation:

$\delta_{^{18}\mathrm{O}} = +8, \delta_{\mathrm{D}} + 10.$

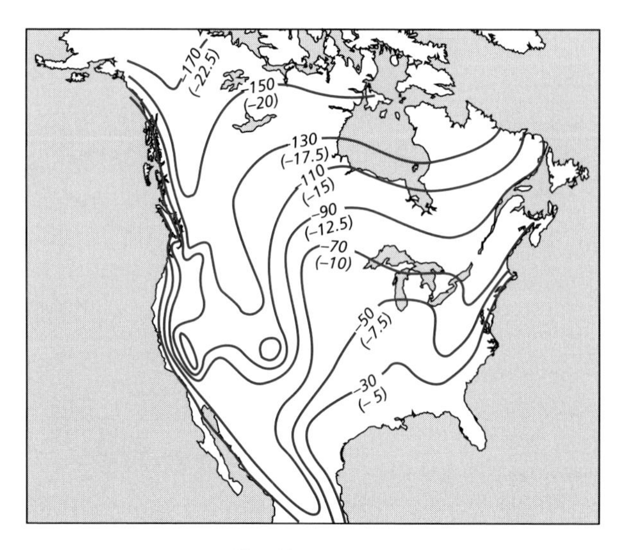

Figure 7.20 Distribution of $^{18}O/^{16}O$ and D/H in rainfall in North America. The $^{18}O/^{16}O$ ratios are in brackets. After Taylor (1974).

Answer

Clouds obey a Rayleigh law:

$$\delta_D \approx \delta_{D,0} + 10^3(\alpha_D - 1)\ln f$$

$$\delta_O \approx \delta_{^{18}O,0} + 10^3(\alpha_O - 1)\ln f.$$

This simplifies to:

$$\frac{\delta_D - \delta_{D,0}}{\delta_{^{18}O} - \delta_{^{18}O,0}} \approx \left(\frac{\alpha_D - 1}{\alpha_O - 1}\right).$$

At 20 °C, as seen in the previous problem, $\alpha_D = 1.0850$ and $\alpha_{^{18}O} \approx 1.0098$, hence:

$$\frac{\alpha_D - 1}{\alpha_{^{18}O} - 1} \approx 8.$$

We therefore have the slope. The ordinate at the origin seems more difficult to model because for vapor formed at 20 °C, $\delta_{D,0} - 8\delta_{^{18}O,0} = -6.8$ whereas we should find 10. We shall not go into the explanation of this difference, which is a highly complex problem, as shown by **Jean Jouzel** of the French Atomic Energy Commission. The different aspects of the hydrological cycle, including kinetic effects during evaporation, play a part.

7.5.2 Juvenile water

It is well known that in the water cycle, there is an input from hot water from the depths of the Earth. It was long thought that this hot water was the gradual degassing of water trapped by the Earth when it first formed, as with the primitive ocean. If this were so, this water would progressively increase the volume of the hydrosphere. Water from deep beneath the surface

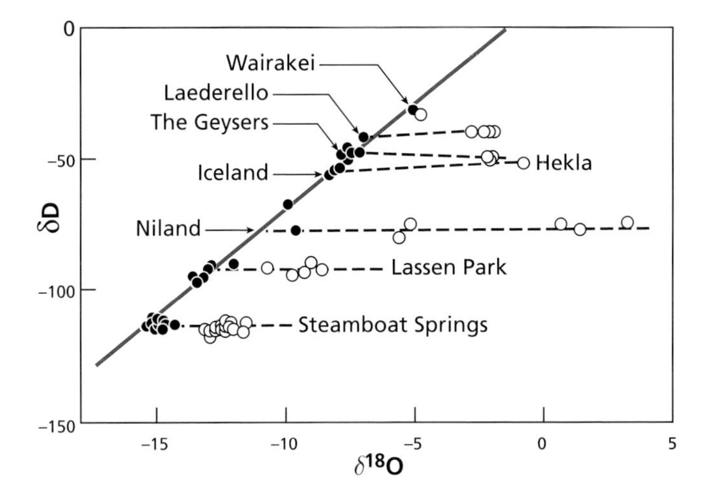

Figure 7.21 Correlation diagram for ($^{18}O/^{16}O$, D/H) in geothermal waters. They form horizontal lines cutting the meteoric straight line at the point corresponding to local rainwater. This is interpreted by saying that water has exchanged its oxygen isotopically with the rock but the hydrogen of water does not change because it is an infinite hydrogen isotope reservoir compared with rocks that are relatively poor in hydrogen. After Craig (1963).

is still referred to as juvenile water. **Harmon Craig** (he again) studied geothermal water to determine the isotope composition of any juvenile water. He showed that the δ values of geothermal waters from the same source can be plotted on a ($\delta D, \delta^{18}O$) diagram on straight lines pretty well parallel to the $\delta^{18}O$ axis and which cut the straight line of precipitation corresponding to the composition of rainwater for the region. And so the composition of geothermal water can be explained by the evolution of meteoric water via isotope exchange of oxygen with the country rock. There is no need to invoke juvenile water from the mantle to explain these isotope compositions (Figure 7.21).

As these relations are systematic for all the geothermal regions studied, Craig concluded that the input of juvenile water into the current water cycle is negligible and that geothermal waters are only recycled surface water. The same goes for water from volcanoes. This hypothesis has been confirmed by more elaborate studies of variations in the isotope composition of geothermal water over time. In many cases, it has been shown that variations tracked those observed in the same place for rainwater, with a time lag corresponding to the transit time which varied from months to years.[12]

EXAMPLE

Iceland's geysers

In some instances, such as the geysers of Iceland, the straight line of ($\delta D, \delta^{18}O$) correlation is not horizontal but has a positive slope (Figure 7.22).

[12] A spa water company signed a research contract with a Parisian professor to study the isotopic composition of the water it sold to prove it was "juvenile" water, a name whose advertising value can be well imagined. As the studies showed the water was not juvenile, the company terminated the contract and demanded that the results should not be published!

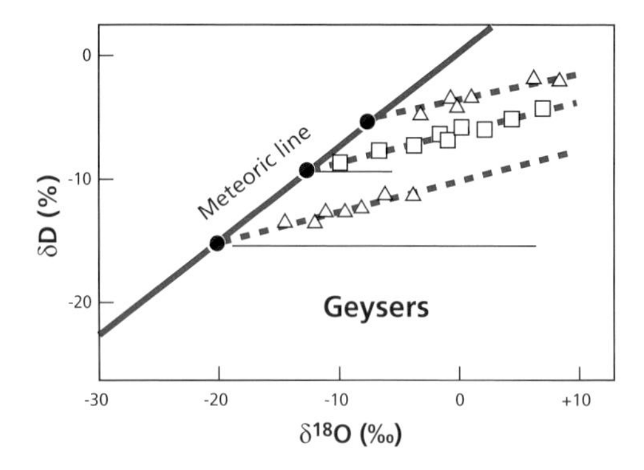

Figure 7.22 Correlation diagram for ($^{18}O/^{16}O$, D/H) in acidic geothermal waters and geysers. The diagram is identical to the previous one except that these are acidic geothermal waters with a high sulfate content whose pH is close to 3 and for which the correlated enrichment in D and ^{18}O results mostly from more rapid evaporation of light molecules with kinetic fractionation into the bargain. After Craig (1963).

Suppose we begin with rainwater of local composition and that this water undergoes distillation by evaporating. Then:

$$\delta_D \approx \delta_{O,D} + 10^3(\alpha_D - 1)\ln f.$$

$$\delta^{18}O \approx \delta^{18}O, O + 10^3(\alpha^{18}O - 1)\ln f.$$

Eliminating ln f gives:

$$\frac{\delta_D - \delta_{O,D}}{\delta_{18}O - \delta_{18}O, O} \approx \left(\frac{\alpha_D - 1}{\alpha_O - 1}\right).$$

We know that at 100 °C, for water–vapor fractionation, $\alpha_D = 1.028$ and $\delta^{18}O = 1.005$. The slope corresponds to 5.6, a lower value than that of equilibrium fractionation (8). The effect is therefore a combination between exchange and distillation.

In fact, in nature, isotopic compositions of geothermal water or vapor are combinations between Rayleigh distillation and the water–rock oxygen isotope exchange, between kinetic fractionation and equilibrium fractionation. A horizontal slope indicates that isotope exchange has been possible and so the transit time is long. When the slope is identical to that of the Rayleigh law, the transit time has been short.

7.6 Oxygen isotopes in igneous processes

Examination shows that the $^{18}O/^{16}O$ isotope composition of unaltered rock of deep origin, whether ocean basalts or ultrabasic rocks, is extraordinarily constant at $\delta^{18}O = +5.5$ (Figure 7.23). This value is analogous to the mean value of meteorites. It has therefore been agreed that this value is the reference value for the mantle. When taking stock of measurements on basic or acid, volcanic or plutonic igneous rocks, the results are found to divide between:

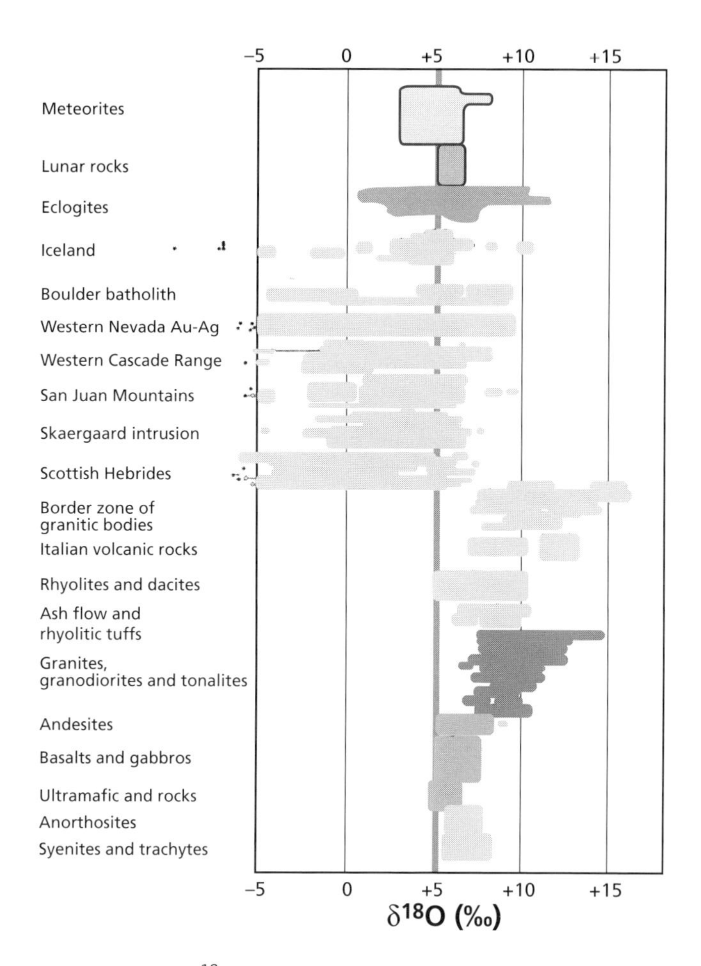

Figure 7.23 Values of $\delta^{18}O$ in rocks and minerals. After Taylor (1974).

- igneous rocks with a $\delta^{18}O$ value greater than 5.5;
- igneous rocks with a $\delta^{18}O$ value less than 5.5, and some with negative values.

These two trends correspond to two types of phenomena affecting igneous rocks: contamination by crustal rocks and *post solidus* exchanges with hydrothermal fluids.

7.6.1 Contamination phenomena

These phenomena are classified under two types: those involving mixing at the magma source where melting affected both acidic and basic metamorphic rocks, and those where contamination occurred when the magma was emplaced. The latter process, known as assimilation, obeys a mechanism already accounted for by **Bowen** (1928). Mineral crystallization in a magma chamber releases **latent heat** of crystallization. This latent heat melts rock around the edges of the magma chamber leading to their assimilation.

$$m \downarrow L = m \uparrow C_P \Delta T,$$

where L is the latent heat, $m\downarrow$ the mass of crystals precipitating per unit time, $m\uparrow$ the mass of rock assimilated, C_P the specific heat of the surrounding rocks, and ΔT the temperature difference between the wall rock and the magma. If we can write $m\downarrow = kM$, then $kML = m\downarrow C_P \Delta T$, therefore:

$$\left(\frac{m\uparrow}{M}\right) = \left(\frac{kL}{C_P \Delta T}\right).$$

The magma is contaminated isotopically too by the mixing law:

$$(\delta^{18}O)_{Hy} = (\delta^{18}O)_{magma}(1 - x) + (\delta^{18}O)_{country\ rock}(x)$$

with $x = (m \downarrow / M)$, because the oxygen contents of the country rock and the magma are almost identical. This was shown by **Hugh Taylor** (1968) of the California Institute of Technology (see also Taylor, 1979).

Exercise

What is the $\delta^{18}O$ value of a basaltic magma whose $\delta^{18}O = 0$ and which assimilates 1%, 5%, and 10% of the country rock whose $\delta^{18}O = +20$?

Answer

	Assimilation		
	1%	5%	10%
$\delta^{18}O$	5.64	6.22	6.95

The contamination effect therefore increases the $\delta^{18}O$ value because sedimentary and metamorphic rocks have positive $\delta^{18}O$ values. An interesting approach to studying the contamination of magmas by continental crust is to cross the studies of oxygen isotopes with those of strontium isotopes. The (O–Sr) isotope diagram can be calculated quite simply because it is assumed that the oxygen content is analogous in the different rocks. The mixing diagram depends only on the Sr contents of the two components of the mixture. Figure 7.24 is the theoretical mixing diagram.

Such combined studies have been made of volcanic rocks of the Japan arcs and the Peninsular Range batholith in California (Figure 7.25).

Exercise

A basaltic magma is emplaced and assimilates 1%, 5%, and 10% of the country rock. The $\delta^{18}O$ values are those of the previous exercise. The $^{87}Sr/^{86}Sr$ values are 0.703 for the magma and 0.730 for the country rock. The Sr content of the magma is 350 ppm and that of the country rock is 100 ppm. Calculate the isotopic compositions of the mixture and plot the ($\delta^{18}O$, $^{87}Sr/^{86}Sr$) diagram.

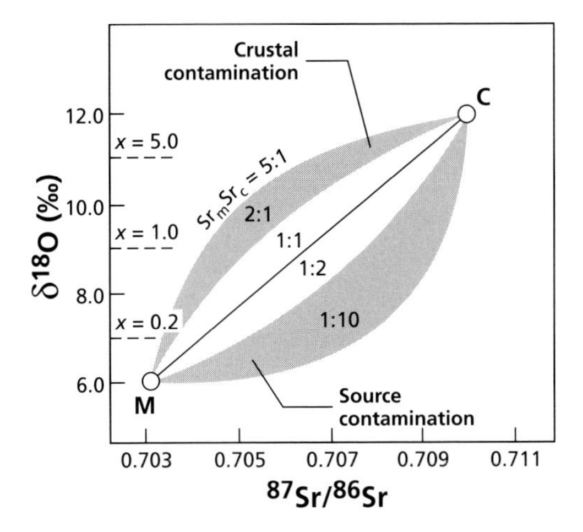

Figure 7.24 Theoretical O–Sr isotope mixing plots. The *x* values show the proportion of country rock relative to the magma. The $Sr_{magma}/Sr_{country\ rock}$ parameter varies from 5 to 0.1. M, magma; C, crust. After James (1981).

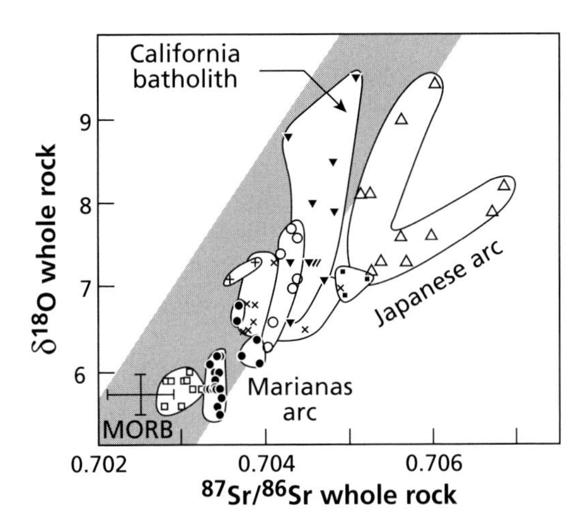

Figure 7.25 Example of an O–Sr isotope correlation diagram showing the Japan arcs and batholith granites in California. After Ito and Stern (1985).

Answer

The $\delta^{18}O$ values have already been given.

	Assimilation		
	1%	5%	10%
$^{87}Sr/^{86}Sr$	0.703 08	0.7034	0.703 94

It is left to the reader to plot the diagram.

7.6.2 Water–rock interaction

As we have said, isotope memory is retained if no exchange occurs after crystallization. When this is not the case, secondary isotopic disturbances can be turned to account. **Hugh Taylor** and his students observed when examining various granite massifs or hydrothermal mineral deposits that the $^{18}O/^{16}O$ isotope compositions had been disrupted after their initial crystallization by water–rock exchanges. The calibration made on water–mineral fractionation was therefore turned directly to account.

Whereas the $\delta^{18}O$ values of minerals and rocks of deep origin are generally positive (between +5 and +8), these rocks had negative $\delta^{18}O$ values of −6 to −7. In the same cases, relative fractionation as can be observed between minerals, such as quartz–potassium feldspar fractionation, was reversed. Taylor remembered Craig's results on thermal waters and postulated that, rather than observing the waters, he was observing rock with which the waters had swapped isotopes. From that point, he was able to show that the emplacement of granite plutons, especially those with associated mineral deposits, involves intense fluid circulation in the surrounding rock. Of course, the existence of such fluids was already known because they give rise to veins of aplite and quartz pegmatite and they engender certain forms of mineralization around granites, but their full importance was not understood.

In a closed system, we can write the mass balance equation:

$$W g_{\mathrm{w}} \delta_{0,\mathrm{W}} + R g_{\mathrm{r}} \delta_{0,\mathrm{R}} = W g_{\mathrm{w}} \delta_{\mathrm{W}} + R g_{\mathrm{r}} \delta_{\mathrm{R}}$$

where W is the mass of water and R the mass of rock, g_{w} is the proportion of oxygen in the water and g_{r} the proportion of oxygen in the rock, $\delta_{0,W}$ and $\delta_{0,R}$ are the initial compositions of water and rock, and δ_W and δ_R are the final compositions thereof.

$$\frac{W}{R} = \left(\frac{g_{\mathrm{r}}}{g_{\mathrm{w}}}\right)\left(\frac{\delta_R - \delta_{0,R}}{\delta_{0,W} - \delta_W}\right),$$

since δ_W and δ_R are related by fractionation reactions $\delta_W = \delta_R - \Delta$. This gives:

$$\frac{W}{R} = \left(\frac{g_{\mathrm{r}}}{g_{\mathrm{w}}}\right)\left(\frac{\delta_R - \delta_{0,R}}{\delta_{0,W} - (\delta_R - \Delta)}\right) \qquad \frac{g_{\mathrm{r}}}{g_{\mathrm{w}}} \approx 0.5.$$

Indeed, $g_{\mathrm{r}} = 0.45$ and $g_{\mathrm{w}} = 0.89$. We estimate $\delta_{0,R}$ from the nature of the rock and the catalog of sound rock (close to +5), and we estimate Δ by calibrating and estimating temperature by fractionation among minerals. This temperature can be compared with the temperature obtained by the heat budget.

A calculation may be made, for example, for a feldspar with $\delta^{18}O = +8$ and $\delta_{W,^{18}O} = -16$ at various temperatures (Figure 7.26a). It shows that in a closed system, the W/R ratios may be extremely variable.

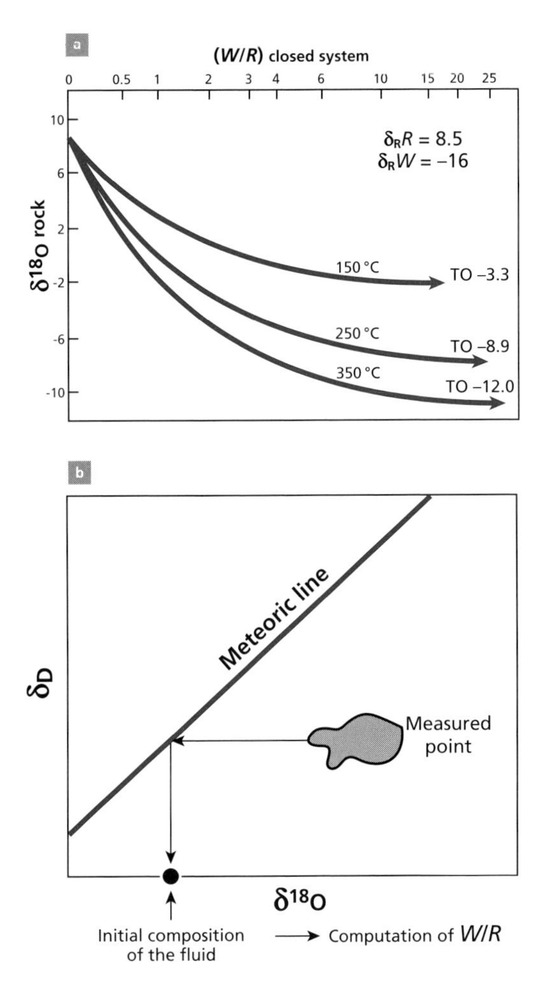

Figure 7.26 Variation in δ^{18}O composition and (δD, δ^{18}O) correlation diagram. (a) The variation in the δ^{18}O composition of a feldspar with an initial composition $\delta = +18$ is calculated as a function of (W/R) for various temperatures, with the initial composition of water being $\delta = -16$. (b) It is assumed the altered rocks are represented by the blue area in the (δD, δ^{18}O) diagram. We can try to determine the initial composition of the fluid by assuming, as a first approximation, that the δD values of the rock and water are almost identical. The intersection between the horizontal and the (δD, δ^{18}O) correlation diagram of rainwater gives the value of water involved in alteration. Reconstructed from several of Taylor's papers.

Exercise

What is the W/R ratio of a hydrothermal system supposedly working in a closed system at 400 °C?

The initial δ^{18}O value of feldspar is +8, that of the water determined by the meteoric straight line is $\delta_{O,W} = -20$. The $\Delta_{\text{feldspar–water}}$ fractionation factor is $3.13 \times 10^6\ T^{-2} - 3.7$. The δ^{18}O of feldspar is measured as $\delta_R = -2$.

Answer

The fractionation factor $\Delta = 3.13 \cdot 10^6/(673)^2 - 3.7 = 3.21 = \delta_R - \delta_W(W/R) = 0.34$.

Exercise

Let us now suppose the W/R ratio = 5, that is, there is much more water. All else being equal, what will be the $\delta^{18}O$ value of the feldspar measured?

Answer

$\delta^{18}O = -14.54$.

The process described in the previous exercise involves a double exchange and it is either the water or the rock that influences the isotopic composition of the other depending on the W/R ratio.

Allowing for the point that Δ varies with temperature, a whole range of scenarios can be generated.

Exercise

Let us pick up from where we left off in the previous exercise. Imagine an exchange between sea water and oceanic crust whose $\delta_{0,R} = +5.5$. The exchange occurs at $W/R = 0.2$. What will the isotope composition of the water and rock be?

Answer

At high temperature $\Delta = 0$; maintaining $g_r/g_w = 0.5$ gives $\delta_{rock} = +3.64$ and $\delta_{water} = +3.64$.

Exercise

Let us imagine now that the water is driven out of the deep rock and rises to the surface and cools to, say, 200 °C. It attains equilibrium with the country rock and its minerals. If the rock contains feldspar, what will the isotope composition of the feldspar be?

Answer

At 100 °C, $\Delta_{feldspar-water} = 10$. Therefore the feldspar of the rock will have a δ value of $13.92 \approx +14$.

It can be seen from the water cycle in the previous exercise that hydrothermal circulation reduces the δ value of deep rocks and increases the δ value of surface rocks. (This is what is observed in ophiolite massifs.)

7.7 Paleothermometry and the water cycle: paleoclimatology

We have just seen how hydrothermalism can be studied by combining information on the isotope cycle of water and that of isotope fractionation. We are going to see how these two effects combine to give fundamental information on the evolution of our planet and its climate. After the initial impetus from **Cesare Emiliani** at Miami University and **Sam Epstein** at the California Institute of Technology, European teams have been the more active ones in this field: for sediment paleothermometry, the teams from Cambridge and Gif-sur-Yvette; for glacial records, those from Copenhagen, Berne, Grenoble, and Saclay.

7.7.1 The two paleoclimatic records: sediments and polar ice

Carbonate paleoclimatology

In order to use oxygen isotopes as a thermometer, we must, strictly, know the $\delta^{18}O$ values of two compounds in equilibrium: water and carbonate. The formula established by **Urey** and his team for the carbonate thermometer draws on $\delta^{18O}_{CaCO_3}$ and $\delta^{18O}_{H_2O}$. In a first approach, the Chicago team had considered that δH_2O, that is the δ of sea water, was constant over geological time and therefore that the $\delta^{18O}_{CaCO_3}$ measurement gave paleotemperatures directly. The discovery of extreme $\delta^{18}O$ values for Antarctic ice challenged this postulate. If the amount of Antarctic ice lost every year into the ocean varies, the $\delta^{18}O$ value of the ocean must vary too, since this ice may have $\delta^{18}O$ values as low as -50. In this case, the hypothesis of constant δH_2O is untenable and it seems that temperatures cannot be calculated simply. On the other hand, if the dissolution of Antarctic ice in the ocean varies in volume, this phenomenon must be related to climate and therefore, to some extent, must reflect the average global temperature.

The first idea developed by **Emiliani** in 1955 was therefore to measure the $\delta^{18}O$ values of carbonate foraminifera in Quaternary sediment cores for which (glacial and interglacial) climatic fluctuations have long been known. Variations in isotope composition are observed (Figure 7.27) and seem to be modulated by glacial and interglacial cycles and more

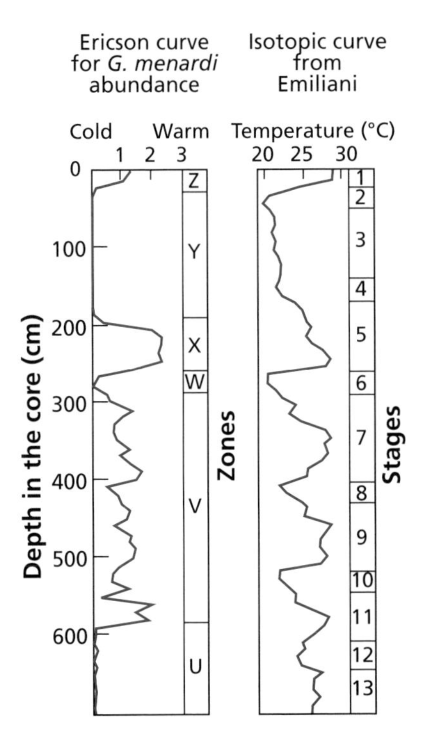

Figure 7.27 The first isotope determination using $\delta^{18}O$ of Quaternary paleotemperatures by Emiliani (1955) compared with glacial–interglacial divisions by **Ericson**. (*Globorotalia menardi* is a foraminifer.)

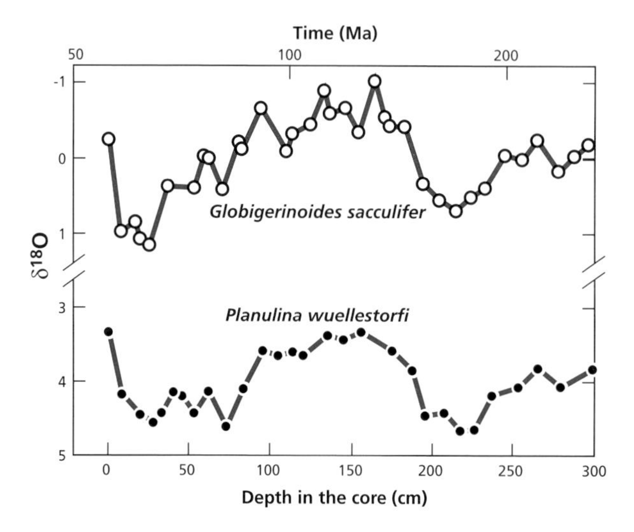

Figure 7.28 Variation in $\delta^{18}O$ in samples of two species of foraminifer. Top: a pelagic (ocean surface) species. Bottom: a benthic (ocean floor) species. Modified after Duplessy *et al.* (1970).

specifically to follow the theoretical predictions of the Yugoslav astronomer **Milankovitch**. Are these variations a direct effect of temperature on (carbonate–water) fractionation or are they the effect of ^{18}O dilution by polar ice? The question remained unanswered.
The formula:

$$T_{^\circ C} = 16.9 - 4.2\left(\delta^{18}O_{CaCO_3} - \delta^{18}O_{H_2O}\right) + 0.13\left(\delta^{18}O_{CaCO_3} - \delta^{18}O_{H_2O}\right)^2$$

shows us that that two effects work in the same direction. When T increases, ($\delta CaCO_3 - \delta H_2O$) fractionation decreases and so $\delta CaCO_3$ decreases if δH_2O remains constant. If, with constant local Δ fractionation, $\delta^{18}O_{H_2O}$ decreases for want of polar ice then $\delta^{18}O_{CaCO_3}$ also declines. **Nick Shackleton** of the University of Cambridge suggested that the $^{18}O/^{16}O$ variations measured in forams were the result of fluctuations in the volume of polar ice, a climate-related phenomenon.

Jean-Claude Duplessy and his colleagues in the Centre National de la Recherche Scientifique (CNRS) at Gif-sur-Yvette had the idea of comparing $\delta^{18}O$ fluctuations of surface-living (pelagic) foraminifera and bottom-dwelling (benthic) foraminifera. It is known that the temperature of the deep ocean varies little around +4 °C.

In conducting their study they realized that $\delta^{18}O$ fluctuations of pelagic and benthic species were very similar (Figure 7.28). At most, extremely close scrutiny reveals an additional fluctuation of 2‰ in the $\delta^{18}O$ of pelagic species, whereas no great difference appears for the same comparison with $\delta^{13}C$. This means, then, that the $\delta^{18}O$ variations in foraminifera reflect just as much variation in the δ value of sea water as variations in local temperature. The signal recorded is therefore meaningful for the global climate (Emiliani, 1972).

In fact, more recent studies have confirmed that, for pelagic species, some 50% of the signal reflects a Urey-type local temperature effect (remember that when temperature rises $\delta^{18}O$ falls), above all in the temperate zones, and 50% the effect of melting of the polar ice caps. For benthic species, the dominant factor is the isotopic fluctuation of the ocean as **Shackleton** (1967b) had surmised. In addition, **Duplessy**'s group and **Shackleton** established that an additional isotope fractionation occurred which was characteristic of each species of foraminifera studied. But those "vital effects" were calibrated and so isotopic measurements on different species could be made consistent with each other.

Exercise

We have just seen that the $\delta^{18}O$ variation of foraminifera was mostly due to $\delta^{16}O$ variation because of melting ice. Let us look more closely at the quantitative influence of melting polar ice on $\delta^{18}O$. Imagine an intense glacial period when the sea level falls by 120 m. What would be the volume of polar ice and the $\delta^{18}O$ value of sea water?

Answer

The ocean surface area is $3.61 \cdot 10^8$ km^2. The volume of the ocean is $1370 \cdot 10^6$ km^3. The volume of present-day polar ice is $29 \cdot 10^6$ km^3. If the sea level is 120 m lower, $46 \cdot 10^6$ km^3 has been stored in the ice caps, corresponding to a mass fraction of the hydrosphere of 3.3%. The polar ice caps were 1.6–2 times larger than today.

If we take the $\delta^{18}O$ value of ice as $-50‰$, then $-50‰ \times 0.033 = -1.65‰$. There is indeed a difference in $\delta^{18}O$ of this order of magnitude between glacial and interglacial periods. Notice that, as with radiogenic isotopes, these effects could not be detected if we did not have a very precise method for measuring $\delta^{18}O$.

Glaciers

Another interesting application of this fractionation was begun on glaciers independently by **Samuel Epstein** of the California Institute of Technology and (more systematically and continuously) by **Willi Dansgaard** of Copenhagen University. When a core of polar glacier ice is taken, it has layers of stratified ice which can be dated by patient stratigraphy and various radiochronological methods. Now, the study of these ice strata reveals variations in $\delta^{18}O$ and δD (Dansgaard, 1964; Epstein *et al.*, 1965; Dansgaard and Tanber, 1969) (Figures 7.29 and 7.30).

For a single region such variations are analogous and mean the sequence of one glacier can be matched with the sequence of a neighboring glacier. An isotope stratigraphy of glaciers can be defined. We can venture an interpretation of these facts in two ways. Either we accept that the origin of precipitation has varied over recent geological time and we then have a way of determining variations in the meteorological cycle of the past. Or we consider that the fractionation factor has varied and therefore the temperature has varied.

Research by **Dansgaard** and his team on the ice first of Greenland and then of Antarctica showed that **the temperature effect is predominant**. By simultaneously measuring isotope composition and temperature, he showed that the $\delta^{18}O$ and δD correlation did indeed correspond to this effect. Moreover, the qualitative rule is the reverse of the carbonate rule: when the temperature rises, both $\delta^{18}O$ and δD increase, because fractionation diminishes with temperature; but the δ values are negative and so move closer to zero (Figure 7.31). A simple empirical rule is that whenever the temperature rises by 1 °C, $\delta^{18}O$ increases by 0.7‰.

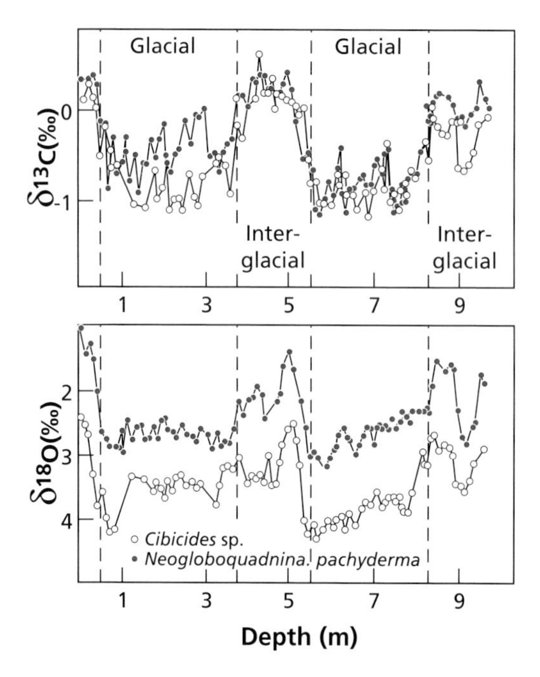

Figure 7.29 Comparison of $\delta^{13}C$ and $\delta^{18}O$ fluctuations in a pelagic (blue circles) and a benthic (white circles) species of foraminifer from the Antarctic Ocean. The $\delta^{13}C$ variation (above) is represented to show there is a shift for $\delta^{18}O$ (below) but not for $\delta^{13}C$.

We can investigate why $\delta^{18O}_{H_2O}$ fluctuation is very important for foraminifera and why it is the local temperature effect that dominates with ice. Because isotope fractionation at very low temperatures becomes very large and dominates isotope fluctuation related to the water cycle. But we shall see that this assertion must be qualified. Modern studies of isotope fluctuations of glaciers use a combination of both effects, local temperatures and isotopic changes in the water cycle, as for foraminifera, but with different relative weightings.

7.7.2 Systematic isotope paleoclimatology of the Quaternary

We have just seen there are two ways of recording past temperatures.

(1) One is based on $\delta^{18}O$ analysis of fossil shells in sedimentary series (marine and continental cores).
(2) The other uses $\delta^{18}O$ analysis of accumulated layers of ice in the ice caps.

Both these methods have progressively converged to allow very precise studies of climatic fluctuations in the Quaternary and more especially for the last million years. **Nick Shackleton** and **Willi Dansgaard** shared the Crafoord Prize in recognition of their complementary achievement. Each method has its limits, and it is only gradually that we have been able to compare and use both types of records in a complementary way to decipher climatic variations that have affected our planet over the last million years and which consist in alternating glacial and warmer interglacial periods.

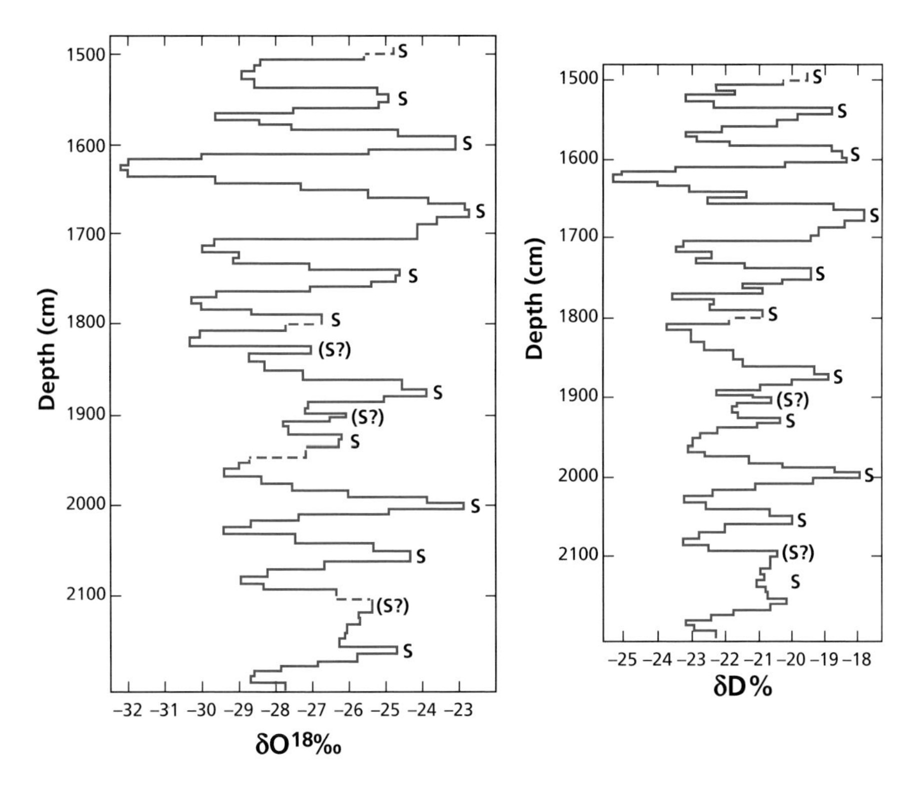

Figure 7.30 Variations in $\delta^{18}O$ and δD with depth in an Antarctic glacier. Summers (S) can be distinguished from winters. (Caution! δD is in percent!) After Epstein and Sharp (1967).

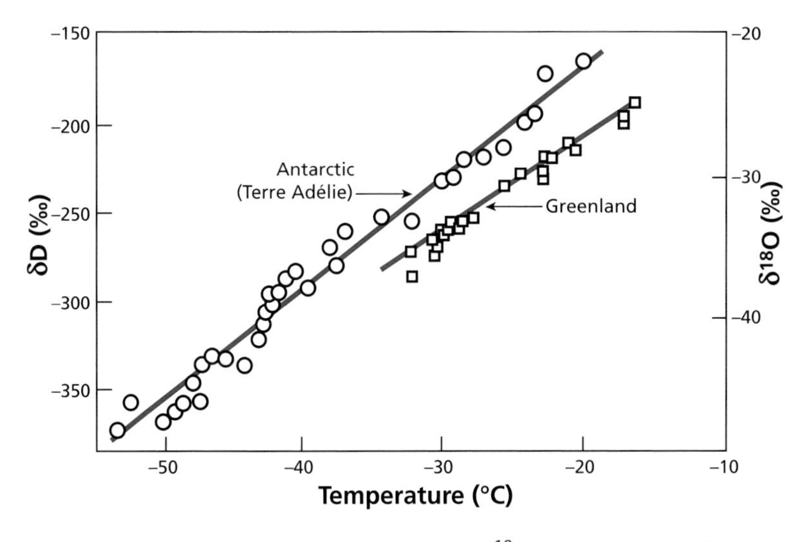

Figure 7.31 Relationship between temperature, $\delta^{18}O$, and δD in snowfall at the poles. After Lorius and Merlivat (1977) and Johnsen *et al.* (1989).

Core sampling (sequential records)

Cores of marine sediments can be taken from all latitudes and longitudes (in the ocean and from continents and lakes); however, two conditions restrict their use. First, sedimentation must have occurred above the "carbonate compensation depth" for there to be any measurable fossil tests left. And second, sedimentation must have been very rapid to provide a record with good time resolution. Sedimentary cores have no time limits other than the lifespan of the ocean floor. Quaternary, Tertiary, and Secondary cores can be studied up to 120 Ma, which is the age of the oldest remnants of oceanic crust that have not been swallowed up by subduction (ancient cores are compacted and transformed into hard rocks and so time resolution is not as good).

For ice caps, the first problem is, of course, their limited geographical and temporal extent. Geographically, records are primarily from the glaciers of Antarctica and Greenland. Mountain glaciers have also recorded climatic events but over much shorter time-spans.[13]

Ice caps are limited in time. For a long time, the longest core was one from Vostok in Antarctica covering 420 000 years. A new core of EPICA has been drilled and covers 700 000 years. Cores from the big mountain glaciers go back a mere 2000 years or so. For both types of record – sediments and ice – precise, absolute dating is essential, but here again many difficulties arise. Especially because as research advances and as studies become ever more refined for ever smaller time-spans, the need for precision increases constantly. There is scope for ^{14}C dating and radioactive disequilibrium methods on sedimentary cores, but their precision leaves something to be desired. Useful cross-checking can be done with paleomagnetism and well-calibrated paleontological methods. In turn, the oxygen isotopes of a well-dated core can be used to date the levels of other cores. Thus, gradually, a more or less reliable chronology is established, which must be constantly improved. Dating is difficult on ice caps except for the most recent periods where annual layers can be counted. Methods based on radioactive isotopes such as ^{14}C, ^{10}Be, ^{36}Cl, ^{87}Kr, and ^{37}Ar are used, but they are extremely difficult to implement both analytically (ice is a very pure material!) and in terms of reliability. Switzerland's **Hans Oeschger** (and his team) is associated with the development of these intricate techniques for dating ice, which, despite their limitations, have brought about decisive advances in deciphering the ice record (Oeschger, 1982).

These clarifications should make it understandable that establishing time sequences of records is a difficult and lengthy job that is constantly being improved. All reasoning should make allowance for this.

Deciphering sedimentary series and the triumph of Milankovitch's theory

Between 1920 and 1930, the Yugoslav mathematician and astronomer **Milutin Milankovitch** developed a theory to account for the ice ages that had already been identified by Quaternary geologists (see Milankovitch, 1941). These periods were thought to be colder. The polar ice extended far to the south and mountain glaciers were more extensive too (Figure 7.32). Alpine glaciers stretched down as far as Lyon in France. These glacial traces can be identified from striated rock blocks forming what are known as moraines.

[13] They have been used by **Lony Thompson** of Ohio State University for careful study of recent temperature fluctuations (see his 1991 review paper).

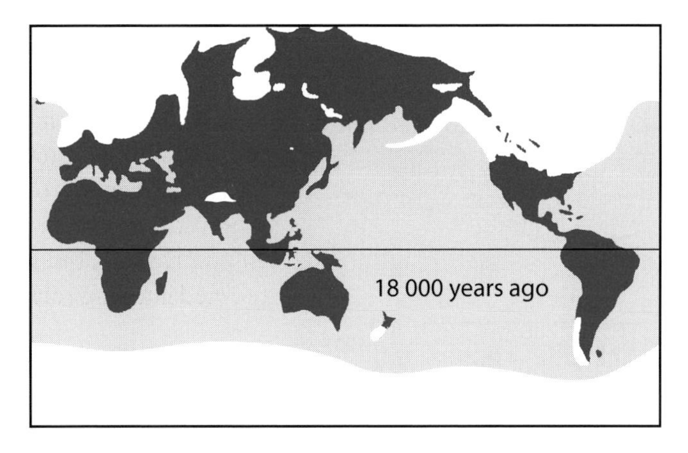

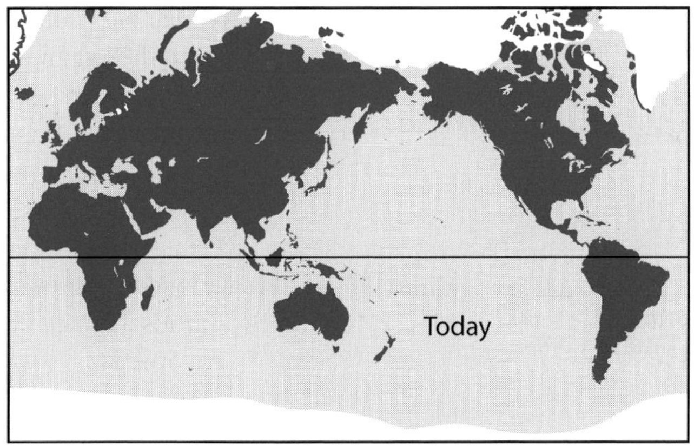

Figure 7.32 Worldwide distribution of ice in glacial and interglacial times.

Milankovitch's theory gave rise to vehement controversy (as vehement as that over **Wegener**'s theory of continental drift[14]). And yet this vision of the pioneers of the 1920s was largely accurate. This is not the place to set out this theory in full. It can be found in textbooks on paleoclimatology, for example, Bradley (1999). However, we shall outline the main principles and the terms to clarify what we have to say about it.

The Earth's axis of rotation is not perpendicular to its plane of rotation around the Sun. It deviates from it by 23° on average. But the axis of this deviation rotates around the vertical over a period of 23 000 years. A further movement is superimposed on these, which is the fluctuation of the angle of deviation between 21.8° and 24.4°. The period of this fluctuation is 41 000 years. The first of these phenomena is termed **precession**, the second is **obliquity**. A third phenomenon is the variation in the **ellipiticity of the Earth's orbit**, with a period of 95 000 years. These three phenomena arise from the influence on the Earth of the Sun, Jupiter, and the other planets and the tides. They not only combine but are

[14] Wegener had been the first, with his father-in-law Köppen, to suggest an astronomical explanation for the ice ages before becoming an ardent defender of Milankovitch.

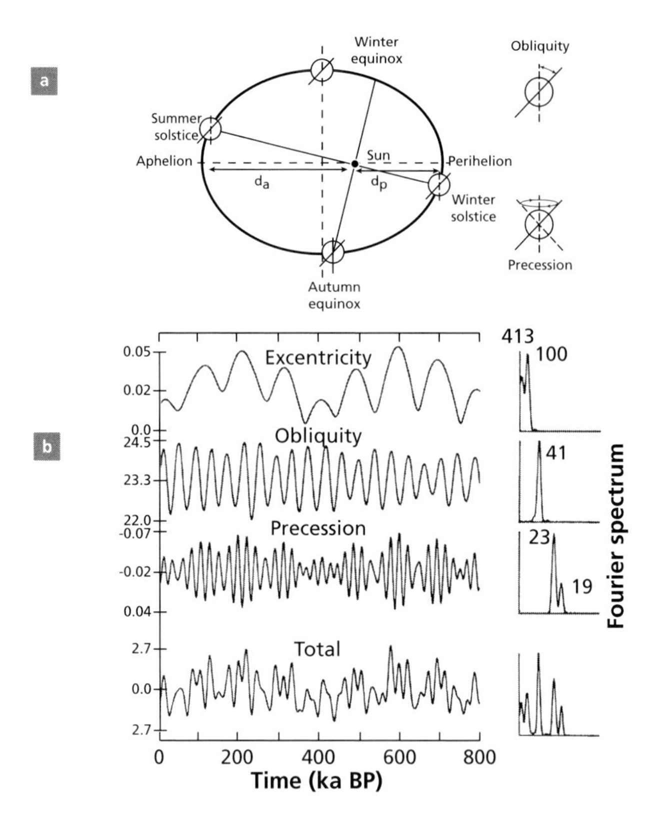

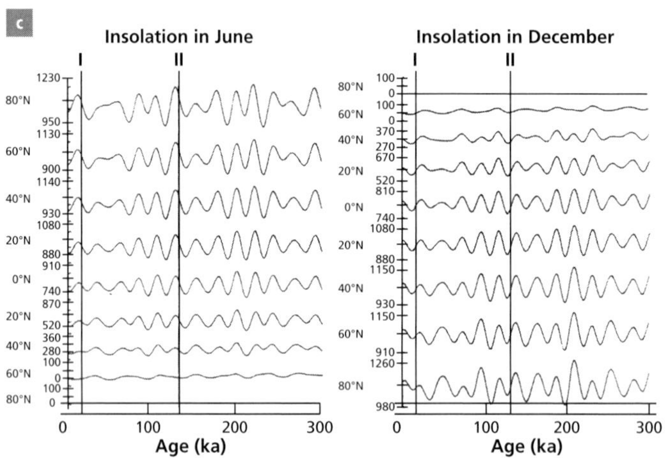

Figure 7.33 The principle of Milankovitch's theory. (a) The three parameters that change: eccentricity of the orbit, obliquity, and precession. (b) Variations in the three parameters calculated by astronomical methods with their Fourier spectrum on the right. (c) Variation in the sunlight curves in June and December with latitude, calculated by the theory.

superimposed, leading to complex phenomena. Thus, at present, the Earth is closest to the Sun on 21 December, but the Earth's axis is aligned away from the Sun, and so, in all, the northern hemisphere receives little sunlight. It is winter there, but other conjunctions also occur. Thus we can calculate the sunshine received during the year at various latitudes. Celestial mechanics mean such calculations can be made precisely (see Figure 7.33 for a simplified summary). As Milankovitch understood, if little sunshine reaches the Earth at high latitudes in summer the winter ice will remain, the white surface will reflect solar radiation, and the cooling effect will be amplified. This is a good starting point for climatic cooling.

What should be remembered is that when we break down the complex signal of sunlight received by the Earth using Fourier analysis methods (that is, when we identify the sine-wave frequencies that are superimposed to make up the signal) we find peaks at 21 000, 41 000, and 95 000 years. When we conduct a similar Fourier decomposition for $\delta^{18}O$ values recorded by foraminifera in sedimentary series, we find the same three frequencies (Figure 7.34).

The $\delta^{18}O$ variations reflect those of the Earth's temperatures. This finding confirms Milankovitch's theory (at least as a first approximation) and so fully bears out the early studies of Emiliani. In complete agreement with the theory, the sedimentary series also showed that climatic variations were very marked at the poles (several tens of degrees), very low in the intertropical zone, and intermediate in the temperate zones (of the order of $\pm 5\,^{\circ}C$).

Figures 7.35 and 7.36 give a fairly complete summary of the essential isotopic observations made from sedimentary cores.

The sedimentary core record (Figure 7.35) also shows in detail how the temperature variations evolved. Cooling is slow, followed by sudden warming. Finer fluctuations are superimposed on these trends but their frequencies match those of the Milankovitch cycles.

Confirmation of Milankovitch cycles by Antarctic isotope records

It was some considerable time before Milankovitch cycles were confirmed in the ice records, for two reasons. There were no ice cores long enough and so covering a long enough time-span and the dating methods were too imprecise.

It was only after the famous Vostok core from Antarctica was studied by the Franco-Russian team that evidence of Milankovitch cycles was found in the ice record. But the core yielded much more than that: it allowed climatic variations to be correlated with variations of other parameters:

- dust content: it was realized that during ice ages there was much more dust and therefore more wind than during interglacial periods.
- greenhouse gas (CO_2 and CH_4) content in air bubbles trapped in the ice – when the temperature increases there is an increase in CO_2 (in the absence of human activity!).

This last question on the debate about the influence of human activity on the greenhouse effect and so on climate is a fundamental one. Which increased first, temperature or CO_2 levels? It is a difficult problem to solve because temperature is measured by δD in ice

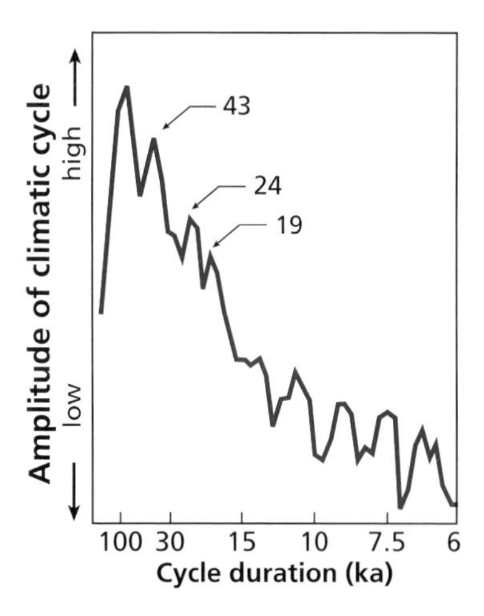

Figure 7.34 Fourier spectrum of paleotemperatures using oxygen isotopes.

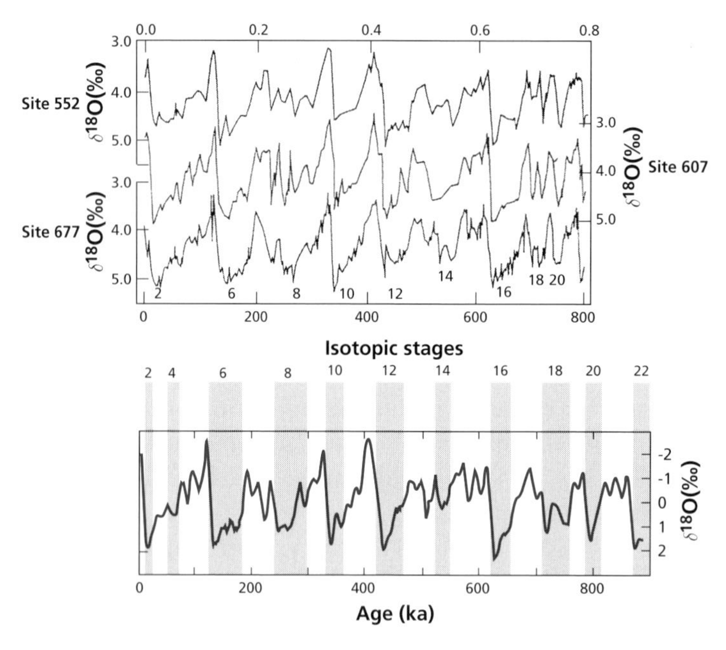

Figure 7.35 Records of $\delta^{18}O$ in foraminifera and the synthetic reference curve. (a) Record of $\delta^{18}O$ for benthic foraminifera at three sites: site 552 – 56° N, 23° W in the North Atlantic; site 607 – 41° N, 33° W in the mid Atlantic; site 667 – 1° N, 84° W in the equatorial Pacific. Correlation between the three cores is excellent. (b) Synthetic reference curve produced by tuning, which consists in defining the timescale so that the Fourier decomposition frequencies of the $\delta^{18}O$ values of the cores match the astronomical frequencies from Milankovitch's theory. The period is then subdivided into isotopic stages. The odd stages are warm periods and the even stages (shaded) glacial periods. (Notice that the interglacials correspond to increased $\delta^{18}O$ values and fluctuations are just a few per mill.) After various compilations from Bradley (1999).

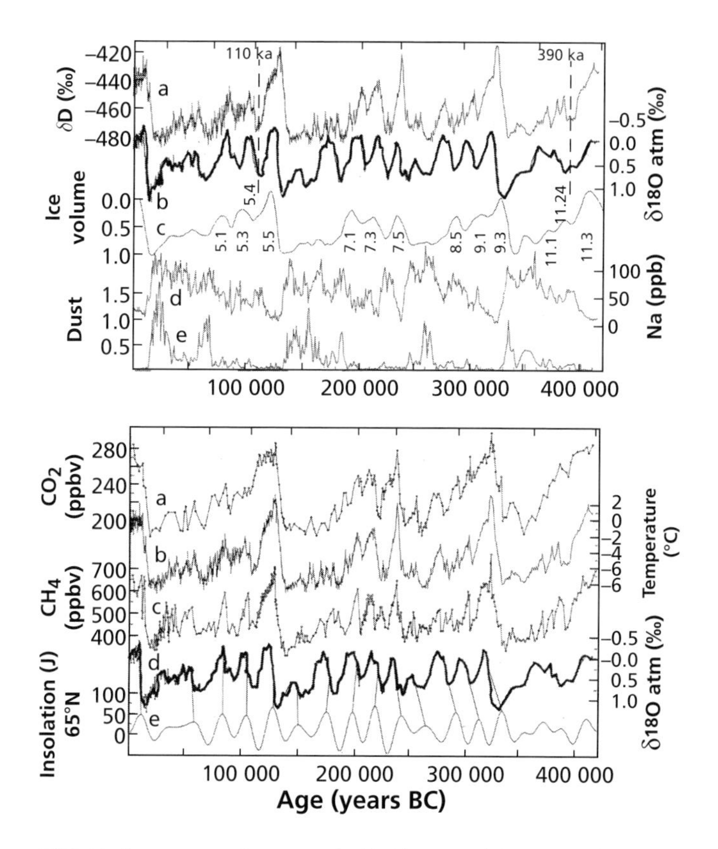

Figure 7.36 Various parameters recorded in the Vostok ice core. After Jouzel *et al.* (1987) and Petit *et al.* (1999).

whereas CO_2 is measured from its inclusion in ice. Now, gaseous inclusions are formed by the compacting of ice and continue to equilibrate with the atmosphere, that is, the air samples are younger than the ice that entraps them (Figure 7.37).

We need, then, to be able to measure the temperature of inclusions directly and compare it with the temperature measured from the δD value of the ice. A method has been developed by **Severinghaus** *et al.* (2003) for measuring the temperature of fluid inclusions using $^{40}Ar/^{36}Ar$ and $^{15}N/^{14}N$ isotope fractionation. After stringent calibration, the teams at the Institut Simon-Laplace at Versailles University and at the Scripps Institution of Oceanography (La Jolla, California) managed to show that the increase in CO_2 lags behind the increase in temperature by 800 years (Figure 7.38) and not the other way round as asserted by the traditional greenhouse-effect model (Severinghaus *et al.*, 1999; Caillon *et al.*, 2003). Now, we know that CO_2 solubility in sea water declines as temperature rises and that the characteristic time for renewal of the ocean water is 1000 years. The first phase of temperature increase followed by the increase in CO_2, with a lag of 800 years, can be readily understood, then, if we invoke the lag because of the thermal inertia of the ocean. There may also be some feedback of the CO_2 effect on temperature.

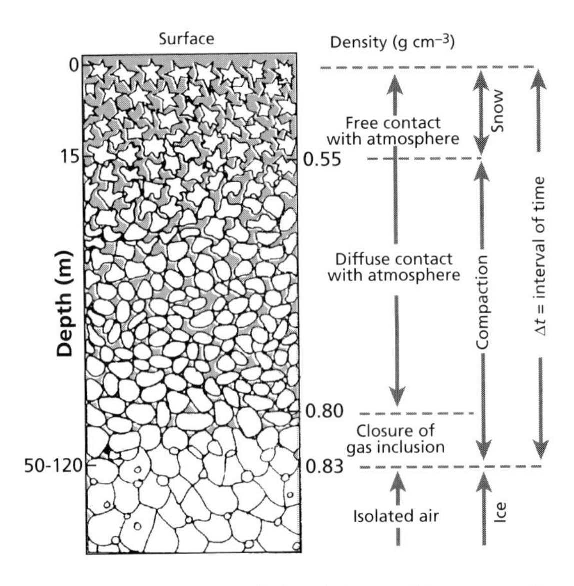

Figure 7.37 As ice is compacted, the air trapped is younger than the snow.

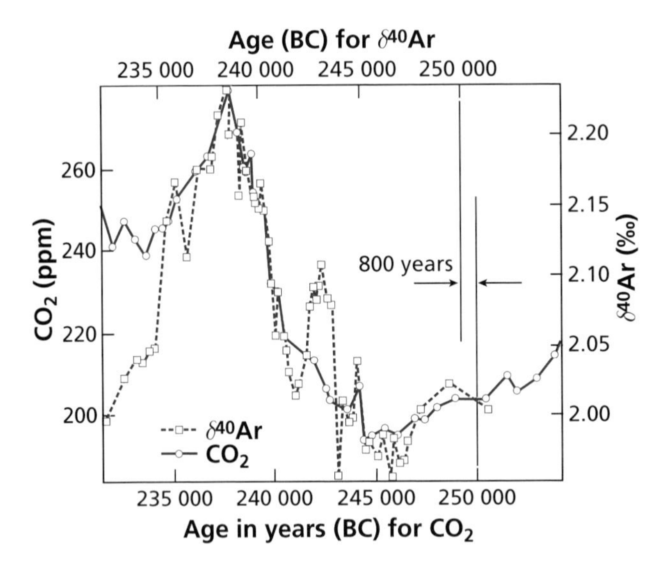

Figure 7.38 Records of δ^{40}Ar and CO_2 from the Vostok core after shifting the CO_2 curve 800 years backwards. After Caillon *et al.* (2003).

Exercise

Argon and nitrogen isotope fractionations are caused by gravitational fractionation in the ice over the poles. Using what has been shown for liquid–vapor isotope fractionation, find the formula explaining this new isotope thermometer.

Answer

If we write the fractionation:

$$\alpha = \frac{C_{m_1} P(1)}{C_{m_2} P(2)}$$

$$\ln \alpha = \ln P_{m_1} - \ln P_{m_2} = \Delta m \frac{gz}{RT}.$$

From the approximation formulae we have already met:

$$\frac{\Delta\delta}{1000} \approx \left(\Delta m \frac{gz}{RT}\right)$$

with $g = 10\ \mathrm{m\ s^{-1}}$, $T = 200\ \mathrm{K}$, for $z = 10\ \mathrm{m}$, and $R = 2$.

This gives $\Delta\delta = 0.25$ for nitrogen and $\Delta\delta = 1.1$ for argon. Once again, extremely precise methods for measuring isotope ratios had to be developed.

Greenland and Antarctica records compared and the complexity of climatic determinants

Although studies of Greenland ice cores pre-dated that of the Vostok core by far, it was only after the Vostok core had been deciphered that the signification of the Greenland cores was fully understood by contrast (Figure 7.39). It was observed that the record of the oxygen and hydrogen isotopes at Vostok was much simpler than in Greenland and that the Milankovitch cycles were clearly recorded.

Things are more complex in Greenland because sudden climatic events are superimposed on the Milankovitch cycles. The first well-documented event is a recurrent cold period at the time of transition from the last ice age to the Holocene reported by **Dansgaard**'s team in 1989. While some 12 800 years ago a climate comparable to that of today set in, it was interrupted 11 000 years ago by a cold episode that was to last about 1000 years and which is known as the Younger Dryas. This event was found some years later in the sedimentary record of the North Atlantic.

Generalizing on this discovery, Dansgaard used oxygen isotopes to show that the glacial period was interrupted by warm periods that began suddenly and ended more gradually. These events, of the order of a few thousand years, correspond to a 4–5‰ change in the $\delta^{18}O$ value of the ice and so to a temperature variation of about 7 °C. Dansgaard's team reported 24 instances of this type of D–O episode, as they are called (D is for Dansgaard and O is for Oeschger), between 12 000 and 110 000 years ago. They have been detected in the sedimentary records of the North Atlantic and as far south as the intertropical zone. Cross-referencing between sedimentary cores and polar ice cores has proved so instructive that both types of record continue to be used to analyze one and the same event.

A second series of events was read this time from the sedimentary records. These were brief events characterized by the discharge of glaciers as far as the Azores. There are 34 of these H events (see Heinrich, 1988) during the last glacial period and nothing comparable has been identified in the ice of Greenland. The relationship between D–O and H events is unclear, but what is clear is that they are not identical happenings.

What compounds the mystery is that these brief events are recorded very faintly in Antarctica, with a time lag. There are therefore one or more mechanisms governing climate

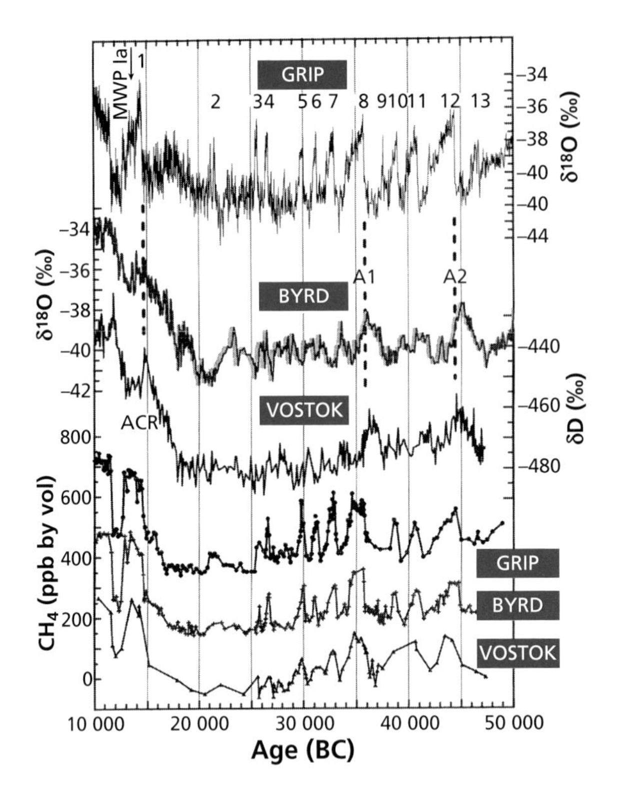

Figure 7.39 Comparison of records from the Antarctic and Greenland. Vostok and Byrd are two stations in the Antarctic; Grip is a borehole in Greenland. Notice that the two CH_4 (methane) peaks are quite comparable in both hemispheres, which is evidence that chemically the atmosphere is broadly homogeneous. By contrast, the $\delta^{18}O$ (and δD) records are very different. Greenland is the site of many sudden climatic events that have been numbered, which events are not seen in Antarctica. After Petit *et al.* (1999).

that are superimposed on the Milankovitch cycles. Why are these events more readily detectable in the northern hemisphere? One of the big differences between the two hemispheres is the asymmetrical distribution of landmasses and oceans (Figure 7.40). This is probably an important factor, but how does it operate? We don't know. This qualitative asymmetry is compounded by a further asymmetry. It seems that the transitions between glacial and interglacial periods occur 400–500 years earlier in the Antarctic. Once again, there is as yet no clear and definite explanation for this.

Exercise

When the $\delta^{18}O$ values of foraminiferan shells are analyzed for glacial and interglacial periods, variations are of 1.5‰ for cores from the intertropical zone but of 3‰ for cores from the temperate zones. How can you account for this phenomenon?

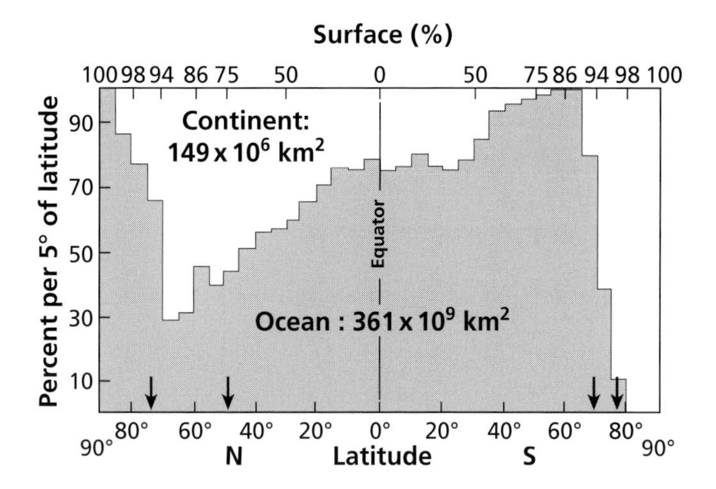

Figure 7.40 Distribution of landmasses and continents by latitude. The difference between northern and southern hemispheres is clearly visible.

Answer

Temperature variations between glacial and interglacial periods are very large at the poles, very low in the intertropical zone, and intermediate in the temperate zones. The variation of 1.5‰ for the intertropical zone is caused by the melting of ice. It must therefore be considered that the additional variation of 1.5‰ of the temperate zones is due to a local temperature effect. Applying the Urey–Epstein formula of the carbonate thermometer

$$T_{°C} = 16.3 - 4.3(\Delta\delta)/0.13(\Delta\delta)^2$$

gives $\Delta T \approx 4.3\ \Delta\delta$. This corresponds to 5–6 °C, the type of temperature difference one would expect in the temperate zone between a glacial and interglacial period.

Very recently, the ice record has been extended to 700 000 years thanks to the core drilled by an international consortium in Antarctica at the EPICA site. As Figure 7.41 shows, this facilitates comparison between sedimentary and ice records. New data can be expected shortly.

7.8 The combined use of stable isotopes and radiogenic isotopes and the construction of a global geodynamic system

Climate is a complex phenomenon with multiple parameters. Temperature, of course, is a cardinal parameter, but the distribution of rainfall, vegetation, mountain glaciers, and winds are essential factors too. Oxygen and deuterium isotopes provide vital information about temperatures and the volume of the polar ice caps. The $^{13}C/^{12}C$ isotopes are more difficult to

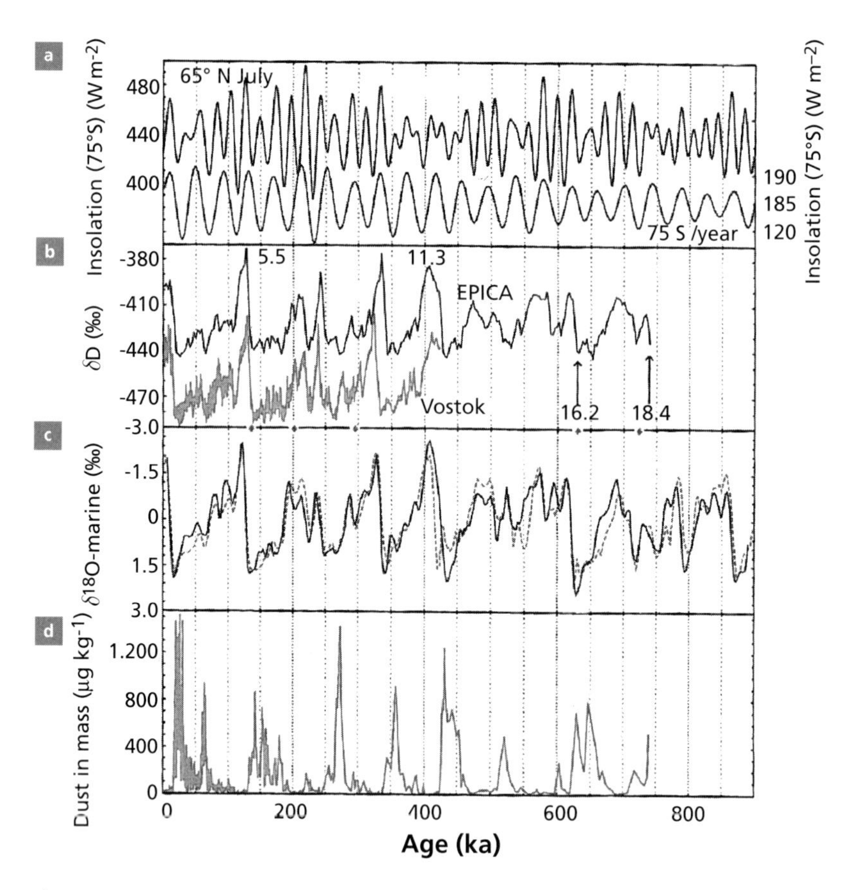

Figure 7.41 Comparisons of (a) insolation and sedimentary (c) and glacial (b, d) records at the EPICA site. After the EPICA Community (2004).

interpret but yield useful pointers, for example, about the type of vegetation (C_3/C_4 plants) and its extent. But such information, which should in principle enable us to construct a biogeochemical picture, is still very difficult to decipher and has been oversimplified in the past.

The use of long-lived radiogenic isotopes has a similar objective, namely to determine how climatic fluctuations are reflected in the planet's erosional system. Erosion is a fundamental surface process. It is what changes volcanoes or mountain ranges into plains and peneplains. The end products of erosion are of two types. Some chemical elements are dissolved as simple or complex ions, while others remain in the solid state. The former are transported in solution, the latter as particles. Whichever state they are in, they are carried by rivers down to the oceans where they form sea water in one case and sediments in the other. The radiogenic isotope ratios are preserved throughout these erosion and transport processes. They are then mixed in the ocean, either as solutions in sea water or as particles in sediments. The erosion sites have characteristic radiogenic isotope signatures which distinguish old landmasses, young continents, and volcanic products of mantle origin. The isotopic compositions of the mixture that makes up sea water thus reflect the proportion of the various sources involved in the erosion processes.

7.8.1 Strontium in the ocean

A first example is the $^{87}Sr/^{86}Sr$ isotopic composition of present-day sea water. The $^{87}Sr/^{86}Sr$ ratio is 0.709 17 and is identical whichever ocean is considered. Where does the Sr come from? Obviously from the erosion of landmasses and subaerial or submarine volcanic activity. Measurement of the isotopic composition of Sr dissolved in rivers yields a mean value of 0.712 ± 0.001. The mean isotopic composition of the various volcanic sources (mid-ocean ridges, island volcanoes, and subduction zones) lies between 0.7030 and 0.7035 (depending on the relative importance attributed to the various sources). From the mass balance equation:

$$\left(\frac{^{87}\mathrm{Sr}}{^{86}\mathrm{Sr}}\right)_{\text{sea water}} = \left(\frac{^{87}\mathrm{Sr}}{^{86}\mathrm{Sr}}\right)_{\text{continental rivers}} x + \left(\frac{^{87}\mathrm{Sr}}{^{86}\mathrm{Sr}}\right)_{\text{volcanic input}} (1-x).$$

The fraction from continental rivers corresponds to $66\% \pm 2\%$ of the Sr in sea water. (Notice this fraction x reflects the mass of chemically eroded continent modulated by the corresponding absolute concentration of Sr.)

$$x = \frac{\dot{m}_c \, C_c}{\dot{m}_c \, C_c + \dot{m}_v \, C_v}$$

where $\dot{m}_c$ is the mass of continent eroded chemically per unit time, $\dot{m}_v$ is the mass of volcanoes eroded chemically per unit time, and C_c and C_v are the Sr contents of rivers flowing from continents and of volcanoes, respectively.

We can go a little further in this breakdown. The isotopic composition of Sr in rivers is itself a mixture of the erosion of silicates of the continental crust, whose mean isotopic composition we have seen is $^{87}Sr/^{86}Sr \approx 0.724 \pm 0.003$,[15] and the erosion of limestones, which are very rich in Sr and are the isotopic record of the Sr of ancient oceans. For reasons we shall be in a better position to understand a little later on, the mean composition of these ancient limestones is $^{87}Sr/^{86}Sr = 0.708 \pm 0.001$. The mass balance can be written:

$$\left(\frac{^{87}\mathrm{Sr}}{^{86}\mathrm{Sr}}\right)_{\text{rivers}} = \left(\frac{^{87}\mathrm{Sr}}{^{86}\mathrm{Sr}}\right)_{\text{limestones}} y + \left(\frac{^{87}\mathrm{Sr}}{^{86}\mathrm{Sr}}\right)_{\text{silicates}} (1-y),$$

which means that limestones make up 71% of the Sr carried to the ocean in solution from the continents. Therefore, all told, the Sr of sea water is made up of 49.5% from reworked limestone, of 16.5% from the silicate fraction of landmasses, and of 34% from volcanic rock of mantle origin. Of course, these figures must not be taken too strictly. The true values may vary a little from these, but not the relative orders of magnitude.

The $^{87}Sr/^{86}Sr$ isotopic composition of present-day marine carbonates is identical to that of sea water in all the oceans. It is assumed, then, that the $^{87}Sr/^{86}Sr$ isotopic compositions measured on more ancient limestones were identical to those of the oceans from which they precipitated (Figure 7.42).

[15] Corresponding to the mean value of detrital particles transported by rivers.

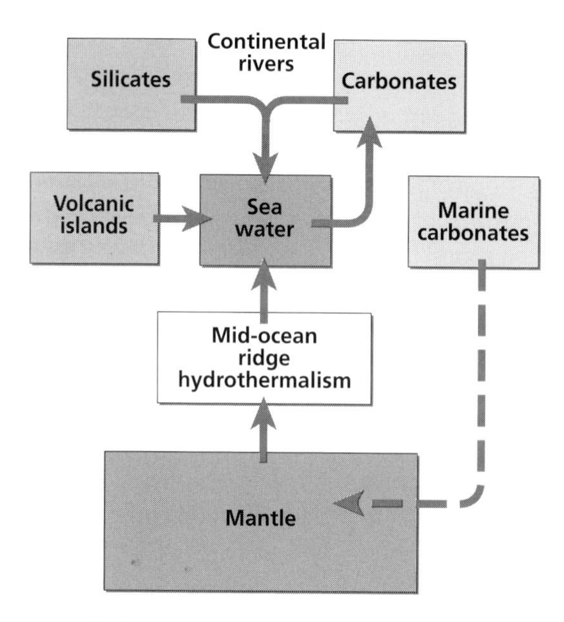

Figure 7.42 The determining factors of the isotope composition of strontium in sea water. The strontium in the ocean comes from alteration of the continents and volcanic arcs. Hydrothermal circulation along the ridges also injects strontium into sea water. Limestones reflect the isotopic composition of the ocean at the time they formed. They are recycled either by erosion or by inclusion in the mantle.

These isotopic compositions have been studied and are found to have varied in the past. The curve of variation of strontium in the course of the Cenozoic has been drawn up with particular care (Figure 7.43).

It shows that the $^{87}Sr/^{86}Sr$ ratio was lower 65 Ma ago than it is today and remained roughly constant from 65 Ma to 40 Ma, from which date it began to rise, at varying rates, up to the present-day value. Why did these variations occur?

Returning to our fundamental mass balance equation, we must conclude that the growth in the ocean's $^{87}Sr/^{86}Sr$ ratio since 40 Ma can be attributed to a variation in the relative input of erosion from the landmasses or the input from the mantle, or both. Initially, a fierce conflict opposed proponents of the growth of continental input with mantle input supposedly remaining constant and their adversaries who thought that it was the mantle input that had varied along with the intensity of erosion. For the former, the predominant phenomenon was the uplift of the Himalayas after India collided with Asia 40 Ma ago (Raymo and Ruddiman, 1992). For the latter, the essential variation was in the activity of mid-ocean ridges in the form of the hydrothermal circulation occurring there (Berner *et al.*, 1983). Now, it was once thought that the total activity of the mid-ocean ridges had varied over geological time and in particular had declined since 40 Ma ago. We no longer think this. The idea today is that both changes are concomitant. The increased erosion because of the uplift of the Himalayas is consistent with reduced input from the mantle, which derives mostly from erosion of subduction zone volcanoes. Such erosion has been partially slowed by the disappearance of a significant source which was swallowed up in the Himalayan collision. All in all, then, it is the formation of the Himalayas that explains the $^{87}Sr/^{86}Sr$ curve as

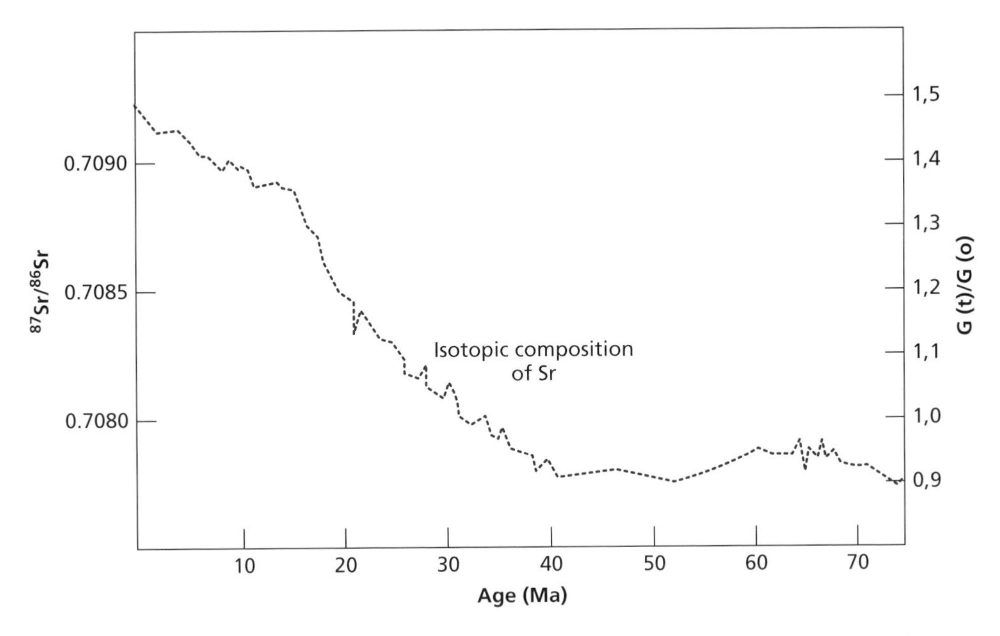

Figure 7.43 Curve of evolution of the $^{87}Sr/^{86}Sr$ ratio in sea water in Tertiary times (Cenozoic).

suggested by the former Columbia team (Raymo and Ruddiman, 1992) and John Edmond (1992) from MIT independently.

Regardless of any causal explanations, the curve is nowadays used to date Cenozoic limestones. This is known as Sr stratigraphic dating. The idea is straightforward enough. Since the curve is identical for all the oceans, it is an absolute marker. As the variation in the $^{87}Sr/^{86}Sr$ curve is all one way, the measurement of an $^{87}Sr/^{86}Sr$ ratio for any limestone can be used to determine its age from the curve. As can be seen, this clock is effective from 0 to 40 Ma, but barely beyond that as the curve flattens out. The precision achieved for an age is ± 1 Ma. This is a useful coupling with micropaleontological techniques.

But, of course, what everyone wants to know is why the $^{87}Sr/^{86}Sr$ ratio curve in limestones is identical whichever the ocean? The answer is that the residence time of Sr in the ocean is very much greater than its mixing time and so it has time to homogenize on a global scale. This will be explained in the next chapter.

The second question relates to climate. As it is observed that $\delta^{18}O$ increased throughout the Cenozoic, corresponding to a general cooling, what relation is there between tectonic activity and climate? This is a fundamental question to which there is as yet no clear answer.

Exercise

There are two inputs to the isotope composition of continental rivers: one from silicates and one from carbonates, in the proportions of 75% from carbonates and 25% from silicates, with average isotope ratios of $(^{87}Sr/^{86}Sr)_{silicates} = 0.708$ for carbonates and $(^{87}Sr/^{86}Sr)_{limestone} = 0.724$ for silicates. Suppose that carbonate recycling falls to 70%. How much will the Sr content of sea water vary assuming that the Sr content of rivers remains the same (which is unrealistic, of course)?

Answer

In the current situation, the Sr isotope ratio of rivers is 0.712 with 75% carbonate and 25% silicate. Recycled limestone has a Sr isotope composition of about 0.708 while that of silicates is 0.724. The composition for rivers with the new proportions becomes 0.7128, which gives a value of 0.709 52 for sea water. The recycling of limestone is clearly an important parameter, then.

Exercise

Assuming the isotope different compositions and inputs remain constant and the flow from volcanic sources remains the same, by how much does the Sr flow from rivers have to vary to change the ocean values from 0.708 to 0.709?

Answer

The calculation is the same as before:

$$\frac{\Delta R}{R} = \frac{\Delta x}{(1 - x)x}.$$

Therefore $\Delta F/F = +0.49$. So the flow from rivers must increase by 50%.

7.8.2 Isotopic variations of neodymium in the course of glacial–interglacial cycles

As with Sr, the isotopic variations of Nd are related to long-period radioactivity. When isotopic variations are measured in Quaternary sedimentary cores, these variations can be attributed to differences in origin alone. The *in situ* decay of ^{147}Sm in the core has virtually no influence. The fundamental difference between the behavior of Sr and of Nd in the ocean is that Sr, having a long residence time (1 or 2 Ma), is isotopically homogeneous on the scale of the world's oceans whereas Nd, having a shorter residence time (500–2000 years), varies isotopically between oceans and even within oceans. For example, the ε_{Nd} value today averages -12 for the Atlantic Ocean, -3 for the Pacific, and -7 for the Indian Ocean. These variations are interpreted by admitting that the Nd of sea water is a mixture between a volcanic source coming from subduction zones ($\varepsilon \approx 0$ to $+6$) and a continental source ($\varepsilon \approx -12 \pm 2$). The ε_{Nd} value varies depending on the degree of volcanic activity in the region relative to continental input (see Goldstein and Hemming, 2003).

Study of a Quaternary core from the Indian Ocean, south of the Himalayas, has allowed the Paris laboratory (Gourlan *et al.*, 2007) to highlight an interesting phenomenon. The sedimentary core is mostly carbonated (more than 70% carbonate). By appropriate chemical treatment, it is possible to extract the Nd of ancient sea water trapped in the small coatings of Mn surrounding foraminifers. Isotopic analysis of the Nd shows that ε_{Nd} varies from -7.5 to -10.5 and that the variations follow the pattern of ^{18}O/^{16}O fluctuation corresponding to the glacial–interglacial pattern. There is an excellent (inverse) correlation between $\delta^{18}O$ and ε_{Nd} (Figure 7.44).

South of the Bay of Bengal, the mixture of outflow from Indonesia and the input from the Rivers Ganga–Brahmaputra (but also the Irrawaddy and the Salween rivers) homogenizes the values of sea water of about $\varepsilon \approx -6 \pm 1$.

So we can suppose that in the Bay of Bengal the fluctuation during glacial–interglacial alternation corresponds to the fluctuation in the impact from the Ganga–Brahmaputra Rivers from the Himalayas. Those variations are linked with variations in intensity of the

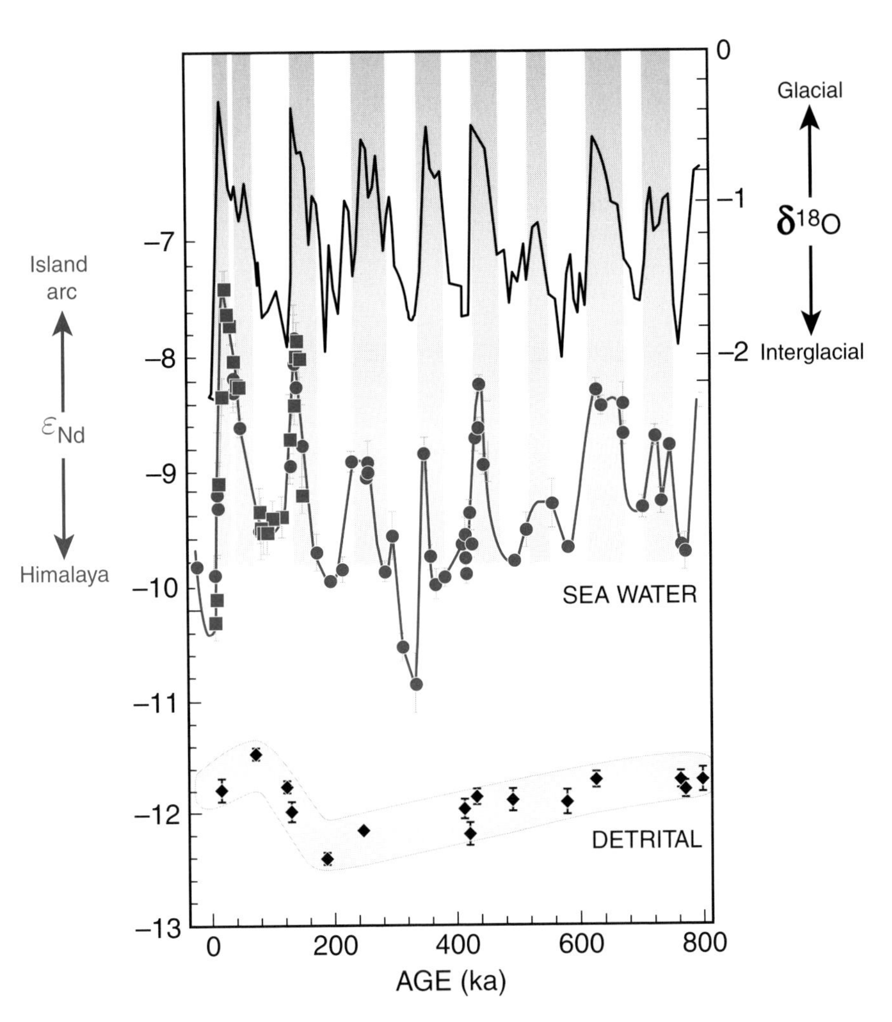

Figure 7.44 Variation in the isotopic composition of neodymium in sea water expressed as ε_{Nd} in a predominantly calcareous marine sedimentary core from the Ninety East Ridge in the Indian Ocean, representing the last 800 ka. Top: the curve of $\delta^{18}O$ variations allowing comparison of climatic fluctuations. Bottom: isotopic composition of the detrital fraction. (a) Oxygen isotopes compared with ε_{Nd} variations. (b) Enlargement before 2 Ma. (c) Oxygen-neodymium maxima and minima correlations.

monsoon and the existence of large glaciers during glacial periods in the high Himalayas. Monsoons are weaker during glacial times and glaciers accumulate snow and then stop (or strongly reduce) the river runoff.

Exercise

Calculate the relative input of the Ganges–Brahmaputra (GB) Rivers between interglacial and glacial if we suppose that $\varepsilon_{ND}^{GB} = -12$ and $\varepsilon_{Nd}^{ocean} = -6$. The measured ratios of concentration are $\varepsilon_{Nd} = -10$ during interglacials and $\varepsilon_{Nd} = -7.5$ during glacials, the concentration of Nd in river and ocean staying the same during all periods.

Answer

We applied the mixing formula. Notating concentration as C_{Nd} and the masses as m, we have:

$$\varepsilon_{Nd}^{measured} = \varepsilon_{Nd}^{GB} x + \varepsilon_{Nd}^{ocean}(1 - x)$$

$$x = \frac{\text{mass of fresh water} \cdot C_{Nd}^{river}}{\text{mass ocean} \cdot C_{Nd}^{ocean} + \text{mass of fresh water} \cdot C_{Nd}^{river}}.$$

With a little manipulation

$$\frac{m_{river}}{m_{ocean}} = \frac{C_{Nd}^{ocean}}{C_{Nd}^{river}} \left(\frac{1}{x^{-1} - 1} \right).$$

So for the ratios between interglacial (i) and glacial (g):

$$\frac{m_{river}^{i}}{m_{river}^{g}} = \left(\frac{x_g^{-1} - 1}{x_i^{-1} - 1} \right) = \frac{3}{0.5} = 6.$$

The river flux from the Himalayas was 6 times higher in interglacial than in glacial times.

This is a simple example in a work in progress in author's laboratory to illustrate the power of investigation of combining O and Nd isotopes.

7.9 Sulfur, carbon, and nitrogen isotopes and biological fractionation

We give two examples of how stable isotope geochemistry can be used in various types of study.

7.9.1 A few ideas on sulfur isotope fractionation

When we examine the $^{34}S/^{32}S$ composition of naturally occurring sulfur isotopes as a function of their geological characteristics, several features stand out. All sulfides associated with basic or ultrabasic rocks have extremely constant compositions close to

$^{34}S/^{32}S = 0.045$ ($\delta = 0$), that is, largely analogous to that of sulfur in meteorites (Nielsen, 1979).

Sulfur mineralization in veins crossing geological structures, with a gangue of quartz, fluorite, or barite, have δ values of about 0 which are very constant. It is therefore legitimate to attribute a deep origin to them or at least an origin related to deep-lying rocks. Cluster mineralization exhibits much more variable compositions, particularly mineralization related to sedimentary strata. Its composition may range from $\delta = +22$ to $\delta = -52$. This observation is tied in with the point that oxidation–reduction reactions $S^{2-} \Leftrightarrow SO_4^{2-}$ are accompanied by equilibrium isotope fractionation which, at low temperatures, is substantial (1.075 at 25 °C) (Tudge and Thode, 1950). Moreover, $S^{2-} \rightarrow SO_4^{2-}$ is an easy reaction at low temperature. However, reduction can only occur through *Desulfovibrio desulfuricans* bacteria. This bacterial reduction is accompanied by an isotopic effect that is weaker than the equilibrium reaction ($\alpha = 1.025$ at 25 °C) (Harrison and Thode, 1958). Remembering that sulfates of sea water and fresh water have $\delta^{34}S$ values that range from +26 to +4, we can explain the dispersion observed by assuming that the sulfides related to strata derive from bacterial reduction of sulfates, but that such reduction exhibits a number of variations. Sometimes reduction may involve sea water, sometimes groundwater circulation. Sometimes it occurs in replenished systems, sometimes in bounded reservoirs (Rayleigh distillation). Sometimes it is followed by isotope exchange leading to equilibrium fractionation, sometimes not. Here we find, but in a different context, variations in scenarios similar to what was calculated for bacterial reduction in sediments.

In any event, case by case, the sulfur isotope composition, associated with metallogenic and geological observations, allows distinctions to be drawn between the various types of deposits (Figure 7.45) and then allows the potential mechanism for the origin of mineralization to be limited. Generally, these data have made it possible to assert the occurrence of sulfur mineralization of exogenous origin, which many workers had contested before, claiming that all mineralization derived from the depths of the planet through mineralizing fluids (Ohmoto and Rye, 1979).

One particularly fascinating observation with sulfur isotope geochemistry relates to mass-independent fractionation (MIF). Such fractionation has been mentioned for oxygen, but it exists for sulfur too. Sulfur has four isotopes: ^{32}S, ^{33}S, ^{34}S, and ^{36}S. In terrestrial sulfur compounds variations in $^{33}S/^{32}S$ ratios account for about half of $^{34}S/^{32}S$ fractionations (0.515 to be precise). If we define $\Delta^{33}S = (\delta^{33}S) - 0.515\,(\delta^{34}S)$, this difference is generally zero. When measuring the isotopic composition of sulfides and sulfates of geologically varied ages, we obtain an unusual result. Between 2.30 Ga and the present day, $\Delta^{33}S = 0$. For samples of 2.30–2.60 Ga, $\Delta^{33}S$ varies with an amplitude of 12‰. For older samples, fluctuations are smaller but around 4‰. Samples of barium sulfate are depleted in ^{33}S (compared with "normal" fractionation, their $\Delta^{33}S$ is negative). Sulfide samples are enriched in ^{33}S (their $\Delta^{33}S$ is positive). This observation cannot be easily interpreted. **James Farquhar** and his team think that there was little oxygen in the atmosphere in ancient periods. The ozone layer surrounding the Earth at an altitude of 30 km and which now filters the Sun's ultraviolet rays did not exist. Sulfur reduction phenomena shifted sulfur from the degree of oxidation −2 (sulfide) to +6 (sulfate) via a cycle of photochemical reactions involving these ultraviolet rays. Now, laboratory experiments show that photochemical reactions (that

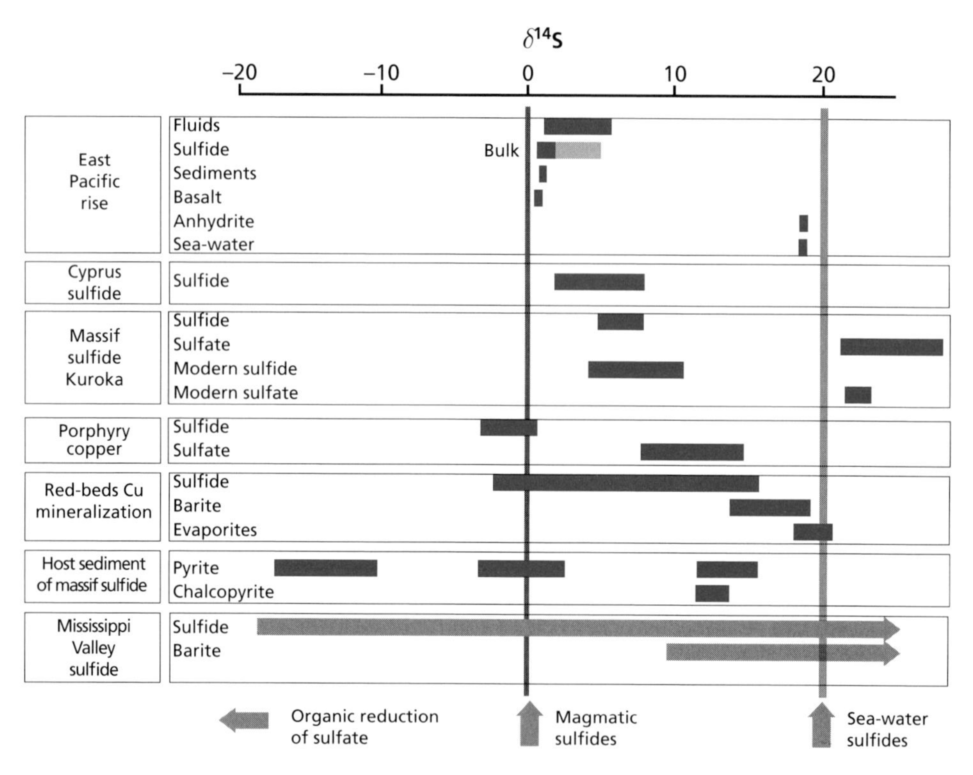

Figure 7.45 Distribution of sulfur isotopes in the main sulfur-bearing deposits.

is, reactions taking place under the influence of light) produce important non-mass-dependent fractionations (Farquhar *et al.*, 2007).

This idea of oxygen being absent from the ancient atmosphere is consistent with many geochemical observations: the presence of detrital uranium in the form of UO_2 in ancient sedimentary series and particularly in the famous Witwatersrand deposits of South Africa. Uranium in its degree of oxidation +4 is insoluble whereas in the +6 form, it forms soluble complex ions. Today uranium is mostly in the +6 state (in solution), but in the Archean it was in the +4 state (as detrital minerals) . Until 2 Ga, very special rich iron deposits are found, known as banded hematite quartzite or banded iron formation (BIF). These are evidence that at that time rivers carried soluble iron in the +2 oxidation state and that it precipitated in the +3 oxidation state on reaching the ocean. Nowadays, surface iron is in the +3 oxidation state and forms insoluble compounds in soils. These iron compounds give tropical soils their characteristic red coloring.

Dick Holland (1984) has long used these observations to argue that the ancient atmosphere was rich in CO_2 and N_2 (as are the atmospheres of Mars and Venus today) and that oxygen, which makes up 20% of our atmosphere today, appeared only 2 Ga ago as a consequence of the superactivity of bacteria or of photosynthetic algae. The appearance of oxygen meant the end of both detrital uranium and chemical iron deposits, which, in fact, are not found after that period. Observations of sulfur isotope

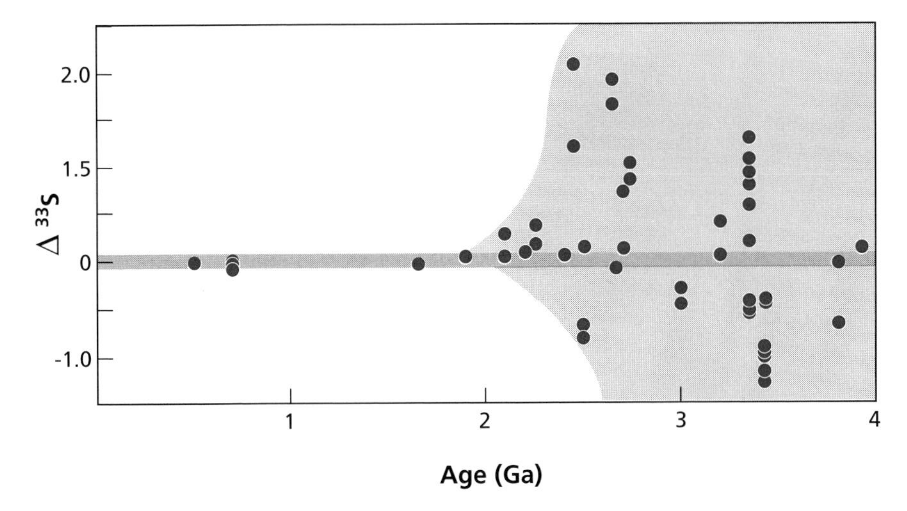

Figure 7.46 $\Delta^{33}S = (\delta^{33}S) - 0.515\ (\delta^{34}S)$. The figure shows ^{33}S variation in $\Delta^{33}S$ of sulfides and sulfates of various ages.

fractionations by **Thiemens**'s team refine this model. They seem to indicate that the growth of oxygen in the atmosphere occurred very quickly, almost suddenly, between 2.5 and 2.1 Ga and that this growth was accompanied by the progressive formation of the ozone layer protecting the Earth's surface from excessive solar ultraviolet radiation (Figure 7.46).

7.9.2 Carbon–nitrogen fractionation and the diet of early humans

Biochemical operators fractionate carbon and nitrogen isotopes. Gradually the mechanisms and the practical rules such fractionation obeys have been determined. Thus, it was correctly predicted that C_3 plants (the first product of photosynthesis with three carbon atoms) (trees, wheat, and rice) fractionate differently from C_4 plants (corn, grass, sugar cane). It has also been shown that marine plants are different again. From these observations **Michael DeNiro** of the University of California at Los Angeles studied the isotope composition of herbivores (eating the various types of plant) and of carnivores eating those herbivores. Oddly enough, a number of regularities were preserved and turned up in the isotope composition of bone (in the mineral matter and also in collagen which withstood decay quite well). He was thus able to determine what early humans ate (Figure 7.47). Those of the Neolithic ate C_3 plant leaves and then people later certainly began to eat corn (C_4). Wheat does not seem to have been grown until much later. This is an example of isotope tracing which is developing in biology and archeology. Stable isotopes measured on bone and tooth remains of extinct animals can be used to answer questions about the type of metabolism of certain dinosaurs (hot or cold blooded), the diet of extinct animals, or the effect of paleoclimate on the cellulose of tree rings. Once again this discipline offers considerable prospects.

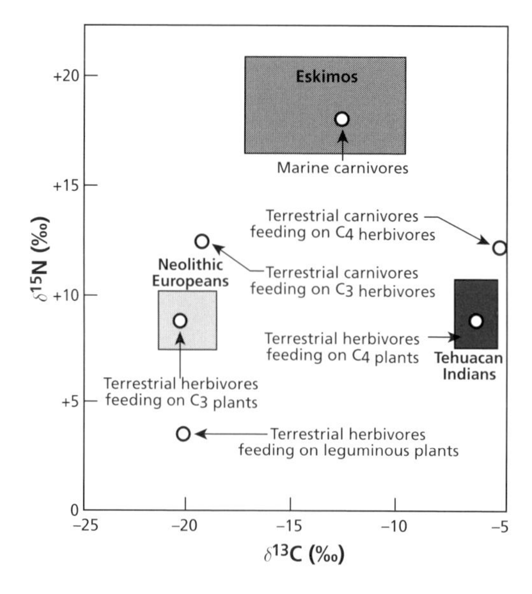

Figure 7.47 Isotopic composition of fossil plants, animals, and humans in the $\delta^{15}N$ and $\delta^{13}C$ diagram. After DeNiro (1987).

7.10 The current state of stable isotope geochemistry and its future prospects

As has been repeated incessantly throughout this book, developments in isotope geology have always tracked advances in measurement methods, which themselves are often the consequence of technological progress. The development of the double-collection mass spectrometer by **Nier** and his collaborators (Nier, 1947; Nier *et al.*, 1947) made it possible to study the effects of very weak isotopic fractionation (oxygen, hydrogen, carbon, and sulfur) in carbonates, water, rock, and living matter.

Since then technical advances have moved in three directions. The first was that of sensitivity. It has become possible to analyze isotopic fractionation on small quantities of material. **Hugh Taylor** managed to analyze D/H in rocks while **Françoise Pineau** and **Marc Javoy** have analyzed $^{13}C/^{12}C$ and $^{15}N/^{14}N$ in basalts.

The second direction was that of precision. In 1950, ratios could be measured to 0.5‰. Now the figure is 0.05‰. This has made it possible to analyze sedimentary cores with precision and to highlight Milankovitch cycles. **Robert Clayton** was able to discover paired $^{17}O/^{16}O$ and $^{18}O/^{16}O$ fractionations of meteorites, which had many consequences for the study of meteorites even if the initial interpretations have been modified. **Mark Thiemens** has been able to move on from there to open up the study of mass-independent fractionation.

The third advance has been the automation of analytical procedures which has enabled large numbers of samples to be studied both in sedimentary carbonates and in polar ice for O, C, and H isotopes. Climatology has gained enormously from this.

Today two new technical advances have occurred: multicollection ICPMS and the development of *in situ* probes, ion probes, or ICPMS laser ionization. In addition, advances in computing and electronics have brought progressive gains in precision, sensitivity, and measurement time for all conventional techniques, TIMS, or double-collection gas spectrometry.

What will come of all this? It is probably too early to answer this question but the trends as perceived can be set out. The most spectacular trend is probably the rush to study isotopic fractionation of "non-classical" elements that are often present in terrestrial materials. These include some major elements (Si, Mg, Fe, or Ca) for which physicochemical fractionations have been identified and then minor light elements like B and Li and minor heavy elements like Cr, Cu, Zn, Cd, Se, Mo, or Tl (the list is not exhaustive) (see the review edited by Johnson *et al.*, 2004). It is undeniable that some interesting results have been obtained for the major elements Mg, Fe, Ca, and Si as well as for B, Li, Cu, Mo, Tl, and Cl. For trace elements, no result has as yet allowed new tracers of geological phenomena to be introduced, as is the case for the isotopes of the major elements H, O, C, and S. Analyses are difficult, the results are often uncertain, and approaches are not systematic enough. These attempts have not achieved the results expected. The present author thinks, but this is open to question, that the most interesting processes are:

- first in biogeochemistry. It seems that living organisms fractionate some isotopes: Ca for the food chain ending with shells, Si for the food chain ending with diatoms, Cu for cephalopods. This, combined with C, N, and S geochemistry, may be the advent of the famous biogeochemistry we have been waiting for since Vernadsky's 1929 book!
- then, for pH conditions boron is a hope, provided the hypothesis of constant $\delta^{11}B$ for the ocean over geological time is eliminated. The degree of oxidation–reduction with the use of iron isotopes and molybdenum isotopes is also relevant. We shall review this if this book runs to a new edition!

The other trend is illustrated by **John Eiler**'s program at the California Institute of Technology. He is trying to take advantage of the improvement in techniques of analysis of traditional elements to develop new and original methods, the most spectacular of which is intercrystalline order–disorder fractionation, which we have spoken of, but also for $^{18}O/^{16}O$ or D/H fractionations in high-temperature phenomena.

The study of non-mass-dependent fractionation by **Mark Thiemens**'s team has probably still not yielded all its results but perhaps requires a more structured approach.

Problems

1 Take a cloud that evaporates at the equator with a mass M_0 $\delta_{H_2O} = 0$ for D and ^{18}O (to simplify). It moves polewards and when the temperature is $+10\,^{\circ}C$ loses one-third of its mass as rain and continues in the same direction. In the cold zone, where the temperature is $0\,^{\circ}C$, it loses one-third of its remaining mass. It moves on and loses another one-third of its remaining mass at $-20\,^{\circ}C$. When it reaches temperatures of $-30\,^{\circ}C$ it loses a further one-third of its mass. The fractionation factors at three temperatures are given in Table 7.4 below.

Table 7.4 Fractionation factors

T (°C)	α_D	$\alpha_{^{18}O}$
+20	1.085	1.0098
0	1.1123	1.0117
−20	1.1492	1.0411

(i) Calculate the δD and δ^{18}O composition of the rain and snow.
(ii) Plot the (δD, δ^{18}O) curve and calculate its slope.
(iii) Plot the δD and δ^{18}O curves as a function of the remaining fraction of the cloud.

2 Take a magma chamber whose magma has an initial δ^{18}O isotope composition of +5.5. Some 30% of olivine Mg_2SiO_4 precipitates in the chamber. Then we precipitate a eutectic mixture with equal proportions of olivine–pyroxene. We precipitate 30% of the remaining melt and then the olivine, orthopyroxene, and plagioclase mixture in equal proportions for 20% of the remaining melt. Given the melt–silicate partition coefficients in Table 7.5 below, calculate the isotope evolution of the melt and the minerals.

Table 7.5 Partition coefficients

Plagioclase–melt	Olivine–melt	Pyroxene–melt
− 0.6‰	− 0.2‰	− 0.3‰
$\alpha = 0.9994$	$\alpha = 0.9998$	$\alpha = 0.9997$

3 Consider rainwater with $\delta_D = -70$‰. This water penetrates into the ground and finally reaches a metamorphic zone where it meets a schist whose proportion relative to water is 15% and whose composition is O = 53.8%, Si = 33.2%, Al = 7.8%, Fe = 2.8%, Ca = 7.1%, Na = 0.6%, K = 1.5%, and C = 1.8%. This schist contains the following minerals which equilibrate with water at 550 °C in a closed system. The composition of the rock is: 40% quartz, 4% magnetite, 16% plagioclase, 15% muscovite, 20% alkali feldspar, and 5% calcite. Calculate the oxygen isotope compositions of the minerals and the water in the end.

4 The CO_2 content of the recent atmosphere is 320 ppm, its δ^{13}C value is −7. As a result of burning of coal and oil the δ^{13}C value has shifted from −7 to −10 in 20 years.
(i) Given that $\delta^{13}C_{oil} = -30$, what quantity of carbon has been burned?
(ii) However, a problem arises. The CO_2 content of the atmosphere is 330 ppm. How can you explain this?
(iii) Suppose the $\alpha_{calcite-CO_2}$ fractionation at 20% is 1.0102. What is the variation observed in the δ value in the calcites precipitating in sea water?
(iv) Does the δ^{13}C isotope analysis of limestone seem to you a good way of testing CO_2 degassing in the atmosphere by human activity? Mass of the atmosphere: $5.1 \cdot 10^{21}$ g.

5 Basalt magma contains sulfur in the form S^{2-} (sulfide) and SO_4^{2-} (sulfate), whose proportions vary with oxygen fugacity.

$$S^{2-} + 2\,O_2 \Leftrightarrow SO_4^{2-}$$

$$\frac{[S^{2-}][O_2]^2}{[SO_2]} = K(T).$$

Therefore:

$$\frac{S^{2-}}{SO_4^{2-}} = \frac{K(T)}{[O_2]^2}.$$

When the magma degasses, it loses almost exclusively its SO_2 and the H_2S content is usually negligible (even if it smells). Determine the partition $\Delta_{gas-magma} = (\delta^{34}S)_g - (\delta^{34}S)_m$.
Given that the magma contains S^{2-} and SO_4^{2-}, show that degassing of the magma leads to an increase in the $\delta^{13}S$ value of the solidified magma or to a decrease depending on oxygen fugacity (after Sakaï *et al.*, 1982). (We know that $\Delta_{S^{2-}}^{SO_2} = +3$ and $\Delta_{S^{2-}}^{SO_4^{2-}} = +7$.)

6 Various scenarios are imagined in which the temperature of the Earth reaches extremes. The first scenario, known as the snowball scenario, says that all the landmasses are covered by a layer of ice 100 m thick in addition to the present-day polar ice which has doubled in volume. The second, reverse, scenario says that the Earth has heated and the polar ice caps melted. In the first scenario the $\delta^{18}O$ of continental ice is supposed to reach -30, with the polar ice caps being like today at -50. The ocean is at $\delta = 0$.

(i) What is the δ value of sea water in the snowball scenario?
(ii) In the scorching Earth scenario, what is the δ value of sea water?
(iii) Examine each scenario. Calculate the rate of increase (or decrease) of $\delta^{18}O$ in meters above sea level.
(iv) Does this figure vary with the speed of the process?

CHAPTER EIGHT

Isotope geology and dynamic systems analysis

We have seen that the Earth can be subdivided into five main reservoirs:

(1) the **continental crust**, where elements extracted from the mantle are stored (K, Rb, U, Th, rare earths, etc.) and which is made up of age provinces assembled like a mosaic;
(2) the **upper mantle**, the MORB source, which is mainly a residue of extraction of continental crust and the place where oceanic crust is formed and destroyed;
(3) the **lower mantle**, of which little is known by direct information other than that it is the source of the rare gas isotope signature in OIB;
(4) the **atmosphere**, which is the recipient of rare gases given out by the mantle and is not sealed for some gases (He, Ne); and
(5) the **hydrosphere**, which is the driving force and the potential vector of all transfers of material at the surface, through the cycle of erosion, transfer, and sedimentation.

These reservoirs exchange material with each other. Material is extracted from the upper mantle to form the continental crust, probably during subduction processes. Material from the landmasses is reinjected into the upper mantle, either during subduction or during episodes when the continental crust is delaminated and falls into the mantle (**Dupal province**).

Exchanges probably occur between the lower and upper mantle, as shown by mass balance calculations for the depleted mantle and the results for rare gases, implying the existence of two reservoirs in the mantle, with injection of the lower mantle into the upper mantle. But the exact processes are unknown even if it is obvious that mantle plumes and subduction are related in some way with these exchanges.

Exchanges with the atmosphere are by volcanic outgassing and, for the continental crust, erosion (which destroys rock and releases some of the rare gases in rock) or hydrothermal processes. It seems that gases are reinjected into the mantle through subduction phenomena as shown by **Marc Javoy** and colleagues of the Institut de Physique du Globe in Paris for CO_2 and N_2 (Javoy *et al.*, 1982) when these gases are transformed into chemical compounds. This does not seem to be so for rare gases, for which subduction is apparently a barrier, and which return to the atmosphere through volcanism in subduction zones.

There remain major questions that are the subject of fierce debate. Do subduction processes affect the lower mantle? Are plumes created in the upper mantle, the lower mantle, or both, and by what processes? Does the lower mantle exchange material with the upper mantle? All of these questions are suggested by the findings of seismic tomography

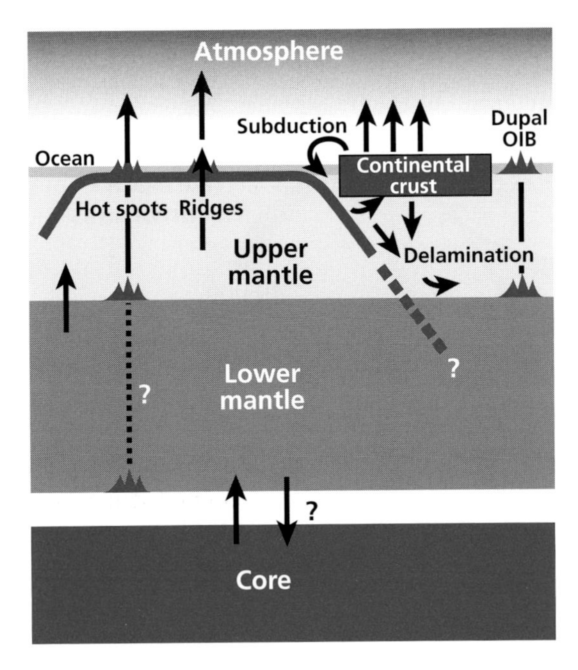

Figure 8.1 The structure and dynamics of the mantle–landmass–atmosphere system. The Earth system is made up of reservoirs exchanging matter and energy with each other.

which provides spectacular color images but which cannot readily be interpreted because it is difficult to distinguish between heat and mass transfers. This is a reality that relates to the present day for geophysicists while isotope geochemistry includes the whole history of the Earth.

These exchanges of material between reservoirs have been modulated by the vagaries of geological history and, for the external reservoirs, by the vagaries of climate. All of this creates what is nowadays termed a dynamic system, or several interlocking dynamic systems if you prefer (Figure 8.1).

We have looked at the Earth's external system where the ocean exchanges with the atmosphere, is fed water charged with ions by the landmasses, precipitates some compounds, and stores water in ice and releases it under the influence of dynamic fluctuations (Figure 8.2). Once again, this is a huge dynamic system. The way in which the system receives energy from the Sun, distributes it, and modulates it triggers the water cycle, modifies the surface temperature and determines the climate.

Quite what the relative influence is of the greenhouse effect, caused by CO_2 or CH_4, and of the hydrological cycle, an extraordinary thermal machine, remains an unresolved issue. Geographical distribution is essential in this dynamic system and, as has been seen, the very different effects between the northern hemisphere with many landmasses and the southern hemisphere, which is largely ocean, are still poorly understood.

The evolution and determining factors of the chemical workings of the oceans and the question of the relative influence of chemical elements extracted from the mantle compared

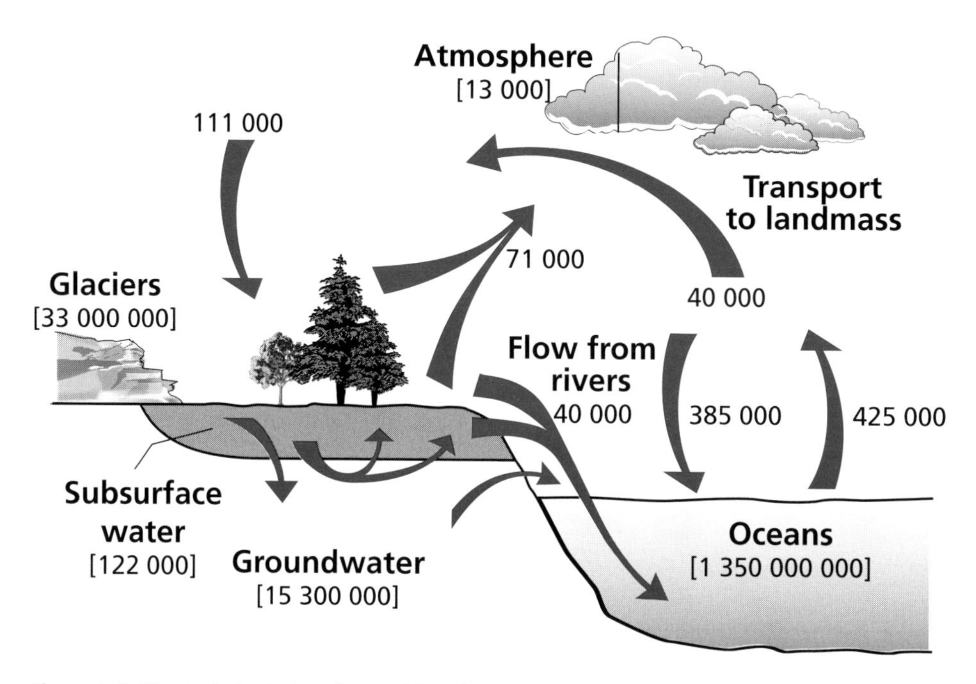

Figure 8.2 The hydrological cycle considered as a system. The masses of the reservoirs are shown in brackets. Flows are marked by arrows. Units are km^3, corresponding to 10^{12} kg, and $km^3\ yr^{-1}$ for flows.

with those from the landmasses on the record of the past chemical or isotopic composition of the oceans depend largely on closely interconnected parameters. This is the example of vast dynamic systems with complex interactions whose determining factors we still do not understand properly.

Future developments of paleo-oceanography or paleoclimatology, like those of chemical geodynamics, are related to our understanding of such complex systems. Accordingly, at the end of this book, it has been thought useful to address, in an admittedly very elementary and succinct way, but prospectively, what is now a huge field of study in the earth sciences but whose formal features exceed our present framework and apply equally to biology and the chemical industry as well as to ecology or to physics. It is this methodology that we call **dynamic systems analysis**. When we say that the Earth is a "chemical plant" it is not just for the pleasure of a new image. The same methods of systems analysis apply to chemical plants as to the Earth and its component parts.

We shall give a few very elementary bases of these methods that we shall apply, of course, to the system that the Earth constitutes, looking at the questions specific to it, but, more than that, opening up future prospects. In this context, isotopes appear as tracers, indicators, like radioactive tracers in biology or industrial chemistry, but they contain a wealth of information as we shall see (for general references see: Jacobsen and Wasserburg, 1981; Beltrani, 1987; Jacobsen, 1988; Albarède, 1995; Haberman, 1997; Lasaga and Berner, 1998; Rodhe, 2000).

8.1 Basic reservoir analysis: steady states, residence time, and mean ages

8.1.1 Well-mixed, simple reservoirs

Let us consider a first simple example. A reservoir of mass M, in a steady state, into which a flow of matter $\dot{M}_{\downarrow}$ enters (and from which a quantity of matter $\dot{M}^{\uparrow}$ exits). The reservoir itself is well mixed and is assumed to be statistically homogeneous.[1] The reservoir might be the Earth's mantle, the ocean or the mass of sediments, etc. For the steady state $\dot{M}_{\downarrow} = \dot{M}^{\uparrow}$, we define the **residence time** as the average time matter spends in the reservoir. The residence time R is defined by

$$R = \frac{M}{\dot{M}_{\uparrow}} = \frac{M}{\dot{M}_{\downarrow}}.$$

The dynamic equation is written $\mathrm{d}M/\mathrm{d}t = \dot{M}_{\downarrow} - \dot{M}^{\uparrow}$, with $\dot{M}_{\downarrow} = \dot{M}^{\uparrow}$ in the steady state. What goes into the reservoir equals what comes out of it, and so residence time can be calculated by considering either the input term or the output term.

To reason in terms of age, we must speak of the **mean age of material** leaving the reservoir and which entered at time $t = 0$. What is the mean age of material in the reservoir?

The output flow $\dot{M}^{\uparrow}$ may be written in the form: $\dot{M}^{\uparrow} = kM$, where k is the fraction of the reservoir exiting per unit time. Then $\mathrm{d}M = kM\,\mathrm{d}t$, which may be written $\mathrm{d}M/\mathrm{d}t = kM$. Thus k may be considered as the probability that an element of the reservoir will exit at a time t. As it is notated, $k = 1/R$ is the inverse of residence time. We can therefore write that the quantity of particles exiting the reservoir is:

$$\mathrm{d}M/\mathrm{d}t = -kM,$$

hence $M = M_0 \mathrm{e}^{-kt}$.

This equation translates the evolution at time t of the number of particles that entered the reservoir at time $t = 0$. It is also the age distribution of the "particles" of matter present in the reservoir.

The proportion of particles of age t still in the reservoir is written:

$$M(t)/M_0 = \mathrm{e}^{-kt}.$$

The mean age of particles in the reservoir is written:

$$\langle T \rangle = \int_0^{\infty} t\mathrm{e}^{-kt}\,\mathrm{d}t = \frac{1}{k}\int_0^{\infty} kt\,\mathrm{e}^{-kt}\,\mathrm{d}t.$$

Integrating by parts, $\int_0^{\infty} kt\,\mathrm{e}^{-kt}\,\mathrm{d}t = 1$, therefore $\langle T \rangle = 1/k = R$.

In this well-mixed reservoir, the **mean age** of particles is equal to the **residence time**. This is a fundamental result.

[1] We often refer to such reservoirs as **boxes** (a **single box** or **multiple boxes**).

Exercise

The mantle may be thought of as a reservoir from which mantle escapes at the ocean ridges and which it enters at the subduction zones. Accepting that it operates in a steady state, what is the residence time of a lithospheric plate in the upper mantle? We ask the same question for the whole mantle.

It is taken that the rate of formation of ocean floor is 3 $km^2\ yr^{-1}$, the thickness of a plate is 8 km, and its mean density 3.5.

Answer

The area subducted (swallowed up by the mantle) equals the area created, therefore the mass of the lithosphere subducted is $8.4 \cdot 10^{14}$ kg yr^{-1}.

The mass of the upper mantle is $1.05 \cdot 10^{24}$ kg, therefore the corresponding residence time is $R = 1.25$ Ga. The mass of the total mantle is $4.02 \cdot 14^{24}$ kg, therefore for the whole mantle $R = 4.78$ Ga.

Exercise

Suppose we wish to calculate the residence time of the ocean crust alone, which would tend to suggest the ocean lithosphere is made up of oceanic crust and a piece of average mantle attached to it. What is the residence time of such an ocean crust in the upper mantle?

Answer

Oceanic crust is 6 km thick, the mass of oceanic crust subducted is therefore $8.4 \cdot 10^{13}$ kg yr^{-1}. Hence $R = 12.4$ Ga for the upper mantle alone.

Exercise

The quantity of ^3He in the atmosphere is $1.26 \cdot 10^9$ moles. The degassing rate of ^3He at the mid-ocean ridges is 1100 moles yr^{-1}. Atmospheric ^3He is lost to space. Accepting that the atmosphere is in a steady state for helium, what is the residence time of ^3He in the atmosphere?

Answer

R is about 1 million years, 1.18 Ma to be precise.

Let us now try to calculate the residence time of a chemical element i in a well-mixed reservoir (**Galer** and **O'Nions**, 1985). We reason as before. Let $C^i_\downarrow$ be the chemical concentration of the element entering the reservoir and C^i_M that present in the reservoir:

$$R^i = \frac{M\, C^i_M}{\dot{M}_\downarrow\, C^i_\downarrow} = R\, \frac{C^i_M}{C^i_\downarrow}$$

where M and $\dot{M}_\downarrow$ are the overall mass and the mass of the flow entering the reservoir, respectively, and R is the residence time of the material.

Exercise

Suppose the ocean is a system in a steady state. The mass of the ocean is $M = 1.39 \cdot 10^{21}$ kg, and the mass of water entering from rivers $\dot{M}_\downarrow = 4.24 \cdot 10^{16}$ kg yr^{-1}.

Given the Sr and Nd concentrations in sea water and rivers in the table below, what is the residence time of these elements in sea water?

	Sr (ppm)	Nd (ppt)
Sea water	7.65	3.1
Rivers	0.06	40

Answer

For Sr, $R=4\cdot10^6$ yr. For Nd, $R=2500$ yr.

Exercise

It is considered that most of the chemical elements in sea water disappear into pelagic sediments by various processes (absorption, precipitation, biochemical reaction, etc.), that the sedimentation rate for such sediments is $3\cdot10^{-3}$ kg m^{-2}, that the ocean area, excluding continental margins is $3.1\cdot10^{14}$ m^2, and that the average concentrations of Sr and Nd of these sediments are $C_{Sr}=2000$ ppm and $C_{Nd}=40$ ppm. Calculate the residence time of Sr and Nd in sea water and compare this result with the previous one.

Answer

For Sr, $R=5.5\cdot10^6$ yr. For Nd, $R=1162$ yr.
The figures are similar to the previous ones, which confirms the hypothesis that the ocean is in a steady state. (These values are controversial and not hard and fast, but they do give a good order of magnitude.)

Exercise

What is the residence time of U and Ni in the upper mantle if we consider the ocean crust is "independent" of its underlying lithosphere? The enrichment of U in the oceanic crust is given as 10 and that of Ni as 0.1.

Answer

$R_U=12.4/10=1.24$ Ga and $R_{Ni}=12.4/0.1=124$ Ga. They differ by a factor of 100.
The average age is calculated using exactly the same process and leads to the various elements being differentiated by their geochemical properties. The ocean lithosphere structure is probably intermediate between the two extremes evoked. It is composed of an oceanic crust, a part of the depleted mantle, the residue of partial melting, and perhaps also a thermally added part which is just accreted averaged mantle.

8.1.2 Segmented reservoirs

Let us now look at a very different kind of reservoir formed by the juxtaposition of cells (or boxes) which do not mix but are adjacent and passed through in turn.

This reservoir is segmented. It takes a time T to go from the first to the last segment. The maximum age of the reservoir is the exit age from the reservoir, which is the residence time $R = T$. The mean age of the reservoir, supposing it is created in a uniform way, is:

$$\langle T \rangle = \frac{1}{\dot{M}^{\uparrow} T} \int_{0}^{T} t \, \dot{M}^{\uparrow} \, dt = \frac{T}{2}.$$

An example is that of the human population. The residence time is 80 years, the mean age 40 years. Of course, if $\dot{M}^{\uparrow}$ is not constant, $\langle T \rangle$ is different from $T/2$ and varies between 0 and T.

Reasoning like this can be applied to the continental crust, supposing the survival of a portion of the continent depends on its age and therefore that the pieces of continents are destroyed by the geodynamic processes (erosion, metamorphic recycling, etc.) and that their maximum age is 4.2 Ga and their mean age 2.2 Ga, as calculated for the (Sr, Nd) isotope balances. The relation between residence time and mean age of a reservoir reflects the internal dynamics of this reservoir and its internal structure.

Exercise

Calculate the mean age of a continental reservoir made up of four segments aged 3.5, 2.5, 1.5, and 0.5 Ga whose masses in arbitrary units are in a first instance 10, 6, 3, and 1 and in a second instance 1, 3, 6, and 10.

Answer

The proportions of the four segments are 0.5, 0.3, 0.15, and 0.05 respectively, and the inverse for the second case. The mean ages are $\overline{T} = 2.75$ Ga in the first case and $\overline{T} = 1.25$ Ga in the second.

Exercise

Let us suppose regular mantle convection (without mixing with the environment) with, for example, subduction, a loop, and partial melting at a mid-ocean ridge (Figure 8.3). Suppose the subduction rate is 4 cm yr^{-1}, that the ridge is 20 000 km from the subduction zone (Pacific), and the depth is 400 km. What is the residence time of this portion of the mantle?

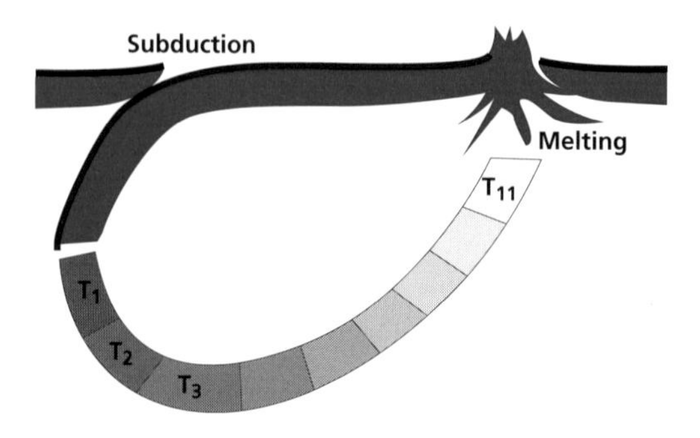

Figure 8.3 Regular mantle convection. $T_1, T_2, \ldots, T_{11}$ are a set of subduction dates with T_{11} the earliest.

Answer

The length of time spent in the mantle is written:

$$T = \frac{\text{distance}}{\text{speed}} = \frac{400 \times 2 + 20\,000 \cdot 10^5}{4} = 502\ \text{Ma}.$$

This is just under half the residence time measured in the upper mantle. But such a process is very similar to that of a segmented reservoir!

8.2 Assemblages of reservoirs having reached the steady state

A whole series of complex (multi-box) systems can be imagined by combining the two types (well mixed or segmented) of simple system. We shall give a few examples which could apply to the upper mantle but also to the ocean, considered as a series of separate reservoirs (Atlantic, Indian, Pacific, surface water, deep water, etc.).

8.2.1 Model 1

Let us consider a mantle made up of two layers which exchange material (Figure 8.4). It may be the upper mantle–lower mantle system or even the upper mantle separated into two layers by the 400-km seismic discontinuity.

Let M_1 and M_2 be the masses of the two mantles. Let $\dot{S}$ and $\dot{D}$ be the flow of matter in the subduction zone and at the mid-ocean ridge. We have $\dot{S} = \dot{D}$. In addition, let $\dot{M}_{1-2}$ be the flow of material from mantle 1 to mantle 2 and $\dot{M}_{2-1}$ the flow from mantle 2 to mantle 1. Supposing a steady state (therefore $\dot{M}_{1-2} = \dot{M}_{2-1}$), what are the residence times (and therefore the mean ages)?

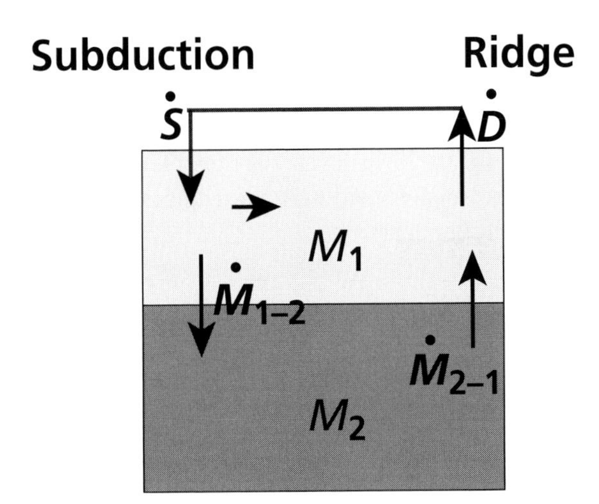

Figure 8.4 Model of mantle with two layers exchanging matter (and energy). M_1 and M_2 are the reservoir masses; $\dot{M}_{1-2}$ and $\dot{M}_{2-1}$ are the mass fluxes.

$$R_1 = \frac{M_1}{\dot{S}} \text{ and } R_2 = \frac{M_2}{\dot{M}_{1-2}}.$$

The total residence time τ is such that:

$$R = \frac{M_1 + M_2}{\dot{S}} = \frac{M_1}{\dot{S}} + \frac{M_2}{\dot{S}} = \frac{M_1}{\dot{S}} + \frac{M_2}{\dot{M}_{1-2}} \cdot \frac{M_{1-2}}{\dot{S}}.$$

Positing $M_{1-2}/\dot{S} = f$, we get:

$$R = R_1 + R_2 f.$$

The mean ages are equal to the residence times.

Exercise

Calculate R, R_1, and R_2 assuming a division between upper and lower mantle with $f = 0.1$.

Answer

$R = 4.75$ Ga; $R_1 = 1.28$ Ga; $R_2 = 34.7$ Ga.

Exercise

Let us now suppose the upper mantle is divided in two by the 400-km discontinuity. Assuming $f = 0.3$, calculate R, R_1, and R_2.

Answer

Rounding the figures, the mass of the mantle above 400 km is $0.60 \cdot 10^{24}$ kg and from 400 to 670 km it is $0.4 \cdot 10^{24}$ kg.
$R = 1.28$ Ga; $R_1 = 0.769$ Ga; $R_2 = 1.7$ Ga.

8.2.2 Model 2

The upper mantle is assumed to be divided into two reservoirs: an upper layer (asthenosphere) and a lower layer (transition zone) of masses M_1 and M_2 (Figure 8.5). Subduction injects 50% into the upper layer and 50% into the lower layer. The upper layer is assumed to be well mixed, while the lower layer is segmented and matter advances horizontally in sequence and is then finally injected into the upper layer.

We have:

$$R_1 = \frac{M_1}{\dot{S}}$$

because the input is both direct and indirect, and

$$R_2 = \frac{M_2}{\dot{M}_{1\to 2}},$$

and $\langle T_1 \rangle = \tau_1$ and $\langle T_2 \rangle = \frac{\tau_2}{2}$.

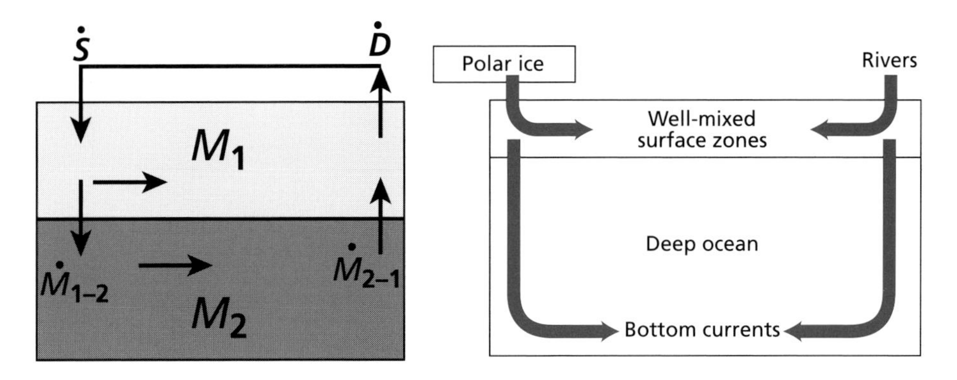

Figure 8.5 Model 2 of the mantle (left) and ocean (right) showing the two layers that exchange material.

With $M_1 = 0.6 \cdot 10^{24}$ kg and $M_2 = 0.4 \cdot 10^{24}$ kg, we get: $R_1 = 0.714$ Ga, $R_2 = 0.952$ Ga, $\langle T_1 \rangle = 0.714$ Ga, and $\langle T_2 \rangle = 0.47$ Ga.

It can be seen that the mean age of the deep reservoir is less than that of the upper reservoir, while the opposite applies for residence time.

In these examples, information can be deduced about the convective structure of the reservoirs by juggling with **residence times** and **mean ages**. Such a model can be constructed for ocean dynamics with different time constants. The upper layer is the well-mixed surface layer, and the lower layer is traversed by slow moving, deep currents, similar to a segmented reservoir.

8.3 Non-steady states

8.3.1 Simple reservoirs

General equations

Let us go back to the simple case of a reservoir fed by an influx J, with an outflow considered proportional to the quantity of material M in the reservoir. Let k be the kinetic constant. The dynamic equation is written:

$$\frac{dM}{dt} = J - kM,$$

with at $t = 0$, $M = M_0 = 0$.

Integrating gives:

$$M = \frac{J}{k}\left(1 - e^{-kt}\right).$$

It is confirmed that when $t \to 0$, $M \to 0$. When $t \to \infty$, M tends towards J/k, that is, the steady state, since J/k is the solution to the differential equation at equilibrium when $dM/dt = 0$.

But what is the significance of k? At equilibrium, the residence time is $R = M/kM = 1/k$, therefore $R = 1/k$. So k is the inverse of residence time, as already said.

Let us rewrite the solution to the equation:

$$M = JR\left(1 - e^{t/R}\right).$$

The equilibrium solution for the full reservoir is therefore: M = flow × residence time. The "speed" to attain equilibrium is proportional to $1/R = k$.

Suppose now that the reservoir is full, and so its mass is $M = JR$. We decide to empty it. What is the drainage law?

$$\frac{dM}{dt} = -kM$$

hence $M = JR\, e^{t/R}$.

The constant of the drainage time is the residence time. Likewise, R controls the time the system takes to attain the state of equilibrium.

Exercise

How long does the system take to attain 99% of its equilibrium mass?

Answer

$t = 4R$.

Exercise

What is the mean age of the reservoir in this case? Remember that

$$T = \frac{1}{m}\int_0^\infty t\,\frac{dm}{dt}\,dt.$$

Answer

$$\frac{dM}{dt} = J - k\,\frac{J}{k}\left(1 - e^{-kT}\right).$$

Hence, as with the simple case of the well-mixed reservoir, $T = R$.

Such a model could apply to the growth of the continents where it is considered that at each period the same amount of material is formed and that the material is destroyed over time and reinjected into the mantle. As we shall see, this is a very general equation.

The case of chemical elements

The problem of the evolution of concentrations of a chemical element in a reservoir is treated analogously following Galer and O'Nions (1985). Let us consider the evolution of the mass of an element i:

$$\frac{d(C^iM)}{dt} = C^i_{ex}\,J - k^iC^iM$$

where C^i is the concentration in the reservoir, C^i_{ex} the concentration in the flow of material entering from outside, J the flux of incoming material, and M the mass of the reservoir.

If the mass is constant, we can divide by M. This gives:

$$\frac{dC^i}{dt} = C^i_{ex}\left(\frac{J}{M}\right) - kC^i.$$

The residence time of the chemical element is written:

$$R = \frac{MC^i}{JC^i_{ex}} = R_M\,\frac{C^i}{C^i_{ex}} = \frac{1}{k},$$

where R_M is the residence time of the mass making up the reservoir, a formula we have already established in the steady state.

The case of a radioactive isotope

Let us write the equation for the evolution of the radioactive isotope:

$$\frac{dC^*}{dt} = Q^* - kC^* - \lambda C^*$$

with $Q^* = C^*_{ex}\,(J/M)$ (λ is the radioactive constant). The constant k is replaced in this equation by $(k+\lambda) = k^*$. The residence time is $R^* = 1/k^*$.

As can be seen, radioactivity is involved in residence time. If λ is very large, it reduces the residence time.

Exercise

The residence time of carbon in the ocean is 350 years. What is the residence time of ^{14}C?

Answer

About the same since for ^{14}C, $\lambda = 1.209 \cdot 10^{-4}\ yr^{-1}$.
$k^* = 1.2 \cdot 10^{-4} + 2.8 \cdot 10^{-3} = 2.92 \cdot 10^{-3}$.
$R_{^{14}C} = 342$ years.

8.3.2 Creation–destruction processes

The Earth is a living planet; all terrestrial structures are created by some processes and destroyed by others. What we observe is merely the outcome of antagonistic processes: birth and death. Thus, magmatic volcanic and metamorphic processes create portions of continents. These landmasses are then destroyed by erosion or by recycling of their materials in a new metamorphic cycle. The oceanic crust is formed at the mid-ocean ridges. It is swallowed up in the mantle in the subduction zones. There it is destroyed either by being thoroughly mixed with the middle mantle or by being fed back through an ocean ridge where it is again melted and reformed. These problems are analogous to those of birth and death in demography, as can be seen by consulting books on the subject.

Let us first consider the question of the geodynamic cycle. Let D be the quantity of mantle differentiated at the mid-ocean ridges. We have:

$$\frac{d(D)}{dt} = J - k(D)$$

where J is the rate of formation at the ocean ridge, and $k(D)$ is the destruction of quantity D which is fed back through an ocean ridge. But $J = kV$, if V is the total volume of the mantle and if the residence time R is such that $R = 1/k$.

$$\frac{dD}{dt} = k(V - D).$$

The solution of this differential equation where V is constant is:

$$D = V\left(1 - e^{-kt}\right).$$

Time is set to $t = 0$, 4.55 Ga ago. So $D/V = (1 - e^{-kt})$ tends towards 1 when t is large, and $D = 0$ when $t = 0$. The question is, of course, what value to take for k.

Let us take the value for the residence time of the upper mantle recycled by plate tectonics and assumed to be in a steady state: $R = 1.2 \cdot 10^9$ yr and $k = 0.83 \cdot 10^{-9}$ yr^{-1} (Figure 8.6). It may also be that there was more recycling in the past because more heat was produced in the mantle and the time may have been 0.7 Ga, that is, $k = 1.4 \cdot 10^{-9}$ yr^{-1}.

On this basis then, let us look at two issues. The first is to determine the quantity of virgin upper mantle, that is, mantle that has not been melted at a mid-ocean ridge. This part of the virgin mantle is the complement of the mantle that has been recycled through ocean ridges:

$$1 = V\left(1 - e^{-kt}\right) + \text{virgin mantle } (V_0)$$

$$\frac{V_0}{V} = e^{-kt}.$$

It can be seen therefore that the mass of the virgin mantle decreases with time, of course. What of the upper mantle? For $k = 0.83 \cdot 10^{-9}$ yr^{-1}, it is 2.3‰. For $k = 1.4 \cdot 10^{-9}$ yr^{-1} it is 1.8‰ (these are per mil values!).

Exercise

What proportion of virgin mantle would be preserved if the flow at the ocean ridges were identical but the system were the entire mantle?

Answer

Let us now try to calculate the age spectrum of this recycled mantle. It is the age distribution at the time the mantle went through the ridge-subduction cycle. It follows the law:

$$N(t_n) = kVe^{-kt}.$$

As can be seen, the mathematical equations are analogous to those of the previous example, but their physical meaning is very different. It is an example to remember.

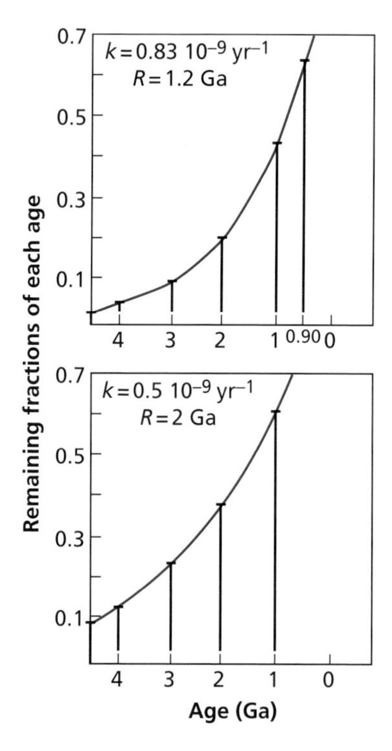

Figure 8.6 Spectrum calculation for $k = 0.83 \cdot 10^{-9}\ \mathrm{yr}^{-1}$ and $0.5 \cdot 10^{-9}\ \mathrm{yr}^{-1}$. Notice the very rapid growth towards recent times. This is because the probability of double or triple recycling increases very rapidly with time.

8.3.3 Time-dependent formation

Let us now consider a slightly more complex case where the formation term decreases with time. This is geologically fairly logical as we know that the creation of radioactive energy has decreased with time. The equation becomes:

$$\frac{\mathrm{d}M}{\mathrm{d}t} = J_0\, \mathrm{e}^{-qt} - kM.$$

This is a very general equation found in many cases (Allègre and Jaupart, 1985). For example, for continental growth, where the equations found can be modified as a consequence or also for degassing of the mantle, which we have looked at.

Integrating gives:

$$M = \frac{J_0}{k-q}\left(\mathrm{e}^{-qt} - \mathrm{e}^{-kt}\right).$$

Therefore M depends on the two constants k and q.

To get our ideas straight, let us take the case of degassing of $^{40}\mathrm{Ar}$ from the upper mantle, and study the $^{40}\mathrm{Ar}$ content of the upper mantle:

$$\frac{\mathrm{d}\,^{40}\mathrm{Ar}}{\mathrm{d}t} = \lambda_{\mathrm{e}}\,^{40}\mathrm{K}_0\, \mathrm{e}^{-\lambda t} - G\,^{40}\mathrm{Ar}$$

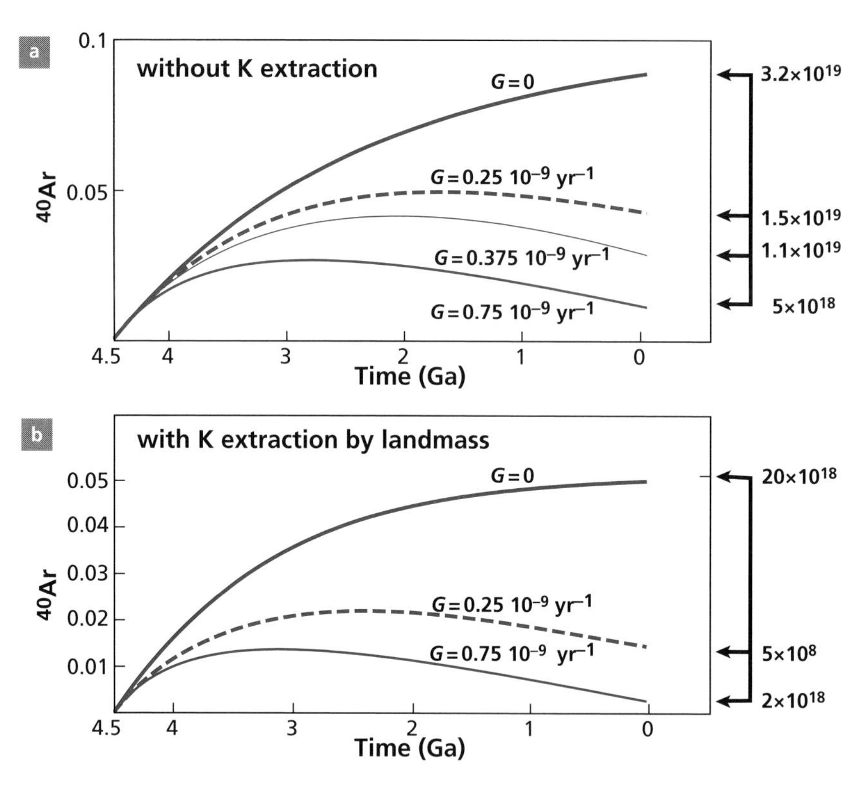

Figure 8.7 Mantle degassing model. (a) Evolution of ^{40}Ar in the mantle over time, assuming that ^{40}K varies by radioactive decay alone. The scale on the right is drawn assuming $^{40}K_0 = 1$. The scale on the left is in 10^{19} g of ^{40}Ar and corresponds to the total mass of ^{40}Ar on the hypotheses made in the main text. (b) Evolution of ^{40}Ar in the mantle assuming ^{40}K decreases both by extraction from the continental crust and by radioactive decay. The scale on the left is in 10^{18} g of ^{40}Ar. On the right, it is assumed that $^{40}K_0 = 1$.

where λ_e is the fraction of ^{40}K giving ^{40}Ar; λ is the total decay constant of ^{40}K; G is the constant of outgassing (Figure 8.7). The equation can be easily integrated:

$$^{40}\mathrm{Ar} = \frac{\lambda_e\, ^{40}\mathrm{K}_0}{G - \lambda}\left(e^{-\lambda t} - e^{-Gt}\right).$$

We are going to study the behavior of ^{40}Ar for the different values of *G*, arbitrarily setting $G_0 = 1$.

The qualitative behavior is understandable. When *G* is quite large, the quantity of ^{40}Ar reaches a maximum and then declines. So let us try to apply this information to the actual upper mantle.

The present-day potassium level in the non-depleted mantle is K = 250 ppm, and $^{40}\mathrm{K} = 1.16 \cdot 10^{-4}$ or 116 ppm. The decay constant is $\lambda = 0.554 \cdot 10^{-9}\ \mathrm{yr}^{-1}$. Some $4.55 \cdot 10^9$ years ago, the ^{40}K level was 12 times higher and so equal to 0.35 ppm.

Given that the mass of the upper mantle is $1 \cdot 10^{27}$ g, 4.55 Ga ago, there was therefore $0.35 \cdot 10^{21}$ g of ^{40}K. From the previous calculation we obtain the following values of ^{40}Ar today:

	$G(10^{-9}\text{yr}^{-1})$			
	0.25	0.375	0.75	0
^{40}Ar(g)	$1.5 \cdot 10^{19}$	$1.1 \cdot 10^{19}$	$0.49 \cdot 10^{19}$	$3.2 \cdot 10^{19}$

The present-day mass is estimated at $2\text{–}3 \cdot 10^{18}$ g, which is not bad!

There is one assumption in the model that is not very satisfactory because it was considered that ^{40}K decreased in the upper mantle by radioactive decay alone. Now, we know that K has also decreased in the upper mantle because it has been extracted at the same time as continental crust in which it is enriched. Let us therefore consider that ^{40}K decreases by the law:

$$^{40}\text{K} = {^{40}\text{K}_0}\, \text{e}^{-(\lambda+\beta)t}$$

where β is the constant for extraction of K from the mantle to the continental crust. (Such extraction does not obey an exponential law exactly, but it is a first approximation.)

We take $\beta = 0.35 \cdot 10^{-9}$ yr^{-1} because, with this constant, total K evolves from 250 ppm to 50 ppm in $4.5 \cdot 10^9$ years, which is about the value estimated for the degree of depletion of the upper mantle in K. The change in ^{40}Ar content of the mantle can therefore be calculated by replacing the value of $\lambda = 0.5 \cdot 10^{-9}$ yr^{-1} by $(\lambda + \beta) = 0.85 \cdot 10^{-9}$ yr^{-1}.

$$^{40}\text{Ar} = \frac{\lambda_e}{G - (\lambda + \beta)} \left(\text{e}^{-(\lambda+\beta)t} - \text{e}^{-Gt}\right).$$

As can be seen from the curves, we obtain an acceptable value: $2.77 \cdot 10^{18}$ g of ^{40}Ar in the upper mantle for $G = 0.75 \cdot 10^{-9}$ yr^{-1}. Now, what does $0.75 \cdot 10^{-9}$ yr^{-1} represent? The value $1/G = 1.3 \cdot 10^9$ is about the residence time of plates subducting into the mantle. If it is taken that the oceanic lithosphere degasses entirely by being fed through the mid-ocean ridges, a plausible scenario can be reconstructed for degassing the upper mantle. Which is no guarantee at all that the model is a proper reflection of reality!

8.3.4 Effects of cyclic fluctuations

General equations

We saw when examining climatic variations that the Earth's temperature varied cyclically (Milankovitch cycles). We can readily imagine that the Earth's tectonic activity varies cyclically (Wilson's plate tectonic cycles). How will such variations affect a dynamic system, a reservoir which receives and exchanges matter?[2]

Let us consider the equation for the dynamic evolution of a reservoir as we have already examined it:

$$\text{d}M/\text{d}t = J - kM,$$

but in which we write that the input flux varies periodically with the formula $J = J_0 + b \sin \omega t$, where $\omega = 2\pi/T$, T being the oscillation period and $k = 1/R$, R being the residence time.

[2] This problem has been addressed by Richter and Turekian (1993) and then by Lasaga and Berner (1998). The latter presentation has inspired the one given here.

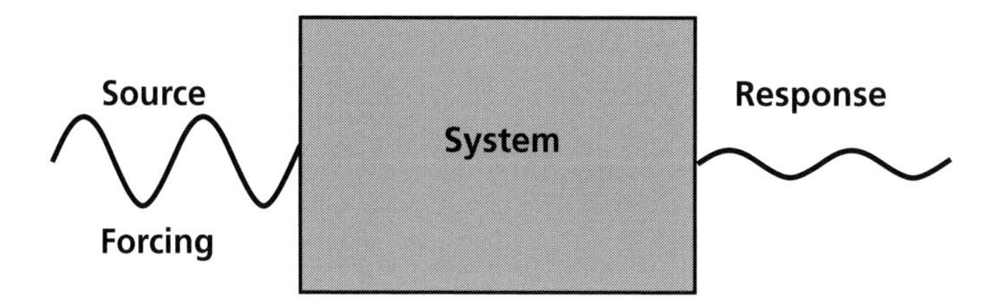

Figure 8.8 Model of response of a system subjected to external action.

The equation for mass evolution M (or concentration in an element) is therefore:

$$\frac{dM}{dt} = J_0 + b \sin \omega t - kM.$$

The equation can be written in the form:

$$kM + \frac{dM}{dt} = J_0 + b \sin \omega t.$$

The first term of this equation depends on M alone, the second term being the forcing imposed on the system from outside (we shall also call this the source term). The question is, of course, how does the system react and what is its response to external stimulus (Figure 8.8).

We integrate this first-order differential equation with constant coefficients in the standard manner: integrate the equation without the second member dependent on t, then calculate the first integration "constant." In the course of calculation, we get on the integral:

$$\int e^{kt} \sin \omega t \, dt.$$

This is the only minor mathematical difficulty.

$$\int e^{x} \sin x \, dx$$

is integrated by parts, integrating twice. This leads to:

$$\int e^{x} \sin x \, dx = e^{x} \sin x - e^{x} \cos x - \int e^{x} \sin x \, dx,$$

hence the calculation of $\int e^{x} \sin x \, dx$. Integrating this general equation therefore gives:

$$M = M_0 \, e^{-kt} + \frac{J_0}{k}\left(1 - e^{-kt}\right) + \frac{b\omega}{k^2 + \omega^2} \, e^{-kt} + \left(\frac{bk}{k^2 + \omega^2} \sin \omega t\right) - \left(\frac{b\omega}{k^2 + \omega^2} \cos \omega t\right).$$

It can be seen that if $M_0 = 0$ and $b = 0$, we find the equation in Section 8.3.1:

$$M = \frac{J_0}{k}\left(1 - e^{-kt}\right).$$

Whenever t is quite large (say more than $3/k$), the equation is simplified and can be written:

$$M = \frac{J_0}{k} + \frac{bk}{k^2 + \omega^2}\sin \omega t - \frac{b\omega}{k^2 + \omega^2}\cos \omega t.$$

If $\omega \ll k$, that is, if $R \ll T$ (the residence time is very much less than the oscillation period), the cosine term is negligible. The equation can then be written:

$$M = \frac{J_0}{k} + \frac{b}{k}\sin \omega t$$

where J_0/k is the equilibrium value around which oscillation occurs.

The system response is in phase with the source oscillation. The amplitude of fluctuation around equilibrium depends on the ratio J_0/b, that is, on the relative value between the equilibrium value and the pulse value, but if this ratio is not too large, the amplitude remains large.

Conversely, if $\omega \gg k$, that is, $R \gg T$ (the residence time is very much greater than oscillation period), the equation becomes:

$$M = \frac{J_0}{k} - \frac{b}{w}\cos \omega t.$$

But we have the trigonometric equality $\cos \omega t = -\sin(\omega t - \pi/2)$. Therefore the system oscillation is delayed by $(\pi/2)$, that is, by a quarter of a period relative to the source oscillation. Moreover, the amplitude is subtracted from a term J_0/k, which is very large since k is very small, which greatly damps the signal. Between the two extremes we find intermediate behavior.

Let us examine the consequences of this mathematical result. Suppose first that the source oscillation period is set, say at 100 ka. Chemical elements whose residence time in the oceans is much less than 100 ka, for example neodymium ($R = 1000$ years), will be subjected to fluctuations (if the geochemical processes governing them are involved) in phase with the source oscillations and which may be of large amplitude depending on the corresponding J_0/b ratio (Figure 8.9).

The chemical elements with very long residence times like strontium ($R = 2$–4 Ma)) will be subjected to fluctuations with a time lag of half a period compared with the source, but that are greatly damped too. If, for example, the disturbance reflects the glacial–interglacial cycles, the strontium fluctuations will be offset and damped.

Let us examine the effect of the oscillation period. We stay with strontium with its residence time of 2–4 Ma. If the fluctuations are of the order of 100 or 1000 ka (climate, for example) the results will be as already described. If the fluctuations are from some geological source of the order of 20–30 Ma (tectonic), the strontium fluctuations will follow the source fluctuations, in phase and without any attenuation. Of course intermediate cases arise between these two extremes.

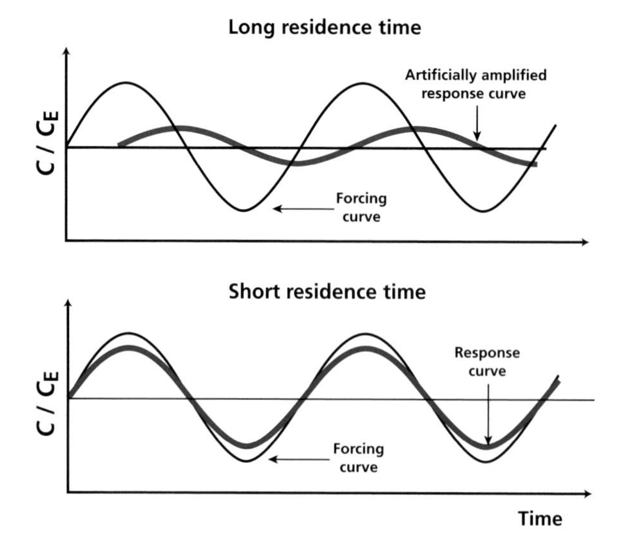

Figure 8.9 Relationship between residence time and lag in the response curve. The response curve at the top has been greatly amplified vertically so the effect can be seen. C, concentration; C_E, equilibrium concentration.

The relative importance of the sine and cosine terms depends on the ratio k/ω. We shall see that these properties extend to the fluctuations of isotopic compositions.

Exercise

We consider sea water and the disturbance to its chemical composition as a result of climatic disruptions introducing different concentrations in rivers. We consider a sinusoidal disturbance with a period of 10 000 years.

We assume the disturbance follows the law: $J = a + b \sin \omega t$.

For four chemical elements Pb, Nd, Os, and Sr whose residence times in sea water are $R_{Pb} = 10^3$ years, $R_{Nd} = 3 \cdot 10^3$ years, $R_{Os} = 3 \cdot 10^4$ years, and $R_{Sr} = 2 \cdot 10^6$ years, respectively, we represent the fluctuations by changing the values of a and b so that $a/R = 1$ and $b = 1$ (this greatly amplifies the Os and Sr fluctuations).

Draw the response curves for the different elements and estimate the phase-shift values. Try to identify any quantitative relationship between residence time and phase shifts. Next try to evaluate damping, taking $a = 1$ and $b = 1$ for all of the elements.

Answer

The equation describing the response curve is written:

$$\frac{dC}{dt} = a + b \sin \omega t - kC.$$

Integrating gives:

$$C = \frac{a}{k} + \frac{bk}{k^2 + \omega^2} \sin \omega t - \frac{b\omega}{k^2 + \omega^2} \cos \omega t.$$

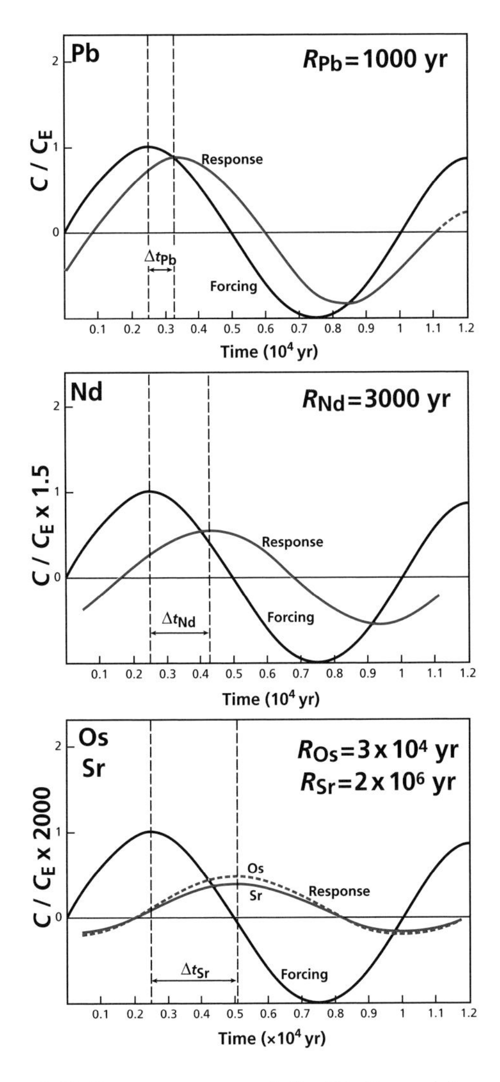

Figure 8.10 Response curves of Pb, Nd, Os, and Sr concentrations in the ocean, if the sources vary sinusoidally with the characteristics given in the main text. ΔC = concentration – equilibrium concentration.

We take 10^3 years as the time unit and $\omega = 2\pi/T = 0.628$.

For Pb, $k = 1$; for Nd, $k = 0.033$; for Os, $k = 0.033$; and for Sr, $k = 5 \cdot 10^{-4}$.

The curves are illustrated in Figure 8.10 ignoring damping. They show the offset in terms of residence time. When the logarithm of R (residence time) is plotted against the phase shift, we get a curve similar to a hyperbola[3] with two asymptotes corresponding to the two limiting cases where k/ω is very large and k/ω is very small (Figure 8.11).

To evaluate damping we take $a = 1$ for all the elements, therefore a/k is 1, 3, 30, and 2000 for Pb, Nd, Os, and Sr, respectively. We evaluate the percentage variation:

[3] In fact, it is the curve $tg\ \omega.\Delta t = \omega R$.

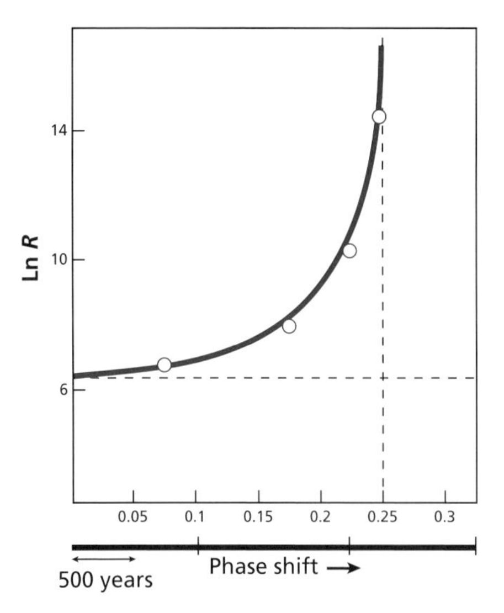

Figure 8.11 Relationship between residence time as a logarithm and phase shift, expressed in intervals of 10 000 years.

$$\frac{\Delta C}{\bar{C}} = \Delta C / \left(\frac{a}{k}\right).$$

We find 90%, 65%, 0.2%, and 0.15% for the fluctuations of Pb, Nd, Os, and Sr, respectively.

We shall see that these properties extend to fluctuations in isotope composition.

Generalization

Extension to all forcing functions

The great interest of this study of response to a sinusoidal source is that a response can be obtained for any source function by decomposing it into its Fourier sine functions and then finding the response for each sine function and summing them.

A LITTLE HISTORY

Joseph Fourier

When he was prefect of France's Isère department under Napoleon, Joseph Fourier (after whom Grenoble's science university is named) proposed one of the most important theorems in mathematics. He showed that any periodic function could be represented by the sum of the sine and cosine functions with appropriate amplitudes and phases. He then showed that when a function was not periodic "by nature," it could be made periodic by truncating it and then repeating the sampled portion. On this basis, any function can be separated into its Fourier components and so represented as a spectrum (frequency, amplitude). Milankovitch cycles are an example that we have already mentioned.

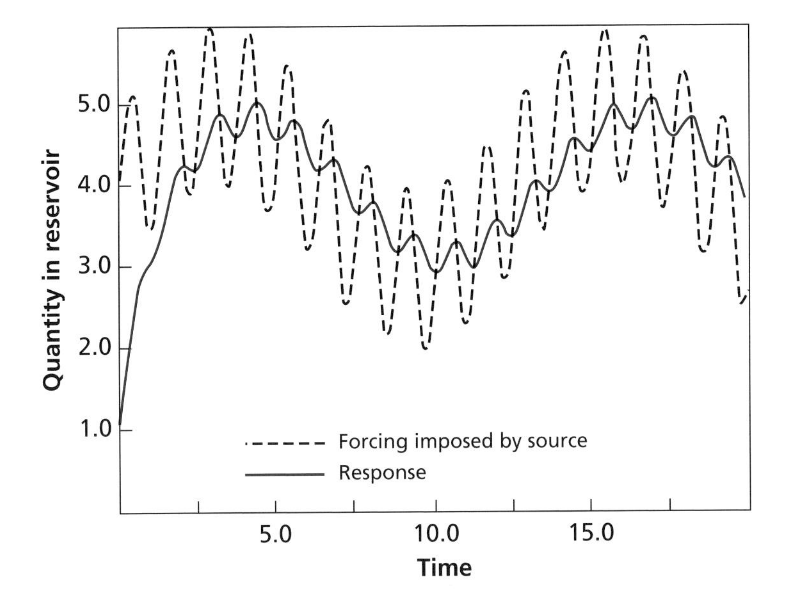

Figure 8.12 Quantity in the reservoir as a function of time. After Lasaga and Berner (1998).

Exercise

We take a system (borrowed from Lasaga and Berner) obeying the previous equations with an imposed fluctuation of $F(t) = 4 + \sin(5t) + \sin(t/2)$. The residence time constant is taken as 1 in the arbitrary time units (t). Calculate and represent the function at the reservoir outlet.

Answer

See Figure 8.12. Notice the phase shift, the damping of high frequencies, and the slight alteration of low frequencies.

The case of radioactive isotopes

We have said that for radioactive systems, the residence time should take account of the mean life of the radioactive element, which is a sort of intrinsic residence time. The dynamic equations we have seen are identical, except for this modification of residence time. In some cases, when the residence time of the reservoir is very large, it is the isotope's lifetime that determines the overall residence time.

The same reasoning applies depending on the value of the λ/ω ratio, where ω is the frequency of the periodic disturbance and λ is the decay constant. The fluctuations imposed by the source shall be taken into account differentially with or without a phase shift and with or without amplitude damping.

EXAMPLE

Cosmogenic isotopes in the atmosphere

There are three types of cosmogenic isotope in the atmosphere we can look at: ^{10}Be, ^{3}H, and ^{14}C. For ^{3}H, $\lambda = 5.57 \cdot 10^{-2}\ yr^{-1}$, for ^{10}Be, $\lambda = 4.62 \cdot 10^{-7}\ yr^{-1}$, and for ^{14}C, $\lambda = 1.209 \cdot 10^{-4}\ yr^{-1}$.

When we observe fluctuations in the abundance of these isotopes, ^{3}H reflects rapid fluctuations in the atmosphere whereas ^{14}C largely damps these variations. However, ^{10}Be, with a much smaller decay constant, should damp them even more than ^{14}C. In fact, it reflects rapid oscillations. Why? Because its residence time in the atmosphere of 1–3 years is very short and it is this constant that prevails.

For similar reasons when it comes to deciphering the Earth's early history, the information provided by extinct forms of radioactivity is not the same as that provided by long-lived forms, as we have seen.

8.3.5 Non-linear processes

In all the examples examined so far, the basic differential equation has been linear, even if the forcing term was non-linear over time, as in the previous example. Let us now consider a model of growth of the Earth's core. Let the variable N represent the quantity of iron in the core. Then:

$$\frac{dN}{dt} = K(1 - N)$$

where $(1 - N)$ is the mass fraction of iron in the core, the total quantity of iron being normalized to unity. It can be considered that the attraction of this iron dispersed in the form of small minerals is proportionally stronger when the quantity of iron in the core is high (gravitational attraction). To characterize the quantity N, we can therefore take $K = QN$.

This gives the equation:

$$\frac{dN}{dt} = QN\,(1 - N).$$

This is a well-known equation, especially in population dynamics. It is the logistic equation (see Haberman, 1997):

$$\frac{dN}{N\,(1 - N)} = Q\,dt.$$

By separating out the simple elements, we can write:

$$\left[\frac{1}{N} + \frac{1}{(1 - N)}\right] dN = Q\,dt.$$

Integrating gives:

$$\ln|N| - \ln|1 - N| = Qt + C,$$

hence:

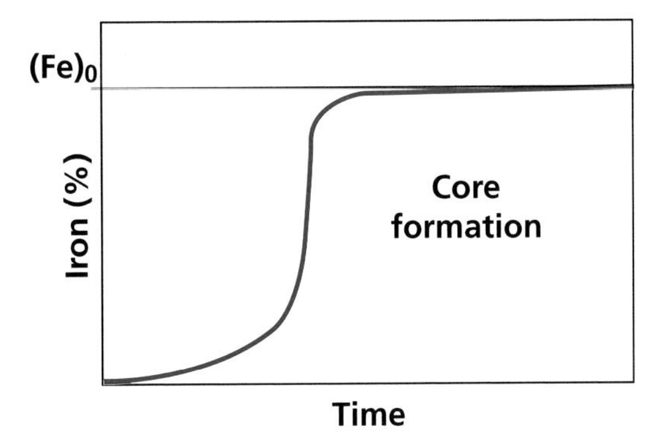

Figure 8.13 Model curve of core growth. To simplify, it is taken that all the iron is in the core although a fraction (approx. 6%) is in the mantle.

$$N = \frac{N_0}{N_0 + (1 - N_0)\, e^{-Qt}}$$

where N_0 is the mass of iron in the core that may be considered as the mass of iron located at the center of the Earth and pooled by melting. This can be shown graphically as an S-shaped curve beginning slowly then accelerating suddenly and ending as an asymptote, with value 1, which is the total quantity of iron (Figure 8.13).

An attempt can then be made to connect time and the constant Q. Assuming that $N_0 = 0.1$, the time taken to form 95% of the core is written:

$$t = 5/Q.$$

If this time is estimated at 50 Ma, we get $Q = 1 \cdot 10^{-7}$. (These calculations are to give an order of magnitude of plausible processes.[4])

This logistic law is probably as general in naturally occurring processes as the exponential law is in fundamental physical processes.

8.4 The laws of evolution of isotope systems

The various examples we have developed can be used to model the Earth's geodynamic systems using isotope systems as tracers, just as radioactive isotopes can be used to see how the human body or a complex hydrological system behave. This model is made much easier by using general equations describing the evolution of isotope systems in complex dynamic systems. These equations can be used to solve problems directly. But, of course, one of the aims is to obtain a formulation for calculating the system parameters from direct observations.

[4] In actual processes, allowance must be made for oxidation of iron, which allows a fraction of the iron to remain in the mantle.

8.4.1 Equation for the evolution of radiogenic systems: Wasserburg's equations

Let us begin with a straightforward case. A reservoir exchanges radioactive and reference isotopes ^{87}Rb, ^{87}Sr, ^{86}Sr, notated r, i, j, with the outside. We notate what enters the reservoir $()_\downarrow$ and what exits $()_\uparrow$. The decay constant is notated λ. The fluxes are notated $\dot{r}, \dot{i}, \dot{j}$. We can then write the equations:

$$\frac{dr}{dt} = -\lambda r - (Hr)_\uparrow + (\dot{r})_\downarrow.$$

$$\frac{di}{dt} = \lambda r - (Gi)_\uparrow + (\dot{i})_\downarrow.$$

$$\frac{dj}{dt} = -(Gj)_\uparrow + (\dot{j})_\downarrow.$$

We have written what leaves the reservoir as proportional to what is in the system, assuming the factor is the same for (i) and (j), that is, assuming no isotope fractionation occurs. Let us combine these equations to bring out the ratios (r/j) and (i/j), that is, by our notations $(^{87}Rb/^{86}Sr)$ and $(^{87}Sr/^{86}Sr)$:

$$\frac{d\left(\frac{r}{j}\right)}{dt} = \left(\frac{r}{j}\right)\left[\frac{dr}{r\,dt} - \frac{dj}{j\,dt}\right] = \left(\frac{r}{j}\right)\left[-\lambda - H + \frac{(\dot{r})_\downarrow}{r} + G\frac{(\dot{j})_\downarrow}{j}\right].$$

We have notated dr/dt and dj/dt as $\dot{r}$ and $\dot{j}$ respectively, as is often done. By bringing out $(r/j)_\downarrow$, that is what leaves the system, and by writing:

$$\frac{(\dot{r})_\downarrow}{r} = \frac{(\dot{r})_\downarrow}{(\dot{j})_\downarrow}\frac{(\dot{j})_\downarrow}{(r)},$$

then by multiplying the bracketed term by (r/j), and noting $r/j = \mu$, as is our standard practice, we get:

$$\frac{d\mu}{dt} = [(G - H) - \lambda]\,\mu + (\mu_{ex} - \mu)\frac{(\dot{j})_\downarrow}{j}.$$

Doing the same calculations for $(i/j) = \alpha$ gives:

$$\frac{d\alpha}{dt} = \lambda\mu + (\alpha_{ex} - \alpha)\frac{(\dot{j})_\downarrow}{j}.$$

By noting $(\dot{j})_\downarrow/j = L(t)$ and $G - H = F$, we finally obtain:

$$\begin{cases} \dfrac{d\mu}{dt} = [F - \lambda]\,\mu + (\mu_{ex} - \mu)\,L(t) \\[2ex] \dfrac{d\alpha}{dt} = \lambda\mu + (\alpha_{ex} - \alpha)\,L(t). \end{cases}$$

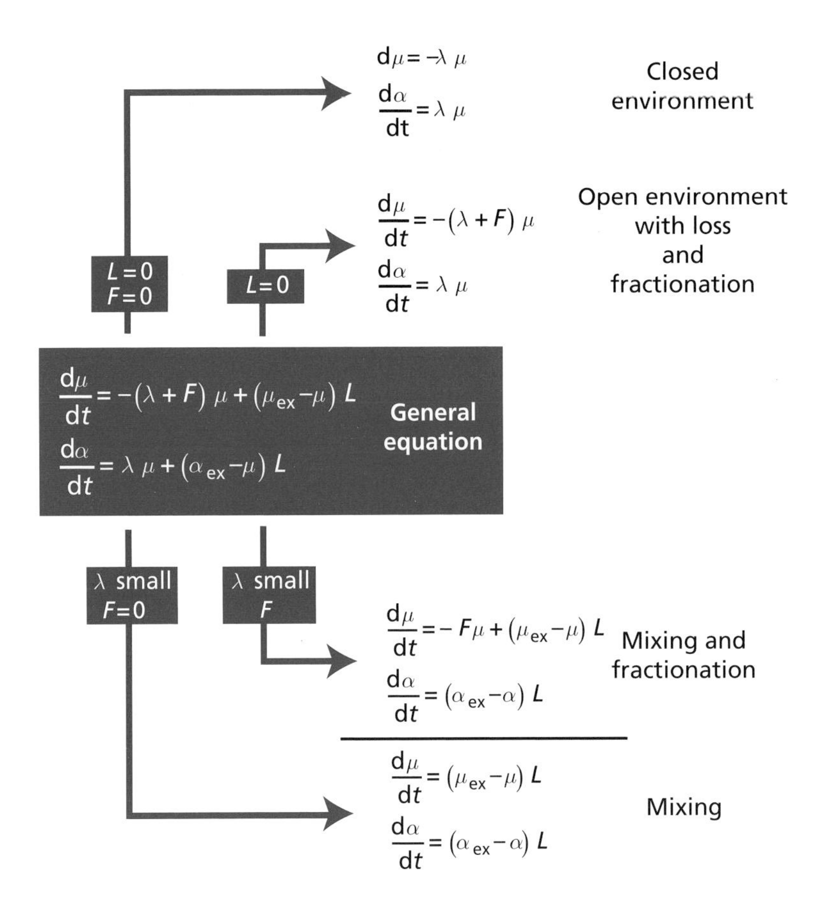

Figure 8.14 Schema explaining the generality of Wasserburg's equations. Most of the basic equations we have used for radioactive–radiogenic systems can be found in his figure.

These are **Wasserburg's equations** (see Wasserburg, 1964) (Figure 8.14).

Let us make a few remarks about them. Notice that parameter F may be either positive or negative depending on whether the daughter element "leaks" from the reservoir faster ($G > H$) or more slowly ($G < H$) than the parent element. These are two concatenated (and not paired) equations. One (μ) describes the system's chemical evolution. It therefore involves a chemical fractionation factor for the material leaving the reservoir. The other (α) involves two terms, one for radioactive decay $\lambda\mu$ and the other for mixing ($\alpha_{ex} - \alpha$). Solving the problem entails integrating the first and then the second. These are very general equations applying to all systems – minerals, rocks, atmosphere, mantle – and thus can be used in both geochronology and isotope geology. Notice that $L(t)$ is the inverse of residence time $(1/R(t))$.

8.4.2 The steady-state box model

This model is developed here through two examples.

Exercise

Take a reservoir into which flows material from a single external source with a constant chemical and isotopic composition over time (μ_{ex} and α_{ex}) and a constant flow rate. Let us assume that the reservoir attains a steady state.

What is the residence time if we suppose the reservoir is well mixed (that is, having a homogeneous isotope composition α)?

Answer

We find: $-\lambda\mu + (\mu_{ex} - \mu)\, 1/R = 0$ and $\lambda\mu + (\alpha_{ex} - \alpha)\, 1/R = 0$.
This gives:

$$R = \frac{1}{\lambda}\left(\frac{\mu_{ex}}{\mu} - 1\right).$$

We find quite simply the residence time calculated for the $^{14}C/C$ ratio of the deep ocean if $\mu = {}^{14}C/C$ and μ_{ex} is the $^{14}C/C$ ratio of the surface water in equilibrium with the atmosphere. The time can be computed from the simple radioactive decay $\mu = \mu_{ex}\, e^{-\lambda t}$ or $\mu_{ex} = \mu\, e^{\lambda t}$. If we develop the Taylor series and keep the first two terms:

$$\mu_{ex} = \mu(1 + \lambda t)t = \frac{1}{\lambda}\left(\frac{\mu_{ex}}{\mu} - 1\right).$$

As for the isotope ratio:

$$R = \frac{1}{\lambda}\left(\frac{\alpha - \alpha_{ex}}{\mu}\right) = \tau.$$

This is the expression of the model age of the reservoir, so we are back to the equality:

residence time = model age.

(This is not so for ^{14}C because the product ^{14}N is drowned in normal ^{14}N.)

Exercise

The $^{87}Sr/^{86}Sr$ isotope composition of sea water is the result of erosion of the continents and of exchange at the mid-ocean ridges where Sr from the mantle is injected into sea water, but also of alteration by volcanoes in subduction zones and oceanic islands. We denote the isotope ratios of the continental crust $()_{cc}$ and of the mantle $()_{m}$: $(\alpha^{Sr})_{cc} = 0.712$, $(\alpha^{Sr})_{m} = 0.703$, $(\alpha^{Sr})_{sea\ water} = 0.709$.

What are the relative flows L_{cc} and L_m, given that the residence time of Sr is $R_{Sr} = 4 \cdot 10^6$ years and that a steady state is attained in the ocean? If we know the ratio of the mass of river water inflow to the mass of the ocean is $3 \cdot 10^{-5}$, calculate the river/ocean Sr concentration ratio.

Answer

We write the simplified Wasserburg equations, as there is no radioactive decay or growth term:

$$\left(\alpha^{Sr}_{cc} - \alpha^{Sr}_{water}\right) L_{cc} + \left(\alpha^{Sr}_{m} - \alpha^{Sr}_{water}\right) L_m = 0.$$

We obtain:

$$\frac{L_m}{L_{cc}} = \left(\frac{\alpha_{cc}^{Sr} - \alpha_{water}^{Sr}}{\alpha_{water}^{Sr} - \alpha_m^{Sr}}\right) = 0.5.$$

We also have:

$$L_m + L_{cc} = 1/R = 2.5 \cdot 10^{-7}\,\text{yr}^{-1},$$

hence $L_{cc} = 1.66 \cdot 10^{-7}\,\text{yr}^{-1}$ and $L_m = 0.83 \cdot 10^{-7}\,\text{yr}^{-1}$.
We then have:

$$L_{cc} = \frac{\text{river water flow} \times \text{Sr concentration in rivers}}{\text{ocean mass} \times \text{Sr concentration in ocean}},$$

hence

$$\frac{\text{Sr concentration in rivers}}{\text{Sr concentration in sea water}} = 0.005.$$

Exercise

Suppose the variation in continental Sr input into sea water follows the tectonic cycle with a period of 100 Ma. Suppose also that this fluctuation occurs with constant isotope composition for the continental crust and mantle. By how much would the $^{87}Sr/^{86}Sr$ isotope ratio of sea water as indicated by limestone be offset?

Answer

It would not be offset. See Section 8.4.

8.4.3 The non-steady state

Naturally enough, in the general case, all of the parameters are a function of time: $F(t)$, $\mu_{ex}(t)$, and $L(t)$. It is not generally very easy to integrate these equations when we are unaware a priori of the form of variation of the various parameters. It can easily be seen that if we write $L(t) = 0$, that is, if we are in the case of evolution with no input from outside, without mixing, we come back to the equations developed in Chapter 3 for open geochronological systems with $F(t)$ taking the form of an episodic or continuous loss. Conversely, if we cancel the terms of radioactive decay, we are dealing with pure mixing. The mixing equation we are used to is:

$$\alpha_M = \alpha_1 x + \alpha_2(1 - x).$$

Let us assume that $\alpha_M = \alpha(t+1)$ and $\alpha = \alpha(t)$ are the two values of the reservoir at $(t+1)$ and (t). If we modify the value α of the reservoir by adding a corresponding quantity to Δx from the outside, then:

$$\alpha_2(t+1) = (\alpha_1 - \alpha(t))\Delta x + \alpha(t),$$

therefore

$$\alpha(t+1) - \alpha(t) = \Delta\alpha = (\alpha_1 - \alpha)\Delta x.$$

This is the expression for differential mixing already established.

Wasserburg's equations may be extended to several reservoirs exchanging material and with different isotope ratios. For each reservoir, we write:

$$\begin{cases} \dfrac{\mathrm{d}\mu}{\mathrm{d}t} = \left[\sum_{i=1}^{i=n} Fi - \lambda\right]\mu + \sum_{i=1}^{i=n} (\mu_i - \mu)\, L_i \\ \dfrac{\mathrm{d}\alpha}{\mathrm{d}t} = \lambda\mu + \sum_{i=1}^{i=n} (\alpha_i - \alpha)\, L_i \end{cases}$$

where *Fi* is the total of fractionation factors.

There are as many pairs of equations as there are reservoirs and we switch therefore to a matrix system whose solution is complex and generally difficult to solve because there are many unknown, time-dependent parameters. However, in some cases it can be approximated.

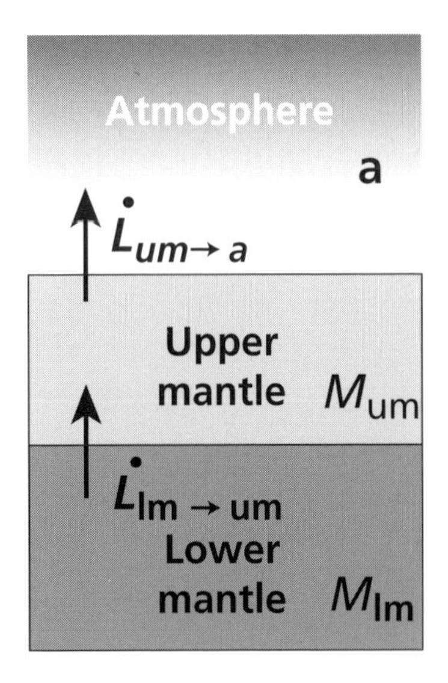

Figure 8.15 The simplified three-reservoir system.

Exercise

A simplified three-reservoir system is considered: the lower mantle (lm), upper mantle (um), and atmosphere (a) with transfers of radiogenic rare gases described by the transfer coefficients shown in Figure 8.15, plus a chemical fractionation (*F*) during the transition from the upper mantle to the atmosphere.

Write the matrix equation describing the dynamic evolution of the system. It is assumed there is no reverse transfer either from the atmosphere to the upper mantle or from the upper to the lower mantle and that transfer from the lower to upper mantle does not involve chemical fractionation.

Keep the α and μ notations as used in this book.

Answer

The subscripts to denote the reservoirs are a = atmosphere, um = upper mantle, lm = lower mantle.

$$\mu = \begin{bmatrix} \mu_a \\ \mu_{um} \\ \mu_{lm} \end{bmatrix} \text{ and } \alpha = \begin{bmatrix} \alpha_a \\ \alpha_{um} \\ \alpha_{lm} \end{bmatrix}.$$

$$\frac{d\mu}{dt} = [T_\mu]\mu$$

where $[T_\mu]$ is the transfer matrix of μ.

$$\frac{d\alpha}{dt} = [T_\alpha]\,\alpha + \lambda\mu$$

where $[T_\alpha]$ is the transfer matrix of α.

$$[T_\mu] = \begin{vmatrix} 0 & 0 & 0 \\ 0 & -(\lambda - F + L_{lm\to um}) & +L_{lm\to um} \\ 0 & 0 & -\lambda \end{vmatrix} \text{ and } [T_\alpha] = \begin{vmatrix} -L_{um\to a} & +L_{um\to a} & 0 \\ 0 & -L_{lm\to um} & +L_{lm\to um} \\ 0 & 0 & 0 \end{vmatrix}.$$

This exercise is designed to show how complex the problems are, but also to prepare readers for the mathematical processing used in research work.

The cases of stable isotopes may also be covered by generalizing from these equations somewhat, but in this case, we must consider a combined formula because while there is no radioactive decay to be considered there is isotope fractionation.

By positing α_s = stable isotope ratio, we can write:

$$\frac{d\alpha_s}{dt} = F_i\,\alpha_s + \sum_i \left(F_{ex}\,\alpha_s^i - \alpha_s\right) L^i(t)$$

where F_i is fractionation internal to the system and F_{ex} is fractionation for elements entering the system. This equation shows how worthwhile but how difficult it is to use stable isotopes in balance processes, because it involves an extra parameter – fractionation.

EXAMPLE

The crust–mantle system

Consider the continental crust-mantle system. Let us take the example of exchange between the continental crust and the mantle. Let us keep our conventional use of μ and α; let us take the $^{87}Rb/^{86}Sr$ system to clarify things. To simplify, we shall assume that the μ values for the mantle are much lower than those for the continental crust.

The evolution of the crust is written:

$$\frac{d\mu_{cc}}{dt} = -\lambda\mu_{cc} + \left(\mu_{lm} - \mu_{cc}\right) L$$

$$\frac{d\alpha_{cc}}{dt} = +\lambda\mu_{cc} + \left(\alpha_{\downarrow m} - \alpha_{cc}\right) L$$

where $\mu_{\downarrow m}$ and $\alpha_{\downarrow m}$ indicate the transfer from the mantle to the continental crust.

It is assumed, to simplify matters, that L is constant over time. Integrating the equation in μ gives:

$$\mu_{cc} = \frac{\mu_{m\downarrow}\, L}{\lambda + L}\left[1 - e^{-(\lambda+L)t}\right] + \mu_{0cc}\, e^{-(\lambda+L)t},$$

μ_{0cc} being the initial value for continental crust. Notice straight away that since $1 - e^{(\lambda+L)t}\ e^{(\lambda+L)t} = 1$, this is a mixing-type equation whose proportions are time dependent. If $x(t) = e^{(\lambda+L)t}$, we get:

$$\mu_{cc} = \frac{\mu_{m\downarrow}\, L}{\lambda + L}\left(1 - x(t)\right) + \mu_{0cc}\, x(t),$$

from which L is of the order of $0.3 \cdot 10^{-9}\ \mathrm{yr}^{-1}$ and $\lambda = 0.0142 \cdot 10^{-9}\ \mathrm{yr}^{-1}$. Therefore $\frac{\mu_{m\downarrow} L}{\lambda + L} \approx \mu_{m\downarrow}$, which is much less than μ_{cc} and is ignored.

$$\frac{d\alpha_{cc}}{dt} = \lambda\, \mu_{0cc}\, e^{-(\lambda+L)t} + \left(\alpha_{m\downarrow} - \alpha_{cc}\right) L.$$

Integrating gives:

$$\alpha_{cc} = \left[\alpha_{cc} + \lambda\mu_{0cc}\, t\right] e^{-(\lambda+L)t} + \alpha_{m\downarrow}\left[1 - e^{-(\lambda+L)t}\right].$$

Notice that this is a mixing equation between the evolution of isolated continental crust (whose evolution is written $\alpha_{0cc} + \lambda\mu_{0cc}\, t$) and the $\alpha_{m\downarrow}$ coming from the mantle, which we have taken to be constant, the terms of the mix being weighted by e^{-Lt} (Figure 8.16).

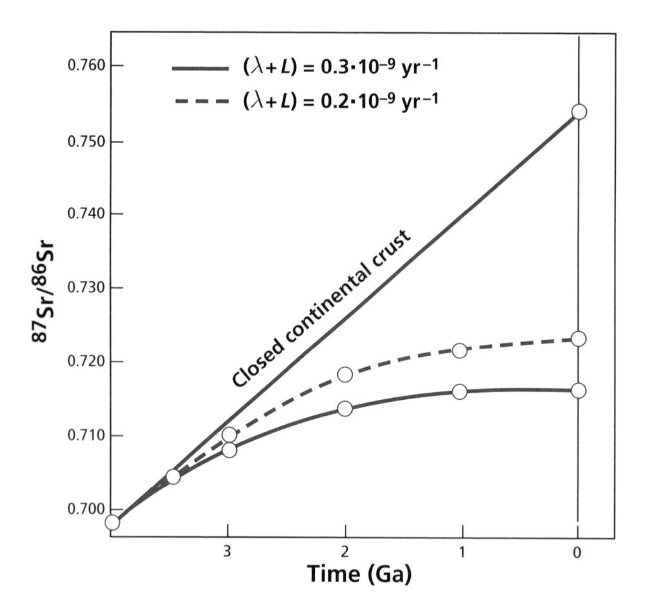

Figure 8.16 Results of the exercise above with $\mu_{cc} = 0.1$.

Exercise

Write the Wasserburg equation for stable isotopes (e.g., $^{18}O/^{16}O$) in δ notation.

Answer

$$\frac{d\delta}{dt} = (\delta - \Delta_{s\uparrow}) + \sum (\delta_i - \Delta_{i\downarrow})\, L^i(t).$$

The $\Delta_{i\downarrow}$ are isotopic fractionations during transfer from sources to the exterior, and $\Delta_{s\uparrow}$ is the isotopic fractionation towards the reservoir, the $L^i(t)$ are identical to those already defined.

Exercise

Consider the sea water reservoir. The concentration of elements is determined by the influx of products of erosion from rocks of continental and mantle origin.

We assume the equation governing this composition is written for element *i*:

$$\frac{dC^i}{dt} = J_i(t) - k_i\, C^i$$

and that $1/k_i = R_i$ is the residence time of element *i*. It is assumed that:
$J_i\,(t) = J_{0i} + bi \sin \omega t$
where $\omega = 2\pi/10^4\ yr^{-1}$.
For all elements we take $J_0 = Gi = 1$.

Calculate the curves of variation of C^i for the elements Nd: $R_i = 1000$ years, Os: $R_i = 30\,000$ years, and Sr: $R_i = 2 \cdot 10^6$ years.

Answer

See Figure 8.10.

8.4.4 Statistical evolution of radiogenic systems: mixing times

Here, we shall again take the example of the upper mantle, but our approach is more general and could apply to other convective geochemical reservoirs like the ocean, the atmosphere, or a river. We are now going to look at not just the average values although they are essential, but also the statistical distributions that can be described summarily by their different statistical parameters: a mean, a dispersion, an asymmetry, etc. (Allègre and Lewin, 1995).

As said, the upper mantle is subjected to two types of antagonistic processes. First, **chemical fractionation** (extraction of oceanic crust, extraction of continental crust, reinjection of material via subduction phenomena). These processes result in chemical and isotopic heterogeneity with the help of time. Second, it is also subjected to **mixing processes** related to mantle convection, which stretch the rocks, break them, fold them, and mix them. There are also melting processes which also tend to homogenize the isotope ratios of the source zone.

Thus two types of phenomena are opposed: those producing chemical and isotopic dispersion and those which mix, homogenize, and tend to destroy the heterogeneities (see McKenzie, 1979; Allègre *et al.*, 1980; Allègre and Turcotte, 1986).

Exercise

Suppose that unaltered oceanic crust, altered oceanic crust, and fine sediments plunge into the mantle in a subduction zone. The extreme compositions for the ^{87}Rb–^{86}Sr system for unaltered oceanic crust are $\alpha_0^{Sr} = 0.7025$ and $\mu^{Rb/Sr} = 0.1$ and for the sediments $\alpha_S^{Sr} = 0.712$ and $\mu^{Rb/Sr} = 0.4$. The altered crust has intermediate values: $\alpha_{0A}^{Sr} = 0.705$ and $\mu^{Rb/Sr} = 0.2$.

What will be the dispersion of the α isotope ratios if this subducted oceanic crust is in the mantle without mixing for 1 billion years?

Answer

Dispersion can be evaluated by calculating the two extreme values. We obtain 0.703 92 and 0.717 68 respectively for the pieces of unaltered oceanic crust and for the sediments. The difference $\Delta\alpha$ is 0.0137, while the difference in α^{Sr} values during subduction was 0.0095.

We can also calculate it directly:

$$\Delta\alpha = (\Delta\alpha)_{\text{initial}} + \lambda(\Delta\mu)T.$$

Hence Δ indicates the dispersion: $\Delta\alpha = 0.0095 + 0.004\,26 = 0.0137$. (Notice that given the value of λ, the term $-\lambda\mu$ is negligible in the evolution of μ.)

Suppose now that these subduction products are subjected to multiple mixing processes in the mantle. Suppose the result is that $\Delta\mu_i$, which was 0.3 initially, becomes $\Delta\mu = 0.1$, and that $\Delta\alpha$ becomes 0.003.

The effectiveness of this isotopic mixing can be measured by the ratio $30/137 = 0.21$ whereas chemical mixing, measured by $0.1/0.3 = 0.33$, is not quite as effective. (This reasoning probably fails to allow for isotope exchange, but it is a first approximation.)

Let us try to generalize these simple examples. We are looking to write an equation derived from Wasserburg's, but dealing with distributions. So we take as variables not the average values, which has already been done, but the dispersions. Dispersion can be measured by the standard deviation (or by the standard deviation of two extreme values, which, as we know, are about three times the standard deviation) (Allègre and Lewin, 1995).

The following equation can be derived from Wasserburg's equations by noting dispersion $\langle\rangle$. Thus we note the dispersion of isotope ratios $\langle\alpha\rangle$ and the dispersion of chemical ratios $\langle\mu\rangle$ with subscripts α_{ex} (external), α_i (internal), μ_{ex} (external), and μ_i (internal).

$$\frac{d\langle\mu_i\rangle}{dt} = [\langle\mu_{ex}\rangle - \langle\mu_i\rangle]\,L - M\langle\mu_i\rangle$$

where the term $-\lambda\langle\mu_i\rangle$ is ignored (Figure 8.17).

The new term that has been introduced is $M\langle\mu_i\rangle$, the term of homogenization of the mixture; M has the dimension of the inverse of a time we shall call **mixing time** (τ). It is the time it takes to reduce the dispersion $\langle\mu\rangle$ by a factor (e) exponential. Suppose a steady state is attained, then:

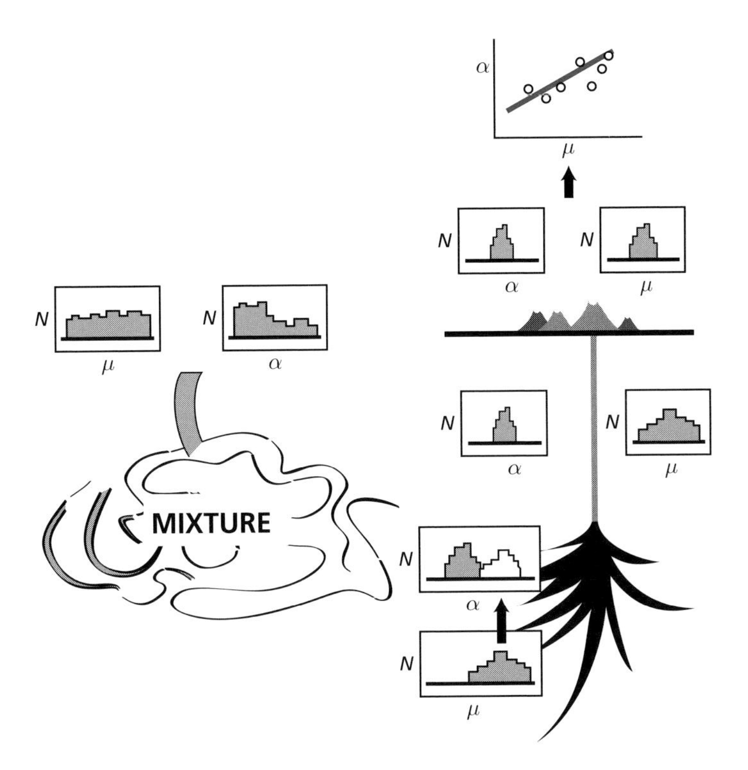

Figure 8.17 Diagram explaining how the histograms for parameters μ and α evolve during the geodynamic cycle. Values of α and μ are represented by histograms. We start on the left with a dispersed histogram for α and μ. In the mantle the spread is reduced by mixing but supplementary α is heated by radioactivity (white in histogram). Then melting reduces the spread of α but not so much as for μ.

$$\langle \mu_i \rangle = \langle \mu_{ex} \rangle \frac{L}{M + L}.$$

Replacing L by its expression 1/residence time $= 1/R$ and $M = 1/\tau$, gives:

$$\langle \mu_i \rangle = \langle \mu_{ex} \rangle \frac{\tau}{R + \tau}.$$

If $\tau \ll R$, mixing is very rapid and $\langle \mu_i \rangle \rightarrow 0$ when $\tau \rightarrow 0$. This is intuitive enough. If mixing time is short, homogenization is vigorous and therefore the standard deviation is zero. If, conversely, $\tau \gg R$, homogenization is poor and $\langle \mu_i \rangle = \langle \mu_{ex} \rangle$: dispersion in the reservoir is the same as that introduced.

For isotope ratios, the equation for standard deviation is written:

$$\frac{d\langle \alpha_i \rangle}{dt} = \frac{\langle \alpha_{ex} \rangle - \langle \alpha \rangle}{2R} + \frac{\lambda}{2} \langle \mu \rangle - \frac{\langle \alpha \rangle}{2\tau}.$$

Let us make two remarks. The first, a purely arithmetical one, is that the factor $^1/_2$ is found everywhere because the calculation made rigorously with variance is $d\langle \alpha \rangle^2/dt = 2\langle \alpha \rangle\, d\langle \alpha \rangle/dt$. The second and more important remark is that in the terms of creation

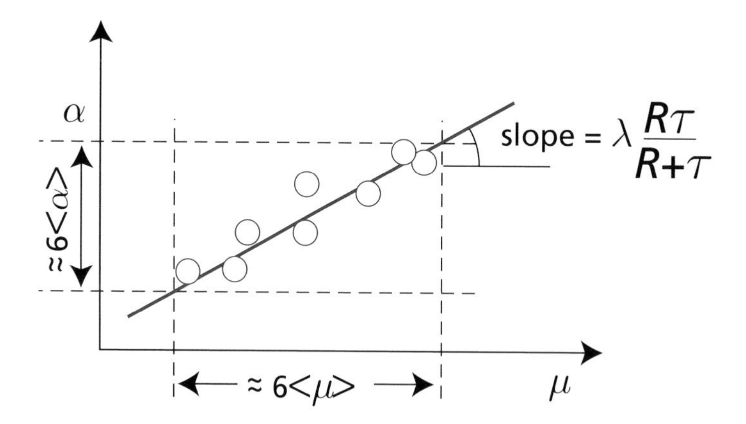

Figure 8.18 Mantle isochrons constructed from dispersion of (μ) and (α). The slope gives a pseudo-age $(R+\tau)$.

of heterogeneity, there is the term of external inputs in $\langle \alpha_{ex} \rangle$ but also a term from the dispersion of $\langle \mu \rangle$ values.

If λ is small enough for us to speak of a steady state, we get:

$$\langle \alpha_i \rangle = \lambda \langle \mu_i \rangle \left(\frac{R\tau}{R + \tau} \right) + \langle \alpha_{ex} \rangle \left(\frac{\tau}{R + \tau} \right).$$

If τ is small compared with R (very active mixing):

$$\langle \alpha_i \rangle \approx \lambda \langle \mu_i \rangle \tau + \langle \alpha_{ex} \rangle \left(\frac{\tau}{R} \right).$$

If τ is very small, $\langle \alpha_i \rangle$ in the mixture is also very small. The mantle is therefore isotopically very homogeneous (small standard deviation).

If τ is much larger than R, mixing is poor, and the standard deviation is equal to the standard deviation of μ multiplied by residence time, plus the deviation of input from outside.

Let us examine the relationship there may be between the dispersion of $\langle \alpha \rangle$ and that of the $\langle \mu \rangle$ values. We saw when calculating the least squares that in an (α, μ) plot, the statistical slope equals $\langle \alpha \rangle / \langle \mu \rangle$ (the ratio of standard deviations multiplied by the correlation coefficient of approximately 1).

If we plot the points representing mantle rock on an (α, μ) diagram, e.g., $(^{143}Nd/^{144}Nd, ^{147}Sm/^{144}Nd)$ or $(^{87}Sr/^{86}Sr, ^{87}Rb/^{86}Sr)$, the slope of the straight line yielding an apparent age is equal to the ratios of the standard deviations of the (α) and (μ) values.

If it can be considered that $\langle \alpha_{ex} \rangle$ is about constant (this is not the absolute value of α_{ex} but its dispersion!), the slope is equal to $R\tau/(R+\tau)$ (Figure 8.18).

If τ is small compared with R, the slope is about equal to (τ), the mixing time. Unfortunately it is not easy to estimate $\langle \mu \rangle$, the "chemical" dispersion ratio, because chemical fractionation in basalt formation greatly increases $\langle \mu \rangle$ dispersion even if it can be assumed that $\langle \alpha \rangle$ remains the same.

The slope of the correlation diagram obtained for basalts is then written:

$$D \left(\frac{R\tau}{R + \tau} \right)$$

where D is a coefficient greater than unity and rather difficult to estimate.

This difficulty makes the exercise somewhat hazardous. Let us attempt it none the less to get an order of ideas. An isochron has been obtained on oceanic basalt by the $^{147}Sm/^{144}Nd$ method with an age of 350 Ma.

Let us admit a value for D between 2 and 1.5 and a value of $R = 1$ Ga for the residence time of the upper mantle. This gives a mixing time τ of 530 Ma.

Another way of addressing the issue is to use the $^4He/^3He$ ratios (Allègre *et al.*, 1995). The ratios measured in MORB are the outcome of mixing of OIB ratios which represent an unmixed mantle. As Figure 6.18 shows, the $^4He/^3He$ ratios of MORB seem to be slightly dispersed around the average for OIB.

Suppose, as a first approximation, that the dispersion caused by the decay of uranium is faithfully reflected by the OIB. We can then write:

$$\frac{\langle \alpha_{\text{MORB}} \rangle}{\langle \alpha_{\text{OIB}} \rangle} = \frac{\tau}{R + \tau}.$$

With $\langle \alpha_{\text{OIB}} \rangle = 45 \cdot 10^3$ and $\langle \alpha_{\text{MORB}} \rangle = 9$, and still taking $R = 1$ Ga, we find $\tau = 0.25$ Ga. We are still dealing with the same order of magnitude but this time it is for the mantle MORB source alone.

There are two important points to remember from this section. **If we have an (α, μ) relation for present-day rocks from the mantle, therefore from a convecting reservoir, the slope does not measure the age of some sudden past event but is related to the physical characteristics of the reservoir: mixing time and residence time.** We must henceforth set about describing geochemical reservoirs not by average parameters but by distributions and even by regionalized distributions. Work is under way in this direction.

Notice too that a number of conclusions about the case where external dispersion fluctuates with $\sin \omega t$ can be applied to the dispersion equation. The decisive parameter in this case is $L/(M + L)$, that is: $\tau/(R + \tau)$. Dispersion will be in phase or out of phase, damped or undamped, depending on the value of this parameter. We leave this matter to readers who wish to investigate this area further and who, for this purpose, may transpose the calculations already set out.

Exercise

It is supposed that the $^4He/^3He$ ratios of the upper mantle vary as a result of dispersion introduced by variable ratios of the OIBs and of mantle convection (see Allègre *et al.*, 1995). The values measured on OIB by $^4He/^3He$ ratios are: mean 93 390, dispersion 45 330.

The values measured for MORB of the North Atlantic ridges are: mean 8938, dispersion 9330.

Calculate the upper mantle mixing time in the North Atlantic, given that the residence time for the Atlantic lithosphere is 1.3 Ga.

The residence time of the North Pacific upper mantle is 582 Ma and the dispersion 3000. Calculate the mixing time of these two zones of the upper mantle.

Answer

The mixing times for the two upper mantle zones are 300 Ma for the North Atlantic and 40 Ma for the North Pacific.

Problems

1 Consider a reservoir whose concentrations evolve in accordance with the equation with standard notation: $dC/dt = J - kC^2$, which is therefore a non-linear evolution equation. What is the residence time of the element in question? Can you imagine a geochemical process for which such a formula might apply? What is the system's response law if a flux J_0 is suddenly injected and then left to evolve by itself?

2 It is assumed that erosion fluctuates with glacial cycles. These cycles are supposedly modeled by the superimposition of three frequencies: 100 ka, 40 ka, and 20 ka, with relative amplitudes of 2, 1, and 1, respectively. Uranium has a residence time in the ocean of 3 Ma. Supposing that the $^{234}U/^{238}U$ ratios injected into rivers vary with climate, draw the $^{234}U/^{238}U$ response curve of the ocean (without calculating).

3 The residence time of oceanic lithosphere in the primitive upper mantle is 1 Ga. It can be supposed that the corresponding 70 km of mantle are fully degassed when they go through the mid-ocean ridge. The ^4He in the upper mantle is the sum of two terms: the radiogenic part formed *in situ* over 1 Ga and the part coming from the lower mantle at the same time as the ^3He.

(i) Calculate how much ^4He accumulated in 1 Ga in the upper mantle with U = 5 ppm and Th/U = 2.5.

(ii) Given that the degassing of ^3He from the mid-ocean ridge is $1 \cdot 10^3$ moles yr^{-1} and that $^4He/^3He = 10^5$, calculate the residence time of ^4He in the upper mantle.

(iii) What do you conclude about the melting process at the mid-ocean ridges?

4 Suppose that the dispersion of $^4He/^3He$ ratios in the upper mantle is due to the incorporation of a dispersion through the OIB counterbalanced by mantle convection. Dispersion measured in the MORB of various oceans is given in the table below.

	Dispersion	Expansion rate (cm yr^{-1})	Residence time (Ma)
North Atlantic	9 000	2.3	1400
South Atlantic	6 000	3.5	900
South-west Indian Ocean	11 000	1.7	1600
North Pacific	3 000	8.0	580
Central Indian Ocean	4 700	3.6	700

Ocean expansion rates are also given in the table along with the residence time of oceanic lithosphere in the corresponding mantle province. Calculate the mixing time of the various portions of the upper mantle, given that the OIB dispersion is 45 000. Is there a relation with the expansion rate? What is the relation? Draw it.

5 Consider Figure 8.2 showing the hydrological cycle. Construct a system of boxed reservoirs, with four boxes: atmosphere/ocean, atmosphere/landmass, groundwater/runoff, and oceans. Draw the input and output and the corresponding flows for each box. What is the residence time of water in each box? What is the proportion of ocean going through the groundwater/runoff box and that has flowed into the sea as rivers over 1 million years?

REFERENCES

Chapter 1

Aldrich, L. T., Doak, J. B., and Davis, G. (1953). The use of ion exchange columns in universal analysis for age determination. *Am. J. Sci.* **251**, 377–80.

Arriens, P. A. and Compston, W. (1968). A method for isotopic ratio measurement by voltage peak switching and its applicaton with digital input. *Int. J. Spectrom. Ion Phys.* **1**, 471–81.

Aston, F. W. (1919). A positive ray spectograph. *Phil. Mag.* (Series 6) **38**, 707–14.

Bainbridge, K. T. and Jordan, E. B. (1936). Mass spectrum analysis. *Harvard Univ., Jefferson Phys. Lab. Contrib.* (Series 2) **3**, No. 2.

Becquerel, H. (1896a). Sur les radiations invisibles émises par phosphorescence. *Comptes Rend. Acad. Sciences Paris* **122**, 420.

Becquerel, H. (1896b). Sur les radiations invisibles émises par les corps phosphorescents. *Comptes Rend. Acad. Sciences Paris* **122**, 501.

Becquerel, H. (1896c). Sur les radiations invisibles émises par les sels d'uranium. *Comptes Rend. Acad. Sciences Paris* **122**, 689.

Beiser, A. (1973). *Concepts of Modern Physics*. New York: McGraw-Hill.

Burchfield, J. D. (1975). *Lord Kelvin and the Age of the Earth*. London: Macmillan.

Curie, P. (1902a). Sur la constante de temps caracteristique de la disparition de la radioactivité induite par le radium dans une enceinte fermée. *Comptes Rend. Acad. Sciences Paris* **135**, 857.

Curie, P. (1902b). Nobel Lecture (6 June 1902). In *Nobel Lectures*, Stockholm: Royal Swedish Academy.

Curie, P. and Laborde, A. (1903). Sur la chaleur dégagée spontanément par les sels de radium. *Comptes Rend. Acad. Sciences Paris* **136**, 673–5.

Ingram, M. G. and Chupka, W. A. (1953). Surface ionisation source using multiple filaments. *Rev. Sci. Instrum.* **24**, 518–20.

Mattauch, R. H. (1934). Über eine neuen Masspektrographen. *Z. Physik* **89**, 786–95.

Nier, A. O. (1940). A mass spectrometer for routine isotope abundance measurements. *Rev. Sci. Instrum.* **11**, 212–16.

Rutherford, E. and Soddy, F. (1902a). The cause and nature of radioactivity. Pt. I. *Phil. Mag.* (Series 6) **4**, 370–96.

Rutherford, E. and Soddy, F. (1902b). The cause and nature of radioactivity. Pt. II. *Phil. Mag.* (Series 6) **4**, 569–85.

Rutherford, E. and Soddy, F. (1902c). The radioactivity of thorium compounds. Pt. II. The cause and nature of radioactivity. *J. Chem. Soc. Lond.* **81**, 837–60.

Rutherford, E. and Soddy, F. (1903). Radioactive change. *Phil. Mag.* (Series 6) **5**, 576–91.

Thomson, J. J. (1914). Rays of positive electricity. *Proc. Roy. Soc. Lond.* (Series 5) **89**, 1–20.

Wasserburg, G. J., Papanastassiou, D., Nerow, E.V., and Bauman, C. A. (1969). A programmable magnetic field spectrometer with online data processing. *Rev. Sci. Instrum.* **40**, 288–95.

Chapter 2

Aldrich, L. T. and Nier, A. O. (1948a). Argon 40 in potassium minerals. *Phys. Rev.* **74**, 876–7.

Aldrich, L. T. and Nier, A. O. (1948b). The occurrence of ^{3}He in natural sources of helium. *Phys. Rev.* **74**, 1590–4.

Aldrich, L. T., Herzog, L., Holve, W., Witting, F., and Ahrens, L. (1953). Variation in isotopic abundance of strontium. *Phys. Rev.* **86**, 631–4.

Barbo, L. (2003). *Les Becquerel: Une Dynastie des Scientifiques*. Paris: Belin.

Bateman, H. (1910). Solution of a system of differential equations occurring in the theory of radioactive transformations. *Proc. Cambridge Phil. Soc.* **15**, 423–7.

Birck, J. L. and Allègre, C. J. (1985). Evidence for the presence of ^{53}Mn in the early solar-system. *Geophys. Res. Lett.* **12**, 745–8.

Harper, C. L. and Jacobsen, S. B. (1994). Investigation of ^{182}Hf–^{182}W systematics. *Lunar Plan. Sci.* **25**, 509–10.

Hirt, B., Tilton, G. R., Herr, W., and Hoffmeister, W. (1963). The half-life of ^{187}Re. In Geiss, J. and Goldberg, E. D. (eds.) *Earth Science and Meteorites*, Amsterdam: North-Holland, pp. 273–80.

Ivanovich, M. (1982). Uranium series disequilibria application in geochronology. In Ivanovich, M. and Harmon, R. (eds.) *Uranium Series Disequilibrium: Application to Environmental Problems*, Oxford, UK: Oxford University Press, pp. 56–78.

Jeffrey, P. M. and Reynolds, J. H. (1961). Origin of excess ^{129}Xe in stone meteorites. *J. Geophys. Res.*, **66**, 3582–3.

Kelly, W. R. and Wasserburg, G. J. (1978). Evidence for the existence of ^{107}Pd in the early Solar System. *Geophys. Res. Lett.* **5**, 1079–82.

Kuroda, P. K. (1960). Nuclear fission in the early history of the Earth. *Nature* **187**, 36–8.

Lee, D.-C. and Halliday, A. N. (1995). Hafnium–tungsten chronometry and the timing of terrestrial core formation. *Nature* **378**, 771–4.

Lee, T., Papanastassiou, D. A., and Wasserburg, G. J. (1977). Aluminium-26 in the early solar-system: fossil or fuel? *Astrophys. J. (Lett.)* **211**, L107–10.

Leighton, R. B. (1959). *Principles of Modern Physics*. New York: McGraw-Hill.

Lin, Y., Guan, Y., Leshin, L. A., Ouyang, Z., and Wang, D. (2005). Short-lived chlorine-36 in a Ca- and Al-rich inclusion from the Ningqiang carbonaceous chondrite. *Proc. Natl Acad. Sci. USA* **102**, 1306–11.

Luck, J. M., Birck, J. L., and Allègre, C. J. (1980). ^{187}Re–^{187}Os systematics in meteorites: early chronology of the solar system and the age of the galaxy. *Nature* **283**, 256–9.

Lugmair, G. W. and Marti, K. (1977). Sm–Nd–Pu timepieces in the Angra dos Reis meteorite. *Earth Planet. Sci. Lett.* **35**, 273–84.

Lugmair, G. W., Scheinin N. B., and Marti, K. (1975). Search for extinct ^{146}Sm. Pt. I. The isotopic abundance of ^{143}Nd in the Juvinas meteorite. *Earth Planet. Sci. Lett.* **27**, 9479–84.

McKeegan, K. D., Chaussidon, M., and Robert, F. (2000). Incorporation of short-lived ^{10}Be in a calcium–aluminium-rich inclusion from the Allende meteorite. *Science* **289**, 1334–7.

Nakaï, S., Shimizu, H., and Masuda, A. (1986). A new geochronometer using lanthanum-138. *Nature* **320**, 433–5.

Nielsen, S. G., Rehkämper, M., and Halliday, A. N. (2006). Large thallium isotopic variation in iron meteorites and evidence for lead-205 in the early solar system. *Geochim. Cosmochim. Acta* **70**, 2643–57.

Notsu, K., Mabuchi, H., Yoshioka, O., Matsuda, J., and Ozima, M. (1973). Evidence of the extinct nuclide ^{146}Sm in "Juvinas" achondrite. *Earth Planet. Sci. Lett.* **19**, 29–36.

Patchett, P. J. and Tatsumoto, M. (1980a). Lu–Hf total rocks isochron for eucrite meteorites. *Nature* **288**, 263–7.

Patchett, P. J. and Tatsumoto, M. (1980b). A routine high-precision method for Lu–Hf isotope geochemistry and chronology. *Contrib. Mineral. Petrol.* **75**, 263–7.

Price, P. B. and Walker, R. M. (1962). Observation of fossil particle tracks in natural micas. *Nature* **196**, 732–4.
Ramsay, W. and Soddy, F. (1903). Gases occluded in radium bromide, *Nature* **68**, 246.
Reynolds, J. H. (1960). Determination of the age of the elements. *Phys. Rev. Lett.* **4**, 8–10.
Rutherford, E. (1906). *Radioactive Transformations*. New York: Charles Scribner's Sons.
Schonbachler, M., Rehkämper, M., Halliday, A. N., *et al.* (2002). Niobium–zirconium chronometry and early Solar System development. *Science* **295**, 1705–8.
Shukolyukov, A. and Lugmair, G.W. (1993). Live iron-60 in the early solar system. *Science* **259**, 1138–42.
Srinivasan, G., Ulyanov, A. A., and Goswami, J. N. (1994). ^{41}Ca in the early Solar System. *Astrophys. J. (Lett.)* **431**, L67–70.

Chapter 3

Ahrens, L. H. (1955). Implications of the Rhodesia age pattern. *Geochim. Cosmochim. Acta* **8**, 1–15.
Aldrich, L. T. and Nier, A. O. (1948a). Argon-40 in potassium minerals. *Phys. Rev.* **74**, 876–7.
Aldrich, L. T. and Nier, A. O. (1948b). The occurrence of ^{3}He in natural sources of helium. *Phys. Rev.* **74**, 1590–4.
Aldrich, L. T. and Wetherill, G.W. (1958). Geochronology by radioactive decay. *Ann. Rev. Nuclear Sci.* **8**, 257–98.
Allègre, C. J. (1964). Géochronologie: de l'extension de la méthode de calcul graphique concordia aux mesures d'âges absolus effectués à l'aide du déséquilibre radioactif – cas des minéralisations secondaires d'uranium. *Comptes Rend. Acad. Sciences Paris* **259**, 4086–9.
Allègre, C. J. (1967). Méthode de discussion géochronologique concordia généralisée. *Earth Planet. Sci. Lett.* **2**, 57–66.
Allègre, C. J. (1968). ^{230}Th dating of volcanic rocks: a comment. *Earth Planet. Sci. Lett.* **5**, 209–10.
Allègre, C. J. and Condomines, M. (1976). Fine chronology of volcanic processes using ^{238}U/^{230}Th systematics. *Earth Planet. Sci. Lett.* **28**, 395–406.
Allègre, C. J. and Dars, R. (1966). Chronologie au rubidium–strontium et granitologie. *Geol. Rundschau* **8–55**, 226–37.
Allègre, C. J. and Michard, G. (1964). Sur les discordances des âges obtenus par les méthodes au strontium et à l'argon. *Comptes Rend. Acad. Sciences Paris* **259**, 4313–16.
Allègre, C. J., Albarède, F., Grünenfelder, M., and Köppel, V. (1974). ^{238}U/^{206}Pb–^{235}U/^{207}Pb–^{232}Th/^{208}Pb zircon geochronology in Alpine and non-Alpine environments. *Contrib. Mineral. Petrol.* **43**, 163–244.
Allègre, C. J., Birck, J.-L., Capmas, F., and Courtillot, V. (1999). Age of the Deccan Traps using ^{187}Re–^{187}Os systematics. *Earth Planet. Sci. Lett.* **170**, 197–204.
Berger, G.W. and York, D. (1981). Geothermy from ^{40}Ar/^{39}Ar dating experiments. *Geochim. Cosmochim. Acta* **45**, 795–811.
Blichert-Toft, J., Albarède, F., and Kornpolst, J. (1999). Lu–Hf isotope systematics of garnet pyroxenites from Beni-Bousera, Morocco: implication from basalt origin. *Science* **283**, 1303.
Brévart, O., Dupré, B., and Allègre, C. J. (1986). Lead–lead age of the komatiitic lavas and limitations on the structure and evolution of the Precambrian mantle. *Earth Planet. Sci. Lett.* **77**, 293–303.
Castaing, R. and Slodzian, G. (1962). Optique corpusculaire: premiers essais de microanalyse par émission ionique secondaire. *J. Microsp.* **395**, 185–9.
Compston, W. and Jeffrey, P. M. (1959). Anomalous common strontium in granite. *Nature* **184**, 1792–3.
Compston, W., Williams, I. S., and Meyer, C. (1984). U–Pb geochronology of zircons from lunar breccia 73217 using a sensitive high mass-resolution ion microprobe. *J. Geophys. Res.* **89** (Suppl.), B525–34.

Davis, G. and Aldrich, L. T. (1953). Determination of the age of lepidolites by the method of isotope dilution. *Bull. Geol. Soc. Amer.* **64**, 379–80.

De Sigoyer, J., Chavagnac, V., Blichert-Toft, J., *et al.* (2000). Dating the Indian continental subduction and collisional thickening in the north-west Himalaya multichronology of the Morai eclogites. *Geology* **28**, 487–90.

Hamilton, P. J., Evensen, N. M., O'Nions, R. K., and Tarney, J. (1979). Sm–Nd systematics of Lewisian gneisses: implications for the origin of granulites. *Nature* **277**, 25–8.

Hanson, G. N. and Gast, P. W. (1967). Kinetic studies in contact metamorphic zones. *Geochim. Cosmochim. Acta* **31**, 1119–53.

Harrison, T. M. and McDougall, I. (1981). Excess ^{40}Ar in metamorphic rocks from Broken Hill. *Earth Planet. Sci. Lett.* **55**, 123–49.

Hart, S. R. (1964). The petrology and isotopic–mineral age relations of a contact zone in the Front Range, Colorado. *J. Geol.* **72**, 493–525.

Hirt, B., Tilton, G. R., Herr, W., and Hoffmeister, W. (1963). The half life of ^{187}Re. In Geiss, J. and Goldberg, E. (eds.) *Earth Science and Meteorites*, Amsterdam: North-Holland, pp. 273–80.

Hohenberg, C. M., Podosek, F., and Reynolds, J. H. (1967). Xenon–iodine: sharp isochronism in chondrites. *Science* **156**, 233–6.

Hohenberg, C. M., Brazzle, R. H., Pravdivtseva, O., and Meshik, A. P. (1998). Iodine–xenon. *Proc. Indian Acad. Sci.* **107** (no. 4), 1–11.

Kober, B. (1986). Whole-grain evaporation for ^{207}Pb/^{206}Pb-age investigations on single zircons using a double-filament thermal ion source. *Contrib. Mineral. Petrol.* **93**, 482–90.

Köppel, V. and Grünnenfelder, M. (1971). A study of inherited and newly formed zircons from paragneisses and granitized sediments of the Strona-Ceneri zone (Southern Alps). *Schweiz. Miner. Petrog. Mitt.* **51**, 387–411.

Koztolanyi, C. (1965). Nouvelle méthode d'analyse isotopique des zircons à l'état naturel. *Comptes Rend. Acad. Sciences Paris* **260**, 5849–51.

Lancelot, J. R. and Allègre, C. J. (1974). Origin of carbonatitic magma in the light of the Pb–U–Th isotope system. *Earth Planet. Sci. Lett.* **22**, 233–8.

Lanphere, M. A., Wasserburg, G. J., Albee, A. L., and Tilton, G. R. (1964). Redistribution of strontium and rubidium isotopes during metamorphism, World Beater complex, Panamint Range, California. In Craig, H., Miller, S. L., and Wasserburg, G. J. (eds.), *Isotopic and Cosmic Chemistry*, Amsterdam: North-Holland, pp. 269–320.

Lee, D.-C. and Halliday, A. N. (2000). Hf–W internal isochron for ordinary chondrites and the initial ^{182}Hf^{180}Hf of the solar system. *Chem. Geol.* **169**, 35–43.

Luck, J. M., Birck, J. L., and Allègre, C. J. (1980). ^{187}Re–^{187}Os sytematics in meteorites: early chronology of the solar system and the age of the galaxy. *Nature* **283**, 256–9.

Lugmair, G. W. and Marti, K. (1977). Sm–Nd–Pu timepieces in the Angra dos Reis meteorite. *Earth Planet. Sci. Lett.* **35**, 273–84.

Manhès, G., Allègre, C. J., Dupré, B., and Hamelin, B. (1980). Lead isotopic study of basic–ultrabasic layer complexes: speculations about the age of the Earth and primitive mantle characteristics. *Earth Planet. Sci. Lett.* **47**, 370–82.

Merrihue, C. and Turner, G. (1966). Potassium–argon dating by activation with fast neutrons. *J. Geophys. Res.* **71**, 2852–7.

Michard, A. and Allègre, C. J. (1979). A study of the formation and history of a piece of continental crust by the ^{87}Rb–^{87}Sr method: the case of the French oriental Pyrenees: *Contrib. Mineral Petrol.* **50**, 257–85.

Nicolaysen, L. O. (1961). Graphic interpretation of discordant age measurements on metamorphic rocks. *Ann. N. Y. Acad. Sci.* **91**, 198–206.

Nier, A. O. (1939). The isotopic constitution of radiogenic leads and the measurement of geologic time. Pt. II. *Phys. Rev.* **55**, 153–63.

Notsu, K., Mabuchi, H., Yoshioka, O., Matsuda, J., and Ozima, M. (1973). Evidence of the extinct nuclide ^{146}Sm in "Juvinas" achondrite. *Earth Planet. Sci. Lett.* **19**, 29–36.

Patchett, P. J. (1983). Importance of the Lu Hf isotopic system in studies of planetary chronology and chemical evolution. *Geochim. Cosmochim. Acta* **47**, 81–91.

Patchett, P. J. and Tatsumoto, M. (1980). A routine high-precision method for Lu–Hf isotope geochemistry and chronology. *Contrib. Mineral. Petrol.* **75**, 263–7.

Reynolds, J. H. (1960). Determination of the age of the elements. *Phys. Rev. Lett.* **4**, 8–10.

Shukolyukov, Y., Ashkinadze, G., and Komarou, A. (1974). A new X_{es}–X_{en} neutron induced method of mineral dating. *Dokl. Akad. Nauka USSR* **219**, 952–4. (in Russian)

Shukolyukov, A., Meshik, A., Meshik, D., Krylov, D., and Pravdivtseva, O. (1994). Current status of X_{es}–X_{en} dating. In Matsuda, X. (ed.) *Noble Gas Geochemistry and Cosmochemistry*, pp. 125–00.

Steiger, R. H. and Wasserburg, G. J. (1966). Systematics in the ^{208}Pb–^{232}Th, ^{207}Pb–^{235}U, ^{206}Pb–^{238}U systems. *J. Geophys. Res.* **71**, 6065–90.

Teitoma, A. W., Clarke, C. J., and Allègre, C. (1974). Spontaneous fission – neutron fission in xenon: a new technique for dating geological events. *Science* **189**, 878–80.

Tilton, G. R. (1960). Volume diffusion as a mechanism for discordant lead ages. *J. Geophys. Res.* **65**, 2933–45.

Tilton, G. *et al.* (1958). Isotopic composition and distribution of lead, uranium and thorium in a Precambrian granite. *Bull. Geol. Soc. Amer.* **66**, 1131–48.

Turner, G. (1968). The distribution of potassium and argon in chondrites. In Ahrens, L. H. (ed.) *Origin and Distribution of the Elements*, New York: Pergamon, pp. 387–97.

Vervoort, J. D., Patchett, P. J., Gehrels, G. E., and Nutman, A. P. (1996). Constraints on early Earth differentiation from hafnium and neodymium isotopes. *Nature* **379**, 412–14.

Wasserburg, G. J. (1985). Short-lived nuclei in the early solar system. In Black, D. C. and Matthews, M. S. (eds.) *Protostars and Planets II*, Tucson, AZ: University of Arizona Press, pp. 703–37.

Wasserburg, G. J. and Hayden, R. J. (1955). Time interval between nucleogenesis and the formation of meteorites. *Nature* **176**, 130–1.

Wetherill, G. W. (1956). Discordant uranium–lead ages. *Trans. Am. Geophys. Union* **37**, 320–7.

Wetherill, G. W., Tilton, G. R., Davis, G. L., Hart, S. R., and Hopson, C. A. (1966). Age measurements in the Maryland Piedmont. *J. Geophys. Res.* **71**, 2139–55.

Wetherill, G. W., Davis, G. L., and Lee-Hu, C. (1968). Rb–Sr measurements on whole rocks and separated minerals from the Baltimore Gneiss, Maryland. *Bull. Geol. Soc. Amer.* **79**, 757–62.

York, D., Hall, C. M., Yanese, Y., Hanese, J. A., and Kenyon, W. J. (1981). ^{40}Ar/^{39}Ar dating of terrestrial minerals with a continuous laser, *Geophys. Res. Lett.* **8**, 1136–8.

Zinner, E. (1996). Presolar material in meteorites. In Bernatowicz, T. J. and Zinner, E. (eds.) *Astrophysical Implications of the Laboratory Study of Presolar Material*, New York: American Institute of Physics, 59–72.

Chapter 4

Arnold, J. R. (1956). Beryllium-10 produced by cosmic rays. *Science* **124**, 584–5.

Atkinson, R. and Houtermans, F. G. (1929). Zur Frage der Aufbaumöglichkeit der Elemente in Sternen. *Z. Physik* **54**, 656–65.

Bard, E., Hamelin, B., Fairbanks, R. G., and Zindler, A. (1990). Calibration of the ^{14}C timescale over the past 30 000 years using mass spectrometric U–Th ages from Barbados corals. *Nature* **345**, 405–10.

Broecker, W. S. and Li, Y. H. (1970). Interchanges of water between the major oceans. *J. Geophys. Res.* **75**, 354–55.

Broecker, W. S., Gerard, R., Ewig, M., and Heezen, B. C. (1960). Natural radiocarbon in the Atlantic Ocean *J. Geophys. Res.* **65**, 2903–31.

Burbidge, E. M., Burbidge, G. R., Fowler, W. A., and Hoyle, F. (1957). Synthesis of the elements in stars. *Rev. Mod. Phys.* **29**, 547–647.

Gamow, G. (1946). Expanding Universe and the origin of elements. *Phys. Rev.* **70**, 572–5.

Harrison, T. M. and McDougall, I. (1981). Excess ^{40}Ar in metamorphic works from Broken Hill. *Earth Planet. Sci. Lett.* **55**, 123–49.

Honda, M. and Arnold J. R. (1964). Effects of cosmic rays on meteorites. *Science* **143**, 203–212.

Kieser, W. E., Beukens, R. P., Kilius, L. R., Lee, H. W., and Litherland, A. E (1986). Isotrace radiocarbon analysis: equipment and procedures. *Nucl. Instrum. Meth. Phys. Res. B* **15**, 718–21.

Lal, D. (1988). *In situ*-produced cosmogenic isotopes in terrestrial rocks. *Ann. Rev. Earth Planet. Sci.* **16**, 355–88.

Lal, D. and Peters, B. (1967). Cosmic ray-produced radioactivity on the Earth. In *Handbook of Physics*, vol. 46/2, Berlin: Springer-Verlag, pp. 551–612.

Lal, D., Malhotra, K., and Peters, B. (1958). On the production of radioisotopes in the atmosphere by cosmic radiation and their application to meteorology. *J. Atmos. Ten. Phys.* **12**, 306–28.

Lee, D.-C. and Halliday, A. N. (2000). Hf–W internal isochron for ordinary chondrites and the initial ^{182}Hf/^{180}Hf of the solar system. *Chem. Geol.* **169**, 35–43.

Libby, W. F. (1946). Atmospheric helium-3 and radiocarbon from cosmic radiation. *Phys. Rev.* **69**, 671–672.

Libby, W. F. (1970). Ruminations on radiocarbon dating. In Olsson, I. U. (ed.) *Radiocarbon Variations and Absolute Chronology*, New York: John Wiley, pp. 629–40.

Libby, W. F., Anderson, E. C., and Arnold, J. R. (1949). Age determination by radiocarbon content: world-wide assay of natural radiocarbon. *Science* **109**, 227–8.

Lin, Y., Guan, Y., Leshin, L. A., Ouyang, Z., and Wang, D. (2005). Short-lived chlorine-36 in a Ca- and Al-rich inclusion from the Ningqiang carbonaceous chondrite. *Proc. Natl Acad. Sci. USA* **102**, 1306–11.

Marti, K. (1982). Krypton-81 dating by mass spectrometry. In Lloyd, A. (ed.) *Nuclear and Chemical Dating Techniques in Interpreting the Environmental Record*, American Chemical Society, pp. 129–00.

McKeegan, K. D., Chaussidon, M., and Robert, F. (2000). Incorporation of short-lived ^{10}Be in a calcium–aluminium-rich inclusion from the Allende meteorite. *Science* **289**, 1334–7.

Merrill P. W. (1952). Technetium in the stars. *Science* **115**, 484–5.

Oeschger, H. (1982). The contribution of radioactive and chemical dating to the understanding of the environmental system. In Lloyd, A. (ed.) *Nuclear and Chemical Dating Techniques in Interpreting the Environmental Record*, American Chemical Society, pp. 5–12.

O'Nions, R. K., Franck, M., von Blanckenburg, F., and Ling, H.-F. (1998). Secular variation of Nd and Pb isotopes in ferromanganese crusts from the Atlantic, Indian and Pacific Oceans. *Earth Planet. Sci. Lett.* **155**, 15–28.

Paneth, F., Raesbeck, P. and Mayne, K., (1952). Helium-3 content and age of meteorites. *Geochim. Cosmochim. Acta* **2**, 300–3.

Staudacher, T. and Allègre, C. J. (1993). Ages of the second caldera of Piton de la Fournaise volcano (Réunion) determined by cosmic ray produced ^{3}He and ^{21}Ne. *Earth Planet. Sci. Lett.* **119**, 395–404.

Stuiver, M. (1965). Carbon-14 content of 18th- and 19th-century wood: variations correlated with sunspot activity. *Science* **149**, 533–5.

Voshage, H. and Hintenberger, H. (1960). Cosmogenic potassium in iron meteorites and cosmic ray exposure age. In *Summer Course on Nuclear Geology*, Pisa: Laboratorio di Geologica Nucleare, pp. 81–235.

Chapter 5

Allègre, C. J., Manhès, G., and Göpel, C. (1995). The age of the Earth. *Geochim. Cosmochim. Acta* **59**, 1445–56.

Barbo, L. (1999). *Pierre Curie: Le Rêve Scientifique*. Paris: Belin.

Barrell, J. (1917). Rhythms and the measurement of geologic time. *Bull. Geol. Soc. Amer.* **28**, 745–50.

Berger, G.W. (1975). 40Ar/39Ar step heating of thermally overprinted biotite, hornblende and potassium feldspar from Eldora (Colorado). *Earth Planet. Sci. Lett.* **26**, 387–90.

Bevington, P. R and Robinson, K. (2003). *Data Reduction and Error Analysis for Physical Sciences*. New York: McGraw-Hill.

Boltwood, B. B. (1907). On the ultimate disintegration products of the radioactive elements. Pt. II. The disintegration products of uranium. *Am. J. Sci.* **23**, 78–88.

Brévart, O., Dupré, B., and Allègre, C. J. (1986). Lead–lead age of komatiitic lavas and limitations on the structure and evolution of the Precambrian mantle. *Earth Planet. Sci. Lett.* **77**, 293–303.

Burchfield, H. P. (1975). *Lord Kelvin and the Age of the Earth*. London: Macmillan.

Clayton, D. D. (1968). *Principles of Stellar Evolution and Nucleosynthesis*. New York: McGraw-Hill.

Crumpler, T. B. and Yeo, J. H. (1940). *Chemical Computation and Errors*. New York: John Wiley.

Dalrymple, B. (1991). *The Age of the Earth*. Stanford, CA: Stanford University Press.

Dupré, B. and Arndt, N. T. (1987) Komatiites: témoins précieux pour retracer l'évolution du manteau. *Bull. Soc. Géol. Fr.* **III**, 1125–32.

Eve, A. S (1939). *Rutherford*. New York: Macmillan.

Gerling, E. K. (1942). Age of the Earth according to radioactivity data. *Dokl. Akad. Nauka USSR* **34–9**, 259–72.

Göpel, C., Manhès, G., and Allègre, C. J. (1994). U–Pb systematics of phosphates from equilibrated ordinary chondrites. *Earth Planet. Sci. Lett.* **121**, 153–71.

Hallam, A. (1983). *Great Geological Controversies*. Oxford, UK: Oxford University Press.

Hamilton, P. J., Eversell, N. M., O'Nions, R. K., Smith, H. S., and Erlank, A. J. (1979). Sm–Nd dating on the Ouverwacht group of volcanoes, South Africa. *Nature* **279**, 298.

Hart, S. R., Davis, G. L., Steiger, R. H., and Tilton, G. R. (1968). A comparison of the isotopic mineral age variations and petrological changes induced by contact morphism. In Hamilton, E. I. and Farquhar, R. M. (eds.) *Radiometric Dating for Geologists*, New York: Wiley Interscience, pp. 73–110.

Hohenberg, C. M. (1969). Radioisotopes and the history of nucleosynthesis in the galaxy. *Science* **166**, 212–15.

Holmes, A. (1911). The association of lead with uranium rock minerals and its application to the measurement of geological time. *Proc. Roy. Soc. Lond. A* **85**, 248–50.

Holmes, A. (1927). *The Age of the Earth: An Introduction to Geological Ideas*. London: Ernest Benn.

Holmes, A. (1946). An estimate of the age of the Earth. *Nature* **157**, 680–4.

Houtermans, F. G. (1946). Die Isotopen-Häufigkeiten im natürlichen Blei und das Alter des Urans. *Naturwissenschaften* **33**, 185–7.

Hutchinson, R. (2004). *Meteorites*. Cambridge, UK: Cambridge University Press.

Kanber, B. S. and Moorbath, S. (1998). Initial Pb of the Amitsoq gneiss revisited: Implication for the timing of early Archean crustal evolution in West Greenland. *Chem. Geol.* **15**, 19–41.

Lee, D.-C. and Halliday, A. N. (2000). Hf–W internal isochron for ordinary chondrites and the initial $^{182}Hf/^{180}Hf$ of the solar system. *Chem. Geol.* **169**, 35–43.

Luck, J. M., Birck, J. L., and Allègre, C. J. (1980). ^{187}Re–^{187}Os sytematics in meteorites: early chronology of the Solar System and the age of the Galaxy. *Nature* **283**, 256–9.

Ludwig, K. R. (1999). *Using Isoplot/Ex Version 2.01: A Geochonological Toolk it for Microsoft Excel*. Berkeley, CA: Geochronological Center.

Michard-Vitrac, A., Lancelot, J., Allègre, C. J., and Moorbath, S. (1977). U–Pb age on single zircons from Early Precambrian rocks of West Greenland and the Minnesota River Valley. *Earth Planet. Sci. Lett.* **35**, 449–53.

Moorbath, S. and Taylor, P. N. (1981). Isotopic evidence from continental growth in the Precambrian. In Kröner, A. (ed.) *Precambrian Plate Tectonics*, Amsterdam: Elsevier, pp. 49–62.

Moorbath, S., Taylor, P. N., and Jones, N. W. (1986). Dating the oldest terrestrial rocks: fact and fiction. *Chem. Geol.* **57** (1–2), 63–86.

Moorbath, S., Whitehouse, M. J., and Kanber, B. S. (1997). Extreme Nd isotope heterogeneity in the early Archean: fact or fiction? Illustrations from northern Canada and West Greenland. *Chem. Geol.* **135**, 213–31.

Nier, A. O. (1938). The isotope constitution of calcium, titanium, sulphur and argon. *Phys. Rev.* **53**, 282–6.

Nier, A. O., Thompson, R. W., and Murphy, B. (1941). The isotopic composition of lead and the measurement of geologic time Pt. III. *Phys. Rev.* **60**, 112–16.

Nutman, A. P., McGregor, V. B., Friend, C. R. L., Benson, V. C., and Kinny, P. D. (1996). The Itsaq gneiss complex of southern West Greenland: the World's most extensive record of early crustal evolution, 3900–3600 Ma. *Precamb. Res.* **78**, 1–39.

Patterson, C. C. (1953). The isotopic composition of meteoritic–basaltic and oceanic lead and the age of the Earth. *Proc. Conf. Nuclear Processes in Geology*, William Bay, 36–40.

Patterson, C. C. (1956). Age of meteorites and the Earth. *Geochim. Cosmochim. Acta* **10**, 230–7.

Reid, M., Coath, C., Harrison, M., and McKeegan, K. (1997). Prolonged residence times for the youngest rhyolites associated with Long Valley Caldera: ^{230}Th–^{238}U ion microprobe dating of young zircons. *Earth Planet. Sci. Lett.* **150**, 27–39.

Strutt, R. J. (1908). On the accumulation of helium in geological time. *Proc. Roy. Soc. Lond. A* **76**, 88–101.

Tatsumoto, M., Knight, R. J., and Allègre, C. J. (1973). Time differences in the formation of meteorites as determined from the ratio of lead-207 to lead-206. *Science* **180**, 1279–83.

Taylor, R. S. (1982). *Planetary Science: A Lunar Perspective*. Houston, TX: Lunar and Planetary Institute.

Wasson, J. (1984). *Meteorites*. New York: W. H. Freeman.

Wasserburg, G. J., Busso, M., Gallino, R., and Nollett, K. M. (2006). Short-lived nuclei in the early solar system: possible AGB sources. *Nucl. Phys A*. **777**, 5–69.

Wendt, J., and Carl, C. (1991). The statistical distribution of the mean squared weighted deviations. *Chem. Geol.* **86**, 275–85.

Wood, J. (1968). *Meteorites and the Origin of the Planets*. New York: McGrow-Hill.

York, D. (1969). Least squares fitting of a straight line with correlated errors. *Earth Planet. Sci. Lett* **5**, 320–4.

Zindler, A. (1982). Nd and Sr isotopic studies of komatiites and selected rocks. In Arndt, N. T. and Nisbet, E. G. (eds.) *Komatiites*, London: George Allen and Unwin, pp. 103–22.

Chapter 6

Aldrich, L. T. and Nier, A. O. (1948). The occurrence of He-3 in natural sources of helium. *Phys. Rev.* **74**, 1590–4.

Allègre, C. J. (1982). Chemical geodynamics. *Tectonophysics* **81**, 109–32.

Allègre, C. J. (1987). Isotope geodynamics. *Earth Planet. Sci. Lett.* **86**, 175–203.

Allègre, C. J. (1997). Limitation on the mass exchange between the upper and lower mantle: the evolving regime of the Earth. *Earth Planet. Sci. Lett.* **150**, 1–6.

Allègre, C. J. and Ben Othman, D. (1980). Nd–Sr isotopic relationship in granitoid rocks and continental crust development: a chemical approach to orogenesis. *Nature* **286**, 335–342.

Allègre, C. J. and Condomines, M. (1976). Fine chronology of volcanic processes using ^{238}U–^{230}Th systematics. *Earth Planet. Sci. Lett.* **28**, 395–406.

Allègre, C. J. and Condomines, M. (1982). Basalt genesis and mantle structure studied through Th-isotopic geochemistry. *Nature* **299**, 21–4.

Allègre, C. J. and Jaupart, C. (1985). Continental tectonics and continental kinetics. *Earth Planet. Sci. Lett.* **74**, 171–86.
Allègre, C. J. and Luck, J. M. (1980). Osmium isotopes as petrogenetic and geological tracers. *Earth Planet. Sci. Lett.* **48**, 148–54.
Allègre, C. J. and Moreira, M. (2004). Rare gas systematics and the origin of oceanic islands: the key role of entrainment at the 670 km boundary layer. *Earth Planet. Sci. Lett.* **228**, 85–92.
Allègre, C. J. and Rousseau, D. (1984). The growth of the continents through geological time studied by Nd isotope analysis of shales. *Earth Planet. Sci. Lett.* **67**, 19–34.
Allègre, C. J. and Turcotte, D. L. (1985). Geodynamical mixing in the mesosphere boundary layers and the origin of oceanic islands. *Geophys. Res. Lett.* **12**, 207–10.
Allègre, C. J. and Turcotte, D. L. (1986). Implications of a two-component marble-cake mantle. *Nature* **323**, 123–7.
Allègre, C. J., Ben Othman, D., Polvé, M., and Richard, P. (1979). The Nd–Sr isotopic correlation in mantle materials and geodynamic consequences. *Phys. Earth Planet. Int.* **19**, 293–306.
Allègre, C. J., Brévard, O., Dupré, B., and Minster, J. F. (1980). Isotopic and chemical effects produced in a continuously differentiating convecting Earth mantle. *Phil. Trans. Roy. Soc. Lond A* **297**, 447–77.
Allègre, C. J., Hart, S. R., and Minster, J. F. (1983a). The chemical structure of the mantle determined by inversion of isotopic data. I. Theoretical method. *Earth Planet. Sci. Lett.* **66**, 177–90.
Allègre, C. J., Hart, S. R., and Minster, J. F. (1983b). The chemical structure of the mantle determined by inversion of isotopic data. II. Numerical experiments and discussion. *Earth Planet. Sci. Lett.* **66**, 191–213.
Allègre, C. J., Staudacher, T., Sarda, P. and Kurz, M. (1983c). Constraints on evolution of Earth's mantle from rare gas systematics. *Nature* **303**, 762–6.
Allègre, C. J., Manhès, G., and Göpel, C. (1995). The age of the Earth. *Geochim. Cosmochim. Acta* **59**, 1445–56.
Allègre, C. J., Hofmann, A., and O'Nions, R. K. (1996). The argon constraints on mantle structure. *Geophys. Res. Lett.* **23** (24), 3555–7.
Alvarez, L. and Cornog, R. (1939). Helium and hydrogen of mass 3. *Phys. Rev.* **56**, 613–15.
Armstrong, R. L. (1981). Radiogenic isotopes: the case for crustal recycling on a near steady-state continental growth earth. *Phil. Trans. Roy. Soc. Lond. A* **301**, 443–72.
Bennett, V. C. and DePaolo, D. J. (1987). Proterozoic crustal history of the Western United States as determined by neodymium isotopic mapping. *Bull. Geol. Soc. Amer.* **99**, 674–85.
Ben Othman, D., Polvé, M., and Allègre, C. J. (1984). Nd–Sr isotope composition of granulite and constraints on the evolution of the lower continental crust. *Nature* **307**, 510–15.
Blichert-Toft, J. and Albarède, F. (1997). The Lu–Hf isotope geochemistry of chondrites and the evolution of the mantle–crust system. *Earth Planet. Sci. Lett.* **148**, 243–58.
Bowen, N. (1928). *The Evolution of the Igneous Rocks*. Princeton, NJ: Princeton University Press.
Boyet, M. and Carlson, R. (2005). ^{142}Nd evidence for early (>4.53 Ga) global differentiation of the silicate Earth. *Science* **309**, 576–80.
Caro, G., Bourdon, B., Birck, J.-L., and Moorbath, S. (2003). ^{146}Sm–^{142}Nd evidence from Isua metamorphosed sediments for early differentiation of the Earth's mantle. *Nature* **423**, 428–32.
Clarke, W., Beg, M., and Craig, H. (1969). Excess ^{3}He in the sea: evidence for terrestrial primoridal helium. *Earth Planet. Sci. Lett.* **6**, 213–30.
Compston, W. and Williams, I. S. (1984). U–Pb geochronology of zircons from lunar brecchia 73217 using a sensitive high mass resolution ion probe. *J. Geophys. Res.* **89** (suppl. B), 525–34.
Condomines, M. and Sigmarson, O. (1993). Why are so many arc magmas close to ^{238}U–^{230}Th radioactive equilibrium? *Geochim. Cosmochim. Acta* **57**, 4491–7.
Condomines, M., Hemond, C., and Allègre, C. J. (1988). U–Th–Ra radioactive disequilibria and magmatic processes. *Earth Planet. Sci. Lett.* **31**, 369–85.
Craig, H. and Lupton, J. (1976). Primordial neon, helium and hydrogen in oceanic basalts. *Earth Planet. Sci. Lett.* **31**, 369–85.

Damon, P. E. (1954). An abundance model for lead isotopes based on the continuous creation of the Earth's sialic crust. *Trans. Am. Geophys. Union.* **35**, 631–42.

Damon, P. E. and Kulp, J. L. (1958). Inert gases and the evolution of the atmosphere. *Geochim. Cosmochim. Acta* **13**, 280–300.

DePaolo, D. J. (1981a). A neodymium and strontium isotopic study of the Mesozoic calcalkaline granitic batholith of the Sierra Nevada and Peninsular range, Calif. *J. Geophys. Res.* **86**, 10470–88.

DePaolo, D. J. (1981b). Neodymium isotopes in the Colorado Front Range and crust–mantle evolution in the Proterozoic. *Nature* **291**, 193–6.

DePaolo, D. J. (1981c). Implication of correlated Nd and Sr isotopic variations for the chemical evolution of the crust and the mantle. *Earth Planet. Sci. Lett.* **43**, 201–11.

DePaolo, D. J. (1988). *Neodymium Isotope Geochemistry.* Heidelberg: Springer-Verlag.

DePaolo, D. J. and Wasserburg, G. J. (1976a). Nd isotopic variations and petrogenetic models. *Geophys. Res. Lett.* **3**, 249–52.

DePaolo, D. J. and Wasserburg, G. J. (1976b). Inferences about magma sources and mantle structure from variations of neodymium-143/neodymium-144. *Geophys. Res. Lett.* **3**, 743–6.

Dietz, R. S. (1963). Continent and ocean evolution by spreading of the sea floor. *Nature* **190**, 854–7.

Doe, B. R. (1970). *Lead Isotopes.* Heidelberg: Springer-Verlag.

Doe, B. R. and Zartman, R. E. (1979). Plumbotectonics I: the Phanerozoic. In Barnes, H. L. (ed.) *Geochemistry of Hydrothermal Ore Deposits*, New York: John Wiley, pp. 22–70.

Dupré, B. and Allègre, C. J. (1980). Pb–Sr–Nd isotopic correlation and the chemistry of the North Atlantic mantle. *Nature* **286**, 17–22.

Dupré, B. and Allègre, C. J. (1983). Pb–Sr isotope variation in Indian Ocean basalts and mixing phenomena. *Nature* **303**, 142–6.

Elliot, T., Plank, T., Zindler, A., White, W., and Bourdon, B. (1997). Element transport from slab to volcanic front at the Mariana arc. *J. Geophys. Res.* **12**, 1491–4.

Farmer, G. L. and DePaolo, D. J. (1983). Origin of Mesozoic and Tertiary granite in the Western United States and implications for pre-Mesozoic crustal structure. I. Nd and Sr isotopic studies in the geocline of the Northern Great Basin. *J. Geophys. Res.* **88**, 3379–402.

Fukao, Y., Obayashi, H., Inoue, H., and Neuberg, M. (1992). Subduction slabs stagnant in the mantle transition zone. *J. Geophys. Res.* **97**, 4809–22.

Gangarz, A. J. and Wasserburg, G. J. (1977). Initial Pb of the Amitsoq gneiss, West Greenland, and implications for the age of the Earth. *Geochim. Cosmochim. Acta* **41**, 1283–301.

Gast, P. W. (1960). Limitations on the composition of the upper mantle. *J. Geophys. Res.* **65**, 1287–90.

Gast, P. W. (1968). Trace element fractionation and the origin of tholeiitic and alkaline magma types. *Geochim. Cosmochim. Acta* **32**, 1057–87.

Gast, P. W., Tilton, G. R., and Hedge, C. E. (1964). Isotopic composition of lead and strontium from Ascension and Gough Island. *Science* **145**, 1181–88.

Gaudette, H., Vitrac-Michard, A., and Allègre, C. J. (1981). North American Precambrian history recorded in a single sample: high resolution U–Pb systematics of the Potsdam sandstone detrital zircons, New York State. *Earth Planet. Sci. Lett.* **54**, 248–60.

Geiss, J. (1954). Isotopic analysis of ordinary lead. *Z. Naturforsch.* **4/9**, 218.

Göpel, C., Allègre, C. J., and Rong Hua Xu. (1984). Lead isotopic study of the Xiagaz ophiolite (Tibet): the problem of the relationship between magmatites (gabbros, dolerites, lavas) and tectonites (harzburgites). *Earth Planet. Sci. Lett.* **69**, 301–10.

Grand, S., van der Hilst, R., and Widiyautaro, S. (1997). Global seismic tomography: a snapshot of convection in the Earth. *GSA Today* **7**, 1–17.

Hamelin, B. and Allégre, C. J. (1985). Large scale regional units in the depleted upper mantle revealed by an isotope study of Southwest Indian Ridge. *Nature* **315**, 52–5.

Hamelin, B., Dupré, B., and Allègre, C. J. (1984). Lead–strontium isotopic variations along the East Pacific Rise and the Mid Atlantic Ridge: a comparative study. *Earth Planet. Sci. Lett.* **67**, 340–50.

Hamelin, B., Dupré, B., and Allègre, C. J. (1986). Pb–Sr–Nd isotopic data of the Indian Ocean Ridge: new evidence of large-scale mapping of mantle heterogeneities. *Earth Planet. Sci. Lett.* **76**, 288–98.

Harper, C. L. and Jacobsen, S. B. (1992). Evidence from coupled $^{147}Sm-^{143}Nd$ and $^{146}Sm-^{142}Nd$ systematics for very early (4.5 Gyr) differentiation of the Earth's mantle. *Nature* **360**, 728–32.

Harrison, T. M., Blichert-Toft, J., Müller, W., *et al.* (2005). Heterogeneous Hadean hafnium: evidence of continental crust at 4.4 to 4.5 Ga. *Science* **310**, 1950–70.

Hart, S. R. (1984). A large-scale isotope anomaly in the southern hemisphere mantle. *Nature* **309**, 753–7.

Hart, S. R. (1988). Heterogeneous mantle domains: signatures, genesis and mixing chronologies. *Earth Planet. Sci. Lett.* **90**, 273–96.

Hart, S. R. and Zindler, A. (1986). In search of a bulk Earth composition. *Chem. Geol.* **57**, 247–67.

Hart, S. R., Schilling, J. G., and Powell, J. L. (1973). Basalts from Iceland and along the Reykjanes Ridge: strontium isotope geochemistry. *Nature* **246**, 104–7.

Hauri, E. H. and Hart, S. R. (1993). Re–Os isotope systematics of HIMU and EMII oceanic island basalts from the South Pacific Ocean. *Earth Planet. Sci. Lett.* **114**, 353–71.

Hawkesworth, C. J. (1979). $^{143}Nd/^{144}Nd$, $^{87}Sr/^{86}Sr$ and trace element characteristics of magmas along destructive plate margins. In Atherton, M. P. and Tarney, J. (eds.) *Origin of Granite Batholiths*, Sevenoaks, UK: Shiva, pp. 76–89.

Hawkesworth, C. J. and Kemp, A. I. (2006). Using hafnium and oxygen isotopes in zircons to unravel the record of crustal evolution. *Chem. Geol.* **226**, 144–7.

Hofmann, A. (1988). Chemical differentiation of the Earth: the relationship between mantle, continental crust and oceanic crust. *Earth Planet. Sci. Lett.* **90**, 297–304.

Hofmann, A. and Hart, S. R. (1978). An assessment of local and regional isotopic equilibrium in the mantle. *Earth Planet. Sci. Lett.* **38**, 44–62.

Hofmann, A. and White, W. M. (1982). Mantle plumes from ancient oceanic crust. *Earth Planet. Sci. Lett.* **57**, 421–36.

Hofman, A., Jochum, K., Seufert, M., and White, W. M. (1986). Nd and Pb in oceanic basalts: new constraints on mantle evolution. *Earth Planet. Sci. Lett.* **79**, 33–45.

Holmes, A. (1946). An estimate of the age of the Earth. *Nature* **157**, 680–4.

Houtermans, F. G. (1946). Die Isotopenhäufigkeiten im natürlichen Blei und das Alter des Urans. *Naturwissenschaften* **33**, 185–7.

Hurley, P. M. and Rand, R. (1969). Pre-drift continental nuclei. *Science* **164**, 1229–42.

Hurley, P. M., Hughes, H., Faure, G., Fairbairn, H. W., and Pinson, W. H. (1962). Radiogenic strontium-87 model of continent formation. *J. Geophys. Res.* **67**, 5315–34.

Jacobsen, S. (1988). Isotopic constraints on crustal growth and recycling. *Earth Planet. Sci. Lett.* **90**, 315–29.

Jacobsen, S. and Wasserburg, G. J. (1979). The mean age of mantle and crustal reservoirs. *J. Geophys. Res.* **84**, 7411–27.

Jacobsen, S. and Wasserburg, G. J. (1980). Sm–Nd isotopic evolution of chondrites. *Earth Planet. Sci. Lett.* **50**, 139–55.

Kleine, T., Munster, C., Mezger, K., and Palme, H. (2002). Rapid accretion and early core formation on asteroids and the terrestrial planets from Hf–W chronometry. *Nature* **418**, 952–5.

Kuroda, P. K. (1960). Nuclear fission in the early history of the Earth. *Nature* **187**, 36–8.

Kurz, M. and Jenkins, W. J. (1981). The distribution of helium in oceanic basalt glasses. *Earth Planet. Sci. Lett.* **53**, 41–54.

Kurz, M., Jenkins, W. J., and Hart, S. R. (1982). Helium isotopic systematics of oceanic islands and mantle heterogeneity. *Nature* **297**, 43–7.

Lambert, D. D., Morgan, W. J., Walker, R. J., *et al.* (1989). Rhenium–osmium and samarium–neodymium isotopic systematics of the Stillwater Complex. *Science* **244**, 1169–74.

Ledent, D., Patterson, C., and Tilton, G. R. (1964). Ages of zircon and feldspar concentrates from North American beach and river sands. *J. Geol.* **72**, 112–22.

Lupton, J. E. and Craig, H. (1975). Excess ^{3}He in oceanic basalts: evidence for terrestrial primordial helium. *Earth Planet. Sci. Lett.* **26**, 133–9.

Mamyrin, B. A. and Tolstikhin, I. (1984). *Helium Isotopes in Nature*. Amsterdam: Elsevier.

Mamyrin, B. A., Tolstikhin, I., Anufriev, G. S., and Kamensky, I. L. (1969). Anomalous isotopic composition of helium in volcanic gases. *Dokl. Akad. Nauka USSR* **184**, 1197–9.

McCulloch, M. and Wasserburg, G. J. (1978). Sm–Nd and Rb–Sr chronology of continental crust formation. *Science* **200**, 1003–11.

McKenzie, D. (1979). Finite deformation during fluid flow. *Geophys. J. Roy. Astron. Soc.* **58**, 689–705.

McKenzie, D. (1985). ^{230}Th–^{238}U disequilibrium and the melting process beneath the ridge axis. *Earth Planet. Sci. Lett.* **72**, 149–57.

Montelli, R., Nolet, G., Dahlen, A., *et al.* (2004). Finite frequency tomography reveals a variety of plumes in the mantle. *Science* **303**, 338–43.

Moorbath, S. and Taylor, P. N. (1981). Isotopic evidence from continental growth in the Precambrian. In Kröner, A. (ed.) *Precambrian Plate Tectonics*, Amstendam: Elsevier, pp. 491–525.

Moreira, M. and Allègre, C. J. (1998). Helium–neon systematics and the structure of the mantle. *Chem. Geol.* **147**, 53–9.

Moreira, M., Kunz, M., and Allègre, C. J. (1998). Rare gas systematics in popping rock: isotopic and element compositions in the upper mantle. *Science* **279**, 1178–81.

Morgan, W. J. (1971). Convection plumes in the lower mantle. *Nature* **230**, 42–3.

Nolet, G., Karato, S., and Montelli, R. (2004). Flux estimates from tomographic plume images yield evidence for chemical stratification in the mantle. *EOS (Trans. Am. Geophys. U.)* **85**, 117–26.

O'Nions, R. K. and Oxburgh, E. R. (1983). Heat and helium in the Earth. *Nature* **306**, 429–36.

O'Nions, R. K., Hamilton, P. J., and Evensen, N. M. (1977). Variations in ^{143}Nd/^{144}Nd and ^{87}Sr/^{86}Sr in oceanic basalts. *Earth Planet. Sci. Lett.* **34**, 13–22.

O'Nions, R. K., Hamilton, P. J., and Evensen, N. M. (1980). Differentiation and evolution of the mantle. *Phil. Trans. Roy. Soc. Lond. A* **297**, 479–93.

O'Nions, R. K. and Hamilton, P. J. (1983). A Nd isotope investigation of sediments related to crustal development in the British Isles. *Earth Planet. Sci. Lett.* **63**, 229–40.

Patchett, P. J. (1983). Importance of the Lu–Hf isotopic system in studies of planetary chronology and chemical evolution. *Geochim. Cosmochim. Acta* **47**, 81–91.

Patchett, P. J. and Tatsumoto, M. (1980). Hafnium isotope variations in oceanic basalts. *Geophys. Res. Lett.* **7**, 1077–80.

Patterson, C. (1963). Characteristics of lead isotope evolution on a continental crust. In Craig, H., Miller, S., and Wasserburg, G. J. (eds.) *Isotopic and Cosmic Chemistry*, Amsterdam: North Holland, pp. 244–68.

Patterson, C. and Tatsumoto, M. (1964). The significance of lead isotopes in detrital feldspar with respect to chemical differentiation within the Earth's mantle. *Geochim. Cosmochim. Acta* **28**, 1–22.

Reynolds, J. (1960). Determination of the age of the elements. *Phys. Rev. Lett.* **4**, 5–9.

Richard, P., Shimizu, N., and Allègre, C. J. (1976). ^{143}Nd/^{146}Nd, a natural tracer: an application to oceanic basalts. *Earth Planet. Sci. Lett.* **3**, 269–78.

Roy Barman, M. and Allègre, C. J. (1994). ^{187}Os/^{186}Os ratios of mid-ocean ridge basalts and abyssal peridotites. *Geochim. Cosmochim. Acta* **58**, 53–84.

Russell, R. D. (1972). Evolutionary model for lead isotopes in conformable ore and oceanic volcanoes. *Rev. Geophys. Space Phys.* **10**, 529–36.

Russell, R. D. and Farquhar, R. (1960). *Lead Isotopes in Geology*. New York: Wiley Interscience.

Sarda, P., Staudacher, T., and Allègre, C. J., (1985). ^{40}Ar/^{36}Ar in MORB glasses: constraints on atmosphere and mantle evolution. *Earth Planet. Sci. Lett.* **72**, 357–75.

Sarda, P., Staudacher, T., and Allègre, C. J. (1988). Neon isotopes in submarine basalts. *Earth Planet. Sci. Lett.* **91**, 73–88.

Sarda, P., Moreira, M., Staudacher,T., Schilling, J. G., and Allègre, C. J. (2000). Rare gas systematics on the southernmost Mid-Atlantic Ridge: constraints of the lower mantle and Dupal source. *J. Geophys. Res.* **83**, 5973–96.

Schilling, J. G. (1973). Iceland mantle plume: geochemical study of the Reykjanes Ridge. *Nature* **242**, 565–571.

Schilling, J. G. (1992). In Duthous, X. (ed.) *Les Isotopes Radiogéniques en Géologie*, Paris: Société française de Mineralogie et Cristallographie, pp. 1–34.

Sobolev, A., *et al.* (2007). The amount of recycled crust in sources of mantle-derived melt. *Science* **316**, 412–17.

Staudacher,T. and Allègre, C. J. (1982). Terrestrial xenology. *Earth Planet. Sci. Lett.* **60**, 389–406.

Sun, S. S. (1980). Lead isotopic study of young volcanic rocks from mid-ocean ridges, ocean islands and island arcs. *Phil. Trans. Roy. Soc. Lond. A* **297**, 409–45.

Sun, S. S. and Hanson, G. N. (1975). Evolution of the mantle: geochemical evidence from alkali basalt. *Geology* **3**, 297–302.

Tatsumoto, M. (1966). Genetic relations of oceanic basalts as indicated by lead isotopes. *Science* **153**, 1088–94.

Tatsumoto, M., Hedge, C. E., and Engel, A. E. J. (1965). Potassium, rubidium, strontium, thorium, uranium and the ratio of strontium-87 to strontium-86 in oceanic tholeiitic basalt. *Science* **150**, 886–8.

Tatsumoto, M., Knight, R., and Allègre, C. J. (1973). Time differences in the formation of meteorites as determined from the ratio of lead-207 to lead-206. *Science* **180**, 1278–83.

Tolstikhin, I., Mamyrin, B., Khabarin, L. U., and Erlich, E. N. (1974). Isotope composition of helium in ultrabasic xenoliths from volcanic rocks of Kamchatka. *Earth Planet. Sci. Lett.* **22**, 75–84.

Treuil, M. (1973). Critères petrologiques, géochimiques et structuraux de la génèse et de la différentiation des magmas basaltiques, exemple de l'Afar. Ph.D. thesis, University of Orleans.

Treuil, M. and Joron, J. L. (1975). Utilisation des elements HYB en la simplification de la modelisation quantitative des processus magmatiques. *Soc. Ital. Mineral. Petrol.* **31**, 125–40.

Trieloff, M., Kunz, M., Clague, D., Harrison, D., and Allègre, C. J. (1998). The nature of pristine noble gases in mantle plumes. *Science* **288**, 1036–8.

Turcotte, D. and Schubert, G. (2002). *Geodynamics*. Cambridge, UK: Cambridge University Press.

Turekian, K. K. (1959). The terrestrial economy of helium and argon. *Geochim. Cosmochim. Acta* **17**, 37–43.

Turner, S., Bourdon, B., and Gill, J. (2003). Insights into magma genesis at convergent margins from U-series isotopes. *Rev. Mineral. Geochem.* **52**, 255–315.

van der Hilst, R., Engdahl,W., Spakman,W., and Nolet, G. (1991). Tomographic imaging of subducted lithosphere below the Northwest Pacific Island Arc. *Nature* **353**, 37–43.

Vervoort, J. D. and Blichert-Toft, J. (1999). Evolution of the mantle Hf isotope evidence from juvenile rocks through time. *Geochim. Cosmochim. Acta* **63**, 533–56.

Wasserburg, G. J. (1964). *Geochronology and Isotopic Data Bearing on the Development of the Continental Crust*. Cambridge, MA: MIT Press.

Wetherill, G.W. (1954). Isotopic variations of neon and argon extracted from radioactive materials. *Phys. Rev.* **96**, 679–83.

White,W. (1985). Sources of oceanic basalts: radiogenic isotopic evidence. *Geology* **13**, 115–22.

White,W. and Hoffman, A. (1982). Sr and Nd isotope geochemistry of oceanic basalts and mantle. *Nature* **296**, 821–5.

Winkler, H. (1974). *Petrogenesis of Metamorphic Rocks*. New York: Springer-Verlag.

Zindler, A. and Hart, S. R. (1986). Chemical geodynamics. *Ann. Rev. Earth Planet. Sci.* **14**, 493–510.

Chapter 7

Berner, R. A., Lasaga, A. C., and Garrels, R. M. (1983). The carbonate–silicate geochemical cycle and its effects on atmospheric carbon dioxide over the past 100 million years. *Am. J. Sci.* **283**, 641–83.

Bigeleisen, J. (1965). Chemistry of isotope science. *J. Chem. Phys.* **147**, 463–71.

Bigeleisen J. and Mayer, M. (1947). Calculation of equilibrium constant for isotope exchange reactions. *J. Chem. Phys.* **15**, 261–7.

Bottinga, Y. and Javoy, M. (1975). Oxygen isotope partitioning among the minerals in igneous and metamorphic rocks. *Rev. Geophys. Space Phys.* **13**, 401–18.

Bowen, N. L. (1928). *The Evolution of Igneous Rocks*. Princeton, NJ: Princeton University Press.

Bradley, R. S. (1999). *Paleoclimatology.* New York: Academic Press.

Burton, K. and Vance, D. (1999). Glacial–interglacial variations in the neodymium isotope composition of seawater in the Bay of Bengal recorded by planktonic foraminifera. *Earth Planet. Sci. Lett.* **76**, 425–46.

Caillon, N., Severinghaus, J., Jouzel, J., *et al.* (2003). Timing of atmospheric CO_2 and Antarctic temperature changes across termination. III. *Science* **299**, 172–82.

Clayton, R. N., Grossman, L., and Mayeda, K. (1973). A component of primitive nuclear composition in carbonaceous chondrites. *Science* **182**, 485–7.

Craig, H. (1961). Isotopic variations in meteoric waters. *Science* **133**, 1702–3.

Craig, H. (1963). The isotopic geochemistry of water and carbon in geothermal areas. In: *Proceedings of Conference on Isotopes in Geothermal Waters*, Spoleto, Sept. 9–13, Pisa: Laboratorio di Geologia Nucleare, pp. 53–70.

Craig, H. (1965). The measurement of oxygen isotope paleotemperatures. In: *Stable Isotopes in Oceanographic Studies and Paleotemperature*, Spoleto July 26–27, Pisa: Laboratorio di Geologia Nucleare, pp. 1–24.

Craig, H. and Boato, G. (1955). Isotopes. *Ann. Rev. Chem.* **6**, 403–20.

Craig, H., Boato, G., and White, D. (1956). Isotopic geochemistry of thermal water. *Procs. 2nd Conf. Nuclear Processes in Geology*, pp. 29–42.

Dansgaard, W. (1953). The abundance of ^{18}O in atmospheric water and water vapour. *Tellus* **5**, 461–9.

Dansgaard, W. (1964). Stable isotopes in precipitation. *Tellus* **16**, 436–68.

Dansgaard, W. and Tauber, H. (1969). Glacier oxygen-18 content and Pleistocene ocean temperatures. *Science* **166**, 499.

Dansgaard, W., White, J. W., and Johnsen, S. J. (1969). The abrupt termination of the Younger Dryas climatic event. *Nature* **339**, 532–4.

De Niro, M. J. (1987) Stable isotopes in archeology. *Am. Scientist* **75**, 182–8.

Duplessy, J. C., Lalou, C., and Vinot, A. (1970). Differential isotopic fractionation in benthic foraminifera and paleotemperature re-assessed. *Science* **168**, 250–1.

Edmond, J. (1992). Himalayan tectonics, weathering processes and the strontium isotope record in marine limestone. *Science* **258**, 1594.

Emiliani, C. (1955). Pleistocene temperature. *J. Geol.* **63**, 538–78.

Emiliani, C. (1972). Quaternary paleotemperatures. *Science* **154**, 851–78.

EPICA Community (2004). Eight glacial cycles from an Antarctic ice core. *Nature* **429**, 623.

Epstein, S. (1959). The variation of $^{18}O/^{16}O$ ratio in nature and some geological applications. In Abelson, P. (ed.) *Research in Geochemistry*, New York: John Wiley, pp. 1217–40.

Epstein, S. and Mayeda, T. (1953). Variation of ^{18}O content of waters from natural sources. *Geochim. Cosmochim. Acta* **4**, 213–24.

Epstein, S. and Sharp, R. (1967). Oxygen and hydrogen isotope variations in a firm core, Eight Station, Western Antarctica. *J. Geophys. Res.* **72**, 5595–618.

Epstein, S. and Taylor, H. P. (1967). Variation of $^{18}O/^{16}O$ in minerals and rocks. In Abelson, P. (ed.) *Research in Geochemistry*, New York: John Wiley, pp. 229–62.

Epstein, S., Buchsbaum, R., Lowenstam, H., and Urey, H. (1953). Revised carbonate–water isotopic temperature scale. *Bull. Geol. Soc. Amer.* **64**, 1315–26.

Epstein, S., Sharp, R., and Gow, A. J. (1965). Six-year record of oxygen and hydrogen isotope variation in South Pole firn. *J. Geophys. Res.* **70**, 1809–14.

Farquhar, J., Bao, H., and Thiemens, M. (2001). Atmospheric influence of Earth's earliest sulphur cycle. *Science* **289**, 757.

Farquhar, J., Peters, H., Johnston, D., *et al.* (2007). Isotopic evidence for mesoarchean anoxia and changing atmospheric sulfur chemistry. *Nature* **449**, 706–9.

Fourcade, S. (1998). Les isotopes : effets isotopiques bases de la radiochimie. In Hageman, J. and Treuil, M. (eds.) *Introduction à la Géochimie*, Paris: CEA.

Galimov, E. (1985). *The Biological Fractionations of Isotopes*, New York: Academic Press.

Gao, Y. Q. and Marcus, R. A (2001). Strange and non-conventional isotope effect in ozone formation. *Science* **293**, 259.

Garlick, G. D. and Epstein, S. (1967). Oxygen isotope ratios in coexisting minerals from regionally metamorphosed rocks. *Geochim. Cosmochim. Acta* **31**, 181–214.

Ghosh, P., Garzione, C., and Eiler, J. (2006a). Rapid uplift of the Altiplano revealed through $^{13}C-^{18}O$ bonds in paleosol carbonates. *Science* **311**, 2093–4.

Ghosh, P., Adkins, J., Affek, H., *et al.* (2006b). $^{13}C-^{18}O$ bonds in carbonate minerals: a new kind of paleothermometer. *Geochim. Cosmochim. Acta* **70**, 1439–56.

Goldstein, S. L. and Hemming, S. R. (2003). Long-lived isotopic tracers in oceanography, paleo-oceanography and ice-sheet dynamics. In Elderfield, H. (ed.), *Treatise on Geochemistry*, vol. 6, London: Elsevier, pp. 625–000.

Harrison, A. G. and Thode, H. G. (1957). The kinetic isotope effect in the chemical reduction of sulfate. *Trans. Faraday Soc.* **53**, 1–4.

Harrison, A. G. and Thode, H. G. (1958). Mechanism of the bacterial reduction of sulfate from isotope fractionation studies. *Trans. Faraday Soc.* **54**, 84–92.

Heinrich, M. (1988). Origin and consequences of cyclic ice rafting in the northeast Atlantic Ocean during the past 130 000 years. *Quatern. Res.* **29**, 143–52.

Holland, H. (1984). *The Chemical Evolution of the Atmosphere and Oceans*. Princeton, NJ: Princeton University Press.

Ito, E. and Stern, R. J. (1985). Oxygen and strontium isotopic investigation of subduction zone volcanism: the case of the volcano arc and the Marianas island arc. *Earth Planet. Sci. Lett.* **76**, 312–20.

James, D. (1981). The combined use of oxygen and radiogenic isotopes as indicators of crustal contamination. *Ann. Rev. Earth Planet. Sci.* **9**, 311–44.

Javoy, M. (1977). Stable isotopes and geothermometry. *J. Geol. Soc. Lond.* **133**, 609–36.

Javoy, M., Fourcade, S., and Allègre, C. J. (1970). Graphical method for examination of $^{18}O/^{16}O$ fractionation in silicate rocks. *Earth Planet. Sci. Lett.* **10**, 12–16.

Johnsen, S., Dansgaard, W., and White, J. W. (1989). The origin of Arctic precipitation under present-day glacial conditions. *Tellus* **41**, 452–69.

Johnson, C., Beard, B., and Albarède, F. (eds.) (2004). *Reviews in Mineralogy and Geochemistry*, vol. 55, *Geochemistry of Non-Traditional Stable Isotopes*, Mineralogical Society of America.

Jouzel, J. (1986). Isotopes in cloud physics: multistep and multistage processes. In *Handbook of Environmental Isotope Geochemistry*, vol. 2, Amsterdam: Elsevier, pp. 61–112.

Jouzel, J., Lorius, C., and Petit, J. R. (1987). Vostok ice core: a continuous isotopic temperature record over the climatic cycle (160 000 years). *Nature* **329**, 403–8.

Labeyrie, L. (1974). New approaches to surface sea-water paleotemperatures using $^{18}O/^{16}O$ ratios in silica of diatom fructules. *Nature* **248**, 40–2.

Lasaga, A. (1997). *Kinetic Theory*. Princeton, NJ: Princeton University Press.

Longinelli, A. and Nutti, S. (1973). Revised phosphate–water isotopic temperature scale. *Earth Planet. Sci. Lett.* **19**, 373–6.

Lorius, C. and Merlivat, L. (1977). Distribution of mean surface stable isotope values in east Antarctica: observed changes with depth in coastal areas. In *Isotopes and Impurities in Snow and Ice*, IAHS Publ. **118**, pp. 127–37.

Milankovitch, M. M. (1941). *Case of Insolation and the Ice-Age Problem*. Washington, DC: US Department of Commerce and National Scientific Foundation.

Nielsen, H. (1979). Sulfur isotopes. In Jäger, E. and Hunziker, J. C. (eds.) *Lectures in Isotope Geology*, Berlin: Springer-Verlag, pp. 283–312.

Nier, A. O. (1947). A mass spectrometer for isotopes and gas analysis. *Rev. Sci. Instrum.* **18**, 398–411.

Nier, A. O., Ney, E. P., and Inghram, M. (1947). A new method for the comparison of two ion currents in a mass spectrometer. *Rev. Sci. Instrum.* **18**, 294–7.

O'Neil, J. R. (1986). Theoretical and experimental aspects of isotopic fractionation. In Valley, J. M, Taylor, H. P., and O'Neil, J. R. (eds.) *Review of Mineralogy*, vol. 16, *Stable Isotopes*, pp. 1–40.

O'Neil, J. R. and Clayton, R. N (1964). Oxygen isotope geochemistry. In Craig, A., Miller, S., and Wasserburg, G. J. (eds.) *Isotopic and Cosmic Chemistry*, Amsterdam: North-Holland, pp. 157–68.

Oeschger, H. (1982). The contribution of radioactive and chemical dating in the understanding of environmental systems. In Lloyd, A. (ed.) *Nuclear and Chemical Dating Techniques: Interpreting the Environmental Record*, American Chemical Society, pp. 5–18.

Ohmoto, H. and Rye, R. O. (1979). Isotopes of sulfur and carbon. In Barnes, H. L. (ed.) *Geochemistry of Hydrothermal Ore Deposits*, New York: John Wiley, pp. 509–67.

Park, M. and Epstein, S. (1960). Carbon isotope fractionation during photosynthesis. *Geochim. Cosmochim. Acta* **27**, 110–26.

Petit, J. R., Jouzel, J., Raynaud, D., *et al.* (1999). Climate and atmospheric history of the past 420 000 years from the Vostok ice core, Antarctic. *Nature* **399**, 429–36.

Pineau, F. and Javoy, M. (1983). Carbon isotopes and concentrations in mid-oceanic ridge basalts. *Earth Planet. Sci. Lett.* **62**, 239–57.

Raymo, M. and Ruddiman, W. (1992). Tectonic forcing of late Cenozoic climate. *Nature* **259**, 117.

Severinghaus, J. P. and Brook, E. (1999). Abrupt climate change at the end of the glacial period inferred from trapped air in polar ice. *Science* **286**, 930–4.

Severinghaus, J. P., Grachev, A., Luzon, B., and Caillon, N. (2003). A method for precise measurement of argon 40/36 and krypton/argon ratios in trapped air in polar ice. *Geochim. Cosmochim. Acta* **67**, 325–43.

Shackleton, N. J. (1967a). Oxygen isotope analyses and the Pleistocene temperature re-assessed. *Nature* **215**, 15–17.

Shackleton, N. J. (1967b). Oxygen isotopes, ice volume and sea-level Quaternary. *Science Rev.* **6**, 183–90.

Taylor, H. P. (1968). The oxygen isotope geochemistry of igneous rocks. *Contrib. Mineral. Petrol.* **19**, 1–71.

Taylor, H. P. (1974). Oxygen and hydrogen isotope evidence for large-scale circulation and interaction between groundwaters and igneous intrusion with particular reference to the San Juan volcanic field. In Hoffman, A. W., Giletti, B., Yoder, H., and Yund, R. (eds.) *Geochemical Transport and Kinetics*, Washington, DC: Carnegie Institution Press, pp. 299–324.

Taylor, H. P. (1979). Oxygen and hydrogen isotope relationships in hydrothermal mineral deposits. In Barnes, H. L. (ed.) *Geochemistry of Hydrothermal Ore Deposits*, New York: John Wiley, pp. 236–77.

Taylor, H. P. (1980). Oxygen effects of assimilation of country rocks by magmas on $^{18}O/^{16}O$, $^{87}Sr/^{86}Sr$ systematics in igneous rocks. *Earth Planet. Sci. Lett.* **47**, 243–54.

Thiemens, M. H. (1999). Mass-independent isotope effects in planetary atmospheres and the early solar system. *Science* **280**, 341.

Thiemens, M. H. and Heidenreich, J. E. (1983). The mass independent fractionation of oxygen: a novel effect and the possible cosmochemical implication. *Science* **219**, 1073.

Thompson, L. (1991). Ice core records with emphasis on the global record of the last 2000 years. In Bradley, R. S. (ed.) *Global Change of the Past*, Boulder, CO: University Corporation for Atmospheric Research, pp. 201–24.

Tudge, A. P. and Thode, H. G. (1950). Thermodynamic properties of isotopic compounds of sulfur. *Can. J. Res. B* **28**, 567–78.

Urey, H. C. (1947). The thermodynamic properties of istopic substances. *J. Chem. Soc. Lond.*, 562–81.

Urey, H. C., Lowenstam, H., Epstein, S., and McKinney, C. R. (1951). Measurement of paleotemperatures of the Upper Cretaceous of England, Denmark and the south-eastern United States. *Bull. Geol. Soc. Amer.* **62**, 399–426.

Chapter 8

Albarède, F. (1995). *Introduction to Geochemical Modeling*. Cambridge, UK: Cambridge University Press.

Allègre, C. J. and Jaupart, C. (1985). Continental tectonics and continental kinetics. *Earth Planet. Sci. Lett.* **74**, 171–86.

Allègre, C. J. and Lewin, E. (1995). Isotopic systems and stirring of the Earth's mantle. *Earth Planet. Sci. Lett.* **136**, 629–46.

Allègre, C. J. and Turcotte, D. (1985). Implications of a two-component marble-cake mantle. *Nature* **323**, 123–7.

Allègre, C. J., Brévard, O., Dupré, B., and Minster, J. F. (1980). Isotopic and chemical effects produced in a continuously differentiating convecting Earth mantle. *Phil. Trans. Roy. Soc. Lond. A* **297**, 447–77.

Allègre, C. J., Moreira, M., and Staudacher, T. (1995). $^4He/^3He$ dispersion and mantle convection. *Geophys. Res. Lett.* **22**, 2325–8.

Beltrani, E. (1987). *Mathematics for Dynamic Modeling*. New York: Academic Press.

Galer, S. and O'Nions, R. K. (1985). Residence time of thorium, uranium and lead in the mantle with implications for mantle convection. *Nature* **316**, 778–82.

Haberman, R. (1977). *Mathematical Models*. New York: Prentice Hall.

Jacobsen, S. (1988). Isotopic constraints on crustal growth and recycling. *Earth Planet. Sci. Lett.* **90**, 315–29.

Jacobsen, S. and Wasserburg, G. J. (1981). Transport models for crust and mantle evolution. *Tectonophysics* **75**, 163–79.

Javoy, M., Pineau, F., and Allègre, C. J. (1982). Carbon geodynamic cycle. *Nature* **300**, 171–3.

Lasaga, A. and Berner, R. (1998). Fundamental aspects of quantitative models for geochemistry cycles. *Chem. Geol.* **145**, 161–74.

McKenzie, D. (1979). Finite deformation during fluid flow. *Geophys. J. Roy. Astron. Soc.* **58**, 689–705.

Richter, F. and Turekian, K. K. (1993). Simple models for the geochemical response of the ocean to climatic and tectonic forcing. *Earth Planet. Sci. Lett.* **119**, 121–8.

Rodhe, H. (2000). Modelling biogeochemical cycles. In Jacobson, M., Charlson, R. Rodhe, H. and Orians, G. (eds.) *Earth System Science*, New York: Academic Press, pp. 62–92.

Wasserburg, G. J. (1964). Pb–U–Th evolution models for homogeneous systems with transport. *EOS (Trans. Am. Geophys. U.)* **45**, 111–18.

APPENDIX

Table A.1 Symbols and orders of magnitude

Prefix	Factor
exa (E)	10^{18}
peta (P)	10^{15}
tera (T)	10^{12}
giga (G)	10^{9}
mega (M)	10^{6}
kilo (k)	10^{3}
hecto (h)	10^{2}
deca (da)	10^{1}
deci (d)	10^{-1}
centi (c)	10^{-2}
milli (m)	10^{-3}
micro (μ)	10^{-6}
nano (n)	10^{-9}
pico (p)	10^{-12}
femto (f)	10^{-15}
atto (a)	10^{-18}

Table A.2 Constants

	Symbol	Value
Speed of light	c	$2.997\,924\,58 \cdot 10^{8}\,\mathrm{m\,s^{-1}}$
Electron charge	e	$-1.602\,177\,33 \cdot 10^{-19}\,\mathrm{C}$
Planck constant	h	$6.620\,607\,55 \cdot 10^{-34}\,\mathrm{J\,s}$
Boltzmann constant	k	$1.380\,658 \cdot 10^{-23}\,\mathrm{J\,K^{-1}}$
Gravitational constant	G	$6.6726 \cdot 10^{-11}\,\mathrm{N\,m^{2}\,kg^{-2}}$
Electron rest mass	m_e	$0.910\,938\,97 \cdot 10^{-30}\,\mathrm{kg}$
Atomic mass unit	u	$1.660\,540\,2 \cdot 10^{-27}\,\mathrm{kg}$
Avogadro constant	N_A	$6.022\,136\,7 \cdot 10^{23}\,\mathrm{mole^{-1}}$
Ideal gas constant	R	$8.314\,510\,\mathrm{J\,mole^{-1}\,K^{-1}}$
		$1.989\,\mathrm{cal\,mole^{-1}\,K^{-1}}$

Table A.3 Geological data

Mass of the Earth	$5.9737 \cdot 10^{24}$ kg
Volume	$1.08320 \cdot 10^{21}\,m^3$
Mean radius assuming spherical Earth	6 371 000 m
Mean density of Earth	$5515\ kg\,m^{-3}$
Mass of atmosphere	$5.1 \cdot 10^{18}$ kg
Mass of oceans	$1.37 \cdot 10^{21}$ kg
Mass of ice caps	$2.9 \cdot 10^{19}$ kg
Mass of fresh water (rivers and lakes)	$3 \cdot 10^{16}$ kg
Mass of continental crust	$2.36 \cdot 10^{22}$ kg
Mass of whole mantle	$4 \cdot 10^{24}$ kg
Mass of upper mantle (above 670 km)	$1 \cdot 10^{24}$ kg
Mass of core	$1.950 \cdot 10^{24}$ kg
Mass of outer core	$1.85 \cdot 10^{24}$ kg
Mass of inner core	$9.7 \cdot 10^{22}$ kg
Area of Earth	$5.100\,655 \cdot 10^{8}\ km^2$
Area of oceans	$3.62 \cdot 10^{8}\ km^2$
Area of continents (and continental margins)	$2 \cdot 10^{8}\ km^2$
Area of exposed continents	$1.48 \cdot 10^{8}\ km^2$
Area of Atlantic Ocean	$9.8 \cdot 10^{7}\ km^2$
Area of Indian Ocean	$7.7 \cdot 10^{7}\ km^2$
Area of Pacific Ocean	$1.7 \cdot 10^{8}\ km^2$
Hydrothermal flux at ocean ridges	$1-2.3 \cdot 10^{14}\ kg\,yr^{-1}$
Flux of rivers to ocean	$4.24 \cdot 10^{4}\ km^3\,yr^{-1} = 4.24 \cdot 10^{16}\ kg\,yr^{-1}$
Flux of river sediment load	$1.56 \cdot 10^{13}\ kg\,yr^{-1}$
Flux of oceanic crust formed	$8.4 \cdot 10^{13}\ kg\,yr^{-1}$
Flux of oceanic lithosphere created	$8.4 \cdot 10^{14}\ kg\,yr^{-1}$
Length of ocean ridges	50 000 km
Flux from hot spots	$2.5 \cdot 10^{14}\ kg\,yr^{-1}$
Average sea-floor spreading rate	$3\ cm\,yr^{-1}$
Flux of lithospheric subduction	$8.4 \cdot 10^{14}\ kg\,yr^{-1}$
Average altitude of landmasses	875 m
Average depth of oceans	3794 m
Average thickness of continents	35 km
Average thickness of oceanic crust	6.0 km
Average heat flow	$87\ mW\,m^{-2}$
Total geothermal flow	44.3 TW
Average continental heat flow	$65\ mW\,m^{-2}$
Average oceanic heat flow	$101\ mW\,m^{-2}$
Solar flux	$1373\ W\,m^{-2}$

Table A.4 Long-lived radioactivity

Percentage of element	Decay constant, λ (yr^{-1})	Mean life, T (yr)	Half-life, $T_{1/2}$ (yr)	Product
^{40}K	β^- $4.962 \cdot 10^{-10}$	$2.015 \cdot 10^9$	$1.397 \cdot 10^9$	^{40}Ca
$1.167 \cdot 10^{-4}$ K_{total}	e^- cap $0.581 \cdot 10^{-10}$	$17.21 \cdot 10^9$	$11.93 \cdot 10^9$	^{40}Ar
	Total $5.543 \cdot 10^{-10}$	$1.80 \cdot 10^9$	$1.25 \cdot 10^9$	
^{87}Rb β^- 0.25	$1.42 \cdot 10^{-11}$	$70.42 \cdot 10^9$	$48.8 \cdot 10^9$	^{87}Sr
^{138}La [β^-, e^- cap] 0.089	β^- $2.25 \cdot 10^{-12}$	$4.44 \cdot 10^{11}$	$3.08 \cdot 10^{11}$	[^{138}Ce, ^{138}Ba]
La_{total}	e^- cap $4.4 \cdot 10^{-12}$	$2.2 \cdot 10^{11}$	$1.57 \cdot 10^{11}$	
	Total $6.65 \cdot 10^{-12}$	$1.5 \cdot 10^{11}$	$1.04 \cdot 10^{11}$	
^{147}Sm α(15.07)	$6.54 \cdot 10^{-12}$	$152.88 \cdot 10^9$	$1.06 \cdot 10^{11}$	^{143}Nd
^{176}Lu β^-(2.6)	$2 \cdot 10^{-11}$	$50 \cdot 10^9$	$3.5 \cdot 10^{10}$	^{176}Hf
^{187}Re β^- (63.93)	$1.5 \cdot 10^{-11}$	$66.66 \cdot 10^9$	$4.6 \cdot 10^{10}$	^{187}Os
^{190}Pt α (0.0127)	$1.16 \cdot 10^{-12}$	$862 \cdot 10^9$	$6 \cdot 10^{11}$	^{186}Os
^{232}Th (100) ($6\alpha, 4\beta^-$)	$4.9475 \cdot 10^{-11}$	$20.21 \cdot 10^9$	$1.4010 \cdot 10^{10}$	^{208}Pb
^{235}U (0.73) ($7\alpha, 5\beta^-$)	$9.8485 \cdot 10^{-10}$	$1.015\,38 \cdot 10^9$	$0.703\,809\,9 \cdot 10^9$	^{207}Pb
^{238}U (99.27) ($8\alpha, 6\beta^-$)	$1.551\,25 \cdot 10^{-10}$	$6.44 \cdot 10^9$	$4.4683 \cdot 10^9$	^{206}Pb

Table A.5 Extinct radioactivity

Percentage of element	Decay constant, λ (yr^{-1})	Mean life, T (Ma)	Half-life, $T_{1/2}$ (Ma)	Product
^{26}Al β^+	$9.7 \cdot 10^{-7}$	1.03	0.714 58	^{26}Mg
^{10}Be β^-	$4.6 \cdot 10^{-8}$	$2.16 \cdot 10^7$	$1.5 \cdot 10^7$	^{10}B
^{36}Cl β^+	$2.25 \cdot 10^6$	0.44	0.308	^{36}Ar (98.1%)
^{36}Cl β^-				^{36}S (1.9%)
^{41}Ca β^+	$6.7 \cdot 10^{-6}$	0.15	0.1	^{41}K
^{53}Mn β^+	$1.886 \cdot 10^{-7}$	5.3	3.3867	^{53}Cr
^{60}Fe $2\beta^-$	$4.761 \cdot 10^{-7}$	2.2	1.456	^{60}Ni
^{92}Nb β^+	$2.777 \cdot 10^{-8}$	36	25.67	^{92}Zr
^{107}Pd β^-	$1.063 \cdot 10^{-7}$	9.4	6.538	^{107}Ag
^{129}I β^-	$4 \cdot 10^{-8}$	25	17.327	^{129}Xe
^{146}Sm α	$6.849 \cdot 10^{-9}$	146	101.19	^{142}Nd
^{182}Hf $2\beta^-$	$7.692 \cdot 10^{-8}$	13	9.01	^{182}W
^{205}Pb β^+	$4.62 \cdot 10^{-8}$	21.64	15	^{205}Tl
^{244}Pu fission	$8.264 \cdot 10^{-9}$	121	83.91	$Xe_{fission}$
^{247}Cm $2\alpha, 3\beta^+$	$4.4 \cdot 10^{-8}$	22.5	15.6	^{235}U

Table A.6 Half-lives of radium isotopes used in geology

Parent	Ra isotope	Half-life
^{238}U	^{226}Ra	1622 years
^{235}U	^{223}Ra	11.435 days
^{232}Th	^{228}Ra	6.7 years
^{232}Th	^{224}Ra	3.64 days

Table A.7 Half-lives and decay constants of disintegration reactions for radioactive chains used in geology

Nuclide	Half-life (years)	Decay constant λ (yr^{-1})
$^{234}_{92}U$	$2.48 \cdot 10^5$	$2.794 \cdot 10^{-6}$
$^{230}_{90}Th$	$7.52 \cdot 10^4$	$9.217 \cdot 10^{-6}$
$^{226}_{88}Ra$	$1.622 \cdot 10^3$	$4.272 \cdot 10^{-4}$
$^{210}_{82}Pb$	22.26	$3.11 \cdot 10^{-2}$
$^{231}_{91}Pa$	$3.248 \cdot 10^4$	$2.134 \cdot 10^{-5}$

Table A.8 Mass ratios of selected elements

Chemical mass ratio	Isotope ratio
(Rb/Sr)	0.341 ($^{87}Rb/^{86}Sr$)
(Sm/Nd)	1.645 ($^{147}Sm/^{144}Nd$)
(Lu/Hf)	1.992 ($^{176}Lu/^{177}Hf$)
(Re/Os)	0.212 ($^{187}Re/^{188}Os$)
(Re/Os)	0.0252($^{187}Re/^{186}Os$)
(U/Pb)	$\approx$ 70 ($^{238}U/^{204}Pb$)[a]

[a]This is indicative only because the figure varies with the lead isotope composition. Care must be taken with chemical values in the literature for rocks when compared with, say, carbonaceous meteorites, which have primitive isotopic compositions.

Table A.9 Values of selected isotope ratios

	Bulk silicate earth	Initial values
$^{143}Nd/^{144}Nd$	0.512 638	0.505 83
$^{147}Sm/^{144}Nd$	0.1966	
$^{176}Hf/^{177}Hf$	0.282 95	0.279 78
$^{176}Lu/^{177}Hf$	0.036	
$^{87}Sr/^{86}Sr$	0.7047	0.698 998
$^{87}Rb/^{86}Sr$	0.031	
$^{187}Os/^{186}Os$	1.06	0.805
$^{187}Os/^{188}Os$	0.130	0.0987
$^{187}Re/^{186}Os$	3.3	
$^{187}Re/^{188}Os$	0.412	
$^{206}Pb/^{204}Pb$	18.426	9.307
$^{207}Pb/^{204}Pb$	15.518	10.294
$^{208}Pb/^{204}Pb$	39.081	29.476
$^{238}U/^{204}Pb$	9.2	
$^{232}Th/^{238}U$	4.25	

Table A.10 Atomic number (Z), chemical symbol, and atomic mass (A) of the natural elements

Atomic number	Element	Atomic mass
1	Hydrogen (H)	1.0079
2	Helium (He)	4.002 60
3	Lithium (Li)	6.941
4	Beryllium (Be)	9.012 18
5	Boron (B)	10.81
6	Carbon (C)	12.011
7	Nitrogen (N)	14.0067
8	Oxygen (O)	15.9994
9	Fluorine (F)	18.998 40
10	Neon (Ne)	20.179
11	Sodium (Na)	22.9898
12	Magnesium (Mg)	24.305
13	Aluminum (Al)	26.981 54
14	Silicon (Si)	28.086
15	Phosphorus (P)	30.973 76
16	Sulfur (S)	32.06
17	Chlorine (Cl)	35.453
18	Argon (Ar)	39.948
19	Potassium (K)	39.098
20	Calcium (Ca)	40.08
21	Scandium (Sc)	44.9559
22	Titanium (Ti)	47.90
23	Vanadium (V)	50.9414
24	Chromium (Cr)	51.996
25	Manganese (Mn)	54.9380
26	Iron (Fe)	55.847
27	Cobalt (Co)	58.9332
28	Nickel (Ni)	58.71
29	Copper (Cu)	63.545
30	Zinc (Zn)	65.38
31	Gallium (Ga)	69.72
32	Germanium (Ge)	72.59
33	Arsenic (As)	74.9216
34	Selenium (Se)	78.96
35	Bromine (Br)	79.904
36	Krypton (Kr)	83.80
37	Rubidium (Rb)	85.468
38	Strontium (Sr)	87.63
39	Yttrium (Y)	88.9059
40	Zirconium (Zr)	91.22
41	Niobium (Nb)	92.9064
42	Molybdenum (Mo)	95.94
43	Technetium (Tc)	(97)
44	Ruthenium (Ru)	101.07
45	Rhodium (Rh)	102.9055
46	Palladium (Pd)	106.4
47	Silver (Ag)	107.868
48	Cadmium (Cd)	112.40
49	Indium (In)	114.82

Table A.10 (cont.)

Atomic number	Element	Atomic mass
50	Tin (Sn)	118.69
51	Antimony (Sb)	121.75
52	Tellurium (Te)	127.60
53	Iodine (I)	126.9045
54	Xenon (Xe)	131.30
55	Cesium (Cs)	132.9054
56	Barium (Ba)	137.34
57	Lanthanum (La)	138.9055
58	Cerium (Ce)	140.12
59	Praseodymium (Pr)	140.9077
60	Neodymium (Nd)	144.24
61	Promethium (Pm)	(145)
62	Samarium (Sm)	150.4
63	Europium (Eu)	151.96
64	Gadolinium (Gd)	157.25
65	Terbium (Tb)	158.9524
66	Dysprosium (Dy)	162.50
67	Holmium (Ho)	164.9304
68	Erbium (Er)	167.26
69	Thulium (Tm)	168.9342
70	Ytterbium (Yb)	173.04
71	Lutetium (Lu)	174.97
72	Hafnium (Hf)	178.49
73	Tantalum (Ta)	180.9479
74	Tungsten (W)	183.85
75	Rhenium (Re)	186.2
76	Osmium (Os)	190.2
77	Iridium (Ir)	192.2
78	Platinum (Pt)	195.09
79	Gold (Au)	196.9665
80	Mercury (Hg)	200.61
81	Thallium (Tl)	204.37
82	Lead (Pb)	207.2 (variable)
83	Bismuth (Bi)	208.9804
84	Polonium (Po)	(209)
85	Astatine (At)	(210)
86	Radon (Rn)	(222)
87	Francium (Fr)	(223)
88	Radium (Ra)	226.0254
89	Actinium (Ac)	227.0278
90	Thorium (Th)	232.0381
91	Protactinium (Pa)	231.0359
92	Uranium (U)	238.029
93	Neptunium (Np)	237.048
94	Plutonium (Pu)	(244)
95	Americium (Am)	(243)
96	Curium (Cm)	(247)

FURTHER READING

Albarède, F., *Introduction to Geochemical Modeling*, Cambridge University Press, 1995. A fine book on modeling although a little hard going for those not keyed up in mathematics.

Bourdon, B., Turner, S., Henderson, G., and Lundstrom, C. C., *Introduction to U-Series Geochemistry*, Reviews in Mineralogy and Geochemistry vol. 52, Washington, DC, Geochemical Society and Mineralogical Society of America, 2003. A very good book for the different applications of radioactives series.

Bradley, R., *Paleoclimatology*, New York, Academic Press, 1999. An excellent clarification of isotopic methods and problems in climatology.

Broecker, W. S. and Peng, T.-H., *Tracers in the Sea*, New York, Eldigio Press of Columbia University, 1982. The reference book for students and anyone interested in marine isotope geochemistry.

Clayton, D. D., *Principles of Stellar Evolution and Nucleosynthesis*, University of Chicago Press, 1968. A clear and well-written book for anyone wanting to go into the nitty-gritty of nuclear astrophysics and nucleosynthesis.

Dalrymple, B., *The Age of the Earth*, Stanford University Press, 1991. An outstanding book on the age of the Earth and a good introduction to isotope geology.

De Paolo, D., *Neodymium Isotope Geochemistry*, Heidelberg, Springer-Verlag, 1988. The reference book for neodymium isotope geochemistry.

Dickin, A., *Radiogenic Isotope Geology*, 2nd edn, Cambridge University Press, 2005. The book for advanced students to move on to after this one. Excellent.

Encyclopedia of Earth Systems Science, London, Academic Press, 2005. An exhaustive introduction to studying external geodynamic systems.

Faure, G., *Principles of Isotope Geology*, New York, John Wiley, 1986. A primer for a generation now.

Muséum national d'histoire naturelle, *Les Météorites*, Paris, Bordas, 1996. A clear and simple review of meteorites accessible to all. (In French.)

Roth, E. and Poty, B., *Nuclear Methods of Dating*, New York, Springer-Verlag, 1989. A collection of papers reviewing the main methods of nuclear dating.

Rowlinson, H., *Using Geochemical Data*, Harlow, UK, Longman, 1993. A book to initiate students in the use of geochemical data.

Turcotte, D. and Schubert, G., *Geodynamics*, Cambridge University Press, 2002. A reference book for geophysics.

Valley, J. M., Taylor, H. P., and O'Neil, J. R., *Stable Isotopes in High Temperature Geological Processes*, Reviews in Mineralogy vol. 16, Washington, DC, Mineralogical Society of America. A good round-up of research on high-temperature stable isotope geochemistry, for advanced students.

Walker, J., *Numerical Adventures with Geochemical Cycles*, Oxford University Press, 1990. An excellent introduction to geochemical cycles using simple computer technology.

SOLUTIONS TO PROBLEMS

Chapter 1

(1) The mass of one atom of ^{17}O is $28.455\,75 \cdot 10^{-27}$ kg. Since the atomic mass unit is $1.660\,540\,2 \cdot 10^{-27}$ kg, the mass of ^{17}O in mass unit is 17.136 4415.

The mass of each of the two molecules $^{12}CDH_3$ and $^{13}CH_4$ is 17.133 367 132 in mass units. The difference is 0.002 77, which in relative mass is about 1.610^{-4} which corresponds to a resolution power of 6100 for the interference of separation.

(2) There are 31.7 ppm of lithium in the rock

(3) The measured isotopic ration $\frac{^{87}Sr}{^{86}Sr} = R_{mes}$ is equal to the real ratios + the pollution (estimated by the blank). The supposed isotopic ratio of the blank is R_{bl}. The real isotopic ratio of the sample is R_s. So:

$$R_{mes} = R_s(1 - x) + R_{bl}(X),$$

where (X) is the mass fraction of the blank in the mixture (this is the same formula as isotope dilution).

If the precision is measured at 1.10^{-4}, $R_{mes} - R_s$ should be greater than 10 times this value:

$$R_{mes} - R_s < 1.10^{-5}$$

which translates to

$$(R_{bl} - R_s)X < 10^{-5},$$

since $R_{bl} - R_s \sim 0.006$ and

$$x < 1.6\,10^{-3} \qquad x \simeq \frac{\text{blank}}{\text{sample}}.$$

So if the sample is 10^{-6}g of Sr, $x < 1.6\,10^{-9}$g. If we increase the accuracy by 10 times the blank should be $0.16\,10^{-9}$g.

(4) The radius must be 63 meters if the angle of incidence is at 90° to the electromagnet's input faces, 31 meters if the angle is 27%.

(5) Present day: 13.59 dps; $4.5 \cdot 10^9$ years ago: 77.15 dps.

(6) Production of radioactive heat for a mantle of primitive composition: $19.9 \cdot 10^{12}$ W. For a mantle analogous to the upper mantle: $2.6 \cdot 10^{12}$ W.

Urey ratios: (i) 47%, (ii) 6%. The second ratio makes the production of radioactive heat virtually negligible. All the internal heat would therefore be related to the Earth's early history.

Chapter 2

(1) Answers on pp. 288 and following.

(2) Apparent ages in Ma

$^{206}Pb/^{238}U$	$^{207}Pb/^{235}U$	$^{207}Pb/^{206}Pb$	$^{208}Pb/^{232}Th$
473	510	677	502
472	489	572	469
442	460	548	471
439	457	547	492

(3) 5.14 Ma; 4.2 Ma. They correspond to lava from different eruptions.

(4) In activity:

$$\left(\frac{^{234}U}{^{238}U}\right) = \left(\frac{^{234}U}{^{238}U}\right)_0 e^{-\lambda_4 t} + \left(1 - {}^{-\lambda_4 t}\right)$$

$$\left(\frac{^{230}Th}{^{238}U}\right) = \frac{\lambda_{230}}{\lambda_{230} - \lambda_{234}}\left(\frac{^{234}U}{^{238}U} - 1\right)\left(1 - e^{-(\lambda_{230} - \lambda_{234})t}\right) + \left[1 - e^{\lambda_{230} t}\right].$$

(5) (i) $\left(\frac{^{230}Th}{^{231}Pa}\right)_{\text{excess}} = \left(\frac{^{230}Th}{^{231}Pa}\right)_{\substack{\text{initial}\\ \text{excess}}} e^{-(\lambda_{230} - \lambda_{231})t}$

(ii) About 300 000 years.

Chapter 3

(1) 0.75 Ga.

(2) With T being the temperature, we have:

$$^{40}Ar = (^{40}AR)_0 \exp\left[-\left(\frac{aT^2}{2} + GT\right)\right]$$

(3) (i) The ages are: Pyke Hill: 2.75 Ga; Fred's Flow: 2.58 Ga; Theo's Flow: 2.46 Ga.
(ii) The sulfides not containing uranium lie on the isochrons and give roughly the initial values of the lead isotope compositions.

(4) Lu–Hf ages: M 101 = 24.7 ± 1.2 Ma, M 214 = 30.6 ± 2 Ma.
Sm–Nd ages: M 101 = 23.6 ± 4.3 Ma, M 214 = 20.0 ± 7 Ma.
The Sm–Nd ages seem younger than the Lu–Hf ages, but the pyroxenite beds are probably 24 Ma years old.

(5) *Ages of three populations of zircons*

Little Belt Mountains	1935–2000 Ma	and	2630–2650 Ma
Finland	2700–2800 Ma	and	2000 Ma
Maryland			600 ± 60 Ma

It can be seen that two of the zircon populations are double.

(6) A straight line.

Chapter 4

(1) C/N = 14.4 million! Plants have C/N ratios of 100–200. Draw your own conclusions!

(2) About 1350 years! It is therefore not the shroud in which Christ was wrapped.

(3) $$\left(\frac{^{41}K}{^{40}K}\right)_{cosm} = \left(\frac{^{41}K}{^{39}K}\right)_{cosm} \cdot \left(\frac{^{39}K}{^{40}K}\right)_{measured} \left[\frac{\left(\frac{^{41}K}{^{39}K}\right)_{measured} - \left(\frac{^{41}K}{^{39}K}\right)_{normal}}{\left(\frac{^{41}K}{^{39}K}\right)_{cosm} - \left(\frac{^{41}K}{^{39}K}\right)_{normal}}\right].$$

(4) $8.9 \cdot 10^5$ years.

(5) $a = 1.62 \pm 0.1\,\text{mm Ma}^{-1}$.

Chapter 5

(1) (i) Basalt >1.6 mg, granite >3 mg.

(ii) Yes: basalt (2 Ga) = 0.7072; basalt (0.5 Ga) = 0.700 98; granite (2 Ga) = 0.7829; granite (0.5 Ga) = 0.7197.

(iii) 3‰ for Sr, 2% for the Rb/Sr ratio.

(iv) Five 200-g pieces are better than one 1-kg piece. The result is cross-referenced and there is a hope of constructing an isochron.

(2) 110 ± 16 ka.

(3) 277 ± 5 Ma.

(4) (i) (a) 1.12 ions per minute.

(b) The uncertainty is ±11 540 years, or 20%, compared with 0.7% uncertainty on the present-day measurement.

(c) Minimum age: 311 ± 50 years.

(d) Counting for 8 days, 300 mg of carbon would have to be extracted.

(ii) The error by the ^{230}Th–^{234}U method is 420 years for an age of 55 000 years, which corresponds to 0.76%. It is therefore an excellent method for this half-life, far superior to ^{14}C in theory. Geochemical conditions must hold for it to apply, that is: closed system and sufficient abundance levels.

(5) (i) The ^{87}Rb/^{87}Sr age of the gneisses is 1.05 Ga.

(ii) The age of the granite determined by the concordia diagram is 2.11 Ga with an intercept less than about 1 Ga (see figure). The apparent ages and the geometric relations established by geology are therefore contradictory. A granite that cross-cuts a gneiss cannot be older than the gneiss! Either of two hypotheses may hold. The first (A) is that both gneiss and granite are about 1 Ga old (given by the concordia intercept and the Rb–Sr isochron). The second (B) is to accept that the gneiss is 2 Ga old and the granite a little younger, but that the whole was subjected to a large tectonic crisis 1 Ga ago, making the zircons discordant and partially re-homogenizing the ^{87}Rb–^{87}Sr system.

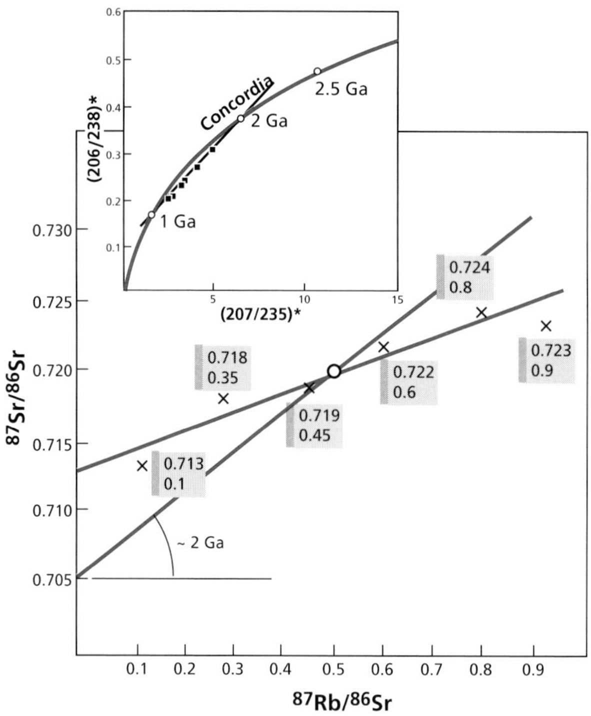

How do we choose between A and B? Hypothesis B is the more likely for two reasons. First, it is the concordia diagram that gives two ages of 2 Ga and about 1 Ga. Now, the 1 Ga corresponds to the ^{87}Rb–^{87}Sr age of the gneiss which has a poor alignment and above all a $(^{87}\mathrm{Sr}/^{86}\mathrm{Sr})_{\mathrm{initial}}$ ratio of about 0.713, which is very clear and indicates isotope re-homogenization. If we take the average of the ^{87}Sr/^{86}Sr and ^{87}Rb/^{86}Sr ratios, we get an average point. Taking an initial ratio of 0.705, we find an age of 1.95 Ga. Everything seems to be coherent therefore. However, of course, a series of U–Pb measurements on the zircon of the granite would be needed to confirm this.

(6) Initial age: 2.7 Ga, Grenville orogeny 1.1 Ga. The zircon ages of 2.8 Ga show that some of the zircon is inherited.

(7) 235 ± 5 Ma.

(8) (i) All the time intervals are mathematically possible. Allowing for diffusion $T < 30$ Ma.

(ii) Rock of interest would contain zircon, apatite, and sphene.

(9) The rhyolite is the result of partial melting (whether followed by differentiation or not) of the ancient basalt crust. This buried hydrated crust is heated and melts a little to give rise to the rhyolite.

(10) See Figure 5.11 for how to construct the diagram.

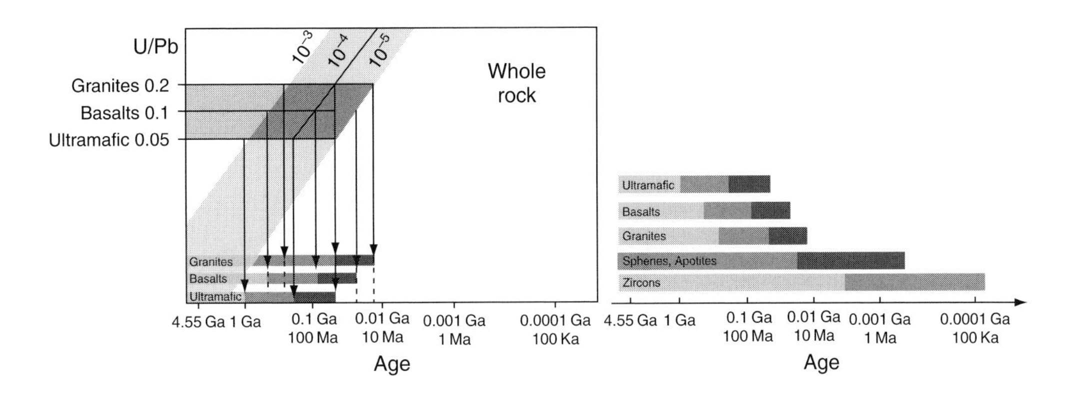

Chapter 6

(1) (i) The initial two proportions in the two volcanogenic sediments are 0.576 and 0.423, respectively, giving:

$$({}^{87}\mathrm{Sr}/{}^{86}\mathrm{Sr})_{\mathrm{J.Sedian}} = 0.709\,46 \text{ and } ({}^{87}\mathrm{Rb}/{}^{86}\mathrm{Sr})_{\mathrm{Sedian}} = 0.1239.$$

(ii) Initial ratios: $({}^{87}\mathrm{Sr}/{}^{86}\mathrm{Sr})_{\mathrm{granite}} = 0.718\,65$, $({}^{87}\mathrm{Rb}/{}^{86}\mathrm{Sr})_{\mathrm{granite}} = 3.885$.
Final ratio: $({}^{87}\mathrm{Sr}/{}^{86}\mathrm{Sr})_{\mathrm{granite}} = 0.7738$.

(2) (i) Mass of ${}^{40}\mathrm{Ar}$ assumed to be contained in the core approx. $20 \cdot 10^{18}$ g, mass of ${}^{40}\mathrm{Ar}$ in the lower mantle $60 \cdot 10^{18}$ g. This putative evaluation does not alter the general idea behind the ${}^{40}\mathrm{Ar}$ balance, a missing part of which must be in the mantle.

(ii) Concentration of non-radiogenic lead in the core $C_{\mathrm{N}}^{\mathrm{Pb}} = 1.81$ ppm.

(iii) For $T = 3$ Ga, $({}^{206}\mathrm{Pb}/{}^{204}\mathrm{Pb})_{\text{lower mantle}} = 17.40$, $({}^{207}\mathrm{Pb}/{}^{204}\mathrm{Pb})_{\text{lower mantle}} = 14.74$. The values for the closed system are 17.35 and 14.53, respectively. The point is slightly to the right of the geochron at 4.5 Ga.

(iv) Taking $T = 4$ Ga, we have $({}^{206}\mathrm{Pb}/{}^{204}\mathrm{Pb})_{\text{lower mantle}} = 17.90$ and $({}^{207}\mathrm{Pb}/{}^{204}\mathrm{Pb})_{\text{lower mantle}} = 16.10$.

(v) This phenomenon places the representative points in the J domain but is insufficient to give lead values with an isotope signature like the island of St. Helena. For this, 50% of the lead in the mantle would have had to pass into the core after 4.4 Ga, which seems a lot. The idea of explaining OIBs with high μ values by this mechanism does not seem to be corroborated by the data.

(3) (i) For the mantle, the reinjection of continental crust plays virtually no part. This is because of the mass difference. Only the initial differentiation is seen.

For the continental crust, it does play a role, but its evolution does not correspond to observation. The concavity observed with time is the opposite of what is observed, which seems on the contrary to involve increasing recycling of continental crust.

(ii) Neither the Nd nor Sr curves of mantle evolution show the evolution really observed, which see a progressive onset of primitive mantle evolution.

(iii) Overall, with the simplifying assumptions made here, this model does not account for the observations very well (see figure).

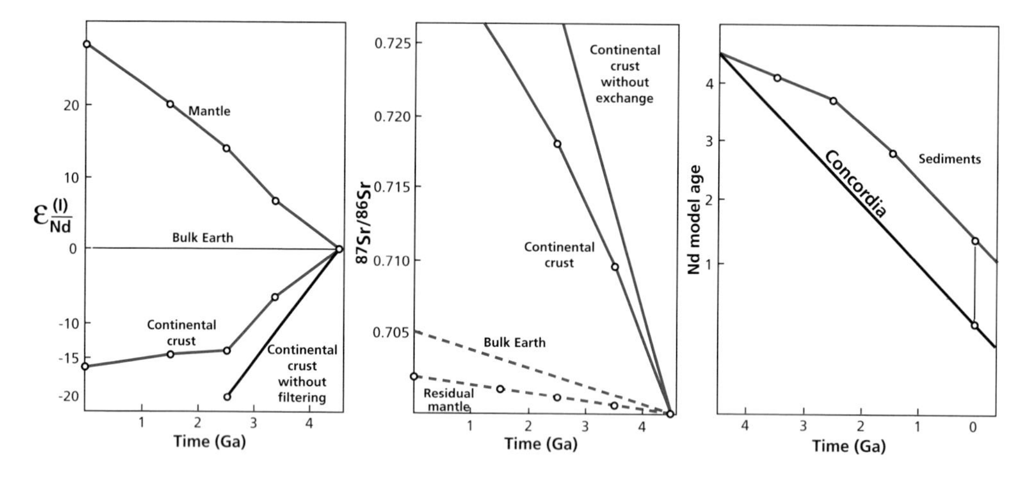

Armstrong's model of continent growth.

(4) (i) The upper mantle has a mass of $1 \cdot 10^{27}$g with 5 ppb uranium, corresponding to $2 \cdot 10^{16}$ moles.

(ii) Production of ^{4}He in 1 Ga is 2.12 times that of uranium in moles, therefore $4.45 \cdot 10^{16}$ moles. The quantity of ^{4}He outgassed is about 10^8 moles yr^{-1}. Therefore, the residence time ^{4}He = 440 Ma is equal to half the residence time of the lithosphere, which means that when the oceanic lithosphere forms at the mid-ocean ridges, there is enrichment in ^{4}He towards the melting zone.

(5) (i) Values of W^{Nd} are calculated in both hypotheses $\alpha_{\mathrm{cc}}^{\mathrm{Nd}} = 0.5110$ and $\alpha_{\mathrm{cc}}^{\mathrm{Nd}} = 0.5120$, which are the extremes using the balance equation μ. Next the T^{Nd} model ages are calculated using the two extreme data by the age formula deduced from continental crust data. This gives a pair of values (W^{Nd}, T^{Nd}) which are (0.2465; 2.84 Ga) and (0.4608; 1.08 Ga). Values of W^{Nd} are calculated with the μ balance equation under both extreme conditions $\mu_{\mathrm{dm}}^{\mathrm{Sm/Nd}} = 0.227$ and $\mu_{\mathrm{dm}}^{\mathrm{Sm/Nd}} = 0.28$.

Then T^{Nd} is calculated using the formula for the depleted mantle. We get a pair of values (W^{Nd}, T^{Nd}) of (0.256; 2.6 Ga) and (0.4882; 0.964 Ga). We then take a (W^{Nd}, T) plane and draw the two straight lines corresponding to the pair of values (W, T)

taken two by two (see figure). Their point of intersection gives $W^{Nd} = 0.335$, $T^{Nd} = 2.05 \cdot 10^9$ years.

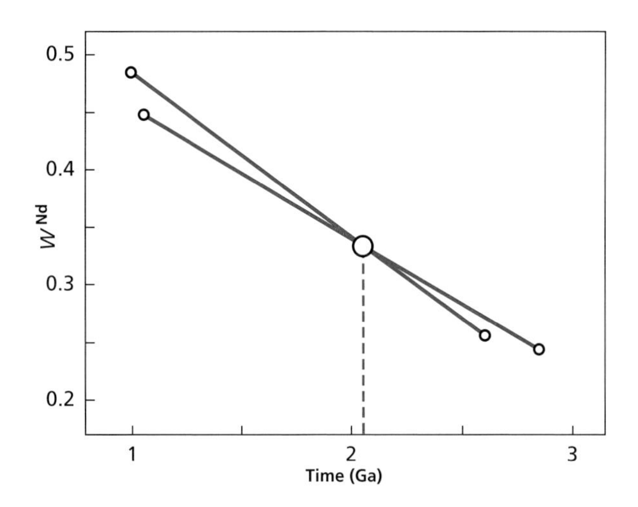

Returning then to the balance equations, we get: $\alpha_{cc}^{Nd} = 0.511\,567$ and $\mu_{dm}^{Sm/Nd} = 0.241$.

(ii) Using W^{Nd}, W^{Sr} is calculated as 0.22. From the balance equation for μ values we deduce $\mu_{cc} = 0.409$. From the balance equation for the α values, we deduce $\alpha_{cc}^{Sr} = 0.713\,28$. We check by calculating the T^{Sr} model age on the depleted mantle values $T^{Sr} = 2.1$ Ga; with continental crust values $T^{Sr} = 1.90$ Ga, for an average of 2 Ga. The overall picture is fairly coherent.

Chapter 7

(1) The first job is to estimate the fractionation factors at the missing temperatures +10 °C and −30 °C.

(i) We plot the two curves $\alpha_D, \delta_{^{18}O}$ against temperature and interpolate and extrapolate linearly, which is warranted because they seem to vary linearly. The complete table is given below.

	α_D	$\alpha_{^{18}O}$
+20	1.085	1.0098
+10	1.10	1.0106
0	1.1123	1.0117
−20	1.1492	1.0141
−30	1.175	1.0155

(ii) The results are as given below.

	δ^{D}_{vapor}	δ^{D}_{rain}	$\delta^{18}O_{vap}$	$\delta^{18}O_{rain}$
1	–133	–33	–16.3	–6.5
2	–208	–96	–24.13	–12.4
3	–307	–158	–33.57	–19.47
4	–424	–249	–43.95	–28.38

(iii) See figure.

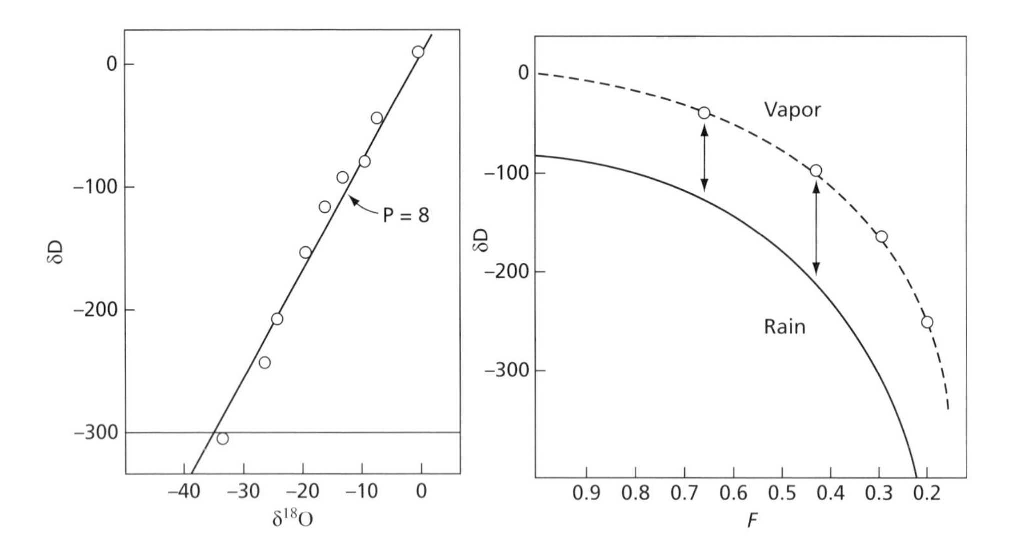

(2) The equation is:

$$\delta = \delta_0 + 10^3(\alpha - 1)\ln f.$$

For the first phase, f varies from 1 to 0.7. The results are given in the table and figure below.

	$\alpha_{-1} = 0.0002$		$\alpha_{-1} = 0.00025$		$\alpha_{-1} = 0.00035$
	$f = 0.9$	$f = 0.7$	$f = 59, \Delta f = 0.84$	$f = 0.49, \Delta f = 0.7$	$f = 0.388, \Delta f = 0.79$
Melt	5.52	5.571	5.614	5.66	5.74
Olivine	5.32	5.37	5.414	5.46	5.54
Pyroxene			5.31	5.36	5.44
Plagioclase					5.14

As can be seen, great precision is required to bring out these variations.

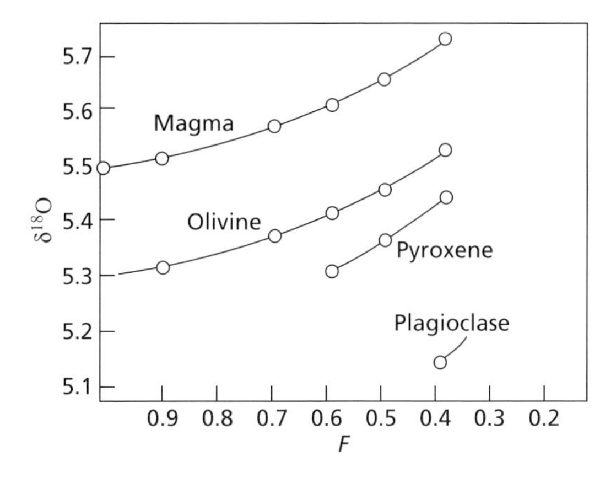

(3) With $\delta_D = 70\%$ and Craig's straight line of precipitation, we deduce $\delta_O = 10\%$. We calculate the partition coefficient at 550 °C.

$\Delta_{A-B} = A \cdot 10^6 T^{-2} + B$

	A	B	$\Delta_{\text{mineral-water}}$
Quartz	4.10	−3.7	2.35
Magnetite	−1.47	−3.7	−5.8
Muscovite	1.9	−3.10	−0.41
Feldspar	3.13	−3.41	1.20
Calcite	2.78	2.89	1.21
Plagioclase	2.15	−2.0	1.149

The balance equation is written:

$$(\delta_Q - \delta_{H_2O})\,x_1 + (\delta_{Mg} - \delta_{H_2O})x_2 + (\delta_{Mg} - \delta_{H_2O}) - x_3 +$$
$$(\delta_{feld} - \delta_{H_2O})x_4 + (\delta_{cal} - \delta_{H_2O})x_5 + (\delta_{plag} - \delta_{H_2O})x_6 = Q.$$

From this we get:

$$\delta_{silicate} - \delta_{H_2O} = 0.99\ (2),$$

but we also know the initial balance:

$$\delta_{\text{initial silicate}} \times y_1 + \delta_{\text{initial water}} y_2 = \delta_{\text{initial}}$$

where y_1 and y_2 are calculated allowing for the fact that by mass:

$H_2O = 15\%$ and silicate $= 85\%$.

In the silicates, oxygen = 54%; in the water, oxygen = 88.8%. Therefore the initial overall value of the system, if assumed closed, is:

$$\delta_{\text{initial}} = 7 \times 0.77 \pm (-10) \times 0.23 = 5.39 - 2.3 = 3.09.$$

This is preserved, so we have:

$$\delta_{\text{silicate}} + \delta_{H_2O} = 3.09\ (2).$$

We then deduce δ_{H_2O} by eliminating δ_{silicate} between (1) and (2):

$$\delta_{H_2O} = 1.05$$

All that is left is to calculate the mineral δ values:

$$\delta_{\text{quartz}} = +3.4$$

$$\delta_{\text{magnetite}} = -4.7$$

$$\delta_{\text{muscovite}} = 0.68$$

$$\delta_{\text{feldspar}} = +2.29$$

$$\delta_{\text{calcite}} = +2.28$$

$$\delta_{\text{plagioclase}} = +2.39.$$

(4) (i) The quantity of carbon burnt is $5.9 \cdot 10^{16}$ g, which corresponds to 410 ppm of CO_2.
(ii) Now, the content is 330 ppm. The remainder has been dissolved in the ocean after homogenization.
(iii) The $\delta^{13}C_{\text{calcite}}$ has gone from +3 to 0.
(iv) If δ can be measured with a precision of 0.1 δ or 0.05 δ, this criterion can be used as a pollution control.

(5) $\Delta_{g-m} = \delta_g - \delta_{\text{maj}} = \delta_{SO_2} - \left[x\delta_{S^{2-}} + (1-x)\delta_{SO_4^{2-}}\right]$

$$x = \frac{S^{2-}}{S^{2-} + SO_4^{2-}}$$

$$x = \frac{K(T)}{(\alpha^2 + K(T))}$$

$$\Delta_{g-m} = 7.5x - 4.5.$$

Therefore if $x < 0.6$ Δ_{g-m} it is negative, otherwise it is positive.

(6) (i) $\delta = 0.344$ and 1.386, respectively.
(ii) $\delta = -1.05$.
(iii) 0.0081 δ m^{-1} and 0.011 08 δ m^{-1} for the two snowball scenarios and for the scorched Earth scenario 0.0148 δ m^{-1}.
(iv) No.

Chapter 8

(1) We find:

$$R = \frac{1}{\sqrt{Jk}}.$$

A decay reaction, for example, decay of organic matter, obeys a second-order kinetic equation:

$$x = \frac{J_0}{J_0 kt + 1}.$$

(2) The frequency 100 ka will be continued unchanged. The 40 ka and 20 ka frequencies will be phase-shifted and damped and therefore the 100 ka frequency will predominate.

(3) (i) The upper mantle $1 \cdot 10^{27}$ g with 5 ppm of U, corresponding to $2 \cdot 10^{16}$ moles. Production of ^{4}He in 1 Ga is 2.2 times that of uranium in moles, therefore $4.4 \cdot 10^8$ yr. The quantity of ^{4}He degassed is about 10^8 moles per year.

(ii) Therefore the residence time of ^{4}He of 440 Ma is twice the residence time in the lithosphere.

(iii) This means that when the oceanic lithosphere forms at the ocean ridges, there is enrichment of ^{4}He towards the melting zones.

(4) Mixing times in the upper mantle

	Mixing time (Ma)
North Atlantic	350
South Atlantic	138
South-west Indian Ocean	533
North Pacific	42
Central Indian Ocean	77

Yes, there is a relation with the expansion rate (see figure).

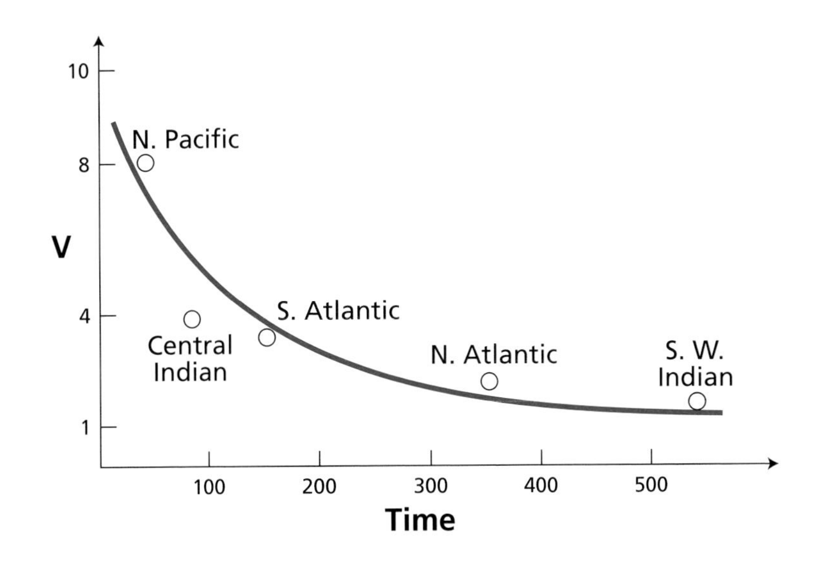

INDEX OF NAMES

SUBJECT INDEX